HUMAN MICROSCOPIC ANATOMY

Human Microscopic Anatomy

SEONG S. HAN
Professor of Anatomy, Medical School
Professor of Oral Biology and
Head, Cell Biology Laboratory, School of Dentistry
Director, Program in Biology of Aging
Institute of Gerontology
The University of Michigan

JAN O. V. HOLMSTEDT
Associate Research Scientist
Institute of Gerontology
The University of Michigan

McGRAW-HILL BOOK COMPANY
New York St. Louis San Francisco
Auckland Bogotá Hamburg
Johannesburg London Madrid Mexico
Montreal New Delhi Panama Paris
São Paulo Singapore Sydney Tokyo Toronto

HUMAN
MICROSCOPIC
ANATOMY

1 2 3 4 5 6 7 8 9 0 VHVH 8 9 8 7 6 5 4 3 2 1 0

This book was set in Palatino by Monotype Composition Company, Inc.
The editors were Richard W. Mixter and Abe Krieger;
the designer was Nicholas Krenitsky;
the production supervisor was Robert A. Pirrung.
The drawings in Chapter 12 were done by Carol Bardolph;
all other drawings were done by Fine Line Illustrations, Inc.
Von Hoffmann Press, Inc., was printer and binder.

Library of Congress Cataloging in Publication Data

Han, Seong S date
Human microscopic anatomy.

Includes index.
1. Histology. I. Holmstedt, Jan O. V.,
joint author. II. Title.
QM551.H148 611'.018 79-22640
ISBN 0-07-025961-5

TO
The Late Professors
Burton L. Baker and Kim Young Chang

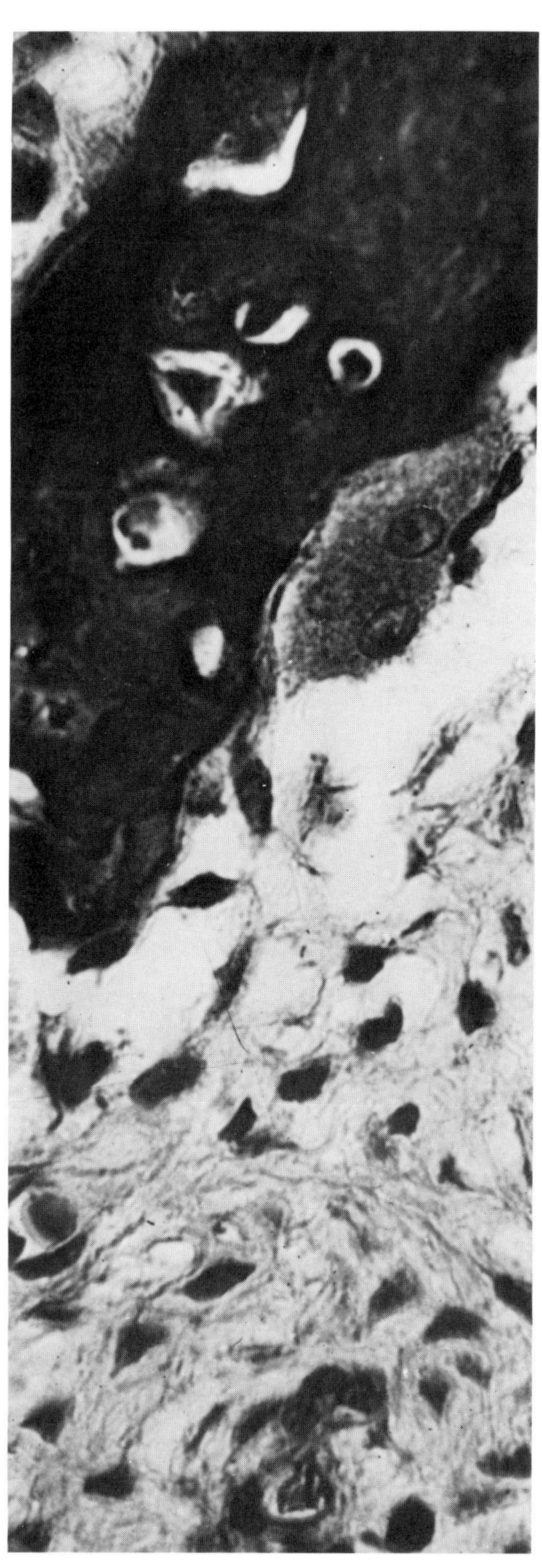

CONTENTS

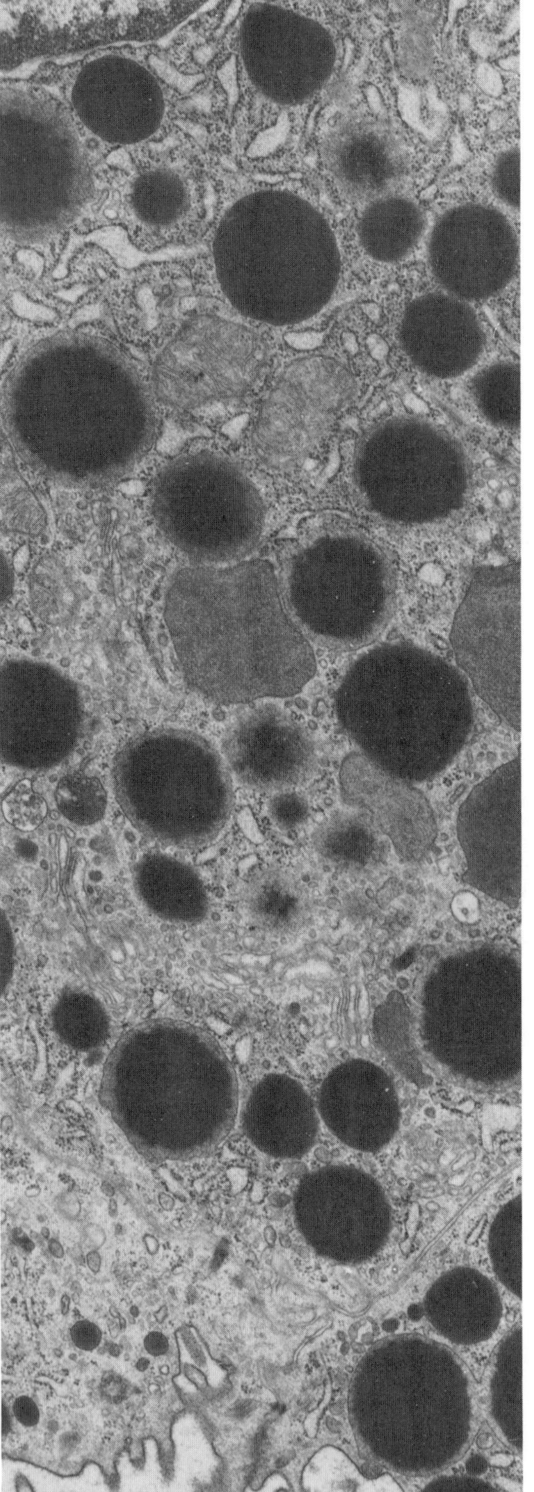

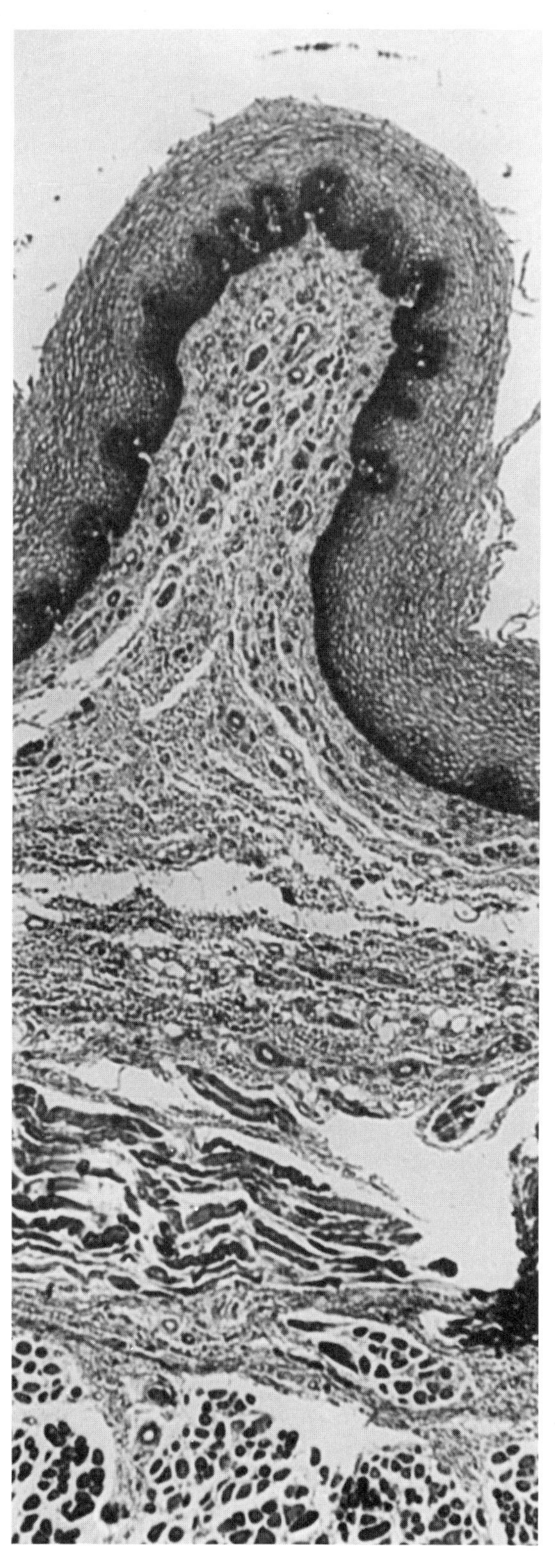

PREFACE

For the past 25 years, I have been deeply involved in the teaching of, and preparation of teaching materials for, courses in microscopic anatomy and cell biology. In these endeavors, I have always attempted to employ the latest available aides to inculcate in my students the nature of the orderly evolution of macromolecules, subcellular organelles, and the birth cells and their eventual differentiation into the building blocks of the human body as we understand it today. Although this has been a period of privilege and enlightenment for me, I have also been troubled by the shortcomings of the traditional methods used to teach human microscopic anatomy. Some of the greatest teachers and scientists have written on this subject but still the problem remains. The problem is that the crowded curricula in medicine, dentistry, and related health science fields allow only so many hours for the exploration of any particular subject area, making it necessarily difficult for the student to extract the essential knowledge needed from the overwhelmingly comprehensive literature presently available.

What the student needs, therefore, is not another reference text, but a book written with the singular purpose of helping him or her assimilate the necessary information. For that reason, I have tried to present the essential knowledge of the subject area in straightforward and simple terms, employing an objective approach which I hope will not only expedite but facilitate the learning process. In fact, because of the many photomicrographs and reviews, this textbook is in itself a complete autotutorial matrix which may be used in conjunction with classroom lectures, or, if the professor so chooses, will allow the student to progress at his or her own rate. Each chapter is headed by a list of behavioral objectives, followed by a description of the subject topic. This, in turn, is followed by a Review section. The Bibliography is divided into two sections: one which deals with directly related books and articles from recent publications, and the other in which pertinent but somewhat more detailed reference materials are found.

One of the basic questions I have had was whether or not the substance of this book could be taught without the aid of slides and a microscope. It is my current belief that in light of the various pressures imposed upon health science curricula, teaching of microscopic anatomy to medical, dental, and other health science students can be carried out equally well either with or without the aid of a microscope, provided the textbook presents adequate illustrations to go along with the particular structure that is being described. This statement should not be misconstrued as a categorical condemnation of the value of laboratory exercises involving the use of the microscope in the teaching of microscopic anatomy. On the contrary, I will confess my fundamental position that looking into a set of slides with ample time and developing three-dimensional concepts and animated functional features of cellular and subcellular elements under observation is a most valuable experience for students in anatomical sciences. The question at hand is just a matter of available time versus what is needed in training a health science professional. For this reason, graduate students in anatomy are strongly encouraged to make use of actual slides and a good light microscope in conjunction with this textbook whenever possible. In the meantime, every effort has been made in writing this book so that it can be used as a self-sufficient textbook by whomever is interested in understanding the basics of human microscopic anatomy without having to spend a large amount of time for laboratory exercises.

Dr. Jan O. V. Holmstedt joined me after the first draft of this book had been written. Since then, Dr. Holmstedt has spent numerous, painful hours with me in revising the manuscript and has been responsible for the selection of bibliographic materials. His efforts have been essential in completing the project, and I feel very fortunate to have him as coauthor of this book.

I am indebted to a number of colleagues and former students at the University of Michigan, without whose critical remarks, thorough reviews of manuscripts, and laborious support in developing illustrative materials the birth of this volume would not have been possible. My appreciations are due to David K. Chang, William E. Check, Yong J. Kang, Jon Arden, and Drs. Moon I. Cho, Marvin Mark, Carol Drinkard, Gerald D. Geisenheimer, David Howard, Yong G. Kim, Booe I. Ma, Arnold P. Morawa, Masseeh Pirbazari, and Lloyd Straffon.

I am grateful to Dr. I. J. Bak of the University of California at Los Angeles, Dr. M. B. Gilula of Rockefeller University, Dr. Schroeder of the University of Zurich, Dr. Steven Wissig of the University of California at San Francisco, and Drs. Kent Christiansen, Joseph Hawkins, Lars Johnsson, S. K. Kim, and M. Ross of the University of Michigan who have kindly provided valuable photomicrographs. I am also grateful to Dr. Richard Cardell of the University of Virginia, Dr. Susumu Ito of Harvard University, and Dr. Russell T. Woodburne of the University of Michigan whose original drawings have been adapted in certain illustrative materials used in chapters on the stomach, intestines, and heart, respectively.

Special thanks are due to Sookja Auh, Eun D. Pyen, and John R. Viery for their wide-ranging assistance in the preparation of the book; to Sarah L. Bernhard and Marion Trainor for their skillful typing and proofreading; to Alice Macnow, Richard W. Mixter, Abe Krieger, and their associates at McGraw-Hill; and to colleagues in the Department of Educational Resources and the Center of Research on Learning and Teaching who have influenced my thinking on the value of new instructional technology in science education. They include Drs. Warren Seibert, Dave Starks, and Pat O'Conner as well as Ruth Ashley, Tom Green, and Karen Schaeffer. I also thank the Office of the Vice-President for Research at the University of Michigan for allowing me to publish the manuscript through McGraw-Hill, and the U. S. Public Health Service whose grant has been helpful in the early phase of content development (No. 5D08 PE0120). Last, but not least, I wish to express my sincere gratitude for their friendship and support to members of my laboratory and to those colleagues in the Department of Anatomy—Medical School, Department of Oral Biology—School of Dentistry, and Division of Biological Sciences—College of Literature, Sciences and the Arts who have spent valuable hours in critically reviewing various chapters.

Seong S. Han

HUMAN MICROSCOPIC ANATOMY

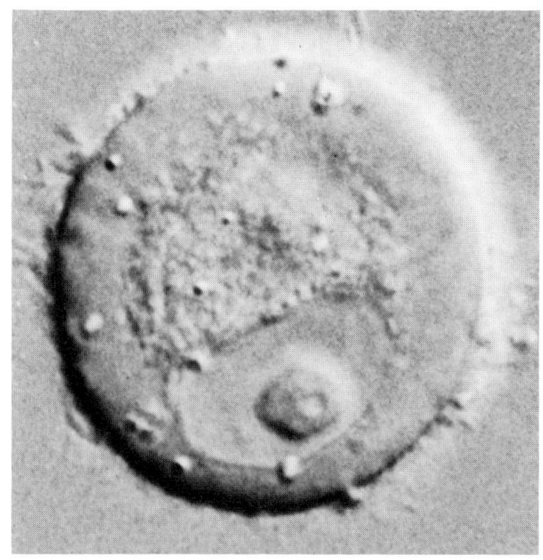

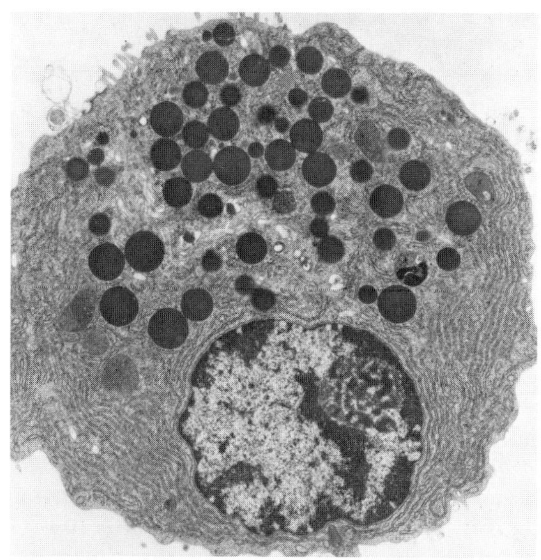

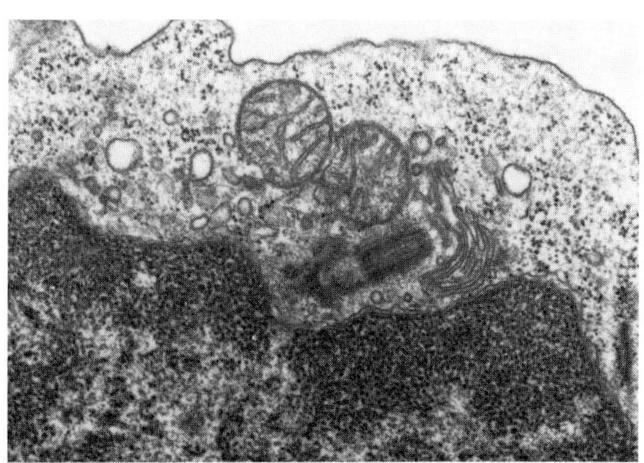

1

INTRODUCTION TO THE CELL

OBJECTIVES

Upon completion of this chapter, the student will be able to:

1 Identify the advantages or disadvantages of the use of the electron and light microscopes with regard to the following:
(a) resolution and magnification (b) staining (c) viewing live tissues

2 Explain the principle of histologic staining, giving hematoxylin and eosin (H and E) as an example.

3 Identify one example of an eosinophilic structure and one example of a basophilic structure.

4 Identify the following structures in a light micrograph (LM) and/or in an electron micrograph (EM), and state the function of each.
Cytoplasmic organelles:
(a) ribosomes (EM) (b) rough-surfaced endoplasmic reticulum/smooth-surfaced endoplasmic reticulum (EM) (c) Golgi complex (LM, EM) (d) lysosomes (EM) (e) centriole (EM) (f) mitochondria (mostly EM) (g) cytoskeletal filaments (LM, EM) (h) microtubules (EM) (i) plasma membranes (EM)
Nuclear structures:
(a) nucleus (LM, EM) (b) nuclear pores (EM) (c) nucleolus (LM, EM)

Every living thing is made up of cells. This is not a startling statement today, but it was the center of scientific debates among biologists during the early nineteenth century. Matthias Schleiden, a German botanist, advocated in 1838 that the cell was the basic structure of plants. A year later, Theodor Schwann also asserted that "cells are organisms and arranged according to definite laws." Thus the groundwork for the modern cell theory has been laid for less than 150 years.

Every scientific advance predates developments of appropriate technology. In the case of microscopic anatomy, it had to wait for the light microscope. This microscope uses visible light rays to form images. The resolution of a microscope is inversely proportional to the square of the wavelengths of the illuminating source, that is, $R = \frac{1}{\lambda^2}$. *Resolution* is the relative ability of a viewer to discern between two close dots when using the device. With the naked eye, one can resolve spaces of 0.1 mm (= 100 μm = 100,000 nm). The light microscope has a resolving power of 0.1 μm, or 100 nm.

Therefore, greater resolution of images depends on the use of shorter wavelengths. In the 1920s ultraviolet light was introduced as a light source because it has a shorter wavelength; but the UV microscope was never widely used because of the hazards inherent in the use of ultraviolet light, as well as the requirement for special, expensive quartz lenses. The discovery in the 1920s that beams of electrons could be focused led to the development of lenses which eventually led to the production of electron microscopes. With an electron microscope, a resolution of less than 0.2 nm can be obtained today. The first useful electron microscope came into being after World War II.

THE LIGHT MICROSCOPE

The light microscope is capable of magnification of up to 2000 times. If specimens have been properly prepared, this magnification allows observation of most bacteria and cellular components.

In the standard light microscope, the light from a tungsten source is taken up by a substage condenser, passed through or around the specimen, and collected by the objective lens (Fig. 1-1). The objective lens spreads the light and sends it through the ocular, which focuses the light on the retina of the eye. The image produced can also be captured by a camera.

Before a specimen is viewed, it must be chemically fixed, embedded in paraffin, sectioned, and stained. In fixation, a piece of tissue is placed into a fixative, such as formaldehyde. Then it is dehydrated and embedded in paraffin, which provides the necessary consistency for sectioning with a microtome blade at 2 to 10 μm in thickness. Numerous biologic dyes are used to stain the sections in order to demonstrate different cellular structures. For general purposes, hematoxylin (H) and eosin (E) staining is the most commonly used procedure. Like other biologic dyes, the H and E stain takes advantage of acidic or basic properties of all structures. Nucleic acids and other acidic structures of a specimen are stained by hematoxylin, the basic dye. For this reason blue-staining structures are referred to as being *basophilic*. Eosin, the acid dye, stains basic components of the cytoplasm an orange to red color; these structures are called *eosinophilic*. Many carbohydrates can be stained with a special technique called the *periodic acid Schiff* (PAS) method. This technique identifies all carbohydrate moieties that release aldehyde groups following periodic acid treatment of the specimen. The free aldehyde groups liberated by the acid treatment in originally reduced form are colorless. The Schiff reagent then combines with the reduced aldehydes resulting in a red-colored product. Silver stains are commonly used to demonstrate such structures as Golgi complex (apparatus) and certain

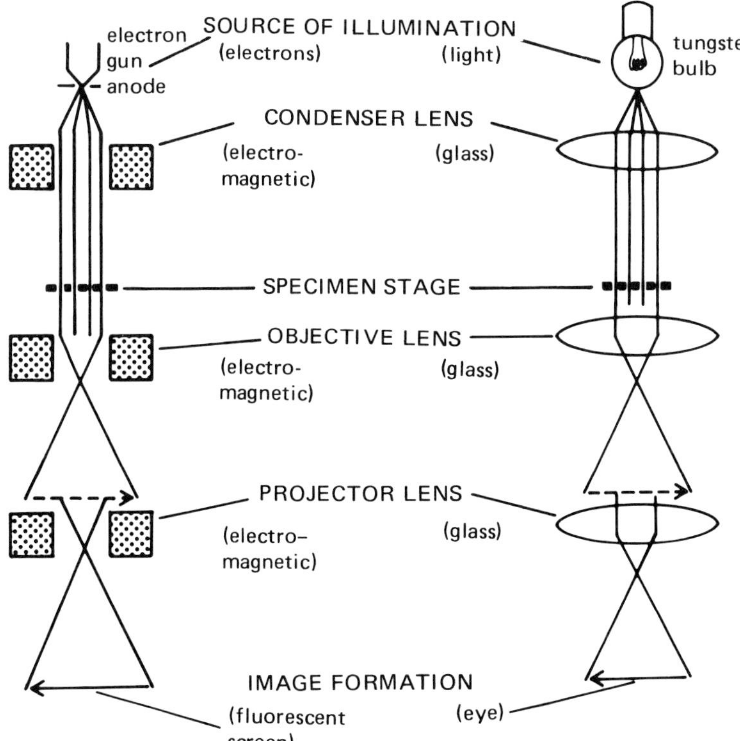

FIGURE 1-1 Comparison between the electron microscope and the optical microscope.

connective tissue fibers. Structures that stain darkly with $AgNO_3$ are called *argyprophilic.* Numerous specific stains other than these mentioned have also been developed.

Certain types of light microscopes, for example, phase and interference, do not require the specimen to be fixed and stained. Figure 1-2 is an interference micrograph of a cell in culture. The tonal variation distinguishing various cell structures reflects the different masses of molecular particles that make up the subcellular components. The phase microscope operates by a similar principle and therefore also allows observations of living cellular structures.

THE ELECTRON MICROSCOPE

The electron microscope (EM) can magnify an original specimen a million times or more. Therefore, tissue preparation for EM viewing is usually more complex than for light microscopy. The fixation is more elaborate and a totally different method of embedding and sectioning is used. Specimens are fixed in an aldehyde or osmium tetroxide and embedded in a plastic resin. The specimen must be very thin (less than 0.1 μm in thickness) and should have minimum distortion from procedural artifacts. Special ultramicrotomes equipped with a glass or diamond knife are available for making very thin sections, which are stained with a heavy metal (osmium, uranium, lead, or tungsten). Some of these metals react preferentially with certain structures, providing a differential deflection of electrons necessary to form the image. These preparatory steps normally prevent the viewing of living materials in the electron microscope, although certain high-voltage electron microscopes have been used to observe small living objects such as bacteria. Because of the greater resolution of electron

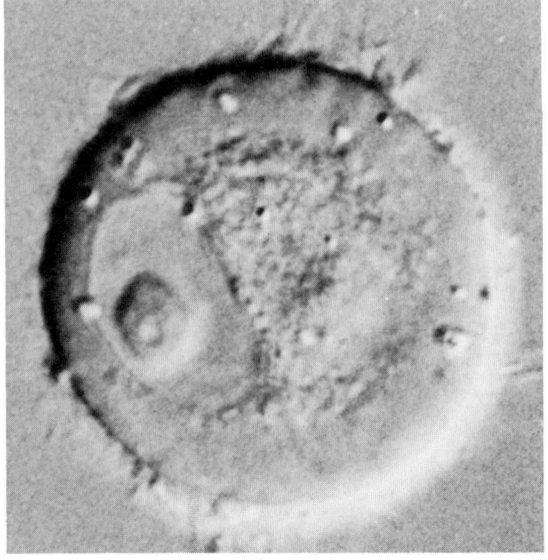

FIGURE 1-2 Interference micrograph of cell in culture.

microscopes, knowledge of the nature of the many artifacts that are invariably introduced in the course of specimen preparations for electron microscopy is essential.

The electron microscope uses a stream of electrons emitted from a tungsten filament called an *electron gun,* energized with an average accelerating voltage of 50 kV as its power source. Since electrons cannot be propelled in air, the electron gun is enclosed in a vacuum column. A series of electromagnetic or electrostatic lenses arranged in a way similar to the light microscope (Fig. 1-1) focus the electrons through the column. A pointed electron gun generates the electrons, and they pass down through the specimen chamber, where the tissue section is positioned on an extremely fine screen above the objective lens. There certain electrons are deflected, while others are collected and focused (by projector and intermediate lenses) onto a phosphorescent screen and form the image. By replacing the screen with a film, an electron photomicrograph is taken. These electron micrographs provide permanent records of the specimen; the specimen itself deteriorates rapidly because of the bombardment by the high-energy beam, and therefore can be viewed only for a short period of time. Figure 1-1 compares the optical path of an electron microscope with that of a light microscope. The light microscope is inverted to facilitate comparison.

UTILIZATION OF RADIOISOTOPES

World War II brought further technological advances that have been essential to the birth of modern cell biology. Among these is the use of radioactive isotope tracers to "tag" molecules so they can be traced through various life processes. Radioactively labeled precursors have been essential to clarification of biochemical processes that occur inside the cell. By combining the radioactive isotope labeling and the photomicrographic techniques, it is possible to localize isotope tracers in histologic specimens. In this technique, called *radioautography,* molecules of living cells are labeled with radioactive markers or precursors and then prepared for histologic observation in a routine manner. After sections are made, a layer of photographic emulsion is laid in total darkness over the specimen, which contains radioactive isotopes. The specimen is kept in darkness for varying periods of time. The radioactivity continues to expose the photographic emulsion, producing small exposed foci over the radioactive sources. Following the exposure, the covering emulsion is developed like a photographic negative and viewed under the microscope, where small dark silver grains appear on or near the cellular structures labeled with the radioactive isotope.

Radioautography can be applied at both the light and electron microscopic levels. Figure 1-3 is an example of a light microscopic radioautograph in which the incorporation of ^{3}H-uridine, a precursor for RNA synthesis, is localized over nuclei (arrows) of large steroid-secreting cells of the ovary. Other

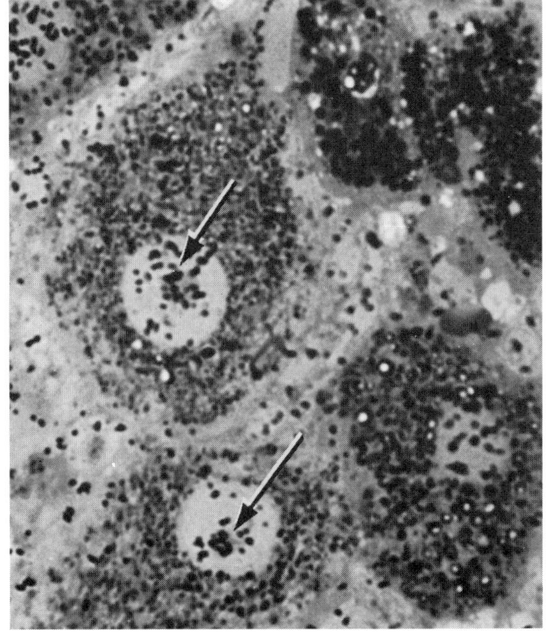

FIGURE 1-3 Light microscope radioautograph. Silver grains are localized over the nuclei (arrows).

dark granules present throughout the cytoplasm are largely lipid droplets and mitochondria. A typical radioautograph at the electron microscope level is seen in Fig. 1-4, in which the incorporation of a radioactive amino acid, ^{3}H-proline, into secretory proteins (arrows) is depicted. Note that the radioautographic grains appear as small dots at light microscope level, whereas in the electron micrograph (EM) the grains show an irregularly coiled appearance. The reason for this unique grain structure in the EM is the physical characteristics of radioactive energy dissipating through the photographic emulsion.

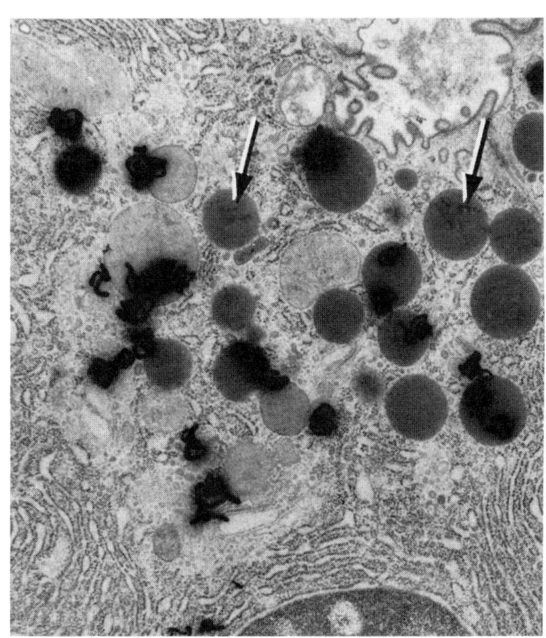

FIGURE 1-4 Electron microscope radioautograph. The incorporation of ^{3}H-proline into secretory proteins (arrows) is illustrated.

CELL STRUCTURE

All living cells are products of evolution. Natural selection has taken place in cells throughout evolution, causing both elimination and preservation of structural components, and thereby reflecting their utilitarian value. Simply put, those structures that are essential to maintain the life of the cell have persisted throughout evolution. Fossils of cells, for example, show that the basic structure of membranes has remained the same. Functional economy characterizes the arrangement of macromolecules that make up all cell structures; thus, the simplest, most effective structures are maintained.

Each cell is fundamentally a sac of protoplasm in balance with its environment. The body is a highly organized collection of individual cells, each of which carries on specific life processes. The cells may be mobile, as are blood cells and macrophages, or fixed, as are nerve and muscle cells. The diagram at the center of Fig. 1-5 shows what could be seen in an EM taken at very low magnification; the structures at the periphery are drawn

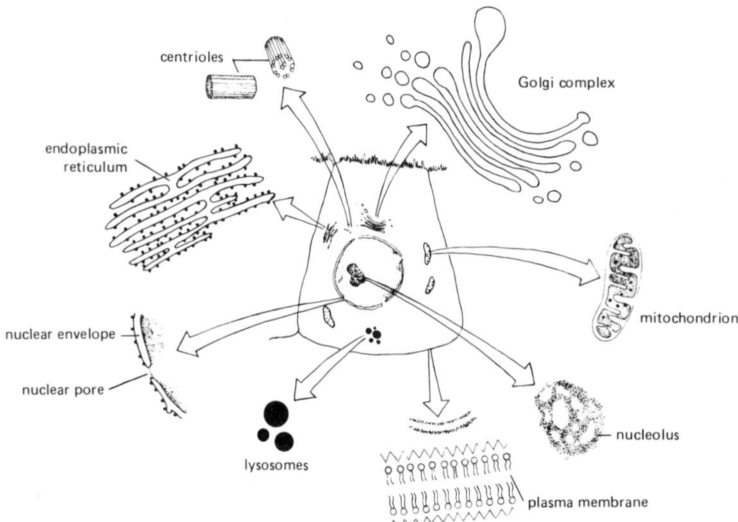

FIGURE 1-5 Diagram of a cell seen at low magnification in the electron microscope. The structures at the periphery are cell organelles seen at higher magnification.

as they might appear at a higher level of magnification.

LIGHT MICROSCOPIC VIEW

The plasma membrane of a cell is too thin to be seen by the light microscope (7.5 nm), but its location has long been evident. In most light microscope preparations the cell boundary may be visualized by the artifactual collection of cytoplasmic materials against the plasma membrane. The membrane separates the cytoplasm from the surrounding environment, yet allows its interaction with the latter.

Cytoplasmic organelles

Many metabolic processes do not occur in the ground cytoplasm but are performed by differentiated subcellular components called *organelles.* Organelles compartmentalize the cytoplasm and provide regions where specialized metabolic functions take place. Examples of organelles visible under the light microscope are ergastoplasm (or basophilic cytoplasm), mitochondria, Golgi apparatus, centrioles, and lysosomes.

Cytoplasm

The basophilic *cytoplasm* is the region of the cytoplasm that stains with basic dyes such as hematoxylin. Basic dyes react with the nucleic acids in ribosomes that make up the basophilic cytoplasm. In Fig. 1-6, the basophilic portion of the cytoplasm (b) occupies much of the base of the cell.

Mitochondria

The light, ovoid- to rod-shaped structures that are interspersed in the basophilic cytoplasm are negative images of *mitochondria* (m). Mitochondria are small ovoid- to rod-shaped organelles which convert nutrients into the form of energy that the cell can use.

Golgi Complex

The Golgi apparatus, or *Golgi complex* (G, Fig. 1-6), is generally located next to the nucleus. One of its known functions is to package and condense secretory materials. Near the Golgi complex, lysosomes can be visualized by certain cytochemical staining. Lysosomes function in *hydrolysis,* the breakdown of unnecessary intracellular structures or foreign materials.

Intracellular inclusions

Depending on the type of cell, various *inclusion* bodies may be present. Secretory cells produce secretion granules. In Fig. 1-6, secretion granules appear as numerous small granules filling the apical cytoplasm; the

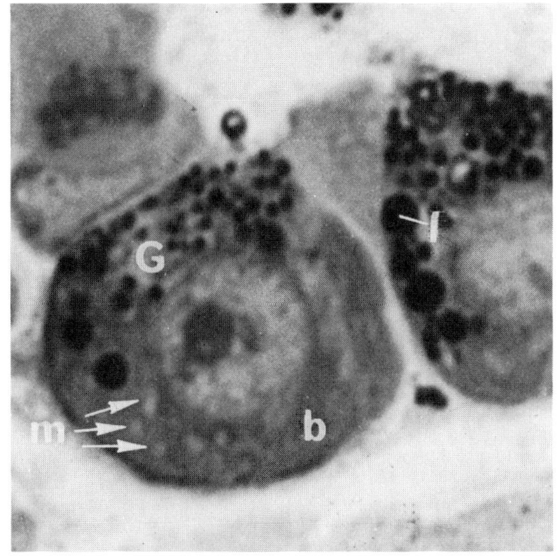

FIGURE 1-6 Light micrograph of secretory cells illustrating cell organelles: basophilic cytoplasm (b), Golgi apparatus (G), mitochondria (m), and inclusions; lipid droplets (l), and secretory granules.

larger granules are lipid (fat) droplets (l). In addition, many cells contain glycogen granules or pigment granules. Lipid droplets and glycogen granules are readily converted into energy when necessary. Frequently lipid droplets are lost in the process of fixation; there are a number of special techniques used to preserve them. Although glycogen granules normally appear in many cells, physiologic and pathologic conditions in a given cell type influence the number and size of lipid droplets and glycogen granules.

Nucleus

The nucleoplasm contains chromatin, one or more nucleoli, and nuclear sap. So named because it stains intensely with basic dyes, the chromatin includes *heterochromatin* and *euchromatin*. Heterochromatin, a condensed form of nuclear material generally located along the periphery of the nucleus, is thought to represent a relatively inactive portion of the chromatin. Euchromatin, on the other hand, a lighter, less dense nuclear material, is active in the production of either deoxyribonucleic acid (DNA) by replication or ribonucleic acid (RNA) by transcription. A relatively quiescent cell is predominantly heterochromatic, while the nucleus of an active cell is predominantly euchromatic. Figure 1-7, which represents a collection of peritoneal lymphocytes and macrophages, shows examples of the active macrophage nucleus (m) and the quiescent, dense nucleoplasm of lymphocytes (l). Note the prominent nucleolus of the macrophage. A similarly large nucleolus is found in the active secretory cell appearing in Fig. 1-6. The nucleolus is associated with the portion of chromatin known as the *nucleolar organizer* and contains precursors of cytoplasmic RNA. It is known to be the cell's major center for the production of ribosomal RNA, which will be discussed later.

ELECTRON MICROSCOPIC VIEW

While the electron microscope has brought some new features of cells into view, it has also clarified the structural details of many individual components previously studied under the light microscope.

Ribosomes

The basophilic portion of the cytoplasm contains ribonucleic acid (RNA). Most cells contain at least three different types of cytoplasmic RNA: ribosomal RNA (*rRNA*), messenger RNA (*mRNA*), and transfer RNA (*tRNA*). These RNA molecules possess distinct physical and chemical characteristics and therefore carry on essential functions unique to different steps in protein synthesis. Ribosomal RNA is responsible for the assembly of amino acids into peptide bonds, while tRNA identifies and facilitates the movement of particular amino acids in the formation of proteins. The mRNA is the direct replica of the genetic code transcribed from DNA and therefore dictates the sequencing of amino acids in protein synthesis. Although all three types of RNA may be responsible for cyto-

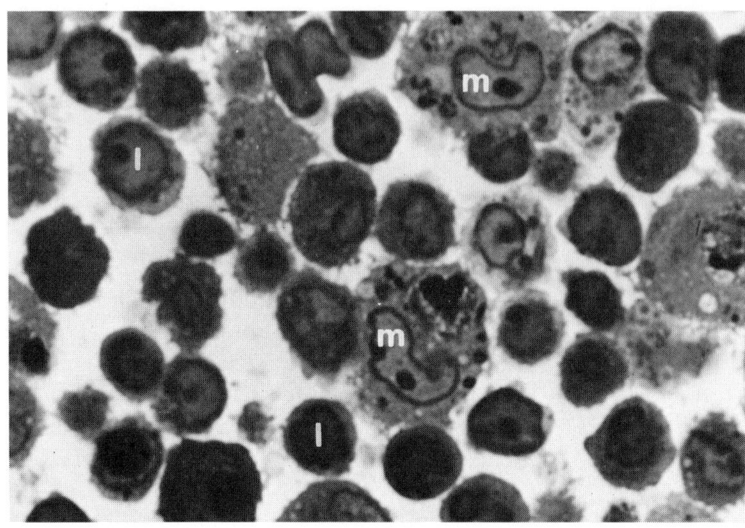

FIGURE 1-7 Light micrograph of peritoneal macrophages (m) and lymphocytes (l).

plasmic basophilia in light micrographs, the rRNA comprises up to 95 percent of all cytoplasmic RNA. Therefore, a deep cytoplasmic basophilia could be regarded as a reflection of a large quantity of rRNA present in the cell.

Ribosomes are made up of a roughly one-to-one mixture of rRNA and proteins. They may be attached to a system of membranes called *endoplasmic reticulum*, or they may be present freely in the cytoplasm. These are called *attached* and *free ribosomes*, respectively. The free ribosomes occur in two forms: *monosomes* are individual ribosomes, and *polysomes*, or polyribosomes, are aggregates of a number of ribosomes. The two types of ribosomes are found in varying proportions, depending on the kind of cell and its func-

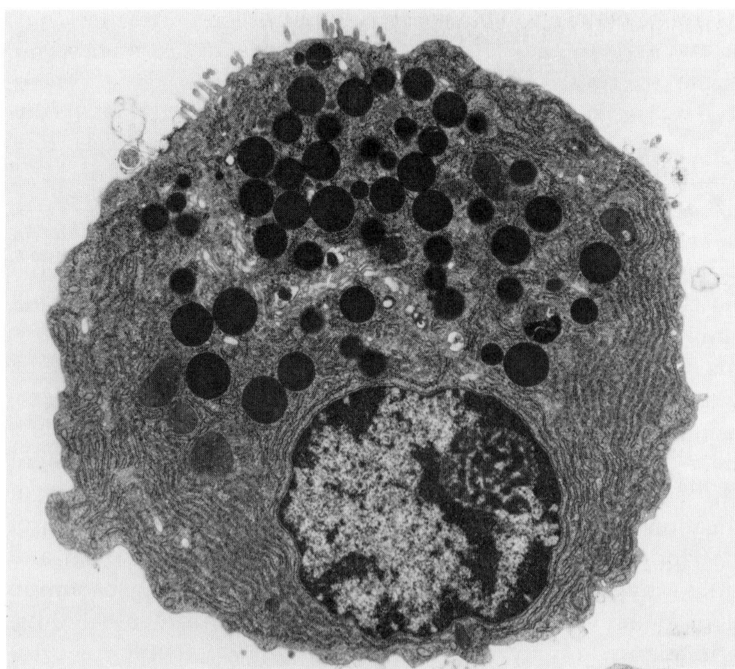

FIGURE 1-8 Low-power electron micrograph of a pancreatic acinar cell.

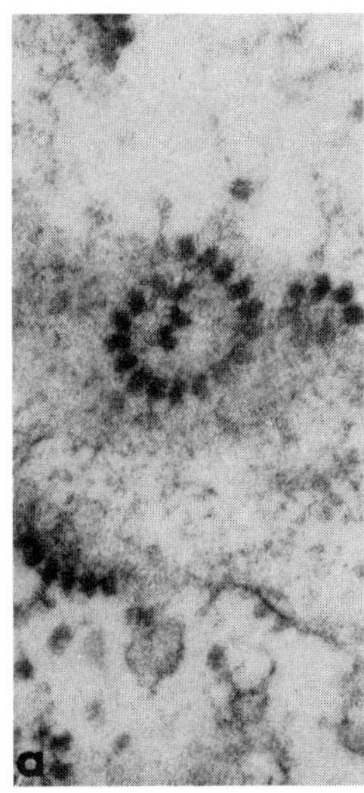

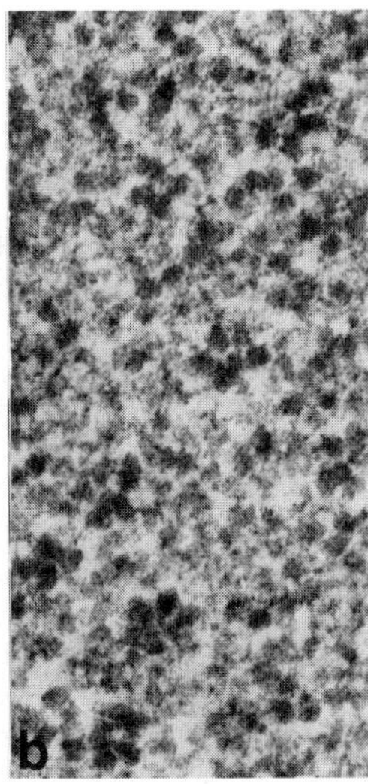

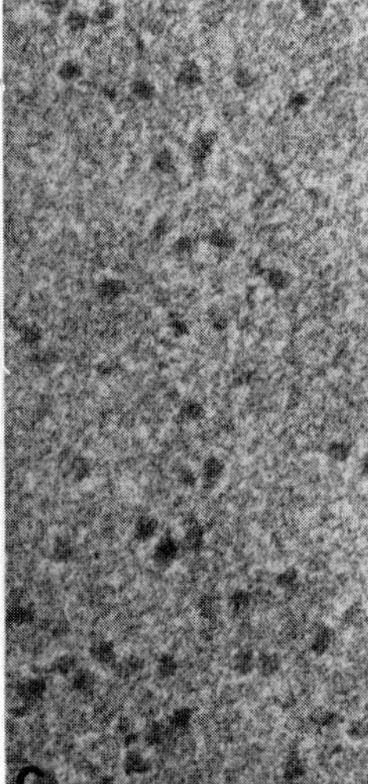

FIGURE 1-9*a* Spiral polyribosome found in a developing fibroblast.

FIGURE 1-9*b* Polyribosomes formed by random clusters of monosomes.

FIGURE 1-9*c* Sparse ribosomes seen in nearly mature red blood corpuscles arranged in a monosomal pattern.

tional state. Figure 1-8 is a low-power EM of a pancreatic acinar cell in which parallel arrays of the endoplasmic reticulum, studded by ribosomes, occupy the basal cytoplasm.

Figure 1-9 gives examples of the arrangement of ribosomes in different cells and in different phases of development. The spiral polyribosomes shown in Fig. 1-9*a*, representing those found in the developing fibroblast, indicate that individual polysomes are made up of varying numbers of monosomes. The arrangement of monosomes in making polysomes could occur either in a spiral shape (Fig. 1-9*a*) or in a random cluster as shown in Fig. 1-9*b*, which is a portion of a developing erythroblast of the bone marrow. In further differentiated cells in which protein synthesis is about to cease, one sees fewer ribosomes than in earlier stages of active differentiation. The nearly mature red blood corpuscle in Fig. 1-9*c* contains fewer ribosomes than the immature erythroblast. Most of the ribosomes appearing in Fig. 1-9*c* are in monosomal arrangement, since they have nearly completed the synthesis of cytoplasmic protein, resulting in hemoglobin, which occupies the dark ground cytoplasm.

These examples support the theory that polysomes are produced when the genetic information on mRNA is being translated by ribosomes which move along the entire length of an mRNA strand. Thus, the number of ribosomes making up a polysome is generally proportional to the size of the mRNA molecule, and hence the length of the peptide produced. There are exceptions to this, however; polycistronic messengers code for more than one peptide molecule. In the polysomal region, where ribosomes pass along the mRNA, tRNA brings amino acids to the ribosomes in accordance with the coded sequence prescribed by the structure of mRNA.

Most polysomes attached to membranes are arranged in aggregates or spiral strings, and therefore are active in protein synthesis. However, the usual plane of section through the attached ribosomes results in an EM in which a polysome may appear as an isolated monosome. Where such membrane has been sectioned tangentially, however, it is clear that the ribosomes are not in monosomal configuration but rather in spiral clusters (Fig. 1-10).

Endoplasmic reticulum

The endoplasmic reticulum (ER) in EM is usually seen as a parallel array of lacy, flattened membranes. In three dimensions, these membranes separate the cytoplasm into two phases: the ground cytoplasm or cytosol outside the ER, and the cytoplasm inside the ER. The space within the ER is called its *cisterna*.

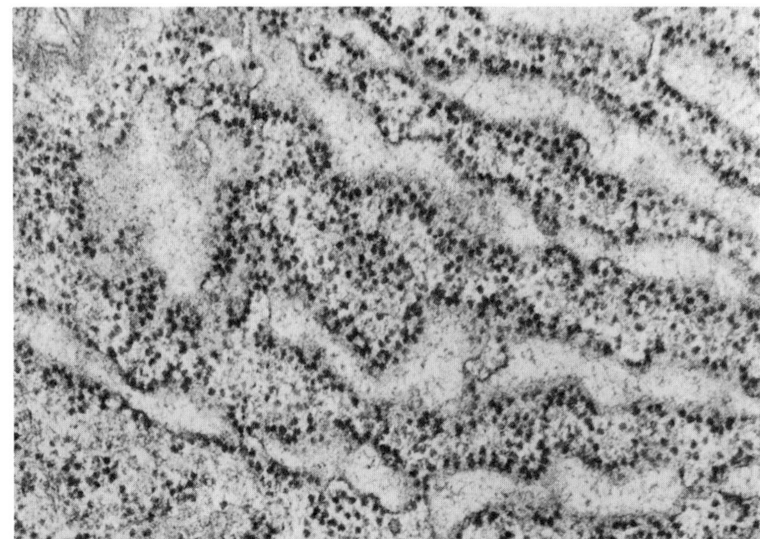

FIGURE 1-10 Electron micrograph of attached polyribosomes.

The ER is differentiated to varying degrees even among the same type of cells, and its profiles are commonly seen in EM throughout the cytoplasm. Two forms of the ER are recognized: rough-surfaced ER (*RER*) and smooth-surfaced ER (*SER*). Figure 1-11*a* shows cisternae of RER, while Fig. 1-11*b* depicts a portion of SER. The RER obtains its name from the presence of attached polysomes along its outer surface. The three-dimensional appearance of a typical RER is diagramed in Fig. 1-12. Note that the cisternae of RER can appear flat or tubular, depending on the way it is sectioned. Often the RER is in continuity with the outer membrane of the nuclear envelope, as seen in Fig. 1-13 (arrows).

It has been hypothesized that the nuclear envelope is derived from ER, or vice versa. The appearance of RER varies according to cell type or physiologic conditions in a cell. In Fig. 1-14*a*, a profile of the RER of a normal pancreatic cell demonstrates the typical lamellated appearance. In contrast, a disturb-

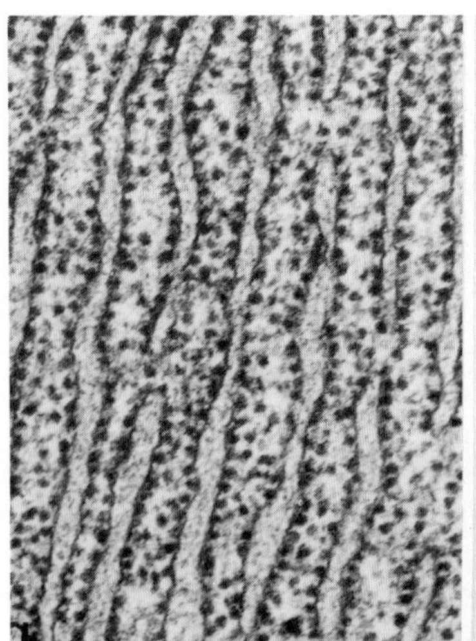

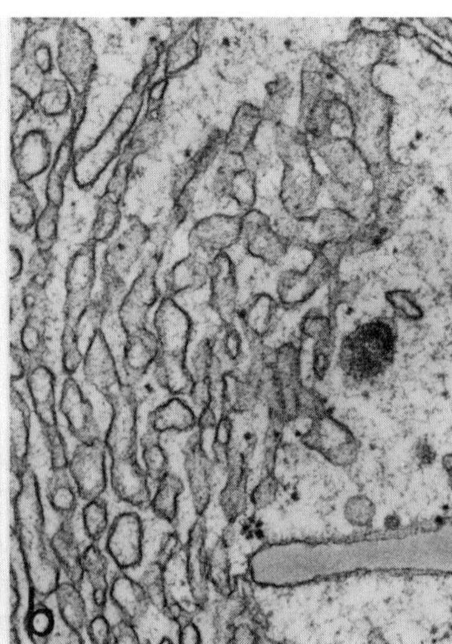

FIGURE 1-11*a* Electron micrograph of rough-surfaced endoplasmic reticulum (RER).

FIGURE 1-11*b* Electron micrograph of smooth-surfaced endoplasmic reticulum (SER).

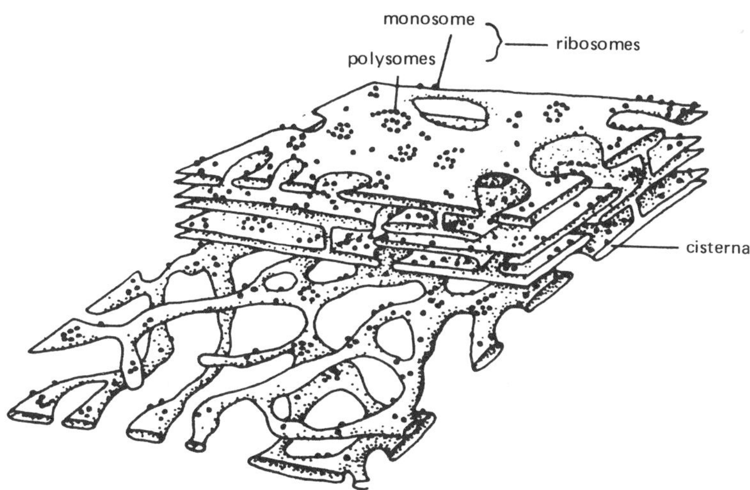

FIGURE 1-12 Diagram of the three-dimensional arrangement of RER.

ance which may inhibit secretory activity of this cell type produces fragmented, swollen profiles (Fig. 1-14*b*). In cells which have fewer profiles of RER, one finds small tubular or irregular profiles, as shown in Fig. 1-14*c*.

SER lacks ribosomes and its profiles frequently appear tubular and more irregularly arranged than RER. The EM in Fig. 1-11*b* shows SER throughout the field. The SER is typically prevalent in cells that produce steroids rather than proteins, since it is known to serve as reservoirs of steroid precursors such as cholesterol.

Mitochondria

The electron microscopic view of a mitochondrion has revealed details of its structure which were not possible to visualize with the light microscope. In the EM, outer and inner *limiting membranes* are visualized in every mitochondrion. The inner membrane usually protrudes into the internal space of the mitochondrion and forms a series of shelf or fingerlike structures. In most human cells these projections, called *mitochondrial cristae*, are usually shelflike in appearance. Cytochromes and other enzymes located on the

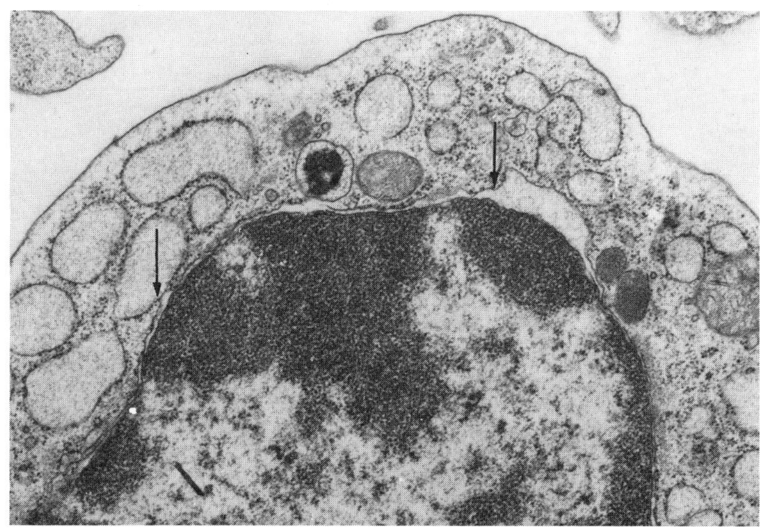

FIGURE 1-13 Electron micrograph showing the continuity between the nuclear envelope and RER (arrows).

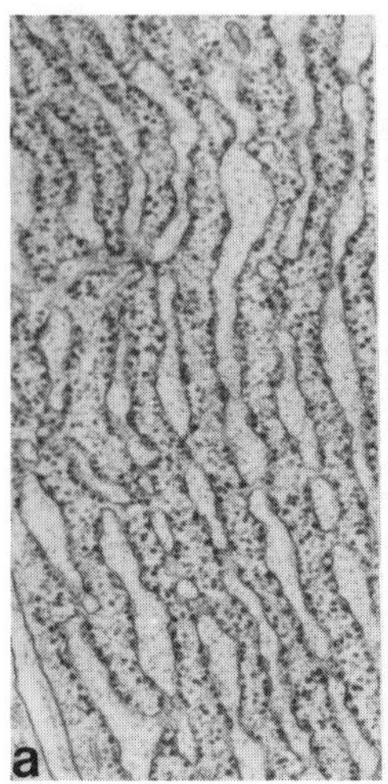

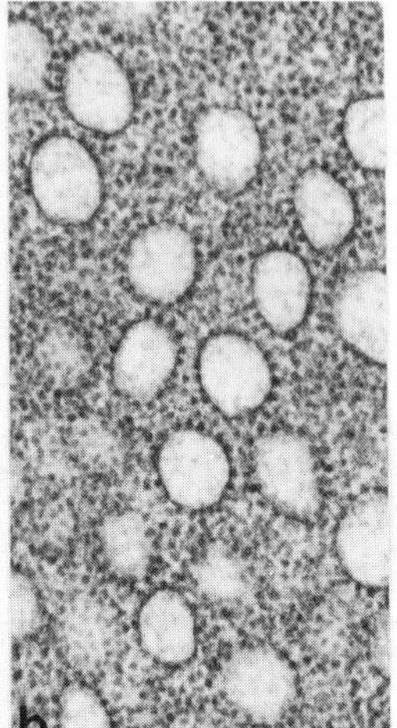

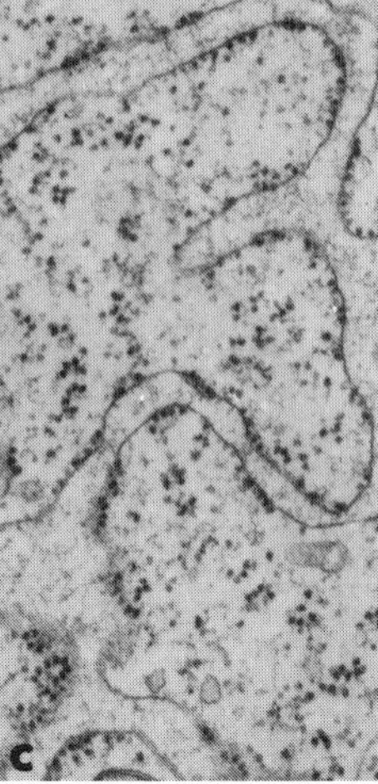

FIGURE 1-14*a* Lamellated RER in a pancreatic acinar cell.

FIGURE 1-14*b* Swollen fragmented profiles of RER in a pancreatic acinar cell, the secretory activity of which has been inhibited.

FIGURE 1-14*c* Tubular and irregular profiles of RER in a cell with sparse RER.

cristae enable the mitochondria to carry out cellular respiration in the presence of oxygen. Since these key enzymes are exclusively located at the surface of cristae, the more active mitochondria have more highly developed cristae. An EM of typical mitochondria sectioned at various angles is depicted in Fig. 1-15. Note the outer and inner limiting membranes, as well as the inward extensions of the inner limiting membrane or cristae. Figure 1-16 is a three-dimensional drawing of a mitochondrion with a portion cut away. In the enlarged circle, the subunits of the inner cisternal membrane are illustrated. They are called *electron transport particles* (ETP).

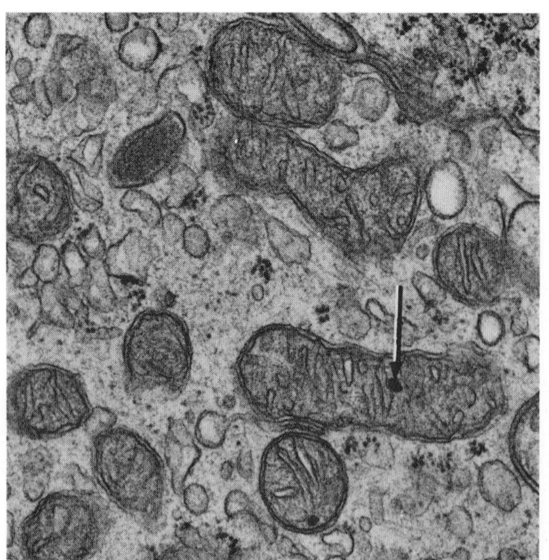

FIGURE 1-15 Electron micrograph of mitochondria. The matrix contains mitochondrial granules (arrow).

Found within the space bordered by the inner membrane, the *mitochondrial matrix* contains a number of proteins, nucleic acids, and small granules, called *(intra) mitochondrial granules.* These granules, 30 to 40 nm in diameter, contain calcium phosphate and are concerned with the regulation of the intracellular calcium concentrations (arrows, Figs. 1-15 and 1-16). Many of the matrix proteins in mitochondria are contractile proteins and are believed to play a role in the intracellular mobility of the organelle. Within the mitochondria are small amounts of RNA and extrachromosomal DNA. The extrachromosomal DNA of the mitochondria differs from the nuclear DNA in that it has a circular configuration. It has been found to produce mRNA coded for subunits of the respiratory enzymes. These subunits combine with complementary enzyme components which are coded by the nuclear DNA and synthesized elsewhere in the cytoplasm. Thus mitochondrial enzymes are thought to be hybrid prod-

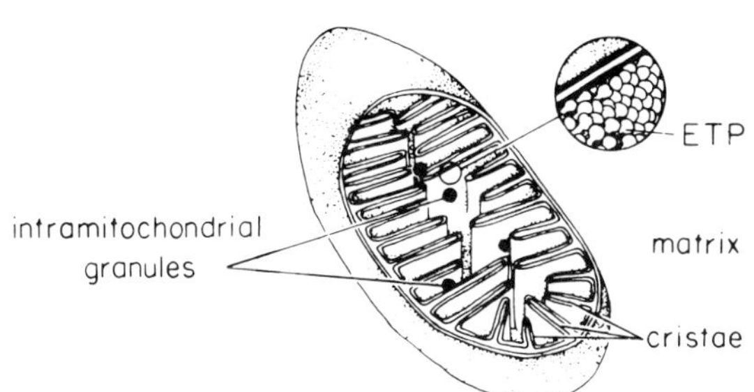

FIGURE 1-16 Three-dimensional drawing of a mitochondrion with a portion cut away illustrating the arrangement of cristae. The matrix contains mitochondrial granules (arrow).

ucts of both the cytoplasm and the mitochondria themselves.

One function common to all mitochondria is the production of adenosine triphosphate (ATP) molecules at the expense of oxygen and nutrients such as simple sugars. Although they appear to be scattered throughout the cytoplasm, mitochondria are not randomly located in the cell. They are usually found near those cytoplasmic organelles which require energy for differentiated cellular functions. For example, mitochondria may be found in close proximity to RER in cells concerned with protein synthesis.

Golgi apparatus

Viewed under the electron microscope, the Golgi apparatus or complex generally appears to be composed of a stack of flattened saccules. Small vesicles, often called *transport vesicles* in secretory cells, are thought to arise from the nearby ER. In such cells these vesicles tend to fuse with flattened vacuoles at the proximal end of the stack, while condensing vacuoles are produced by fission of larger vacuoles at the distalmost lamellae of the Golgi complex. The EMs in Figs. 1-17 to 1-19 show examples of the Golgi complex found in cells of different functions in which the involvement of Golgi complex is valuable.

Figure 1-17, an electron microscopic view of an exocrine cell, shows the relationship of lamellar components of the Golgi apparatus with RER and condensing vacuoles. Note also the number of small transport vesicles (arrows) between the RER and Golgi lamellae. The Golgi complex seen in Fig. 1-18 is from a fully mature white blood cell in which the participation of the Golgi complex in the formation of granules has been completed. Therefore, the Golgi complex in this particular figure is small. In this regard, the struc-

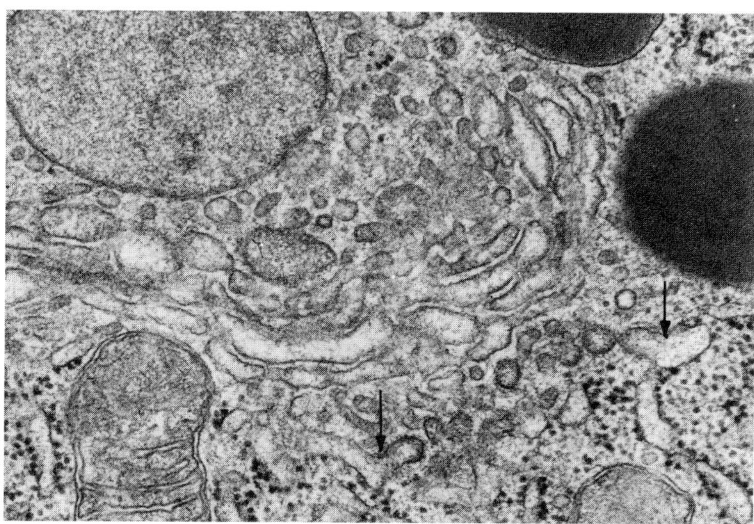

FIGURE 1-17 Golgi apparatus in an exocrine cell. Note transport vesicles (arrows) originating in the RER.

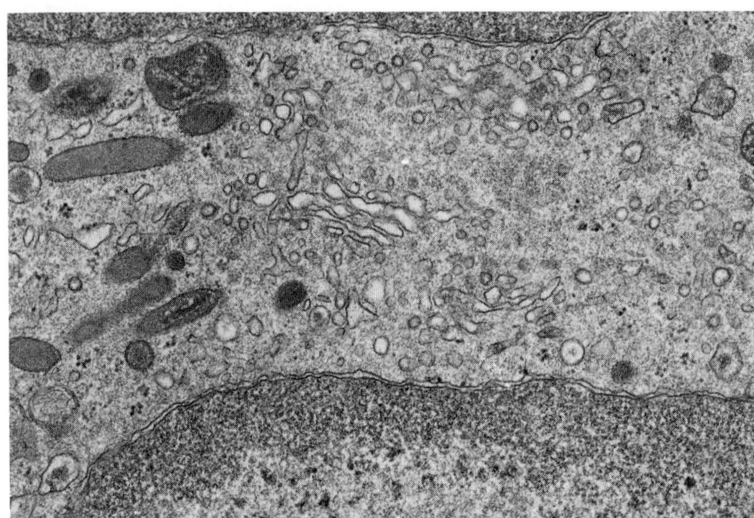

FIGURE 1-18 Small Golgi complex in a fully mature white blood cell.

ture is not unlike the Golgi complex of a lymphocyte (Fig. 1-19), which is not concerned with secretory protein synthesis or intracellular storage granules.

In addition to the commonly known function of the Golgi complex in condensing secretory proteins, recent studies have shown that it plays an important role in the synthesis of polysaccharides. Since most, if not all, of the secretory proteins are glycoproteins, processing the biosynthetic products of RER by adding carbohydrate moieties appears to be a universal function of the Golgi complex. Evidence for this has come from radioautographic studies of sugar incorporation as well as cytochemical demonstration of certain enzymes that are important in polysaccharide synthesis. Figure 1-20 is an example of cytochemical localization of a nucleoside diphosphatase activity, *uridine diphosphatase*, in the distal lamellae of the Golgi complex.

To summarize the function of cellular organelles in exportable protein synthesis (Fig. 1-21), the polysomes lining the RER produce peptides that are transported across the membrane into its inner cavity, or cisterna

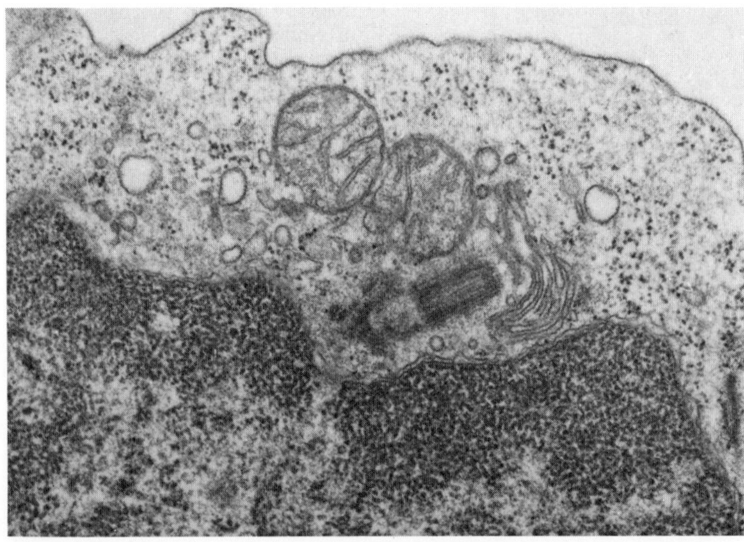

FIGURE 1-19 Golgi apparatus in a lymphocyte.

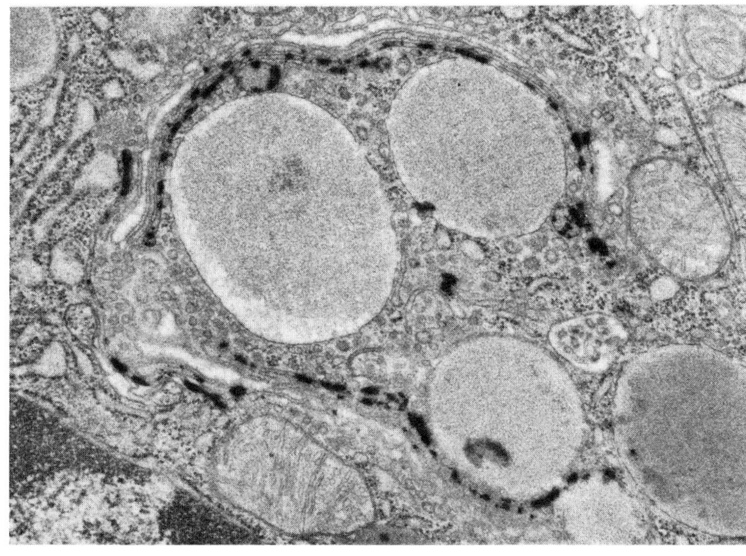

FIGURE 1-20 A portion of an acinar cell showing deposits of reaction product for uridine diphosphatase in the Golgi apparatus.

(steps 2 and 3). This synthesized peptide protein passes through the cisternae of the RER to the region of the Golgi apparatus. At the distal end of the RER, transport vesicles pinch off the ER and fuse with each other or with the proximal Golgi saccule. After the proximal lamella reaches a certain size, vesicles are thought to produce a new lamella. The proteinaceous material is condensed in the Golgi complex and carbohydrate moieties are added (step 4). Parts of the distal lamella are pinched off as condensing vacuoles, in which further condensation and packaging of the product occur (step 5). The vacuole becomes a dense, dark *secretory granule* (step 6). This membrane-covered granule freely moves toward the apical cytoplasm and fuses with the plasma membrane, externalizing its contents (step 7). The energy used throughout protein synthesis is produced by mitochondria (step 1). As the secretion granule membrane fuses with the plasma membrane, the plasma membrane is often seen to buckle and fold because of the added membrane mate-

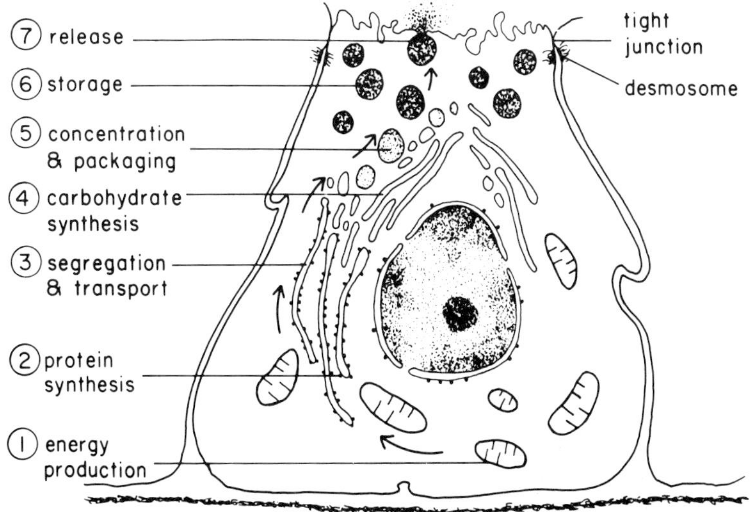

FIGURE 1-21 Drawing summarizing the function of cellular organelles in exportable protein synthesis.

rial. Eventually, this extra membrane is broken down and reutilized by the cell, in this way conserving membrane material.

Lysosomes
Lysosomes are small cytoplasmic bodies that contain hydrolytic enzymes within the cell. Since enzymes are proteinaceous in nature, they are produced by the RER, transported to the Golgi apparatus, and packaged in the same manner as the secretory granules. With special enzyme cytochemical techniques, such as acid phosphatase, the lysosomal enzyme activities can be visualized both in LM and EM. Those lysosomal enzymes that can be identified by cytochemical techniques are often called *lysosomal marker enzymes;* one of the most frequently used is acid phosphatase. Enzymes in the lysosome are typically activated at low pH (about pH 5.0). Several dozen enzymes with low pH optima have been found to exist in lysosomes of different cells. In activated lysosomes, these enzymes can hydrolyze most of the substances brought into the cell. They may even hydrolyze the cell's own components (autophagocytosis) under certain conditions. This autolytic process can break down unnecessary organelles and molecules, recycling their components within the cell. Table 1-1 lists some of the lysosomal enzymes and their specific functions.

SOME LYSOSOMAL ENZYMES AND THEIR SUBSTRATES — TABLE 1-1

ENZYME	SUBSTRATE
Deoxyribonuclease	DNA
Ribonuclease	RNA
Phosphatase	Phosphate ester, phospholipids
Cathepsins	Proteins
Glucuronidase	Carbohydrates

Some lysosomes have dense bodies with irregular contours, as shown in Fig. 1-22. In general, lysosomes are slightly smaller than mitochondria.

Centrioles, microtubules, and microfilaments
The centriole is small and often difficult to resolve when viewed under the light microscope. When visible, it appears as a refractile granule, usually in the vicinity of the Golgi complex. In the EM, centrioles are cylindrical in organization and appear in cross section to be composed of nine tubular structures

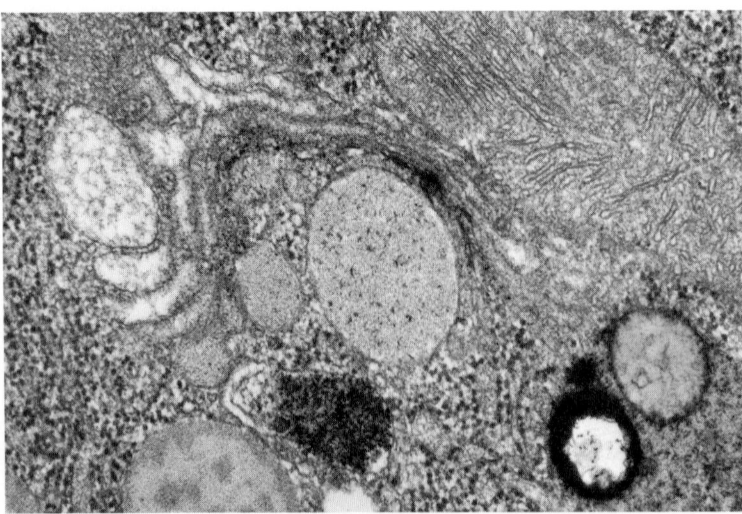

FIGURE 1-22 Portion of a cell containing several lysosomes.

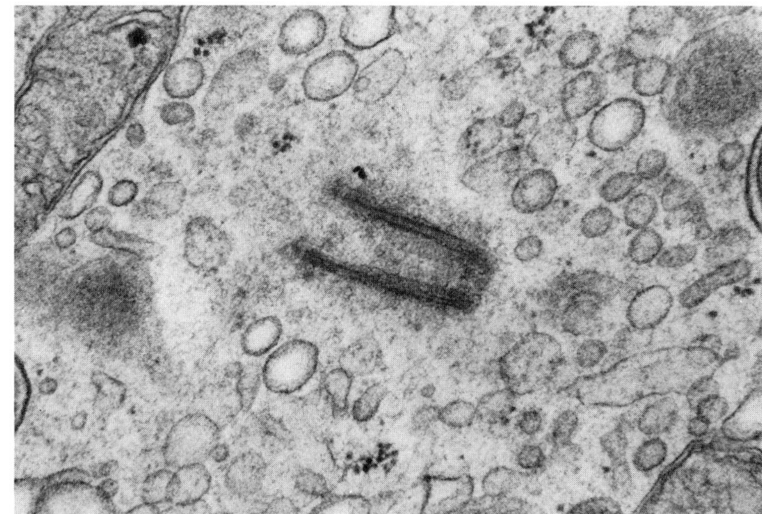

FIGURE 1-23 Longitudinal section of a centriole.

that are circularly arranged. Figure 1-23 is an EM of a centriole in longitudinal section. The centriole is identical in appearance to the basal body of a cilium, as shown in Fig. 1-24. Microtubules extending from the basal body form the ciliary shaft (arrows, Fig. 1-24). When cross-sectioned, the ciliary shaft shows the same ultrastructural arrangement of the microtubules found in the cross-sectioned centriole except that an additional pair of tubules is located in the center of nine pairs of peripheral tubules. Figure 1-25 shows three ciliary shafts so sectioned.

Centrioles play a key role in the process of cellular division, particularly in the organization of spindle fibers. The EM has shown that spindle fibers are also made up of microtubules, which are continuous with the tubules that make up centrioles. It is believed that the centrioles and the associated microtubules provide the basic mechanism for movement of chromosomes away from the equatorial plate during anaphase. Likewise, the centrioles and microtubules form contractile components of motile cilia or flagella. Figure 1-26 is an EM of a portion of motile

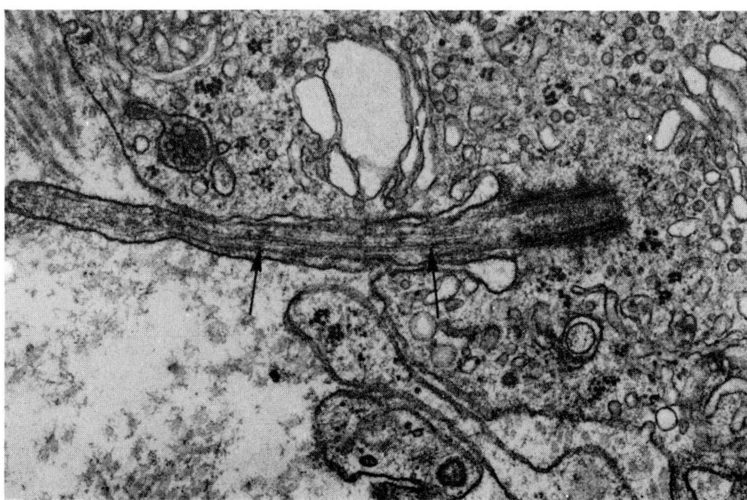

FIGURE 1-24 Longitudinal section of a cilium showing the shaft and the basal body. Microtubules in the shaft are indicated by arrows.

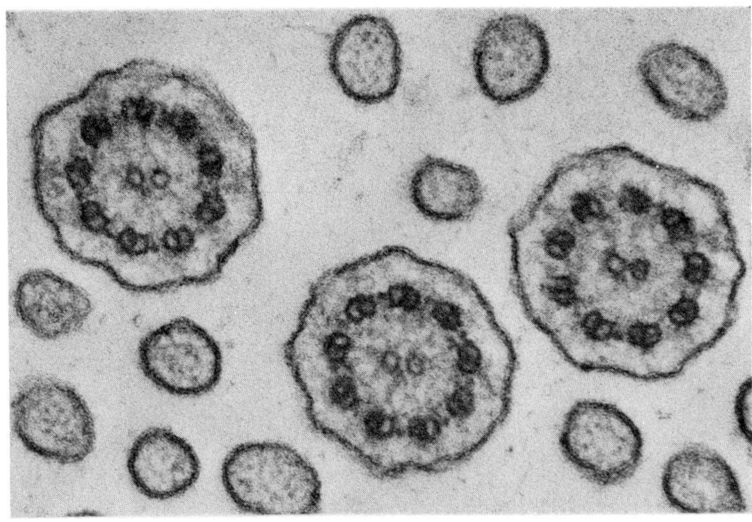

FIGURE 1-25 Cross section of ciliary shafts.

cilia of the respiratory passage. Each one of the ciliary shafts contains axial microtubules that are attached to basal bodies which resemble centrioles. Cilia similar to these are shown in Fig. 1-27, which gives cross-sectional views of cilia and basal bodies at various levels.

Cell membranes

If a portion of cellular membrane is sectioned at right angles to its surface and magnified in the electron microscope beyond 50,000 times or so, its ultrastructure becomes evident (Fig. 1-5). Under these conditions, membranes appear as three-layered structures: two dark layers sandwiching a lucent middle layer. This structure reflects the regular arrangement of macromolecules that make up all cellular membranes. An EM of a plasma membrane of a developing red corpuscle is shown in Fig. 1-28. Because of the irregularity of the cell surface, the trilaminar membrane structure is visible only at portions where the thickness of the membrane is perpendicularly

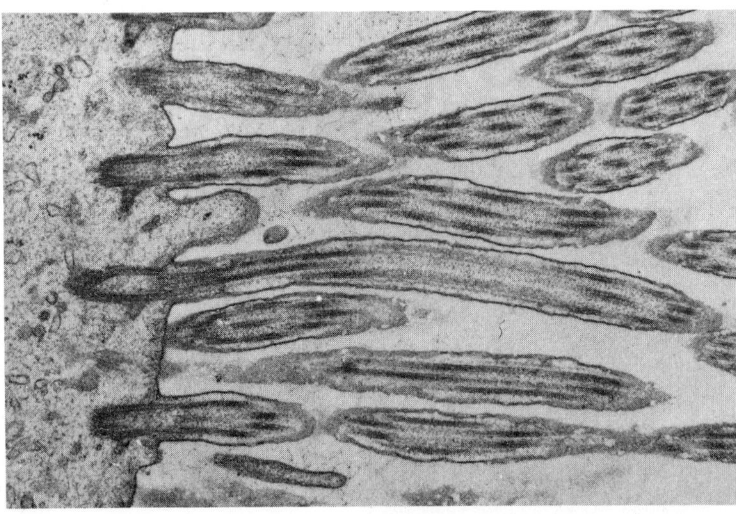

FIGURE 1-26 Apical portion of an epithelial cell from the respiratory passages.

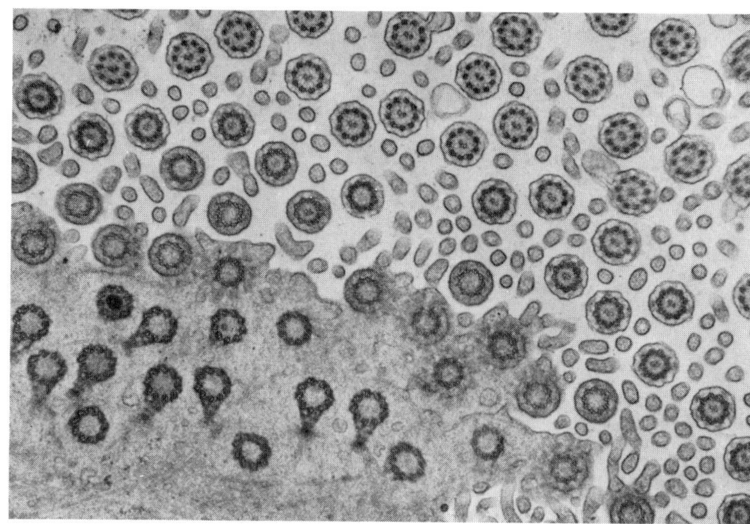

FIGURE 1-27 Cross-sectional views of cilia and basal bodies at various levels.

sectioned (arrows). A few intracytoplasmic vesicles also demonstrate the trilaminar appearance. Note the thickness of the membranes relative to the ribosomes. The two dense lines represent the proteins and the hydrophilic portions of lipid molecules. The inner lucent layer represents the two hydrophobic ends of a phospholipid bilayer facing each other. Thus, all membranes are composed of a lipid bilayer sandwiched by two protein layers. Although there are more current views that better define the molecular architecture of biological membranes in functional terms, since the late fifties the term *unit membrane* has been used to designate the trilaminar appearance of most biological membranes. In addition to the lipoprotein layers, a carbohydrate coat is attached to the outer surface of membranes. This coat influences the recognition of various substances in the immediate environment as well as the permeability characteristics of individual membranes. The carbohydrate coat at the outer surface of the cell membrane appears

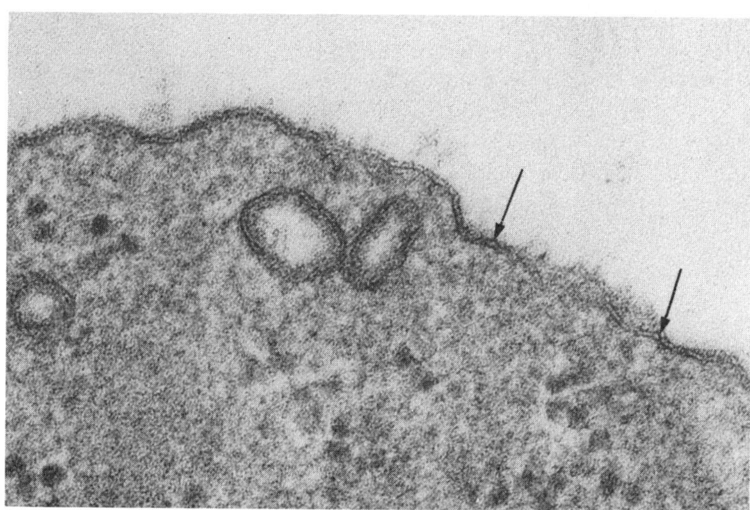

FIGURE 1-28 Electron micrograph demonstrating the trilaminar structure of cell membranes in areas where it is perpendicularly sectioned (arrows).

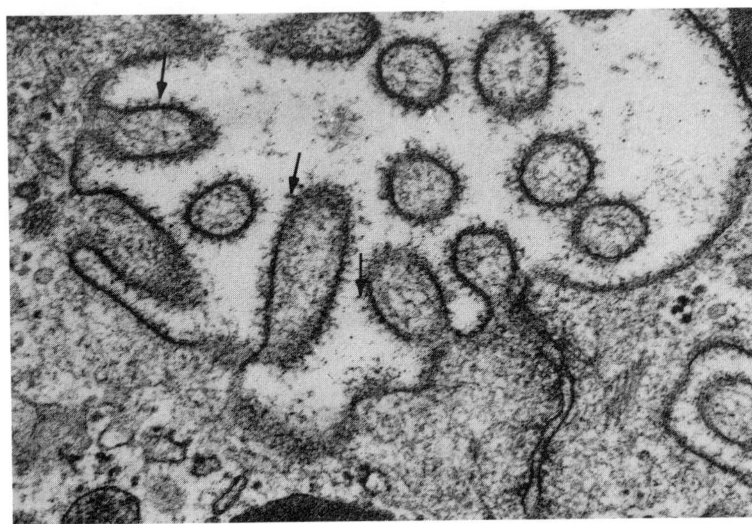

FIGURE 1-29 Portions of cell depicting the cell coat, glycocalyx (arrows).

as a fuzzy substance of irregular thickness which varies depending on the type of cells. Figure 1-29 depicts this; a fuzzy glycoprotein substance appears along the surface of irregular projections in a pancreatic cell. This fuzzy coat, indicated by the arrows, is called *glycocalyx*.

The ability of cells to recognize what is foreign following transplantation of organs or homografts resides in the molecular makeup of the plasma membrane. This biologic quality of plasma membrane is also essential to body defense against invading organisms and noxious substances. The significance of this recognition mechanism will be dealt with in some detail in our discussion of the lymphoid organs and immunity.

In addition to serving to contain the cytoplasmic constituents and segregate cytoplasmic compartments, membranes also create a semipermeable barrier between the environment and the cytosol. This allows differential uptake by the cell and diffusion of substances across the plasma membrane. The nature of transmembrane transport of matters depends on the size, chemical composition, and electric charge of the substances. While diffusion of small molecules is not an energy-dependent phenomenon, energy is expended when a cell participates in an active transport, which occurs when there is a dynamic concentration of substances against higher concentration gradients.

Nucleus

The nucleus is enclosed by two layers of trilaminar membranes which are collectively called the *nuclear envelope* (Fig. 1-30). The outer and inner membranes are periodically penetrated by small openings, *nuclear pores*, where the two membranes become confluent. In each nucleus, nuclear pores can range from several hundred to thousands in number. These pores are covered by a thin diaphragm, but otherwise serve as the sites of linkage between the nucleoplasm and cytoplasm. Therefore, they selectively allow some large molecules, such as RNA, to pass from the nucleus to the cytosol with ease, while excluding other smaller components of the cytoplasm from those in the nucleoplasm or vice versa. For example, the concentration of sodium and potassium ions is different in the cytoplasm and the nucleoplasm, and this difference in concentration gradients is rigidly maintained despite the uninhibited translocation of much larger molecules. Figure 1-31*a* is a portion of the nuclear envelope of a lymphocyte in which a sagittal view of

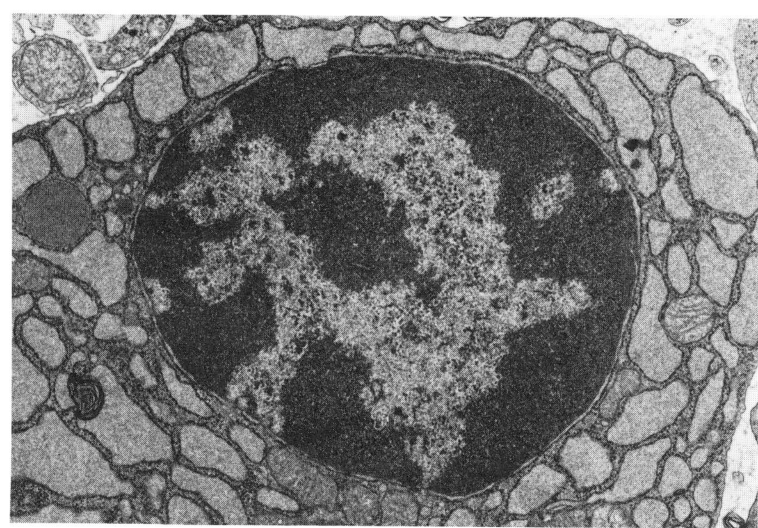

FIGURE 1-30 The nuclear envelope.

the pore is seen (arrow). The pore is closed by a cloudy nonmembranous *diaphragm.* Within the perinuclear cisterna between the two nuclear membranes of the envelope is a small ringlike extension of the diaphragm. As shown at a higher magnification in Fig. 1-31*b*, the ringlike flange resembles a pair of dumbbells (arrows).

As indicated earlier, the dense nuclear chromatin, the *heterochromatin,* is usually seen as peripheral dark regions in the electron microscope. The peripheral heterochromatin is interrupted by *intranuclear channels* that extend from the pore to the inner nucleoplasm (Fig. 1-30). It is presumed to represent supercoiled strands of DNA and associated nuclear proteins such as histones and other nonbasic proteins. For this reason, a nucleus that contains more heterochromatin may be regarded as relatively quiescent. That the

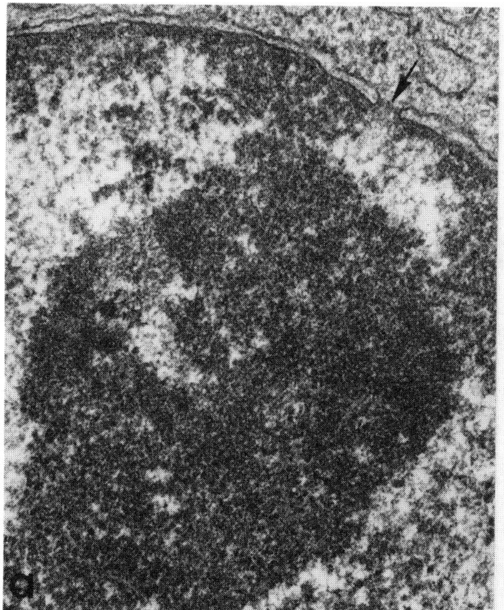

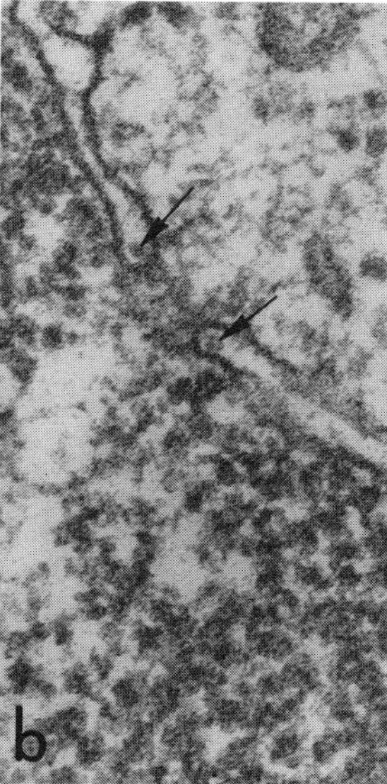

FIGURE 1-31*a* Nuclear pore (arrow).

FIGURE 1-31*b* Electron micrograph of a nuclear pore taken at a higher magnification, depicting the structure of the diaphragm. The ringlike flange resembles a pair of dumbbells (arrows).

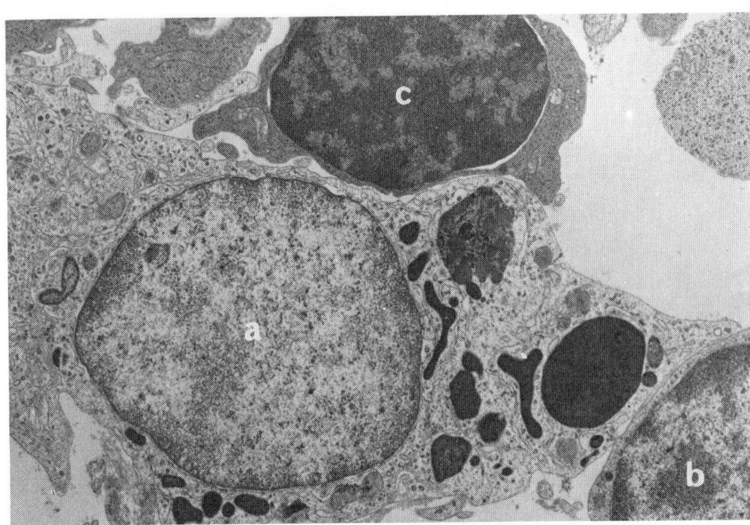

FIGURE 1-32 Electron micrograph of bone marrow cells in varying stages of differentiation and activity. Notice differences in nuclear density.

amount of heterochromatin varies among cells of different functional state can be seen in Fig. 1-32, which is an EM of bone marrow in which most active (a), moderately active (b), and least active (c) cells can be identified based on the relative amounts of heterochromatin. Fine strands seen in the *euchromatin* region denote the area where chromosomal DNA is presumed to be uncoiled and active. This is where either mRNA is being transcribed from DNA or the replication of DNA is taking place. In EM the nucleolus appears as an eccentric, dense structure which is associated with a region of DNA called the *nucleolar organizer.* It can be divided into the following three basic components: (1) the *nucleolonema,* or the fine fibrous portion of high electron density; (2) the *pars amorpha,* or the amorphous, less dense region; and (3) the granular region. The small granules associated with the nucleolus are presumed to contain precursors of rRNA. The nucleolus is the site of synthesis of rRNA which exits through nuclear pores. The EM in Fig. 1-33*a*

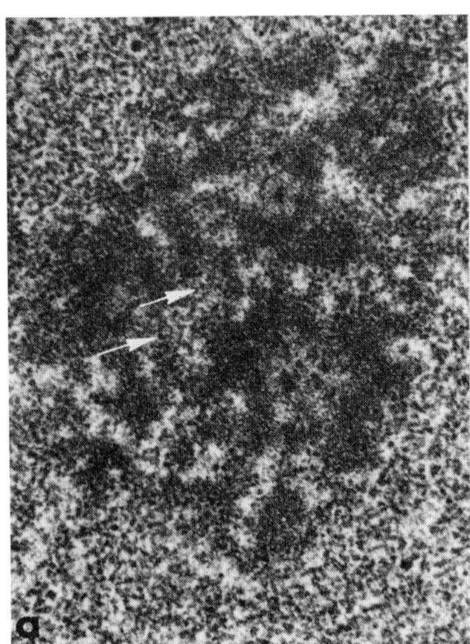

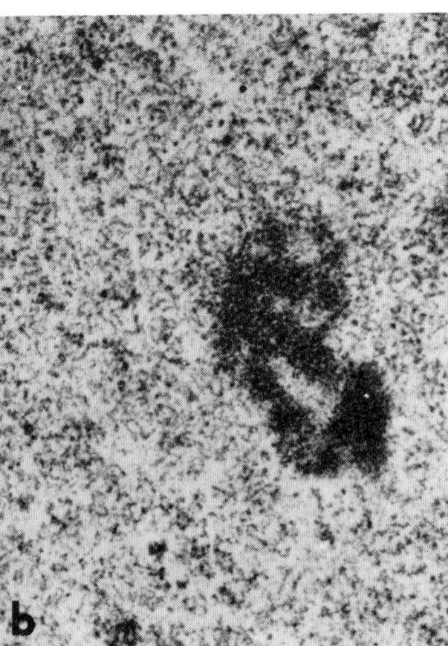

FIGURE 1-33*a* Portion of the nucleus of a pancreatic acinar cell containing a well-developed nucleolus. Ribosome-like granules (arrows) are present between the nucleolonema and the pars amorpha.

FIGURE 1-33*b* Portions of the nucleus from an inactive cell (lymphocyte) containing a poorly developed nucleolus.

shows a portion of the nucleus of a pancreatic acinar cell containing the nucleolus. Note the lacy nucleolonema, which is associated with the less dense pars amorpha. Ribosome-like granules are present in between the above elements (arrows) as well as the periphery of the nucleolus. Figure 1-33*b* is an example of the nucleolus from an inactive lymphocyte. The nucleolonema appears as several small clumps, and the pars amorpha is segregated and occupies the central portion.

CELL DIVISION

The sequence of events involved in the division of somatic cells is called *mitosis.* The stationary period between mitotic divisions is called *interphase.* In dividing cells where continued cell renewal is required, several stages of the cell cycle can be clearly characterized on the basis of biochemical events underlying different phases of a cell's preparation toward the next mitotic division (Fig. 1-34). During the immediate postmitotic period, the centriole in the daughter cell is replicated. This is accompanied by the synthesis of RNA and proteins that are necessary prior to replication of DNA. This period *G phase* has been called the *first-gap phase* (G_1) because of the earlier difficulty in understanding the biochemical events taking place during this phase. It is now known that most of the postmitotic cells which seldom divide (viz nerve and muscle cells) stay in G_1 phase. The second period of interphase nuclei which undergo division is called *S phase;* during this phase the synthesis of DNA, facilitated by the newly produced RNA and proteins during G_1 period, takes place. In addition, nuclei of the nuclear regulatory proteins such as histones are produced during the S period. The last segment of the interphase period, i.e., the time between DNA synthesis and the next mitosis, is called the *second-gap phase,* or *G_2 phase,* which is characterized by the synthesis of RNA and proteins necessary for cytokinesis.

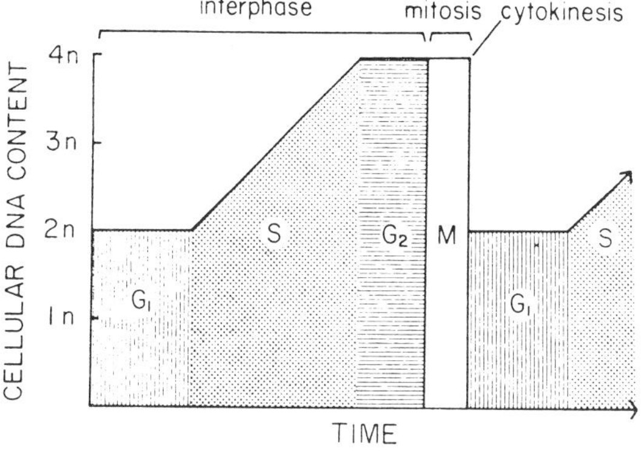

FIGURE 1-34 Diagram illustrating the various phases of the cell cycle.

Since it is believed that most students reading this book have earlier studied the basic process of mitotic division, only a brief review of the subject is given here. The beginning of mitosis, *prophase,* consists of the appearance of heavy chromosome strands and their longitudinal division, the movement of centrioles towards the opposite poles, and disappearance of the nuclear envelope (Fig. 1-35*a* and *b*). During *metaphase* the two sets of spindle "fibers," which are made up of microtubules produced by the centrioles, attach to the *centromeres* or the midpoint of each chromosome. At this time, all chromosomes line up along the equatorial plane with the centromeres making a circle (Fig. 1-35*c*). During *anaphase,* the duplication of centromeres occurs and the two sets of chromatids are drawn toward the opposite poles (Fig. 1-35*d* and *e*).

Figure 1-36 is an electron micrograph of a dividing plasmablast in which an obliquely sectioned chromosome set is visualized. A large number of RER, ribosomes, and mitochondria are present throughout the cyto-

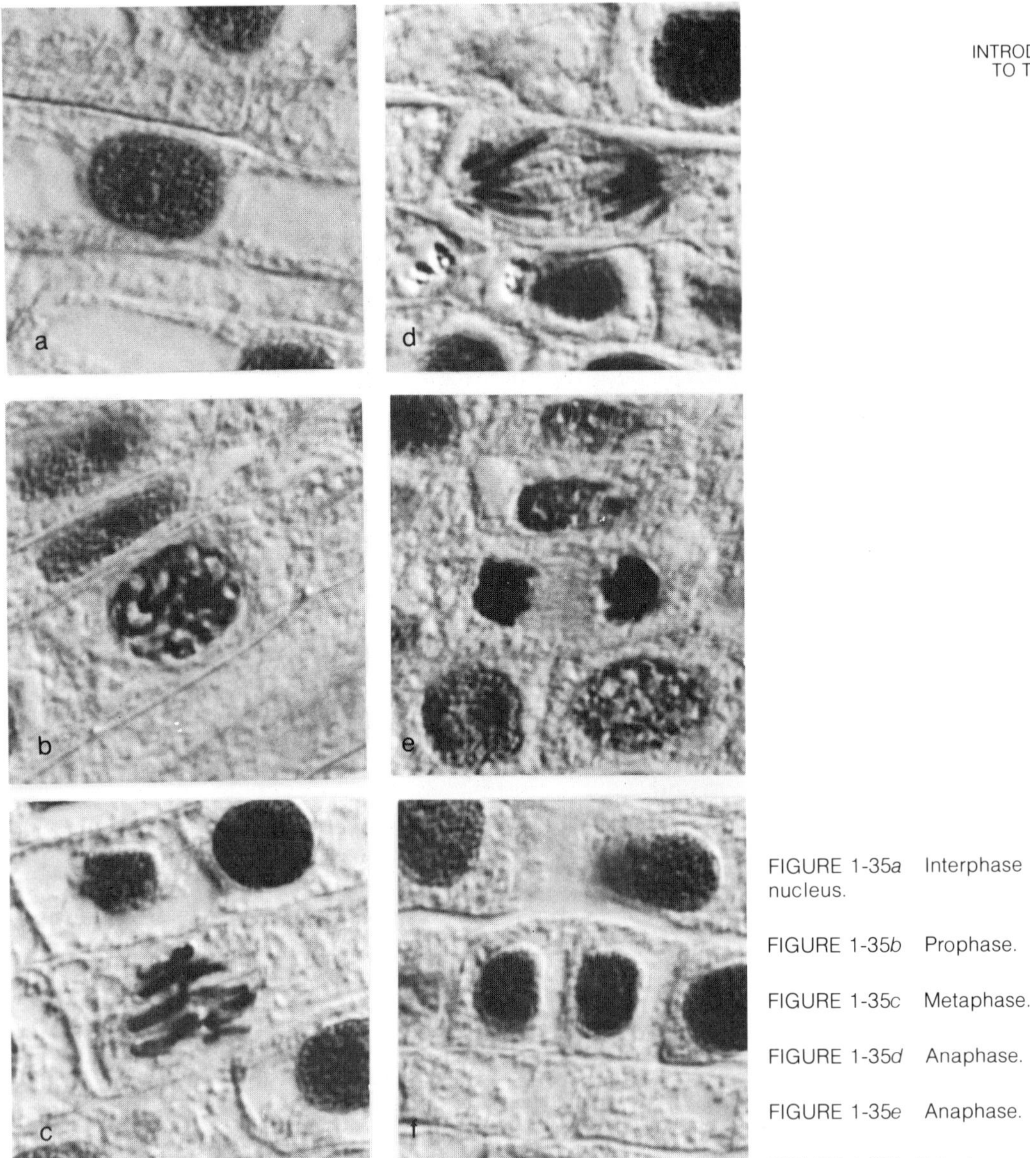

FIGURE 1-35*a* Interphase nucleus.

FIGURE 1-35*b* Prophase.

FIGURE 1-35*c* Metaphase.

FIGURE 1-35*d* Anaphase.

FIGURE 1-35*e* Anaphase.

FIGURE 1-35*f* Telophase.

plasm. *Telophase* is the time when a reversal of prophase characteristics takes place; namely, reappearance of nuclear membranes surrounding the two daughter nuclei, disappearance of spindle fibers, and reappearance of nucleoli. This is followed by the division of cytoplasm (cytokinesis), resulting in two complete daughter cells (Fig. 1-35*f*).

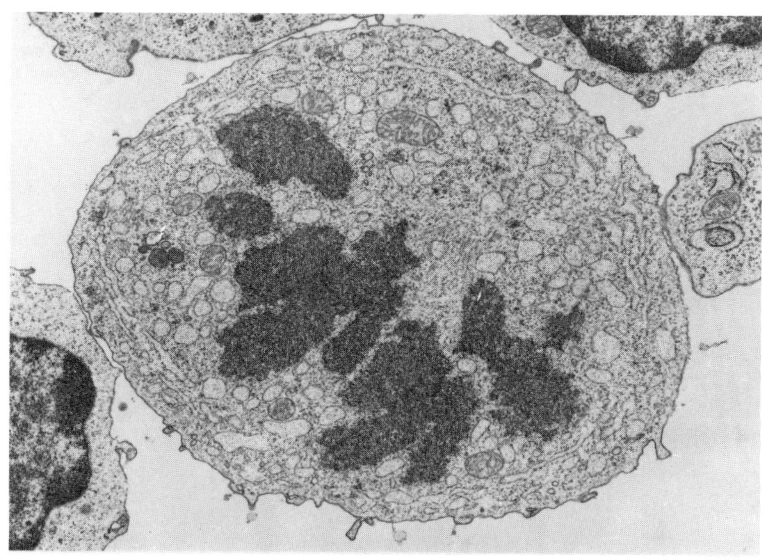

FIGURE 1-36 Electron micrograph of a dividing cell (plasmablast).

REVIEW SECTION

1. Note the two thin, stacked structures indicated by arrows in Fig. 1-37. They are associated with large vacuoles and small vesicles called __________. Collectively, this organelle is called a __________. The large rounded structures labeled S_1 and S_2 are the products of this organelle. They are respectively called __________ and __________.

The stacked lamellae of this structure along with the transport vesicles should tell you that this organelle is a Golgi apparatus. The light and dark granules (S_1 and S_2) in this EM are the products of the Golgi apparatus–condensing vacuole (S_1) and secretory granule (S_2).

2. Figure 1-38 shows a portion of a cell which is typical to a __________-synthesizing cell, as evi-

Numerous polysomes are associated with the ER throughout much of the EM, which shows a portion

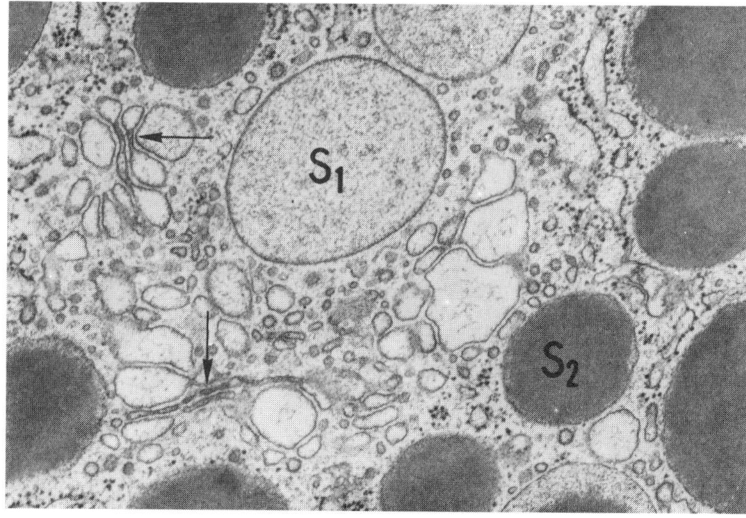

FIGURE 1-37

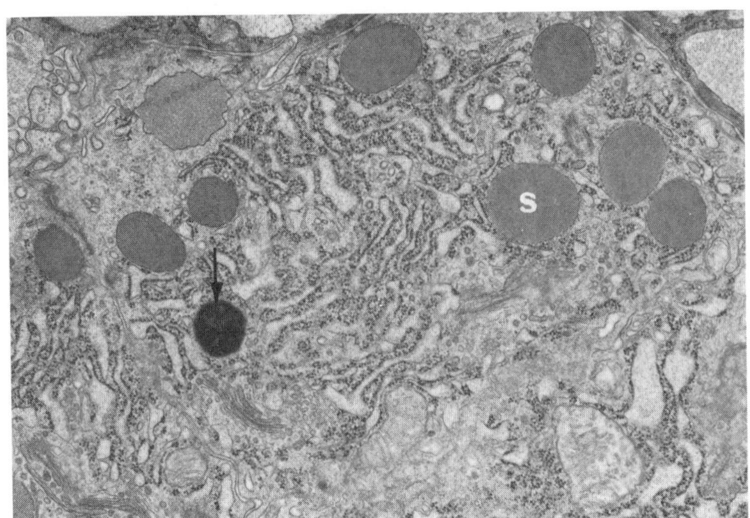

FIGURE 1-38

denced by the many __________ studded on the ER. The dense body indicated by the arrow is called __________.

of the cytoplasm of a pancreatic acinar cell secreting proteinaceous enzymes. The dark, rounded lysosome is thought to contain many acid hydrolases. The lysosome and the Golgi complex are produced by the RER, as is the secretory granule (s).

3. The two panels of EM appearing in Fig. 1-39*a* and *b* represent two different stages of developing red blood corpuscles. Which of the two is from a less developed cell? __________. On what basis did you arrive at this conclusion?

Developing red blood corpuscles produce a large amount of hemoglobin, which darkens the ground cytoplasm as the hemoglobin accumulates in the cytosol. Accordingly, many ribosomes (polysomes) are present during the early phase of their development (Fig. 1-39*a*) and fewer ribosomes are seen as the cell becomes more mature (Fig. 1-39*b*).

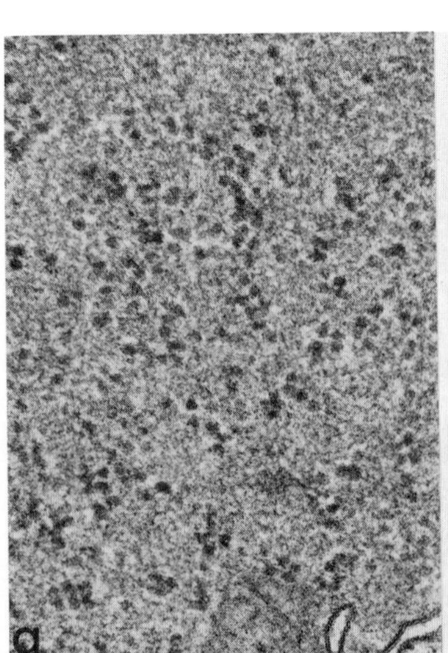

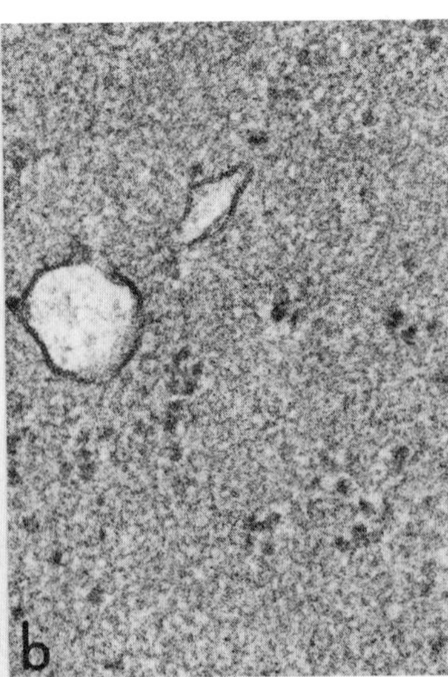

FIGURE 1-39*a*

FIGURE 1-39*b*

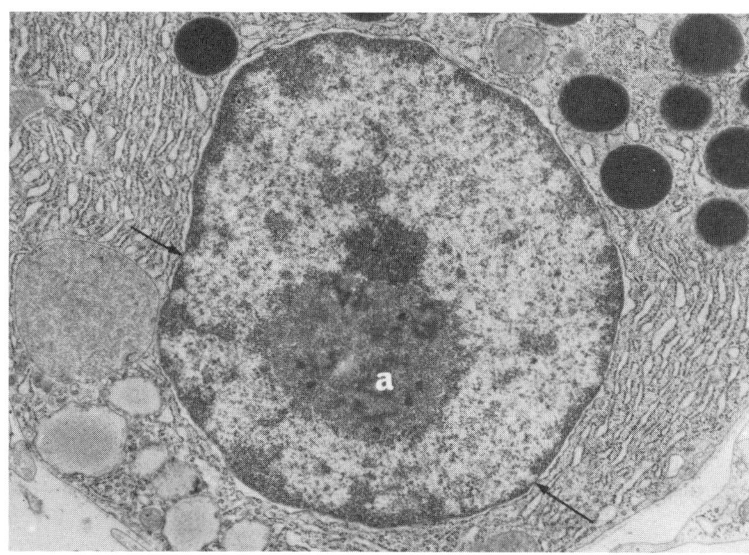

FIGURE 1-40

4. Figure 1-40 shows a nucleus. Around the outside of this nucleus is an abundance of RER. It is seen that the nucleus is surrounded by two membranes which are collectively called __________. The arrows in the EM point to breaks in the nuclear membranes. These breaks are called __________ and are traversed by __________. At the periphery of the nucleoplasm are some dense areas of __________. The dark, finely fibrillar region (a) which is different from chromatin is the __________, which is the site where the synthesis of all __________ takes place.

The arrows in this EM point to discontinuities in the nuclear envelope called *nuclear pores*. The clear areas in the peripherally located heterochromatin of the nucleus, leading to each pore, represent the intranuclear channels which allow for exchange of materials between the cytoplasm and the nucleus. The dark body (a) in the nucleus is the nucleolus, which is composed of the pars amorpha, nucleolonema, and ribosome-like granules. The nucleolus is the exclusive site of rRNA synthesis.

FIGURE 1-41

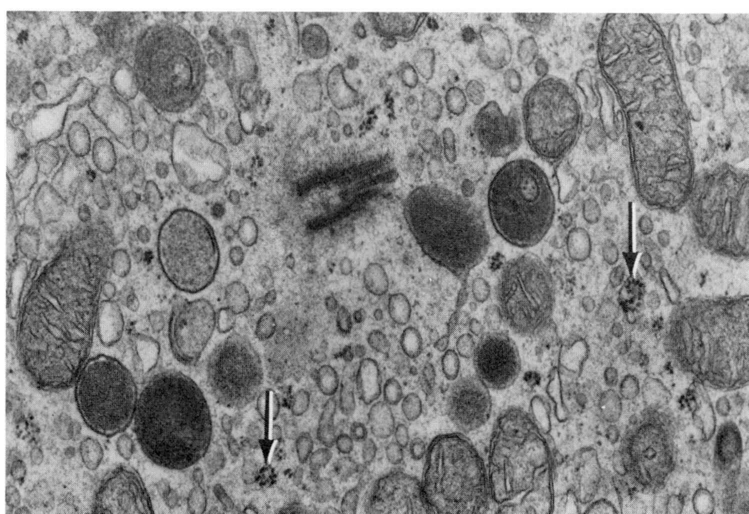

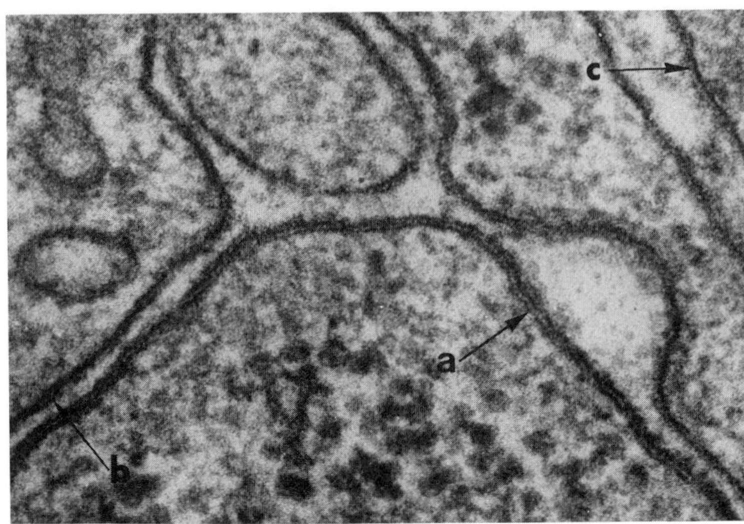

FIGURE 1-42

5. A portion of cytoplasm from a steroid secretory cell is present in this EM (Fig. 1-41). Name as many cytoplasmic organelles and inclusions as you can. You should be able to identify at least five different structures.

The following organelles are present in this EM: (*a*) mitochondria, (*b*) polysomes, (*c*) a few RER profiles (arrows), (*d*) smooth vesicles and SER, (*e*) lysosomes, (*f*) centrioles, and (*g*) portions of lipid droplets.

6. Figure 1-42 is an EM depicting portions of several different cell types at a high magnification. The plasma membrane at (*a*) shows a trilaminar appearance. What chemical components make up the two dark lines? __________ and __________. The plasma membranes at (*b*) and (*c*) are apparently thinner than (*a*), and the trilaminar appearance is difficult to recognize. What could this difference mean?

All plasma membranes tend to show a trilaminar profile which is due to the basic molecular arrangement of biologic membranes—two protein layers sandwiching a lipid bilayer. The electron-dense lines represent the protein layers plus the hydrophilic ends of the lipid bilayer, while the electron-lucent middle line represents the hydrophobic ends of phospholipid molecules. Because of differences in the composition of phospholipids and proteins, plasma membranes of different cells often show different thicknesses.

BIBLIOGRAPHY

Barka, T., and P. J. Anderson: *Histochemistry-Theory, Practice and Bibliography,* Hoeber-Harper, New York, 1963.

Bloom, W., and D. W. Fawcett: *A Textbook of Histology,* 10th ed., Saunders, Philadelphia, 1975.

Copenhaver, W. M., R. P. Bunge, and M. B. Bunge: *Bailey's Textbook of Histology,* 6th ed., Williams & Wilkins, Baltimore, 1971.

Fawcett, D. W.: *The Cell: Its Organelles and Inclusions,* Saunders, Philadelphia, 1966.

Gray, P. (ed.): *The Encyclopedia of Microscopy and Microtechnique,* Van Nostrand Reinhold, New York, 1973.

Ham, A. W.: *Histology,* 6th ed., Lippincott, Philadelphia, 1969.

Hayat, M. A.: *Principles and Techniques of Electron Microscopy: Biological Application,* vol. 1, Van Nostrand Reinhold, New York, 1970.

Leeson, C. R., and T. S. Leeson: *Histology,* 3d ed., Saunders, Philadelphia, 1976.

Lillie, R. D.: *Histopathologic Technique and Practical Histochemistry,* 3d ed., McGraw-Hill, New York, 1965.

Rhodin, J. A. G.: *Histology. A Text and Atlas,* Oxford University Press, New York, 1974.

Weiss, L., and R. O. Greep: *Histology,* 4th ed., McGraw-Hill, New York, 1977.

ADDITIONAL READINGS

Abelson, H. T., and G. H. Smith: "Nuclear Pores: The Pore-Annulus Relationship in Thin Sections," *J. Ultrastruct. Res.* **30:**558–588 (1970).

Allison, A. C.: "The Role of Microfilaments and Microtubules in Cell Movement, Endocytosis and Exocytosis in Locomotion of Tissue Cells," in Ciba Foundation Symposium, *Locomotion of Tissue Cells,* Elsevier, Amsterdam, 1973, p. 109.

Barr, M. L.: "The Significance of the Sex Chromatin," *Int. Rev. Cytol.* **19:**35–95 (1966).

Beams, H. W., and R. G. Kessel: "The Golgi Apparatus: Structure and Function," *Int. Rev. Cytol.* **23:**209–276 (1968).

Behnke, O., and A. Forer: "Evidence for Four Classes of Microtubules in Individual Cells," *J. Cell Sci.* **2:**169–192 (1967).

Bennett, G., C. P. Leblond, and A. Haddad: "Migration of Glycoprotein from the Golgi Apparatus to the Surface of Various Cell Types as Shown by Radioautography after Labeled Fucose Injection into Rats," *J. Cell Biol.* **60:**285–284 (1974).

Buck, F.: "The Variable Condition of Euchromatin and Heterochromatin," *Int. Rev. Cytol.* **45:**25–64 (1976).

Busch, H.: *Histones and Other Nuclear Proteins,* Academic Press, New York, 1965.

Caro, L. G., and G. E. Palade: "Protein Synthesis, Storage and Discharge in the Pancreatic Exocrine Cell, an Autoradiographic Study," *J. Cell Biol.* **20:** 473–495 (1964).

Carr, I.: "The Fine Structure of Microfibrils and Microtubules in Macrophages and Other Lymphoreticular Cells in Relationship to Cytoplasmic Movement," *J. Anat.* **112:**383–389 (1972).

Dallner, G., P. Siekevitz, and G. E. Palade: "Biogenesis of Endoplasmic Reticulum Membranes," *J. Cell Biol.* **30:**73–96 (1966).

de Bruijn, W. C.: "Glycogen: Its Chemistry and Morphologic Appearance in the Electron Microscope," *J. Ultrastruct. Res.* **42:**29–50 (1973).

DePierre, J. W., and M. J. Karnovsky: "Plasma Membranes of Mammalian Cells," *J. Cell Biol.* **56:** 275–303 (1973).

Gosh, S.: "The Nucleolar Structure," *Int. Rev. Cytol.* **44:**1–28 (1976).

Holzman, E.: *Lysosomes: A Survey,* Cell Biology Monograph, vol. 3, Springer-Verlag, New York, 1976.

Jacob, F., and J. Monod: "Genetic Regulatory Mechanisms in the Synthesis of Proteins," *J. Mol. Biol.* **3:**318–356 (1961).

Lehninger, A. L.: *The Mitochondrion,* W. A. Benjamin, Inc., New York, 1964.

Napolitano, L.: "The Differentiation of White Adipose Tissue," *J. Cell Biol.* **18:**663–679 (1963).

Sorokin, S. P.: "Reconstruction of Centriole Formation and Ciliogenesis in Mammalian Lungs," *J. Cell Sci.* **3:**207–230 (1968).

Spiro, A. S., and L. P. Garidova: *The Ribosome,* Springer-Verlag, New York, 1969.

Tandler, B., and C. L. Hoppel: *Mitochondria,* Academic Press, New York, 1972.

Watson, J. D.: *Molecular Biology of the Gene,* 3d ed., W. A. Benjamin, Inc., Menlo Park, Calif., 1976.

Wolfe, J.: "Basal Body Fine Structure," *Adv. Cell Mol. Biol.* **2:**151–192 (1972).

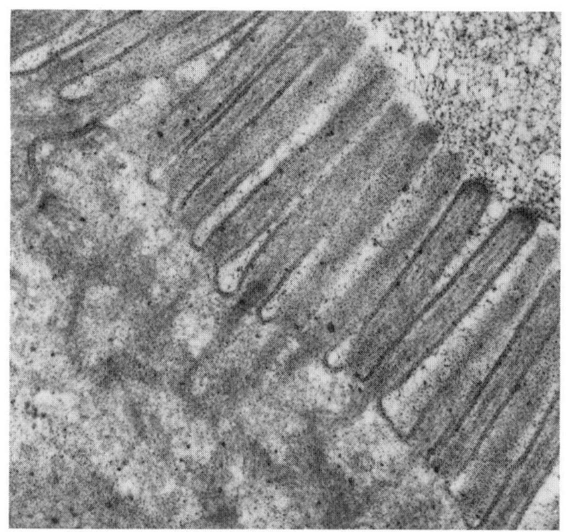

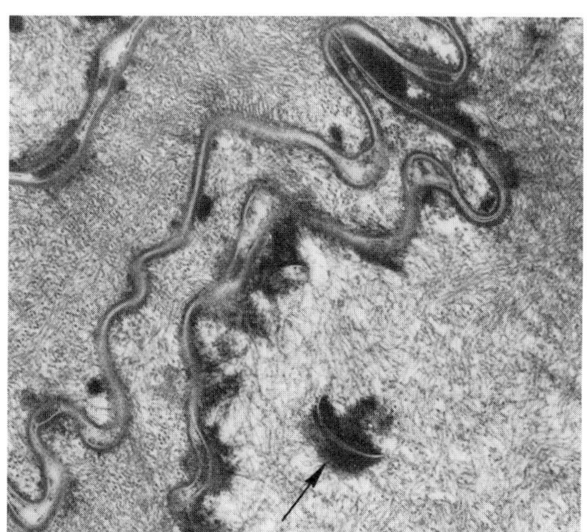

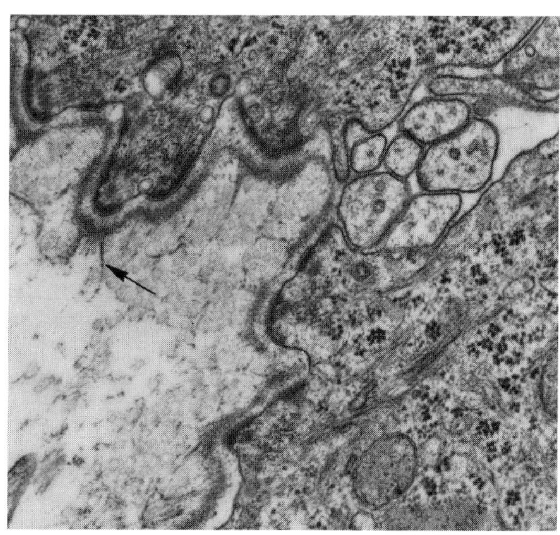

2 EPITHELIUM

OBJECTIVES

Upon completion of this chapter, the student will be able to:

1 Identify the following as characteristics of epithelium:
(a) a tissue of closely adherent cells with sparse intercellular substance (b) arranged in single or multiple layers (c) covering and protecting internal and external surfaces and maintaining an aqueous environment for body cells

2 List the general structural and functional characteristics of epithelium including the following:
(a) location in body (b) intercellular relations (c) relation to underlying connective tissue

3 Identify the structure and state the role of the following specialized attachments between cells:
(a) tight junction (zonula occludens) (b) gap junction (c) zonula adherens (d) desmosomes (macula adherens) (e) tonofilaments (tonofibrils)

4 Identify the structure, function, and at least one location of the following free surface modifications of epithelial cells:
(a) keratinization (b) microvilli (c) cilia (d) stereocilia

5 Identify the structural and physical characteristics and mode of attachment in the major types of epithelium and relate them to the function of the cell.

6 Define, locate, and give the function of connective tissue in relation to epithelium including its role in nutrition.

One theory maintains that human beings began as primitive organisms living under water. Today, reminders of that time are still evident when one observes the phylogenetic relationship of various forms of life. Within the individual organism, almost all the cells of the body live in an aqueous environment. The epithelium on the outer and inner surfaces of the body allows this environment to be maintained.

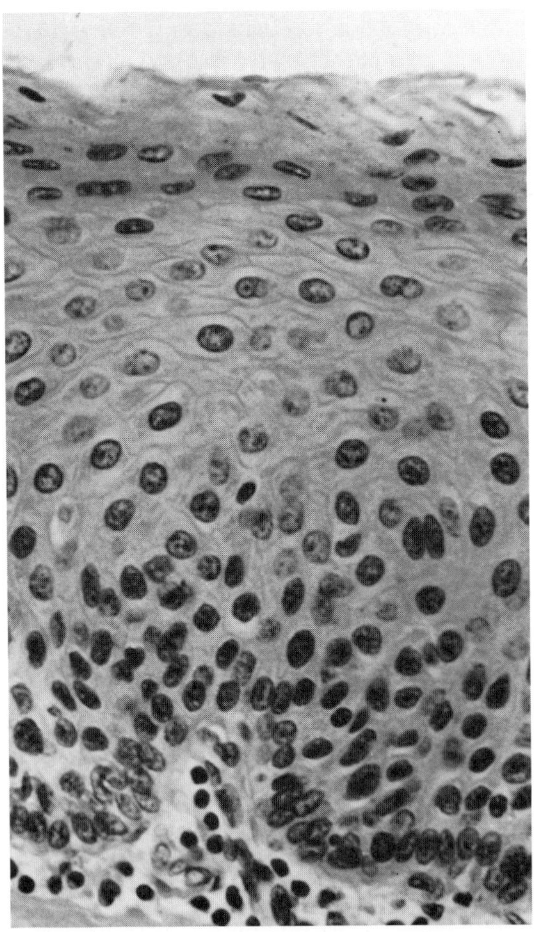

FUNCTIONS

To maintain the aqueous environment for cells, epithelial tissues must be made up of a continuous layer(s) of cells that are attached to each other. They must be leakproof to contain the body fluid.

Protection. Epithelial cells are also responsible for protecting body surfaces from abrasion and other physical abuses. In a single layer, the epithelium does a poor job of protecting; but in multiple or stratified layers, as seen in Fig. 2-1, this protective function is adequately achieved. The cells at the base of this stratified epithelium, stained dark, cover the underlying connective tissue (CT). In areas of the body where there is a great deal of pressure and abrasion, like the sole of the foot, there are more layers of epithelial cells than in a less abused area, such as the eyelid. We will describe the details of the intracellular structures and intercellular attachment complexes that function in maintaining the physical integrity of stratified epithelium later in the chapter.

Lubrication. Epithelial cells aid in lubricating the body surfaces covered by other epithelial cells, using a substance which is a product of different glandular cells. The sweat gland is one example of an invagination of an epithelial element that has differentiated to lubricate the outer surface of the skin. Figure 2-2 shows the spiraling duct of the sweat gland (arrows) as it passes through the

FIGURE 2-1 Stratified nonkeratinized squamous epithelium. The interface between the epithelium and the underlying connective tissue is irregular. Projections of the connective tissue, called *papillae*, allow for a firmer attachment of the epithelium to the connective tissue.

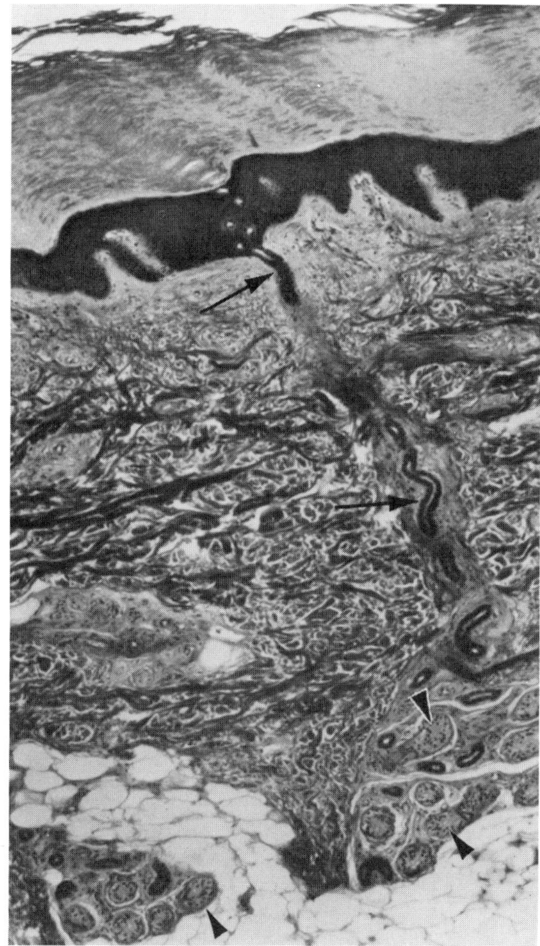

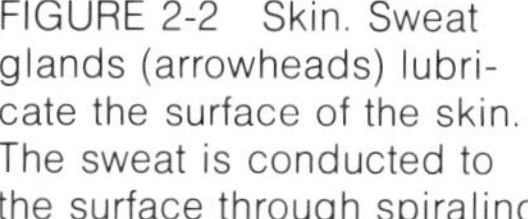
FIGURE 2-2 Skin. Sweat glands (arrowheads) lubricate the surface of the skin. The sweat is conducted to the surface through spiraling ducts (arrows).

many layers of the epithelium from the underlying connective tissue where the glands are located (arrowheads).

Figure 2-3 is a cross section through the base of the glands, seen at a higher magnification. The three circularly arranged groups of darker cells with lumens are sections through the coiled portions of the duct (arrows). The lighter circular areas, two of which appear at the bottom of this micrograph, are made up of a single layer of large cells. These are the secretory portions of the gland. Since the secretory portions of the gland are also coiled, they appear as randomly sectioned tubules mingled with smaller darker profiles of the ducts in cross section.

Another example of a lubricating gland is the sebaceous gland. Found around hair follicles throughout the skin, sebaceous glands help keep the skin "conditioned" and supple. In Fig. 2-4, a cluster of light-stained cells of a sebaceous gland are attached to the hair shaft. The gland's oily product is discharged into the space surrounding the hair through which it reaches the surface. Both sebaceous and sweat glands are found in the stratified squamous epithelia of skin.

Surface lubrication also occurs in single-layered areas of epithelium. In the digestive and respiratory tracts, goblet cells secrete a mucous substance that lubricates the inner surface of the tube. This substance helps food move through the digestive tract and protects the epithelium of the organs against abrasion and against chemical and mechanical damage. The pale-stained goblet cells in Fig. 2-5 are seen interspersed among the regular columnar epithelial cells of the trachea. The mucous products of the goblet cells are secreted directly into its lumen (arrows).

Absorption. The digestive tract epithelium is specialized for food absorption. The epithelial cells are columnar and arranged in a single layer. Note the arrangement of tall cells in Fig. 2-6. The dark border (arrows) of epithelial surfaces along the intestinal lumen marks the location of specialized cell processes that absorb nutrients. The content of a single goblet cell (g) has been stained extremely dark with a special stain for mucous materials.

Similar but lower epithelial cells found in the kidney tubules are specialized for resorption of useful substances from the provisional urine (Fig. 2-7). Note the different heights of epithelial cells that line the various internal surfaces. All epithelial tissues rest on a basement membrane which may be

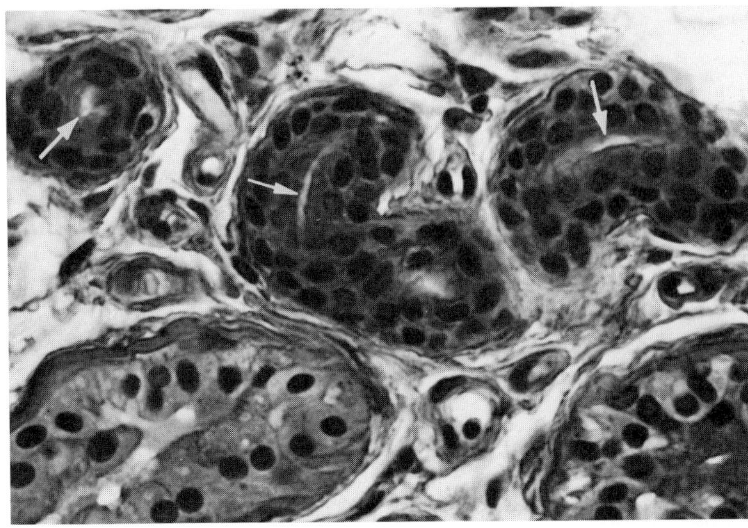

FIGURE 2-3 Cross section of the base portion of a sweat gland. Three portions of the coiled duct (arrows) are seen in the upper part of the micrograph. The lighter-stained circular structures at the bottom of the micrograph which are composed of a single layer of cells are the secretory portions of the gland.

thick as in the respiratory epithelium (arrowheads, Fig. 2-5) or thin as in kidney tubules (arrows, Fig. 2-7).

Gas and Nutrient Exchanges. Extremely thin epithelial cells are found in places where the epithelium separates different phases of body fluid. For example, the epithelium lining blood vessels, called *endothelium,* which separates blood from tissue fluid, consists of a single layer of extremely thin or attenuated cells. These cells (arrows in Fig. 2-8) allow for gas and nutrient exchange to occur.

The EM in Fig. 2-9 shows two flat endothelial cells with the oval nucleus in one of them. Note also the thin cloudy basal lamina (basement membrane of LM) surrounding the endothelial lining (arrows). Similar attenuated endothelial cells occur in many other blood vessels. An example of an extremely attenuated endothelium may be found in air sacs of the lung where oxygen-

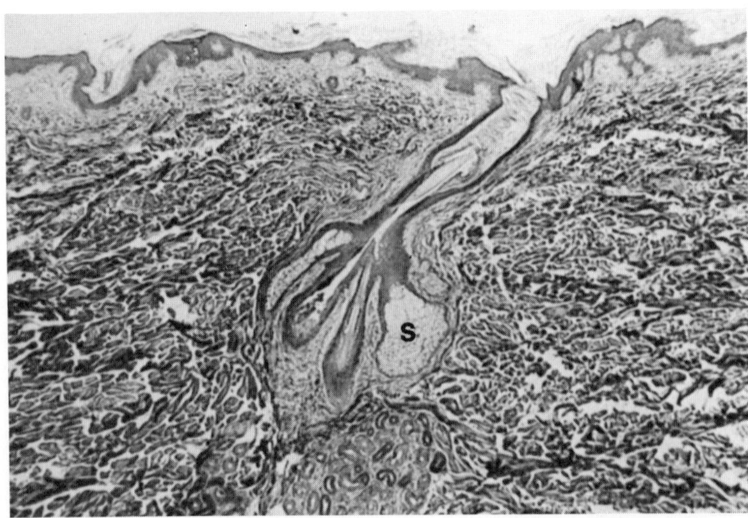

FIGURE 2-4 Sebaceous gland(s) attached to hair shaft in skin.

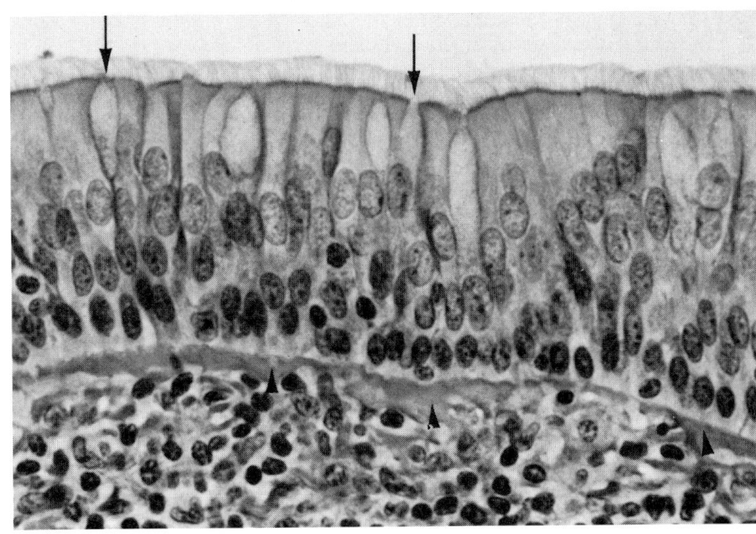

FIGURE 2-5 Columnar epithelium from the respiratory tract (trachea). Pale-staining goblet cells are discharging their mucous secretion product into the lumen (arrows). A thick basement membrane separates the epithelium from the connective tissue (arrowheads).

ation of the blood takes place. Additionally, internal body cavities are lined by single-layered flat cells called *mesothelium,* as seen in Fig. 2-10 (viz pleural, peritoneal, and pericardial cavities).

The preceding examples indicate that there is a large variation in epithelial cell structure. These variations reflect the close relationship between the structure and function of a given epithelial tissue.

INTERCELLULAR JUNCTIONS AND RELATED SPECIALIZATIONS

Although epithelial cells exhibit a wide range of structural variations and arrangements, all are closely attached to each other. This is shown in the section of skin in Fig. 2-11, which is a phase contrast micrograph demonstrating the close relationships of epithelial

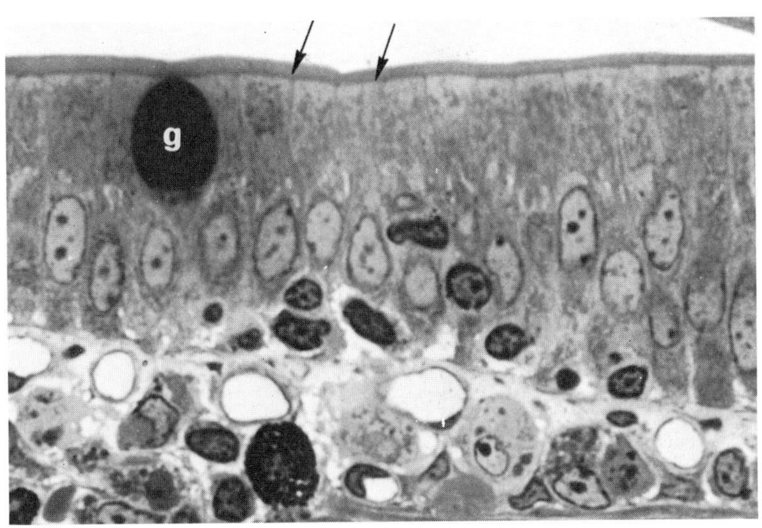

FIGURE 2-6 Columnar cells in the digestive tract have surface specializations (arrows) for enhanced absorption of nutrients. A goblet cell (g) is seen among the absorptive cells.

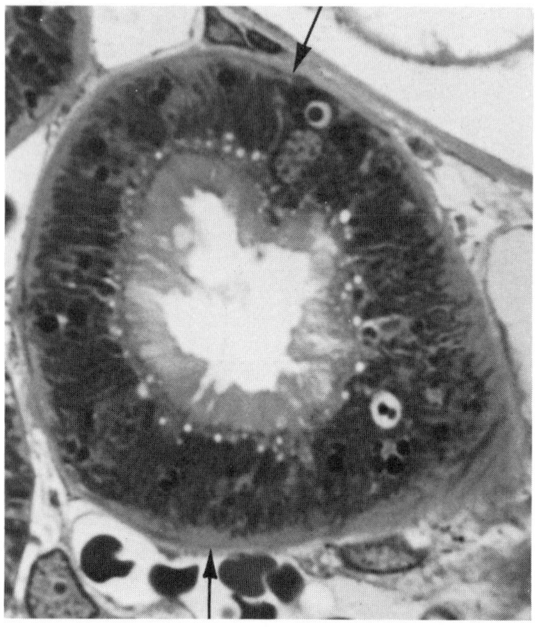

FIGURE 2-7 Light micrograph of kidney tubule. A relatively thin basement membrane separates the tubular epithelium from the connective tissue (arrows).

cells that are attached to one another by one of several junctional complexes, desmosomes, at the light microscopic level. When the close relationship between neighboring cells in different epithelia are examined in the electron microscope, a number of specialized structures that contribute to the maintenance of these relationships can be found. These specializations, along with certain intracellular structures, may reflect the degree of physical stress that a particular epithelium might be exposed to.

Most *epithelial* tissues undergo a continuous process of *cell renewal*. This requires a synchronized regeneration of cells which in turn may necessitate communication between cells in order to regulate the division and differentiation of progenitor cells and preservation of their precursors. In the lining of body cavities, such as the intestines, the single-layered cells must obliterate the intercellular space completely to protect the underlying tissue from the leakage of body fluid or invasion by bacteria. Since most of these cells are continuously being sloughed off in the normal process of epithelial renewal, the intercellular relationships, particularly in stratified squamous epithelium, will also change. This change begins with differentiation of junctional structures in the basal layer and ends with total degeneration when the epithelial cells are exfoliated from the surface.

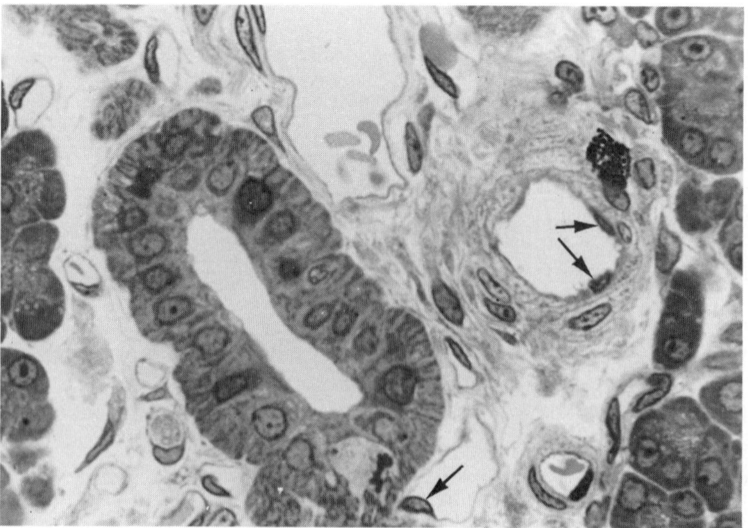

FIGURE 2-8 Photomicrograph of endothelial cells (arrows).

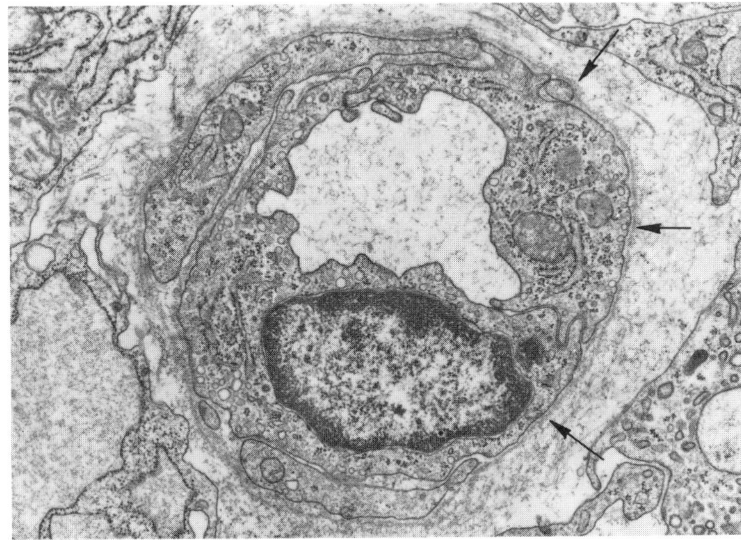

FIGURE 2-9 Electron micrograph of endothelial cells. An amorphous basal lamina surrounds the endothelial lining (arrows).

Several specialized structures have been described which, in certain areas, such as the apical surface of the intestinal epithelium, are called *junctional complexes.* A diagram of a junctional complex, as seen along the apex of intestinal epithelial cells, is shown in Fig. 2-12. This diagram represents a composite figure of the different types of junctional structures as they can be seen at a magnification of about 100,000 times or so. The plasma membrane of the epithelial cells is composed of the trilaminar structure common to all plasma membranes. The junctional complex is made up of three differentiated structures: zonula occludens, zonula adherens, and desmosome (macula adherens).

Zonula Occludens. As indicated in Figs. 2-12 and 2-13, this is a region where the two outer protein layers of adjoining plasma membranes are fused. For this reason, zonula occludens is often called a *tight junction.* The term *zonula* indicates that the tight junction is a continuous zone or belt around the

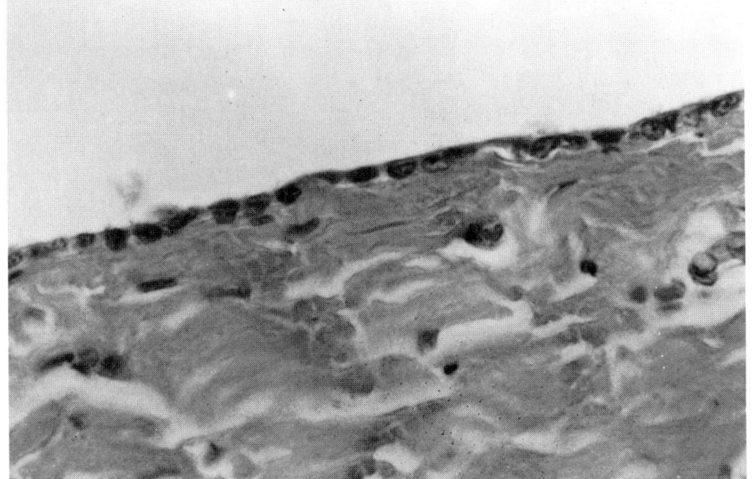

FIGURE 2-10 Mesothelium.

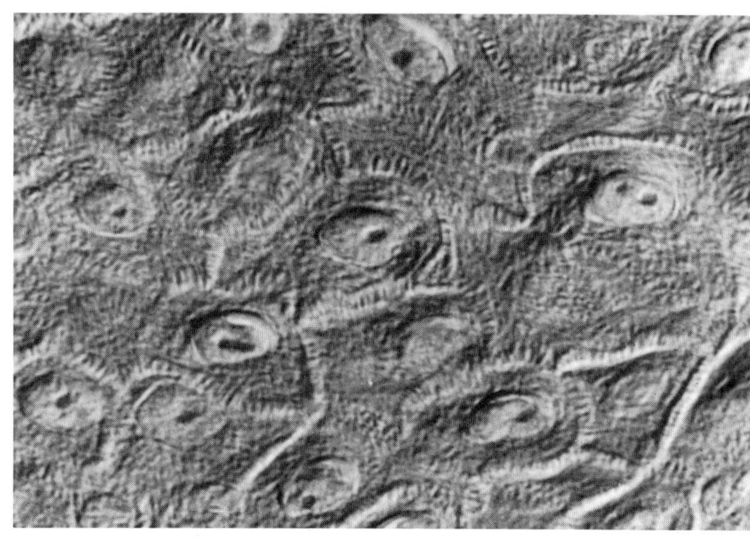

FIGURE 2-11 Phase contrast photomicrograph of skin illustrating the close attachment of epithelial cells to each other.

circumference of the cell, as indicated in Fig. 2-12. When a tight junction in section appears as a small point of fusion, it is often called a *punctate junction.* In recent years, a minute gap of about 2 nm has been observed in certain regions between the apparently fused membranes. Physiological studies have shown that such junctions with gaps, or *gap junctions,* represent areas of low electrical resistance between the two adjacent cells, thereby facilitating an electrical communication between them (Fig. 2-14).

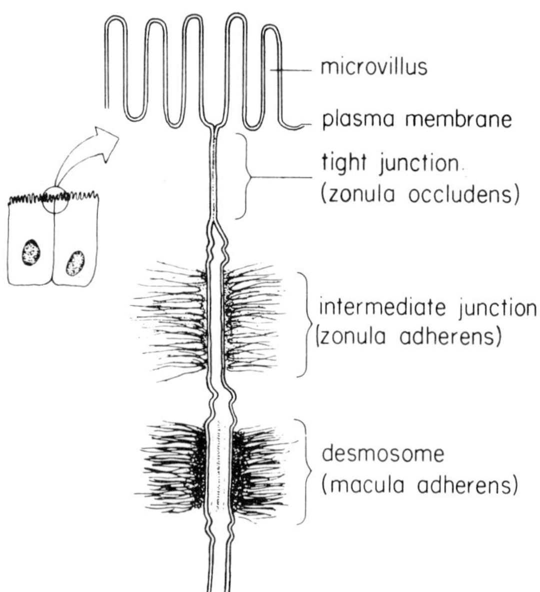

FIGURE 2-12 Diagram of a junctional complex.

Figure 2-14*a* is a transmission EM showing the appearance of gap junction, while Fig. 2-14*b* represents an *en face* view of a freeze-fractured gap junction. *Freeze-fracture* is a technique used to cleave the membrane along the middle of the phospholipid layer. The junction is so split that the extracellular side, or E face (E), shows the typical dips and the cytosol side, or P face (P), demonstrates protein particles that fit the dips on E face.

Since the gap junction can only be visualized by special techniques or high-resolution electron microscopy with appropriately prepared materials, many of the junctions that have been identified as tight junctions in the past might turn out to be gap junctions.

Besides influencing electrical communication, the gap junction is thought to play a role in chemical communication between cells. Where it is absent, the outer protein layer of the cell membrane is coated with polysaccharides which are often negatively charged. This coating may interfere with the passage of any charged chemical messengers

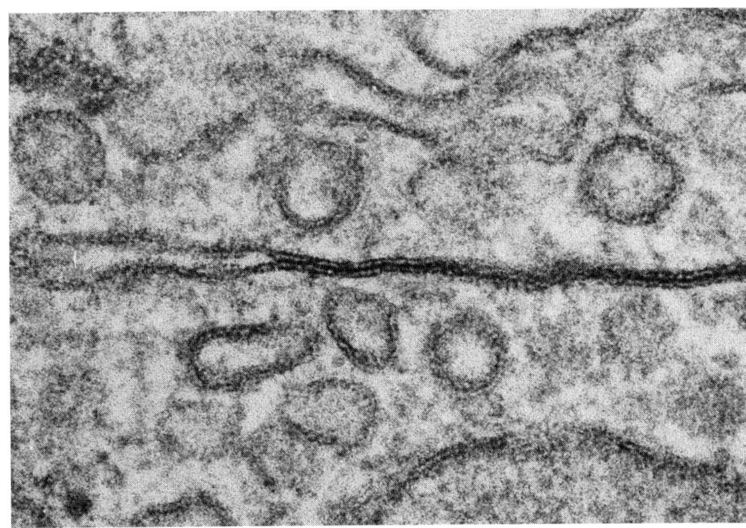

FIGURE 2-13 Electron micrograph of a tight junction.

between cells. Fusion of the membranes, by eliminating the electrostatic interference of the surface with chemical agents, allows the messengers to pass from one cell to the next. Recent advances in techniques, such as the freeze-fracture method described earlier, allow observations of intercellular junctions at the surface view, revealing the topographic appearance in great detail. It is now known that gap junctions occupy small areas of irregular shape. The appearance and the extent of gap, tight, and punctate junctions continually changes among epithelial cells that undergo differentiation or healing. The nature of intercellular communication as related to changing cellular physiology is an area of intensive investigation by biologists from different disciplines.

Zonula Adherens. Figure 2-12 also shows another connection which extends around the cells of the intestinal epithelium, the zonula adherens. This junction differs from the tight junction in that there is a space of about 20 nm between the opposing plasma membranes. Notice the uniform width of the space characteristic of the region. Since the zonula adherens is the least specialized form of physical attachment between cells, only a slight condensation of the subjacent cytoplasm may be seen. A limited number of tonofilaments may also be present in such regions. It is suspected that the zonula occludens and the zonula adherens may vary in size, ranging from a complete zone surrounding the entire circumference of cells as in the small intestine or trachea (Fig. 2-15), to small areas of adherent regions as in the case of endothelium (arrows, Fig. 2-16).

Often the term *intermediate junction* is used to designate plaque-like regions of adherence type where they are presumed to differentiate into desmosomes.

Desmosomes. This type of connection, illustrated in Fig. 2-12 and shown in the EM of Fig. 2-17, is also called *macula adherens.* The terminology reflects the fact that this junction is not a continuous belt, but is found in separated locations (macula, spot), much like a "snap button." In general, desmosomes are ellipsoidal in shape. They are often large, as in Fig. 2-17, but may also be small in size, as illustrated in Fig. 2-18, which is an EM magnified to the same level as Fig. 2-17. As indicated in these figures, the cementing substance of desmosomes shows three electron-dense layers interposed with alternating electron-lucent layers that occupy the intercellular space.

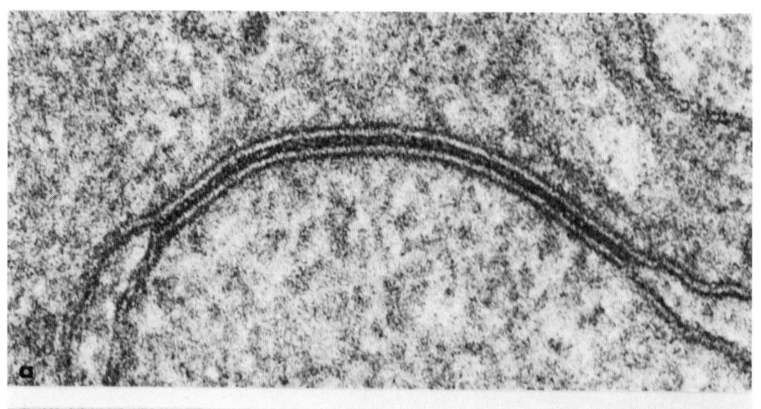

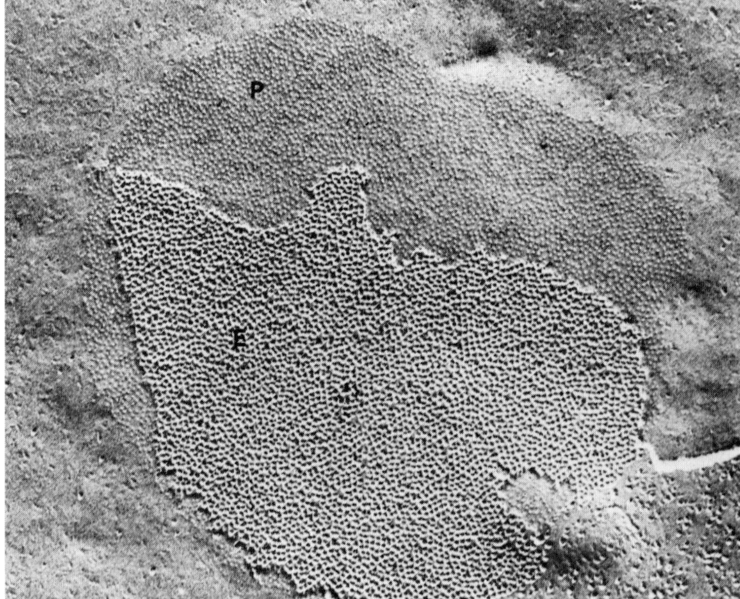

FIGURE 2-14*a* Electron micrograph of gap junction.

FIGURE 2-14*b* Electron micrograph of gap junction after freeze-fracture. Note the projections on the cytoplasmic (P) face and corresponding concavities on the extracellular (E) face. *(Courtesy of N. B. Gilula.)*

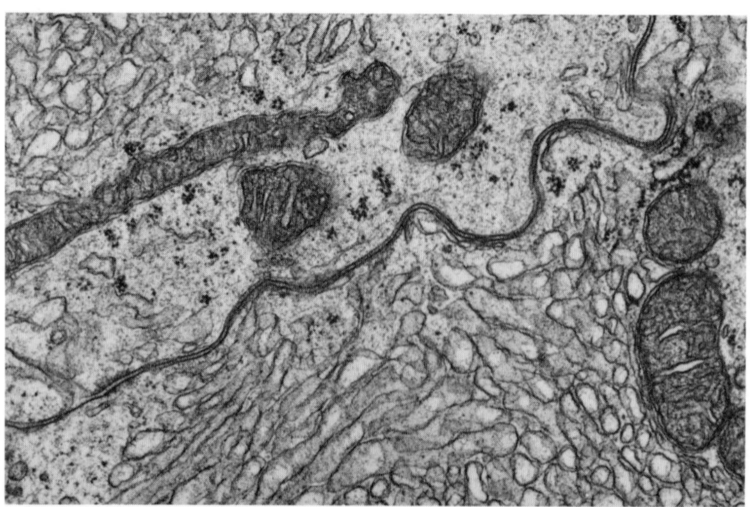

FIGURE 2-15 Electron micrograph of a zonula adherens surrounding columnar cells in the intestine.

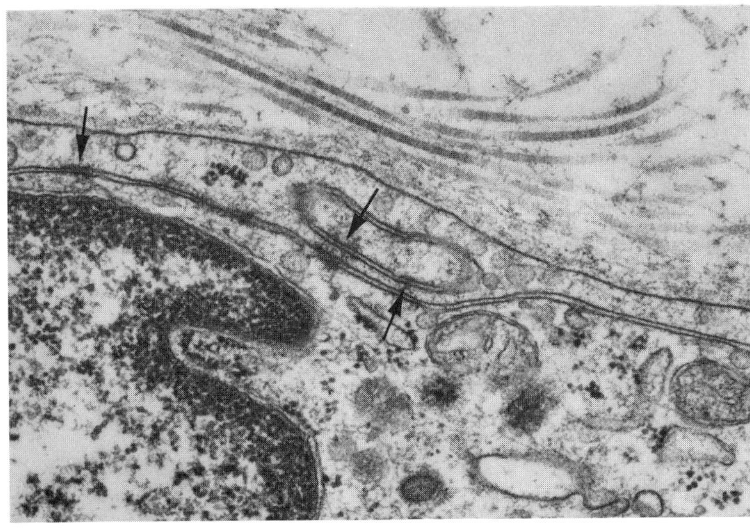

FIGURE 2-16 Electron micrograph of small zonula adherens regions between endothelial cells (arrows).

Desmosomes are the strongest type of intercellular connection and thus are most common where the epithelium is subject to a great deal of physical stress and abrasion. Figure 2-19 shows the large number of tonofilaments which merge into a desmosome creating dark condensations in the subjacent cytoplasm. Because of the sharp angle that tonofibrils take in converging towards the desmosome, their attachment to the plasma membrane is best observed when the desmosome is obliquely sectioned, as shown in Fig. 2-19. Prior to the advent of the electron microscope, the convergence of the tonofilaments seen under a light microscope was interpreted as a bridging of the cytoplasm between cells, and hence the old term *intercellular bridges* was born. It is important to remember that the tonofilaments do not pass out of their cells, and there is no continuity between cells.

Attachment to Connective Tissue

The epithelial cells line both external and internal body surfaces. They must maintain a secure relationship with the underlying connective tissue. This is accomplished by the *basement membrane* which is present under all epithelial tissues. Basement membrane is a light microscopic term that refers to a

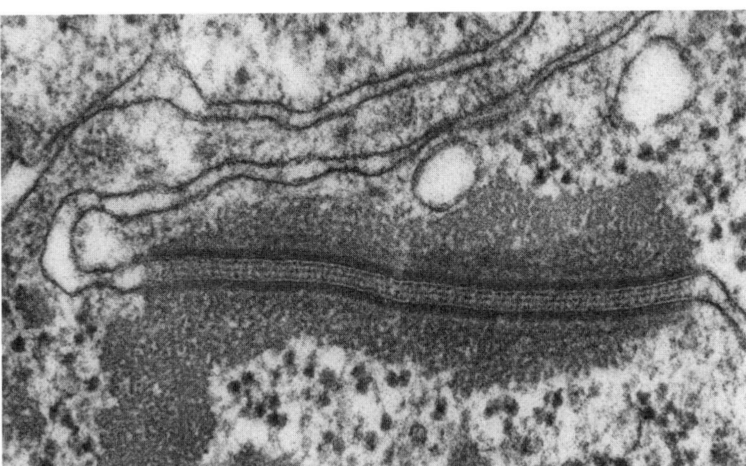

FIGURE 2-17 Electron micrograph of a desmosome showing the thickening of the inner leaflet of the cell membrane, the condensation of tonofilaments in the adjacent cytoplasm, and the lamellar material in the intercellular space.

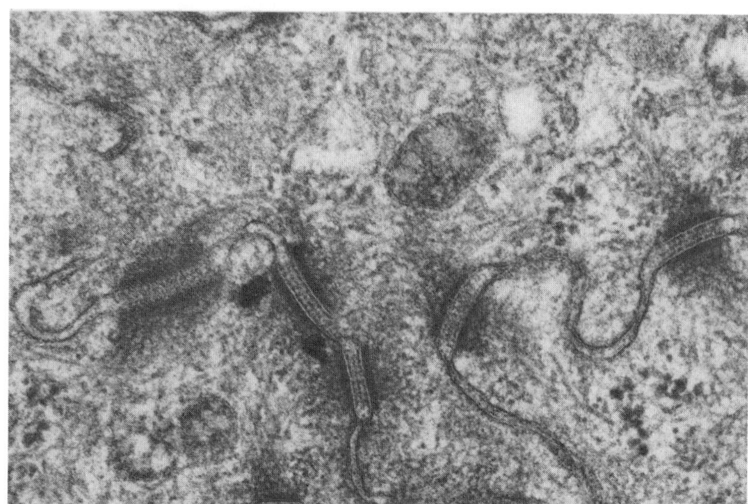

FIGURE 2-18 Smaller desmosomes having the same characteristic structure as the larger one in Fig. 2-17.

polysaccharide complex and some reticular fibers that cement the bases of epithelial cells to the connective tissue. It can be stained with a wide variety of stains; one is the periodic acid Schiff (PAS) technique. The Schiff (PAS) technique stains most polysaccharides that yield free aldehyde groups upon oxidation with periodic acid, giving them a purple to magenta color in sections. Figure 2-20 illustrates the appearance of the basement membrane after PAS staining. In an EM such as Fig. 2-21, the amorphous layer of moderate electron density (arrows) is closely apposed to the base of epithelial cells. Since this layer is universally present and can be distinguished from underlying connective tissue fibers, i.e., reticular fibers, the term *basal lamina* was introduced to designate the EM appearance of this layer. Thus the term *basement membrane* refers to the basal lamina plus other fibrous elements seen in LM and should only be used to designate stainable structures observable under a light microscope. The basement membrane varies in thickness, generally being thicker in the epithelia whose major function is protection

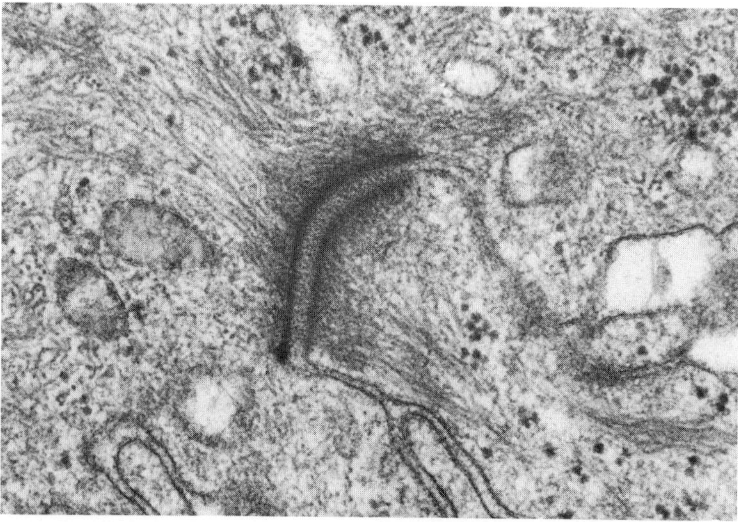

FIGURE 2-19 Electron micrograph of a desmosome. Note the large number of tonofilaments converging on it and merging to create dark condensations in the subjacent cytoplasm.

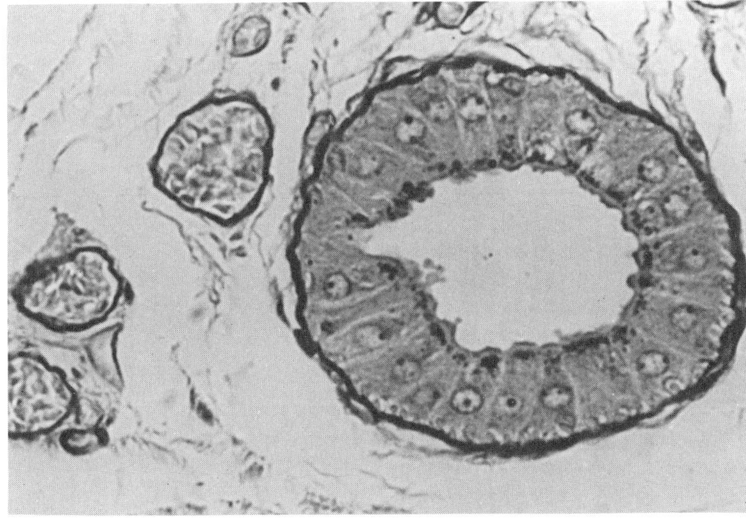

FIGURE 2-20 Photomicrograph of a section stained with the PAS technique showing the reaction of the basement membrane with Schiff reagent.

(viz respiratory epithelia in Fig. 2-5). Along the epithelial–connective tissue interface are specialized attachment apparatuses called *hemidesmosomes,* so named because each approximates half a desmosome, as seen in Fig. 2-22. Hemidesmosomes, shown at low magnification in Fig. 2-23, are characterized intracellularly by the convergence of tonofilaments and increased density of subjacent cytoplasm, as well as the presence of extracellular cementing substances.

The basal lamina in this region might be considered analogous to the central dense line of the desmosome. A second short dense line, called *peripheral density,* appears between the plasma membrane and the basal lamina (arrow, Fig. 2-22), significantly closer to the plasma membrane, as diagramed in Fig. 2-24. Short irregular, but heavy, fibrils with irregular cross striations emanate from the basal lamina of hemidesmosomes, penetrate the underlying connective tissue for varying distances, and often make a loop through which collagen fibrils pass. In highly differentiated stratified epithelium, such as skin, a hundred or more hemidesmosomes

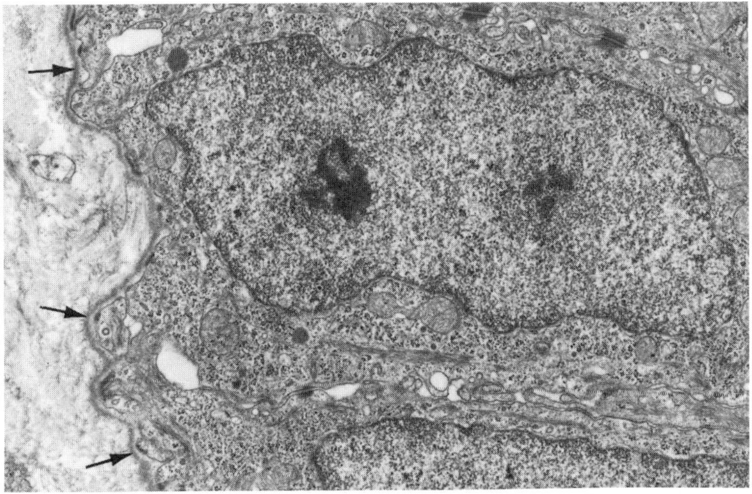

FIGURE 2-21 Electron micrograph of basal cells with the closely apposed basal lamina (arrows).

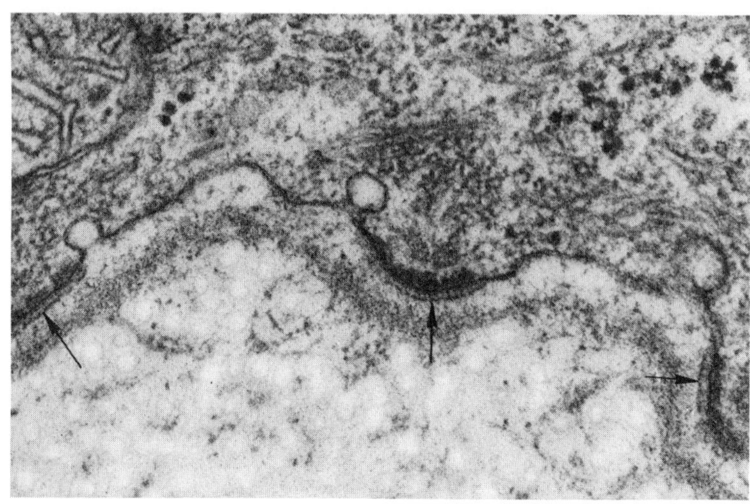

FIGURE 2-22 Electron micrograph of the basal aspect of palatal epithelium. Hemidesmosomes (arrows) aid in anchoring the epithelial cells to the basal lamina.

are present along the basal surface of each cell with many fibrous loops, which help to anchor the cell to the connective tissue through the numerous collagen fibrils. These fibrils with the irregular cross striations are called *anchoring fibrils* (arrows, Fig. 2-23; Fig. 2-24). The high magnification EMs in Fig. 2-25*a* and *b* clearly show the atypical periodicity of anchoring fibrils (arrows) and the cross sections of collagen fibrils (arrowheads) that pass through the loop. Note that there is a considerable range of thickness among anchoring fibrils.

SURFACE SPECIALIZATIONS

There are a number of surface specializations for epithelial cells that can be discussed in terms of their function. The following are some of the more common surface specializations.

Keratinization. Cells subjected to physical abrasion eventually wear out as new cells are generated at the base of the epithelium. Along the way the intercellular connections of more superficial cells that become keratin-

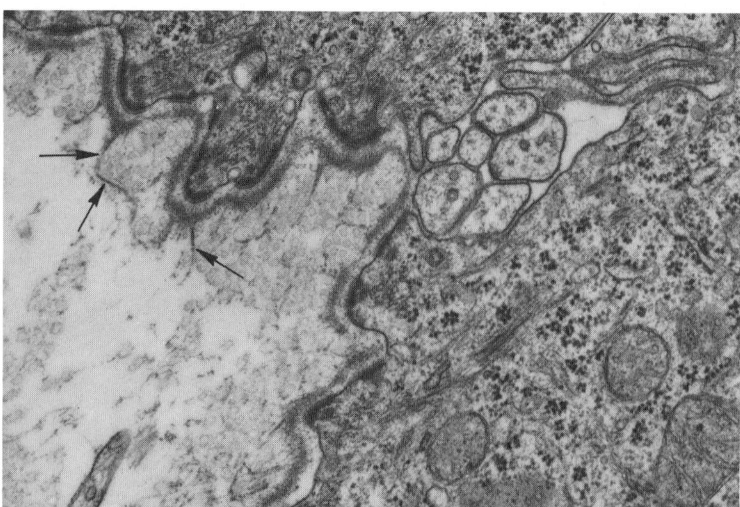

FIGURE 2-23 Hemidesmosomes viewed at lower magnification. They are characterized by an intracellular convergence of tonofilaments on a cytoplasmic density, a peripheral density, and extracellular cementing substance. Anchoring fibrils (arrows) originate in the basal lamina.

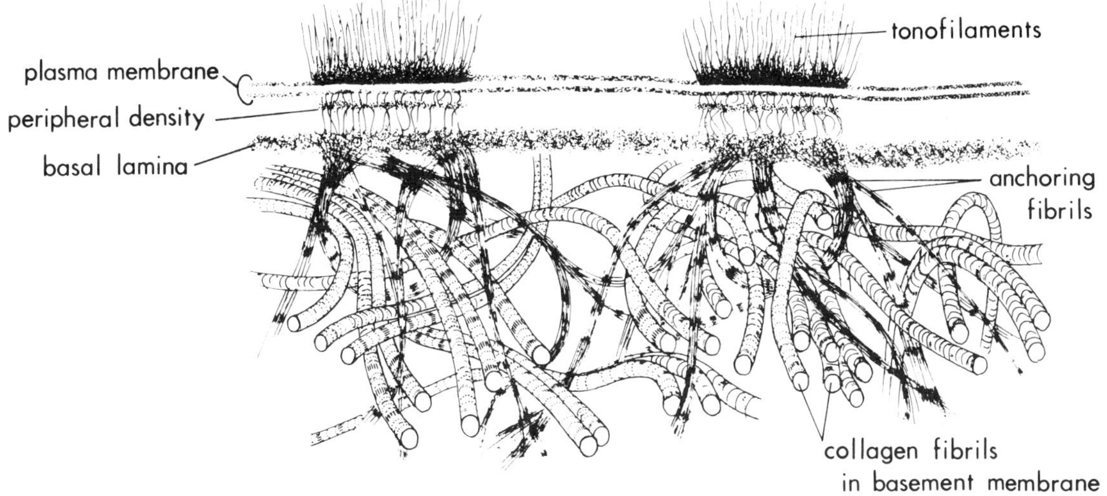

FIGURE 2-24 Diagram of an epithelial–connective tissue junction showing hemidesmosomes, the basal lamina, anchoring fibrils, and the latter's relationship to collagen fibers in the subjacent connective tissue.

ized or cornified are removed as the cells are sloughed off at the surface. Figure 2-26 shows a keratinized layer (k). To replace the lost cells, newly generated cells move up through the layers of the epithelium. As they do so, they begin to develop *keratohyalin granules* in the cytoplasm. Note the appearance of keratohyalin granules in Fig. 2-27. The granules (arrows, Fig. 2-27) are formed in the granular layer bordering the keratinized layer. The figure illustrates the abrupt disappearance of cytoplasmic granules and nuclei at the base of the keratinized layer. Details of changes in the cytoplasm characterized by rearrangement of tonofilaments and dissolution of cytoplasmic organelles, including the nucleus, are depicted in the low-power EM in Fig. 2-28. The differences in electron density of the cytoplasm among adjoining cells of keratinizing epithelia reflect rapid

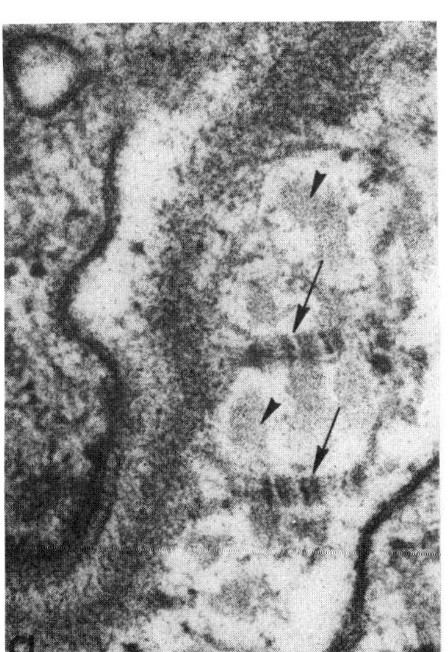

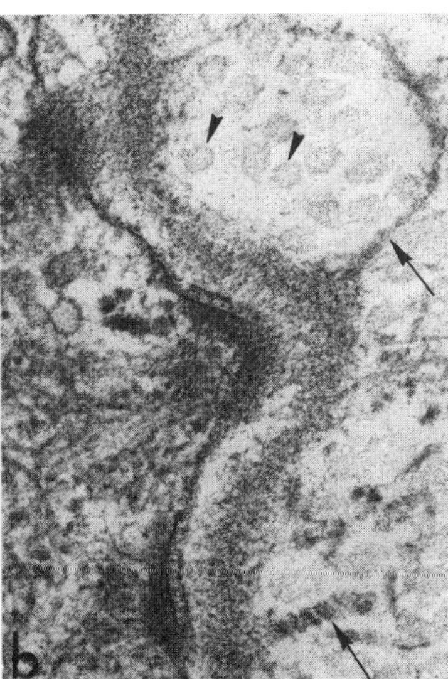

FIGURE 2-25*a–b* High-power view of anchoring fibrils (arrows) showing their atypical periodicity and cross sections of collagen fibrils (arrowheads) that pass through their loops.

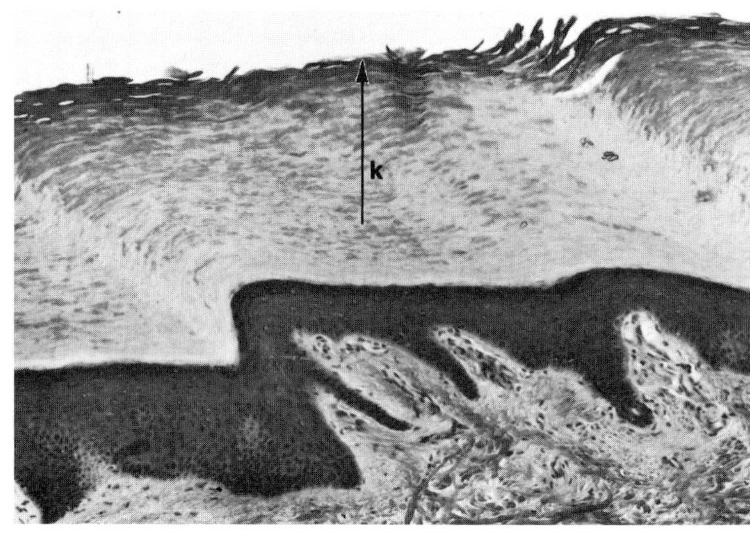

FIGURE 2-26 Photomicrograph of skin. The stratified epithelium has a thick keratinized (k) layer which protects against physical abrasion.

changes in the chemical makeup of the cytoplasm during keratinization.

The structural details of keratinization are seen in the three successive EMs (Figs. 2-29, 2-30, and 2-31). In cells where keratohyalin granules are being formed, one can see aggregations of polysomes, particularly around the developing granules. Figure 2-29 shows a keratohyalin granule that is covered by ribosomes. Note also the tonofilaments which are running irregularly throughout the cytoplasm. The EM in Fig. 2-30 depicts cytoplasmic pieces of five different layers as the keratinization continues from the lower (lower right) to upper layers (upper left). Tonofilaments are more irregular and fill the entire cytoplasm in random tangles. Along with changes in electron density mentioned elsewhere, one can also observe desmosomes that are in the process of breaking down, including some that are apparently endocytosed (arrow, Fig. 2-30). Figure 2-31 illustrates

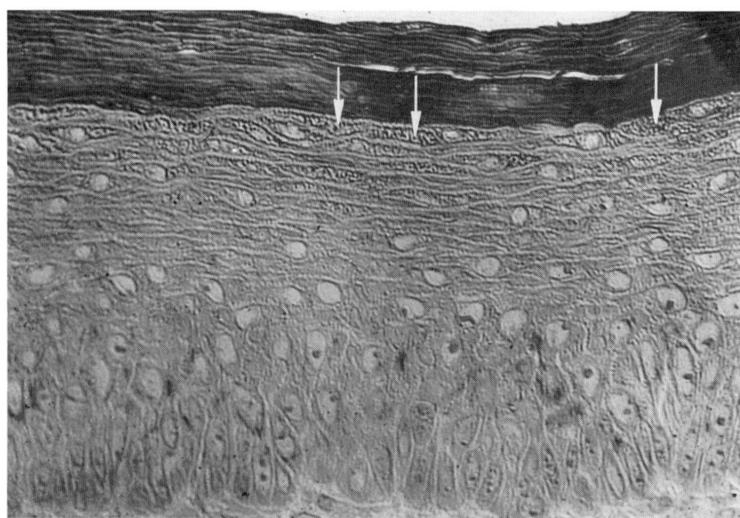

FIGURE 2-27 Photomicrograph of keratinized stratified epithelium. Keratohyalin granules are present in the granular layer (arrows).

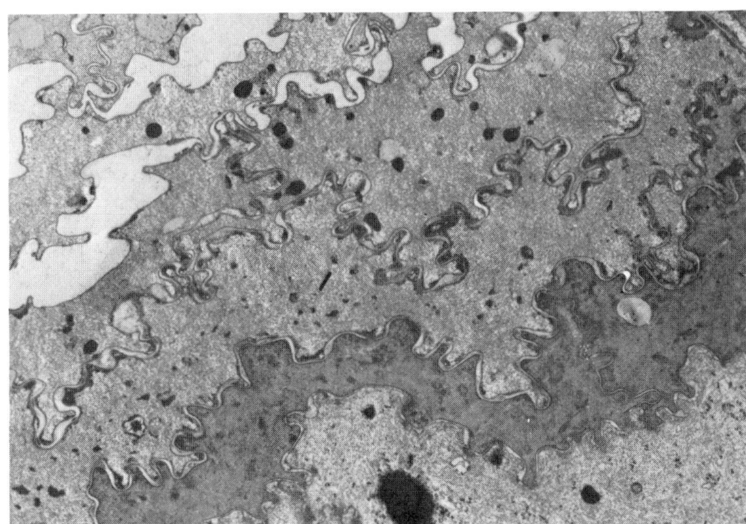

FIGURE 2-28 Low-power electron micrograph depicting structural details of keratinization. Note the abrupt disappearance of nuclei and cytoplasmic granules at the base of the keratinized layer, the rearrangement of tonofilaments, and disruption of desmosomes.

a portion where the final exfoliation is taking place; disruptions of desmosomes are indicated by an arrow.

In areas where there is a relatively small amount of wear and tear of the epithelium, as in some regions of the mucous membrane of the mouth, the epithelial cells are usually not keratinized. The turnover rate of the nonkeratinized cells of the oral mucosa is slower than that of the keratinized cells, but is faster than the turnover rate of the keratinized cells of the epidermis of the cheek.

***Striated Border* (*or Microvilli*).** This surface specialization occurs along the intestines where the process of absorption takes place. Food travels slowly along the digestive tract so that absorption of nutrients from ingested food can occur. A great surface area is needed to accommodate the amount of material to be absorbed. *Microvilli* are so called because they are made up of many uniform cytoplasmic projections which increase the surface area of the intestinal tract and therefore facilitate efficient absorption of food. They

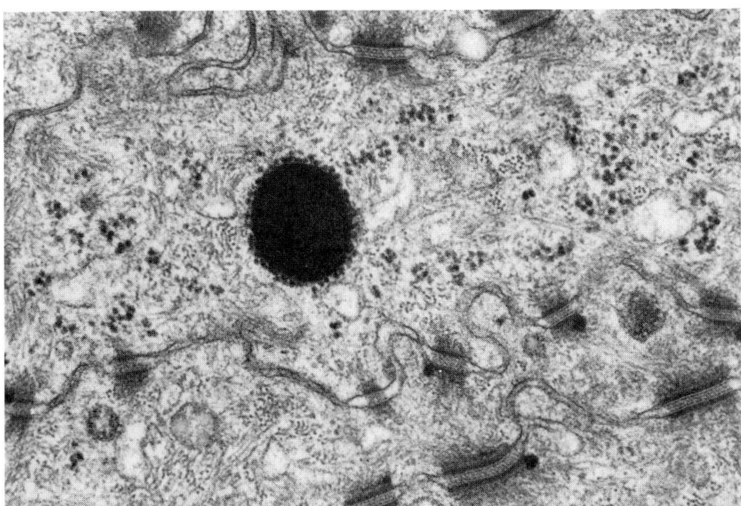

FIGURE 2-29 Portions of cells in the granular layer. One cell contains a keratohyalin granule which is covered by ribosomes. Tonofilaments are running in varying directions in the cytoplasm. Also note aggregations of polysomes in the cytoplasm.

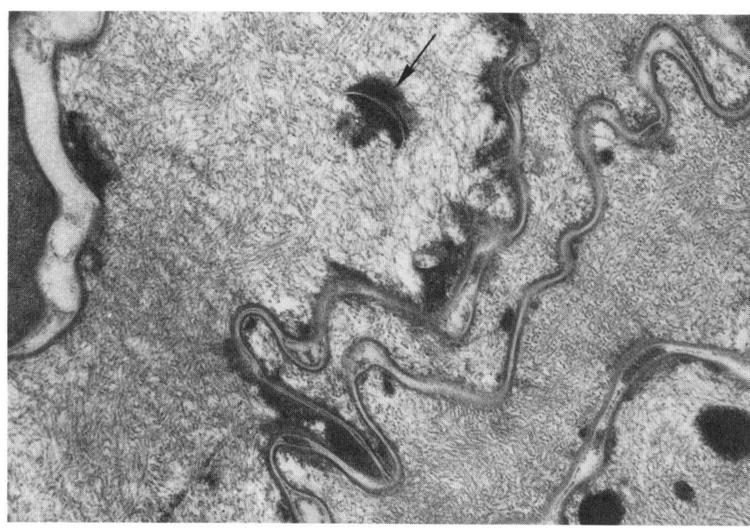

FIGURE 2-30 Electron micrograph of the keratinization process proceeding from the lower right to the upper left of the micrograph. Desmosomes are apparently endocytosed (arrow).

are known by their light microscopic appearance as *striped* or *striated border,* and they appear as a thin, homogeneously stained border along the clear intestinal lumen, as shown in Fig. 2-32. As mentioned previously, the electron microscopic appearance of this border shows numerous fingerlike projections, or microvilli (Fig. 2-33).

The external surface of microvilli has a layer of cloudy substance which is a polysaccharide-protein coat called *glycocalyx,* or "fuzzy coat," augmented by the mucous material produced by goblet cells. Two goblet cells that are unicellular glands are present among the columnar absorptive cells in Fig. 2-32. Chyme (food being digested) moving down the digestive tract comes in contact with a larger surface area by flowing into the valleys or intermicrovillous spaces. Microvilli remain straight and rigid throughout the absorption process despite the tremendous pulsating motion of the intestines and pressure from the ingested food. This is possible because within each microvillus are *axial*

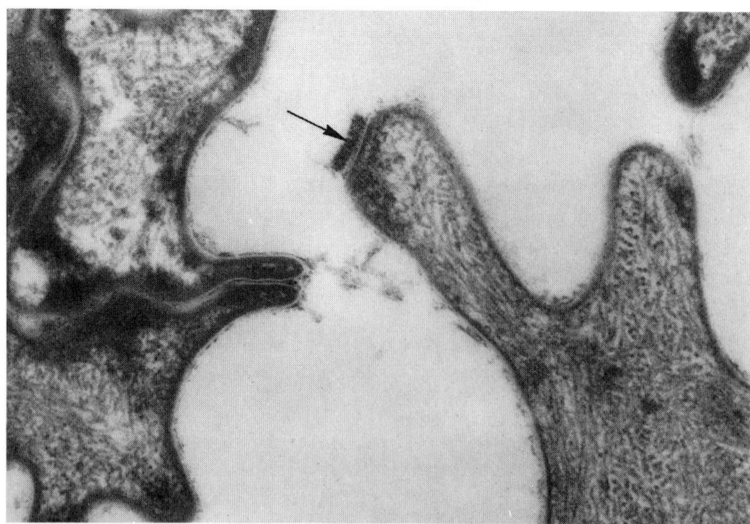

FIGURE 2-31 Exfoliation of fully keratinized cells at the surface of the epithelium. The desmosomes are disrupted (arrow) as cells are sloughed.

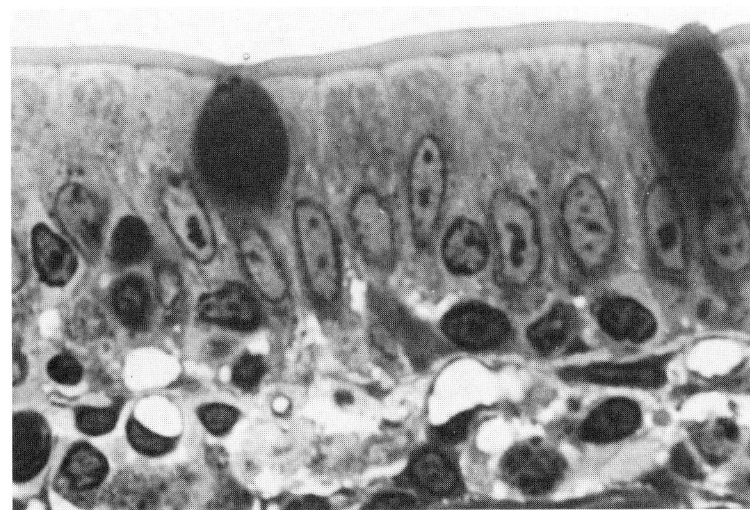

FIGURE 2-32 Normal absorptive epithelium with goblet cells and striated border.

filaments, which provide support, as seen in Fig. 2-33 (arrows). These axial filaments branch off at the base of each microvillus and converge with similar branching filaments originating from neighboring microvilli. This convergence of branched axial filaments creates a weblike structure providing a broad base for the microvilli. For this reason, it is called the *terminal web* (tw in Fig. 2-33).

Brush Border. Villous projections that differ from intestinal microvilli occur in the convoluted tubules of the kidney. Here, too, an increase in surface area is provided by these projections. However, the villous projections are primarily concerned with fluid transport and therefore are not subjected to as much pressure as the digestive microvilli. Because of the large quantity of fluid resorption that occurs in kidney tubule cells during processing of urine, these projections are taller than those of the digestive tract. In addition, they lack the supporting axial filaments and therefore tend to collapse when prepared for ob-

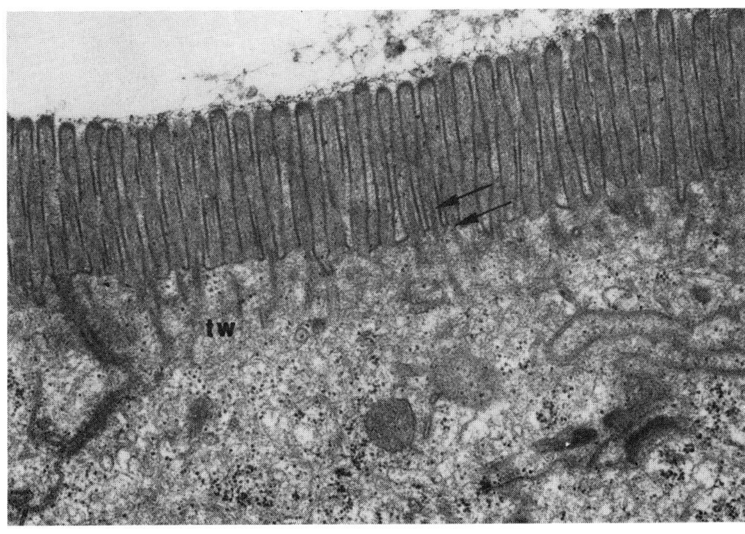

FIGURE 2-33 Microvilli on intestinal absorptive cells (axial filaments, arrows; terminal web, tw).

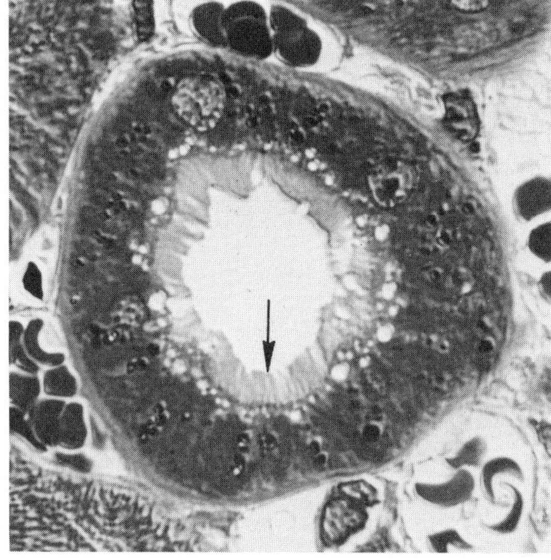

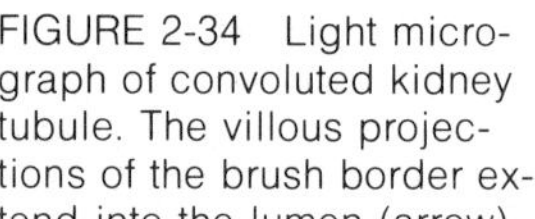

FIGURE 2-34 Light micrograph of convoluted kidney tubule. The villous projections of the brush border extend into the lumen (arrow).

servation. Seen under a light microscope (arrow, Fig. 2-34), these slender projections resemble an old brush and hence have been called *brush border.* Figure 2-35 shows an electron microscopic view of the brush border.

Cilia. Other surface specializations that may look similar to microvilli but are totally different in structure and function are called *cilia.* Cilia occur in large numbers along the respiratory tract at the luminal end of the columnar epithelial cells. Figure 2-36 shows cilia (arrows) of the tracheal epithelium as they appear along the free surface of the lining cells. The three-dimensional effect produced by use of phase contrast optics clearly indicates that the border is made up of individual cilia. An EM view of cilia in Fig. 2-37 shows a dense body called a *basal body* (arrow) at the base of each cilium.

As mentioned in Chap. 1, the ciliary shaft extending from the basal body is covered by a layer of cytoplasm and plasma membrane as it protrudes from the apical cytoplasm. The cross-sectional appearance of cilia as seen in the EM (Fig. 2-38) shows the characteristic arrangement of the microtubules that make up the ciliary shaft. Upon closer observation it is possible to visualize these microtubules as paired structures with nine peripheral pairs surrounding a central pair, as shown in Fig. 2-38. The goblet cells, situated among the ciliated columnar cells (Fig. 2-36), coat the surface with a mucous substance. As inspired air travels along the

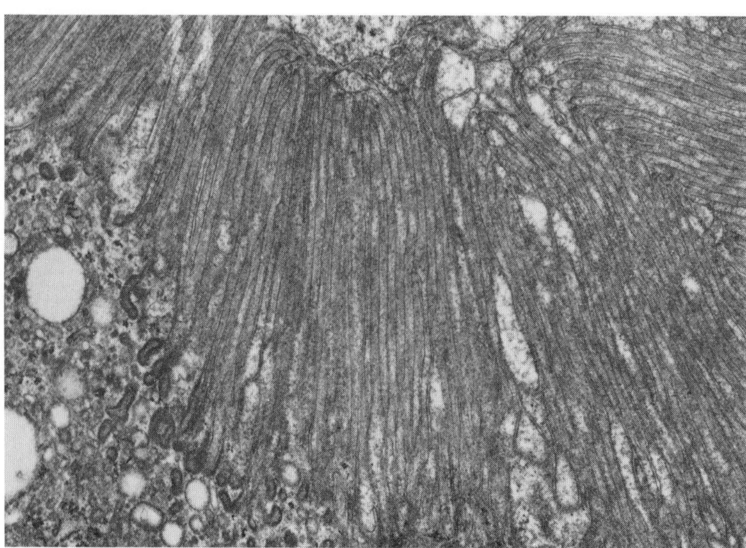

FIGURE 2-35 Electron microscopic view of the brush border. Long projections extend from the apical surface into the lumen.

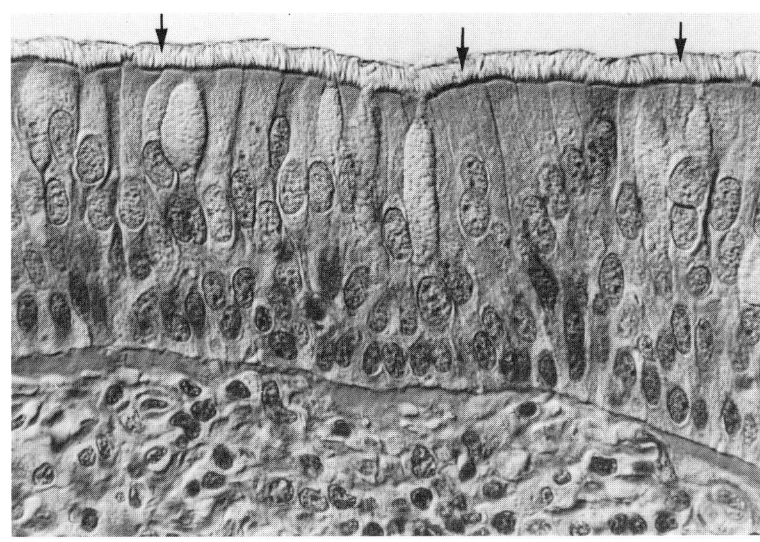

FIGURE 2-36 Phase contrast micrograph of tracheal epithelium. Cilia are seen along the free surface of the columnar cells (arrows).

respiratory tract, bacteria and dust particles are trapped by the sticky surface coating of the ciliated epithelium. The ciliary motion is coordinated and unidirectional, resulting in the upward movement of trapped particles and microbes until they reach the pharynx. The mechanism for continuous intercellular communication necessary among epithelial cells in order to synchronize and coordinate the movement of the many billions of cilia is not fully understood.

Stereocilia. There are other apical projections that resemble cilia. These are found in certain segments of the male genital tract, namely epididymis. Since they do not have the axial microtubules and are therefore nonmotile, they are called *stereocilia*. Figure 2-39 shows a segment of epididymis in which the stereocilia are found along the luminal surface of the epithelium. The stereocilia are taller and more irregularly arranged than the motile cilia shown in Fig. 2-36.

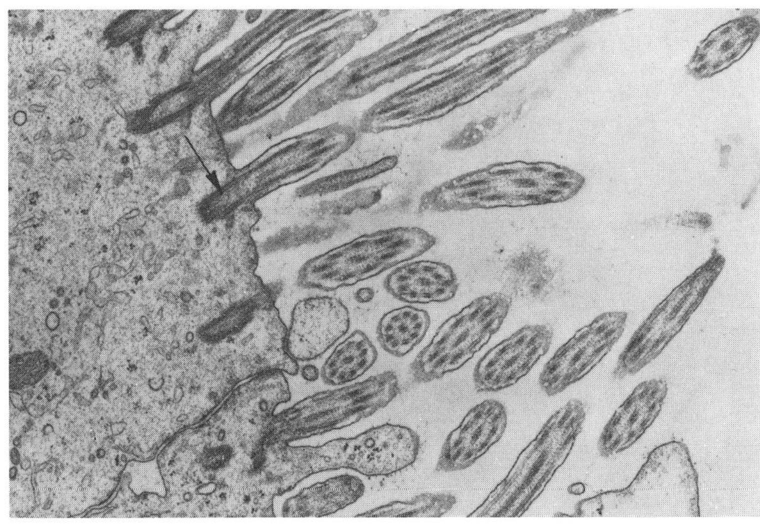

FIGURE 2-37 Electron micrograph of the apical surfaces of tracheal epithelium. Ciliary shafts are extending from basal bodies (arrow) in the apical cytoplasm.

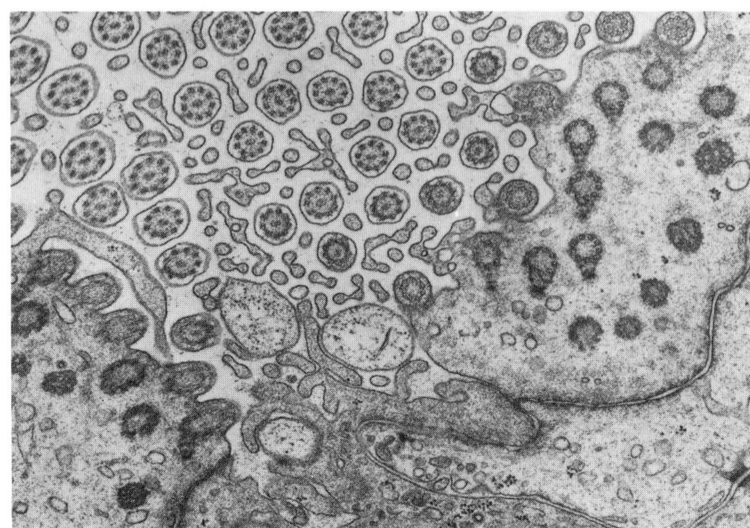

FIGURE 2-38 Cross sectional view of cilia showing the characteristic arrangement of microtubules in the ciliary shafts.

Transitional Epithelium. Lining the bladder and ureter, which undergo drastic changes in luminal size during accumulation and discharge of urine, is the *transitional epithelium.* The inner volume can range in size from 500 mL (distended) to less than 50 mL (empty). This volumetric change necessitates a modulation of the inner surface area, which is accomplished by the sliding over of the superficial cells of the epithelium in such a way that the distended transitional epithelium will be made up of only a few layers of cells, whereas under constricted conditions, it is composed of 10 or more cell layers.

An LM section of a constricted bladder is shown in Fig. 2-40. The more deeply situated cells near the basement membrane are small, while the more superficial cells are larger and umbrella-shaped. As the size of the bladder increases, the number of cell layers is reduced and the cells are generally flatter (Fig. 2-41). Few desmosomes are present between epithelial cells, allowing the free cellular sliding movement. The transition of

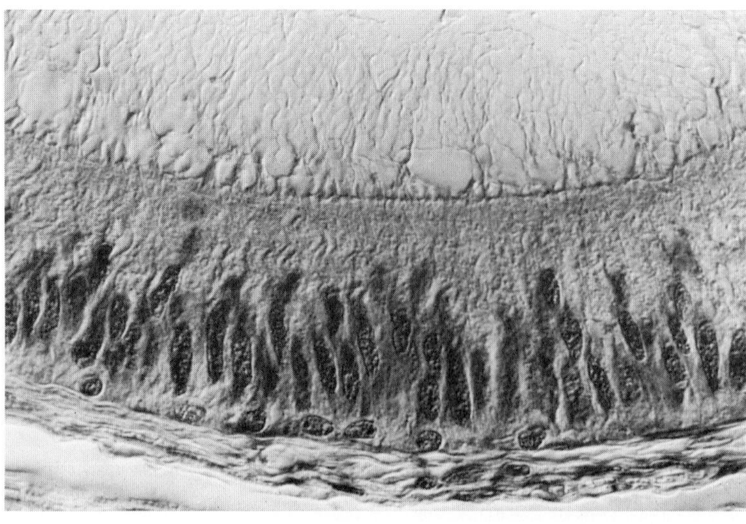

FIGURE 2-39 Phase contrast micrograph of columnar epithelial cells in epididymis with stereocilia along the luminal surface.

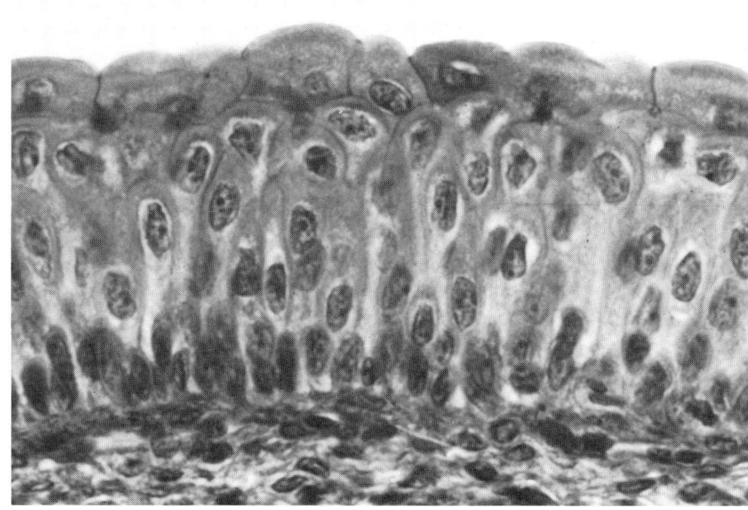

FIGURE 2-40 Appearance of transitional epithelium in the constricted bladder.

the cellular layers from one state to the other, which occurs rapidly as the bladder is emptied, gives the tissue its name.

CLASSIFICATION OF EPITHELIUM

Epithelium is classified according to the shape of the cells and the number of layers of a given cell type it has. Figure 2-42 diagrams the types of epithelium described below.

Epithelial cells may be divided into three basic shapes:

1. *Squamous*—flat platelike cells as in the endothelium
2. *Cuboidal*—cube-shaped cells as in the kidney tubule
3. *Columnar*—tall cells as in the intestine

Each of these cell types may be arranged in single or multiple (stratified) layers. Since the classification of epithelial tissue is usually

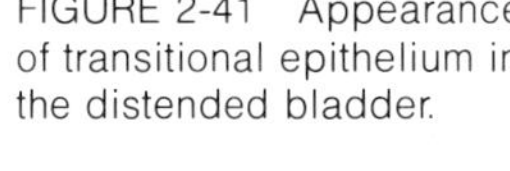

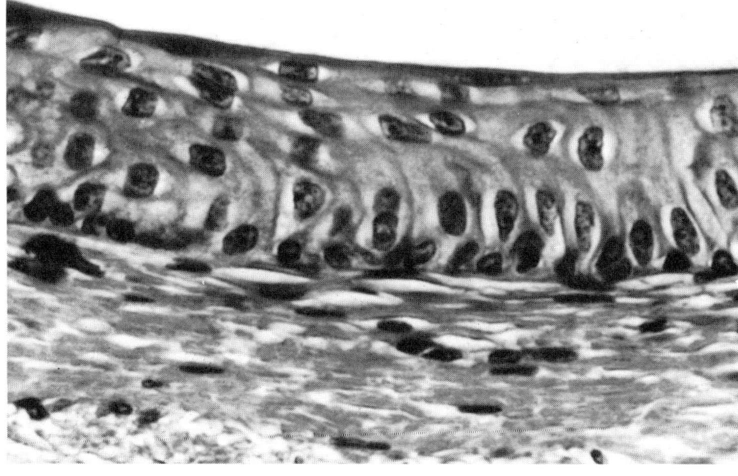

FIGURE 2-41 Appearance of transitional epithelium in the distended bladder.

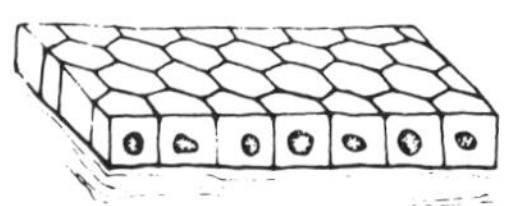

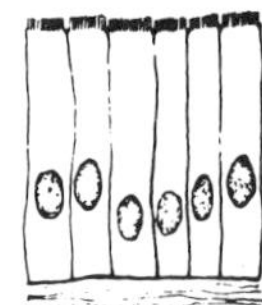

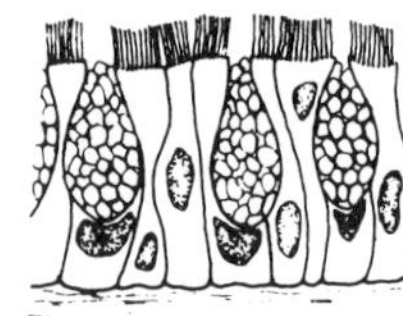

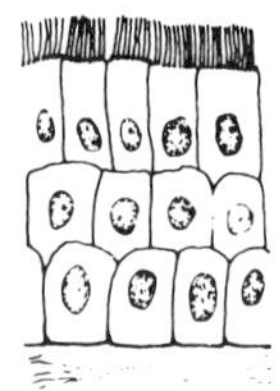

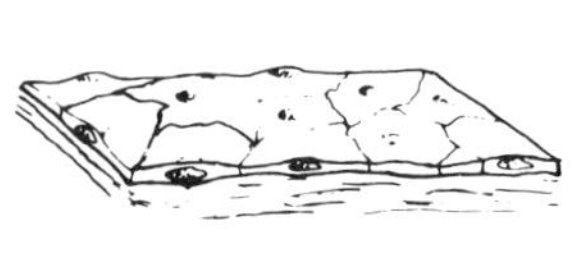

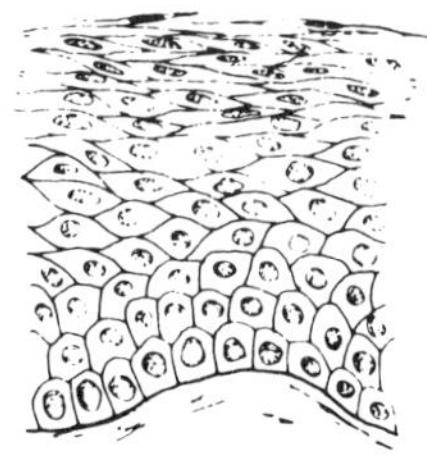

FIGURE 2-42 Diagram of the various epithelial types.

made on the basis of cell type and the manner in which they are arranged, the following combinations are possible:

	Squamous			Squamous
Stratified	Cuboidal	or	Simple	Cuboidal
	Columnar			Columnar

Pseudostratified Epithelium. This type of epithelium is present only in the columnar epithelium of the respiratory and genital tracts in which a single layer of cells are resting on the basement membrane and only some cells reach the free surface. For this reason the nuclei appear at different levels,

EXAMPLES OF THE TYPES OF EPITHELIUM — TABLE **2-1**

TYPE	SQUAMOUS	CUBOIDAL	COLUMNAR
Simple	Endothelium of blood vessels and lymphatics Epithelium of lung Mesothelium of pleural, peritoneal, and pericardial cavities	Some kidney tubules Thyroid gland Sweat gland	Digestive tract
Pseudostratified	None	None	Trachea Epididymis
Stratified	Epidermis Oral mucosa Vagina Anal mucosa	Sweat gland duct (rare)	Trachea Epiglottis (rare)
Transitional		Bladder Ureter	

giving a false stratified appearance. An example of this is shown in Fig. 2-36, in which nuclei of the columnar cells appear somewhat irregularly stratified.

Transitional Epithelium. This type is a special modification of stratified epithelium found in the pelvis of the kidney, the ureter, the bladder, and a portion of urethra. As mentioned previously, the number of cell layers and their shapes change with distention and contraction of the organs.

Epithelial cells may also be named by adding a modifier to the above classifications. Such a modifier describes the surface specializations as follows: keratinized, with microvilli, ciliated, or with stereocilia. In describing epithelia, any of the surface specializations mentioned above should be added prior to the classification. For example, the epithelium of the skin is described as keratinized stratified squamous epithelium, while the tracheal epithelium is called ciliated pseudostratified epithelium with goblet cells. Table 2-1 summarizes major covering epithelia of different organs by types.

REVIEW SECTION

1. Figure 2-43 is a section through the skin of the palm of the hand. The specializations of the epithelium (dark layer) for the protection of the palm include __________. Other places where similar epithelium might be found are __________.

The palm, as well as other locations of stress, such as the sole of the foot or the skin of the elbow, would show the layering of cornified epithelium, or keratinization, to provide protection. There would be a large number of hemidesmosomes connecting the epithelium to the basal lamina which in turn would be literally tied to the underlying connective tissue via the loops produced by anchoring fibrils. Desmosomes would be well-developed and the most frequent type of connections between the epithelial cells.

2. Figure 2-44 shows a gland typically found in the epidermis (arrows). This gland is usually associated with (name of structure) __________. What is the name of the gland?

The sebaceous gland produces sebum to help lubricate the hair and the skin. It opens into the space around the hair shaft; its secretory duct is located alongside the hair shaft.

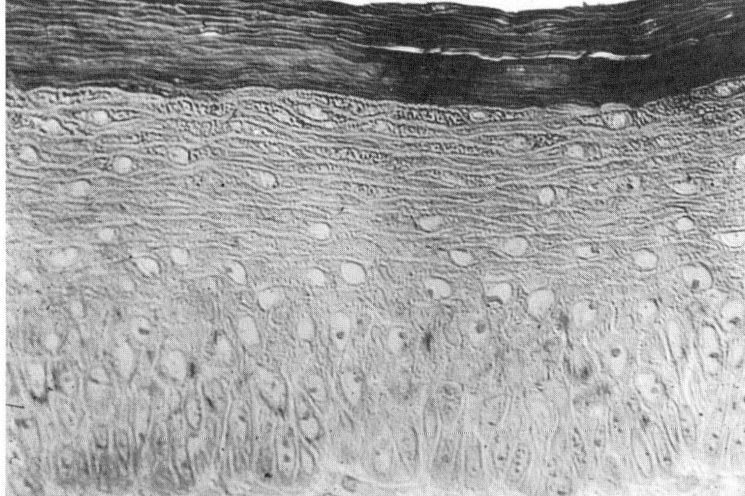

FIGURE 2-43

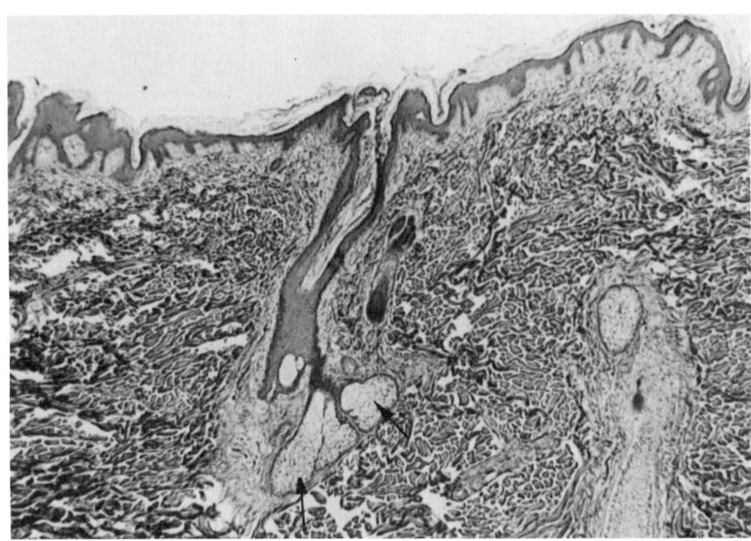

FIGURE 2-44

3. The adaptability of epithelium is shown in the type of epithelium seen in Fig. 2-45. What difference can be seen between the top and bottom layers of cells? ________. This type of epithelium is termed ________. Where can this type of epithelium be found? ________. What kinds of intercellular connection would be found here?

Figure 2-45 shows a section of the bladder. The larger umbrella-shaped cells facing the lumen are characteristic of the transitional epithelium, which is also found in other portions of the urinary tract, such as the ureter. The cells are loosely connected, with a few small desmosomes found in the deeper layers. This aids in the adjustment to rapid volumetric change of the organ.

4. Figure 2-46 is a section through the jejunum of the small intestine. Notice the height of the cells. How would you classify the epithelium lining the lumen? ________. What is the dark line edging the lumen (arrows) called? ________. This border is composed of ________. The function of the border is ________.

Simple columnar epithelium with a striated border lines the small intestine. The microvilli in the border increase the surface area for absorption of nutrients.

FIGURE 2-45

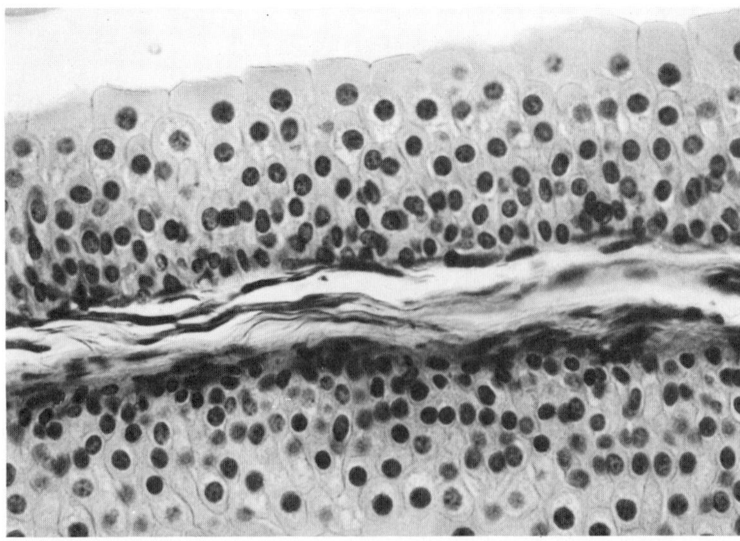

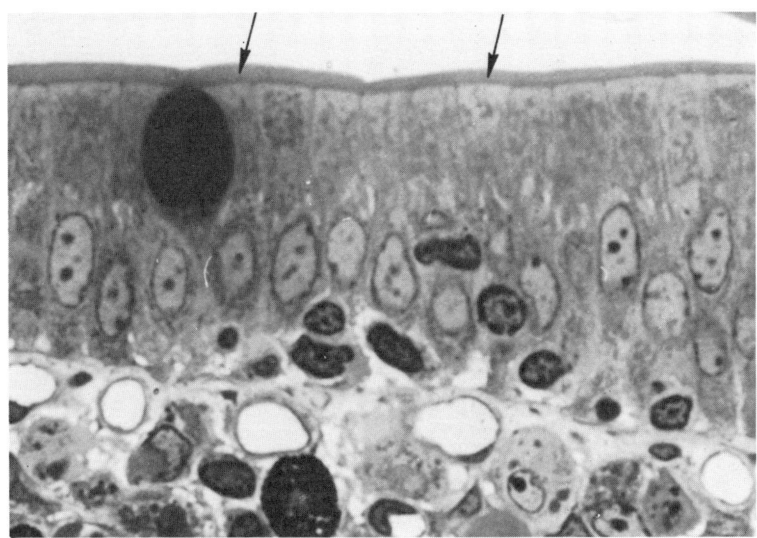

FIGURE 2-46

5. The longitudinal section of the lymphatic vessel shown in Fig. 2-47 is lined with endothelium, a form of epithelium. How would you classify this epithelium? ________. What kind of intercellular junctions would you expect to find in this type of tissue?

The thin cells lining the lymphatic vessel are simple squamous epithelial in nature. As such they are similar to endothelium of small capillaries that are present throughout the field. All types of junctions might be found between endothelial cells. However, fully developed desmosomes are seldom found between endothelial cells.

6. Figure 2-48 is a section through the respiratory tract. How would this epithelium be classified? ________. What forms the somewhat irregular fringe along the lumen? ________. How are the epithelial cell nuclei arranged here? ________. What structure connects the epithelium to the underlying tissues?

Much of the upper respiratory tract is lined with ciliated pseudostratified columnar epithelium. The epithelium makes a fairly impermeable barrier with its regular organization of closely packed cells and a thick basement membrane. The connective tissue beneath the epithelium contains much extracellular fluid and fibers with relatively few cells.

FIGURE 2-47

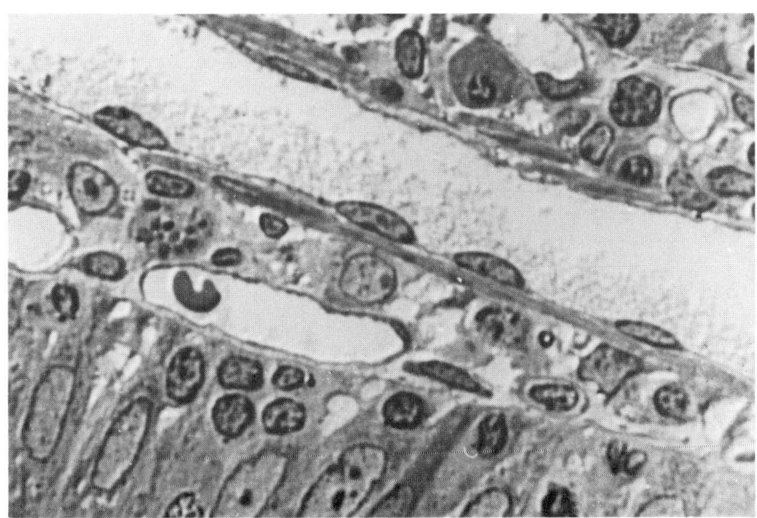

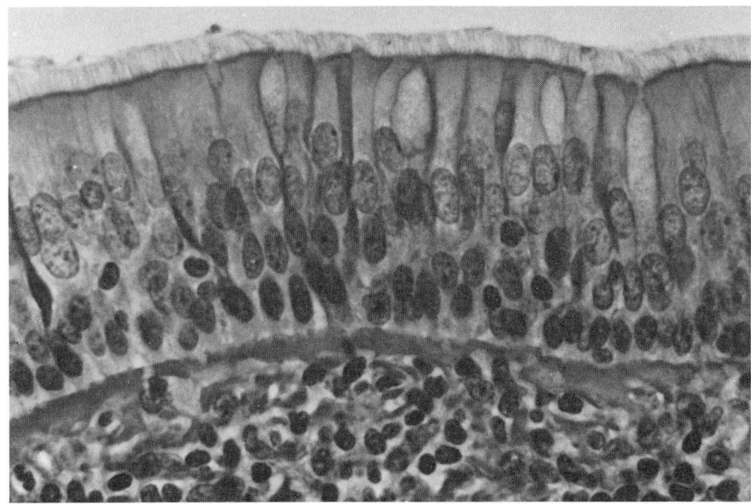

FIGURE 2-48

7. The two large attachment structures in Fig. 2-49 are typical ________ present along the cell membranes between stratified epithelial cells.

These structures are desmosomes, which are the strongest type of intercellular connections. Note the loosely organized but distinct tonofibrils in this field.

8. In this EM (Fig. 2-50) there are several desmosomes sectioned at different angles. What are the structures indicated by the arrow? ________. What are the structures indicated by arrowheads?

In this EM of stratified squamous epithelium, well-developed desmosomes are sectioned at different angles. The dense patches indicated by the arrows are cross-sectioned bundles of tonofibrils, which are much better organized here. It should be noted that, in places where desmosomes are highly developed, tonofibrils make tight and much more clearly delineated bundles. Compare Fig. 2-50 with Fig. 2-49 to confirm this point. Arrowheads indicate occasional tight or gap junctions that are found between the desmosomes.

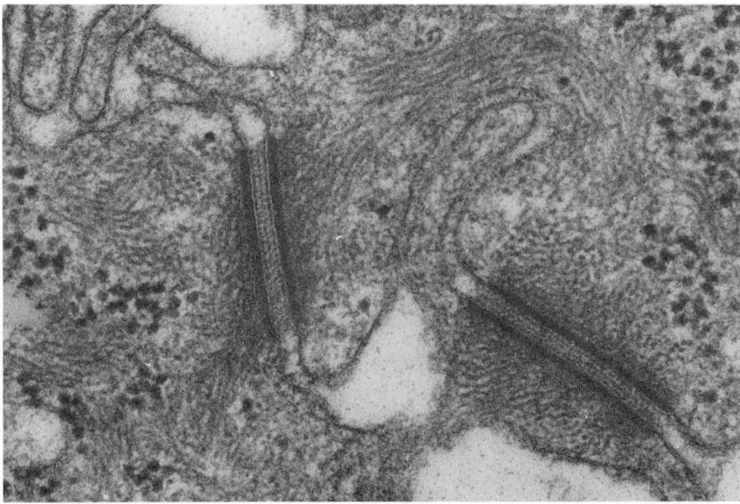

FIGURE 2-49

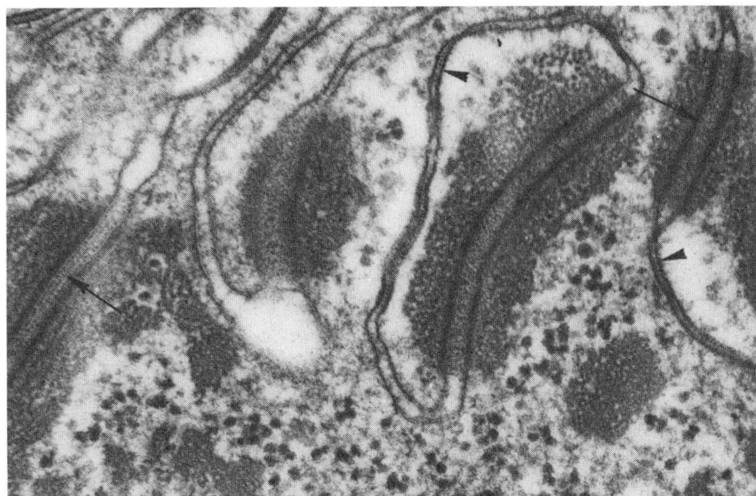

FIGURE 2-50

9. The type of adaptation shown in Fig. 2-51 can be found in __________. What is its function?

Figure 2-51 is a cross section through cilia. The nine regularly organized pairs of microtubules distinguish this cross section from that through a microvillus. Ciliated epithelium occurs in the respiratory tract. It moves the mucus produced by the goblet cells (interspersed among the ciliated cells) to the mouth and nose to be expelled with any trapped foreign matter.

10. The EM of a capillary, shown in Fig. 2-52, demonstrates the appearance of endothelium. How would the epithelium be classified? __________. What element immediately subjacent to the endothelium helps to hold it in place?

The simple squamous epithelium which makes up capillary endothelium is thin and allows for passage of materials and gases between the blood and surrounding tissues. As in all epithelia, these cells are lined with a continuous thin basal lamina, which can barely be seen in this frame as a fuzzy layer on the outside of the capillary.

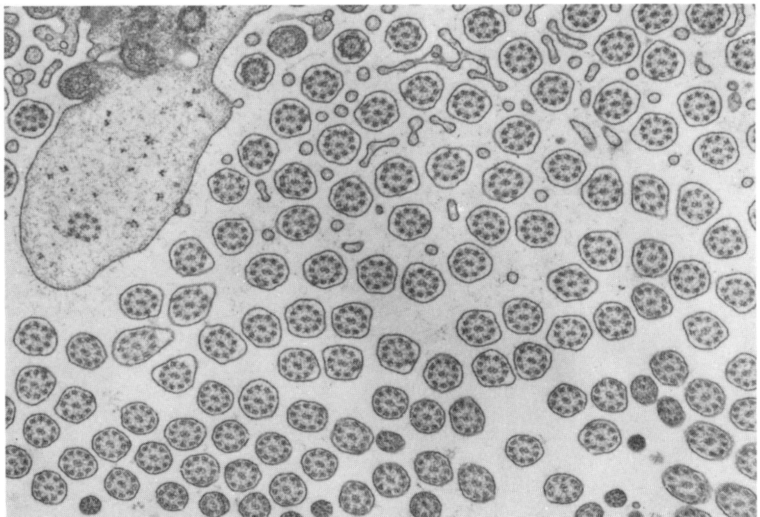

FIGURE 2-51

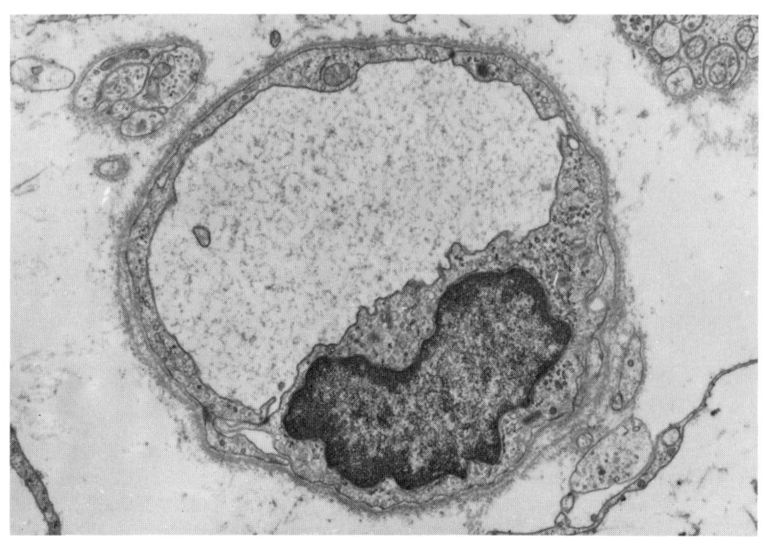

FIGURE 2-52

BIBLIOGRAPHY

Bloom, W., and D. W. Fawcett: *A Textbook of Histology,* 10th ed., Saunders, Philadelphia, 1975.

Brody, I.: "The Ultrastructure of the Tonofibrils in the Keratinization Process of Normal Human Epidermis," *J. Ultrastruct. Res.* **4:**264–297 (1960).

Farquhar, M. G., and G. E. Palade: "Junctional Complexes in Various Epithelia," *J. Cell Biol.* **17:** 375–412 (1963).

Fawcett, D. W.: "Cilia and Flagella," in J. Brachet and A. E. Mirsky (eds.), *The Cell,* vol. II, Academic Press, New York, 1961, pp. 217–297.

Friend, D. S., and N. B. Gilula: "Variations in Tight and Gap Junctions in Mammalian Tissues," *J. Cell Biol.* **53:**758–776 (1972).

Ito, S.: "The Surface Coat of Enteric Microvilli," *J. Cell Biol.* **27:**475–491 (1967).

Matoltsy, A. G., and P. F. Parakkal: "Membrane-Coating Granules of the Epidermis," *J. Cell Biol.* **24:**297–307 (1965).

Montagna, W., and W. C. Lobets, Jr. (eds.): *The Epidermis,* Academic Press, New York, 1964.

Rhodin, J. A. G.: *Histology. A Text and Atlas,* Oxford University Press, New York, 1974.

Snell, R. S.: "The Fate of Epidermal Desmosomes in Mammalian Skin," *Z. Zellforsch.* **66:**471–487 (1965).

Weiss, L., and R. O. Greep: *Histology,* 4th ed., McGraw-Hill, New York, 1977.

Zelickson, A. S. (ed.): *Ultrastructure of Normal and Abnormal Skin,* Lea & Febiger, Philadelphia, 1967.

ADDITIONAL READINGS

Breathnach, A. S.: "Aspects of Epidermal Ultrastructure," *J. Invest. Dermatol.* **65:**2–15 (1975).

Bulger, R. E.: "The Shape of Kidney Tubule Cells," *Am. J. Anat.* **116:**237–255 (1965).

Farbman, A. I.: "Plasma Membrane Changes during Keratinization," *Anat. Rec.* **156:**269–282 (1966).

Fawcett, D. W.: "Surface Specializations of Absorbing Cells," *J. Histochem. Cytochem.* **13:**75–91 (1965).

Goodenough, D. A., and J. P. A. Revel: "A Fine Structural Analysis of Intercellular Junctions in the Mouse Liver," *J. Cell Biol.* **45:**272–290 (1970).

Kelly, D. E.: "Fine Structure of Desmosomes, Hemidesmosomes and an Adepidermal Globular Layer in Developing Newt Epidermis," *J. Cell Biol.* **28:** 51–72 (1966).

Mukherjee, T. M., and L. A. Staehlin: "The Fine Structural Organization of the Brush Border of Intestinal Epithelium," *J. Cell Sci.* **8:**573–599 (1971).

Pierce, G. B., A. R. Midgley, and J. Sri Ram: "Histogenesis of Basement Membrane," *J. Exp. Med.* **117:**339–348 (1963).

Snell, R.: "An Electron Microscopic Study of Keratinization in the Epidermal Cells of the Guinea Pig," *Z. Zellforsch.* **65:**829–846 (1965).

———: "An Electron Microscopic Study of the Human Epidermal Keratinocyte," *Z. Zellforsch.* **79:** 492–506 (1967).

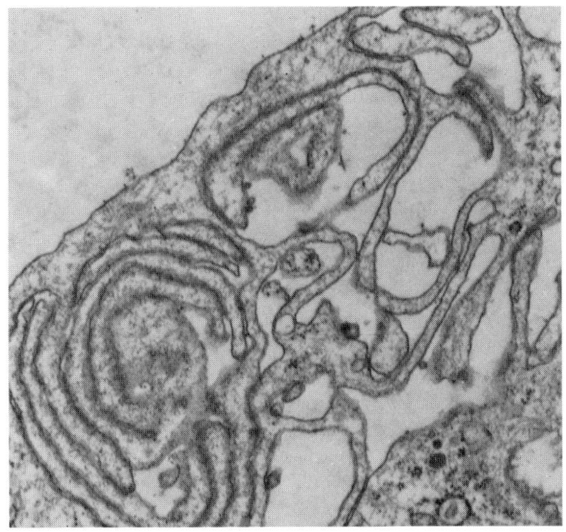

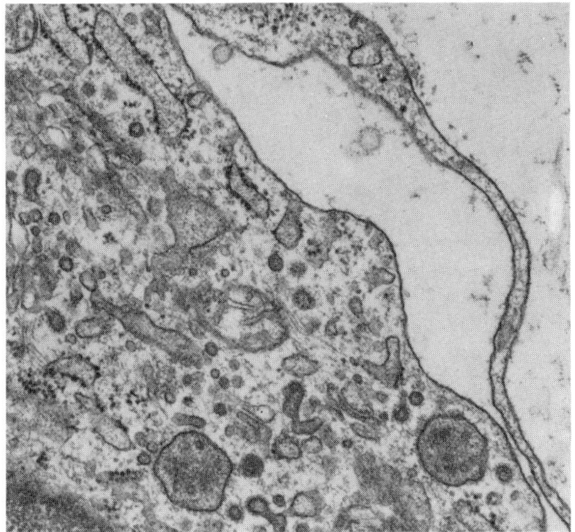

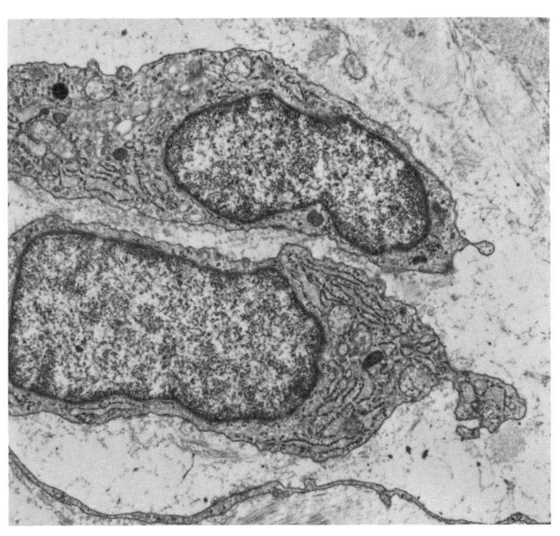

3 CONNECTIVE TISSUE

OBJECTIVES

Upon completion of this chapter, the student will be able to:

1 Identify the origin, morphology, and function of the following connective tissue cells: (a) fibroblasts (b) macrophages (c) plasma cells (d) mast cells (e) adipose cells

2 Identify the origin, morphology, and function of the following connective tissue fibers: (a) collagenous (b) elastic (c) reticular

3 Describe the steps involved in the production of collagenous fibers.

4 Identify the origin, morphology, and function of connective tissue ground substances.

5 Identify the location and the structural and functional characteristics of loose fibroelastic connective tissues.

6 Identify the location and the structural and functional characteristics of dense connective tissues.

7 Name and identify the type of connective tissue found in the embryo that is generally *not* found in the adult.

8 Construct a classification of connective tissues.

Connective tissue, as the term implies, is responsible for maintaining structural interrelationships among various parts of the body, from organs to tissues and cells. The connective tissue also serves as a pathway for substances. Nutrients, gases, wastes, and regulatory chemicals pass through this tissue as they move to and from the blood and various cells of the body. The organization of the extracellular matrix of connective tissues reflects the particular functional requirements of the different types of connective tissues. Even within a given type of connective tissue, such as the *connective tissue proper,* this generalization holds true.

In contrast to other body tissues such as epithelium and muscle, connective tissue has fewer cells in proportion to the intercellular substance. Based on composition of the intercellular substance, adult connective tissue is divided into three main groups: *connective tissue proper, cartilage,* and *bone.* The intercellular substance is soft in connective tissue proper; in cartilage it is often partly calcified but flexible in nature; and in bone it is rigid because of complete calcification of its matrix.

In this chapter the identification and histologic characterization of cells and fibers that make up the connective tissue proper will permit us to develop its subclassification into several categories based on their histologic features.

CELLS

The following cells, present throughout the connective tissue proper, contribute to the synthesis and maintenance of the diverse types of extracellular matrices that best meet the functional needs of each region. For this reason, the number of different cell types in a given region of connective tissue varies widely.

Fibroblasts. This most numerous cell type in the connective tissues produces collagenous and elastic fibers, as shown in Fig. 3-1.

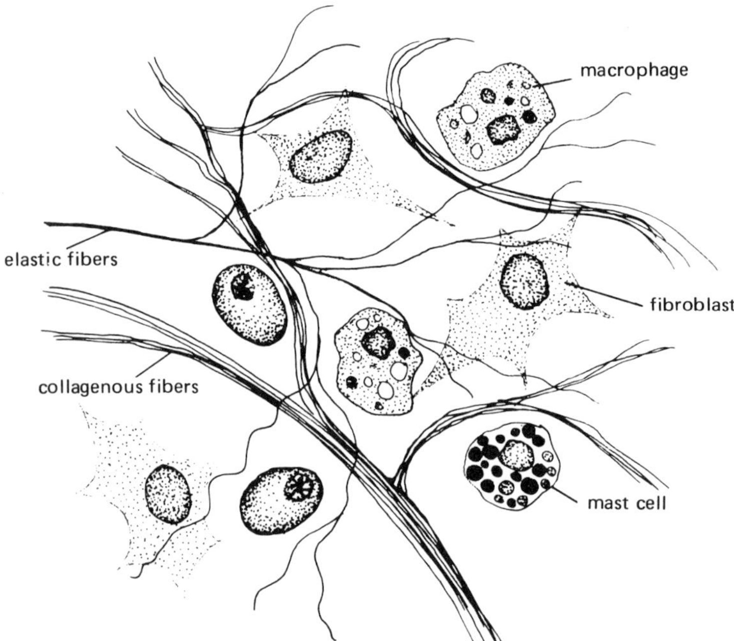

FIGURE 3-1 Drawing of loose fibroelastic connective tissue illustrating its cellular and extracellular components.

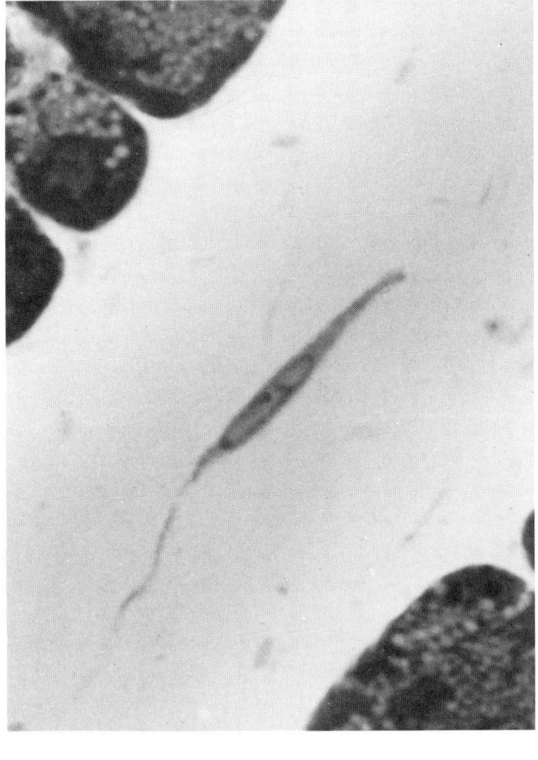

FIGURE 3-2 Light micrograph of fibroblast in side view.

The fibroblasts are also responsible for production of the basic components of the ground substance. Fibroblasts are flat, stellate cells with ovoid nuclei. The fibroblast may appear to have a variety of shapes, depending on the particular plane in which each cell is sectioned and viewed. In Fig. 3-2, for example, the fibroblast appears spindle-shaped because it represents a side view of a flat cell. The cytoplasm generally stains lightly, and it is often difficult to determine the boundary of the cells. An example of the fibroblasts seen from the surface is shown in Fig. 3-3 (arrows). Note again the lightly stained cytoplasm with a large, somewhat angular, ovoid nucleus. The chromatin pattern, when viewed from above, appears dusty with a thin, dense marginal zone of heterochromatin. The nucleus shows occasional nucleoli. Fibroblasts vary in structure according to their functional state. They can be extremely active or quiescent. The appearance of the intracellular organelles in an EM reflects the functional status of the cell. Active fibroblasts, found in healing wounds or developing tissues, produce fibers and ground substances and therefore contain the cytoplasmic machinery necessary for such synthetic activity. A diagram of an EM view of a fibroblast is shown in Fig. 3-4*a*. In this figure, notice the large number of RER profiles and the Golgi complex at the upper left area of the cell. Occasionally, a vestigial cilium is found to arise in the region of the Golgi complex, as indicated by arrows in Fig. 3-4*b*, which is an actual EM of the fibroblast drawn in Fig. 3-4*a*.

An increased number of RER profiles appear in more active fibroblasts (Fig. 3-5). Such fibroblasts will appear more basophilic under the light microscope. In contrast, fi-

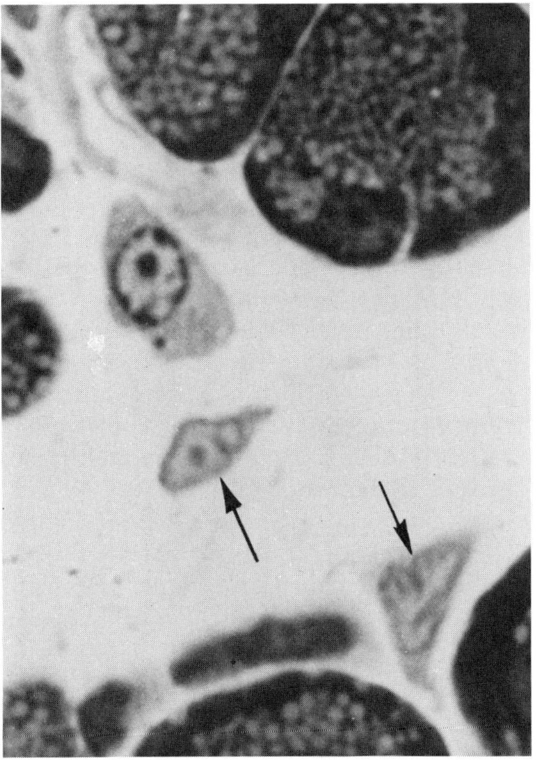

FIGURE 3-3 Surface view of fibroblasts (arrows).

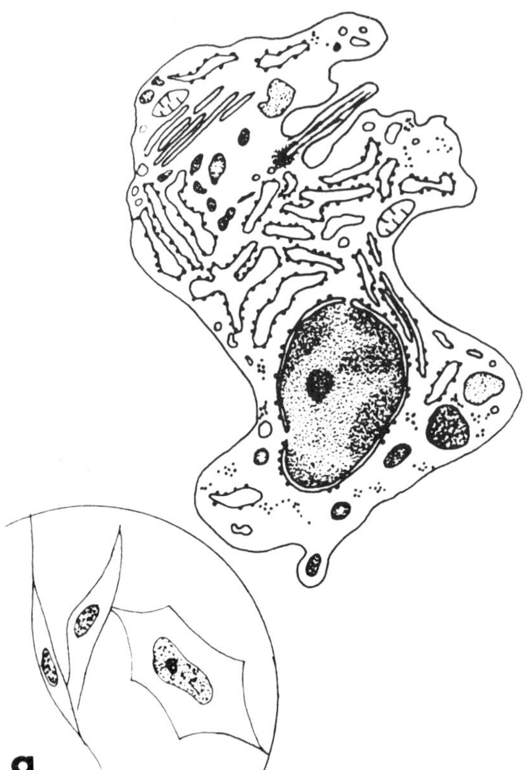

FIGURE 3-4*a* Diagram of a fibroblast seen in the electron microscope.

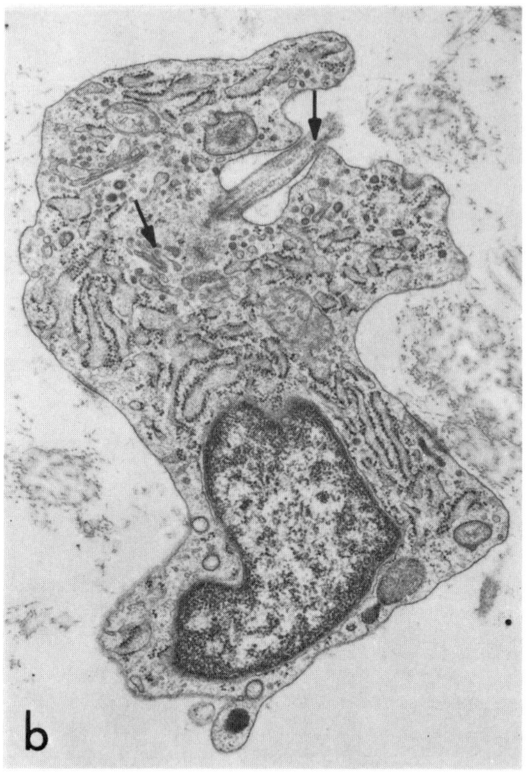

FIGURE 3-4*b* Actual electron micrograph of the fibroblast in Fig. 3-4*a*. Note the cilium originating in the region of the Golgi complex (arrows).

broblasts in a quiescent state become smaller and contain less-differentiated cytoplasm (f in Fig. 3-6). Few of the structures that are actively concerned with the synthesis of intercellular matrix can be distinguished. Quiescent fibroblasts may be found in such regions as the tooth pulp of an individual. The contour of the nucleus is often irregular, as indicated in Fig. 3-7.

Because of the features mentioned above, recognizing fibroblasts under a light microscope is a difficult task; it requires an ability to visualize the cell from different angles and to distinguish relative levels of activity. In an electron microscope the identification of fibroblasts becomes somewhat simpler. In Fig. 3-8, a portion of a fibroblast contains some RER, mitochondria, and a distinct Golgi apparatus. In addition to these cytoplasmic features, the fibroblast often shows regions of the plasma membrane which appear to be specialized for attachment of extracellular microfibrils. These specializations are visible even in a low-power EM (arrows, Fig. 3-8), but can be better visualized at a higher magnification (Figs. 3-9 and 3-10, arrows).

The attachment plaque has not been described in the past. Its structure is not unlike a miniature hemidesmosome in that it involves a densification of subjacent cytoplasm (arrows, Fig. 3-11) and the presence of cloudy, electron-dense material extracellularly. A number of microfibrils are found in association with the attachment plaque (arrows, Fig. 3-10).

Frequently the fibroblast shows a large but diffusely organized Golgi region which is composed of a few to several stacks of

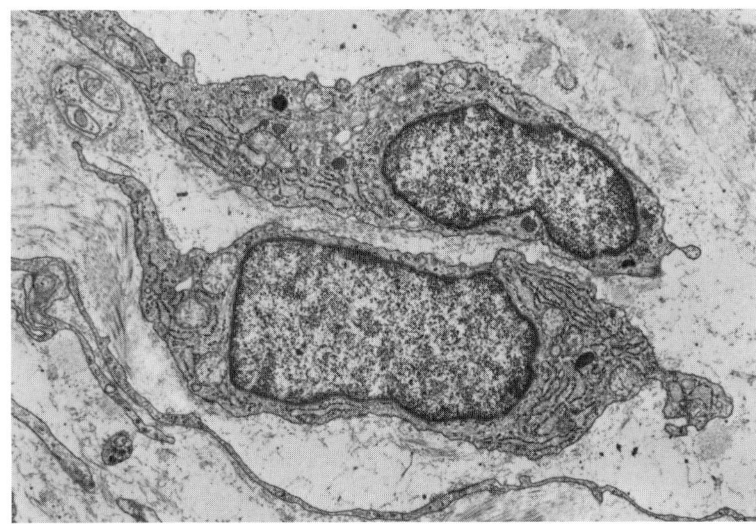

FIGURE 3-5 Electron micrograph of active fibroblasts. Note the numerous profiles of RER.

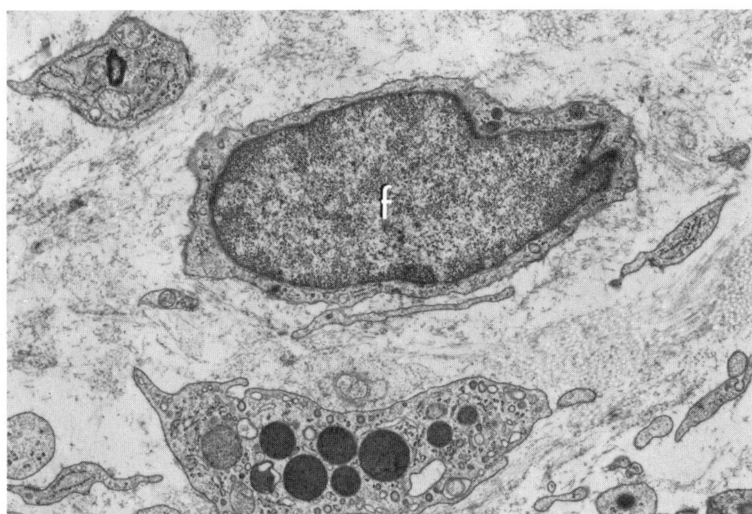

FIGURE 3-6 Electron micrograph of quiescent fibroblast (f).

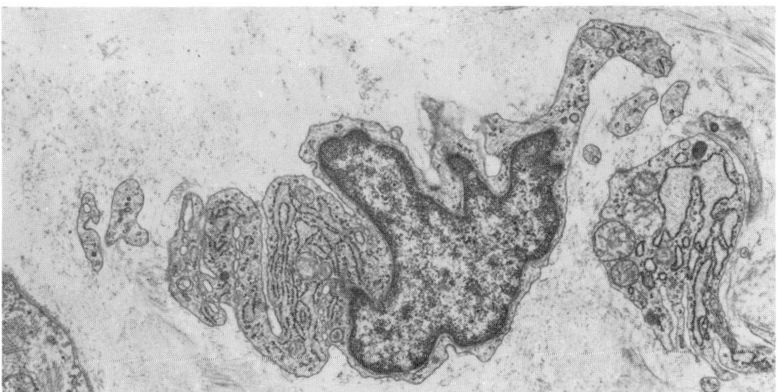

FIGURE 3-7 Electron micrograph of quiescent fibroblast in the tooth pulp. Note the irregular shape of the nucleus.

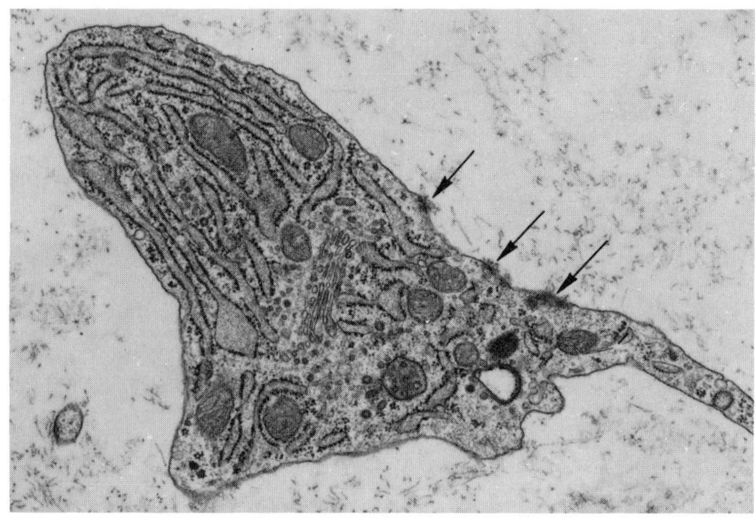

FIGURE 3-8 Portion of a fibroblast containing RER, mitochondria, and a Golgi complex. The plasma membrane shows specializations for the attachment of extracellular microfibrils (arrows).

lamellar elements, transport vesicles, and occasional compound vacuoles (Fig. 3-12). A number of RER profiles, along with several mitochondria, are visualized at the periphery of the Golgi complex. The mitochondria in fibroblasts are generally oblong or rod-shaped, but may appear in EM as rounded structures because of the greater chance of their being sectioned transversely (Figs. 3-8 and 3-9).

Macrophages. As the name implies, these are scavengers that ingest foreign materials they encounter in the tissue space of the body. A prevalent cell type in connective tissue, they are large and usually mobile. The macrophages (m in Fig. 3-13) have a voluminous cytoplasm with an ovoid to round nucleus which shows frequent indentations. Often such a cell would show phagocytized materials along with occasional vacuoles. Macrophages contain a large number of lysosomes to aid in intracellular digestion of phagocytized foreign material (arrows, Fig. 3-14). Figure 3-15 is a portion of a macrophage in which numerous vesicles and vacuoles are

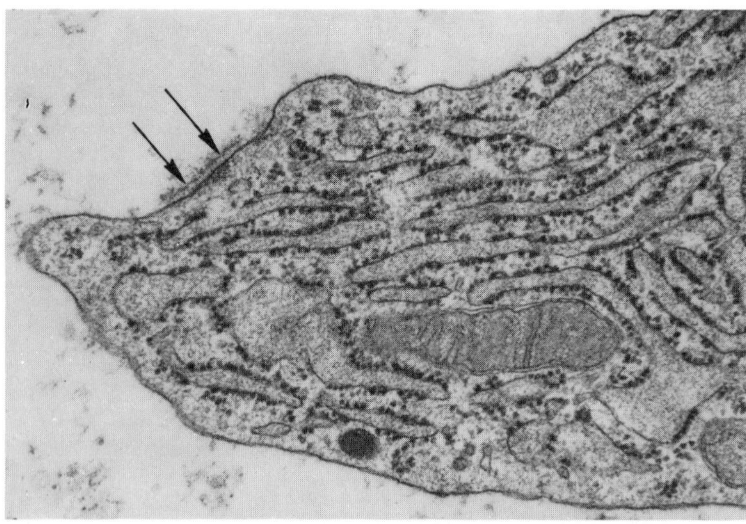

FIGURE 3-9 Portion of a fibroblast showing membrane specialization for attachment of extracellular microfibrils at a higher magnification. An extracellular attachment plaque is indicated by arrows.

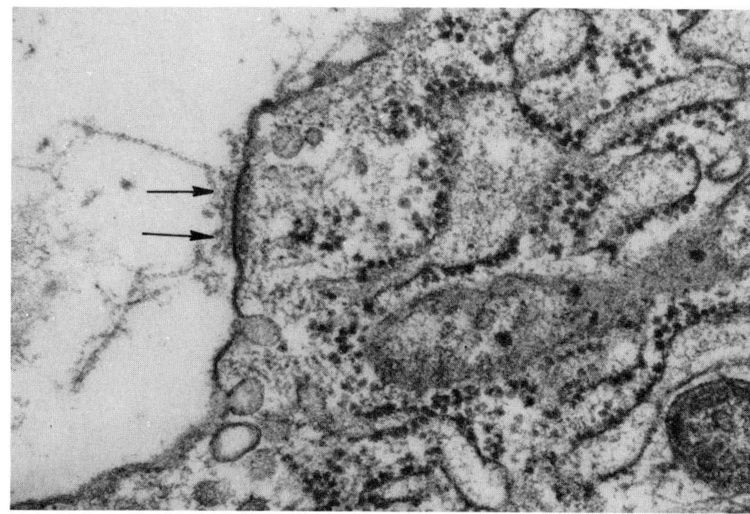

FIGURE 3-10 Hemidesmosome-like attachment specialization on a fibroblast. Microfibrils are associated with the attachment plaque (arrows).

present. Since macrophages are involved in pinocytosis and phagocytosis, the plasma membranes of macrophages often appear extremely ruffled (Fig. 3-16). Figure 3-17 is an EM of a small macrophage ingesting several pneumococci, which are common pathogenic organisms responsible for pneumonia and related respiratory infections.

Figure 3-18 is a diagram illustrating the process of phagocytosis and subsequent intracellular digestion of the microbes. It indicates the initial attachment of the microbe to the plasma membrane (*1*) which then surrounds it (*2*). By constriction of the plasma membrane (*3*), the cell has completely endocytosed the bacterium. Once the cell has engulfed the microbe or a foreign particle, a lysosome fuses with the particle-containing (phagocytic) vacuole (*4*). The joining of a *primary lysosome* of the cell with a phagocytic vacuole produces a *phagosome*, or a secondary lysosome vacuole (*5*). The steps involved in the synthesis of the primary lysosome are indicated in the left half of the diagram by the letters *a*, *b*, *c*, and *d*. Within the phagosome the lysosomal enzymes are activated,

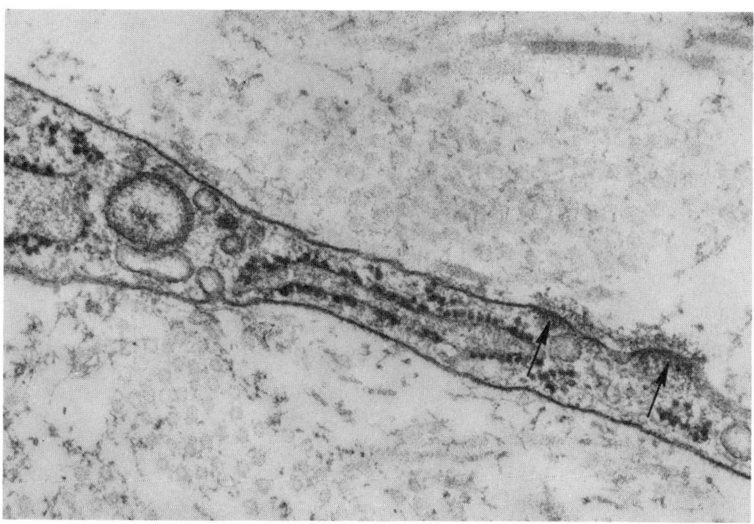

FIGURE 3-11 Cytoplasmic densification associated with attachment specialization of fibroblasts (arrows).

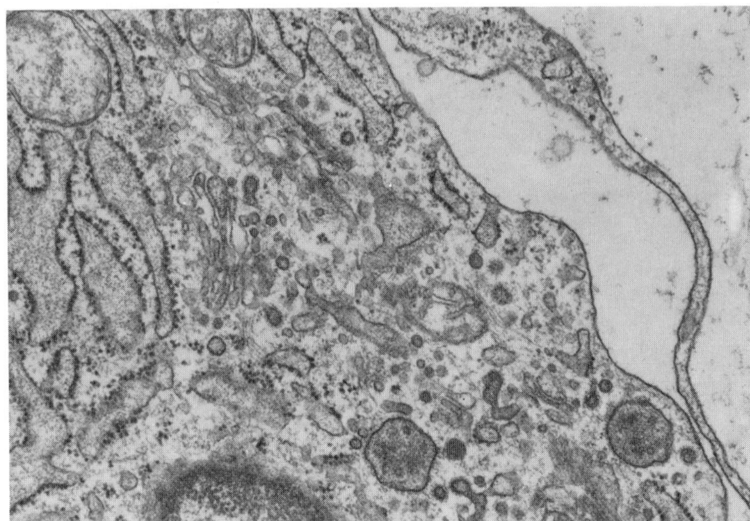

FIGURE 3-12 Portion of a fibroblast having a well-developed Golgi complex.

initiating degradation of the phagocytized substance (*6*). Figure 3-19*a* to *d* are actual EMs depicting the process mentioned above. Note that three lysosomes have fused with a phagocytic vacuole containing two of the microbes (L in Fig. 3-19*d*).

In addition to the phagocytosis and intracellular digestion of matter that is foreign to the organism, a phagosome could result from phagocytosis of unwanted cells such as worn-out red blood corpuscles. Macrophages in the spleen are specialized to eliminate the

FIGURE 3-13 Macrophages (m). Note the irregularly shaped nucleus and dense bodies (lysosomes) and vacuoles in the cytoplasm.

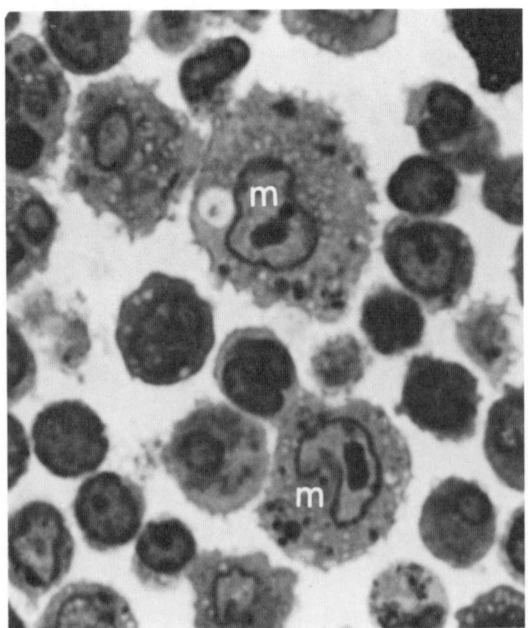

FIGURE 3-14 Electron micrograph of macrophage. Lysosomes (arrows) aid in the digestion of ingested foreign material.

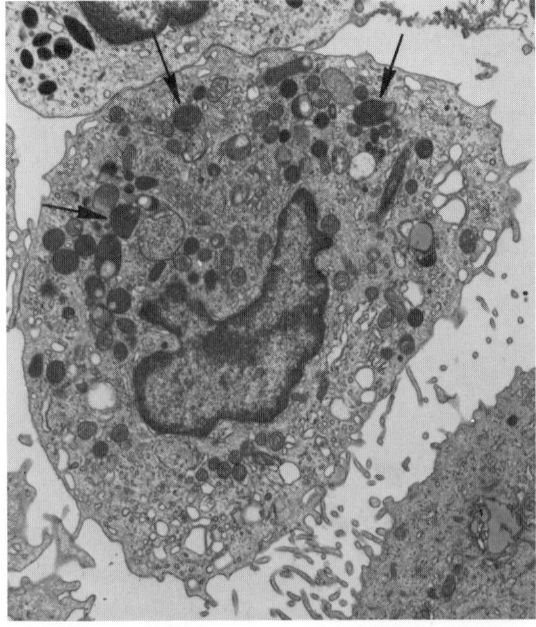

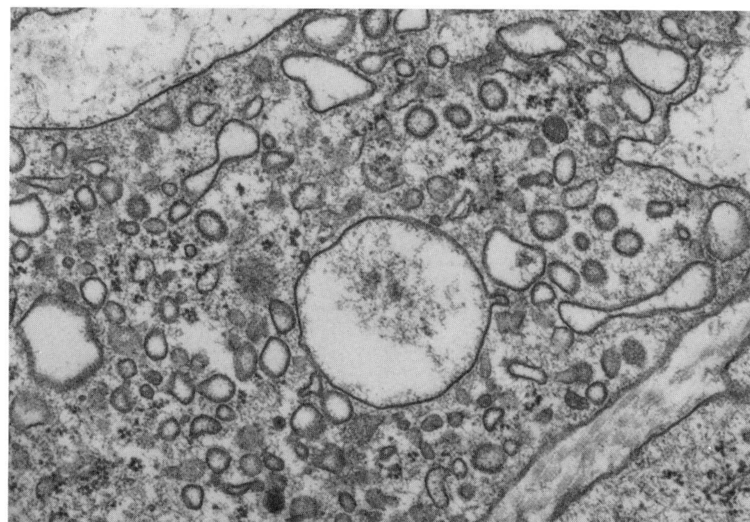

FIGURE 3-15 Portion of a macrophage containing vesicles and vacuoles.

old red blood corpuscles in most mammals. This requires that the macrophage recognizes the aged red corpuscles, engulfs them, and subsequently treats the phagocytized corpuscles as foreign matter.

In the macrophage in Fig. 3-20, several profiles of red blood corpuscles are undergoing different stages of hydrolysis. While trapping foreign material and intracellular digestion are unique functions of macrophages, cells that are not specialized for phagocytosis could segregate portions of their own cytoplasm and digest it. This form of secondary lysosome is called an *autophagic vacuole*. In contrast to a phagosome, the autophagic vacuole formation reflects a cell's response to changing requirements in terms of the quantity of different organelles. Examples of autophagic vacuoles are seen in Fig. 3-21*a* and *b*, in which a number of secretory granules, mitochondria, and RER have been segregated by a membrane.

In summary, a macrophage in an EM can be identified by the presence of an irregular

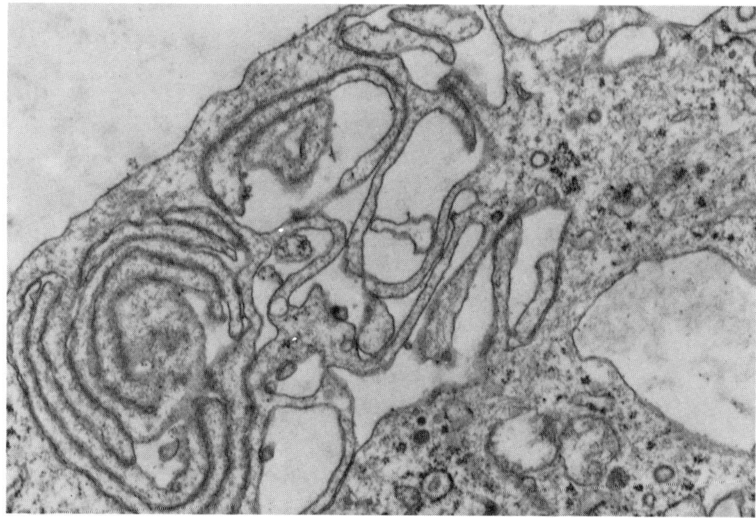

FIGURE 3-16 Electron micrograph of cell membrane ruffles on a macrophage.

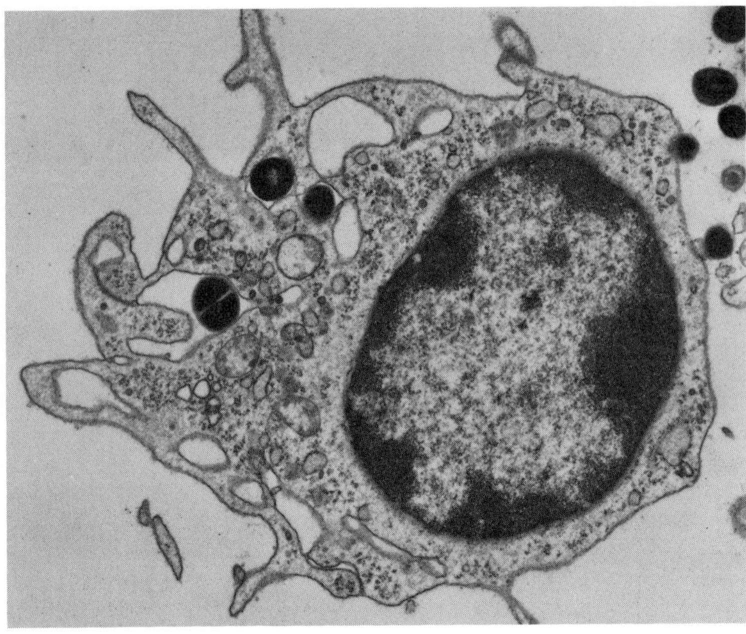

FIGURE 3-17 Electron micrograph of a macrophage phagocytizing pneumococci.

or ruffled cell surface, numerous vesicles and vacuoles, and primary lysosomes and phagosomes.

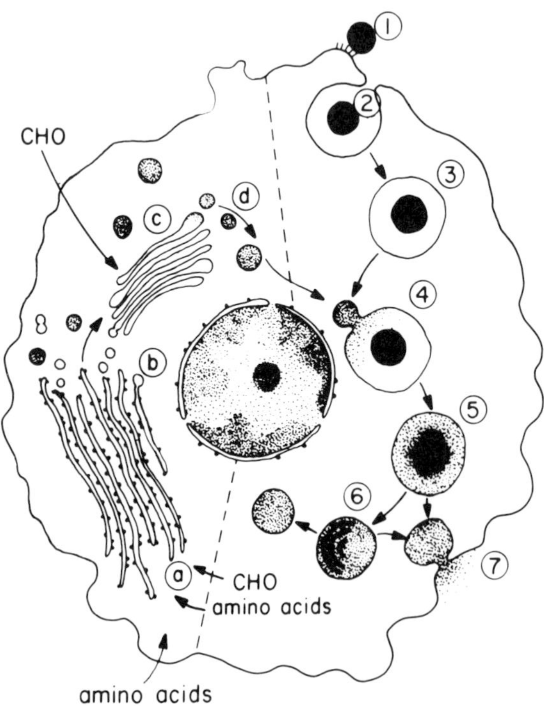

FIGURE 3-18 Diagram illustrating the different steps involved in the process of phagocytosis and intracellular digestion of microbes. CHO stands for carbohydrate components that are also incorporated in the synthesis of primary lysosomes.

Mast Cells. Of the connective tissue cells, these cells can be more readily seen with special staining such as toluidine blue. They are frequently found along small blood vessels throughout the body. These cells resemble basophilic granulocytes (white blood cells) of the peripheral blood and are filled with large granules. Figure 3-22 shows a mast cell (arrow) present between a small artery (a) and vein (b). In Fig. 3-23, four mast cells are present in the connective tissue space. They were specially stained to demonstrate mast cell granules, which appear dark. Compare the size and density of the mast cells with those of the fibroblasts scattered throughout the field (Fig. 3-23). Because they resemble blood basophils, the mast cells have often been called *tissue basophils.* Mast cell

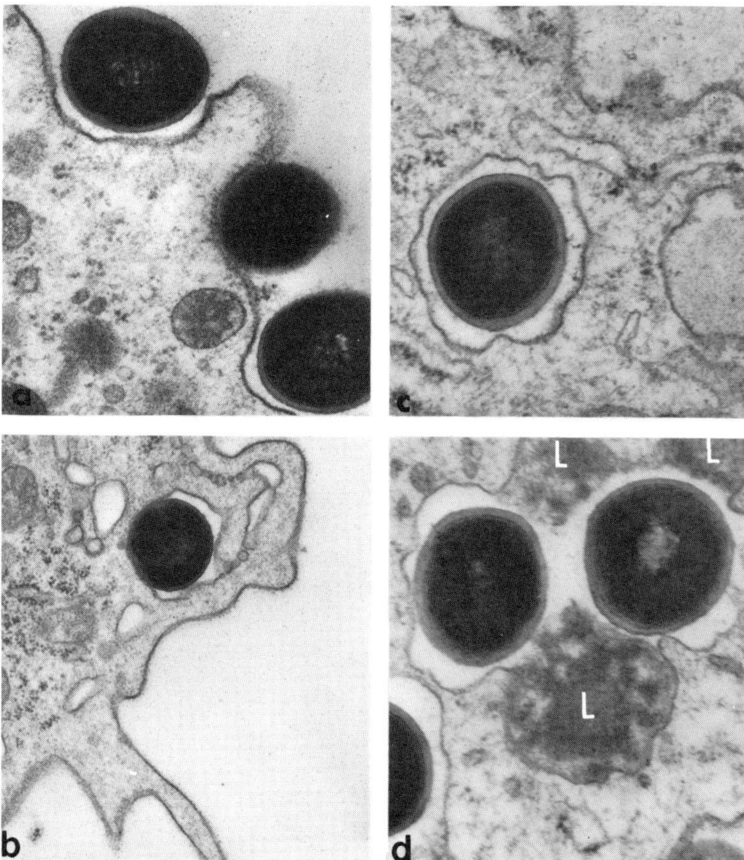

FIGURE 3-19 Electron micrographs depicting the process of phagocytosis. (*a*) Microbes are becoming attached to the cell membrane of the macrophage. (*b*) A microbe is completely surrounded by two cell processes of the macrophage. (*c*) A microbe is completely endocytosed and is located in a vacuole. (*d*) Three lysosomes (L) have fused with a vacuole containing two microbes.

granules contain heparin (an anticoagulant) and histamine (a vasodilator). Heparin changes the physical characteristics of the ground substance to make it more fluid, while histamine is intimately concerned with the regulation of endothelial cell junctions.

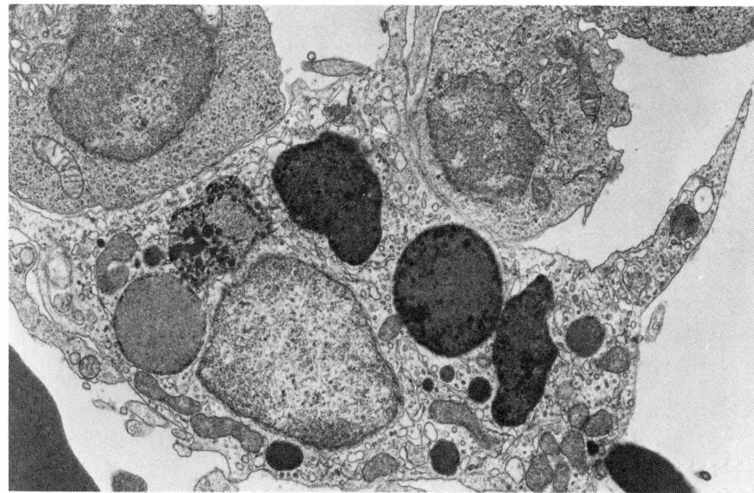

FIGURE 3-20 Macrophage containing several ingested red blood corpuscles.

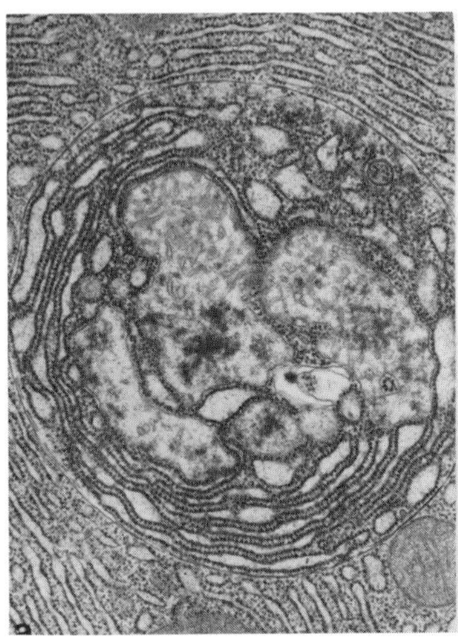

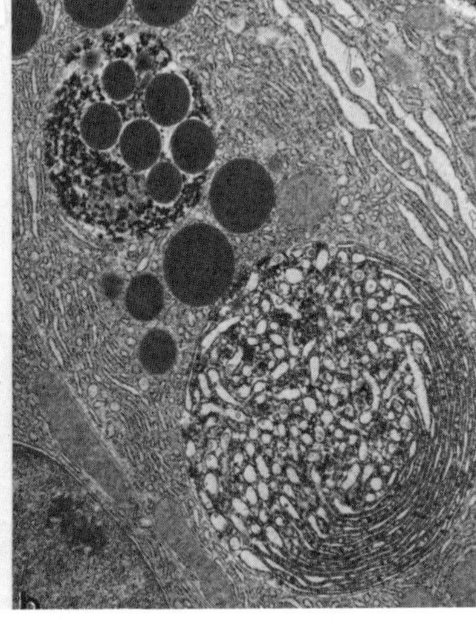

FIGURE 3-21*a* Electron micrograph of autophagic vacuole containing RER, mitochondria, and secretory granules.

FIGURE 3-21*b* Electron micrograph of autophagic vacuoles containing RER, vesicles, and secretory granules.

Hence these cells help to regulate vascular permeability as well as the physical characteristics of the surrounding tissue.

The EM in Fig. 3-24 shows the details of the cytoplasm of mast cells loaded with granules of varying densities. Only a small portion of the nucleus is visible, because the cell is sectioned off the center of the nucleus. In EM the mast cell granules are generally homogeneously electron-dense structures. However, irregularity in contour and lamellated substructures within them have been observed (Fig. 3-25).

Plasma Cells. These cells indicated by arrows in Fig. 3-26*a* are ovoid to rounded cells

FIGURE 3-22 Mast cell (arrow) in loose connective tissue adjacent to small blood vessels (artery (a) and vein (b)).

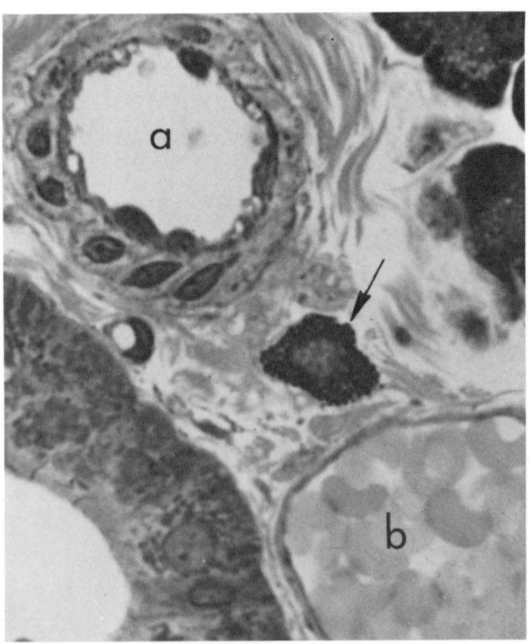

FIGURE 3-23 Mast cells in loose connective tissue.

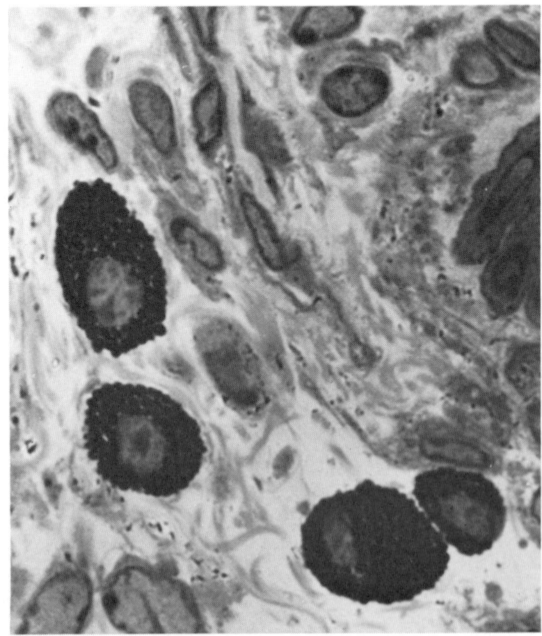

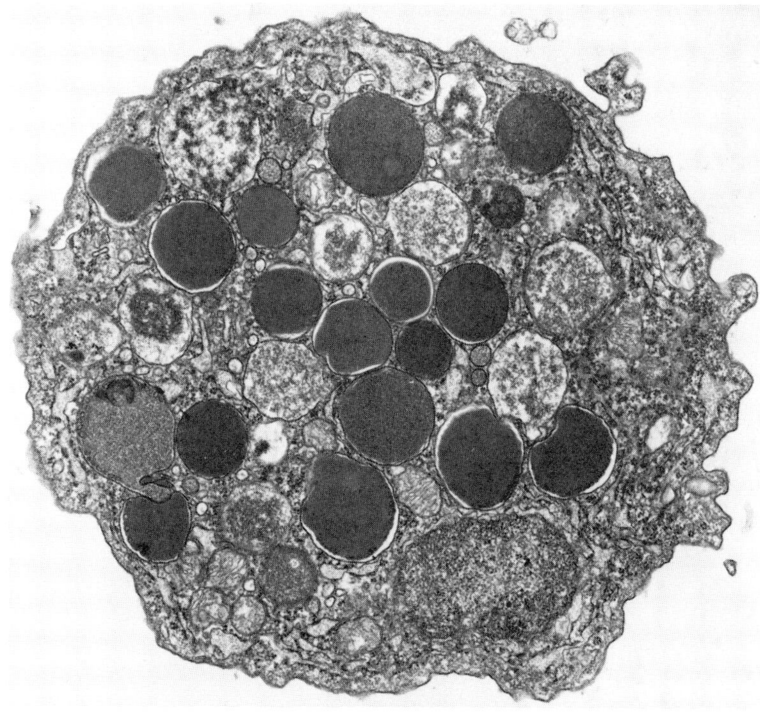

FIGURE 3-24 Electron micrograph of a portion of a mast cell. Note the varying densities of the granules.

with eccentrically located nuclei. The chromatin appears as dark-staining clumps along the periphery of the nucleoplasm, which often presents a cartwheel appearance. Plasma cells and their immediate precursors, plasmablasts, are highly basophilic and the major producers of circulating antibodies which neutralize antigens. Three plasma cells can be seen in Fig. 3-26*a*. Note the clear region next to the nucleus, which is the region where the Golgi complex is located. A technique which employs the fluorescent microscope can demonstrate the presence of antibodies in these cells when the antibodies

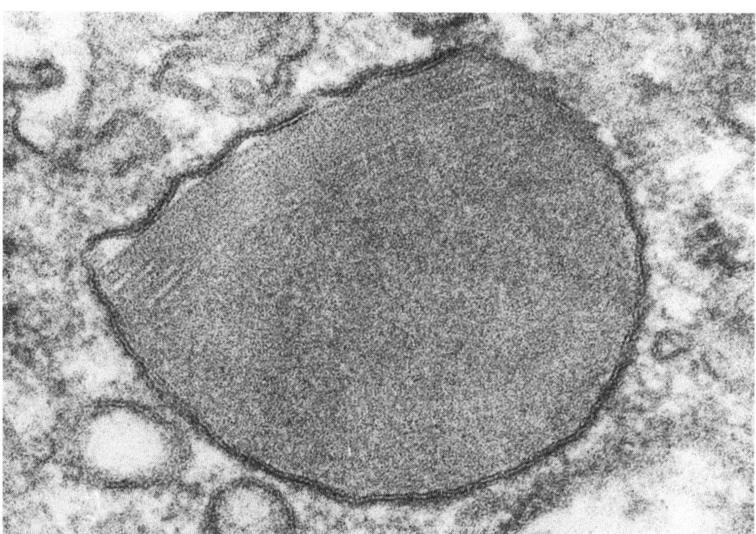

FIGURE 3-25 Electron micrograph of a mast cell granule at higher magnification. Note its irregular contour and the lamellated structure within it.

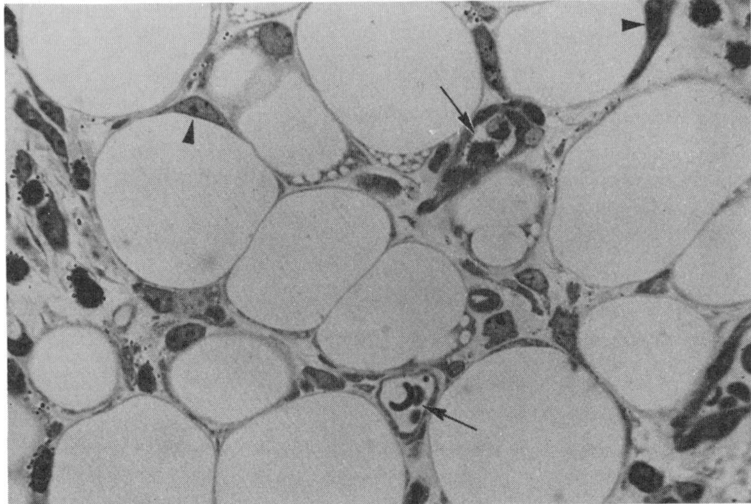

FIGURE 3-26*a* Light micrograph of three plasma cells (arrows). Note the eccentric nuclei with cartwheel arrangement of the chromatin and the negative images of the Golgi apparatuses.

FIGURE 3-26*b* Fluorescent micrograph of plasma cell stained with antigens against antibodies contained in the cytoplasm. The antigens are conjugated with a fluorescein dye which makes them visible in the fluorescent microscope.

are stained with antigens conjugated with fluorescein dyes. An example of this is seen in Fig. 3-26*b*. The light cytoplasm of the cell is where the antibodies are located. Since antibodies are proteins that are chemically identical with serum gamma globulins, the cytoplasm of plasma cells is filled with well-developed RER and a large Golgi complex indicating that they are synthesizing these molecules very actively (Fig. 3-27). These well-developed protein-producing organelles tend to displace the nucleus.

Adipose Cells. These cells are frequently found in loose areas of connective tissue as well as in specialized fat storage regions such as subcutaneous connective tissues and buccal fat pads, and contain a nucleus pushed flat against the cell membrane by stored lipids (Fig. 3-28). Note the peripherally lo-

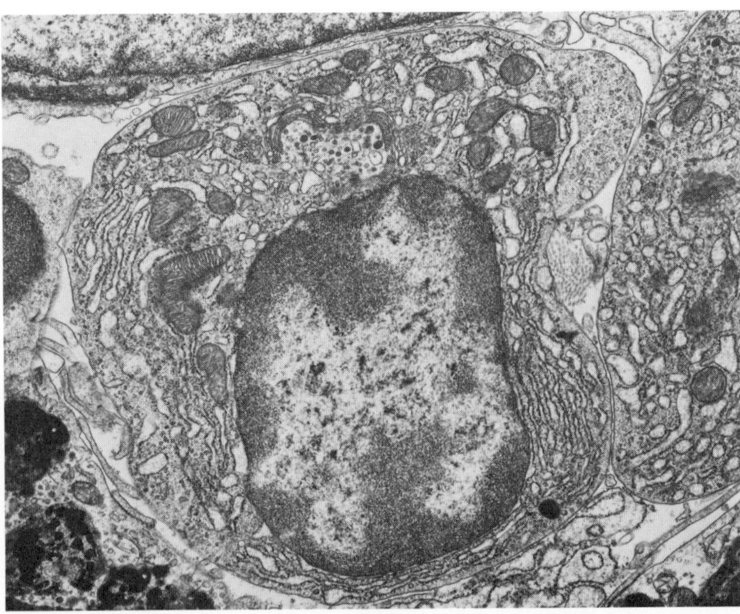

FIGURE 3-27 Electron micrograph of a plasma cell. Note the extensive RER and the well-developed Golgi complex, characteristics of a cell actively synthesizing proteins.

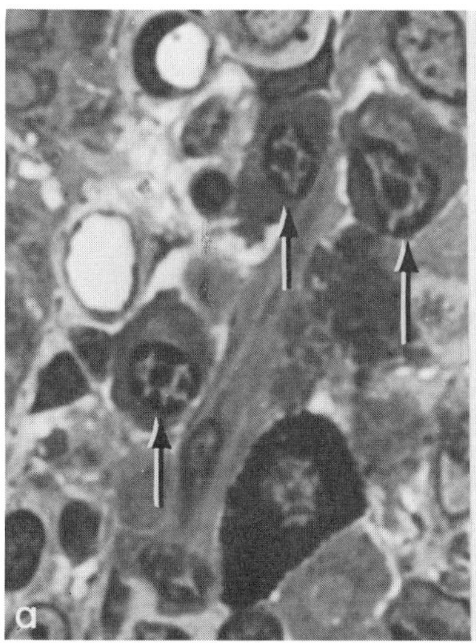

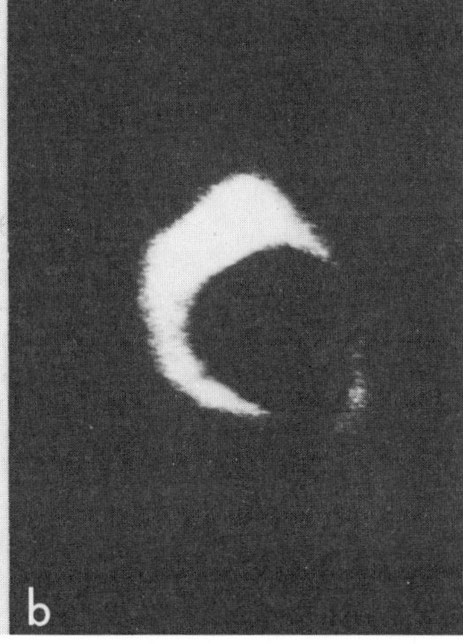

FIGURE 3-28 Adipose cells. Their flattened nuclei are indicated by arrowheads. Interspersed between these cells are capillaries and larger vessels (arrows).

cated nuclei (arrowheads) of the adipose cells. They are often difficult to distinguish from those of fibroblasts.

In microscopic preparations, adipose cells appear empty because the lipid has been dissolved during the process of dehydration. Since the adipose cells represent a storage area of nutrients in the form of lipids and neutral fat, many capillaries and vessels are closely associated with fat cells (arrows, Fig. 3-28).

EXTRACELLULAR MATRIX

Connective tissue proper consists of a scattering of the cell types discussed in the preceding section in a relatively large amount of extracellular matrix. The matrix is made up of varying proportions of fibers and ground substances.

FIBERS

Light microscopic appearance

Fibers in the extracellular matrix can be regarded as ropes which bind the body cells and tissues together. Three major types of fibers are recognized: collagenous, elastic, and reticular. Collagenous fibers, which are most numerous, provide the strength needed to hold cells and tissues together. They occur in various diameters according to the strength needed, and they are capable of withstanding great tensile forces.

Collagenous Fibers

Figure 3-29 contains many *collagenous fibers* of varying diameters with several fibroblasts and a few lymphoid cells and macrophages scattered among them. A collagenous fiber may branch or join a similar fiber to produce smaller or larger fibers. In tendons and ligaments, collagenous fibers could be highly regularly arranged.

Elastic Fibers

Elastic fibers, obviously named for their ability to stretch, can be found in abundance wherever flexibility is required. In Fig. 3-30, which is a spread of mesentery, the dark elastic fibers are thin, stretched, and uniform in diameter. Note the much thicker collagenous fibers that course in a wavy manner. While the elastic fibers are present in almost all connective tissue proper, they require special stains and therefore cannot be readily visualized after H and E staining, as in Fig. 3-29.

Reticular Fibers

Reticular fibers resemble fine collagenous fibers in their chemical composition and elec-

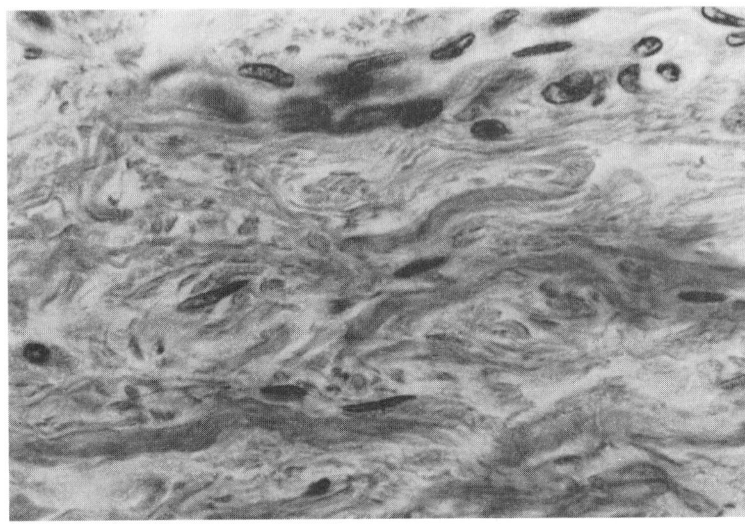

FIGURE 3-29 Light micrograph of collagenous fibers. Several fibroblasts are scattered among them.

tron microscopic appearance. Named for the meshwork-like appearance (reticulum), they are present in many tissues. Reticular fibers form the stroma (or framework) of blood-forming organs and lymphoid tissues. In Fig. 3-31, the reticular fibers from a lymph node are stained black by a silver impregnation technique, indicating the fine, delicate nature of the fibers that are in close contact with the large reticular cells that are associated with them.

A similar area from a lymph node in Fig. 3-32 has not been stained for demonstration of reticular fibers. A number of large cells with light cytoplasm and pale nuclei among the more numerous lymphocytes containing small, dark nuclei are the reticular cells (arrows).

The reticular fibers are singled out as different fiber type despite their similarity to collagenous fibers. This is because they form a delicate meshwork in certain specialized (lymphoid and hemopoietic) organs and contain a ground substance that is richer in carbohydrates than other fibers. Additionally, there are certain differences in fiber-to-

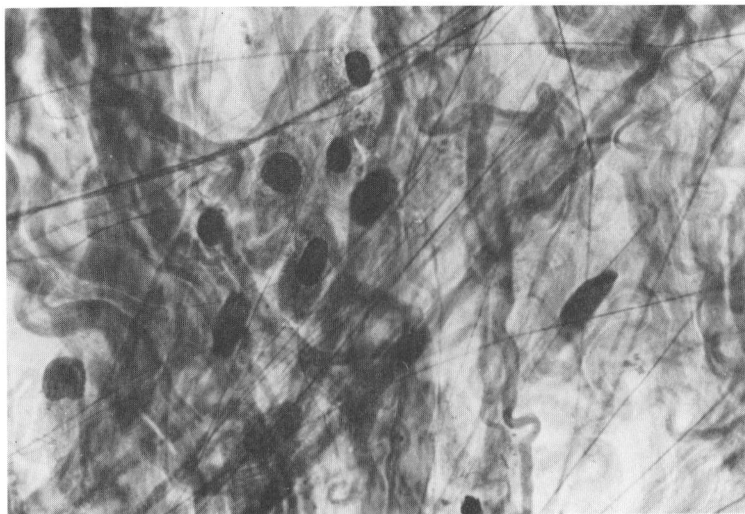

FIGURE 3-30 Elastic fibers seen in a spread of mesentery.

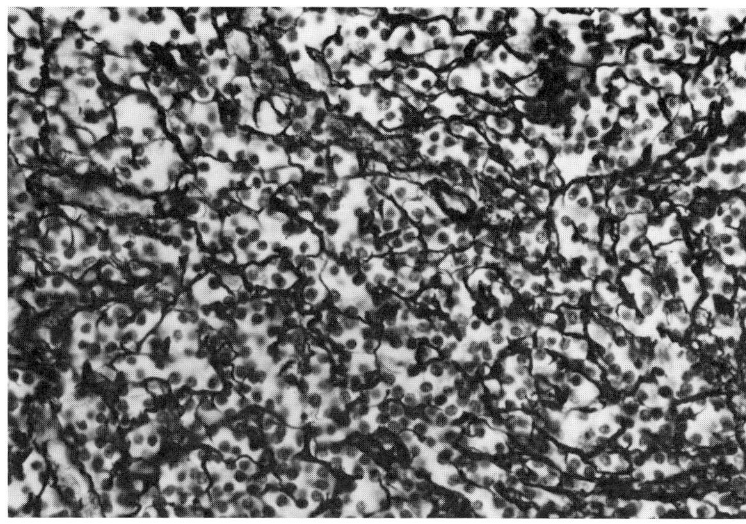

FIGURE 3-31 Light micrograph of a lymph node section. Reticular fibers are stained black by a silver impregnation technique.

cell relationships, which will be discussed in a later chapter dealing with lymphoid organs. Special stains such as PAS readily demonstrate the abundance of polysaccharides. These stains have demonstrated that reticular fibers are also present in other regions of the body, appearing between muscle cells of the digestive tube, among skeletal muscle cells, and in association with basement membranes throughout epithelia of the body (refer to Fig. 2-20). Since reticular fibers are produced by fibroblasts in regions other than blood-forming and lymphoid organs, the rigid distinction between reticular and collagenous fibers may not be justified as was once thought. However, because of the different staining characteristics of the two fibers, it is still useful to make the histologic distinction.

Production and ultrastructure of collagenous fibers

Collagenous fibers are made of collagen and cementing (ground) substances. Originally the term *collagen* was used to refer to a chemically extractable fibrous entity. Recent

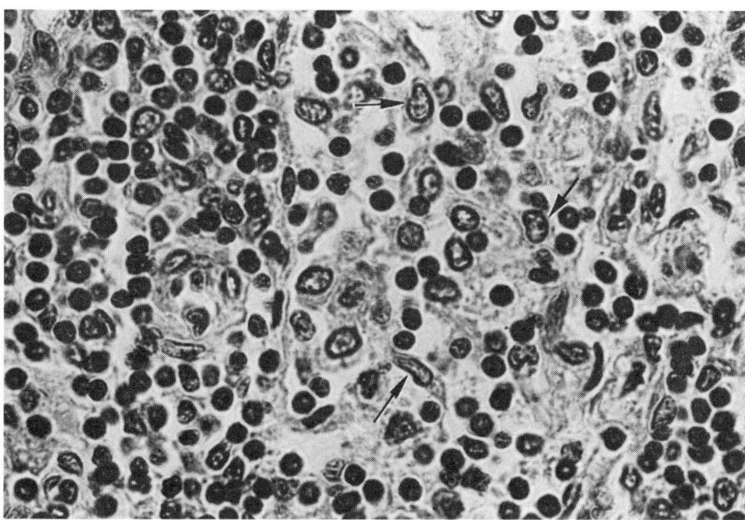

FIGURE 3-32 Light micrograph of a section of a lymph node stained for demonstration of the cellular components. Compare it with Fig. 3-31. Reticular cells (arrows) are seen among the more numerous lymphocytes.

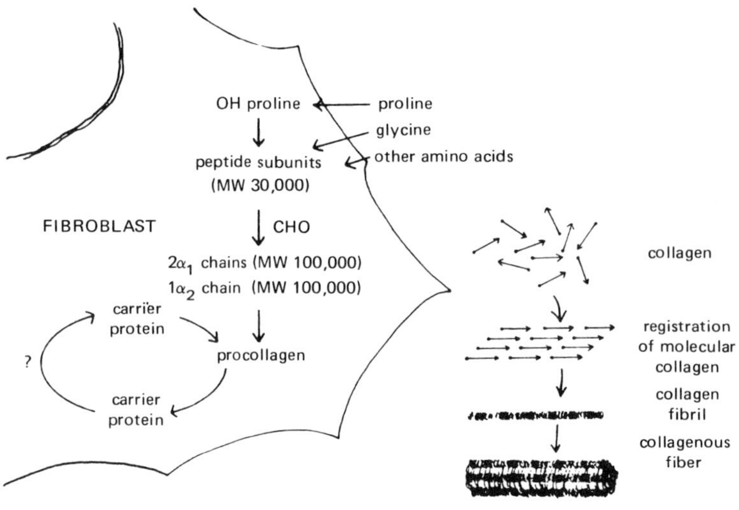

FIGURE 3-33 Diagram of collagen production illustrating steps from the initial intracellular linking of amino acids to the extracellular aggregation of the finished molecules.

advances in our knowledge of the chemistry of collagen and its synthesis have furnished the following understanding of the production of collagenous fibers by the fibroblasts. Collagen, as other proteins, begins to be formed within the fibroblast when the amino acids are linked together into peptide chains by the polysomes located along the cytosol surface of RER (Fig. 3-33). The peptide chains are segregated out of the ground cytoplasm into the RER and then moved through its cisternae to the region of the Golgi complex where their orderly packaging takes place. As you recall from Chap. 1, the Golgi complex is involved in the synthesis of polysaccharides and other carbohydrate moieties. Thus the end products, both peptide chains and polysaccharides, are collected in Golgi (secretory) vacuoles and subsequently exocytosed or secreted out of the cell in the usual way.

The original peptide chains, each having a molecular weight of about 100,000, intertwine in groups of three, as shown in Fig. 3-34, to produce a single needle-shaped collagen molecule with a molecular weight of 300,000 which measures 280 nm in length and 1.5 nm in diameter (Fig. 3-34).

Currently, intensive investigations are under way to determine exactly where these peptide chains combine, although it is generally suspected that the Golgi complex is involved.

Whether produced in or out of the cell, collagen molecules are found in the extracellular matrix adjacent to the cell. The nascent collagen molecule has an extra piece of carrier protein which is necessary in the subsequent registration of the collagen molecules. This combination of the carrier protein and a collagen molecule is called a *procollagen*. The carrier protein of the procollagen is cleaved off when molecular registration is completed.

As a molecular entity, collagen is unique in the proportion of glycine, proline, and hydroxyproline it contains. Its molecular

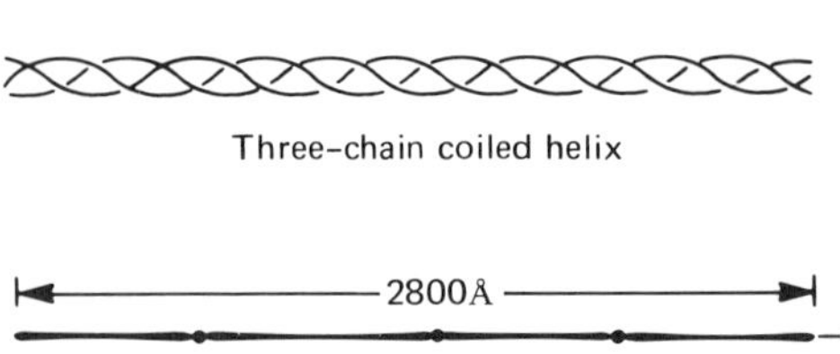

FIGURE 3-34 Diagram of the arrangement of the three coiled polypeptides in the collagen molecule.

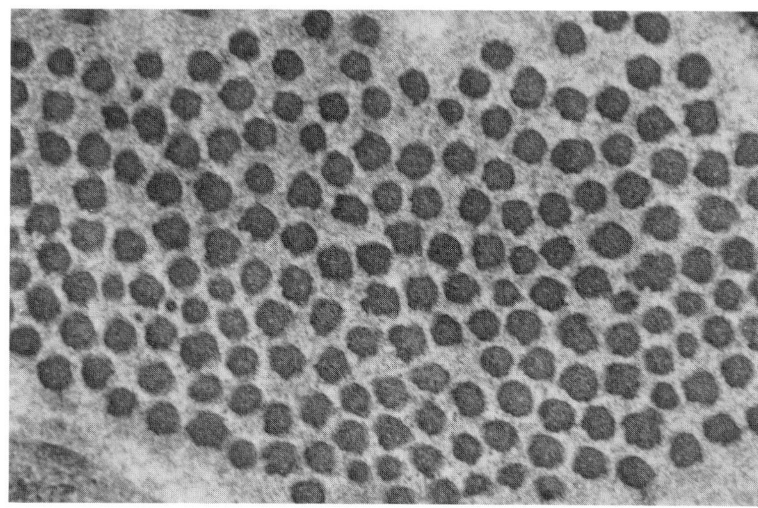

FIGURE 3-35 Electron micrograph of collagen fibrils seen in cross section.

framework includes glycine as every third amino acid and either proline or hydroxyproline as every fourth and fifth amino acid. The importance of glycine is demonstrated by its position in the helical conformation of the three peptides, where it is located at every third amino acid position and extends toward the center of the collagen molecule, producing the bonding of the polypeptide chains to each other in the formation of the three-chained coiled helix.

A *collagen fibril,* about 60 to 80 nm in diameter as seen in electron microscopy, is produced by the quarter-stagger registration of collagen molecules, diagram in Fig. 3-33. An actual EM depicting the cross-sectional view of collagen fibrils is seen in Fig. 3-35. Individual collagen fibrils show the typical cross striations of 64 nm which reflect the major, end-to-end alignment of the quarter-staggered molecules. Figure 3-36 shows a longitudinal section of collagen.

Collagen fibers are composed of bundles of unit fibrils of collagen. Therefore, the

FIGURE 3-36 Longitudinal view of collagen fibrils depicting their typical cross striations.

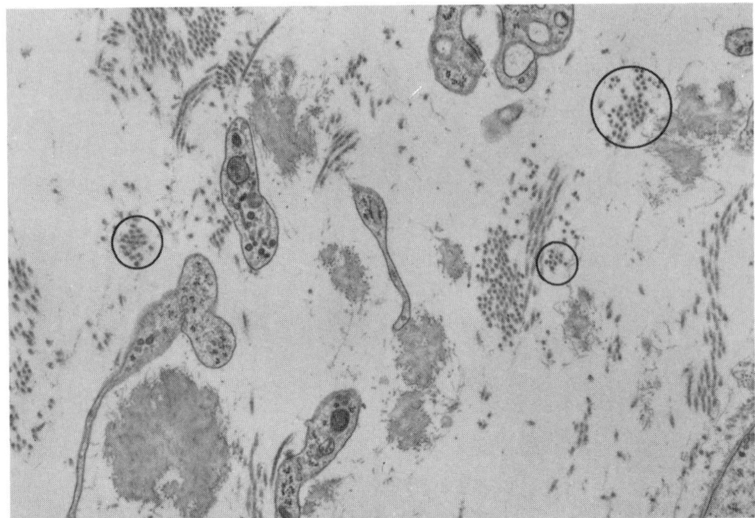

FIGURE 3-37 Bundles of collagen fibrils form larger collagen fibers (circles).

increase in fiber diameter is accomplished by the formation of larger and larger bundles of the individual collagen fibrils, as shown in Fig. 3-37 (circles). A bundle of about 15 fibrils is the smallest *collagenous fiber* that can be seen in LM.

Figure 3-38 shows two small bundles of collagen fibrils in cross section. This is a specimen specially prepared so that the accumulation of flocculent or cloudy ground substance around each individual collagen fibril can be seen (arrows). When these sheaths around individual fibrils fuse, they form a collagenous fiber. Therefore, the ground substance within each bundle serves as a cementing substance in constructing a collagenous fiber. Like a rope, the collagenous fibers gain strength when they are combined in large bundles. It is the combination of greater numbers of individual fibers that determines the tensile strength of the final bundle.

Unlike collagenous fibers, *elastic fibers* are not made up of cross-banded fibrils. But the

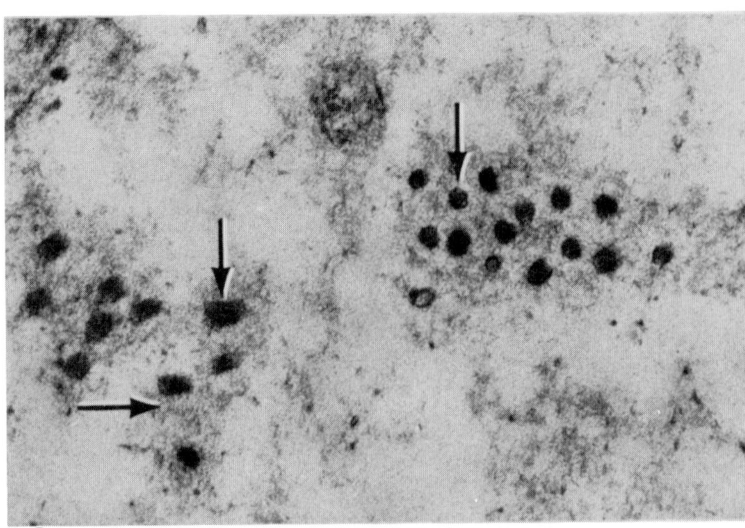

FIGURE 3-38 Bundles of collagen fibrils seen in cross section. This specially prepared specimen illustrates the appearance of the accumulation of ground substance around the fibrils (arrows).

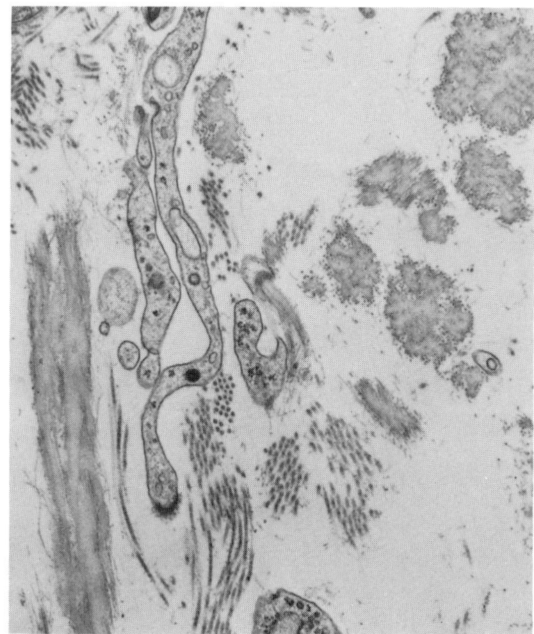

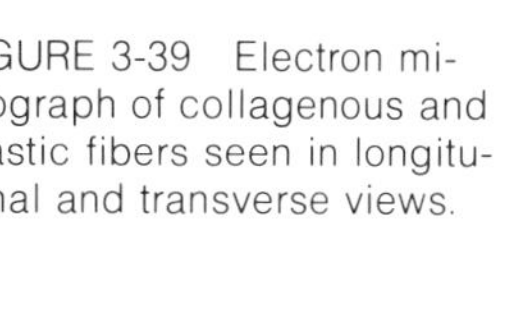

FIGURE 3-39 Electron micrograph of collagenous and elastic fibers seen in longitudinal and transverse views.

fibers do have an organized composition, since slender microfibrils are present embedded in an amorphous material. Chemically composed of elastin, these fibers branch and anastomose freely, forming elaborate networks. They stretch with a high degree of resiliency, are refractile in LM, and appear blue when stained with such special stains as Weigert's. In EM an elastic fiber appears as a homogenous fibrous material of moderate electron density. Compare the appearance of elastic fibers with that of collagenous fibers in Fig. 3-39. Both longitudinal and transverse views of elastic fibers are depicted in this photomicrograph from the soft palate. Elastic fibers are often covered by microfibrils of 11 nm or so in diameter. Figure 3-40 is a cross section of an elastic fiber which shows the close association of microfibrils at its surface. These microfibrils are also found along collagen fibrils.

Ground substance, a term given to the extracellular matrix minus the fibers, contains primarily proteins, complex carbohydrate moieties, ions, and bound and free water. The majority of the carbohydrates are acid mucopolysaccharides (*glycosaminoglycans*), which maintain the physicochemical matrix characteristics of different types of connective tissue. This is accomplished when their highly negatively charged molecular structures act with ions present in the extracellular fluid. The most common mucopolysaccharides include hyaluronic acid and several sulfated forms such as condroitin sulfates A, B, and C. Because the different fibers are present singly and in bundles, the mucopolysaccharides and ions of the matrix create an enormous range of possible interactions, depending on the environmental milieu.

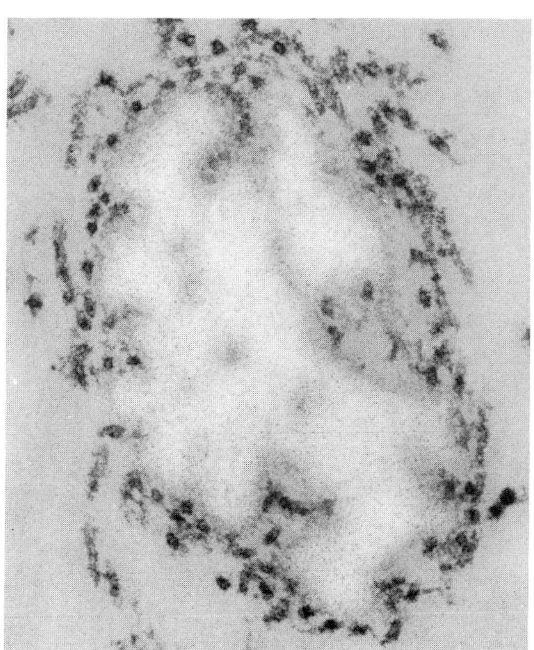

FIGURE 3-40 Cross-sectional view of an elastic fiber showing the close association of microfibrils with its surface.

ADULT CONNECTIVE TISSUE

Cells, fibers, and ground substance combined make up the connective tissue, a functioning unit. The classification of adult connective tissue depends on its functionally related composition.

The most prevalent form of connective tissue is located in the subcutaneous region and the interstices of tissues along blood vessels and nerves. It also forms the stroma of most organs. With cells and fibers loosely packed in the matrix, it is called *loose fibro-elastic connective tissue* (*loose FECT*). Figure 3-41 shows typical loose FECT with many adipose cells. The irregularly arranged collagen and elastic fibers in a highly fluid ground substance produce a fairly soft, displaceable tissue. Loose FECT contains fibroblasts, macrophages, adipose cells, blood vessels, and nerves (arrows, Fig. 3-41). Often it functions in support of associated avascular epithelium as in the case of the mesentery. Notice that there are more fibrous elements around the blood vessels and nerves. Loose FECT with a predominance of adipose cells (Fig. 3-28) is sometimes called *adipose tissue*.

In certain regions of the body, more collagenous fibers of large diameter predominate in the FECT. Such a region is called *dense FECT*. Dense FECT contains more fibers, less ground substance, and fewer cells than loose FECT. Dense FECT can be divided into two groups, depending on fiber orientation: dense irregular and dense regular. Dense irregular FECT has its fibers running in various directions, as shown in Fig. 3-42. It occurs in the dermis of the skin and also makes up the covering of internal organs. Dense regular FECT, in which fibers run in the same direction, is present in areas where increased directional tensile strength is needed. The dense regular FECT, therefore, is found in tendons, ligaments, and aponeuroses where forming secure connections is of primary importance. Figure 3-43 shows a cross section of a tendon. At a higher magnification it is seen to be made up of fibers with tendon cells inserted between them (Fig. 3-44*a*). A longitudinal view of the tendon is visualized in Fig. 3-44*b*. These tissues rarely contain cellular components other than fibroblasts which, in the case of tendon, are called *tendon cells* (arrows, Fig. 3-44*a*) because of their squeezed appearance between fiber bundles (Figs. 3-43 and 3-44). The regular arrangement of the collagenous fibers is more

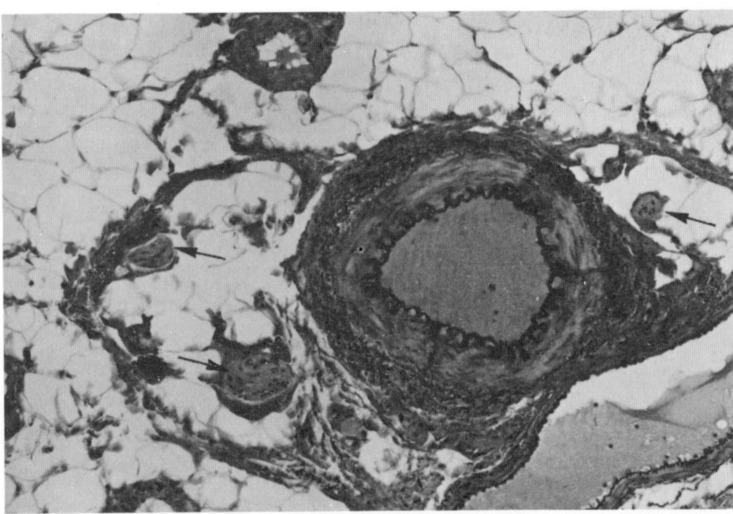

FIGURE 3-41 Light micrograph of loose FECT containing fibroblasts, macrophages, numerous adipose cells, blood vessels, and nerves (arrows).

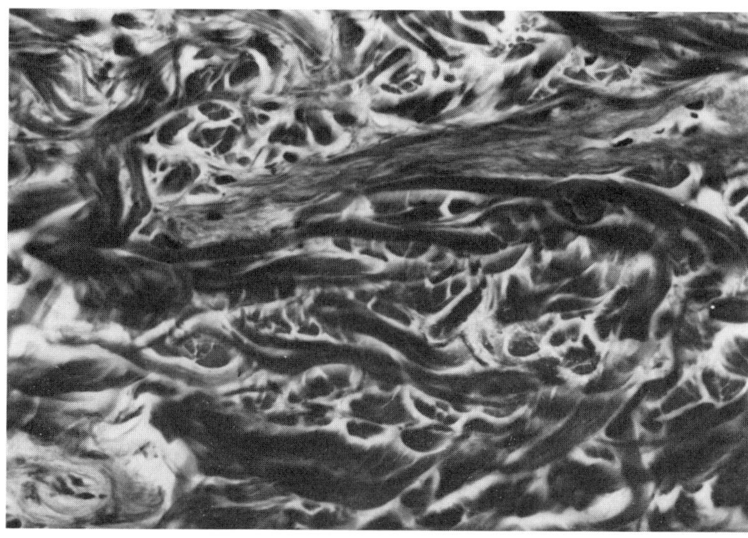
FIGURE 3-42 Dense irregular FECT in skin.

easily discerned in longitudinal sections (Fig. 3-44*b*).

In the tendon, fiber bundles are organized as individual fibers wrapped by a thin layer of collagenous fibers called *endotendineum* (a in Fig. 3-45*b*). Several such bundles are put together to produce a large bundle which is again wrapped by a heavier layer of collagenous *peritendineum* (Fig. 3-45*b*). The even heavier, final wrapping of the entire tendon is the *epitendineum* (Fig. 3-45*a*). Similar organizational patterns may be found in other structures such as muscles and nerves, where bundling of fibrous structures by collagenous fibers is needed.

The presence of *reticular connective tissue* as an adult-type connective tissue has been mentioned in connection with reticular fibers. Although reticular connective tissues are typically present among lymphoid- and blood-forming organs, they are also present in sinusoidal walls of many endocrine tissues. Furthermore, the chemical and structural similarity between reticular and colla-

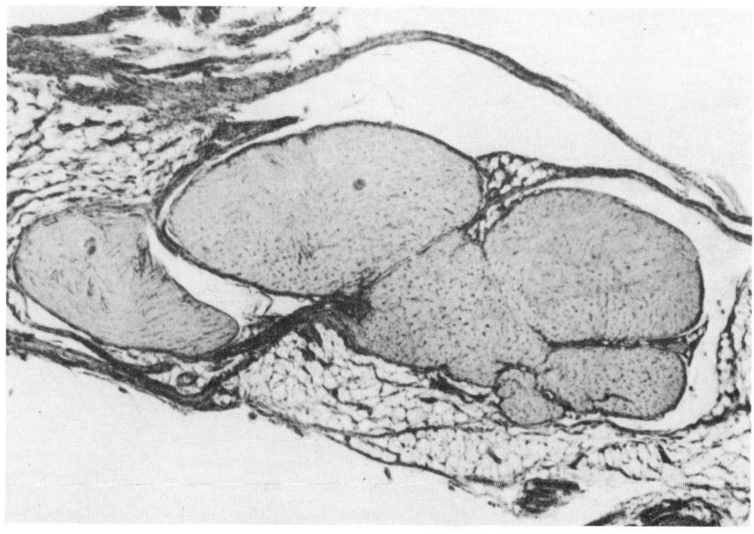
FIGURE 3-43 Cross section of a tendon with dense regular FECT.

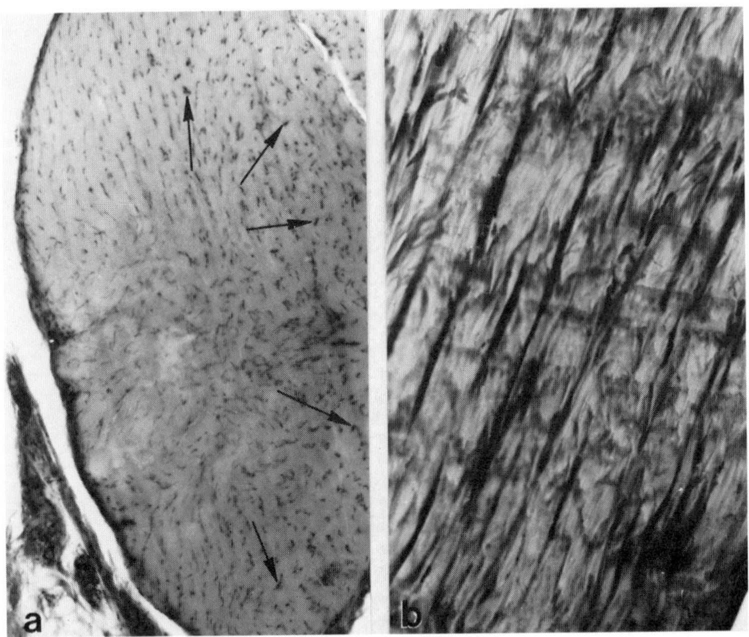

FIGURE 3-44*a* Cross section of a portion of a tendon seen at higher magnification. Tendon cells (arrows) are scattered among the collagenous fiber bundles.

FIGURE 3-44*b* Longitudinal section of tendon at higher magnification.

genous fibers make it difficult to clearly delineate which tissues do or do not contain reticular fibers.

EMBRYONIC CONNECTIVE TISSUE

The connective tissue appearing in embryonic and early fetal development is called *mesenchyme*. This type is composed of large stellate-shaped cells, like those shown in Fig. 3-46, that are separated by abundant intercellular matrix. The intercellular matrix of the mesenchyme lacks definitely formed fiber bundles but is composed of fine collagenous fibrils in a profuse ground substance rich in mucopolysaccharides. Note the empty appearance of the matrix demonstrated by a

FIGURE 3-45*a* Diagram of a cross section of a tendon and its wrappings.

FIGURE 3-45*b* Enlarged diagram of a cross section of tendon; a = endotendineum; b = peritendineum.

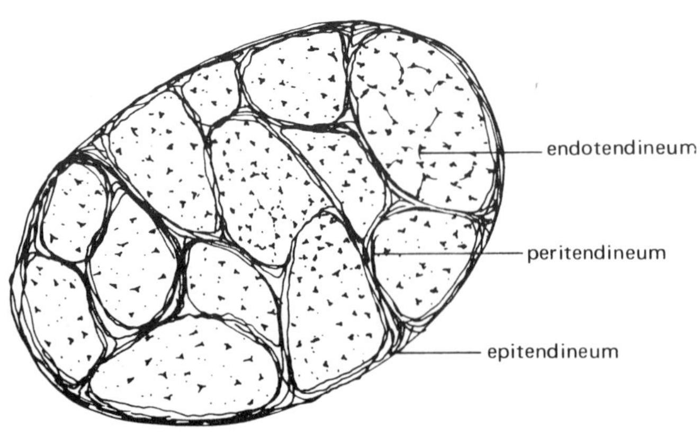

a

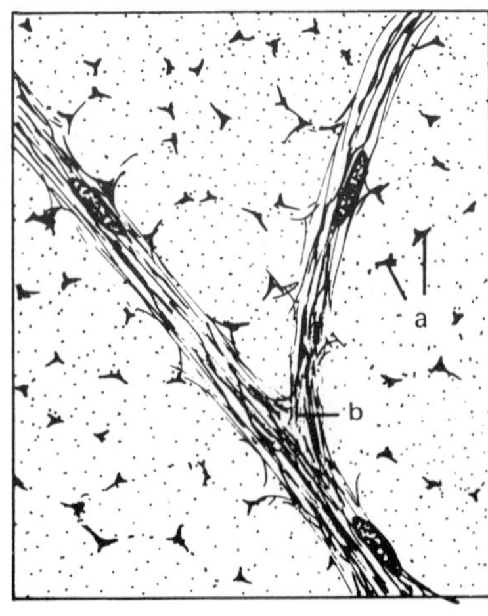

b

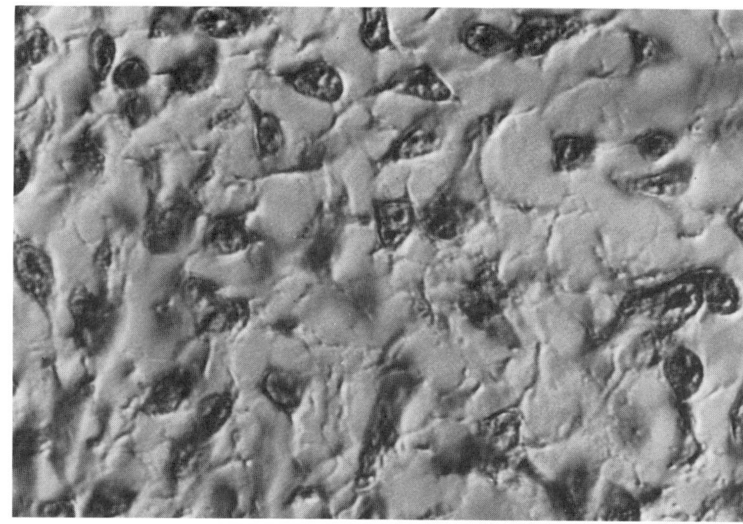

FIGURE 3-46 Embryonic connective tissue or mesenchyme.

phase contrast micrograph of embryonic connective tissue (Fig. 3-46). The cells of the mesenchyme divide and are pluripotential, since they can develop into many different types of connective tissue and related cells during later life. A specialized mesenchyme is present in the umbilical cord connecting a fetus to its mother. Because of the mucoid nature of this ground substance, the mesenchyme of the umbilical cord has been called *Wharton's jelly.*

The considerations of connective tissue organization described in our discussion of adult and embryonic connective tissue have permitted the classification of connective tissue given in Table 3-1.

CLASSIFICATION OF CONNECTIVE TISSUES

TABLE **3-1**

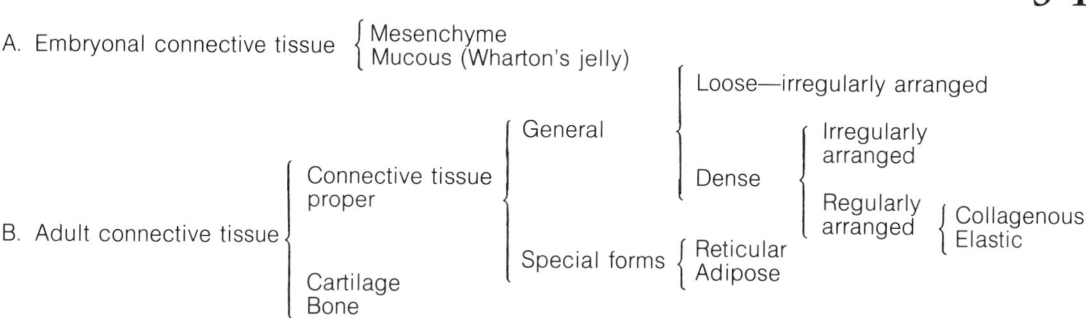

A. Embryonal connective tissue	Mesenchyme				
	Mucous (Wharton's jelly)				
B. Adult connective tissue	Connective tissue proper	General	Loose—irregularly arranged		
			Dense	Irregularly arranged	
				Regularly arranged	Collagenous
					Elastic
		Special forms	Reticular		
			Adipose		
	Cartilage				
	Bone				

REVIEW SECTION

1. Figure 3-47 shows a type of connective tissue. What type of fibers predominate? __________. Would you say the fiber bundles are thick or thin? __________. Are they arranged randomly? __________. Classify this tissue.

The relative predominance and thickness of the fibers help identify this connective tissue as dense. The arrangement helps classify it as regular. This dense regular FECT contains collagenous fibers and is actually a section from a tendon.

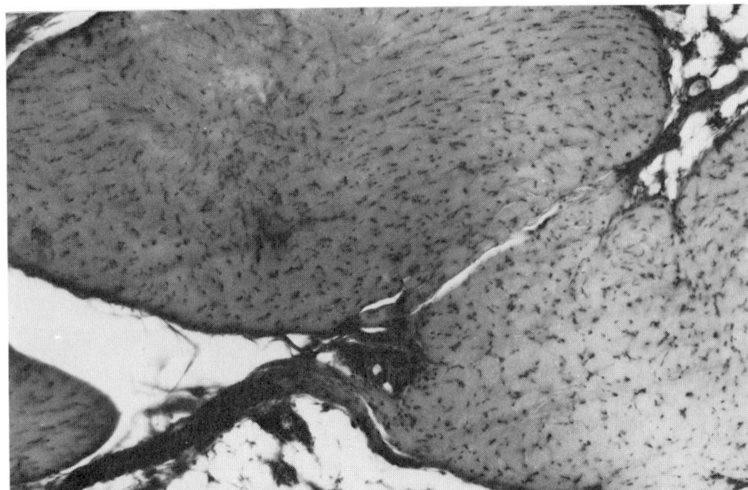

FIGURE 3-47

2. Figure 3-48*a* and *b* show irregularly and regularly arranged fibers and cells. Would you classify this as dense or loose? __________. What type of fibers are they?

The two figures are more or less comparable areas prepared through paraffin embedding (3-48*a*) and epoxy resin preparation (3-48*b*), respectively. Since much of the recent literature, including some of the micrographs used in this book, takes advantage of the greater structural details attained by epoxy-embedded materials, it is important that you become able to identify structures prepared in different ways. The tissue in both figures is loose irregular FECT containing numerous collagenous fibers.

3. Cross sections of several blood vessels can be seen throughout Fig. 3-49. Are blood vessels typically present in connective tissue? __________ Classify the tissue.

Because connective tissue surrounds most vessels, it often shows blood vessels and capillaries in sections. There are irregularly arranged collagenous fibers, fibroblasts, and a few mast cells darkly stained.

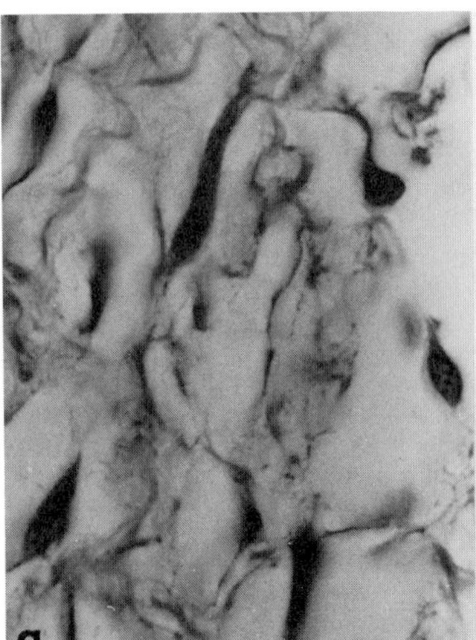

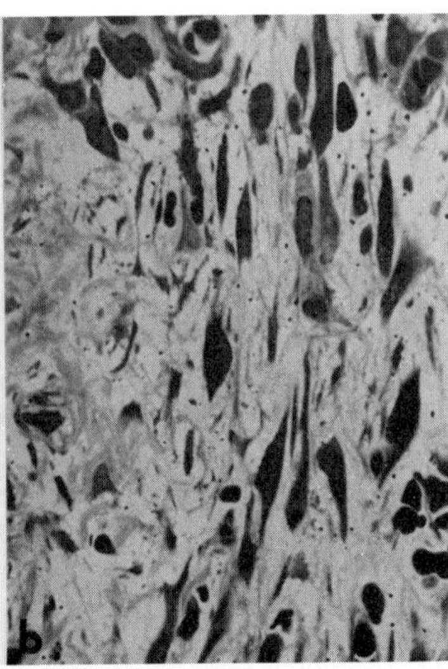

FIGURE 3-48

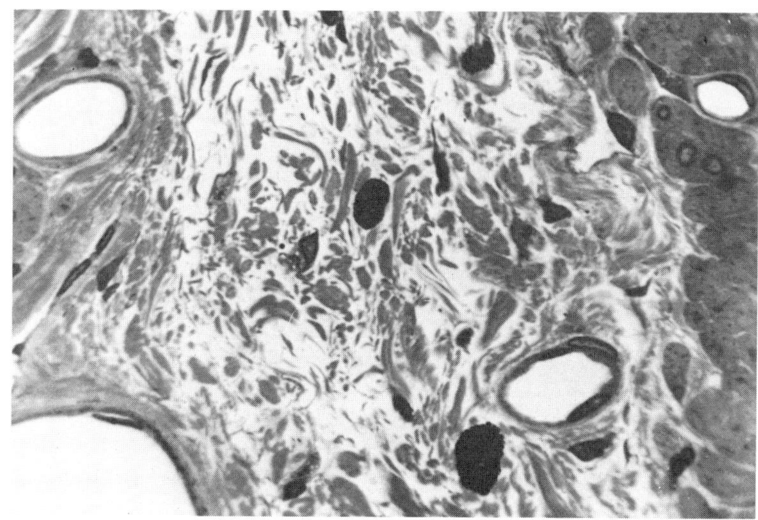

FIGURE 3-49

4. In Fig. 3-50 a cell and a number of cross-sectioned fibers can be seen. Name the fibers labeled "f." __________. What other fibers do you expect to see? __________. The cell represents the most numerous cell type found in connective tissue. Name it. __________. What is the function of the cell? __________. Identify as many as you can of the ground substance components that occupy the rest of the field (i.e., space).

The fibroblast is usually stellate in life. It produces most connective tissue fibers and ground substances. The fibers labeled f are elastic fibers. One can see some collagenous fibers and microfibrils.

5. Figure 3-51 is a region of the connective tissue from the small intestine. Identify the three labeled cell types (a, b, and c) you have studied in this chapter. They are:
(a) __________ (b) __________
(c) __________

The connective tissue appearing in this field contains several different cell types that are responsible for maintenance and defense mechanisms. You should be able to identify (a) macrophages, (b) plasma cells, and (c) fibroblasts.

FIGURE 3-50

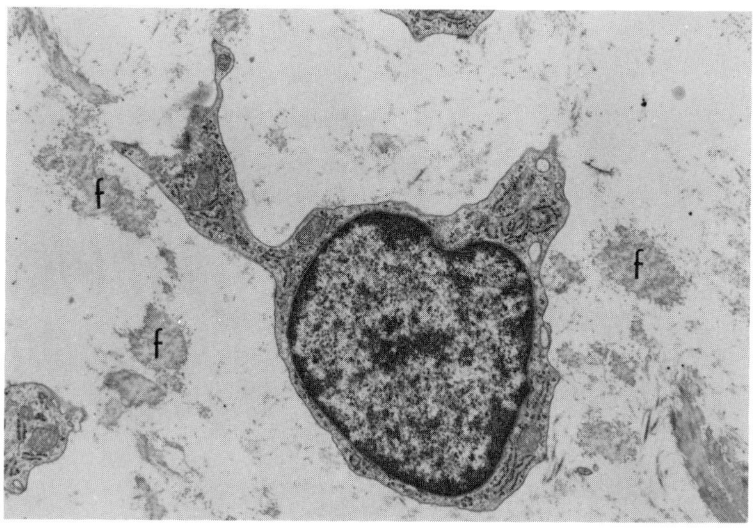

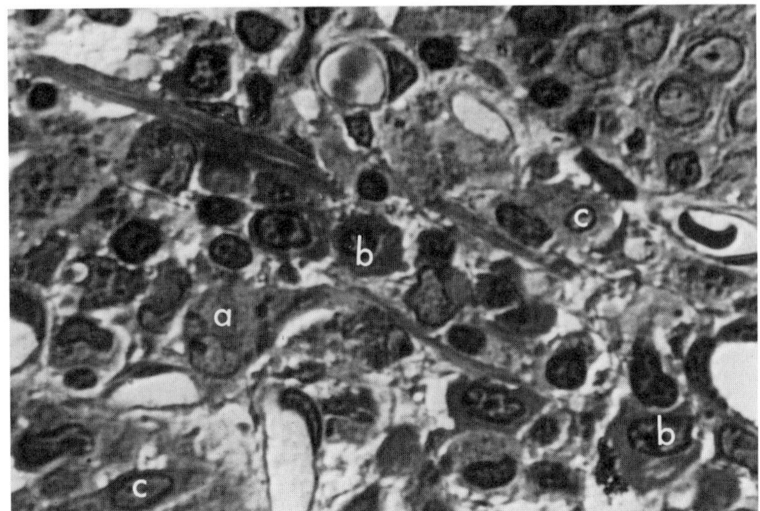

FIGURE 3-51

6. Study this EM (Fig. 3-52) and identify the two different cell types:
(a) __________ (b) __________
How did you identify them? List at least three ultrastructural characteristics for each cell type. In so doing you may include structures that are not present in this particular EM.

In this low-magnification EM of loose FECT one sees a fibroblast and a macrophage. The fibroblast can be identified because of the abundance of RER, the appearance of Golgi complex, and attachment regions at the cell surface. The macrophage contains many vesicles, vacuoles, and lysosomes and shows ruffling of plasma membranes.

7. Figure 3-53 shows a specialized connective tissue cell. Generally, what shape is this cell? __________. Is the nucleus centrally located? __________. How would you describe the nuclear chromatin?

This ovoid cell has an eccentric nucleus with chromatin arranged in packed granular form at the periphery of the nuclear envelope, in a cartwheel shape. It is a typical plasma cell, functioning in the production of circulating antibodies.

FIGURE 3-52

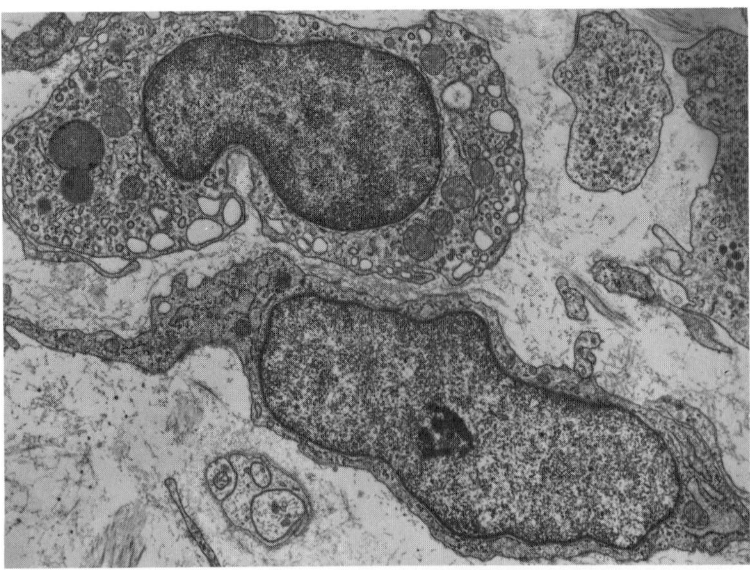

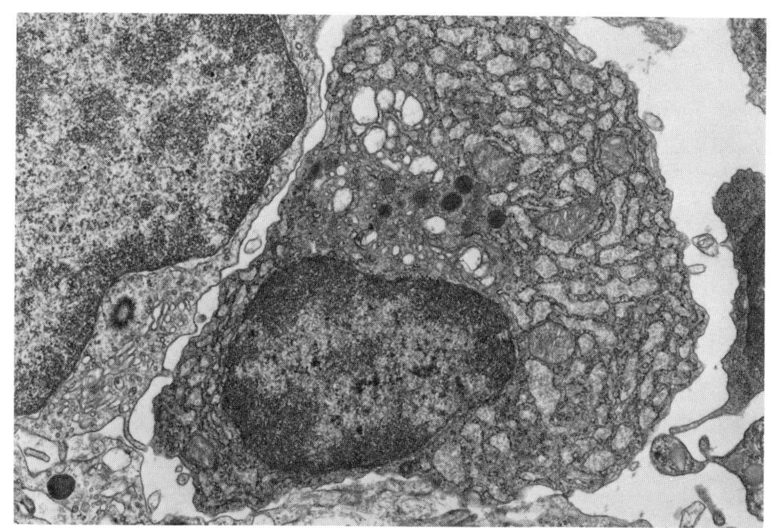

FIGURE 3-53

8. Figure 3-54 shows a preponderance of yet another cell type found in connective tissue. What is the general shape of each cell? __________. Where is the nucleus located? __________. Which are more prevalent, cells or fibers? __________. How would you classify this tissue?

These rounded cells have their nuclei pushed flat against the periphery by the accumulated lipid droplets. The adipose cells are much more in evidence than fibers; thus this is loose FECT, or better yet, adipose tissue.

BIBLIOGRAPHY

Anderson, J. C.: "Glycoproteins of the Connective Tissue Matrix," *Int. Rev. Connect. Tissue Res.* **7:** 251–322 (1976).

Bloom, W., and D. W. Fawcett: *A Textbook of Histology,* 10th ed., Saunders, Philadelphia, 1975.

Carr, I.: *The Macrophage: A Review of Ultrastructure and Function,* Academic Press, New York, 1973.

Fernandez-Madrid, F.: "Collagen Biosynthesis: A Review," *Clin. Orthop. Relat. Res.* **68:**163–181 (1970).

FIGURE 3-54

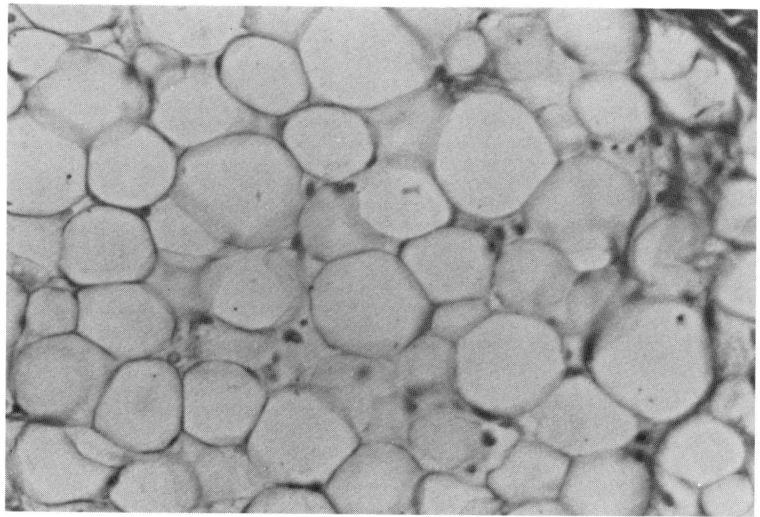

Greenlee, T. K., Jr., R. Ross, and J. L. Hartman: "The Fine Structure of Elastic Fibers," *J. Cell Biol.* **30:**59–71 (1966).

Leduc, E. H., S. Avrameas, and M. Bouteille: "Ultrastructural Localization of Antibody in Differentiating Plasma Cells," *J. Exp. Med.* **127:**109–118 (1968).

Ramachandron, G. N. (ed.): *Treatise on Collagen,* vol. 1, *Chemistry of Collagen,* G. N. Ramachandron, (ed.); vols. 2A and 2B, *Biology of Collagen,* B. S. Gould (ed.); vol. 3, *Clinical Pathology of Collagen,* R. A. Milch (ed.), Academic Press, New York, 1974.

Rhodin, J. A. G.: *Histology. A Text and Atlas,* Oxford University Press, New York, 1974.

Slavin, B. G.: "The Cytophysiology of Mammalian Adipose Tissue," *Int. Rev. Cytol.* **33:**297–334 (1972).

Weinstock, A., and J. T. Albright: "The Fine Structure of Mast Cells in Normal Human Gingiva," *J. Ultrastruct. Res.* **17:**254–256 (1967).

Weiss, L., and R. O. Greep: *Histology,* 4th ed., McGraw-Hill, New York, 1977.

Wilhelm, D. L.: "Inflammation and Healing," in W. D. A. Anderson (ed.), *Pathology,* vol. 1, 6th ed., Mosby, St. Louis, 1971, pp. 14–67.

ADDITIONAL READINGS

Bairoti, A., M. G. Petruccioli, and L. Torri Tarelli: "Studies on the Ultrastructure of Collagen Fibrils. I. Morphological Evaluation of the Periodic Structure," *J. Submicrosc. Cytol.* **1:**113–141 (1969).

Bornstein, P.: "The Biosynthesis of Collagen," *Ann. Rev. Biochem.* **143:**567–603 (1974).

Church, R. L., S. E. Pfeiffer, and M. L. Tanzer: "Collagen Biosynthesis: Synthesis and Secretion of a High Molecular Weight Collagen Precursor (Procollagen)," *Proc. Natl. Acad. Sci. U.S.A.* **68:** 2638–2642 (1971).

Combs, J. W.: "Maturation of Rat Mast Cells: An Electron Microscopic Study," *J. Cell Biol.* **31:** 563–575 (1966).

Gersh, F., and H. R. Catchpole: "The Nature of Ground Substance of Connective Tissue," *Perspect. Biol. Med.* **3:**282–319 (1969).

Lagunoff, D.: "Contributions of Electron Microscopy to the Study of Mast Cells," *J. Invest. Dermatol.* **58:** 296–311 (1973).

Lamberg, S. I., and A. C. Stoolmiller: "Glycosaminoglycans: A Biochemical and Clinical Review," *J. Invest. Dermatol.* **63:**444–449 (1974).

Movat, H. Z., and N. V. P. Fernando: "The Fine Structure of Connective Tissue. I. The Fibroblast," *Exp. Mol. Pathol.* **1:**509–534 (1962).

——— and ———: "Fine Structure of the Connective Tissue. II. The Plasma Cell," *Exp. Mol. Pathol.* **1:**535–553 (1962).

Napolitano, L.: "The Differentiation of White Adipose Tissue Cells: An Electron Microscope Study," *J. Cell Biol.* **18:**663–679 (1963).

Nimni, M. E.: "Collagen: Its Structure and Function in Normal and Pathological Connective Tissue," *Sem. Arthritis Rheum.* **4:**95–150 (1974).

Ross, R.: "Connective Tissue Cells, Cell Proliferation and Synthesis of Extracellular Matrix—A Review," *Philos. Trans. R. Soc. Lond. Biol. Sci.* **271:** 247–259 (1975).

——— and P. Bornstein: "The Elastic Fiber," *J. Cell Biol.* **40:**366–381 (1969).

——— and E. P. Benditt: "Wound Healing and Collagen Formation. V. Quantitative Electron Microscope Radioautographic Observations of Proline-^{3}H Utilization by Fibroblasts," *J. Cell Biol.* **27:** 83–106 (1965).

Slavin, B. G.: "The Cytophysiology of Mammalian Adipose Tissue," *Int. Rev. Cytol.* **33:**297–334 (1972).

Smith, D. E.: "The Tissue Mast Cell," *Int. Rev. Cytol.* **14:**327–386 (1963).

Weinstock, M.: "Collagen Formation. Observations on Its Intracellular Packaging and Transport," *A. Zellforsch.* **129:**455–470 (1972).

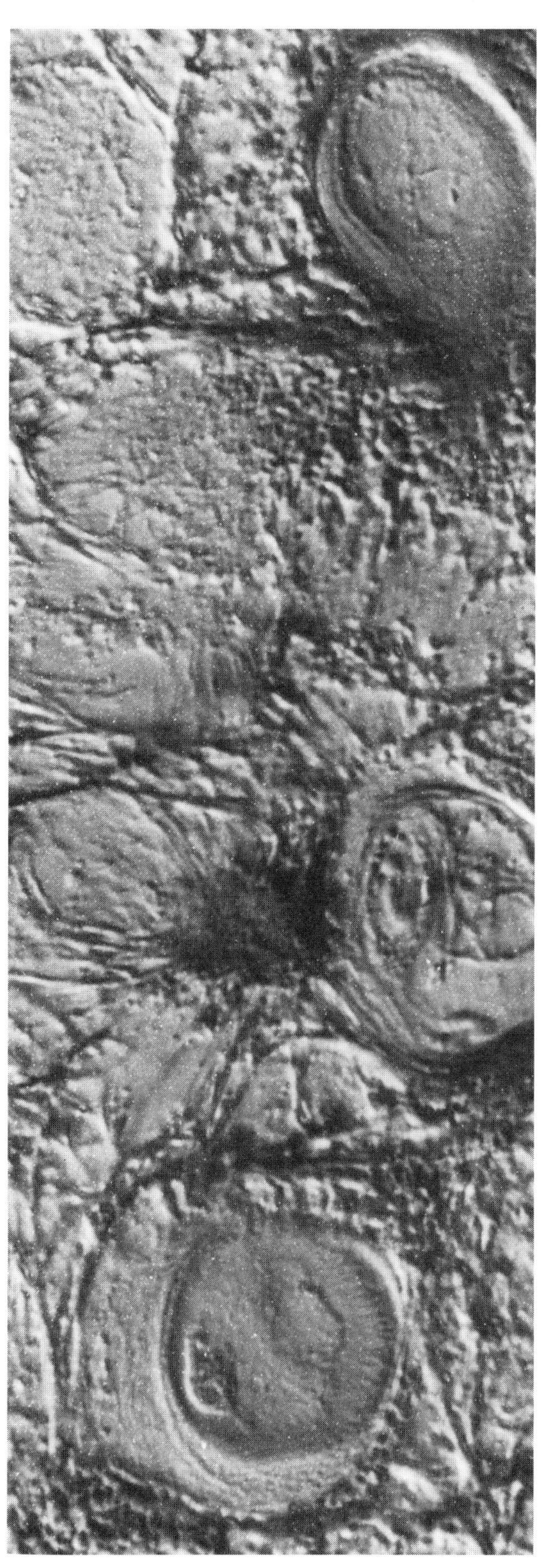

4 CARTILAGE

OBJECTIVES

Upon completion of this chapter, the student will be able to:

1 Identify the locus where each of the following types of cartilage is found:
(a) hyaline (b) elastic (c) fibrous

2 Identify the type and form of fibrous elements characteristic to each type of cartilage.

3 Describe the composition of the intercellular substance of cartilage by identifying the: (a) two major entities found in the substance (b) distribution of cells within the intercellular substance (c) staining characteristics for the substance

4 Describe the morphology of the cell found in cartilage.

5 Identify the characteristics of the following two methods of cartilage development:
(a) interstitial growth (b) appositional growth

6 Describe the nutrition of cartilage.

Cartilage, a specialized type of connective tissue, provides physical rigidity and resilience where there is extreme pressure and abrasive forces. It exists in three forms: hyaline, elastic, and fibrous. These types share certain common characteristics, but each is specialized to meet particular functional requirements unique to the region. Most cartilage, with the exception of articular surfaces, is covered by the dense perichondrium that protects its surfaces and produces cellular progenies.

HYALINE CARTILAGE

Hyaline cartilage, the most prevalent type, is found in the heads of long bones (joints), where its strength and elasticity withstand the constant pressure and stresses applied to the joint. Because of the pressure and constant movement, the cartilaginous head of the joint is usually devoid of perichondrium. Hyaline cartilage is also found in the cartilaginous ends of ribs where they attach to the sternum, and in the nose, larynx, trachea, and bronchi. Nearly all the skeleton in the fetus is formed first by hyaline cartilage that is later replaced by developing bone, a process that will be discussed in the next chapter.

In proportion to the amount of matrix, the number of cells (primarily chondrocytes) in hyaline cartilage is small. Located within small spaces called *lacunae,* the cells are usually round to ovoid. The hyaline cartilage in Fig. 4-1 illustrates these points.

The nucleus is often displaced toward the periphery of the cell, with little detail discernible in a light micrograph. In hyaline cartilage the cells often appear in clusters within the matrix because continual cell division occurs in a rigid matrix which prevents the cells from moving away from each other. These clusters are called *isogenous groups* (arrows). Each isogenous group originates from a single parent cell with its progeny remaining closely associated with each other, often showing four or more cells in a group. Although it appears empty and vacuolated under microscopic observation, the cytoplasm of chondrocytes in life contains a large amount of glycogen and lipid droplets which are lost during histologic processing. Their presence could therefore be demonstrated following special stains. The LMs in Figs. 4-2 and 4-3 were prepared in a special way that

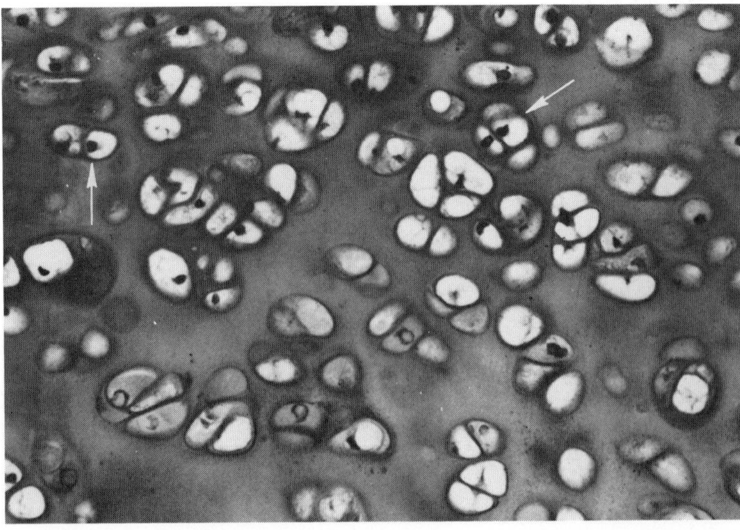

FIGURE 4-1 Light micrograph of hyaline cartilage. Clusters of cells, isogenous groups, are indicated by arrows.

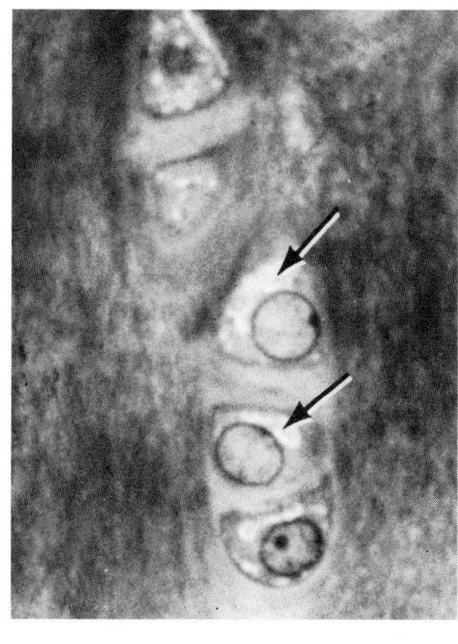

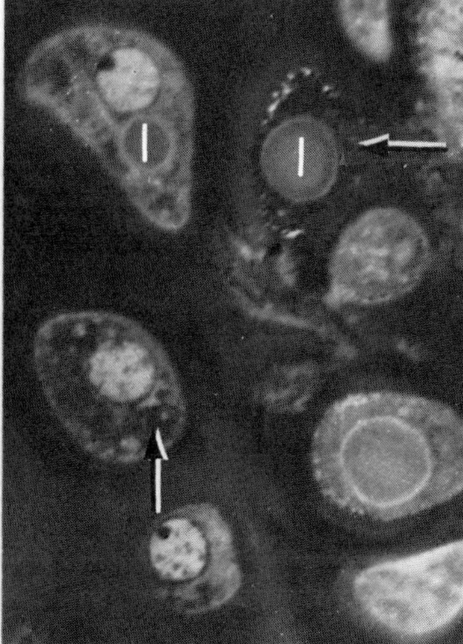

FIGURE 4-2 (*far left*) Hyaline cartilage seen at higher magnification. Note the better preservation of the chondrocytes in this preparation compared with that of the ones seen in Fig. 4-1. The light regions (arrows) in the cytoplasm of the chondrocytes represent areas where glycogen was present.

FIGURE 4-3 (*near left*) Light micrograph of hyaline cartilage prepared with osmium tetroxide for even better preservation of structural details. Note lipid droplets (l) and masses of glycogen (arrows) in the cytoplasm of the chondrocytes.

allowed the preservation of chondrocyte structures. Note the light, irregular-shaped regions in Fig. 4-2 which represent areas where glycogen deposits (arrows) were present. The cells are ovoid to angular in shape and contain a round nucleus in which an eccentric nucleolus is occasionally observed. Figure 4-3 shows the same, except that the lipid droplets (l) are preserved by osmium tetroxide and the deposits of glycogen appear as small dense irregular masses (arrows).

The low-power EM appearing in Fig. 4-4 confirms essentially what was observed under the light microscope. The cell is of somewhat angular ovoid shape. Two large lipid droplets have a moderate electron density. Elsewhere in the cytoplasm are glycogen granules, a few vacuoles, mitochondria, and tracts of intracellular microfibrils (f). Figure 4-5 depicts in detail the large aggregates of glycogen granules. In addition, a couple of dense structures that resemble primary lysosomes, described in Chap. 1, are present. Figure 4-6 is another region of the cytoplasm

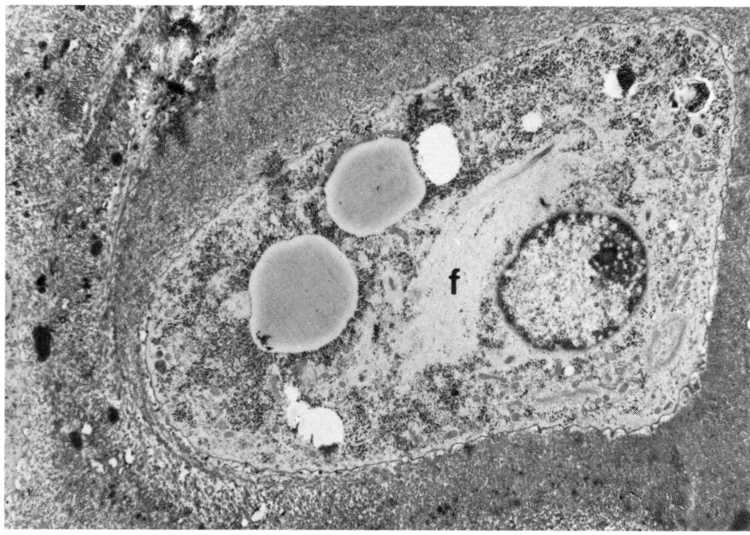

FIGURE 4-4 Low-power electron micrograph of chondrocyte. The cytoplasm contains glycogen granules, a few vacuoles, lipid droplets, mitochondria, and microfibrils (f).

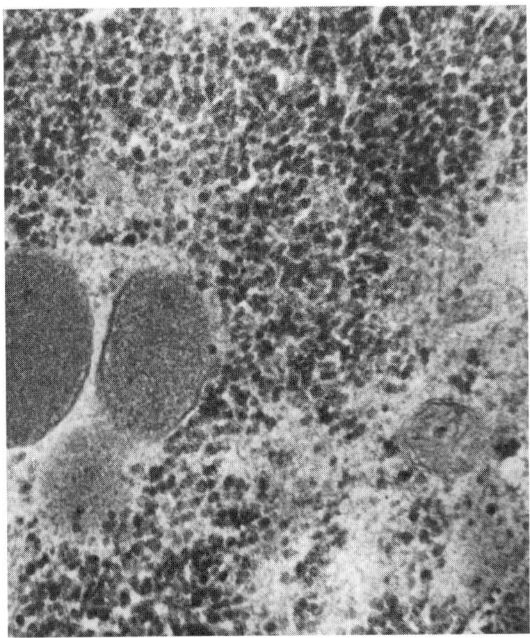

FIGURE 4-5 Portion of a chondrocyte containing large aggregates of glycogen granules and some dense bodies resembling primary lysosomes.

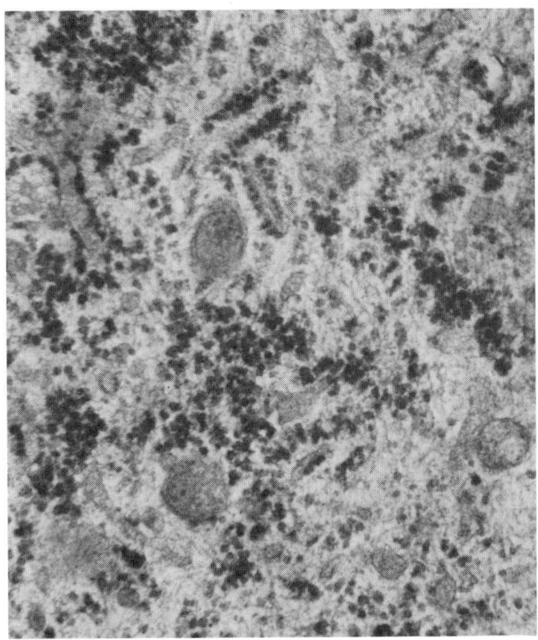

FIGURE 4-6 Portion of chondrocyte cytoplasm containing more sparsely located glycogen granules.

FIGURE 4-7 Portion of a chondrocyte containing a lysosome, several mitochondria, ribosomes, some glycogen granules, and microfibrils. In some areas (arrows) the microfibrils appear condensed.

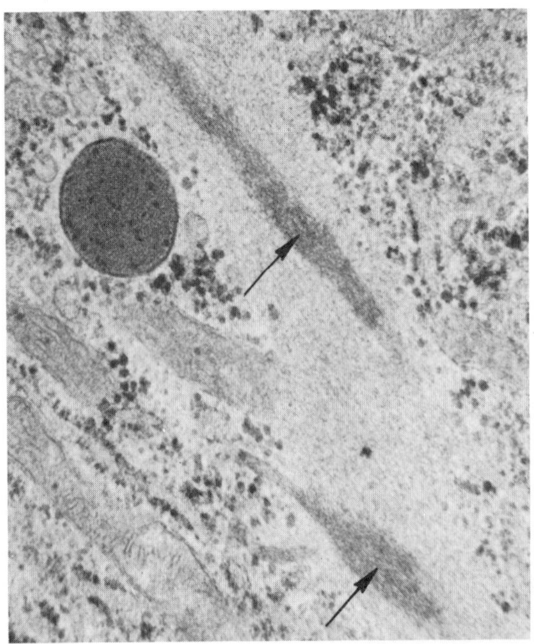

FIGURE 4-8 Another portion of a chondrocyte containing a lipid droplet.

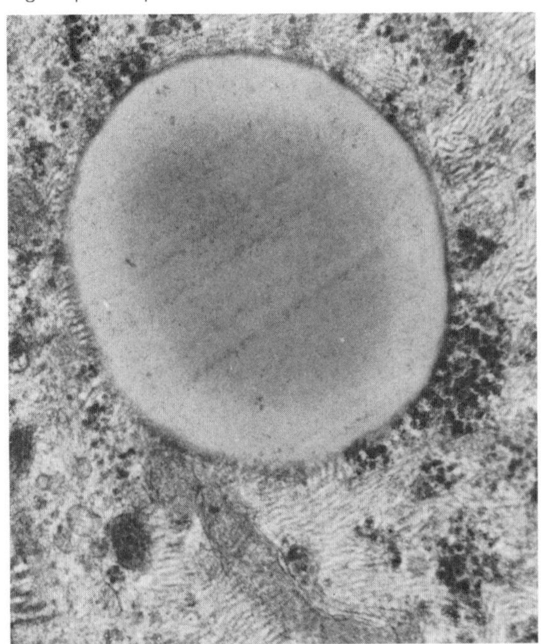

where glycogen is more sparsely located. The EM in Fig. 4-7 shows a lysosome, several small rod-shaped mitochondria, some glycogen granules, ribosomes, and a bundle of microfibrils running diagonally through the field. Note the central region (arrows) where the fibrils appear denser than in the periphery. Another portion of a cartilage cell showing a typical lipid droplet, glycogen, and diffusely oriented fibrils is seen in Fig. 4-8. That the chondrocytes are secretory cells is evidenced by the presence of small but frequently encountered RER (Fig. 4-9) and a well-developed Golgi complex (Fig. 4-10). Both glycogen and lipid droplets that are so abundant in chondrocytes are metabolic reservoirs for the cells. These reservoirs can be mobilized easily during the active phase of cartilage growth.

Extracellular Matrix

In hyaline cartilage the extracellular matrix is composed of fine collagen fibrils immersed in a ground substance which is rich in mucopolysaccharides. Fine individual collagenous fibers in the extracellular matrix run in random patterns among the isogenous groups of cells. These fibers, large enough to be seen with a light microscope with certain staining techniques, can be better visualized in EM, particularly after removal of polysaccharide proteins from the ground substance. In regions where there is relatively little ground substance, the collagen fibrils are clearly visible (Fig. 4-11). With increase in

FIGURE 4-9 Portion of a chondrocyte containing profiles of RER.

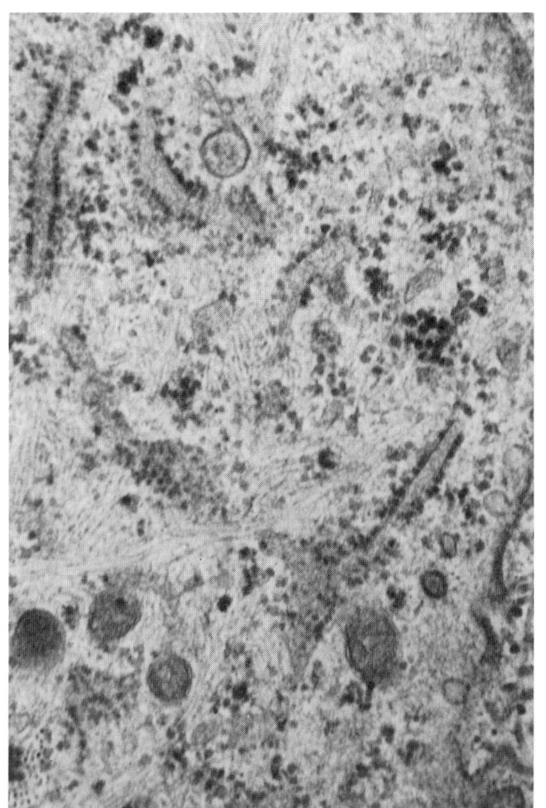

FIGURE 4-10 Portion of a chondrocyte containing a well-developed Golgi complex.

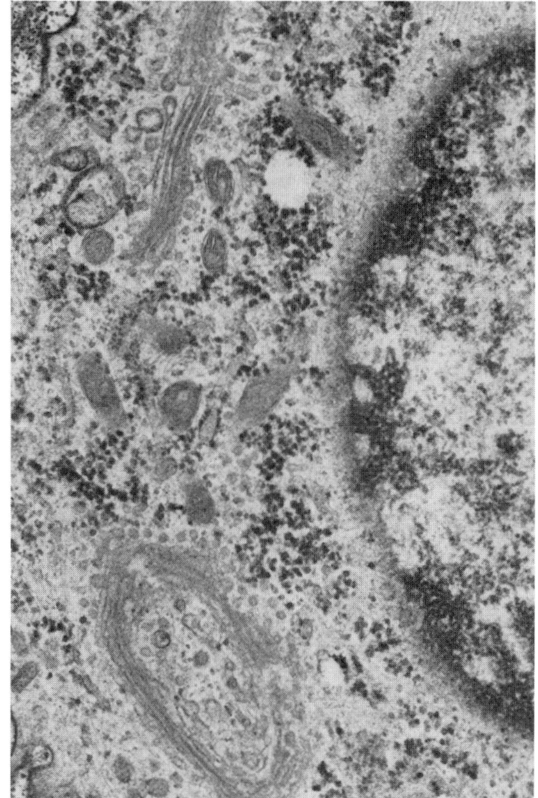

the amount of ground substance, the collagen fibrils in the matrix become covered by electron-dense material (Fig. 4-12) and eventually become totally obscured (Fig. 4-13).

The fibers and ground substance are different in organization in the immediate vicinity of lacunae than in areas between the lacunae and therefore can be stained differently. The *territorial matrix,* including the area closely surrounding the cell and cell groups, stains intensely with special dyes for ground substances (Figs. 4-1 and 4-14). This stain is caused by a large amount of ground substance containing chondroitin sulfate and fewer collagenous fibers in this area. Since the matrix is produced by the cells, the territorial matrix, which is the area closest to the groups, is the youngest. The *interterritorial matrix* is the area further away from the cells, filling the remainder of space between territorial matrices. In general, this area stains less intensely than territorial matrices adjacent to isogenous groups and isolated chondrocytes (Fig. 4-1). This tendency is observed even after staining with routine preparative procedures such as H and E. The interterritorial matrix usually contains less chondroitin sulfate and more collagenous fibers. The character of the cartilaginous matrix changes with age, as more intermolecular cross-linking occurs and water is lost. This causes the different staining properties of the two types of matrices. In a

FIGURE 4-11 Portion of hyaline cartilage matrix with relatively little ground substance. Fine collagen fibers are clearly seen.

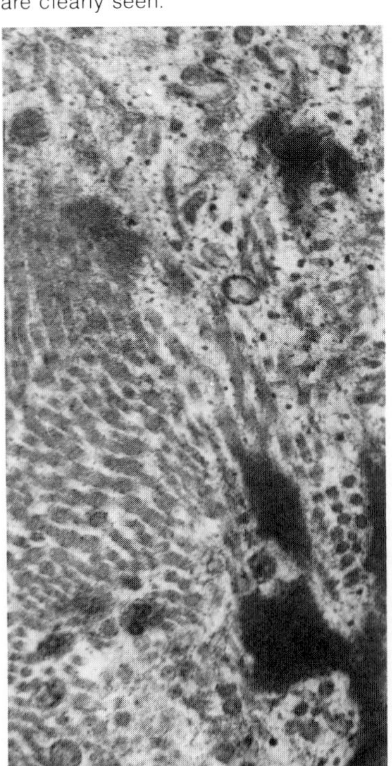

FIGURE 4-12 Portion of hyaline cartilage matrix with increased amount of ground substance. The fine collagen fibers are here partially obscured.

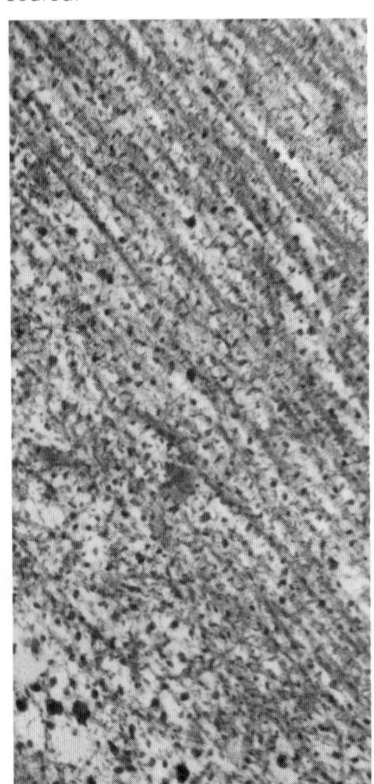

FIGURE 4-13 In this section there is further increase in ground substance. This increase totally obscures the collagen fibers.

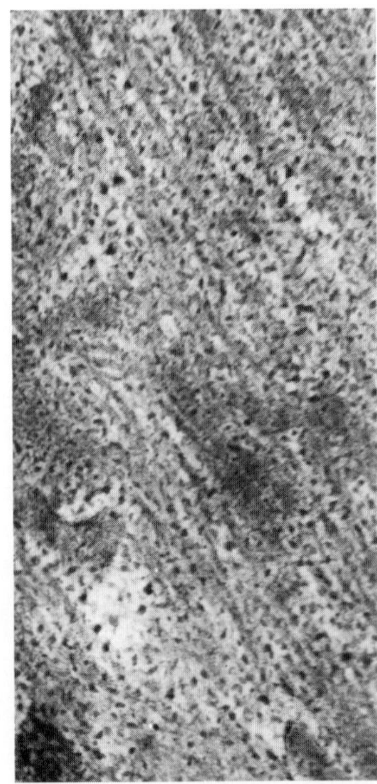

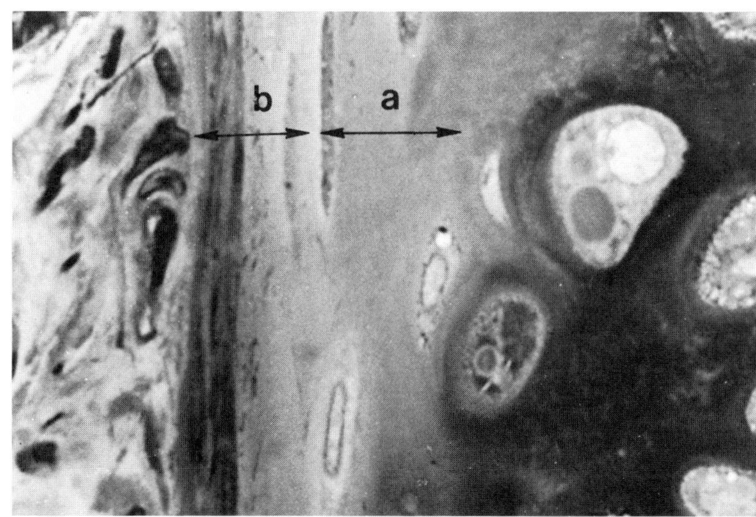

FIGURE 4-14 Light micrograph of hyaline cartilage and its perichondrial covering. The latter is composed of an inner cellular layer (a) and outer fibrous layer (b). Note the intense staining of the territorial matrices.

similar manner, histologic differences may be observed between cartilages of the same type taken from individuals of different ages.

The ground substance in which the fibers are embedded is composed of a number of complex polysaccharide molecules attached to protein skeletons, which are therefore called *proteoglycans.* This negatively charged polysaccharide moiety has a high affinity for water and holds it with such tenacity that water becomes an integral part of the tissue. This gives cartilage its rigidity, and at the same time allows it to maintain a high degree of flexibility. In different areas of the body, differences in the nature of the ground substances cause hyaline cartilage to show wide variances in rigidity and resilience.

Hyaline cartilage grows in two ways. Chondroblasts at the surface of the existing cartilage within the perichondrium may form new cartilage by *apposition,* that is, by adding matrix and fibers to the surface of the cartilage. An example is shown in Fig. 4-14, where the perichondrium is seen on the left. The perichondrium is made up of two layers: the inner cellular layer where chondroblasts proliferate (a, Fig. 4-14) and the outer fibrous layer which is largely collagenous in nature (b, Fig. 4-14). The EM in Fig. 4-15 depicts the two layers of perichondrium. The bottom third or so is the fibrous layer which supports the cellular layer appearing in the upper two-thirds of the micrograph. Because of unidirectional growth during apposition, the chondrogenic cells often show a polarity suggestive of their secretory function. The EMs in Figs. 4-16 and 4-17, which are taken from a rapidly growing cartilage, illustrate this point. Note the scalloped appearance of the plasma membrane facing the cartilage side (C in Fig. 4-16). Apparently, this scalloping is related to the secretory activity of the cell which releases the matrix material through the fusion of secretory granules with the surface plasma membrane (Fig. 4-17). *Interstitial growth* occurs mainly in young cartilage, when chondrocytes within the cartilage mass divide and form isogenous groups of cells and surrounding territorial matrices.

ELASTIC CARTILAGE

Elastic cartilage is found in the external ear and the nose where it helps to retain the shape of mobile structures. Another example of elastic cartilage is in the epiglottis, where a very flexible but elastic flap closes off the respiratory passage from the esophagus during swallowing. Elastic cartilage is more cellular and contains less matrix than hyaline cartilage, and it has a large number of elastic

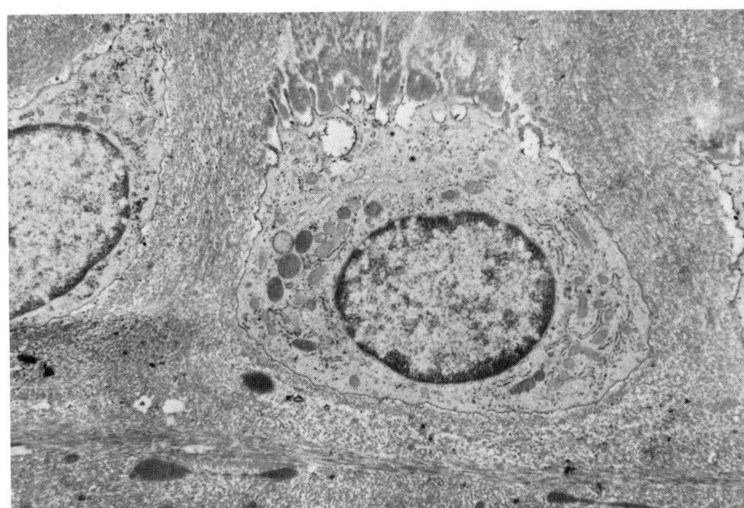

FIGURE 4-15 Electron micrograph of the two layers of the perichondrium. The fibrous layer appears at the bottom and the cellular layer at the top of the micrograph.

FIGURE 4-16 Electron micrograph of rapidly growing hyaline cartilage. Note the polarization of the chondroblast which has a scalloped plasma membrane facing the cartilage matrix (C).

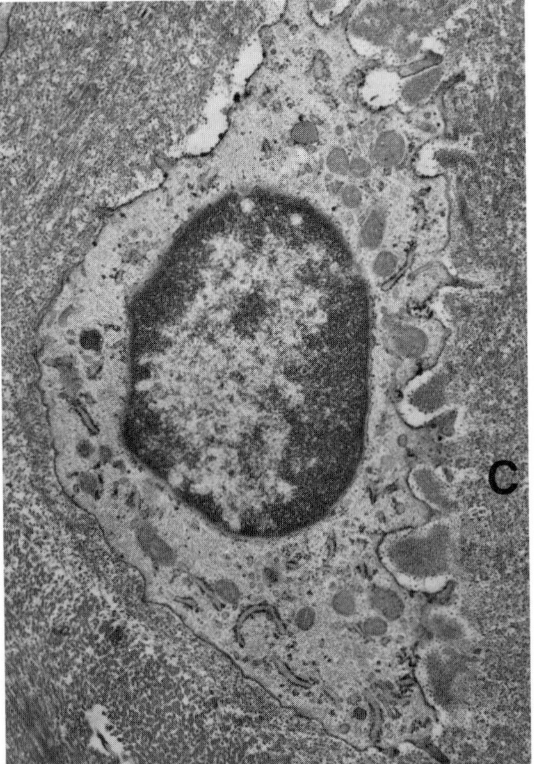

FIGURE 4-17 View of the scalloped plasma membrane at higher magnification. It appears that material is added to the matrix by the fusion of secretory granules with the plasma membrane. It seems that this fusion causes the scalloped appearance of the plasma membrane.

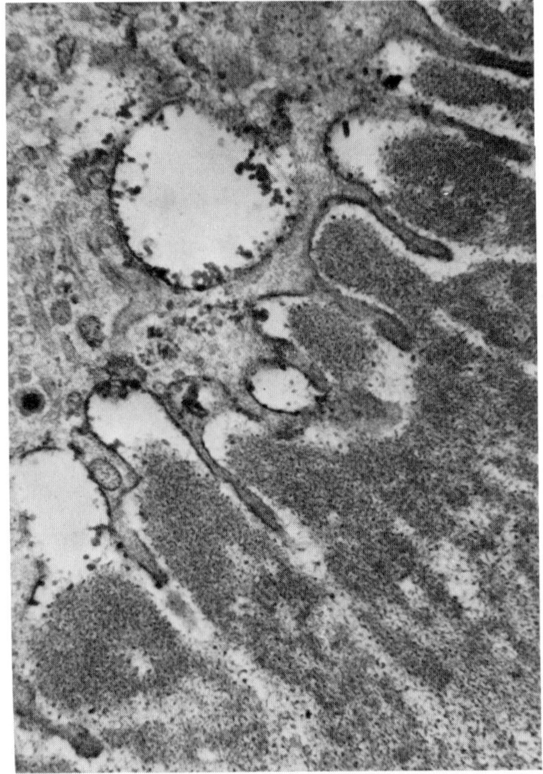

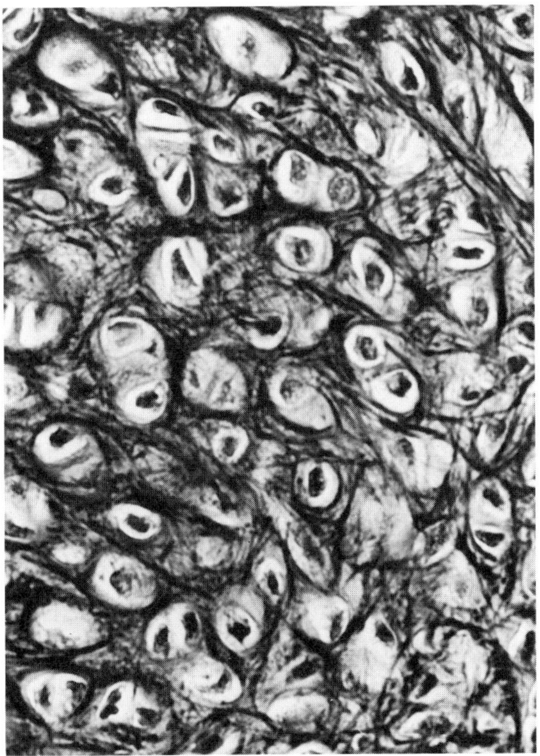

FIGURE 4-18 Light micrograph of a section of elastic cartilage. Note the intense staining of fibers and cells.

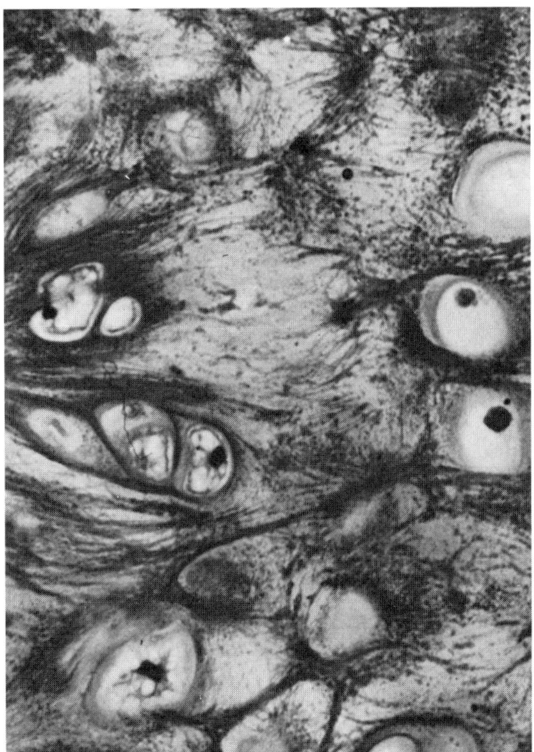

FIGURE 4-19 Fine elastic fibers are running in various directions in this micrograph of elastic cartilage taken at a higher magnification.

fibers. Figure 4-18 is a section from an epiglottis in which the matrix and fibers have been stained. A greater cell density than occurs in hyaline cartilage is observed here. Many fine dark-stained elastic fibers can be seen against the background of the ground substance. In Fig. 4-19, fine elastic fibers of an elastic cartilage running in random directions are better demonstrated. Again, both appositional and interstitial growth occurs in elastic cartilage.

FIBROUS CARTILAGE

Fibrous cartilage, often called *fibrocartilage,* is associated with certain joints in the body where the attachment of a ligament or tendon to bone is adjacent to hyaline cartilage of a joint surface. Intervertebral disks in the spine consist largely of fibrous cartilage. It is continuous above and below with the hyaline cartilage covering of the adjacent vertebrae and continuous peripherally with the spinal ligaments. An example of the fibrous cartilage of an intervertebral disk is seen in Fig. 4-20. The fibrous cartilage consists of cartilage cells interposed between dense collagenous fiber bundles that are abundant between rows of cartilage cells. This can be shown by using a phase contrast microscope, which brings out details of the fibrous matrix (Fig. 4-21). The proportion of fiber bundles, cartilage cells, and matrix varies greatly.

CHANGES DURING AGING

As a person ages, cartilage degenerates because of poor vascularization of the tissue. In many adults there is a mass of degenerating tissue in the middle of the intervertebral disk

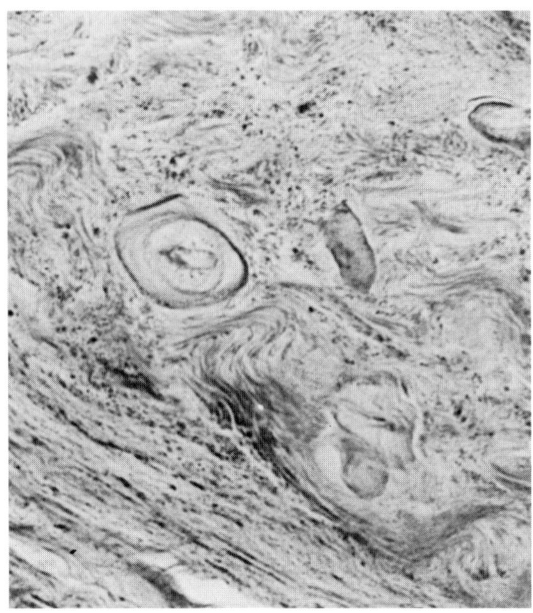

FIGURE 4-20 Light micrograph of fibrous cartilage. Note the rows of chondrocytes interspersed between the thick collagen bundles.

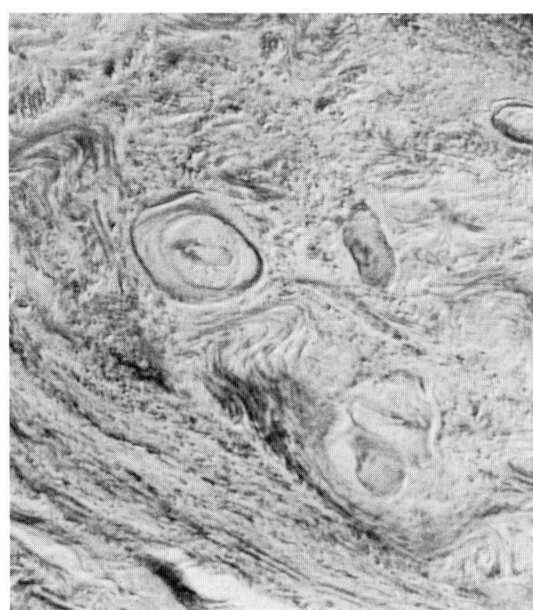

FIGURE 4-21 Phase contrast micrograph illustrating the details of the fibrous matrix of fibrocartilage.

called the *nucleus pulposus*. If this disk ruptures in certain directions, it can cause pressure against the spinal cord and cause great pain. The nutrition of cartilage is completely dependent on diffusion processes since cartilage contains no blood vessels or lymphatics. This lack of blood supply to the cartilage is thought to contribute to the degenerative changes seen in old cartilages.

Several features characterize these age-related changes. Coarse fibers become more predominant in the matrix. An increase in noncollagenous proteins along with a loss of acid mucopolysaccharides results in a cloudy appearance. Occasional calcification occurs slowly and is dependent upon a viable perichondrium to supply nutrients and chondroblasts for repair.

REVIEW SECTION

1. Figure 4-22 shows a section of cartilage tissue in the upper two-thirds of the field. How would you describe the cell-to-matrix ratio? What kind of fibers can you identify in the cartilage? The lower part of the figure shows the covering of the cartilage called the ________. How would you classify the connective tissue of the covering? What type of cartilage is this?

The low cell-to-matrix ratio and the appearance of the cartilage matrix indicates that this is hyaline cartilage. The fine fibers cannot really be identified in this preparation. The perichondrium is made up of dense regular FECT.

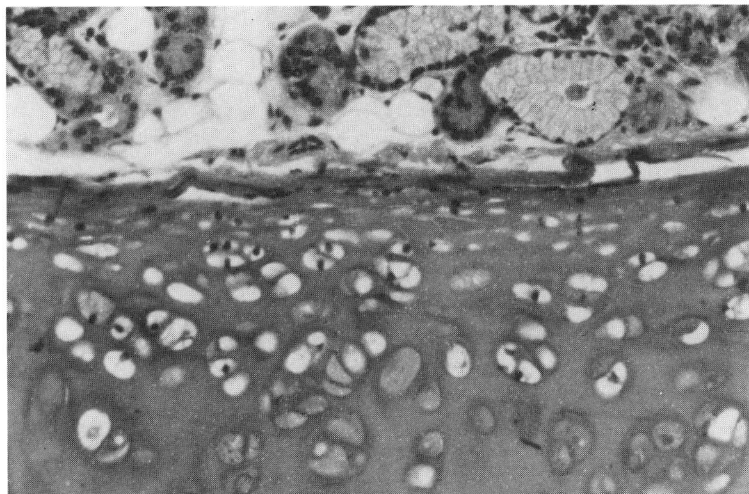
FIGURE 4-22

2. Figure 4-23 is a photomicrograph of another section of cartilage. Does this type of cartilage have more or fewer cells than the hyaline cartilage? What type of fibers can you identify?

The cartilage with a high cell-to-matrix ratio with many fibers in its matrix is elastic cartilage. The density of elastic fibers gives it the needed flexibility. The elastic fibers, which appear dark here, may be stained to show different colors such as blue or red, depending on staining technique. A routine method for staining elastic fibers is Weigert's technique, which renders elastic fibers a dark blue color.

3. Figure 4-24 shows a cartilage stained for collagenous fibers. Chondrocytes here are arranged in __________. The predominance of the collagenous bundles should help you to name this type of cartilage.

The cells in fibrous cartilage tend to be arranged in rows parallel to the fiber bundles that give this tissue its great strength. Note the small territorial matrix between cells.

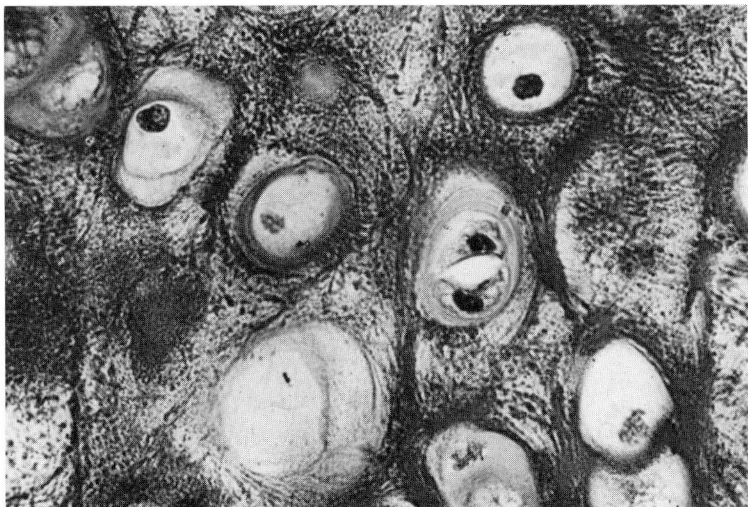
FIGURE 4-23

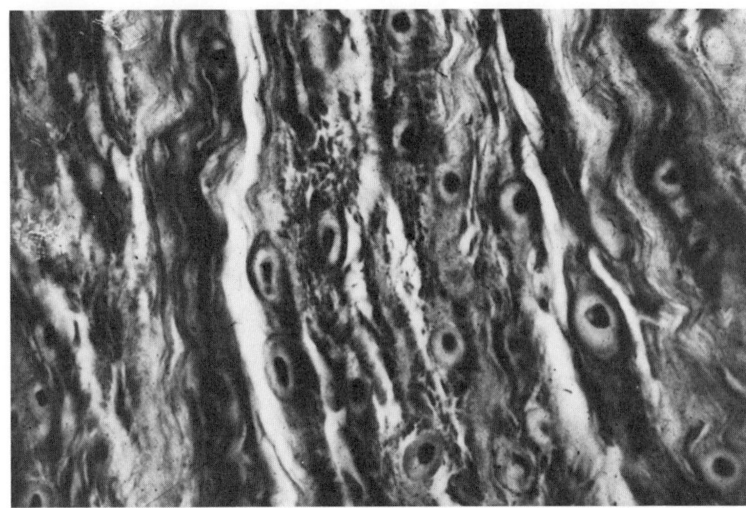

FIGURE 4-24

4. Identify in Fig. 4-25: lipid droplets, lacuna, territorial matrix, and interterritorial matrix.

In this low-magnification EM of a chondrocyte from hyaline cartilage one sees four large lipid droplets, the lacunar wall which appears somewhat irregular, dense territorial matrix immediately surrounding the lacuna, and less dense but ill-defined interterritorial matrix.

BIBLIOGRAPHY

Bloom, W., and D. W. Fawcett: *A Textbook of Histology,* 10th ed., Saunders, Philadelphia, 1975.

Rhodin, J. A. G.: *Histology. A Text and Atlas,* Oxford University Press, New York, 1974.

FIGURE 4-25

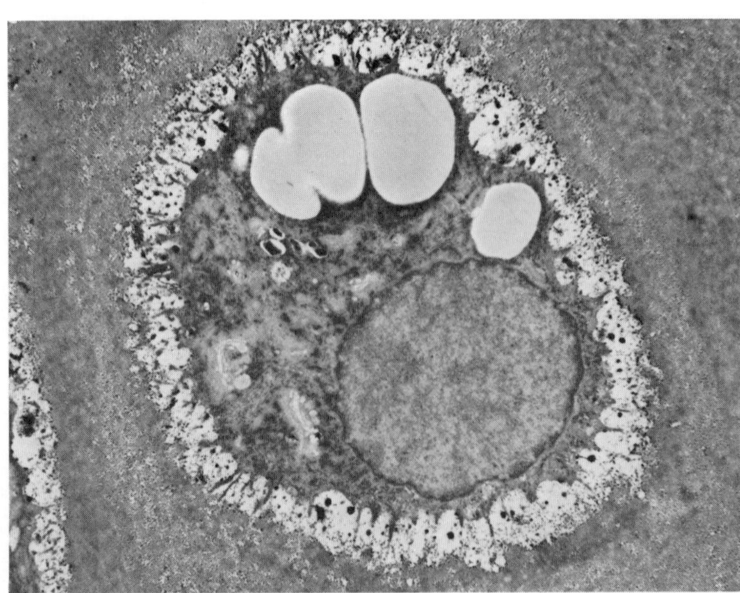

Weiss, L., and R. O. Greep: *Histology,* 4th ed., McGraw-Hill, New York, 1977.

ADDITIONAL READINGS

Anderson, D. F.: "Ultrastructure of Hyaline and Elastic Cartilage of the Rat," *Am. J. Anat.* **114:** 403–434 (1964).

Anderson, H. C.: "Electron Microscopic Studies on Induced Cartilage Development and Calcification," *J. Cell Biol.* **35:**81–101 (1967).

———: "Vesicles Associated with Calcification of the Matrix of Epiphyseal Cartilage," *J. Cell Biol.* **41:**59–72 (1969).

Belanger, L. F., and B. B. Migicovsky: "Comparison between Different Mucopolysaccharide Stains as Applied to Chick Epiphyseal Cartilage," *J. Histochem. Cytochem.* **9:**73–78 (1961).

Clarke, I. C.: "Reticular Cartilage: A Review and Scanning Electron Microscope Study. II. The Territorial Fibrillar Architecture," *J. Anat.* **118:**261–280 (1974).

Cooper, G. W., and D. J. Prockop: "Intracellular Accumulation of Protocollagen and Extrusion of Collagen by Embryonic Cartilage Cells," *J. Cell Biol.* **38:**523–537 (1968).

Fahmy, A., W. Hillman, D. Talley, and V. Long: "Fibrillogenesis in the Epiphyseal Cartilage of Adult Rats," *J. Bone Jt. Surg. Am. Vol.* **51A:**802 (1969).

Godman, G. C., and K. R. Porter: "Chondrogenesis, Studied with the Electron Microscope," *J. Biophys. Biochem. Cytol.* **8:**719–760 (1960).

——— and N. Lane: "On the Site of Sulfation in the Chondrocyte," *J. Cell Biol.* **21:**353–366 (1964).

Goel, S. C.: "Electron Microscopic Studies on Developing Cartilage. I. The Membrane System Related to the Synthesis and Secretion of Extracellular Materials," *J. Embryol. Exp. Morphol.* **23:** 169–184 (1970).

Horowitz, A. L., and D. Dorfman: "Subcellular Sites for Synthesis of Chondromucoprotein of Cartilage," *J. Cell Biol.* **38:**358–368 (1968).

Matukas, V. J., B. J. Panner, and J. L. Orbison: "Studies on Ultrastructural Identification and Distribution of Protein Polysaccaride in Cartilage Matrix," *J. Cell Biol.* **32:**365–378 (1964).

Minor, R. R.: "Somite Chondrogenesis. A Structural Study," *J. Cell Biol.* **56:**27–50 (1973).

Palfrey, A. J., and D. V. Davis: "The Fine Structure of Chondrocytes," *J. Anat.* **100:**213–226 (1966).

Revel, J. P., and E. D. Hay: "An Autoradiographic and Electron Microscopic Study of Collagen Synthesis in Differentiating Cartilage," *Z. Zellforsch.* **61:**110–114 (1963).

Searls, R. L.: "Newer Knowledge of Chondrogenesis," *Clin. Orthop. Relat. Res.* **96:**327–344 (1973).

Sheldon, H., and R. A. Robinson: "Studies on Cartilage: Electron Microscope Observations on Normal Rabbit Ear Cartilage," *J. Biophys. Biochem. Cytol.* **4:**401–406 (1958).

Silverberg, R., M. Silverberg, and D. Feir: "Life Cycle of Articular Cartilage Cells: An Electron Microscope Study of the Hip Joint of the Mouse," *Am. J. Anat.* **114:**17–47 (1964).

Smith, J. W., T. J. Peters, and A. Serafin-Fracassini: "Observations on the Distribution of Protein–Polysaccharide Complex and Collagen in Bovine Articular Cartilage," *J. Cell Sci.* **2:**129–136 (1967).

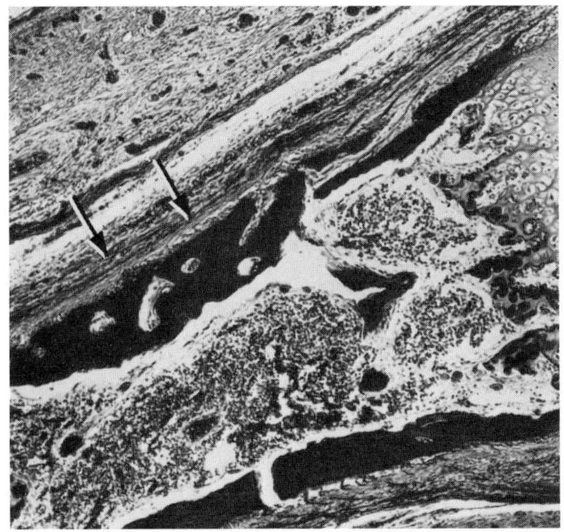

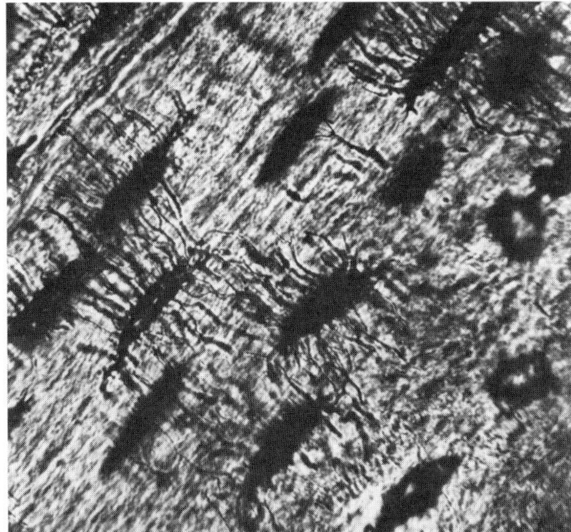

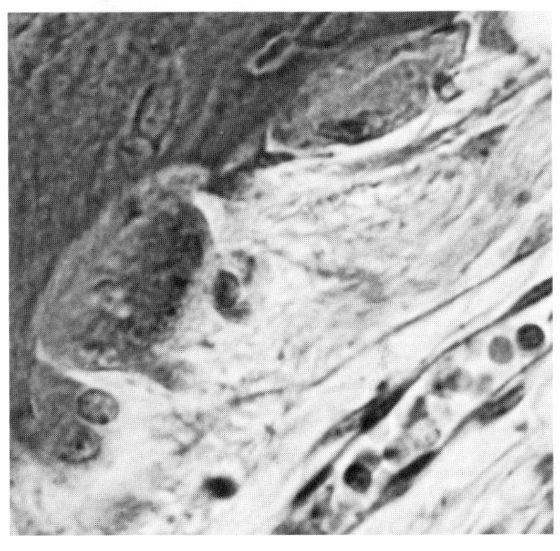

5 BONE

OBJECTIVES

Upon completion of this chapter, the student will be able to:

1 Name the embryonic origin of osteogenic cells.

2 Identify the steps in the process of:
(a) intramembranous bone formation (b) endochondral bone formation

3 Identify the morphology and function of each of the following cells:
(a) osteoblasts (b) osteocytes (c) osteoclasts

4 Identify on a light and electron micrograph the preosseous or osteoid region of bone matrix.

5 Identify the steps involved in the process of:
(a) bone resorption (b) remodeling

6 Identify on a slide or diagram the following structures:
(a) fibrous layer of periosteum (b) cellular layer of periosteum (c) endosteum (d) lacuna (e) haversian lamella (f) haversian canal (g) growth arrest line (h) Sharpey's fibers and bundle bone

Bone is a specialized connective tissue that provides a physical framework for the body and protects various internal organs. For instance, the brain is shielded by the skull, while the ribs encase the vital thoracic organs. Bones also provide space for the bone marrow cells that produce blood elements and contribute to immune functions. The hard cortical bone provides attachment for skeletal muscles. Without the structural characteristics of the long bones of the skeleton, the arms and legs would not have the strength and mobility we depend on.

Bone is more rigid than cartilage because there are a large amount of minerals deposited in the extracellular matrix in the form of apatite crystals. The abundance of calcium salt in the bone matrix causes it to serve as a calcium reservoir in regulating the level of Ca^{2+} in blood and in tissue fluids. The cells present in the bone are therefore highly responsive to hormones and other factors that influence the calcium metabolism.

The two basic types of bone tissue are *spongy* (or cancellous) bone and *dense* (or compact) bone, as shown in Fig. 5-1. Most bones combine these two types in varying proportions. The inner part of most bones, particularly at the ends of long bones, is spongy bone which owes its name to the spongy appearance of the tissue (Fig. 5-1). The space between trabeculae of the cancellous bone (t, Fig. 5-2) is occupied by hemopoietic cells. Developing in accordance with physical and mechanical requirements of the particular region, the trabeculae thus divide the bony space into areas (marrow cavities), which are filled with these hemopoietic elements.

Two types of marrow exist: red and yellow. Red marrow, highly vascularized and active in blood formation, is found in young bones. Fetal bones contain only red marrow, as illustrated in Fig. 5-3. The red marrow becomes gradually replaced by yellow marrow as the bone ages. Yellow marrow develops when hemopoietic cells become fewer in number, allowing the space to be filled by adipose cells. The vacuolar appearance in Fig. 5-4, which is an LM of a bone sample

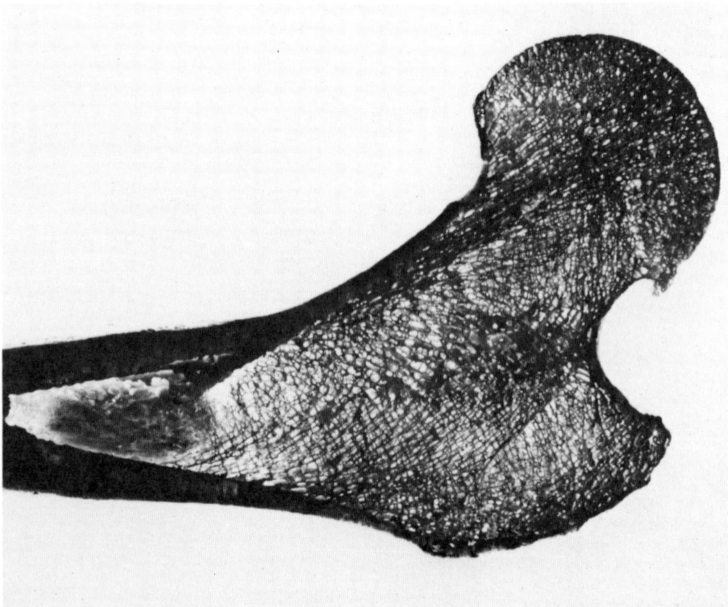

FIGURE 5-1 Photograph of the end of a long bone seen in a ground section. Note the basic two types of bone: compact bone on the surface and spongy bone within it.

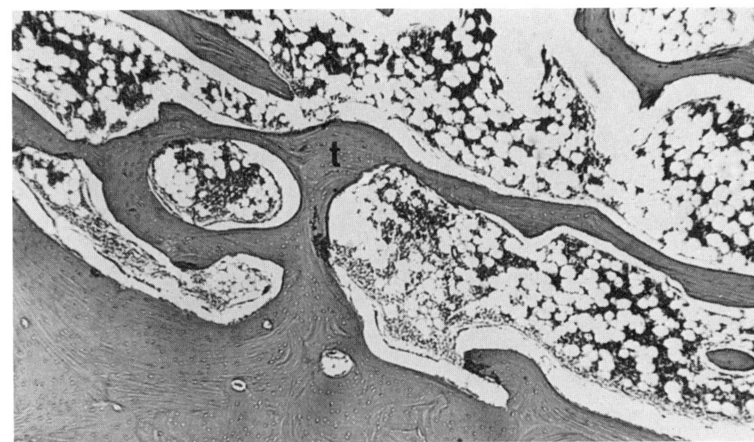

FIGURE 5-2 Light micrograph of bone. Dense bone is at the bottom and spongy bone is at the top. Hemopoietic cells fill the spaces between the trabeculae (t) of the spongy bone.

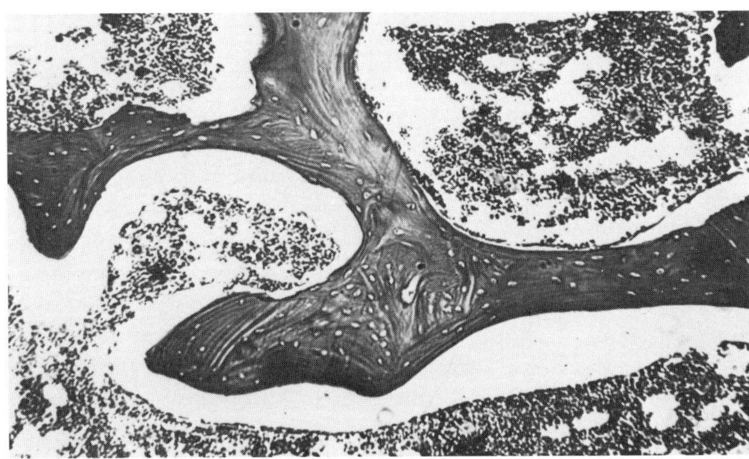

FIGURE 5-3 Light micrograph of red bone marrow in fetal bone. It consists only of hemopoietic cells.

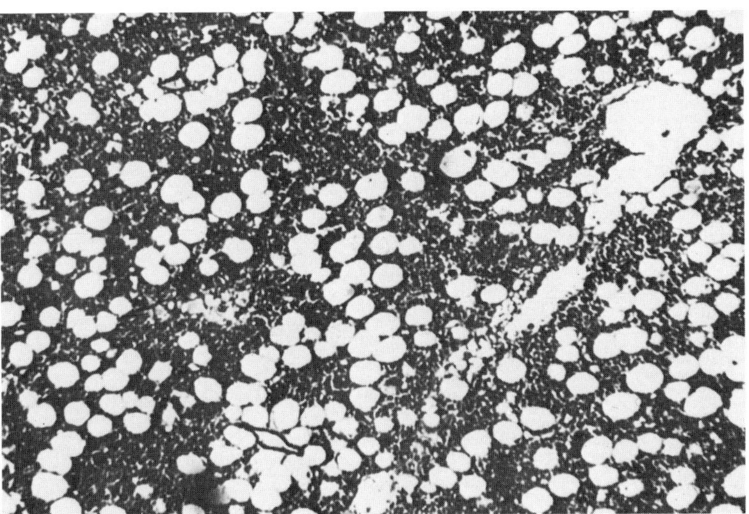

FIGURE 5-4 Yellow bone marrow. An increasing number of adipose cells are interspersed between the red bone marrow cells which increasingly become fewer as the marrow becomes more yellow at maturity.

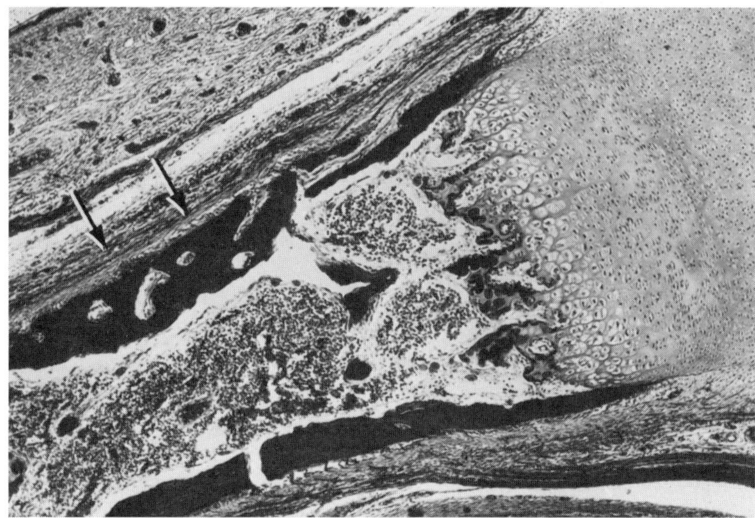

FIGURE 5-5 Light micrograph of a longitudinal section of a developing long bone. The epiphysis is seen in the right third of the micrograph, and the diaphysis extends through the remaining part of the picture to the left. A periosteum (arrows) covers the bone of the diaphysis.

from a middle-aged person, was caused by dissolution of lipid from fat cells in the histologic processing. In the adult, red marrow is restricted to the vertebrae, sternum, ribs, cranial bones, and the ends (epiphyses) of the long bones. Compact bone tissue is a dense, rigid tissue, making up the periphery of long bones. Therefore, the compact bone does not contain marrow spaces, although there are small canals containing vessels throughout the bone.

General Structure of Long Bones. The long bone is commonly said to consist of two parts: the *diaphysis,* which is the body or shaft of a long bone; and the *epiphysis,* which is its enlarged head. The *periosteum,* the vascular fibroelastic connective tissue covering of all bones, is indicated along the diaphysis of the developing long bone in Fig. 5-5 (arrows). As in perichondrium of the cartilage, the periosteum is also made up of two layers (Fig. 5-6). Its outer layer, called a

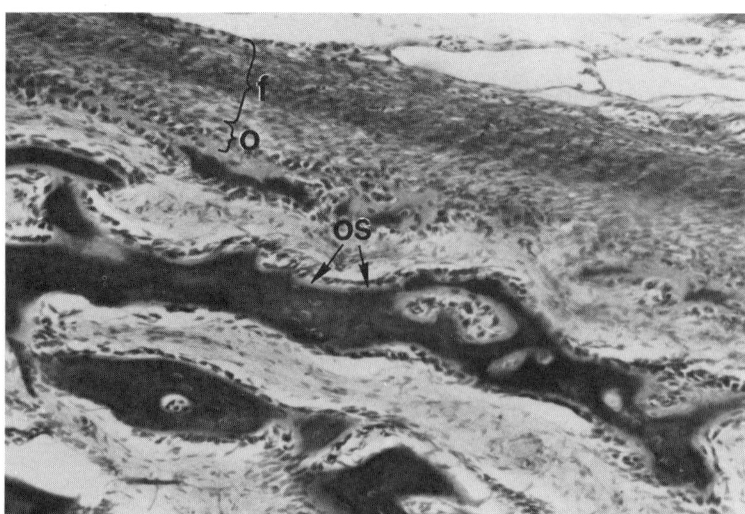

FIGURE 5-6 Developing bone seen at higher magnification. The periosteum covers the outer surface of the developing bone. It is composed of a fibrous layer (*f*) and a cellular layer (*o*) which contains osteoblasts, among other cells. Endosteum lines the marrow spaces. Osteoid (*os*) lines the growing bony trabeculae.

fibrous layer, contains a large number of collagenous fibers, some blood vessels, and nerves. The inner layer, or *cellular layer,* has many connective tissue cells, including precursor cells for osteoblasts; therefore it is also called the *osteogenic layer.* Figure 5-6 is a section through the periosteum in which collagenous fibers and osteoblasts are clearly visible. The fibrous layer is labeled by f, while o marks the osteogenic layer. A very thin layer of connective tissue called *endosteum* lines the marrow spaces of bone.

CELLULAR ELEMENTS OF BONE

Among the different types of cells in and around bone, osteoblasts, osteocytes, and osteoclasts are essential in its formation and maintenance. Figures 5-7 and 5-8 indicate the relative size and locations of these cells in two bony spicules of the growing facial skeleton of the human fetus.

Osteoblasts

In Fig. 5-9, *osteoblasts* are lined up along the subperiosteal surface. At a higher magnification (Fig. 5-10), these cells appear to be large cells with a rounded nucleus, prominent nucleolus, and basophilic cytoplasm due to abundant RER. The clear area on the left half of Fig. 5-10 is the calcified bone matrix which contains the nucleus of an osteocyte (oc). The darker matrix is where a number of rounded cells are in the process of differentiating into osteoblasts (arrow). In general, the nucleus of osteoblasts is eccentric, and there is a juxtanuclear region of the Golgi apparatus, which usually appears as a negative image, situated toward the newly formed matrix (Fig. 5-10). Osteoblasts are thus polarized; they synthesize and secrete the extracellular fibers and ground substances. They are

FIGURE 5-7 Osteoblasts along a developing bony spicula.

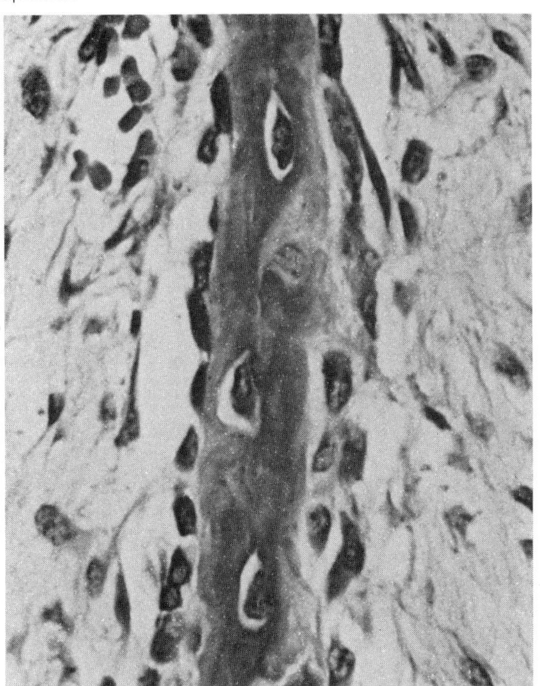

FIGURE 5-8 Photomicrograph of two multinucleated osteoclasts along the surface of a bony spicule.

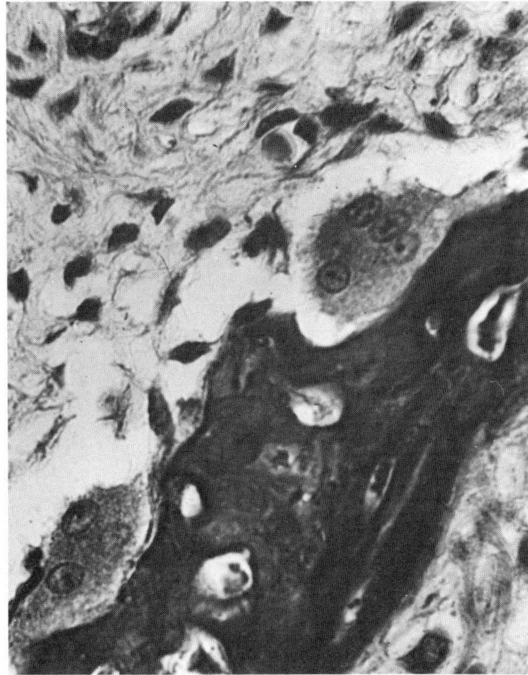

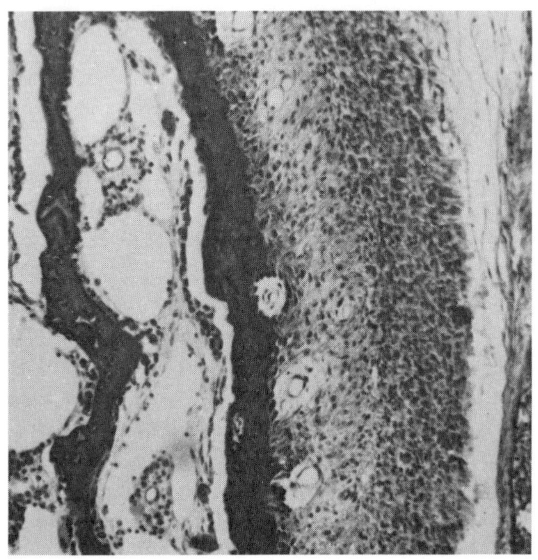

FIGURE 5-9 Osteoblasts aligned along the subperiosteal surface of a developing long bone.

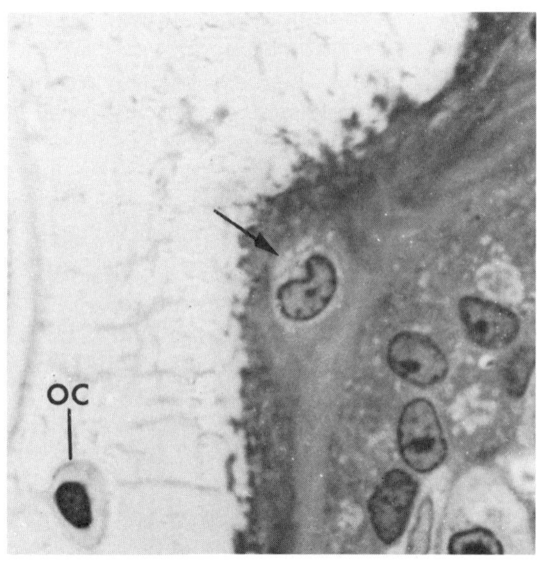

FIGURE 5-10 At higher magnification, osteoblasts (arrow) appear as large cells with a rounded nucleus, prominent nucleolus, and basophilic cytoplasm due to abundant RER. The calcified bone matrix (light-appearing) contains the nucleus of an osteocyte (oc).

known to control the uptake and deposition of minerals by the matrix during ossification. The EM in Fig. 5-11, which represents an osseous surface with a portion of an osteoblast (ob), shows all the features mentioned above. The RER is frequently dilated and contains a moderately electron-dense substance (Figs. 5-11 and 5-12). In Fig. 5-11, the matrix immediately surrounding the cell is mineralized, as shown by the presence of dark hydroxyapatite crystals. The mineralized crystals from the older matrix (o, Fig. 5-11) has been removed by mild decalcification, which left the newer calcified bone

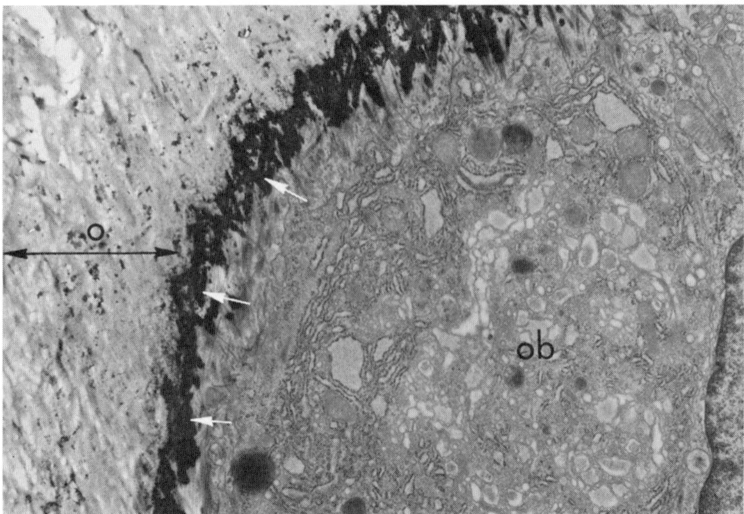

FIGURE 5-11 Electron micrograph of a portion of an osteoblast containing the organelles necessary for the synthesis of proteins and other components of the intercellular matrix. The older matrix (o) has been decalcified slightly. However, crystals in the more newly calcified bone remain intact (arrows).

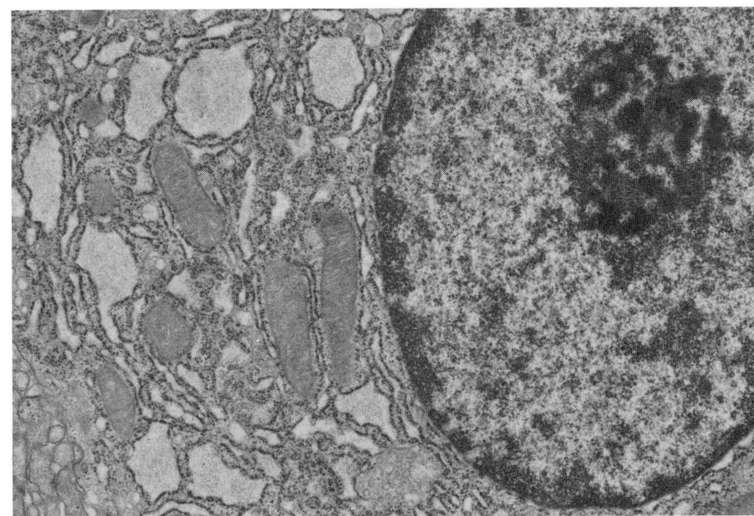

FIGURE 5-12 Portion of an osteoblast. The RER is diluted and contains a moderately electron-dense material.

matrix intact (arrows). However, there is a lag period between the secretion of matrix and calcification. When calcification occurs sufficiently behind the matrix formation, one can see the unmineralized matrix, called *osteoid* or *preosseous matrix,* appearing as a faintly eosinophilic band even in LM (os in Fig. 5-6). The width of the osteoid is highly variable and therefore can be seen at both light and electron microscopic levels. Figures 5-13 and 5-14 are EMs magnified at the same level. Note the difference in the amount of uncalcified matrix (compare the regions indicated by arrows).

Osteocytes

In bone formation, osteoblasts are regularly trapped by the matrix they produce. *Osteocytes* seen within the matrix eventually become encased by the calcified matrix. The EMs in Figs. 5-15 and 5-16 represent the stages of initial trapping and subsequent calcification, respectively.

Because osteocytes are less active than osteoblasts, the cytoplasm and its organelles atrophy, leaving long, attenuated processes. These processes are retained in the calcified matrix and often are in contact with similar processes of adjacent osteocytes. It is known that osteocytes are capable of communicating with each other and therefore quickly respond to such situations as changes in blood calcium level. The space in which an osteocyte resides is called a *lacuna*. The fine tunnels that link neighboring lacunae and contain osteocyte processes are called *canaliculi.* In the LM shown in Fig. 5-17, lacunae and communicating canaliculi are visualized in a ground preparation, which represents a thinly ground, nondecalcified piece of bone. The clear background is the ossified matrix, while the lacunae and canaliculi show up dark because of differences in refractive indices between the calcified matrix and air space.

Osteoclasts

Osteoclasts are cells which break down the ossified matrix of the bone. The large multinucleated cells located along the surface of the bony trabeculae in Fig. 5-8 are osteoclasts. Since they are involved in excavation of bony matrix, osteoclasts are usually situated in a cavity which is called *Howship's lacuna*. The cytoplasm of osteoclasts is often vacuolated and is usually slightly eosinophilic. In the EM, the cytoplasm shows an extremely ruffled border, numerous vesicles and vacuoles, and clusters of ribosomes and lysosomes. A portion of such a cell is diagramed in Fig. 5-18,

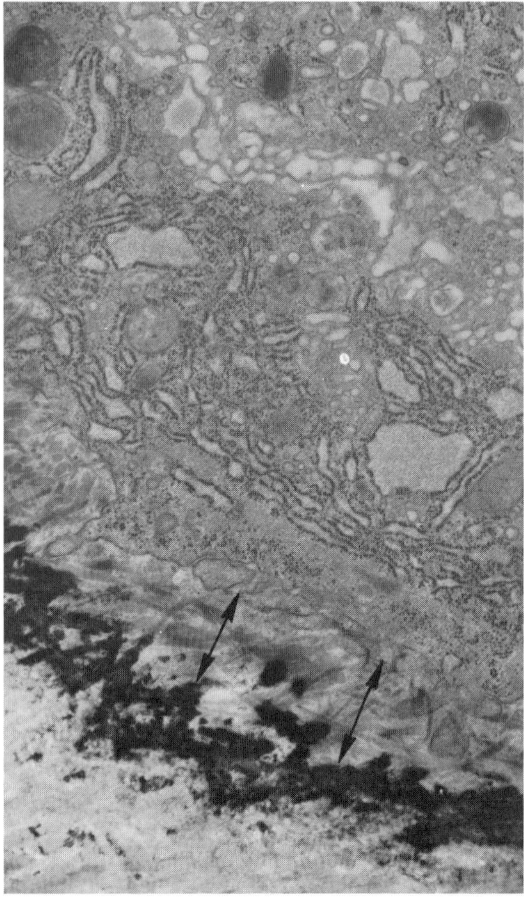

FIGURE 5-13 Electron micrograph of calcifying bone. The width of the uncalcified matrix (arrows), the osteoid, is narrow; therefore mineralization of the matrix has followed shortly after its deposition.

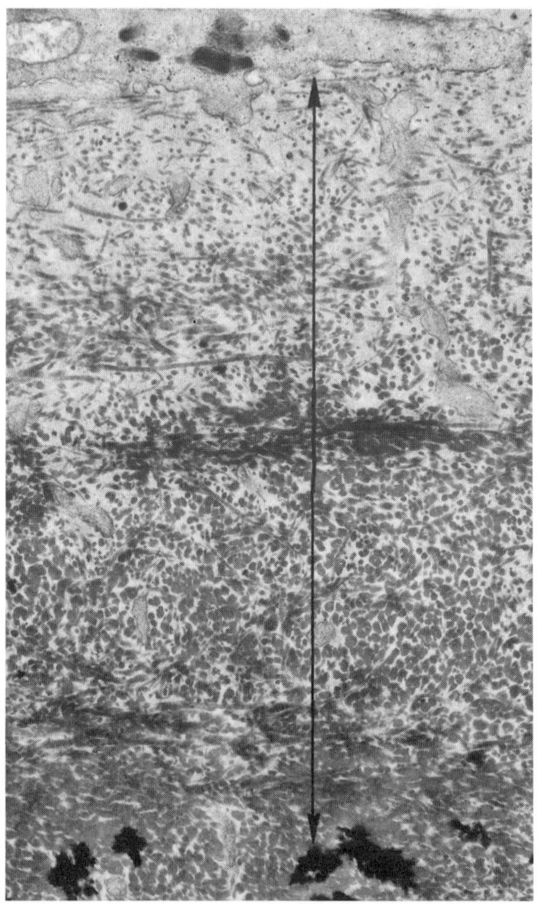

FIGURE 5-14 Electron micrograph of calcifying bone. In this bone the mineralization is delayed. Therefore the width of the uncalcified matrix (arrows) is greater than in more rapidly calcifying bones (compare Figs. 5-13 and 5-14).

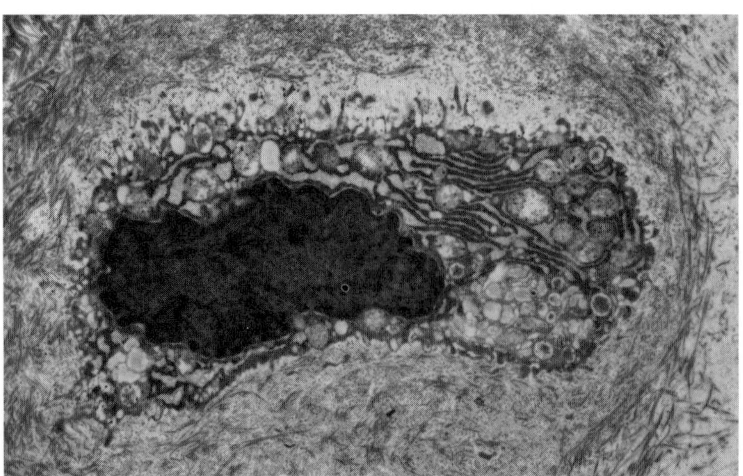

FIGURE 5-15 Electron micrograph of an osteocyte in the initial stage of trapping in the bone matrix. The osteocyte is completely surrounded by osteoid.

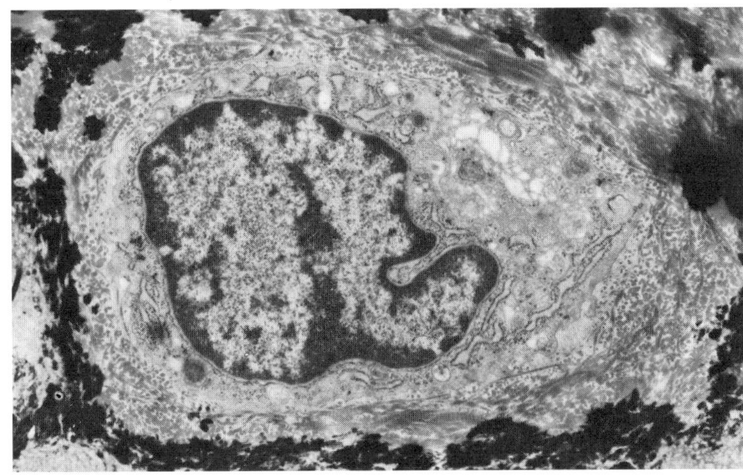

FIGURE 5-16 Electron micrograph of an osteocyte in calcified matrix. It is surrounded by a narrow band of uncalcified matrix in this micrograph. Note the atrophy of cell organelles and the cytoplasm.

which depicts the junction between the osteoclast and bone.

The osteoclasts are present even in the earliest stage of osteogenesis. Bone trabeculae, as will be discussed later in this chapter, must move through tissue space during their formation. Thus an osteoclast may be found in a small isolated bone trabecula during an early stage in embryonic development. This is shown in Fig. 5-19, which is a portion of the developing facial skeleton. Osteoclasts (Fig. 5-19) are usually present on the side of the trabecula opposite from where active osteoblasts are present. This positional relationship indicates the direction in which a bone trabecula moves through the developing tissue space. As a cell specialized for bone matrix removal, the osteoclast is also sensitive to hormones and other factors influencing the calcium metabolism. However, the response is opposite that of the osteoblast and osteocyte, reflecting a complex feedback mechanism between these cell types in maintaining calcium homeostasis.

BONE FORMATION AND GROWTH

Bone formation is a complex process. It involves maintaining skeletal structures from

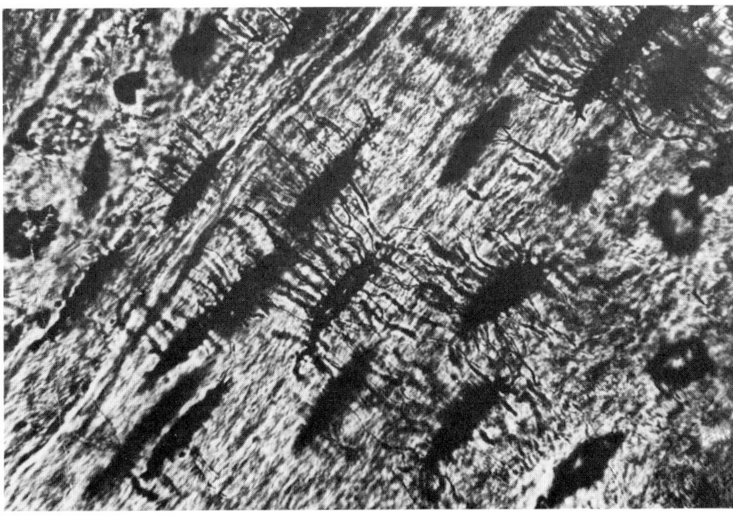

FIGURE 5-17 Ground section of undecalcified bone. Several lacunae where the osteocytes resided in the living bone and their communicating canaliculi are seen in this micrograph.

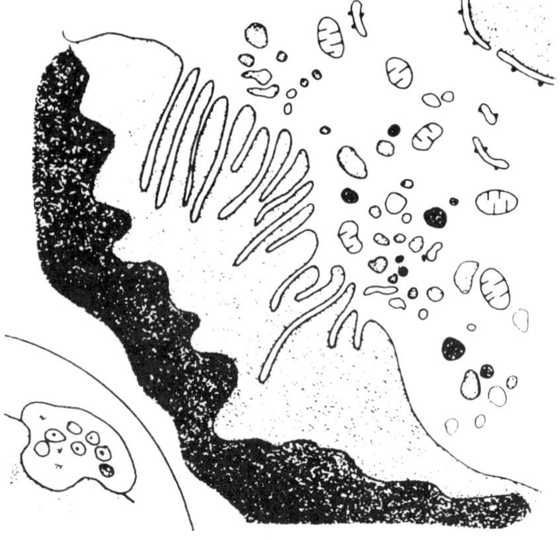

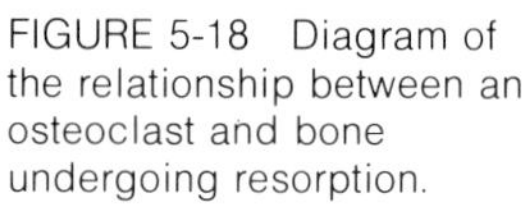

FIGURE 5-18 Diagram of the relationship between an osteoclast and bone undergoing resorption.

a very early stage of fetal development in conjunction with a continuous and orderly growth of the skeleton. During development most bones are necessarily transient, yet physical rigidity is required even during the early development. There are two modes of bone formation: intramembranous (within membranes) and endochondral (within cartilage). Both may occur simultaneously at different locations within a single bone. Because continuous remodeling occurs from the very beginning of osteogenesis, it is important that a student recognize in what way bone formation has occurred in a given histologic picture. However, it is essential that a student recognize the loci and reasons why the two modes occur in different adult bones.

INTRAMEMBRANOUS BONE FORMATION

In the fetus, a well-vascularized mesenchyme develops in presumptive regions where osteogenesis will occur. The stellate-shaped mesenchymal cells proliferate and organize to form a primordial mass of cells, which differentiate into osteoblasts. This is characterized by development of mesenchymal cells into functional osteoblasts, which includes differentiation of intracellular structures necessary for matrix synthesis. An example of such differentiation is seen in Fig. 5-20, where mesenchymal cells condense at the edge of a developing bony trabecula which is covered by osteoblasts. These differentiating cells start to elaborate and secrete

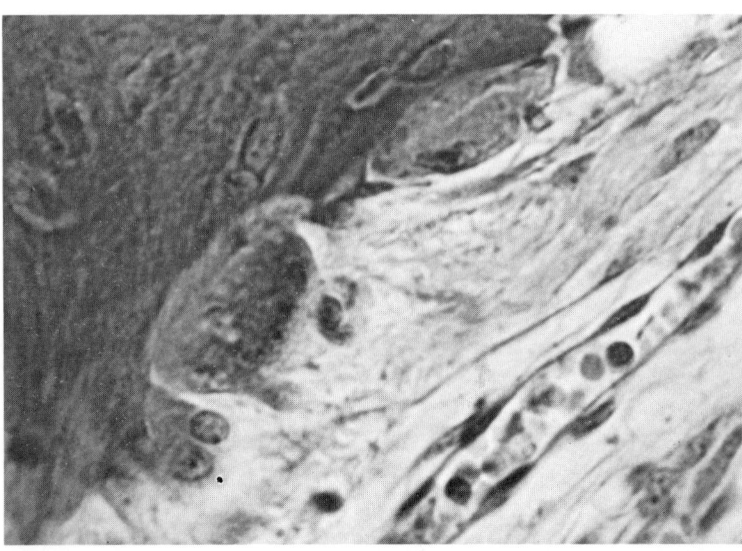

FIGURE 5-19 Light micrograph of an osteoclast on the surface of a bony trabecula in the developing facial skeleton of a human embryo.

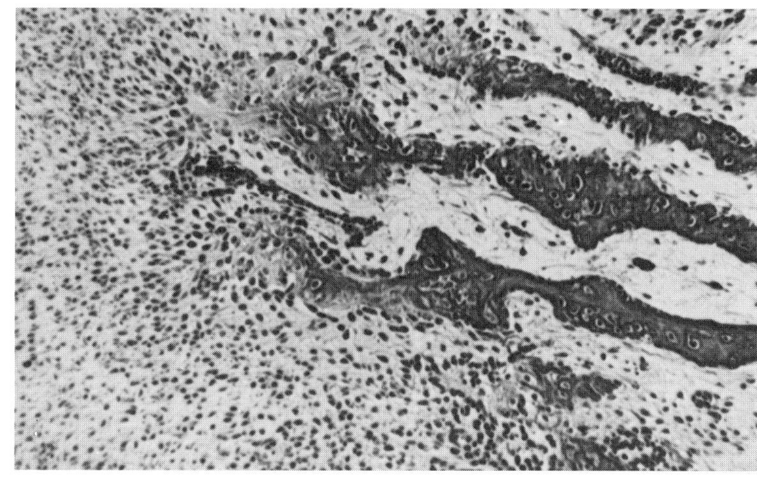

FIGURE 5-20 Mesenchymal cell condensation at the end of a bone trabecula. Functional osteoblasts are differentiating from mesenchymal cells within the cell mass. These young osteoblasts begin to elaborate and secrete the components of the organic constituents of bone.

the substance of organic matrix which includes fine collagenous fibers and much ground substance. As described earlier, this matrix is called *osteoid* or *preosseous matrix*. Osteoid is the organic constituent of the bone which becomes calcified later as inorganic minerals are added to it.

From the beginning of matrix formation, the osteoblasts are polarized. As they continue to elaborate the organic matrix, some osteoblasts become trapped by the matrix material they have secreted. Such trapping, occurring in a rhythmic fashion, involves a large number of osteoblasts at one time and results in the formation of a linear arrangement of trapped osteoblasts that eventually become osteocytes. The transformation of osteoblasts into osteocytes causes a gradual reduction of cytoplasmic activity as the cell runs out of space around them. A lacuna and associated canaliculi are then formed around each osteocyte (Fig. 5-17). Since the osteoblasts are simultaneously trapped, the lacunae from the same generation of cells are lined up in a linear fashion, forming an osseous lamella as indicated in Fig. 5-21 (arrows). As osteoblasts are trapped within the matrix, a new generation of cells develops from division or differentiation of mesenchymal cells along the surface of the matrix to maintain a constant layer of osteogenic cells.

The formation of lamellae at the inner surface of a compact bone takes place along

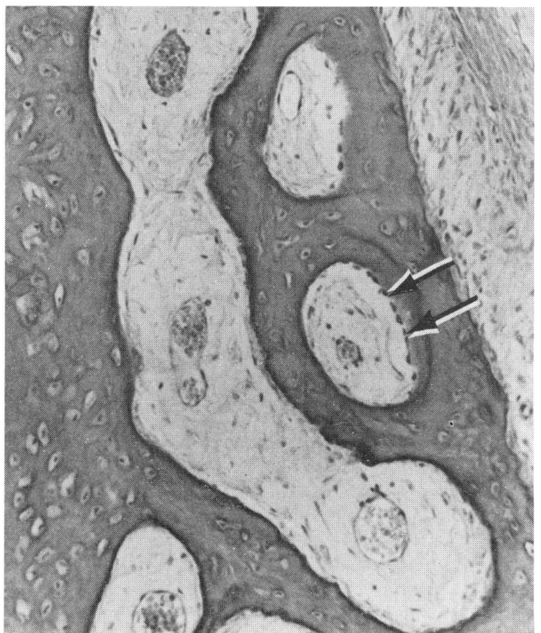

FIGURE 5-21 Primary marrow spaces at the inner surface of compact bone. Osteoblasts (arrows) line these spaces. When they are trapped, circular lamellae are formed, reducing the size of the primary marrow spaces.

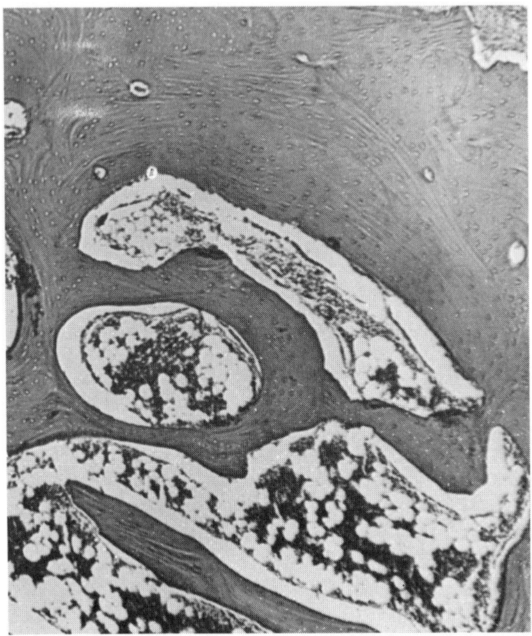

FIGURE 5-22 Primary marrow spaces which have an increased number of circular lamellae. These lamellae are arranged in a concentric manner. Hemopoietic cells now fill the spaces.

the primary marrow spaces that are not unlike the erosion tunnels to be discussed later. Note that each of the primary marrow spaces appearing in Fig. 5-21 is lined by a layer of osteoblasts and contains blood vessels and mesenchymal elements. The micrograph in Fig. 5-22 shows at a lower magnification that a further number of lamellae have been formed. By this time the marrow space has become filled with hemopoietic cells. The trapping of osteoblasts and the successive formation of lamellae occur in a concentric manner. In the process, a certain amount of osseous matrix is deposited before another generation of osteoblasts is trapped, resulting in lamellae that are arranged in concentric layers. The entire structure, i.e., concentric lamellae and central canal with its blood vessels and associated connective tissue, is called a *haversian system* or an *osteon.* The concentric lamellae are often referred to as *haversian lamellae,* while the central canal of the osteon is called a *haversian canal* (Fig. 5-21). This is diagramed in Fig. 5-23, which depicts a cross-sectional view of the shaft of a long bone demonstrating numerous haversian systems. The open cut at the front indicates the longitudinal orientation of haversian systems in long bones.

General nutritional conditions of the body influence the rhythm of lamellar development so that matrix deposition may occur more slowly during a period of malnutrition or disease. Under these diverse conditions, the matrix produced is often hypocalcified and therefore differentially stained. With H and E stain, such a *growth arrest line* appears basophilic in contrast to the eosinophilic matrix of the bone next to it, as seen in Fig. 5-24 (arrows). Because growth arrest lines have different fibrous matrices, a view through the polarizing microscope, which shows dark-bright contrast reflecting fiber orientation, presents an exaggerated picture of the same (Fig. 5-25).

Mineralization. The process of adding minerals to the osteoid matrix to form the fully calcified bone is called *calcification* or *mineralization.* With H and E stain, the mineralized portion of the bone appears darker than osteoid matrix. During calcification, tricalcium phosphate molecules are assembled in the form of *hydroxyapatite* crystals, which are eventually deposited on the collagen fibrils of the matrix. The EM in Fig. 5-26 is a section through mineralizing bone. Hydroxyapatite crystals first appear on the collagen fibril surface (arrows, Fig. 5-26) and then between fibrils. The crystals are aligned with the periodic cross striations of the fibrils. Calcification was previously thought to occur when ideal physicochemical conditions were produced in the microenvironment within the organic matrix, but recent investigations have revealed that the cells may play significant roles in this process. Osteoblasts and

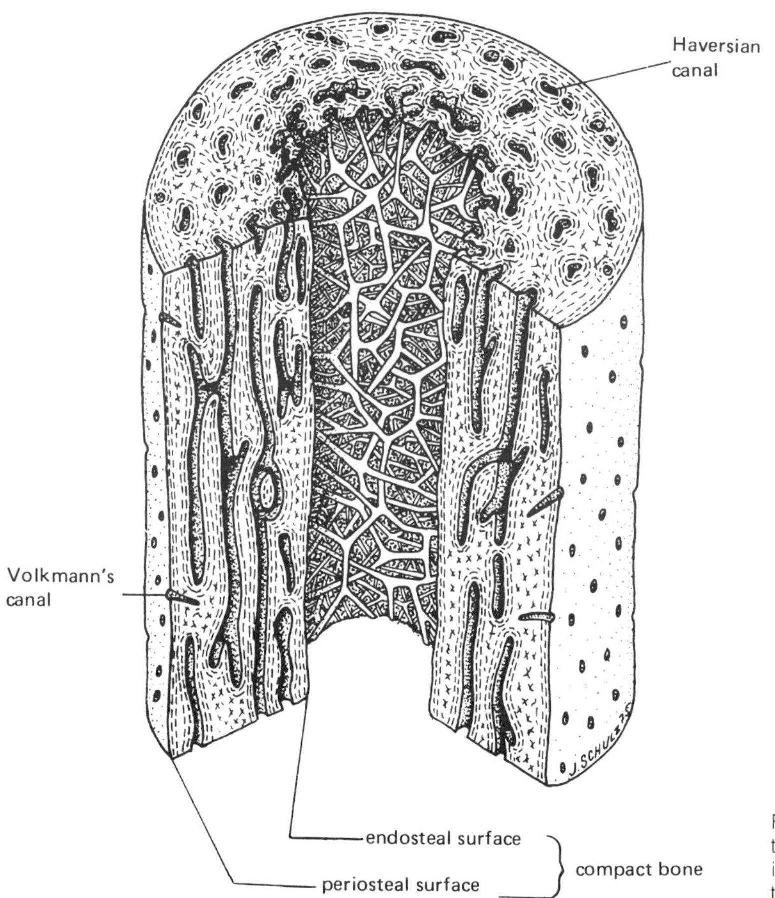

FIGURE 5-23 Diagram of the diaphysis of a long bone illustrating its structural features.

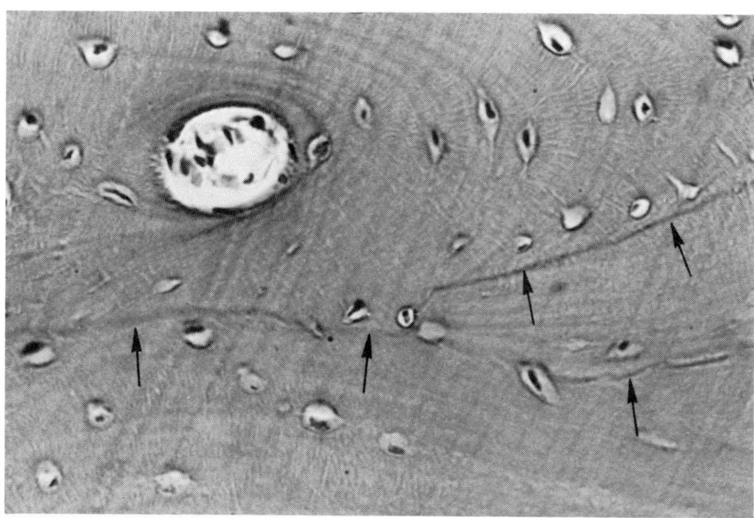

FIGURE 5-24 Section of decalcified bone stained with H and E. An arrest line (arrows) separates two regions of well-mineralized bone.

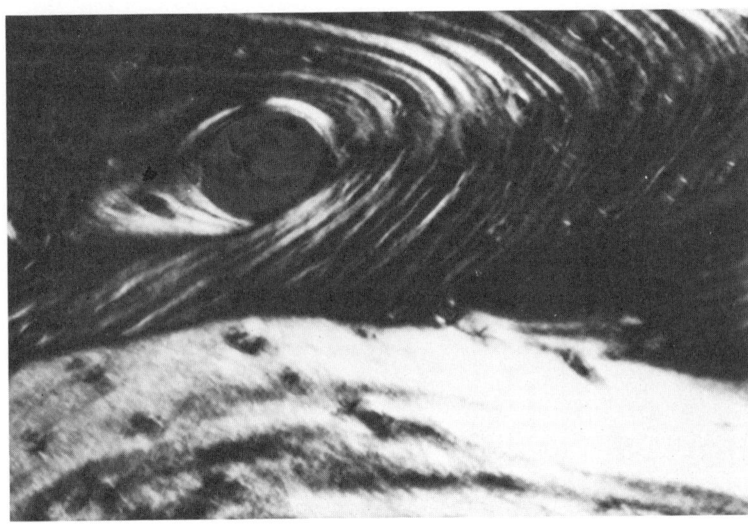

FIGURE 5-25 The same section in Fig. 5-24, viewed through a polarizing microscope. The dark-bright contrast reflects the fiber orientation in the two adjacent bony areas and presents an exaggerated image of the growth arrest line.

osteocytes actively participate through production of insoluble tricalcium phosphate, utilizing a mechanism which involves mitochondria and the extrusion of calcium salt from the cells. As indicated in Chap. 1, mitochondria can trap and concentrate intracellular Ca^{2+} in the form of calcium phosphate crystals, which are located in the intramitochondrial granules. For this reason, mitochondria in mineralizing tissues often have large, numerous intramitochondrial granules (Fig. 5-27).

ENDOCHONDRAL BONE FORMATION

This mode, also called *intracartilaginous bone formation,* replaces preformed cartilage with bone. While certain bones at the base of the skull are formed in this manner, more typical examples are seen in the formation and growth of long bones (Fig. 5-28).

In presumptive areas of long bone formation, aggregations of mesenchymal cells develop into a rod-shaped mass of primordial cartilage cells. As indicated in Fig. 5-28, cartilage cells proliferate and start to produce

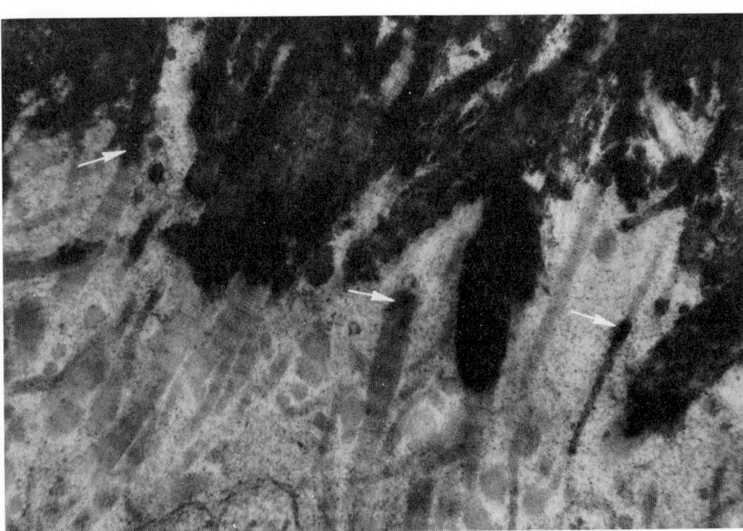

FIGURE 5-26 Electron micrograph of mineralizing bone. Hydroxyapatite crystals first appear on the surface of collagen fibrils in the matrix (arrows) and then between them. Eventually they are equally deposited in fibrils and ground substance.

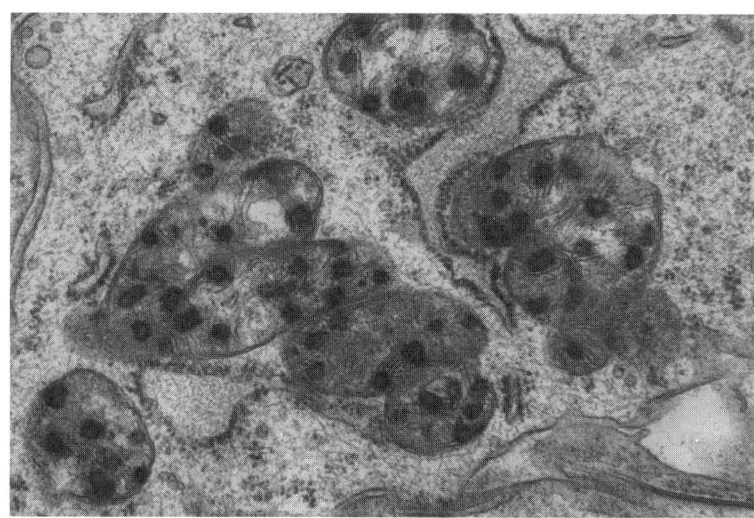

FIGURE 5-27 Mitochondria in cells from a mineralizing tissue. They are involved in the mechanism of calcification. Note the large number of intramitochondrial granules which contain calcium phosphate crystals.

matrix (*a*, *b*, and *c* of Fig. 5-28, interstitial and appositional growths). When the cartilage has reached a certain size, its center is penetrated by developing vascular buds (*d*, Fig. 5-28). While the cartilage continues to exhibit appositional growth in length and width, the invading blood vessels at the middle (presumptive diaphysis) begin to resorb the cartilage matrix. This resorptive process enables initial differentiation of osteoblasts to occur among the primitive mesenchymal cells that enter the space with the invading vascular buds. Continuous excavation in the center and cartilage growth

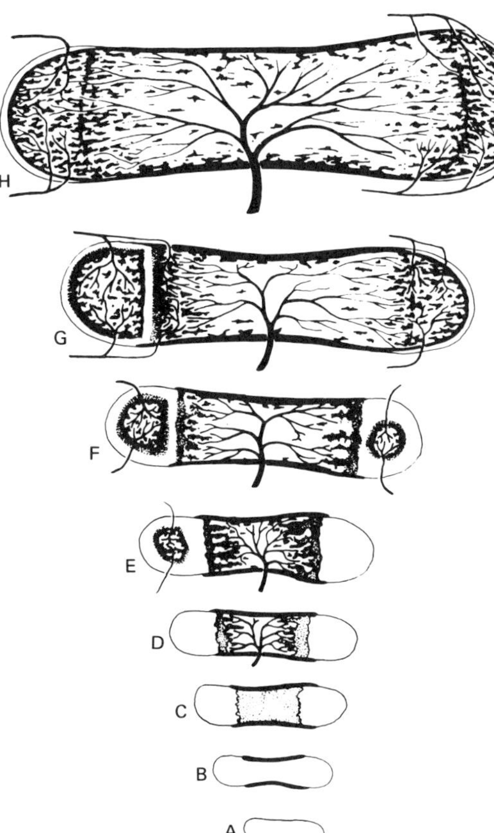

FIGURE 5-28 Diagram of the ossification of a long bone. (a) Cartilage model. (b) Periosteal bone collar appears before any calcification of cartilage. (c) Cartilage begins to calcify. (d) Vascular mesenchyme enters the calcified matrix and divides it into two zones of ossification. (e) Blood vessels and mesenchyme enter the upper epiphyseal cartilage and the epiphyseal ossification center develops in it. (f) A similar ossification center develops in the lower epiphyseal cartilage. (g) As the bone ceases to grow in length, the lower epiphyseal plate disappears first and then the upper. (h) The bone marrow cavity then becomes continuous throughout the length of the bone and the blood vessels of the diaphysis, metaphyses and epiphyses intercommunicate.

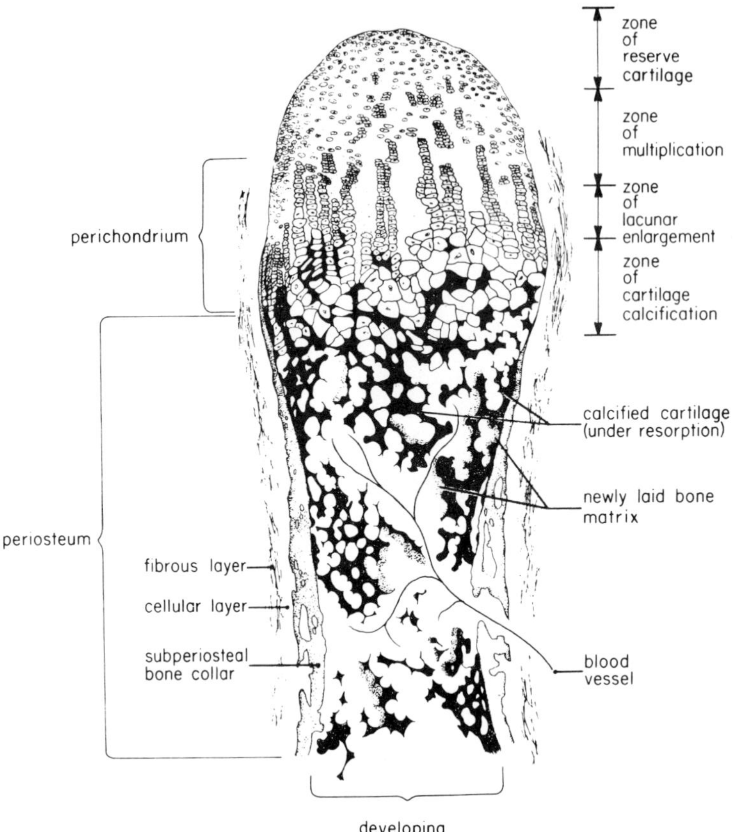

FIGURE 5-29 Diagram of a longitudinal section through the epiphyseal end of a growing long bone.

allow differentiation of various characteristic zones of the epiphyseal plate, which moves away from the diaphysis, as diagramed in Fig. 5-29.

Lengthening of Long Bones. Multiplication and hypertrophy of reserve cartilage cells characterize the growth of cartilage at the ends of long bones. Dividing cartilage cells line up in rows to form the zone of multiplication. As the cells begin to hypertrophy, their lacunae increase in size, reducing the amount of matrix between them. Subsequently, the matrix becomes calcified with mineral salts, producing a more basophilic appearance than the original cartilage. The calcified cartilage is soon removed by chondroclastic cells that differentiate from the mesenchymal cells accompanying the blood vessels entering from the center of the diaphysis. Figure 5-30 is a section through the growing epiphyseal plate, with the cartilage on the right and diaphyseal bone on the left. At higher magnification, as in Fig. 5-31, one can readily identify several different zones. The *zone of reserve cartilage,* barely visible on the extreme right, is the original cartilage tissue of primitive hyaline type. It grows away from the diaphysis. By the time of cessation of long bone growth, the amount of reserve cartilage is drastically reduced.

Adjacent to the reserve cartilage zone is the *zone of cell multiplication,* which is highly cellular and characterized by rows of cartilage cells. Some of these cartilage cells are in isogenous groups. Consult the diagram in Fig. 5-29 and identify the multiplying cartilage cells that have organized into rows (Fig.

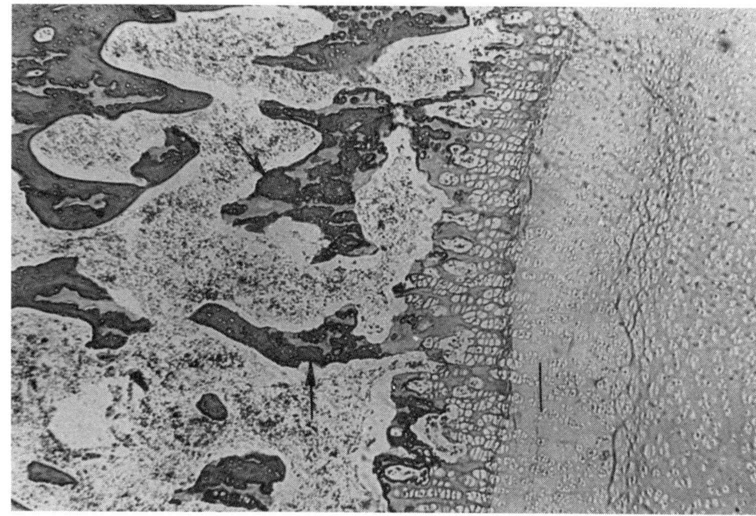

FIGURE 5-30 Light micrograph of a growing epiphyseal plate. The cartilage appears on the right and the growing bone on the left. Bone is being laid down on the surface of the remaining cartilage (arrow) in the diaphysis.

5-31). Where the multiplication of cartilage cells ceases, the cells start to grow in size.

In the next zone, called the *zone of cellular hypertrophy or lacunar enlargement*, cartilage cells show cytoplasmic differentiation accompanied by the enlargement of their size and lacunar space. As the lacunae continue to enlarge, the cartilage matrix undergoes mineralization and is called the *zone of cartilage calcification*. In Figs. 5-30 and 5-31, this zone appears darker than the preceding zones, representing a change in ground substance accompanied by mineralization of the matrix.

Following the calcification of the matrix, chondrocytes give way to the erosive action of the invading blood vessels and associated cells (Fig. 5-29). The removal of cartilage matrix is done by the growing mesenchymal cells of the perivascular region. As mentioned earlier, some of the mesenchymal elements differentiate into osteoblasts which start to lay down an osseous matrix against the calcified cartilage matrix being removed. In Figs. 5-30 and 5-31, the arrows point to the darker osseous matrix that is being laid against the remaining cartilage matrix. The

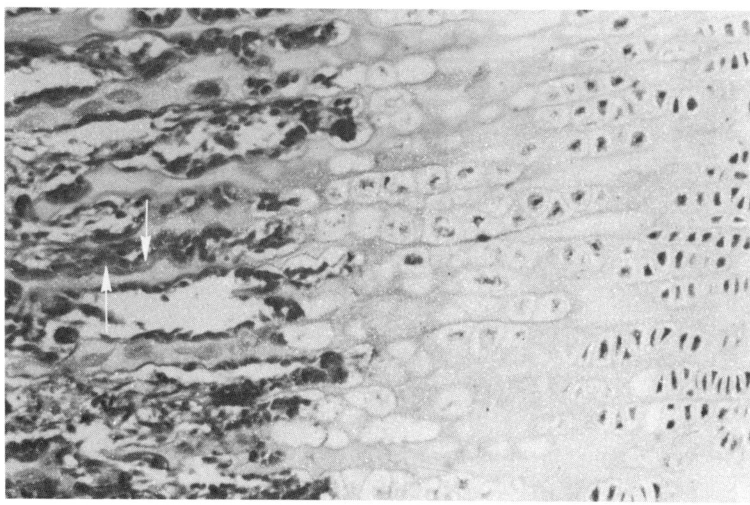

FIGURE 5-31 Epiphyseal plate viewed at higher magnification. Various zones can be identified. From right to left, the zone of multiplication, the zone of cellular hypertrophy, and the zone of cartilage calcification can be seen. Bone is being laid down on the surface of the remaining cartilage in the diaphysis (arrows).

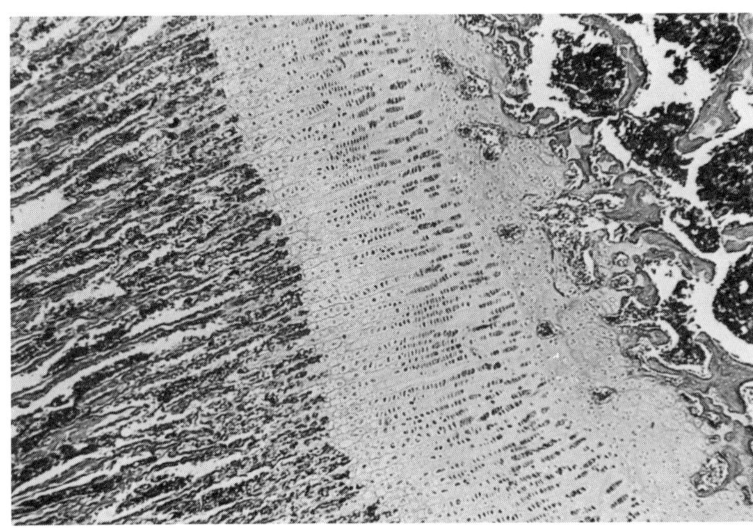

FIGURE 5-32 Light micrograph of the epiphyseal plate of a long bone. The primary ossification center is seen on the left while the secondary epiphyseal ossification center is seen on the right.

cartilage matrix appears light purple in H- and E-stained slides and usually occupies a central portion of the developing bone spicules, while the osteoid matrix is orangish red and is more peripherally located. In three dimensions, however, columns of calcified cartilage matrix are being resorbed by invading mesenchymal elements on one side, while other sides serve as a physical framework against which osteoid matrix is laid down.

The preceding events illustrate the movement of the epiphyseal cartilage away from the diaphyseal center and the continued production of new osseous matrix in the space generated by the movement of the epiphyseal cartilage. This accounts for the lengthwise growth of long bones. The entire histologic region where the endochondral bone formation is taking place is called the *epiphyseal plate.* At a later stage of development, a secondary ossification center develops in the center of the heads at each epiphyseal end of the long bone (*e, f,* and *g,* Fig. 5-28).

Figure 5-32 shows an example of an epiphyseal secondary ossification center, which is seen to the right of the epiphyseal plate. As the zone of reserve cartilage recedes with the progression of osteogenic activities at both primary and secondary ossification centers, they approach each other, i.e., the epiphyseal plate becomes narrower in time. Eventually, the two osteogenic fronts merge as indicated in the diagram in Fig. 5-28 (*g* and *h*). This merger, effectively eliminating the functioning of the epiphyseal plate, occurs at about the time when the long bone stops its longitudinal growth.

Growth in Diameter of Long Bones. At the same time lengthening of long bone is being accomplished by the epiphyseal plate, the diameter of long bones is being achieved by appositional growth at the periosteal surface. In the beginning, the cells of the perichondrium surrounding the primordial long bone differentiate into osteoblasts (Figs. 5-5, 5-28, and 5-29), which deposit a layer of bone called the *bone collar* beneath the periosteum of the shaft. For this reason, the bone collar can be said to result from subperiosteal appositional growth. At this point, the perichondrium has become periosteum, since it now covers bone. The bone collar provides support for the fragile osseous tissue in the epiphyseal plate.

The appositional growth at the subperiosteal surface continues with the lengthening of the bone, and as the marrow cavity becomes enlarged, the removal of bone tissues from the endosteal surface has to take place (Fig. 5-28). Such bone removal continues throughout the appositional growth process. This process involves a dynamic bone remodeling which reorganizes the bone structure so that the fully mature bone will have attained an ideal physical arrangement of matrix proteins and orientation of hydroxyapatite crystals for its function. Since the periosteal growth generates a series of *outer circumferential lamellae,* the blood vessels that run from periosteum into the marrow cavity produce a canal which has no organized osteoblasts around it (see Fig. 5-23). The larger of these canals corresponds to the nutritional canals you have seen in gross specimens, while smaller ones that are seen under the light microscope are called *Volkmann's canals* (see Fig. 5-23). These canals run perpendicular to the long axis of the bone, whereas circumferential lamellae and haversian systems run parallel to it. This orientation of haversian systems is illustrated in Fig. 5-23.

As the bone grows in diameter and length, surrounding soft tissues such as connective tissues and muscle must maintain a functionally meaningful relationship with it. For example, the attachment of collagenous fibers of a tendon to a bone must continually be modified as an individual grows. At the same time, the fibers maintain their tight attachment to the bone surface, which results in an investment of a series of bundles of parallel fibers in newly formed bone matrix. Because of these bundles, such bone is called *bundle bone* (Fig. 5-33). The anchoring fibers that continue into the connective tissue from the bundle bone are called *Sharpey's fibers.* Figure 5-33 shows the periphery of a developing long bone stained for demonstration of collagenous fibers. Note that a number of fiber bundles penetrate the darkly stained bone matrix along its surface (arrows, Fig. 5-33).

BONE REMODELING

Bone is a dynamic tissue, constantly remodeling itself to meet functional changes in stress and strain which may result from growth, changing patterns of exercise, or repair of wounds. The process of remodeling necessarily requires the removal of portions of existing bone to be followed by the addition and calcification of matrix in order to meet evolving characteristics of a particular

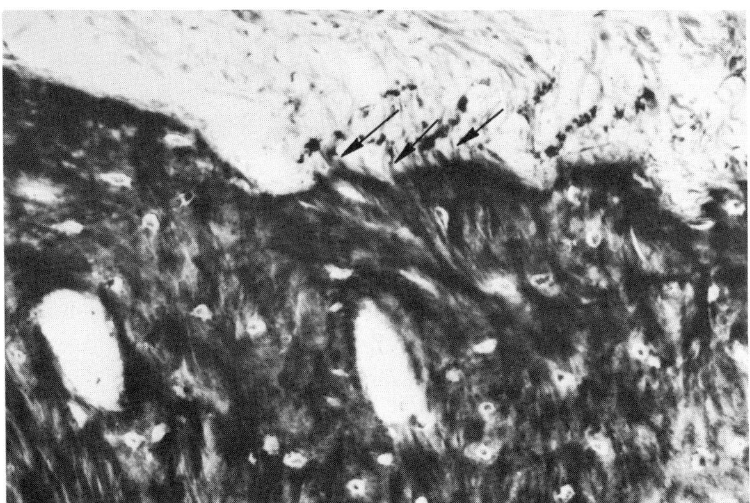

FIGURE 5-33 Light micrograph of bundle bone. Sharpey's fibers (arrows) provide a firm attachment of the periosteum to the bone.

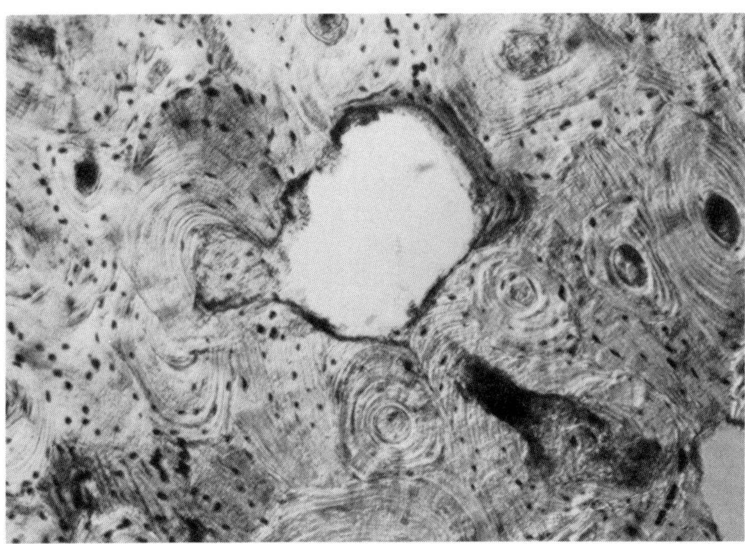

FIGURE 5-34 Erosion tunnel in compact bone (ground section).

region in an ideal scheme. The removal or resorption of bone is usually achieved by invading osteoclasts which are accompanied by blood vessels and perivascular connective tissue elements. As resorption takes place, the advance of these structures into the resorbing cavity creates a channel which is called an *erosion tunnel.* Histologically, an erosion tunnel is characterized by an irregularly shaped lumen and the presence of osteoclasts and accompanying vasculature. The erosion tunnel is totally unrelated to the pattern of haversian systems in a given region (Fig. 5-34). Instead, it presents an excavated appearance of existing haversian systems and eventually forms the boundary for succeeding generations of osteons. Once an erosion tunnel has reached a certain size, the osteoclasts disappear and in their place osteoblasts differentiate from the perivascular connective tissue. These begin to lay down new bone matrix along the inner wall of the erosion tunnel. By periodic trapping of the osteoblasts, concentric haversian lamellae are produced, resulting in a secondary osteon. Since remodeling is a continuous process, second and third generations of erosion tunnels, and hence osteons, may be produced (Fig. 5-35). The partially destroyed haversian lamellae, which no longer maintain a concentric structure, can now be called *intestitial lamellae* (arrows in Fig. 5-35). Since the remodeling takes place largely from the endosteal surface, the circumferential lamellae produced by periosteal apposition tend to remain intact for long periods of time.

The teleologic significance of the life-long process of bone remodeling can be best illustrated by observing the change in fiber orientation of each osteon that is being produced. As mentioned elsewhere, the polarizing microscope provides powerful means to observe molecular orientation of collagenous fiber bundles. Figure 5-36 is a decalcified preparation of a long bone. Although this field is adequate in demonstrating multiple generations of osteons, nothing could be said about the orientation of the matrix fibers. In contrast, the same field seen under a polarizing microscope, seen in Fig. 5-37, shows alternating bright and dark lines in each osteon. Thus the collagenous fibers making up each haversian lamella are oriented at an angle to its neighboring lamellae (usually at 90°). Since there are no organized fibers in haversian canals, they appear as black spaces (Fig. 5-37). Compare the four osteons indicated by arrows in both figures. The alternative angular orientation of fibers between adjacent lamellae provides the strength and

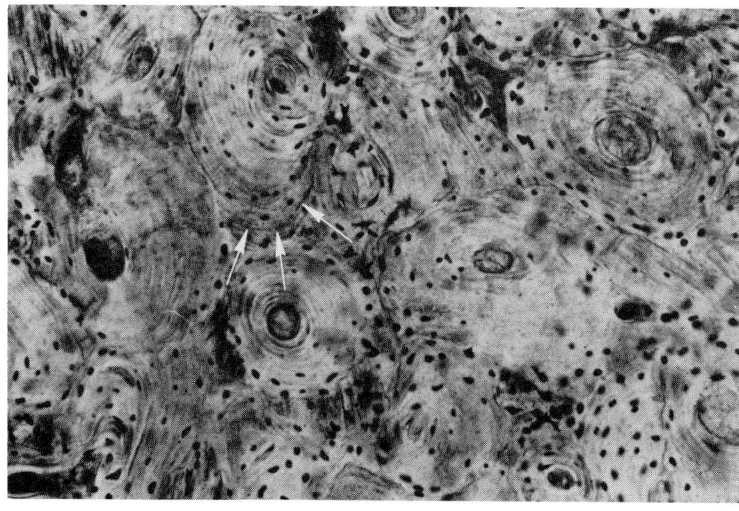

FIGURE 5-35 Light micrograph of compact bone. Interstitial lamellae (arrows) are interspersed among the haversian systems.

resiliency that are important in functional support of long bones.

In summary, an osteon is a dynamic unit of bone tissue which is continuously forming new bone as erosion tunnels excavate old bone matrix. As such, an osteon is fully equipped to support all osteocytes that make up the lamellar system. The osteon has the blood vessels and multipotential reserve cells in the central canal, and therefore it supplies necessary nutrients to each and every osteocyte. An osteon, therefore, can be viewed as a tissue cylinder which responds to changing physiologic and homeostatic conditions, such as changing levels of parathormone and calcitonin, the two hormones that are essential in the maintenance of blood calcium levels. If an injury occurs in blood vessels that supply an osteon, either through trauma or by diseases such as atherosclerosis, the cells that make up the entire osteon may die as a result of the damage. Your observation of microstructure may reveal the complex, dynamic life history of a particular bone. It is estimated that the entire adult skeleton turns over every 7 years or so.

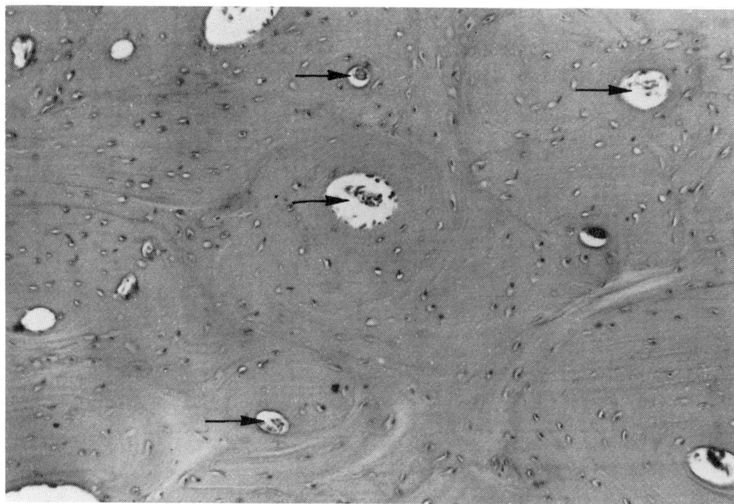

FIGURE 5-36 Light micrograph of a decalcified preparation of long bone. Several haversian systems (arrows) are present.

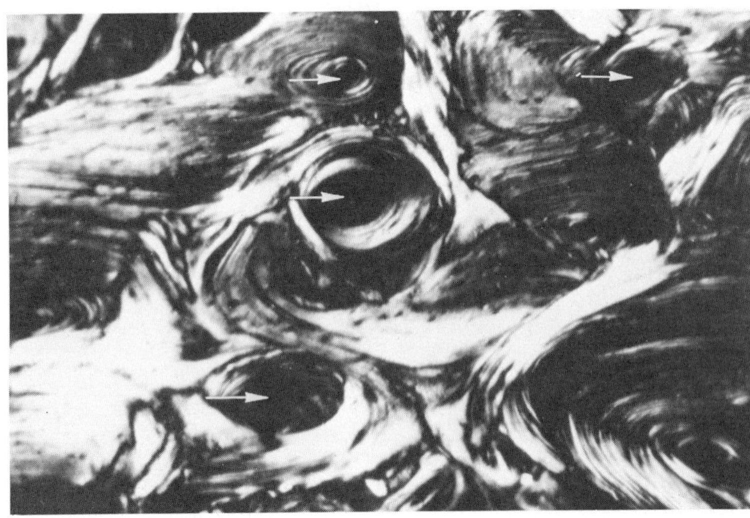

FIGURE 5-37 The same field as in Fig. 5-36 viewed in a polarizing microscope. It illustrates the variation in the orientation of the collagen fibers in each individual haversian system.

REVIEW SECTION

1. Figure 5-38 is a longitudinal section through a developing human long bone. Match the indicated structures in the figure with the following: __________ periosteum __________ spongy bone __________ hyaline cartilage.

The spongy bone (b) is formed by a web of trabeculae in the diaphysis. The periosteum (a) surrounds the developing shaft. Hyaline cartilage (c) covers the end of the bone, protecting the joints and allowing the longitudinal growth. The appositional growth below the periosteum will produce the bone collar, which does not contain a cartilaginous matrix.

2. Figure 5-39 is a cross section through a developing long bone. The numerous small cells

The developing bone is laid down in layers of osteoblasts. Osteoblasts develop from the cellular

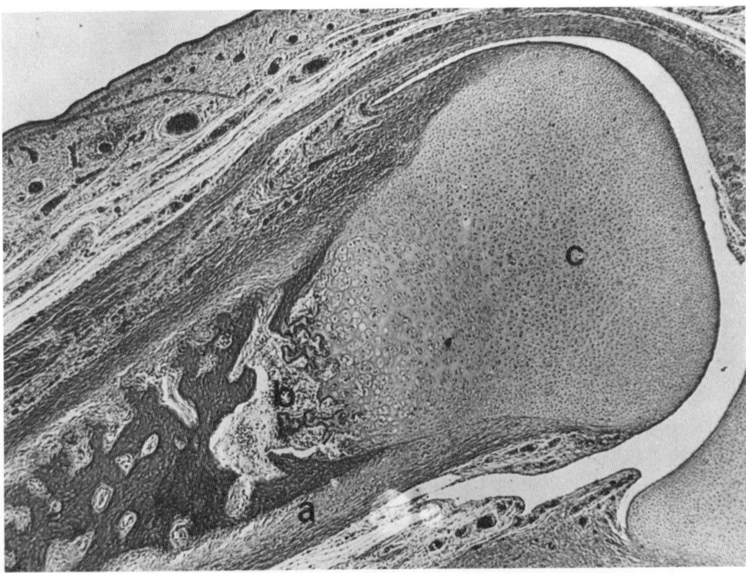

FIGURE 5-38

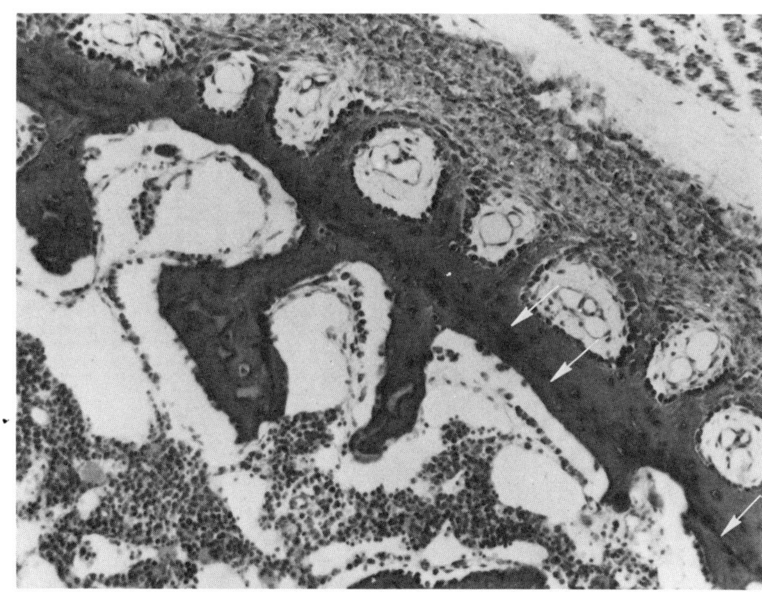
FIGURE 5-39

along the surface of the bone at the bottom of the figure are ________. This tissue is called ________. The dark line indicated by the arrows represents ________. Occasional light spots present in the bone spicules are ________.

layer of the periosteum and lining cells of the marrow cavities. The marrow cavities contain primitive marrow elements and blood vessels. As new bone is laid down, osteoblasts become trapped and remain as osteocytes throughout the compact bone until freed by a new erosion tunnel. The dark line indicated by the arrows is the demarcation between endochondrally formed bone (containing light calcified cartilage matrix) and subperiosteal bone grown through apposition.

3. Figure 5-40 is the growing end of a long bone. The bone being formed is an example of ________ formation. What is the name designating this region of growth in a long bone?

The ends of a long bone grow by endochondral bone formation. The cartilage shows zones of cell hypertrophy and lacunar enlargement as well as calcified cartilage. The dark calcified bone makes

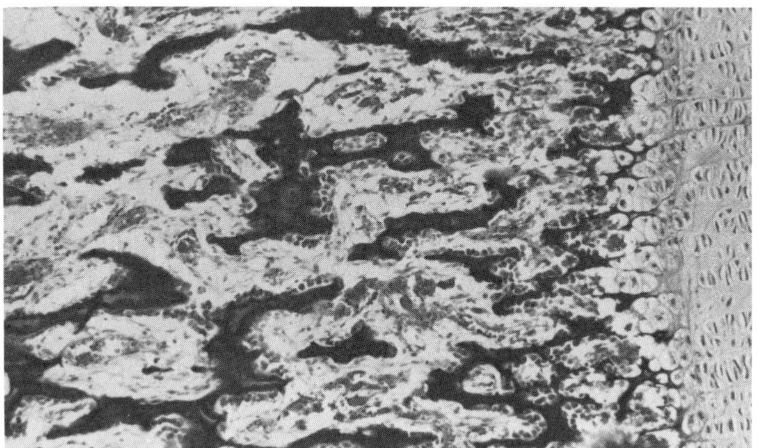
FIGURE 5-40

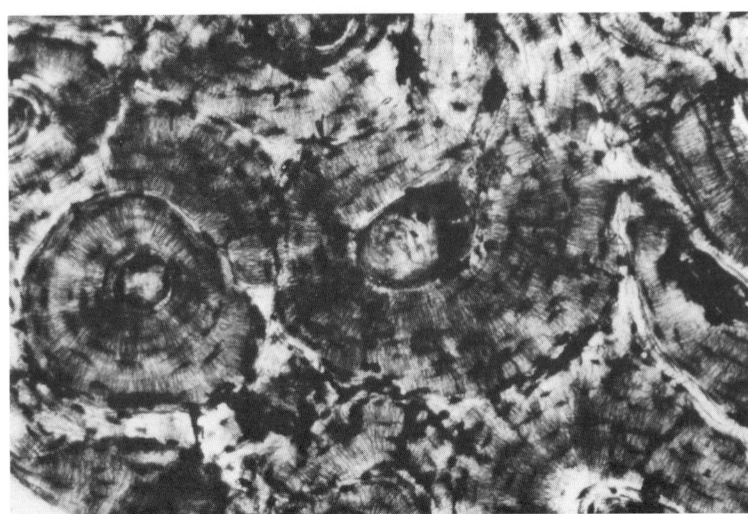
FIGURE 5-41

up the trabeculae of spongy appearance and often contains the calcified cartilage matrix which is yet to be resorbed. The boundary between the epiphyseal plate and the spongy bone is ill-defined.

4. The characteristic groups of circular lamellae shown in Fig. 5-41 are part of the __________ system. How was this section prepared? The dark spots which appear periodically in the lamellar portion of the system are __________. They contain __________. If the body needed more calcium in the blood, how would the haversian systems respond?

In this ground preparation, osteons of several generations are seen. The concentrically located lacunae contained osteocytes which have dried out, leaving an airspace. The canaliculi, which link individual lacunae in this section, reflect the fact that numerous osteocyte processes are in communication with each other in a living state. The extensive surface area produced by the canalicular system is thought to contribute to the rapid mobilization of calcium required by the body. Conversely, the deposition of minerals can take place through the canalicular system, as evidenced by sclerotic regions where bone cell deaths may occur due to pathologic conditions or aging.

5. Refer to Fig. 5-42. In this EM identify (a) osteoid and (b) calcified bone. What structural features of the cell reflect its functional characteristics?

The dark periphery of this EM is the calcified matrix of the bone. Collagenous fibers in the uncalcified matrix show their transverse striations. In the cytoplasm of the osteoblast are a number of RER profiles and a few mitochondria that contain intramitochondrial granules.

BIBLIOGRAPHY

Bourne, G. H. (ed.): *The Biochemistry and Physiology of Bone,* Academic Press, New York, 1972.

Brookes, M.: "The Vascular Architecture of Tubular Bone in the Rat," *Anat. Rec.* **132:**25–41 (1958).

Cohen, J., and J. W. H. Harris: "The Three-Dimensional Anatomy of the Haversian System," *J. Bone Jt. Surg. Am. Vol.* **40A:**419–434 (1958).

Ham, A. W.: *Histology,* 7th ed., Lippincott, Philadelphia, 1974.

Rhodin, J. A. G.: *Histology. A Text and Atlas,* Oxford University Press, New York, 1974.

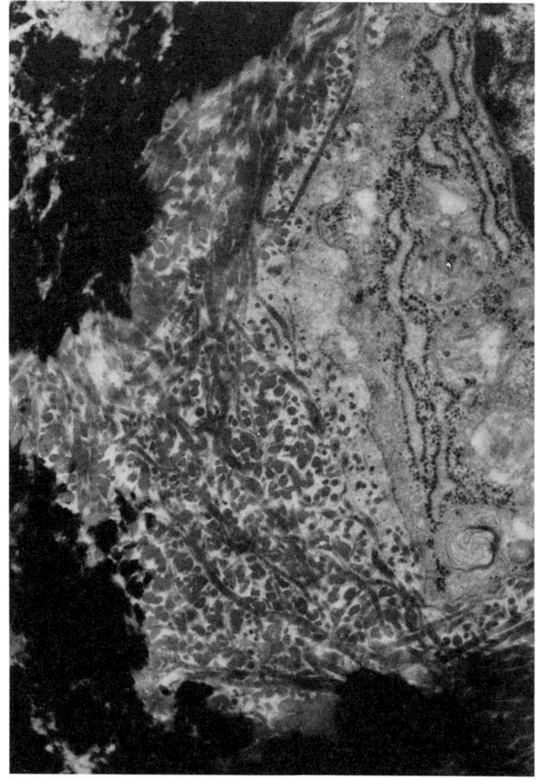

FIGURE 5-42

ADDITIONAL READINGS

Amprino, R.: "On the Growth of Cortical Bone and the Mechanism of Osteon Formation," *Acta Anat.* **52:**177–187 (1963).

Band, C. G.: "Submicroscopic Structure and Functional Aspects of the Osteocyte," *Clin. Orthop. Relat. Res.* **56:**227–236 (1968).

Barnard, G. W.: "The Ultrastructural Interface of Bone Crystals and Organic Matrix in Woven and Lamellar Bone," *J. Dent. Res.* **48:**781–788 (1969).

Baylinck, D., M. Stauffer, J. Wegedal, and C. Rich: "Formation, Mineralization, and Resorption of Bone in Vitamin D–Deficient Rats," *J. Clin. Invest.* **49:** 1122–1134 (1970).

———, J. Sipe, J. Wegedal, and O. J. Whittemore: "Vitamin D-Enhanced Osteocytic and Osteoclastic Bone Resorption," *Am. J. Physiol.* **224:**1345–1357 (1973).

Belanger, L. F.: "Osteocytic Osteolysis," *Calcif. Tissue Res.* **4:**1–12 (1961).

Cooper, R. R., J. W. Milgram, and R. A. Robinson: "Morphology of the Osteon. An Electron Microscopic Study," *J. Bone Jt. Surg. Am. Vol.,* **48A:** 1239–1271 (1966).

Dudley, H. R., and D. Spiro: "The Fine Structure of Bone Cells," *J. Biophys. Biochem. Cytol.* **11:** 627–649 (1961).

Frost, H. M.: "A Model of Endocrine Control of Bone Remodelling," *Henry Ford Hosp. Med. Bull.* **10:**119–170 (1962).

Gonzales, F., and M. J. Karnovsky: "Electron Microscopy of Osteoclasts in Healing Fractures of Rat Bone," *J. Biophys. Biochem. Cytol.* **9:**299–316 (1961).

Hall, B. K.: "The Origin and Fate of Osteoclasts," *Anat. Rec.* **183:**1–12 (1975).

Herring, G. M.: "A Review of Recent Advances in the Chemistry of Calcifying Cartilage and Bone Matrix," *Calcif. Tissue Res.* **4** (Suppl):17–23 (1970).

Jande, S. S., and L. F. Belanger: "Electron Microscopy of Osteocytes and Pericellular Matrix in Rat Trabecular Bone," *Calcif. Tissue Res.* **6:**280–289 (1971).

——— and ———: "Fine Structural Study of Osteocytes and Their Surrounding Bone Matrix with Respect to Their Age in Young Chicks," *J. Ultrastruct. Res.* **37:**279–300 (1971).

Kallio, D. M., P. R. Garant, and C. Minkin: "Evidence of Coated Membranes in the Ruffled Border of the Osteoclast," *J. Ultrastruct. Res.* **37:**169–177 (1971).

Knese, K. H.: "Osteoklasten, Chondroklasten, Mineraloklasten, Kollagenoklasten," *Acta Anat.* **83:** 275–288 (1972).

Luk, S. C., C. Nopajarovonsri, and G. T. Simon: "The Ultrastructure of Cortical Bone in Young Adult Rabbits," *J. Ultrastruct. Res.* **46:**184–216 (1974).

Malkain, K., M. M. Luxembouger, and A. Rebel: "Cytoplasmic Modifications at the Contact Zone of Osteoclasts and Calcified Tissue in the Diaphyseal Growth Plate of Foetal Guinea Pigs," *Calcif. Tissue Res.* **11:**258–264 (1973).

Martin, J. H., and J. L. Matthews: "Mitochondrial Granules in Chondrocytes, Osteoblasts and Osteocytes," *Clin. Orthop. Relat. Res.* **68:**273–278 (1970).

Menczel, J., and A. Harell (eds.): *Calcified Tissue: Structural, Functional and Metabolic Aspects,"* Academic Press, New York, 1971.

Mjor I. A.: "The Bone Matrix Adjacent to Lacunae and Canaliculi," *Anat. Rec.* **144:**327–339 (1962).

Owen, M.: "The Origin of Bone Cells," *Int. Rev. Cytol.* **28:**213–238 (1970).

Rasmussen, H.: "Tonic and Hormonal Control of Calcium Homeostasis," *Am. J. Med.* **50:**567–588 (1971).

Schenk, R. K., D. Spiro, and J. Wiener: "Cartilage Resorption in the Tibial Epiphyseal Plate of Growing Rats," *J. Cell Biol.* **34:**275–291 (1967).

Smith, J. W.: "The Disposition of Protein-Polysaccharide in the Epiphyseal Plate Cartilage of the Young Rabbit," *J. Cell Sci.* **6:**843–864 (1970).

Sognnaes, R. F. (ed.): "Mechanisms of Hard Tissue Destruction," American Association for the Advancement of Science, Washington, D.C., 1963.

Young, R. W.: "Cell Proliferation and Specialization during Endochondral Osteogenesis in Young Rats," *J. Cell Biol.* **14:**357–370 (1962).

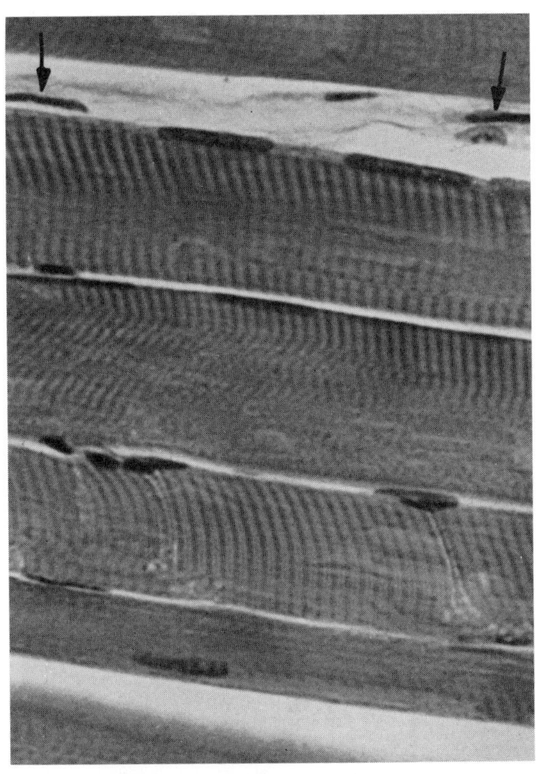

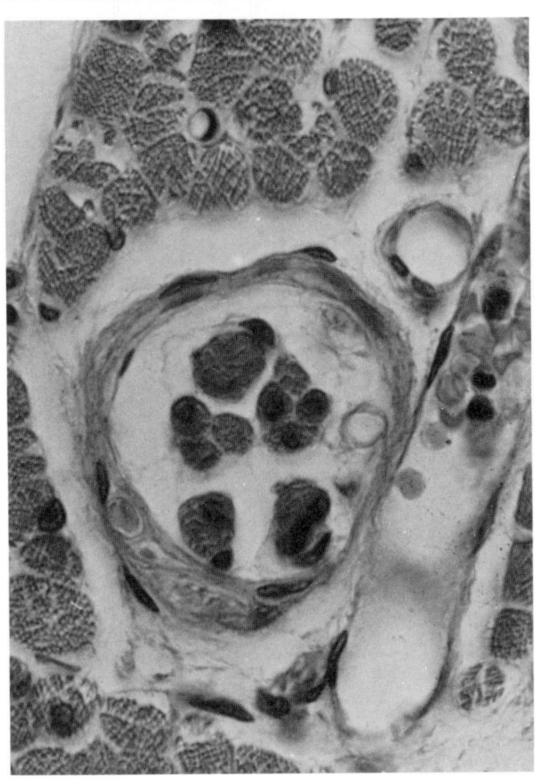

6

MUSCLE

OBJECTIVES

Upon completion of this chapter, the student will be able to:

1 Specify the following for each of the three muscle fiber types:
(a) location in the body (b) location of the nucleus in cross section (c) number of nuclei per fiber (d) classification

2 Identify the following in a cross section of skeletal muscle: (a) epimysium (b) perimysium (c) endomysium

3 Describe the relationships between the following structural elements of muscles:
(a) muscle fibers (b) myofibrils (c) myofilaments

4 Identify the electron microscopic structures which result in the appearance of striations (in longitudinal section) under the light microscope.

5 Label on slides or diagrams of longitudinal sections of sarcomeres the following structures:
(a) A band (b) H band (c) I band (d) Z line (e) thin filament (actin) (f) heavy meromyosin (g) light meromyosin (cross bridges)

6 Identify the location of the trinity (triad) in the sarcomere.

7 Describe the role played by each of the following in skeletal muscle contraction:
(a) motor end plate (b) T tubules (c) calcium ions (identify the structure from which they originate as well as function) (d) cross-bridge formation

8 Identify muscle spindles in cross section and state their function.

9 Identify the intercalated disk and name the types of junctional specialization of which it is composed.

10 Specify the functions served by the intercalated disk.

11 Define *nexus* (gap junction) and identify its function in smooth muscle contraction.

Every animal has some means of locomotion, an ability which distinguishes them from plants. In human beings, skeleton and associated groups of muscles make this function possible. These muscles associated with the skeleton are called *voluntary* or *skeletal muscles*. Muscles responsible for the movement of internal organs, like the respiratory and cardiovascular systems and the gastrointestinal tract, are not consciously controlled by the individual, so these are termed *involuntary* muscles. Muscles, therefore, are classified into two groups on the basis of innervation.

Another classification utilizes their microscopic appearance. Skeletal and cardiac muscle cells, which show cross striations under the microscope, are called *striated* muscle cells. The muscle cells which lack such striations, found in the internal organs, are called *smooth* muscle cells. Because of its long fibrous appearance, a muscle cell is frequently referred to as a *muscle fiber*. Each muscle fiber contains many contractile myofibrils, which are composed of many myofilaments. Figure 6-1 diagrams the structural organization of a skeletal muscle as revealed by successively higher levels of magnification. As will be discussed later, the *contractile filaments* are arranged in a highly ordered manner, producing the basic unit of contraction called the *sarcomere*. A *myofibril* is produced by longitudinal, end-to-end registration of numerous sarcomeres throughout the entire length of a muscle cell. By having many dozens to several hundred myofibrils aligned side by side, a muscle cell or fiber is formed. Although myofibrils are the most prominent

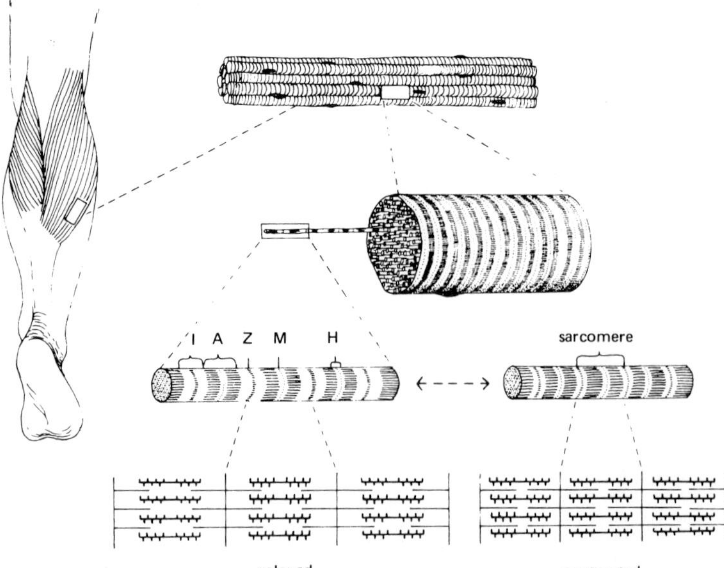

FIGURE 6-1 Diagrams of the structural organization of a skeletal muscle at successively higher levels of magnification.

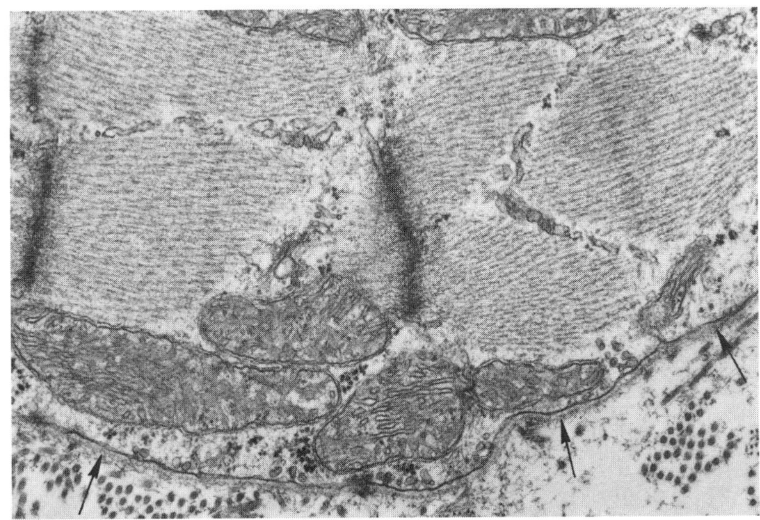

FIGURE 6-2 Electron micrograph of a skeletal muscle fiber. It is separated from the surrounding connective tissue by the sarcolemma (arrows).

structural feature of the cytoplasm of muscle cells (called *sarcoplasm*), it also contains such specialized organelles as mitochondria and ribosomes and a series of specialized ER called the *sarcoplasmic reticulum*. The plasma membrane of all muscles is specialized; it is covered by a basal lamina, an electron-dense structure, and therefore it is referred to as *sarcolemma* (arrows, Fig. 6-2). Within the sarcoplasm are also numerous glycogen and lipid droplets which serve as a metabolic depot.

SKELETAL MUSCLE

Skeletal muscle fibers, which are voluntary and striated, are the largest of the three muscle cell types. Figure 6-3 shows cross striations in a longitudinal section of several muscle fibers. The dark oval bodies seen at the edge of the fibers are the nuclei. Smaller nuclei of the fibroblasts (arrows) are distinguished from the muscle cell nuclei. As the muscle fibers generally extend from the origin of a muscle to its insertion, the length of

FIGURE 6-3 Skeletal muscle fibers seen in longitudinal section. Their characteristic cross striations and their peripheral nuclei are clearly visualized. Smaller nuclei of fibroblasts in the surrounding connective tissue are indicated by arrows.

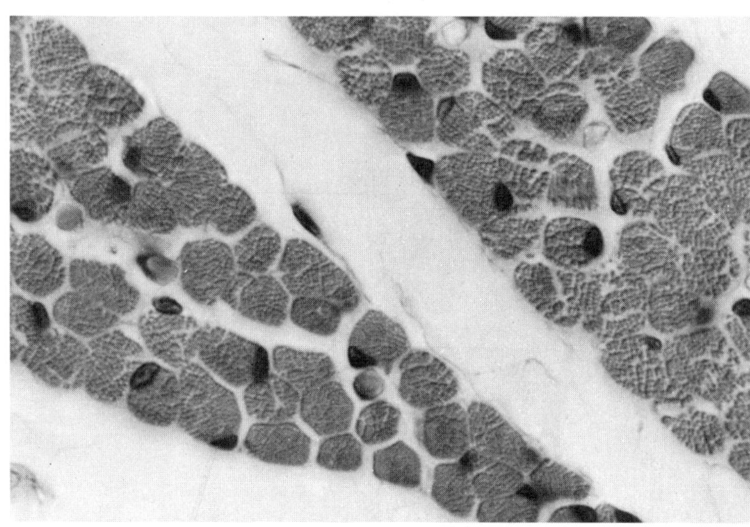

FIGURE 6-4 Cross section of skeletal muscle. Note the organization of the myofibrils within the fibers and the organization of the connective tissue surrounding them. Contrast the appearance of the peripherally located nuclei of the fibers with that of the connective tissue cells.

skeletal muscle varies from several millimeters to half a meter. The longest muscle fibers are found in the sartorius muscle in the thigh, while the muscles that move the eyeball in its orbit are composed of very short fibers.

The cross section of skeletal muscle in Fig. 6-4 indicates the pattern and individual fiber organization as well as the myofibrils within them. Note the peripherally located nuclei contrasted with those in the cells of the connective tissue that surrounds individual fibers.

Skeletal muscles in the adult seldom grow by adding more cells. Individual fibers increase in diameter by synthesizing more myofibrils. The amount of cytoplasm also increases as the cells grow in length. Whether the cell grows in diameter or in length, it produces more nuclei; these nuclei are always present at the periphery of the cell, immediately under the plasma membrane.

STRUCTURE OF SKELETAL MUSCLE

Microscopic Structures of Contractile Elements in a Skeletal Muscle. Each muscle fiber is covered by a plasma membrane which is coated by a glycoprotein layer akin to a basal lamina. As mentioned earlier, this combination of plasma membrane and basal lamina–like substance is called the *sarcolemma.*

The striations of skeletal muscle cells result from the characteristic organization of myofilaments (contractile proteins) which occupy a large portion of the cytoplasm. The high-power LM in Fig. 6-5 shows a longitudinal view of a skeletal muscle cell where cross striations can be seen in some detail. Note that the cells that run horizontally in the field are traversed by alternating light and dark bands. This particular micrograph was taken in a polarizing microscope. The broad light bands are anisotropic and are called *A bands.* In an ordinary light microscope the A band will appear dark. In contrast, the broad dark band is isotropic under a polarizing microscope and hence is called an *I band.* Again, the I bands in your microscope will appear as light bands. The thin line bisecting the I band is called a *Z line* (arrows in Fig. 6-5). Contractile filaments between neighboring Z lines work as units of contraction, and therefore the structure between two Z lines is called a *sarcomere* (see Fig. 6-1). Figure 6-6, an EM of portions of several sarcomeres, shows that the dark A band is bisected by a relatively light band called the *H band.* Through the center of the H band is a moderately dark line called the *M band.*

Under high magnification, as shown in Fig. 6-7, the details of myofibrils making up

FIGURE 6-5 Longitudinal view of skeletal muscle seen under a polarizing microscope. Broad light bands (A bands) alternate with broad dark bands (I bands). The lines bisecting the I bands are the Z lines (arrows).

a sarcomere can be seen. A sarcomere contains two types of filaments or contractile proteins. The thick filaments, along with their small lateral projections, are made up of *myosin,* while the thin filaments are made up of *actin.* These contractile proteins and their interrelationships are diagramed in the lower portion of Fig. 6-1. Chemical isolation and subsequent characterization of myosin have shown the lateral processes to be made up of a subfraction of myosin called *heavy meromyosin* (HMM) and the basic thick filament to be composed of a subfraction called *light meromyosin* (LMM).

A cross section through both sets of filaments indicates that the thick filaments

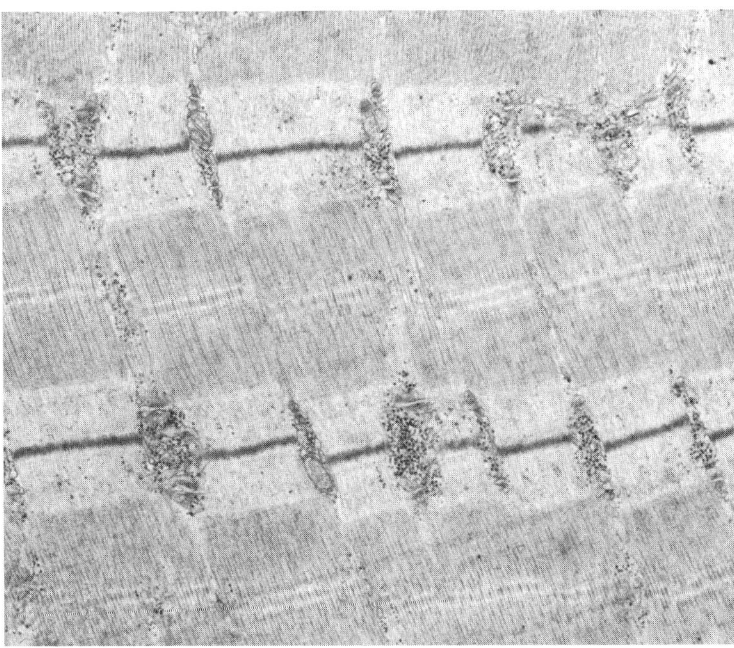

FIGURE 6-6 Electron micrograph of myofibrils which shows several sarcomeres and portions of adjacent ones. It clearly demonstrates the structural features of the sarcomere.

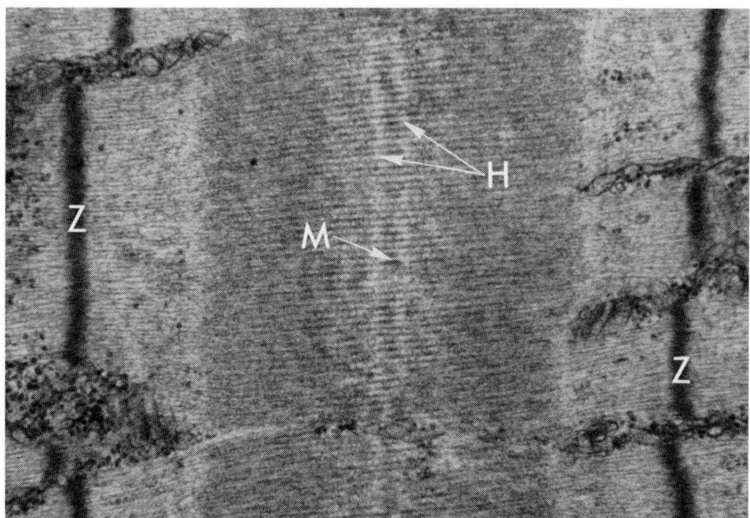

FIGURE 6-7 Sarcomere seen at higher magnification. This micrograph shows the details of the arrangement of myofilaments in the functional unit of skeletal muscle; i.e., the sarcomere which is a segment of a myofibril between two successive Z lines. The dark A band is bisected by the light H band with the darker M line in the center. On either side of the A bands is half an I band limited by a Z line.

(b, Fig. 6-8) are about 45 nm apart, arranged in a triangular pattern. Each thick filament is surrounded by six thin filaments in a corresponding hexagonal arrangement (Fig. 6-8). The thick filaments are continuous throughout the A band. The thin filaments extend from the Z lines into the A band but terminate at the central H band, which appears lighter.

The regular, small lateral projections of HMM along the length of LMM are arranged in a spiral pattern at 60° intervals (Figs. 6-8 and 6-9). This "staircase" arrangement (a, Fig. 6-9) allows contacts between the thick and thin filaments in their hexagonal configuration when viewed from above (b, Fig. 6-9). The movement of these cross bridges between lateral projections and thin filaments draws the thin filaments on either end of the sarcomere together, condensing the myofibrils and causing the muscle to contract. Figure 6-10 shows relationships of filaments during relaxation and contraction. Note how the H band is almost obliterated in the fully contracted state.

Structural Basis of Contraction. An impulse from a motor nerve initiates the contraction of its skeletal muscle. Each muscle fiber is innervated by a *motor end plate*, or myoneural junction. In Fig. 6-11 four different motor end plates present profiles seen from different angles.

Figure 6-12 diagrams the appearance of a motor end plate as viewed in an electron

FIGURE 6-8 Diagram of the arrangement of the two sets of filaments seen in cross section (a = heavy meromyosin, b = thick filament (light meromyosin), c = thin filament).

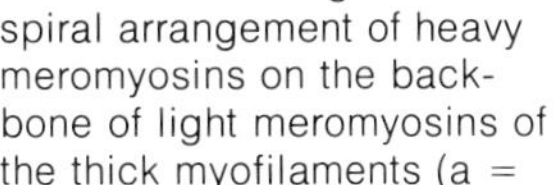

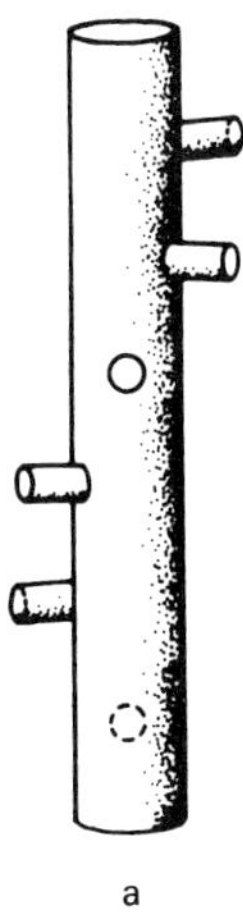

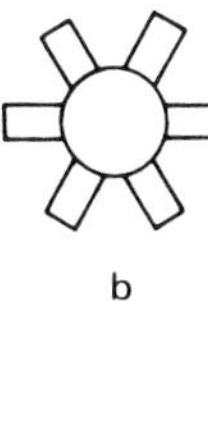

FIGURE 6-9 Diagram of the spiral arrangement of heavy meromyosins on the backbone of light meromyosins of the thick myofilaments (a = longitudinal view; b = cross-sectional view).

microscope. When the impulse arrives at the motor end plate, it induces the release of the active neurotransmitter, *acetylcholine,* from the synaptic vesicles into the synaptic trough. Acetylcholine will depolarize the sarcolemma, which propagates the electric impulse along the entire surface of the cell. After the depolarization has begun, the acetylcholine at the motor end plate is enzymatically broken down by *acetylcholinesterase,* present in the sarcolemma, thus preparing the region to receive the next impulse. The depolarization of sarcolemma results in a sudden increase in the permeability of the muscle cell membrane to Na^+, resulting in a change in the transmembrane potential. This propagation of electric potential is rapid and reaches the entire sarcolemma within milliseconds.

To cause a simultaneous contraction of all contractile filaments within the muscle, the electric message induced by surface membrane depolarization must be immediately communicated to those myofibrils located deep in the center of the muscle cell. Simple diffusion of flowing ions from the sarcolemma to the deeply situated myofilaments would be too slow for such a rapid transmission to occur. Nature has eliminated these problems by producing numerous tubular invaginations which form a system of small transverse tubules, called *T tubules,* which are continuous with the plasma membrane and therefore undergo depolarization as soon as the sarcolemma is depolarized. Within each skeletal muscle fiber there is a system of tubular membranes called the *sarcoplasmic reticulum,* a specialized form of ER. The sarcoplasmic reticulum, lacy in appearance, surrounds the A band of one sarcomere. As diagramed in Fig. 6-13, these longitudinal elements of sarcoplasmic reticulum run parallel to the myofibrils and end at the A-I junction in the form of *terminal cisternae,* which are in close contact with T tubules. Within the I band, the sarcoplasmic reticulum is morphologically different. It appears to be continuous across the Z line and makes two triads within the limit of each I band.

Thus, in EM the T tubules appear to lie between the terminal cisternae of the sarcoplasmic reticulum. When a T tubule consis-

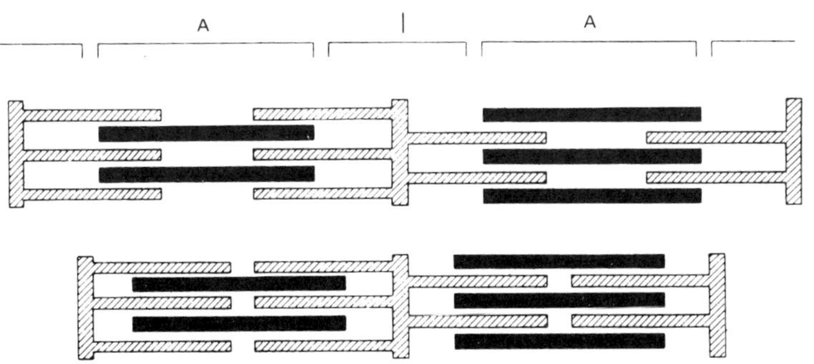

FIGURE 6-10 Diagram illustrating the relationships of the thick and thin filaments during relaxation (top) and contraction (bottom). Note that the H band is almost obliterated in the fully contracted state.

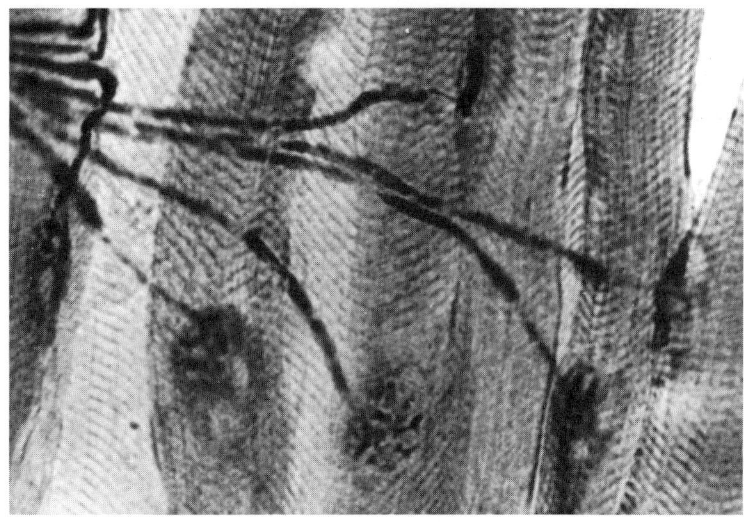

FIGURE 6-11 Light micrograph of motor end plates or myoneural junctions seen at different angles.

tently is associated with two terminal cisternae of sarcoplasmic reticulum along the A band–I band junction, this configuration is called a *triad.* Each sarcomere has two triads running along lines between adjacent sarcomeres. The T tubules and terminal cisternae of the sarcoplasmic reticulum run between and surround myofibrils throughout the entire diameter of the large muscle cell. Through this system of T tubules, an impulse received at the surface is propagated throughout the entire thickness of the cell.

As indicated earlier, a neural impulse delivered by motor end plates generates a wave of depolarization which spreads over the entire surface of the muscle fiber, causing a simultaneous depolarization of the many T tubules present along the length of the cell. The resulting change of potential is transmitted to the terminal cisternae of the sarcoplasmic reticulum, which stores cytoplasmic Ca^{2+} ions. The change in membrane potential causes the sarcoplasmic reticulum to become more permeable to Ca^{2+} ions, allowing them to escape into the sarcoplasm. Since the sarcoplasmic reticulum pervades the muscle fiber around each myofibril, the increase in Ca^{2+} ions in the vicinity of the contractile filaments occurs almost instantaneously. The Ca^{2+} ions are necessary in forming functional bridges between actin and HMM. Dispersed in large numbers through-

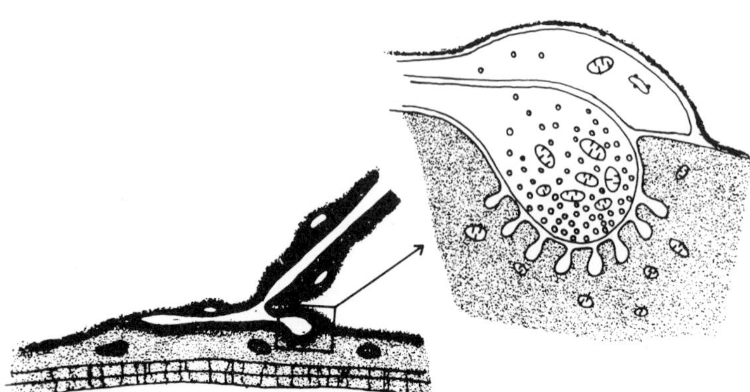

FIGURE 6-12 Diagrams of the structure of a motor end plate seen in the electron microscope. To the left is a longitudinal view of the motor end plate. The diagram to the right is an enlargement of the rectangular area illustrating the intrinsic relationship between the nerve ending and the muscle fiber.

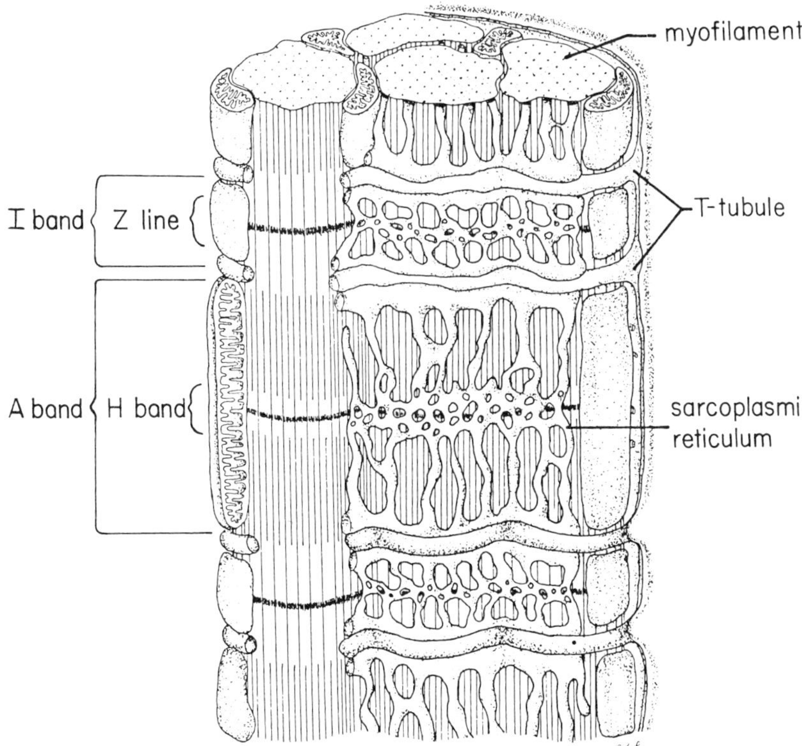

FIGURE 6-13 Diagram illustrating the organization of the sarcoplasmic reticulum and the T tubules, the formation of triads, and the arrangement of the above structures in relation to the bands of the myofibrils.

out the entire fiber, the ions produce the simultaneous contraction of all sarcomeres. With repolarization of the membranes, the Ca^{2+} ions return to the sarcoplasmic reticulum, and the muscle cell begins to relax.

Since the areas surrounding myofibrils appear lighter than the fibrils themselves, the cross-sectional areas in LMs have been referred to as *Cohnheim's fields* (Fig. 6-14). Electron microscopic observations during the past two decades have shown clearly that Cohnheim's fields are produced by the sarcoplasmic reticulum and the numerous mitochondria that are regularly arranged among the myofibrils (Fig. 6-14). The latter structures play an important role in fulfilling energy requirements for contractile processes.

ORGANIZATION OF A SKELETAL MUSCLE

Each muscle fiber is encompassed by the sarcolemma, which is covered by a thin layer of connective tissue called the *endomysium.* The endomysium consists of a few fibroblasts and transversely oriented reticular and fine collagenous fibers. A bundle of several muscle fibers is wrapped by a connective tissue layer called *perimysium.* The perimysium is a loose connective tissue which contains blood vessels and neural elements for the muscle. A number of these muscle fiber bundles are wrapped by another connective tissue layer, the *epimysium.* Because this layer is the outermost covering of muscles and holds the muscles together, it is heavy and thick. These coverings are illustrated in Fig. 6-15.

Muscle spindles are complex sensory units found throughout the skeletal muscle. They are involved in position sense, or *proprioceptive sensation,* and are located in the perimysial connective tissue space along with the blood vessels and nerves. Figure 6-16 shows a cross section of muscle spindle. The small

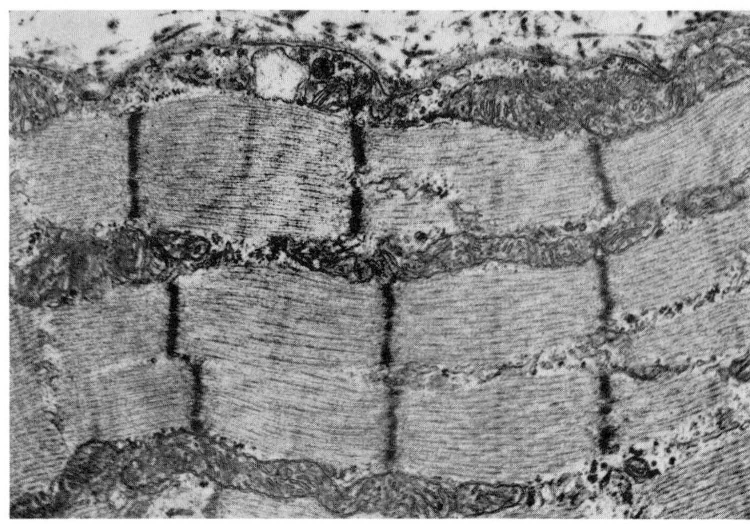

FIGURE 6-14 Electron micrograph of a portion of a muscle fiber. It shows the regular arrangement of the mitochondria and the sarcoplasmic reticulum among the myofibrils. In the light microscope these areas are referred to as Cohnheim's fields.

fibers which contain some contractile elements are called *intrafusal fibers.* Figure 6-17 is a diagram of a muscle spindle which shows intrafusal fibers and nerve fibers within the spindle. The neural elements in the spindle send proprioceptive information to the brain on the magnitude and rate of muscle contraction.

Muscles and bones are connected by tendons. At the muscle-tendon junction, reticular fibers anchored to the sarcolemma become segregated into strands and small bundles, which become continuous with the heavy bundles of collagenous fibers in the tendon. Collagenous fibers of the tendon in turn insert into the bone, where they are called *Sharpey's fibers.*

Because of the high level of activity, the skeletal muscle is richly supplied by blood vessels. Figure 6-18 shows the appearance of capillaries in skeletal muscle after an intravascular injection of a dye. While muscle

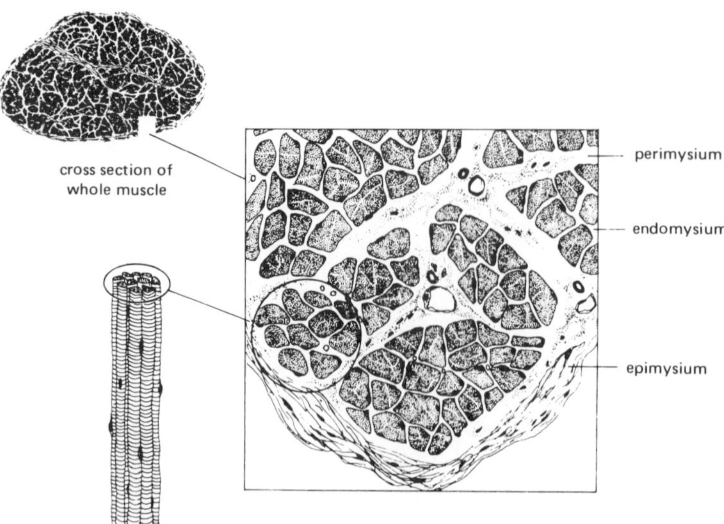

FIGURE 6-15 Diagram of the organization of skeletal muscle illustrating the three levels of connective tissue coverings.

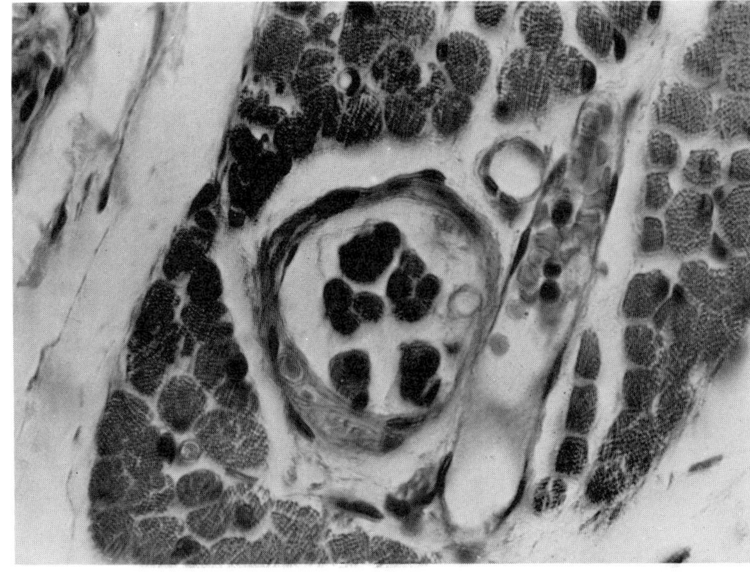

FIGURE 6-16 Light micrograph of a muscle spindle seen in cross section.

cells are not stained, the network of capillaries surrounding individual muscle fibers is clearly shown.

CARDIAC MUSCLE

Smaller than most skeletal muscle fibers, the cardiac muscle cells are branching, often connecting with two or more neighboring cells (Fig. 6-19). Each cell contains only one centrally located nucleus.

The contractile fibrils in the cardiac muscle cell are finer than those of the skeletal muscle. Nevertheless, they can be seen under a light microscope. Figure 6-19 shows the fine cross striations of cardiac muscle cells which are similar to but not as clearly evident as those of skeletal muscle. Myofibrils within cardiac muscle terminate at the *intercalated disks* found between cardiac cells (arrows in Fig. 6-19). The intercalated disks appear as dense boundary lines running perpendicular to the length of the cells. In an EM, the intercalated disks can be distinguished from the straight Z lines of skeletal muscle, since they are much broader and often irregular, as shown in Fig. 6-20. The intercalated disks provide attachment for myofibrils and appear much like exaggerated desmosomes holding the cells together. They also contain gap junctions, which are often called *nexuses*. Formerly, a nexus was thought to be a tight junction, but with improved techniques it has been shown to have a small gap of 2 nm. Each of these junctions serve as an electrical nexus where resistance is low and the im-

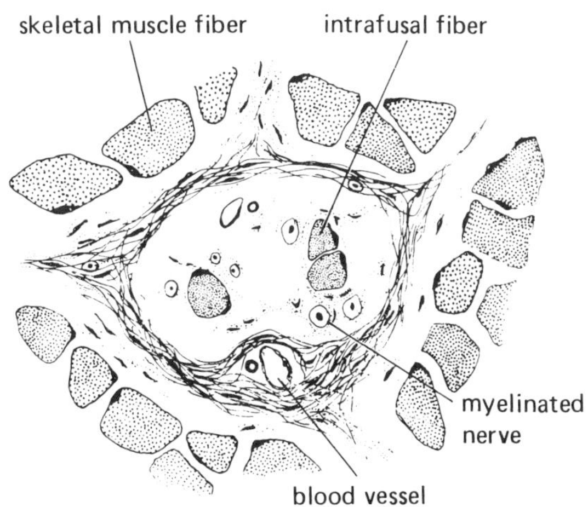

FIGURE 6-17 Diagram illustrating the components of a muscle spindle.

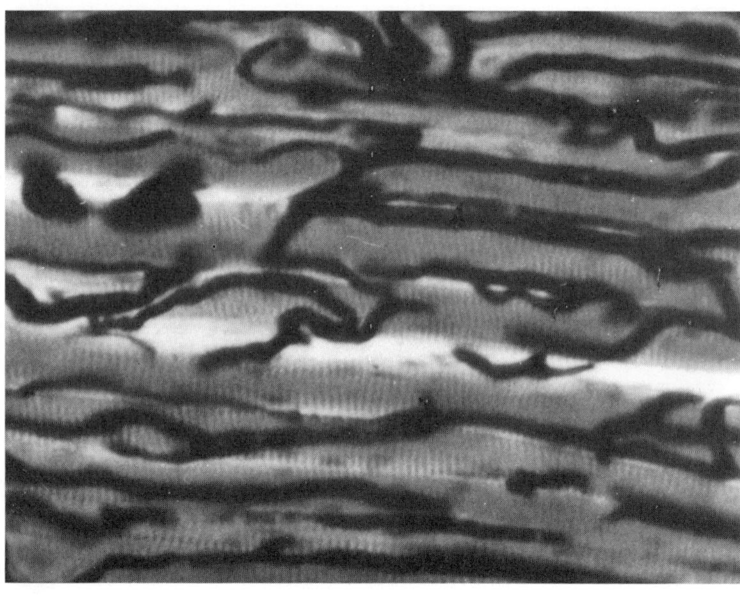

FIGURE 6-18 Light micrograph of capillaries in skeletal muscle. The capillaries have been made demonstrable by the injection of a dye, while the muscle fibers are unstained.

pulse that initiates contraction in one cell can pass to the adjoining cell.

A cross section through cardiac muscle (Fig. 6-21) demonstrates the irregularly shaped profiles of muscle fibers. The centrally located, large nucleus can be seen in some of them. Like skeletal muscle, these cells are invested in a delicate connective tissue stroma that is made up of fine collagenous and reticular fibers. While the muscle fibers are arranged in bundles running in different directions, the connective tissue surrounding the bundles is heavier and is continuous with the cardiac skeleton. Note the rich supply of capillary vessels, which reflects the high level of activities by the cardiac muscle.

A coordinated contraction of the heart depends on the propagation of the impulse from cell to cell. Certain cardiac cells initiate the impulse. A polarization starts at a point called the *sinoatrial* (*SA*) *node* and spreads in a wavelike fashion as the individual fibers contract in sequence. Because of this, cardiac muscle is classifed as involuntary, although

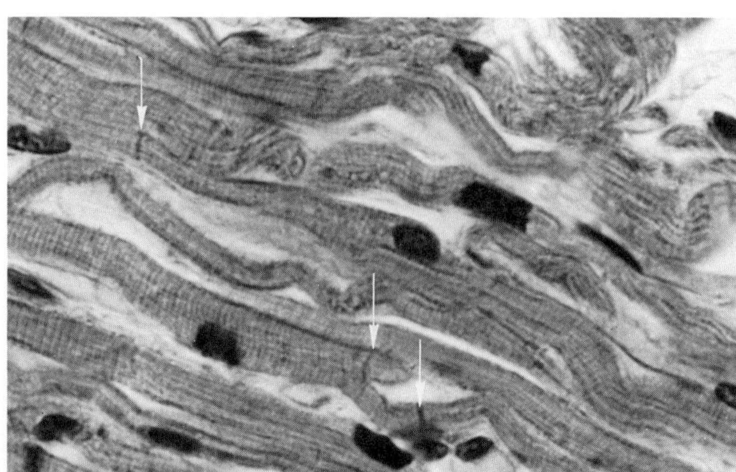

FIGURE 6-19 Light micrograph of cardiac muscle. Note that the cells in cardiac muscle are branching and the cross striations are evident. Intercalated disks are indicated by arrows.

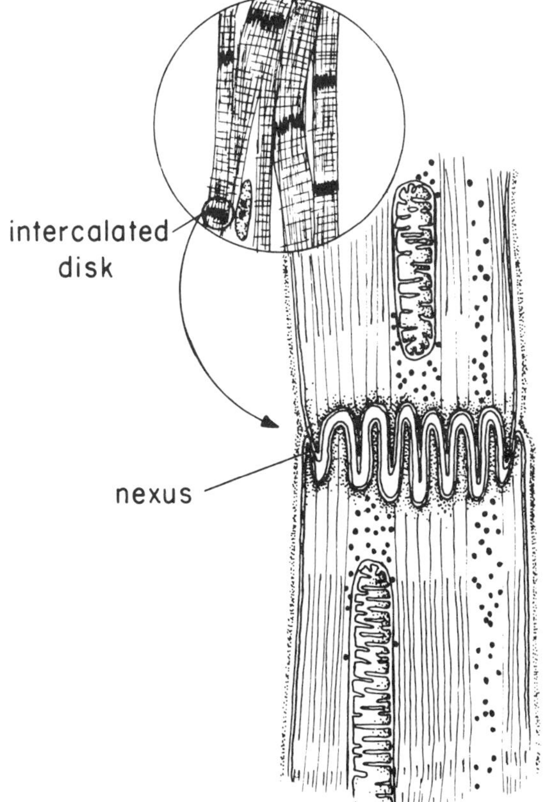

FIGURE 6-20 Diagram of intercalated disk.

it is striated. Like skeletal muscle, cardiac muscle fibers contain well-developed T-tubule and sarcoplasmic reticulum systems that regulate rapid mobilization and recapture of calcium ions. The histologic appearance of specialized muscle cells for impulse conduction within the heart will be described in a later chapter on heart and blood vessels.

SMOOTH MUSCLE

Smooth muscle fibers are the smallest of the muscle cells. However, they vary widely in size depending on the organ and its functional status. In the smallest arterioles, the smooth muscle cells that make up part of the vessel wall may be as small as 10 μm in length, whereas hypertrophied smooth muscle of the uterus during pregnancy may reach a length of several millimeters. In general, they are elongated and somewhat spindle-shaped. Figures 6-22 and 6-23 illustrate lon-

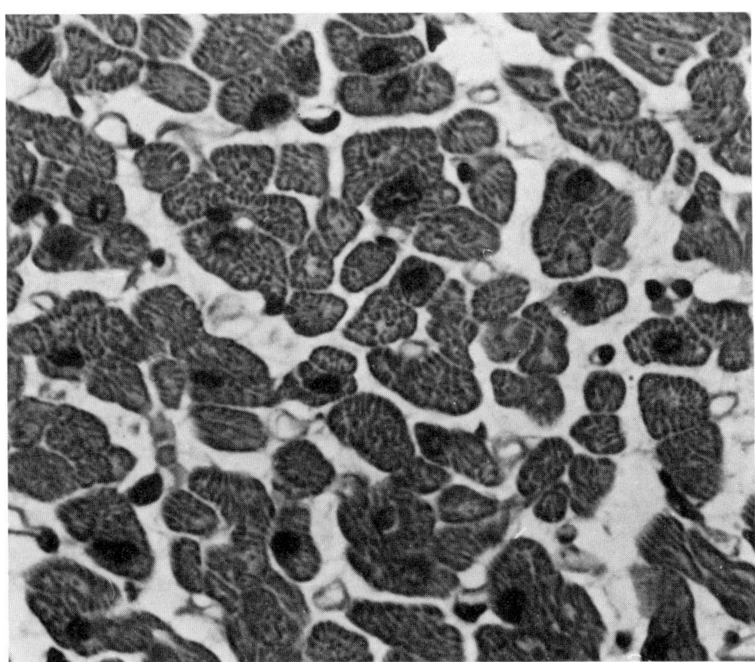

FIGURE 6-21 Cross sections of cardiac muscle. Note the irregularly shaped profiles of muscle fibers with varying sizes and the centrally located nucleus in some of them.

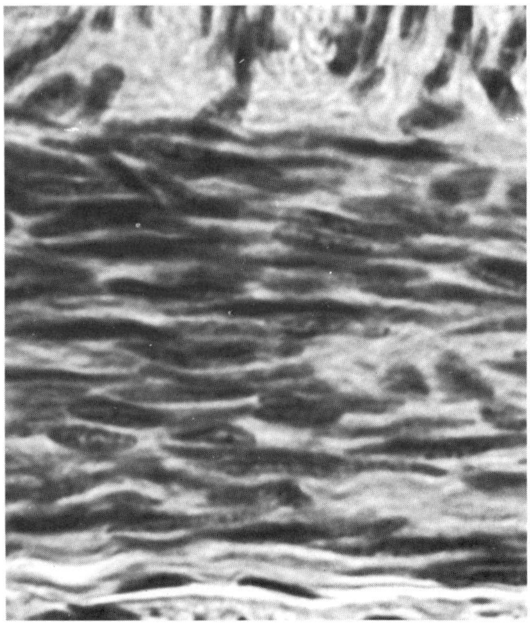

FIGURE 6-22 Longitudinal bundles of smooth muscle fibers found in the wall of the uterus.

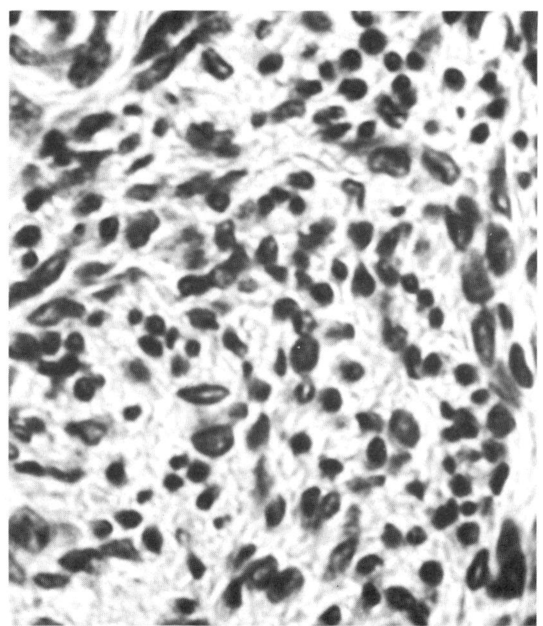

FIGURE 6-23 Cross section profiles of smooth muscle. The diameter of them varies greatly and is dependent upon where the muscle fiber is sectioned. Since the cells are spindle shaped, profiles with the greatest diameter contain the nucleus, while toward the end the profiles gradually decrease in diameter.

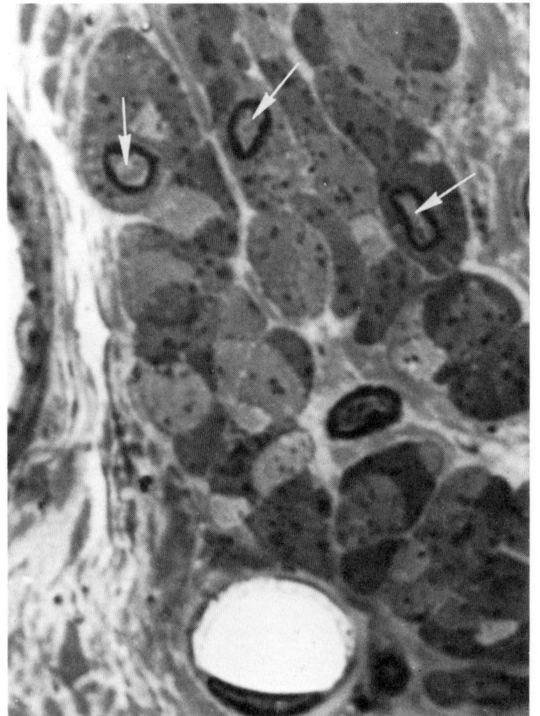

gitudinal and cross sections of smooth muscle. Figure 6-22 shows longitudinal bundles of smooth fibers from a section of the uterus. Cross-sectional profiles of smooth muscle cells from the same organ are seen in Fig. 6-23. Note the many blood vessels found among the muscle cells. In a similar region at a higher magnification (Fig. 6-24) are several centrally located nuclei (arrows) of individual muscle cells. The nucleus, which is club-shaped and shows a fine, dusty pattern of chromatin, can therefore be distinguished from the denser nuclei of fibroblasts present in the connective tissue stroma of the region. Figure 6-25 is a cross section of a small artery in which smooth muscle cells of a considerably smaller size can be seen (arrows).

FIGURE 6-24 Cross section of smooth muscle fibers. Several of them have club-shaped nuclei (arrows) which are centrally located.

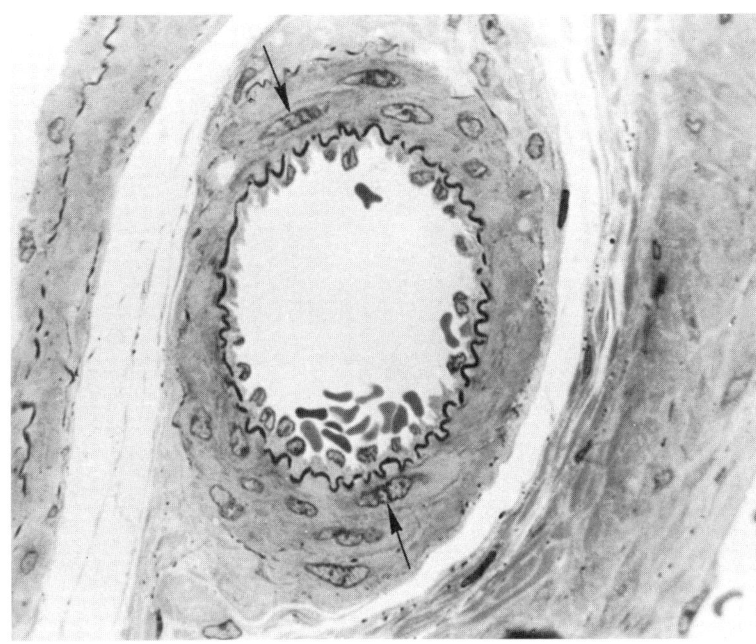

FIGURE 6-25 Cross section of a small artery. Note the smaller size of the muscle cells in the arterial wall (arrows).

Unlike in the striated muscles, the contractile filaments of the smooth muscles are not organized into sarcomeres; therefore no definite striations are evident. For this reason, there is some controversy about the exact mechanism of the contraction of these cells. However, it is known that the cells contain contractile proteins similar to those of striated muscles. In a contracted state, occasional dense and amorphous bands can be seen in LM. These bands are called contractile bands and appear as fusiform densities in an EM (arrow, Fig. 6-27).

Smooth muscle is found in the wall of the gastrointestinal tract, respiratory tubes, blood vessels, and certain bulbous organs, such as the bladder and the uterus, all of which are under involuntary control. The

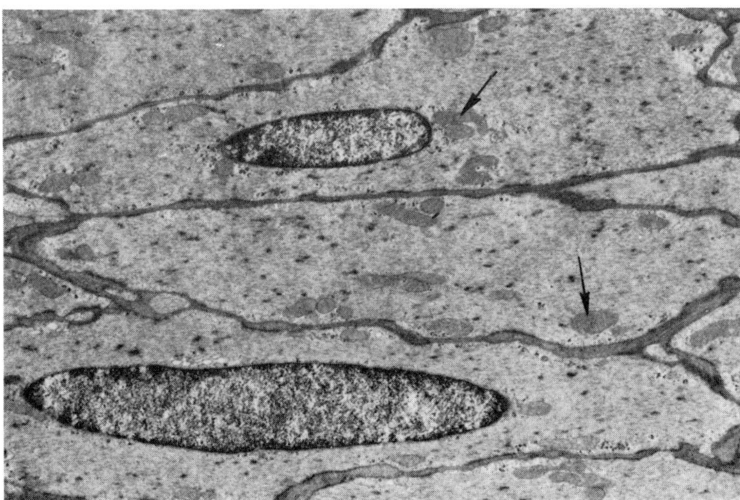

FIGURE 6-26 Low-power electron micrograph of smooth muscle cells. Some of them contain the centrally located elongated nucleus. These cells contain numerous small elongated mitochondria (arrows) in addition to some free ribosomes and RER profiles.

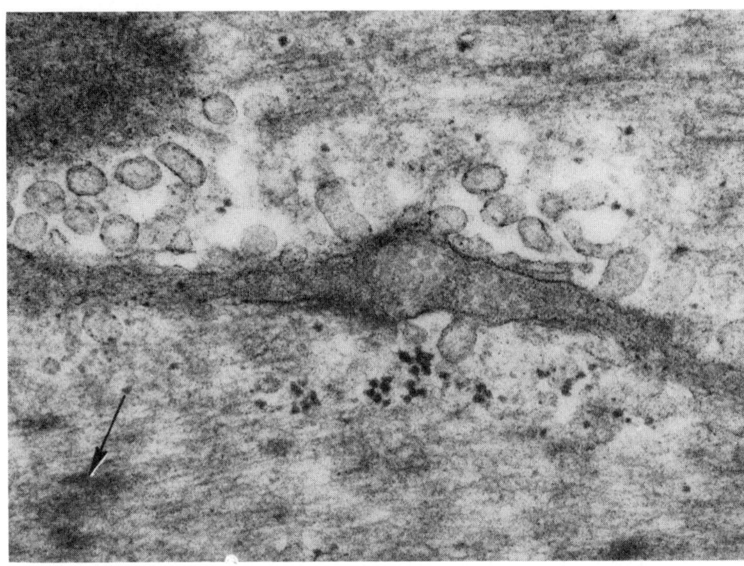

FIGURE 6-27 Portions of smooth muscle cells which contain fusiform densities (arrow) that are called *contractile bands.*

orientation of smooth muscle cells varies depending on the location. The cells may form rings, as in arteries, or run longitudinally, as in the veins. Along the gastrointestinal tract, smooth muscle cells make up two layers with the fibers in each running more or less perpendicularly to each other. Coordinated contractions of the two layers allow

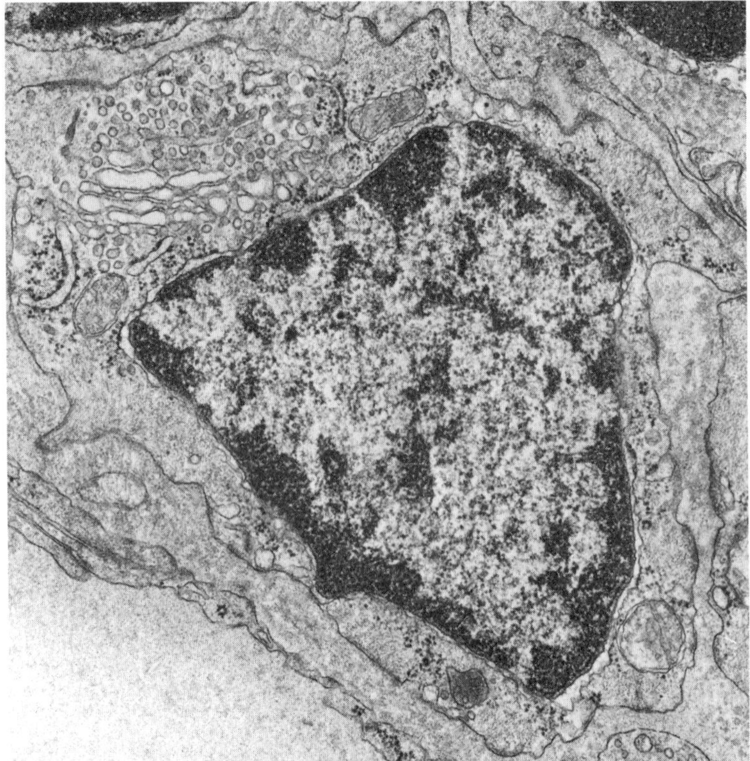

FIGURE 6-28 Smooth muscle cell seen in cross section in the electron microscope. Several cell organelles, Golgi complex, RER, and mitochondria are seen in the juxtanuclear cytoplasm.

for the undulating peristaltic movement throughout the digestive tube. The stomach has an additional obliquely oriented muscle layer which allows for its churning action.

Impulse propagation through smooth muscle is carried out through nexuses, similar to those in cardiac muscle. Since this muscle contracts very slowly, it does not have the T-tubule system, which allows rapid impulse propagation in skeletal and cardiac muscles.

In an EM, smooth muscle cells show parallel arrays of fine contractile filaments that are about 6 nm in diameter and therefore are difficult to clearly resolve at low magnification (Figs. 6-26 and 6-27). Note also the many small vesicles that appear just underneath the plasma membrane (Fig. 6-27).

Next to the nucleus, numerous small elongated mitochondria are present (arrows, Fig. 6-26). A small number of free ribosomes and RER profiles occur in these cells. Occasionally, a small juxtanuclear Golgi complex can also be observed. Figure 6-28 is the cross-sectional appearance of a smooth muscle cell in which a small Golgi complex, a few RER, and mitochondria are present.

The sarcoplasm of smooth muscle cells, like all muscle, is moderately eosinophilic under a light microscope. Here the eosinophilia presents a more pinkish tone than the red-orange of collagen, providing a good clue for distinguishing isolated smooth muscle cells from similar-appearing collagenous fibers of FECT.

REVIEW SECTION

1. Figure 6-29 is a cross section of muscle. The multiple nuclei are seen as small dark dots at the periphery of the cell. The fibers are gathered into bundles. The structure which gathers the fibers into bundles is __________. The structure which surrounds each individual fiber is __________. What elements could be found in the spaces

Skeletal muscles are enclosed by an organized connective tissue stroma which can be broken down to three levels. Each individual muscle fiber is surrounded by endomysium. The fibers are bundled together with perimysium. Structures such as blood vessels, nerves, and muscle spindles can be found in the perimysial spaces in skeletal

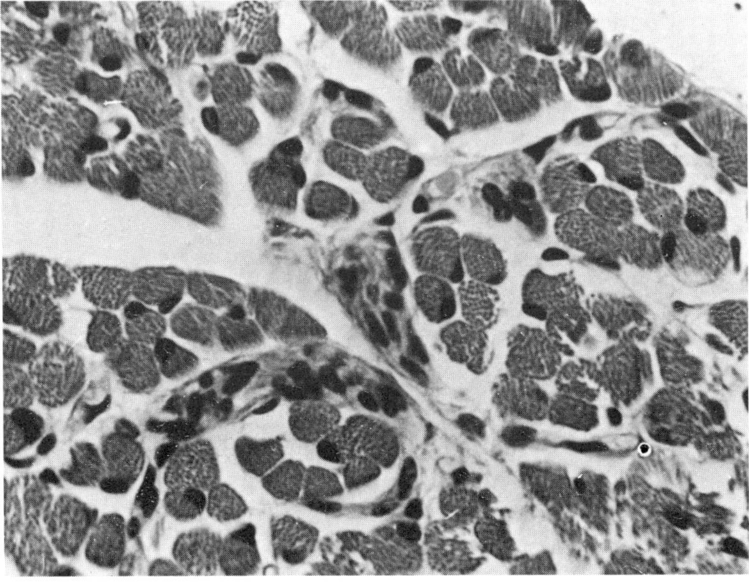

FIGURE 6-29

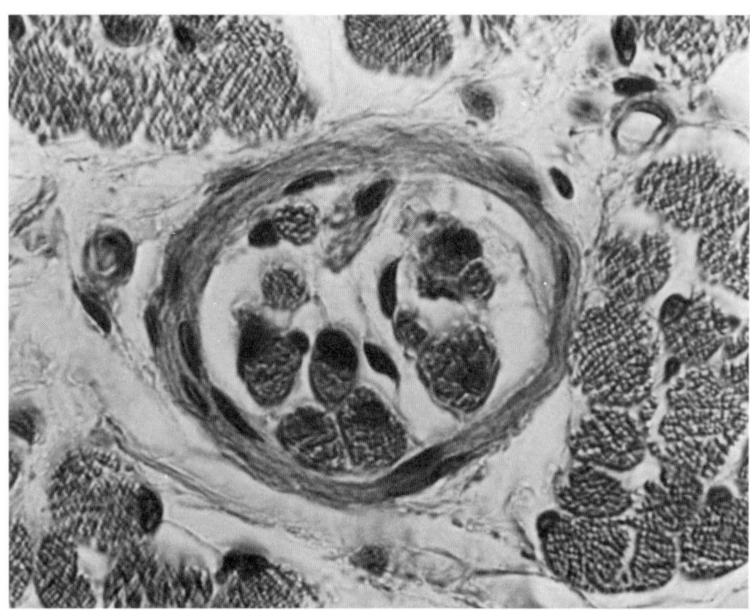

FIGURE 6-30

between major fiber bundles appearing in this section? What type of muscle is this?

2. Figure 6-30 shows a muscle spindle located in the perimysial space of skeletal muscle. What different elements can be found in the spindle? The function performed by the spindle is ________.

muscle. The whole muscle is surrounded by the epimysium, which forms the connective tissue fascia of all skeletal muscles.

The muscle spindle provides proprioceptive information to the central nervous system. It uses the combination of nerve and intrafusal muscle fibers to discern the relative state of contraction of intrafusal and extrafusal fibers which exists at any given time in the muscle.

FIGURE 6-31

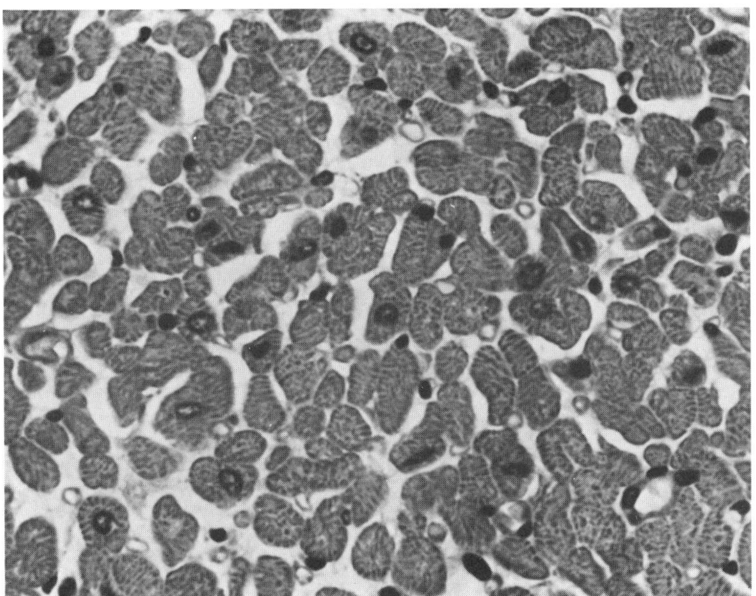

FIGURE 6-32

3. Figure 6-31 is obviously a striated muscle. What type of muscle? The numerous oblong bodies along the fibers' peripheries are __________. The dark flattened structures indicated by arrowheads are __________.

Skeletal muscle can be identified by the large number of nuclei along the periphery of the fiber. They are indicated by the arrows. The arrowheads point to the fibroblast nuclei of the endomysium. Mitochondria are much smaller and are approximately the length of a sarcomere.

4. Figure 6-32 is a cross section through cardiac muscle. How can this cross section be differentiated from smooth muscle? These fibers are surrounded by __________. The clear structure (the size of a fiber) at the right of center in the figure is a __________.

Cardiac muscle is more irregular in its organization and cell size and shape than smooth muscle, when viewed in a cross section. The cardiac muscle has a small, single, centrally located nucleus. Delicate connective tissue corresponding to the endomysium of skeletal muscle is found around each fiber.

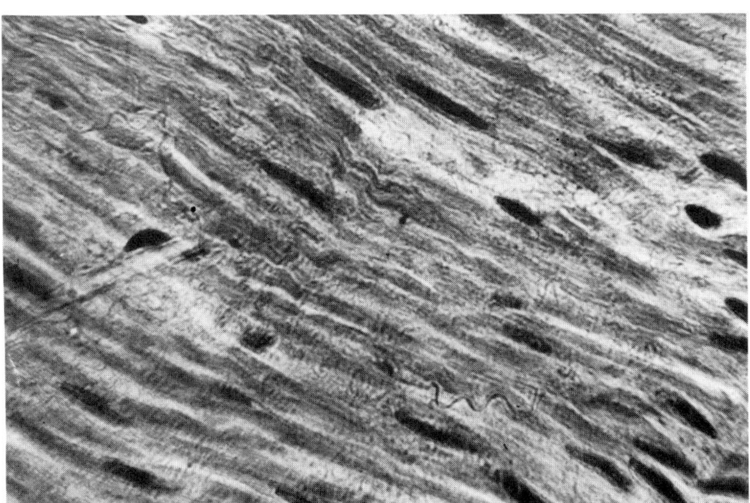

FIGURE 6-33

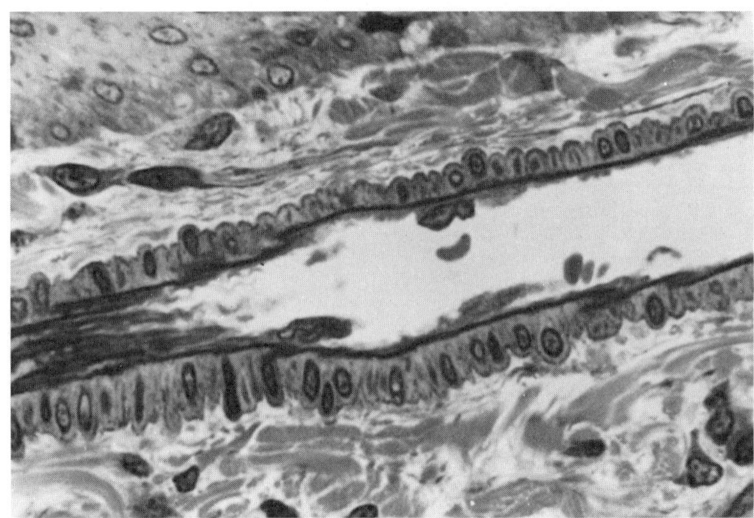

FIGURE 6-34

The lack of bundling in cardiac muscle indicates that no perimysium is found. Capillaries and other blood vessels often appear in a cross section of cardiac muscle.

5. Figure 6-33 is a longitudinal section of muscle from the lower gastrointestinal tract. What type of muscle is characterized by the parallel array of fibers? How many nuclei would be found in each fiber shown in the figure? What types of connective tissue elements would be found here?

Smooth muscle is characterized by this type of parallel arrangement of fibers with fibrous sleeves made up of reticular and elastic fibers. The muscle fibers have only one nucleus.

6. Figure 6-34 is a longitudinal section of an arterial wall. What type of muscle is shown here?

The arterial wall is one of the many sites where smooth muscle is present. As shown in Fig. 6-34, endothelial cells of the blood vessel can be distinguished from the nuclei of muscle cells.

BIBLIOGRAPHY

Bloom, W., and D. W. Fawcett: *A Textbook of Histology,* 10th ed., Saunders, Philadelphia, 1975.

Bourne, G. H. (ed.): *The Structure and Function of Muscle,* 3 vols, Academic Press, New York, 1960.

Langer, G. A., and A. J. Brady (eds.): *The Mammalian Myocardium,* Wiley, New York, 1974.

Podolsky, R. J. (ed.): *Contractility of Muscle Cells and Related Processes,* Prentice-Hall, Englewood Cliffs, N. J., 1971.

Rhodin, J. A. G.: *Histology. A Text and Atlas,* Oxford University Press, New York, 1974.

Timashoff, S. N., and G. D. Fasman (eds.): *Biological Macromolecules Series,* vol. 5, Marcel Dekker, New York, 1971.

Weiss, L., and R. D. Greep: *Histology,* 4th ed., McGraw-Hill, New York, 1977.

ADDITIONAL READINGS

Bois, R. M.: "The Organization of the Contractile Apparatus of Vertebrate Smooth Muscle," *Anat. Rec.* **177:**61–78 (1973).

Bridgman, C. F., E. E. Shumpert, and E. Eldred: "Insertions of Intrafusal Fibers in Muscle Spindles of the Cat and Other Mammals," *Anat. Rec.* **164:** 391–402 (1969).

Coers, C.: "Structure and Organization of the Myoneural Junction," *Int. Rev. Cytol.* **22:**239–268 (1967).

Davies, R. E.: "A Molecular Theory of Muscle Contraction: Calcium-dependent Contractions with H-bond Formation plus ATP-dependent Extensions of Part of the Myosin-Actin Cross-bridges," *Nature (Lond.)* **199:**1068–1074 (1963).

Devine, C. E., and A. P. Somlyo: "Thick Filaments in Vascular Smooth Muscle," *J. Cell Biol.* **49:** 636–649 (1971).

———, ———, and A. V. Somlyo: "Sarcoplasmic Reticulum and Excitation-Contraction in Mammalian Smooth Muscle," *J. Cell Biol.* **52:**690–718 (1972).

Dewey, M. M., and L. Barr: "A Study of the Structure and Distribution of the Nexus," *J. Cell Biol.* **23:** 553–585 (1964).

Fawcett, D. W.: "The Sarcoplasmic Reticulum of Skeletal and Cardiac Muscle," *Circulation* **24:** 336–348 (1961).

——— and N. S. McNutt: "The Ultrastructure of the Cat Myocardium. I. Ventricular Papillary Muscle," *J. Cell Biol.* **42:**1–45 (1969).

Franzini-Armstrong, L.: "Studies on the Triad. I. Structure of the Junction in Frog Twitch Fibers," *J. Cell Biol.* **47:**488–499 (1970).

——— and K. R. Porter: "The Z Disc of Skeletal Muscle Fibrils," *Z. Zellforsch.* **61:**661–672 (1964).

Gauthier, G. F.: "The Motor End Plate," in D. N. London (ed.), *The Peripheral Nerve,* Chapman & Hall, London, 1976.

Huxley, H. E.: "The Mechanism of Muscular Contraction," *Science (Wash. D.C.)* **164:**1356–1366 (1969).

Kelly, D. E.: "Fine Structure of Skeletal Muscle Triad Junctions," *J. Ultrastruct. Res.* **29:**37–49 (1969).

——— and M. A. Cahill: "Filamentous and Matrix Components of Skeletal Muscle Z-Discs," *Anat. Rec.* **172:**623–642 (1972).

Kelly, R. E., and R. V. Rice: "Localization of Myosin Filaments in Smooth Muscle," *J. Cell Biol.* **37:** 105–116 (1968).

Knappeis, G. G., and F. Carlson: "The Ultrastructure of the M Line in Skeletal Muscle," *J. Cell Biol.* **38:**202–212 (1968).

McNutt, N. S., and D. W. Fawcett: "The Ultrastructure of the Cat Myocardium. II. Atrial Muscle," *J. Cell Biol.* **42:**46–67 (1969).

——— and R. S. Weinstein: "The Ultrastructure of the Nexus. A Correlated Thin-Section and Freeze-Cleavage Study," *J. Cell Biol.* **47:**666–688 (1970).

Padykula, H. A., and H. F. Gauthier: "The Ultrastructure of the Neuromuscular Junctions of Mammalian Red, White and Intermediate Skeletal Muscle," *J. Cell Biol.* **46:**27–41 (1970).

Panner, B. J., and C. R. Honig: "Filament Ultrastructure and Organization in Vertebrate Smooth Muscle. Contraction Hypothesis Based on Localization of Actin and Myosin," *J. Cell Biol.* **35:** 303–321 (1967).

Porter, K. R., and G. E. Palade: "Studies on the Endoplasmic Reticulum. III. Its Form and Distribution in Striated Muscle Cells," *J. Biophys. Biochem. Cytol.* **3:**269–300 (1957).

——— and C. Franzini-Armstrong: "The Sarcoplasmic Reticulum," *Sci. Am.* **212:**72–81 (1965).

Rayns, D. G.: "Myofilaments and Cross Bridges as Demonstrated by Freeze-Fracturing and Etching," *J. Ultrastruct. Res.* **40:**103–121 (1972).

Schiaffino, S., V. Hanzlikova, and S. Pieroban: "Relations Between Structure and Function in Rat Skeletal Muscle Fibers," *J. Cell Biol.* **47:**107–119 (1970).

Sommer, J. R., and E. A. Johnson: "Cardiac Muscle: A Comparative Study of Purkinje Fibers and Ventricular Fibers," *J. Cell Biol.* **36:**497–526 (1968).

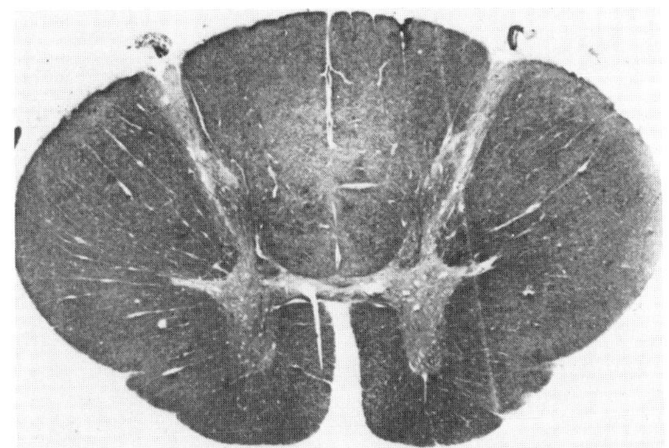

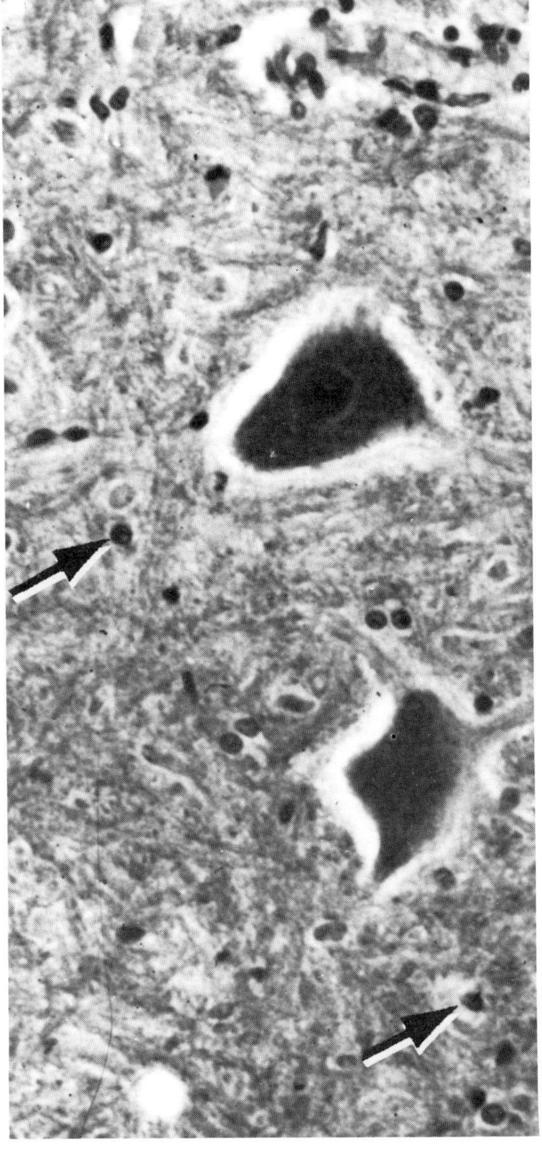

7 THE NERVOUS SYSTEM: ORGANIZATION

OBJECTIVES

Upon completion of this chapter, the student will be able to:

1 Identify the embryonic origin of the nervous system and briefly explain the steps in its development.

2 Draw and identify with labels the structures involved in a spinal reflex arc, including the following:
(a) gray matter (b) ventral root (c) ventral horn (d) dorsal root (e) dorsal root (spinal) ganglion (f) dorsal horn (g) spinal nerve (h) white matter

3 Identify the following in a cross section of a peripheral nerve:
(a) endoneurium (b) perineurium (c) epineurium (d) blood vessels

4 Specify the functions of the sympathetic and parasympathetic divisions of the autonomic nervous system.

5 Identify the location, structure, and function of each of the following, where applicable:
(a) neural ectoderm (b) neural tube (c) central nervous system (CNS) (d) peripheral nervous system (e) spinal nerve (f) ganglion (g) nucleus (h) spinal ganglion (i) sympathetic ganglion (j) parasympathetic ganglion

The nervous system develops primarily from an infolding of ectoderm, called *neural ectoderm,* during the early embryonic life. The infolding fuses dorsally to form the *neural tube,* which eventually gives rise to the spinal cord. Figure 7-1 diagrams the dorsal surface of a developing embryo. The ectoderm infoldings are already evident, fusing to form the neural tube in the center. The cross-sectional appearance of different areas of such a tube is indicated by the four drawings on the right half of the figure. Cells at the cephalic end of the neural tube continue to proliferate, forming the brain. During development, both the brain and spinal cord are ensheathed in a connective tissue investment called the *meninges,* and they become bathed in a clear, sustaining fluid (called *cerebrospinal fluid*).

STRUCTURAL COMPONENTS OF THE NERVOUS SYSTEM

The nervous system is composed of several types of very specialized cells. Nerve cells involved in conducting impulses are called *neurons.* These cells have long processes which may branch and extend as much as a meter or more, as in the sciatic nerve.

Many large neurons have a polygonal cell body called the *perikaryon* (plural *perikaryons*) from which several nerve processes or *fibers* extend. In Fig. 7-2, a section of the spinal cord shows several large motor neurons with processes of varying length. Neurons make up only about 10 percent of the total cell population of the nervous system. The remaining 90 percent are supportive cells called *neuroglia*; they provide nutrition, insulation, and protection from harmful substances. Glial cells are generally smaller than neurons. In Fig. 7-2 are scattered many small nuclei of glial cells (arrows). The remainder of the space is occupied by nerve cell processes or fibers that are too fine to be resolved at this level of magnification. They run in random directions.

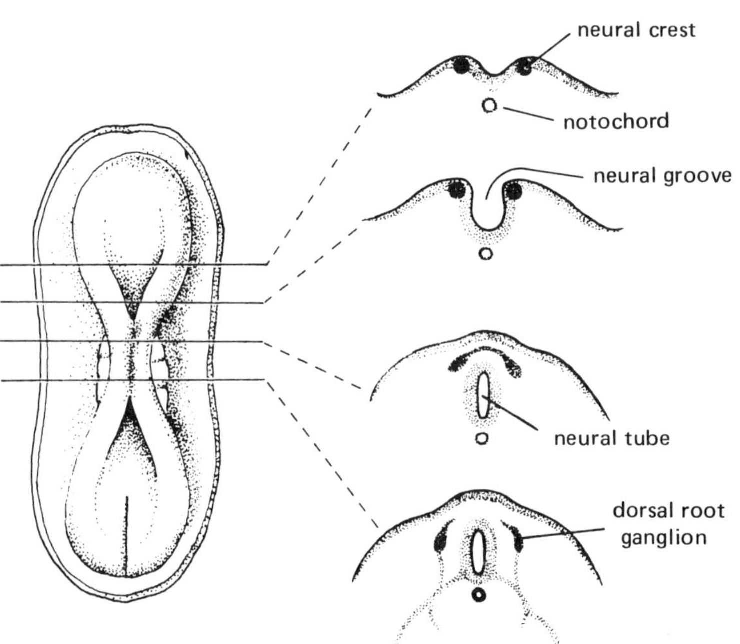

FIGURE 7-1 Diagram illustrating surface and cross-sectional views of the development of the neural tube.

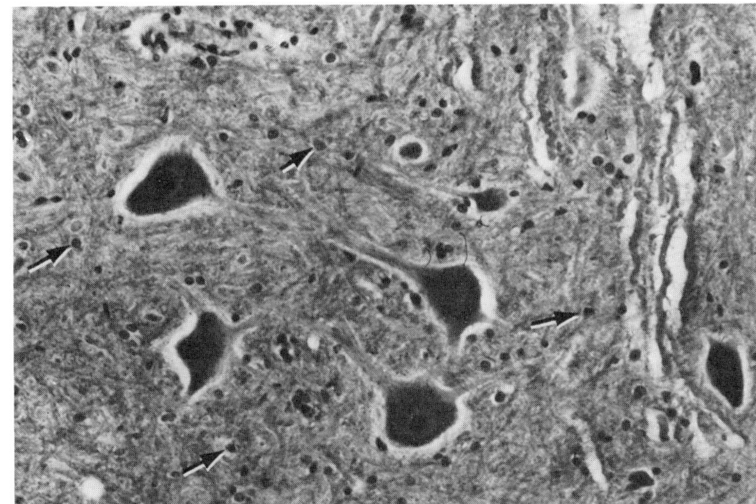

FIGURE 7-2 Light micrograph of motor neurons in the spinal cord. These neurons have large cell bodies—perikaryons—with several processes or nerve fibers; i.e., they are multipolar. Motor neurons contain large vesicular nuclei, each having a prominent nucleolus. Nuclei of glial cells are indicated by arrows.

ORGANIZATION OF THE NERVOUS SYSTEM

The nervous system is divided into two main systems that are anatomically different. The *central nervous system* (CNS), which consists of the brain and spinal cord, collects, integrates, and processes information arriving from the peripheral region of the body via the nerve fibers that make up the *peripheral nervous system*.

CENTRAL NERVOUS SYSTEM

The brain and spinal cord are made up of neurons and neuroglia. Cell bodies or perikaryons of neurons that are densely grouped according to particular functional requirements are called *nuclei* when located within the CNS and *ganglia* when present in the peripheral nervous system. Bundles of nerve fibers running in the same direction are called *tracts* when found within the CNS and *nerves* when located elsewhere in the body.

This chapter will deal only with the spinal cord. This part of the CNS is composed primarily of tracts which transmit impulses to and from the brain as well as nuclei which contain cell bodies of different peripheral nerves and therefore differ in structure depending on the level.

The spinal cord is roughly oval in cross section with a ventral fissure. Histologically, the transverse section of a spinal cord shows the white matter in the periphery (Fig. 7-3). Centrally, the gray matter occupies a region that is roughly H-shaped. Figure 7-4 is a photograph of an actual cross section of a spinal cord.

White matter receives its name from the presence of a large amount of *myelin*, the fatty insulating substance around individual nerve fibers. Gray matter consists of cell bodies and unmyelinated fibers which present a relatively gray appearance compared with the myelinated fibers of the white matter. The gray matter has two *ventral horns* and two *dorsal horns*. The fibers enter the spinal cord via the dorsal horns and exit from the ventral horns in discrete bundles. Since these bundles are present at regular intervals along the spinal column, they are called *spinal nerves*. Thirty-one pairs of such bundles, the spinal nerves, are part of the peripheral nervous system. Depending on the level, these spinal nerves are divided into the cervical, thoracic, lumbar, and sacral nerves (Fig. 7-5).

PERIPHERAL NERVOUS SYSTEM

The peripheral nervous system consists of nervous tissues other than the brain and spinal cord. It includes spinal and cranial nerves from the CNS, with their branches.

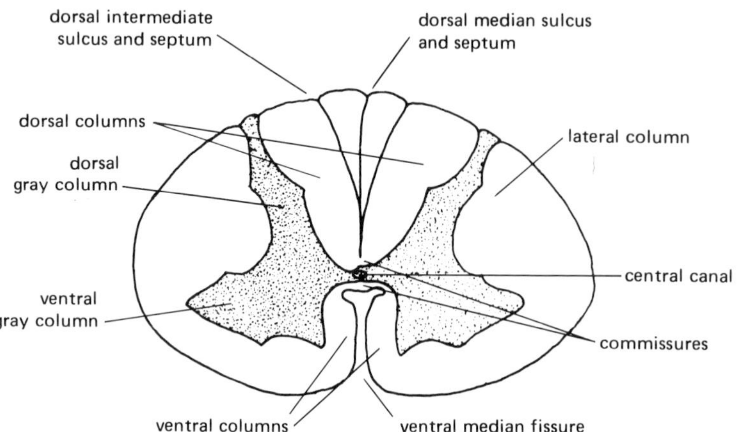

FIGURE 7-3 Diagram illustrating a transverse section of the human spinal cord.

The Spinal Nerves. A spinal nerve consists of both motor and sensory fibers. Figure 7-6 illustrates the motor (or effector) and sensory components of a typical spinal nerve. On the left, motor neurons are located in the ventral horn and sensory fibers can be seen entering the spinal cord via the dorsal root. Their cell bodies are located in the *spinal ganglion,* also called *dorsal root ganglion.* The dorsal root ganglion is a bulbous mass of cell bodies of sensory neurons (Figs. 7-6 and 7-7). The fibers of the dorsal root can be seen as they pass through the groups of large ganglion cells in the upper right of Fig. 7-7. Small cells closely applied to the surface of the large ganglion cells are called *satellite cells* (arrows, Fig. 7-8). They perform a supportive role for ganglion cells in a manner similar to the glial cells.

The Spinal Reflex Arc. A sensory input from the periphery can induce a rapid motor response caused by a local reflex called the *spinal reflex arc.* For example, if a hot object is touched with a finger, the finger recoils quickly. Temperature receptors under the epidermis relay the stimulus through the peripheral nerve to the cell body in the dorsal root ganglion. A series of neurons called *interneurons* relay the stimulus to the motor neuron located in the ventral horn (Fig. 7-6). The large motor neurons located in the ventral

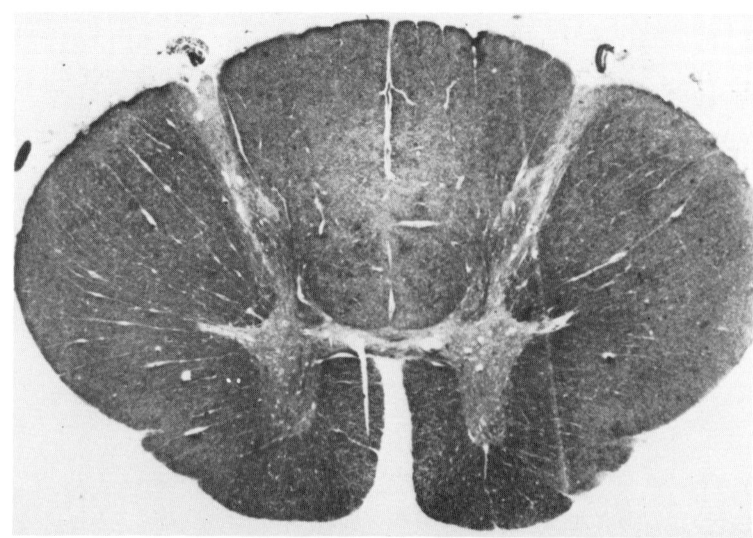

FIGURE 7-4 Photograph of an actual cross section of the human spinal cord. In the center, the gray matter appears approximately H-shaped. The periphery is occupied by white matter. Note the median ventral fissure.

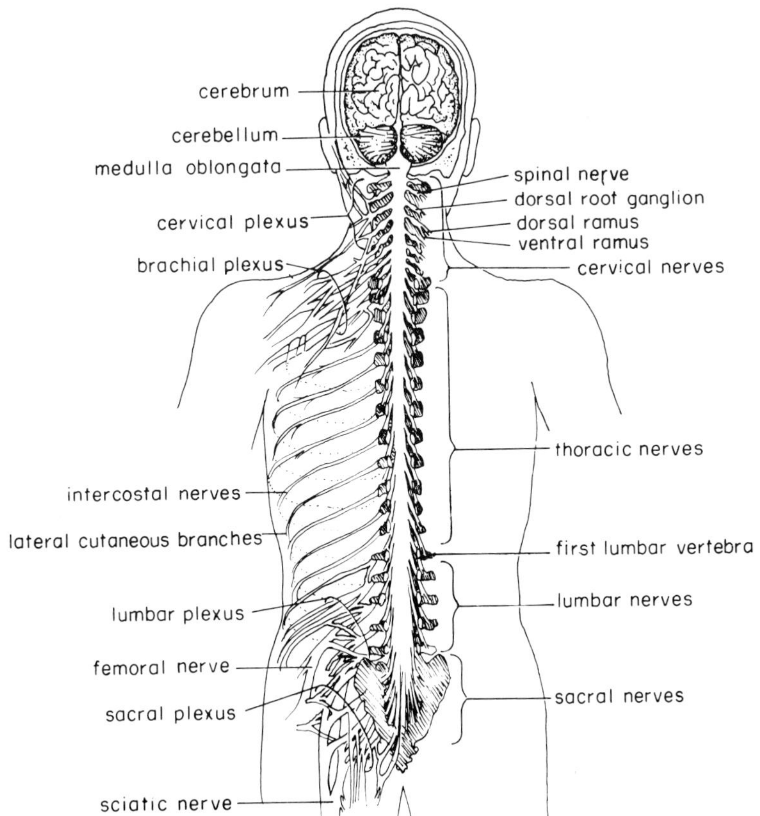

FIGURE 7-5 Diagram illustrating the relationship among components of central and peripheral nervous systems. Note the brain, spinal cord, spinal nerves, and the great limb plexuses, as seen from behind.

gray horn send an impulse toward the periphery through the ventral root. The nerve fibers which supply the appropriate muscles induce them to contract and draw the limb away from the hot object. The sensory impulse also travels up the spinal tracts to the brain so that the subject is made aware of this response.

The Cranial Nerves. Not all peripheral impulses are carried by the 31 pairs of spinal nerves. The peripheral nervous system also

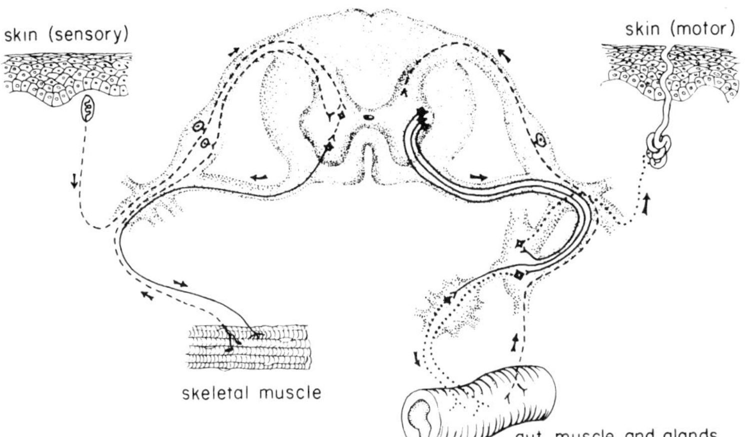

FIGURE 7-6 Diagram illustrating the motor and sensory components of a typical spinal nerve. The spinal reflex arc is shown on the left while the components of the sympathetic autonomic nervous system are illustrated on the right.

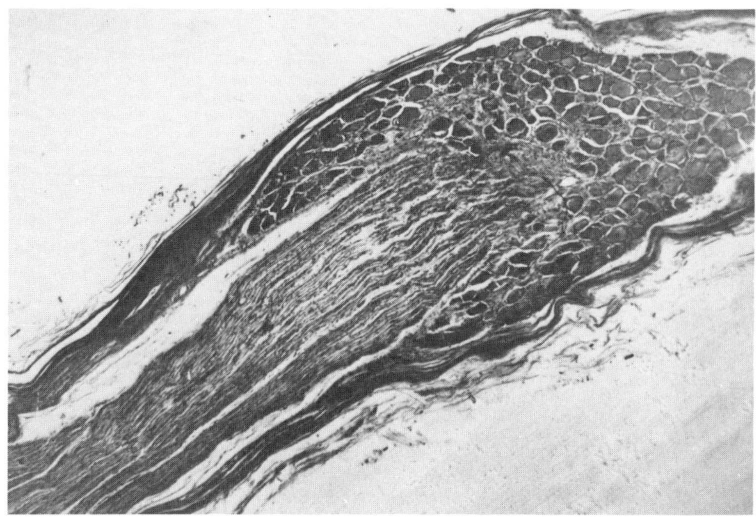

FIGURE 7-7 Light micrograph of a dorsal root ganglion. Numerous nerve fibers are seen passing through the ganglion. The ganglion contains the cell bodies of sensory neurons, which are surrounded by satellite cells.

includes 12 pairs of *cranial nerves.* Cranial nerves arise directly from cell bodies located in the brain and brain stem. These cranial nerves are designated by roman numerals I through XII.

Microstructure of Peripheral Nerves. In all peripheral nerves, the delicate nerve fibers are protected by connective tissue fibers and cells that wrap around them transversely. This relationship, which is established between nerve fibers and connective tissue elements, resembles the organization of skeletal muscle. Figure 7-9 diagrams a cross section of a peripheral nerve in which endoneurium, perineurium, and epineurium are labeled. *Endoneurium* is composed of a small number of connective tissue fibers and cells that surround individual nerve fibers. *Perineurium* is a thin but dense connective tissue layer surrounding a bundle of nerve fibers. The final part of nerve fiber wrapping is the *epineurium,* which contains the usual connective tissue fibers and cells, plus blood vessels that nourish the nerve. Figure 7-10 depicts portions of two nerve fiber bundles

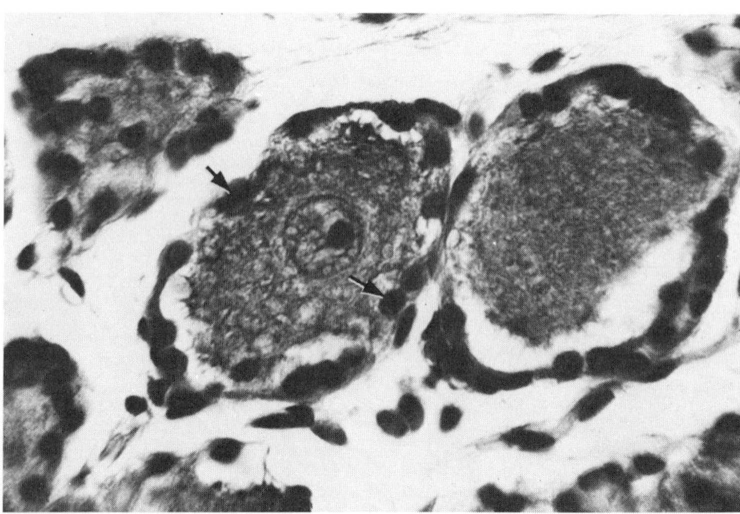

FIGURE 7-8 Light micrograph of sensory neurons viewed at higher magnification. The neurons have a large vesicular nucleus containing a prominent nucleolus. They are surrounded by satellite cells (arrows).

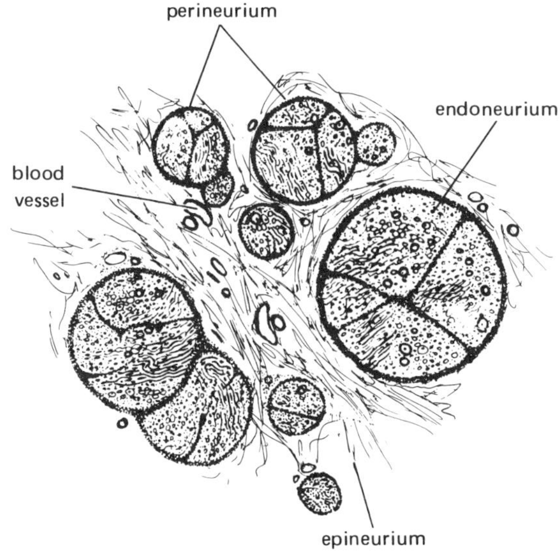

FIGURE 7-9 Diagram illustrating the organization of the connective tissue elements of a peripheral nerve.

surrounded by thin but dense perineurium (arrows) as well as more abundant epineural connective tissue in which several blood vessels are located.

The autonomic nervous system involves all the motor neurons concerned with visceral functions. Autonomic nerve fibers are present in all spinal and in most cranial nerves. They innervate the smooth muscle of visceral organs, as well as a few other structures such as the iris muscle of the eye, sweat glands, salivary glands, and the smooth muscle of blood vessels.

There are two divisions in the autonomic nervous system: the *sympathetic division* and the *parasympathetic division.* Both divisions consist of pre- and postganglionic neurons. In the sympathetic division, *preganglionic neurons* are located in the spinal cord from the first thoracic level to the second or third lumbar level, whereas in the parasympathetic division, they are found in the nuclei of origin of cranial nerves III, VII, IX, and X and in sacral segments II, III, and IV of the spinal cord. The *postganglionic neurons* are located in ganglia outside of the CNS. Based on the loci of the pre- and postganglionic neurons, the two divisions of the autonomic nervous system assume different anatomical positions (see Fig. 7-11).

Pre- and postganglionic neurons give rise to *pre- and postganglionic nerve fibers,* respectively. The preganglionic nerve fibers are myelinated, whereas the postganglionic ones are unmyelinated. Postganglionic fibers of the sympathetic division take their origin in the *chain ganglia* located in the sympathetic

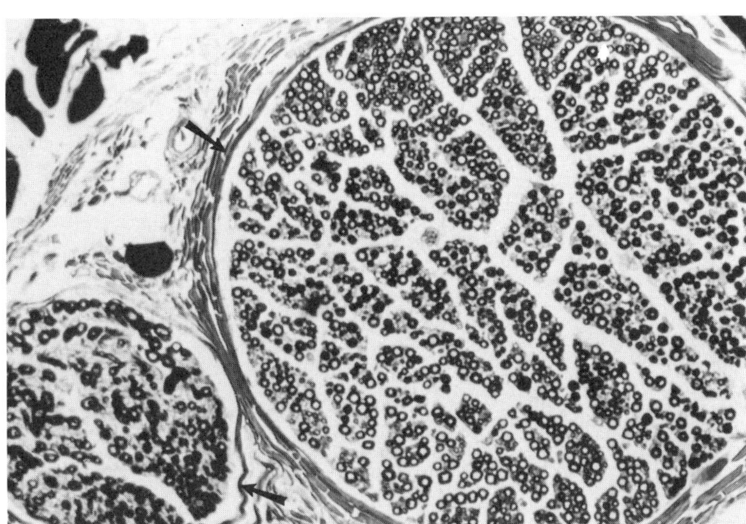

FIGURE 7-10 Light micrograph of a cross section of two nerve fiber bundles. A thin but dense perineurium (arrows) surrounds each bundle.

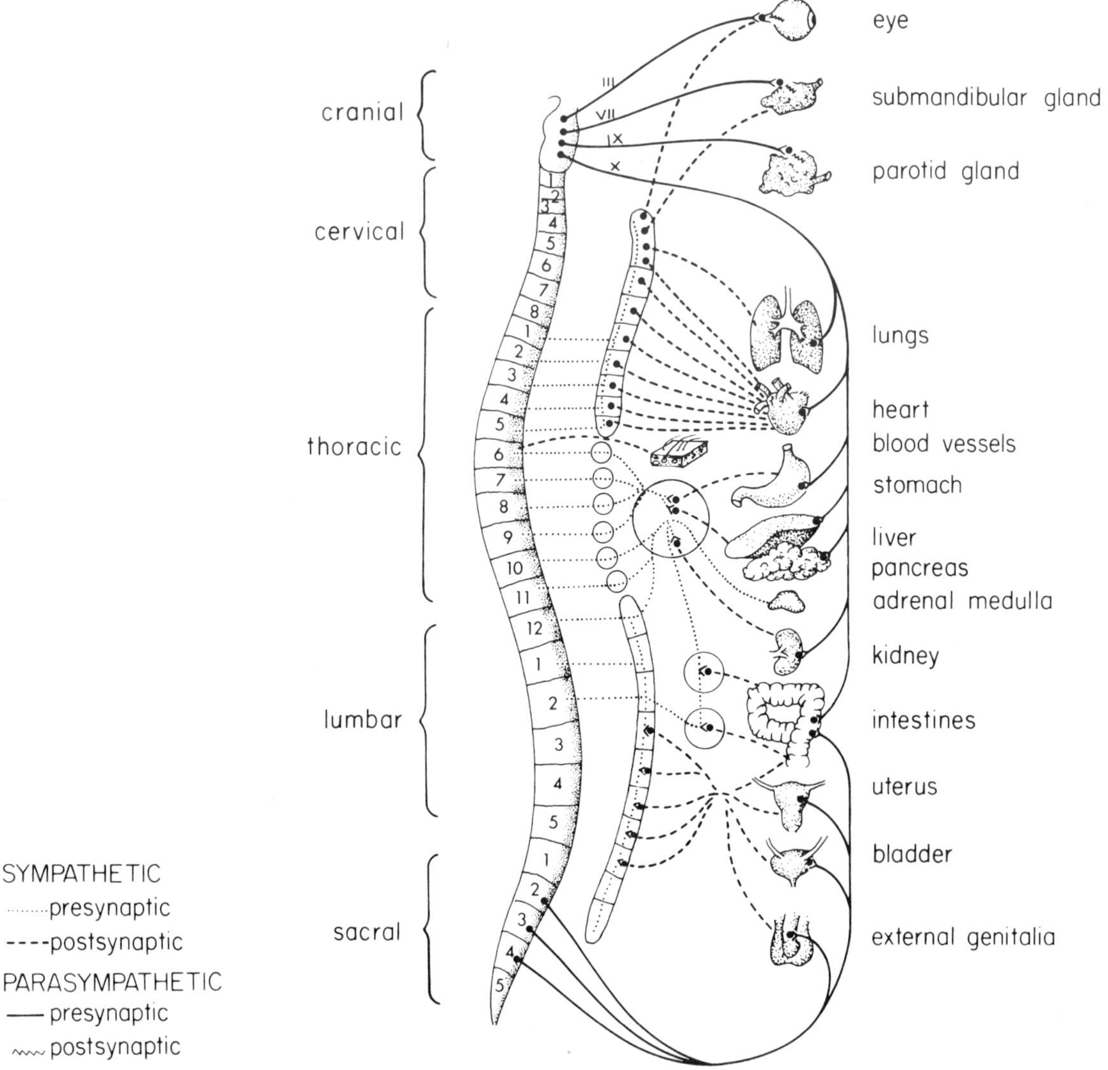

FIGURE 7-11 Diagrammatic illustration of the major pathways of the autonomic nervous system.

trunk along the thoracolumbar division of the spinal cord or in the prevertebral ganglia. Those of the parasympathetic division take their origins in ganglia located near or within structures innervated by them.

The two divisions are generally antagonistic or opposite in function. The sympathetic division prepares the body for crisis. It is responsible for the so-called fight-or-flight response. It accelerates production of energy and other bodily functions such as heart rate, respiration rate, and sweating, and slows such functions as intestinal activity and secretion of digestive enzymes. The parasympathetic division helps conserve or preserve the body in a quiescent state, causing slower heart rate and vasodilatation, resulting in lower blood pressure. Its stimulation also increases digestive processes.

The two divisions of the autonomic nervous system are controlled by different classes of functional neurochemical transmitters, namely adrenaline and acetylcholine. Nerves of the sympathetic division are called *adrenergic* because they act via the release of adrenaline and related substances. Nerves of the parasympathetic system are called *cholinergic* because acetylcholine acts as the main transmitter.

BIBLIOGRAPHY

Bloom, W., and D. W. Fawcett: *A Text Book of Histology,* 10th ed., Saunders, Philadelphia, 1975.

Bourne, G. H. (ed.): *The Structure and Function of Nervous Tissue,* Academic Press, New York, 1968.

Rhodin, J. A. G.: *Histology: A Text and Atlas,* Oxford University Press, New York, 1974.

ADDITIONAL READINGS

Babel, J.: *Ultrastructure of the Peripheral Nervous System,* Mosby, St. Louis, 1970.

Barr, M. L.: *The Human Nervous System,* Harper & Row, New York, 1972.

Bunge, R. P.: "Glial Cells and the Central Myelin Sheath," *Physiol. Rev.* **48:**197–251 (1968).

Everett, N. B.: *Functional Neuroanatomy,* Lea & Febiger, Philadelphia, 1971.

Glees, P.: *Neuroglia: Morphology and Function,* Charles C Thomas, Springfield, Ill., 1955.

Hubbard, J. I. (ed.): *The Peripheral Nervous System,* Plenum, New York, 1974.

Kuntz, A.: *The Autonomic Nervous System,* Lea & Febiger, Philadelphia, 1953.

Rodahl, K., and D. Issekutz (eds.): *Nerve as a Tissue,* Harper & Row, New York, 1966.

Truex, R. C., and M. B. Carpenter: *Strong and Elwyn's Human Neuroanatomy,* Williams & Wilkins, Baltimore, 1969.

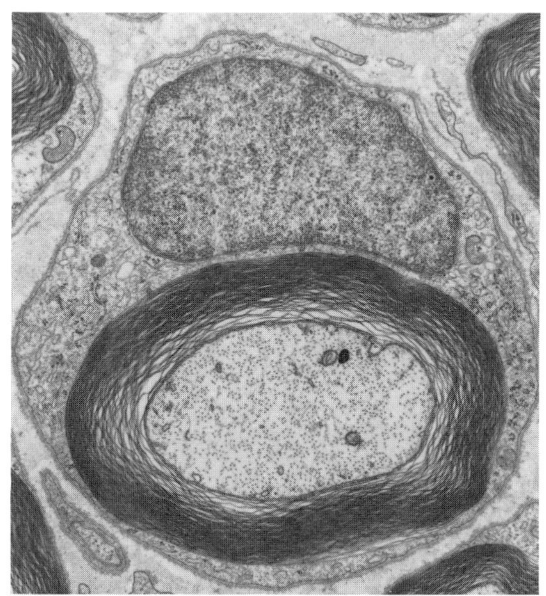

8 THE NERVOUS SYSTEM: CELLULAR COMPONENTS

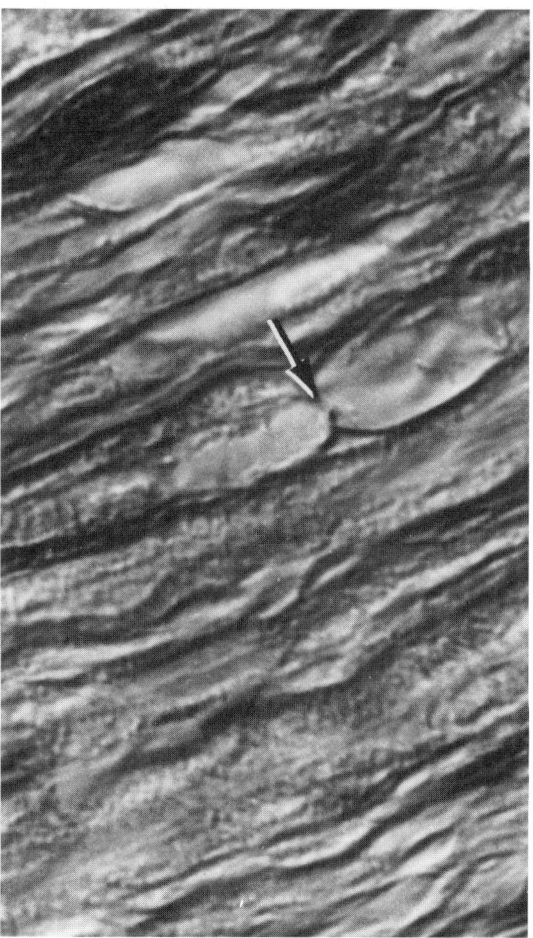

OBJECTIVES

Upon completion of this chapter, the student will be able to:

1 Label a diagram of a light micrograph of a neuron, including the following:
(a) perikaryon (b) nucleus (c) nucleolus (d) Nissl bodies (e) Golgi apparatus (f) dendrite(s) (g) axon hillock (h) myelin sheath (i) node of Ranvier

2 Describe the myelin sheath by identifying:
(a) the cell type of which it is composed (b) its function (c) its chemical composition

3 Identify in cross section in an EM a myelinated nerve and an unmyelinated nerve.

4 Give an example of each of the following types of neurons:
(a) multipolar (b) bipolar (c) unipolar

5 Define the following terms with regard to impulse conduction:
(a) electric potential (b) resting potential (c) depolarization (d) threshold stimulus (e) action potential (f) saltatory conduction

6 Label a diagram of a synapse including the following:

(a) axon bouton (b) mitochondria and synaptic vesicles (c) synaptic cleft (d) postsynaptic membrane

7 Identify the function of the following with regard to synapses:
(a) acetylcholine (b) acetylcholinesterase

8 Draw and label a motor end plate, and describe its function.

9 Name three types of neuroglia.

10 Identify in section the following sensory endings, state where they are found, and give their function:
(a) free nerve endings (b) Meissner's corpuscles (c) pacinian corpuscles (d) muscle spindles

STRUCTURE OF NEURONS

The neuron typically has a large polygonal-shaped cell body, the *perikaryon*. Extending from the perikaryon are a number of processes, or *nerve fibers*, that conduct impulses. The fibers have different lengths and structures and are classified in two basic ways: *Dendrites* collect the impulse and conduct it centripetally towards the perikaryon; and the *axon* conducts the impulse away from the perikaryon.

Figure 8-1 is a diagram of a motor neuron. A long axon and many short, branching dendrites are depicted. Often the nerve fibers are surrounded by a sheath of multilayered cell membranes of Schwann cells, called the *myelin sheath*. Since a Schwann cell produces a myelin sheath of finite size, there is a gap between sheath segments representing the products of individual Schwann cells. These gaps, usually present as constrictions at regular intervals, are called *nodes of Ranvier.*

Figure 8-2 is an LM of a large motor neuron having several processes. Many nuclei of neuroglial cells are also seen. The processes of the neuron extending from the perikaryon reach varying distances and branch frequently. As mentioned earlier, axons of motor neurons located in the lumbar region of the spinal cord must extend a meter or more to innervate the muscles of the toes. Since the neuron must nourish its long fibers, the perikaryon of large neurons is often voluminous, containing many specialized structures. Likewise, the cell processes have certain structural specializations that allow them to conduct impulses at different rates.

Fine Structure of Perikaryon and Nerve Fibers. The neuron has a rounded light-staining nucleus, which contains one or more prominent nucleoli, located in the perikaryon. The perikaryon has dense basophilic cytoplasm. The dense basophilic material is arranged in small patches. These patches are substances known as *Nissl bodies,* which may extend into the dendrites. Nissl bodies are not present in axons or in a small region of the perikaryon near the beginning of the axon, the *axon hillock* (arrow, Fig. 8-2). In the electron microscope, a Nissl body can be identified as a patch of RER, which by virtue of its associated polyribosomes appears basophilic. A region of a Nissl body is seen in

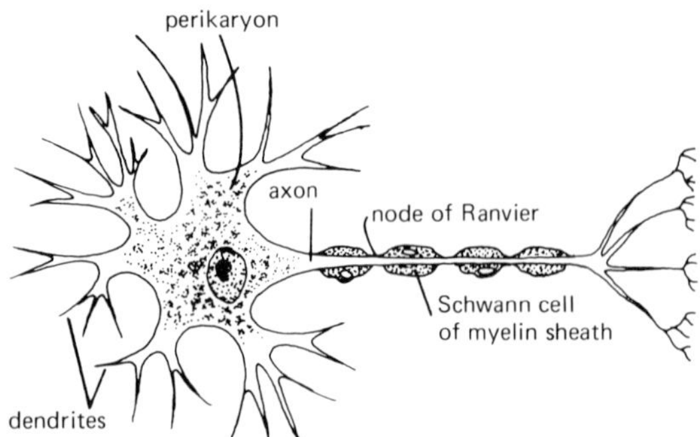

FIGURE 8-1 Schematic drawing of motor neuron. For practical purposes the length of the axon is markedly reduced.

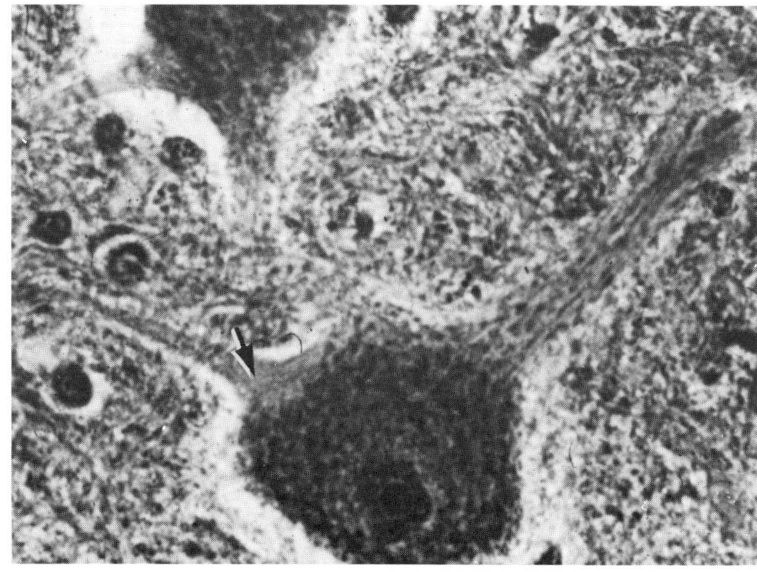

FIGURE 8-2 Light micrograph of motor neuron. The motor neuron is multipolar, i.e., has many processes. The perikaryon contains a large nucleus with a prominent nucleolus and Nissl bodies. Nissl bodies are not present at the beginning of the axon, the axon hillock (arrow).

Fig. 8-3 at the level of EM. Numerous free ribosomes are also found between the cisternae of the RER in the Nissl body region. Elsewhere in the perikaryon, microtubules, mitochondria, and profiles of the Golgi apparatus are also present. The Golgi complex, which is confined to the perikaryon, is extensively developed in most large neurons, and in fact was identified in nerve cell bodies with silver impregnation technique.

In early days of cytology, light microscopists were able to discern what they called *neurofibrils* throughout the perikaryon. Numerous neurofibrils run parallel to one another in the processes. With the advance of electron microscopy, it has been possible to differentiate two structures that account for the traditional neurofibrils of LM. The thinner *neurofilaments* are intracellular filaments similar to the tonofibrils of epithelial cells. These are about 7 nm in diameter. In addition, microtubules resembling those found in

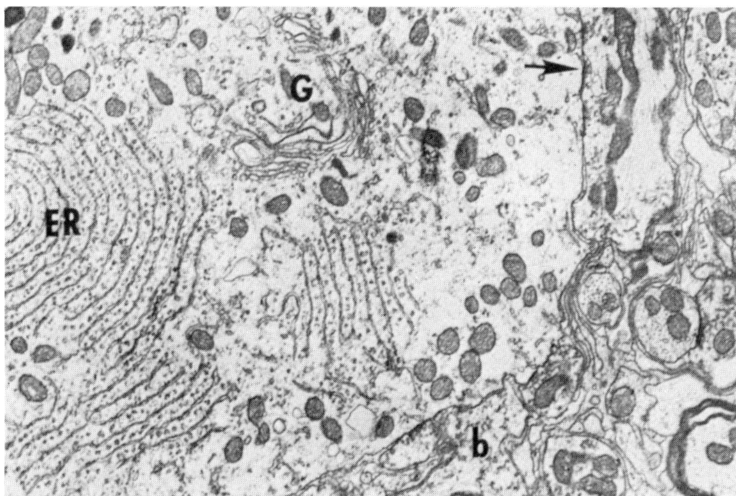

FIGURE 8-3 Electron micrograph of Nissl body. A Nissl body is composed of patches of RER and numerous free ribosomes between the RER profiles (G, Golgi complex; b, bouton). An axosomatic synapse is indicated by arrow. (*Courtesy of Dr. I. J. Bak.*)

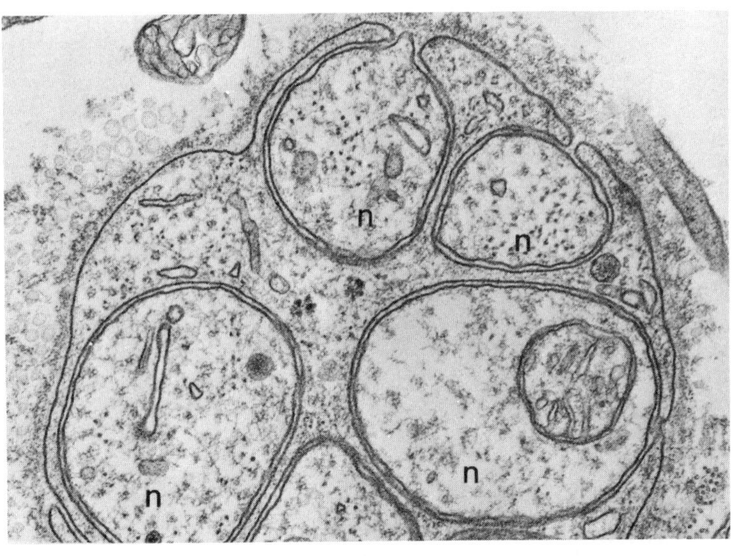

FIGURE 8-4 Electron micrograph of an unmyelinated nerve. Several nerve fibers (n) are invested in the cytoplasm of a Schwann cell. Cross section of neurotubules and neurofilaments are apparent within the nerve fibers. One fiber contains a mitochondrion.

other cell types are present in most neurons and are called *neurotubules.* Neurotubules have a diameter of about 24 nm. Figure 8-4 is a cross-sectional view of several nerve fibers (n) in EM, showing neurofilaments and a few neurotubules.

Mitochondria of the neuron are rather small but numerous. They are ovoid to round and are located throughout the perikaryon as well as its processes. One such mitochondrion is present in Fig. 8-4.

Classification of Neurons. A neuron can be classified on a functional basis as being either

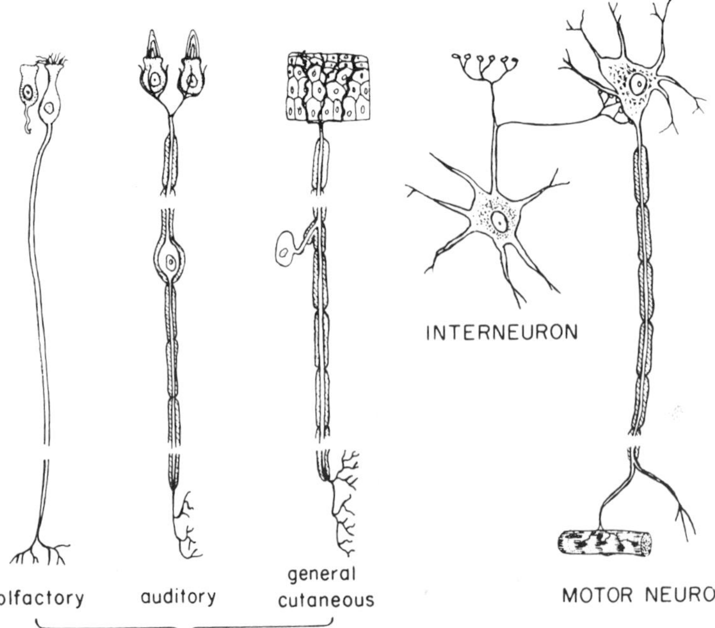

FIGURE 8-5 Schematic drawing illustrating the principal forms of neurons.

a *motor neuron* or a *sensory neuron*. It can also be classified on the basis of its structure (Fig. 8-5). Since most motor neurons have many processes, they are called *multipolar neurons.* The neurons that connect nerve cells within the CNS, called *interneurons,* are also multipolar neurons, but generally they have much shorter processes. Auditory neurons have only two processes; hence they are called *bipolar neurons.* An example of such a neuron is illustrated in Fig. 8-5. Some of the sensory neurons have only one process; hence they are called *unipolar neurons.* Note the structure of the unipolar olfactory neuron. The majority of sensory neurons, such as those specialized for general cutaneous sensation, have two processes which are joined for some distance from the perikaryon; hence they are called *pseudounipolar neurons.*

IMPULSE PROPAGATION

The function of a neuron is to receive electrical or chemical stimuli, generate an electric potential, and propagate it over a distance along the nerve. The process of propagation of an electric potential, which can now be called a *nerve impulse,* moves rapidly along the fiber. It is generated by a sudden change in the electric charge of the plasma membrane of the neuron. Most nerve cells that are in a resting stage have about 70 mV of negative electric charge. Thus the *resting electric potential* (voltage difference) of the neurons is −70 mV.

When a neuron is stimulated at a point along its surface, its membrane (neurilemma) abruptly becomes permeable to sodium ions (Na^+) in that region. The influx of sodium ions into the stimulated region of the neuron raises the electric potential by *depolarization* of the membrane. When the stimulus and hence the depolarization is present in an appropriate amount, an *action potential* is generated, allowing the propagation of the nerve impulse, as shown in Fig. 8-6. As the depolarized region becomes immediately repolarized, the action potential in a neuron passes along its fibers in a wavelike manner. Since the action potential is generated only after the excitatory stimulus has reached a certain level, this level of stimulus necessary for generating an action potential is called a *threshold stimulus.* A good example of this may be seen in the function of synapses which require a predetermined level of chemical transmitter substances before impulse propagation beyond the synapse can occur.

The Synapse. Some neurons, particularly among lower forms of life, relay nerve impulses via direct electrical stimulation, as in nexuses. However, the most common manner in which a nerve impulse is transmitted from one cell to another is across *synapses.* In a synaptic junction, impulses are relayed across a cleft or gap by means of the release of a *neurochemical transmitter* from the axon, as shown in Fig. 8-7. The release of the transmitter substance is accomplished in a manner similar to exocytosis of other secretory substances. The transmitter substance released

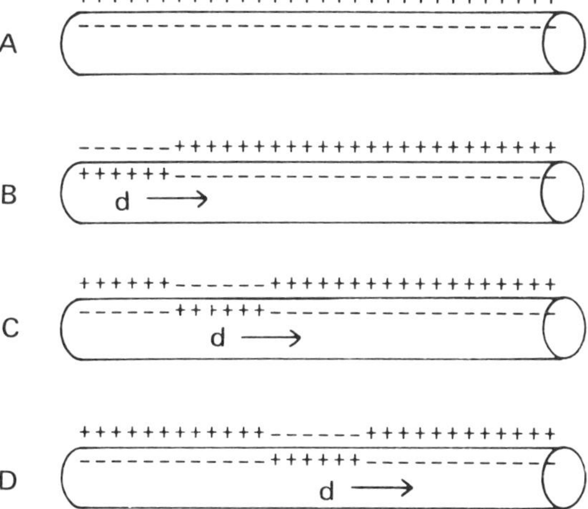

FIGURE 8-6 Diagram illustrating the propagation of a nerve impulse in a nerve fiber (d = depolarization).

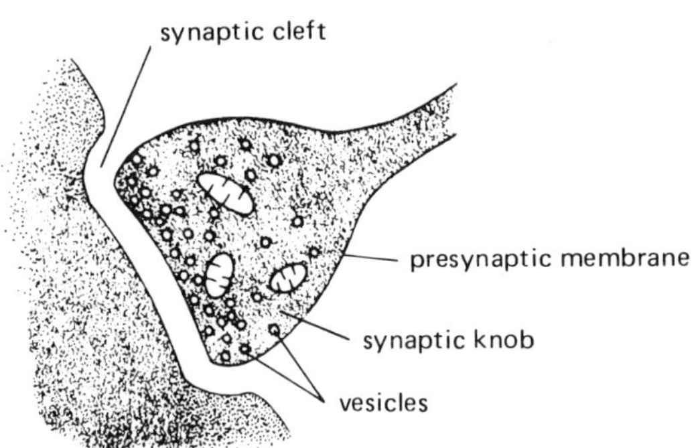

FIGURE 8-7 Drawing illustrating the morphology of a synapse.

into the synaptic cleft diffuses throughout the space and induces depolarization of the second-order nerve cell. Figure 8-8 is a diagram of a mature motor end plate. Although the transmission of impulse at this junction is between nerve and muscle, the region is functionally analogous to a synapse and illustrates the relationship between presynaptic and postsynaptic membranes. Note the irregular infoldings of the postsynaptic membrane. The small vesicles present in the axon terminal contain a neurochemical transmitter, which in this case is *acetylcholine.* As a nerve impulse reaches the end plate, the acetylcholine contained in *presynaptic vesicles* is released into the synaptic cleft. Acetylcholine depolarizes the plasma membrane of a muscle or nerve fiber, causing it to become more permeable to sodium ions. When enough acetylcholine is released to generate an action potential, the impulse is propagated along the cell surface. Following this, acetylcholine is broken down by an enzyme, *acetylcholinesterase,* which is located in the postsynaptic membrane. The synaptic end of an axon is generally enlarged and bulbous and is often called a *bouton.* The bouton has in it fine neurotubules, neurofilaments, mitochondria, and small vesicles containing the transmitter substances, which could be any one of a number of different chemicals. Autonomic elements may be defined as adrenergic or

FIGURE 8-8 Schematic drawing of motor end plate. Details of the relationship between the nerve and the muscle fiber are seen enlarged in the square.

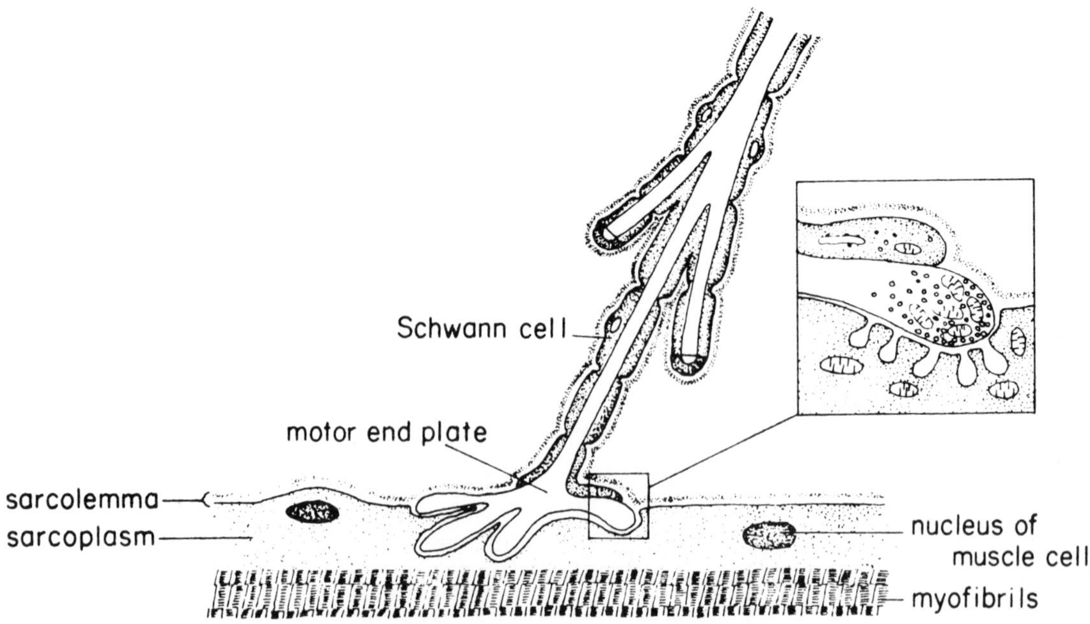

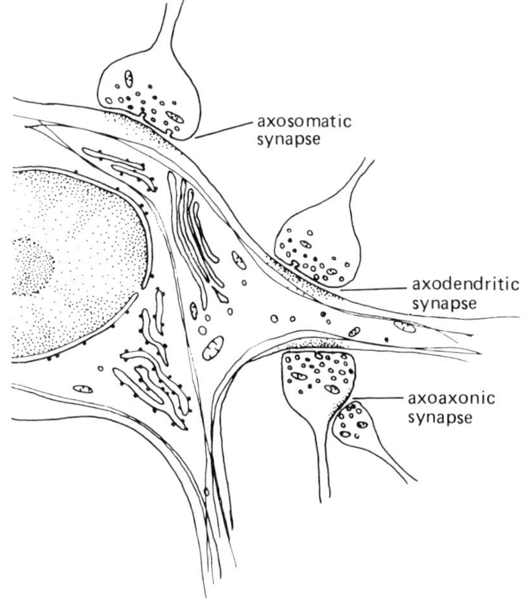

FIGURE 8-9 Drawing illustrating types of synapses occurring on various parts of the neuron.

cholinergic, depending on the type of chemicals contained in its presynaptic vesicles.

Neurons can synapse with other neurons in a variety of ways. Dendrites may receive from one to many hundreds of inputs. Axons, however, generally have only one point of transmission. The axon may form a synapse with dendrites from other neurons, an *axodendritic synapse*. It may form a synapse with an axon from another cell in an *axoaxonal synapse* or with a cell body of another neuron in an *axosomatic synapse*. Figure 8-9 shows examples of the different types of synapses.

THE MYELIN SHEATH

Many nerve fibers are ensheathed by a thick fatty covering, called the *myelin sheath*, that is readily visible under LM. Dendrites of all sensory neurons have a myelin sheath, but dendrites of motor or multipolar neurons do not. Electron microscopy has revealed that the myelin sheath is composed of a spiral layering of plasma membranes of *Schwann cells*, which surround the fiber in a jelly roll–like manner (Fig. 8-10).

During development, Schwann cells differentiate from the neural tube, lining up along the nerve fiber as it grows longer, and start to produce a myelin sheath by wrapping the nerve as shown in Fig. 8-11*a*. In the process of myelinization, the cytoplasmic lip

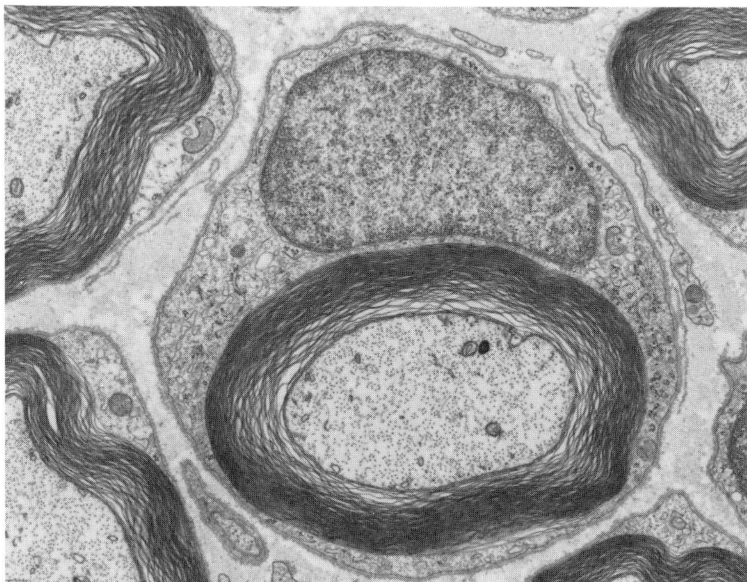

FIGURE 8-10 Electron micrograph of myelin sheath. The cell body of the associated Schwann cell containing the nucleus is apparent.

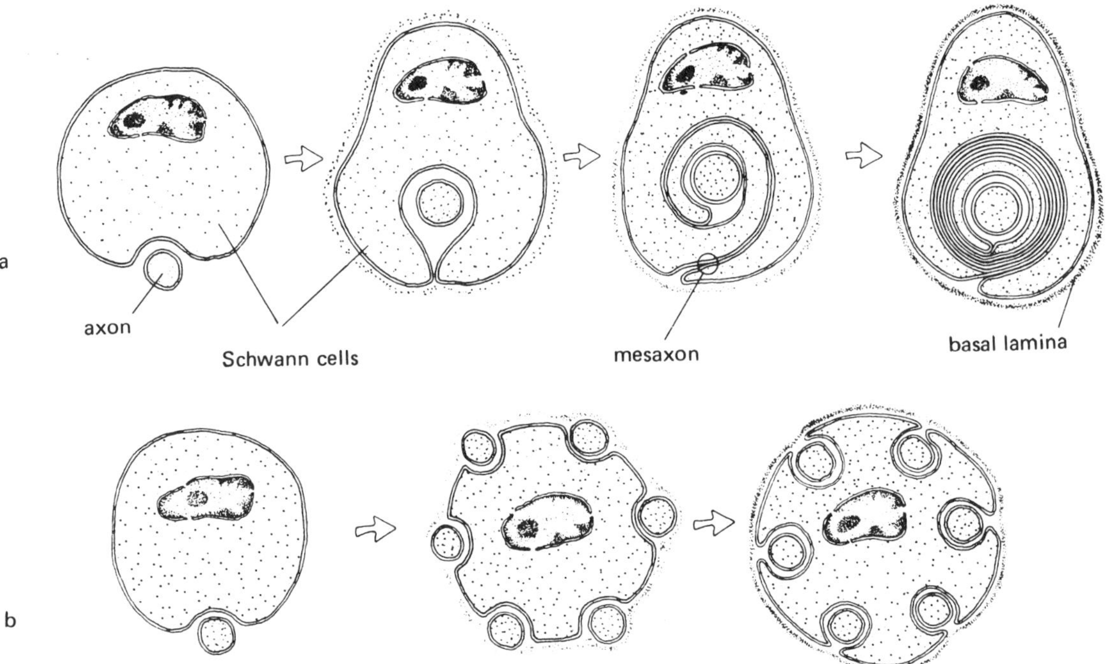

FIGURE 8-11 Diagram illustrating the development of a myelinated nerve (*a*) and an unmyelinated nerve (*b*).

of the Schwann cell, as it moves constantly around the growing fiber, is in close contact with the apposing membrane, resulting in a jelly roll–like sheath. This tight relationship is pertinent, since the primary purpose of the Schwann cell is to protect and insulate the neurilemma of the axon. This relationship is called *mesaxon*. Commonly, 5 to 50 layers of

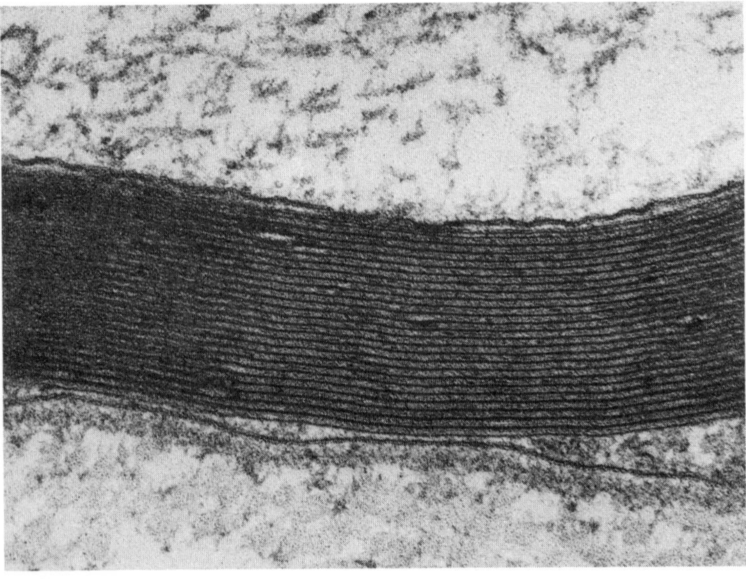

FIGURE 8-12 Electron micrograph of myelin sheath. The myelin sheath is made up by the spiral layering of the Schwann cell membrane.

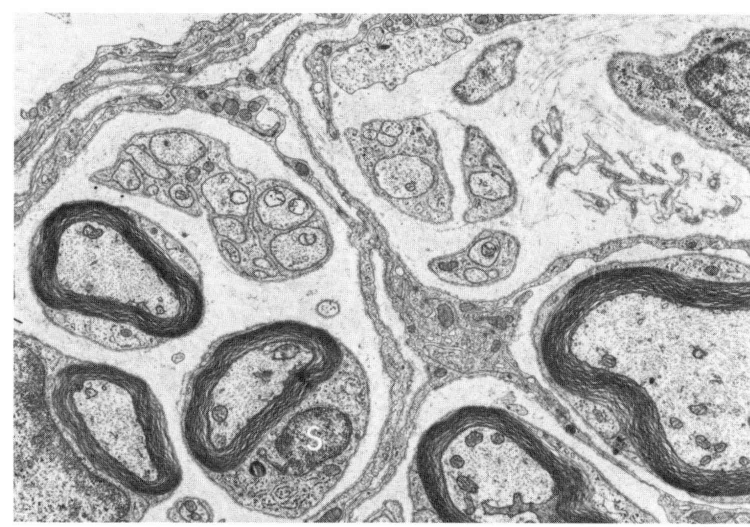

FIGURE 8-13 Electron micrograph of cross sections of myelinated and unmyelinated nerve fibers. The nucleus (s) in one of the Schwann cells can be seen.

Schwann cell membranes make up a myelin sheath. Figure 8-12 is a high magnification EM of a portion of a myelin sheath which is made up of approximately 30 Schwann cell membranes. In general, the larger the diameter of the nerve process, the thicker is the myelin sheath. The large nucleus of the Schwann cell (Fig. 8-13) can occasionally be seen within the outer myelin layer.

In cross section, an LM of a peripheral nerve may show nerve fibers of different sizes that can be preserved by special techniques such as osmium impregnation (Fig. 8-14*a*). Because histologic preparation involves a process of dehydration, the light microscopic appearance of the peripheral nerve often shows a clear rim around the axon from which the lipid-rich myelin substance has been extracted during dehydration (Fig. 8-14*b*). Figure 8-15 is a cross section of an H- and E-stained nerve. Note the centrally located axons in fibers of different sizes.

FIGURE 8-14 Diagram depicting well-preserved (*a*) and poorly preserved (*b*) myelin sheaths.

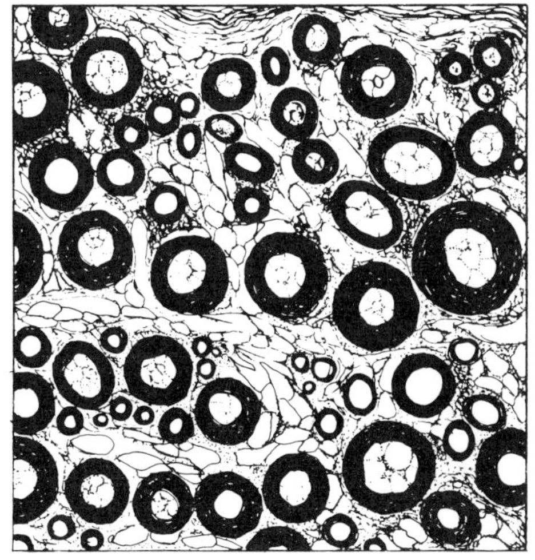

a

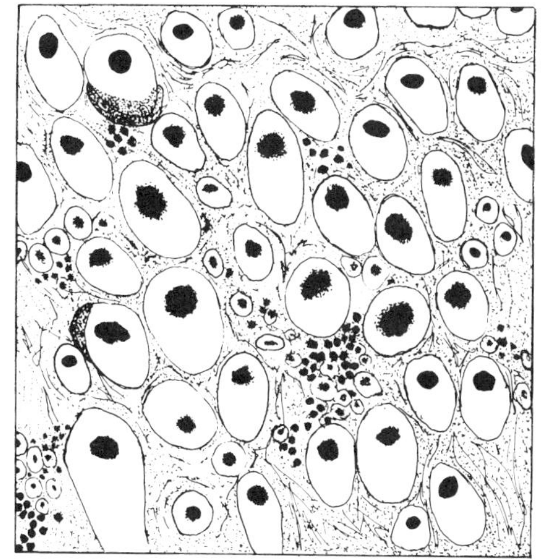

b

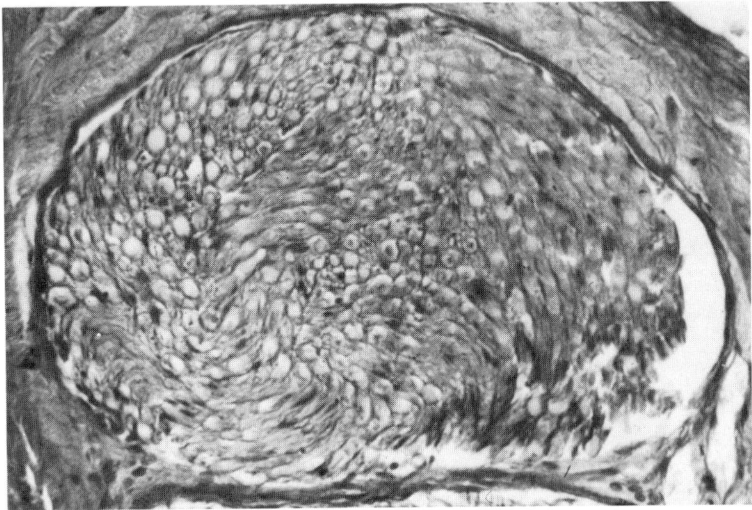

FIGURE 8-15 Light micrograph of cross section of a nerve stained with H and E.

Nodes of Ranvier. In longitudinal sections, the myelinated nerves show node-like areas where the myelin sheath is interrupted. Figure 8-16 is an LM in which an example of this break in the myelin sheath, called *nodes of Ranvier,* is depicted (arrow). Figure 8-17 is a diagram of a longitudinal view of myelinated nerves in which nodes of Ranvier with the axons passing through the middle of the nodes are seen. Note that the Schwann cell in the thicker fiber covers a larger area than in the smaller. Also note that the nuclei of the Schwann cells are clearly within the myelin sheath. These nuclei are club-shaped with a fine chromatin pattern and can be distinguished from those of fibroblasts of the endoneurium which are smaller, denser, and more spindle-shaped. It should be noted that the axon is exposed to the extracellular space at the node.

Unmyelinated Nerves. *Unmyelinated nerves* are groups of fibers that lack a myelin sheath, such as postganglionic nerves. Unmyelinated

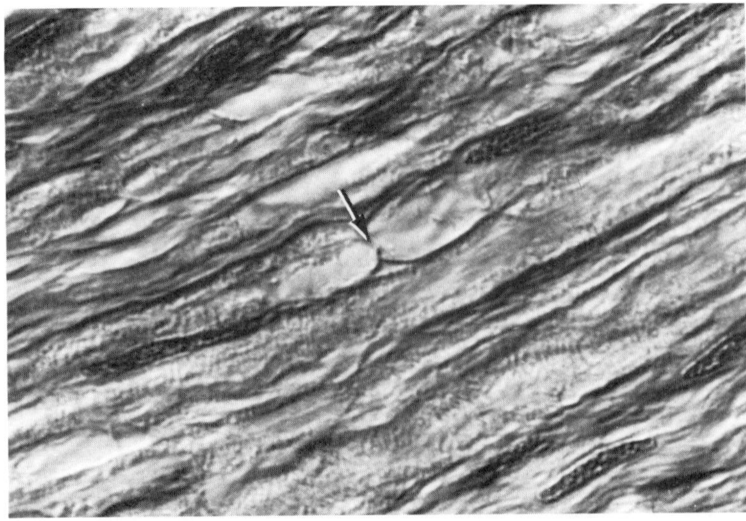

FIGURE 8-16 Light micrograph of myelinated nerve fibers seen in longitudinal section. A node of Ranvier is indicated by arrow.

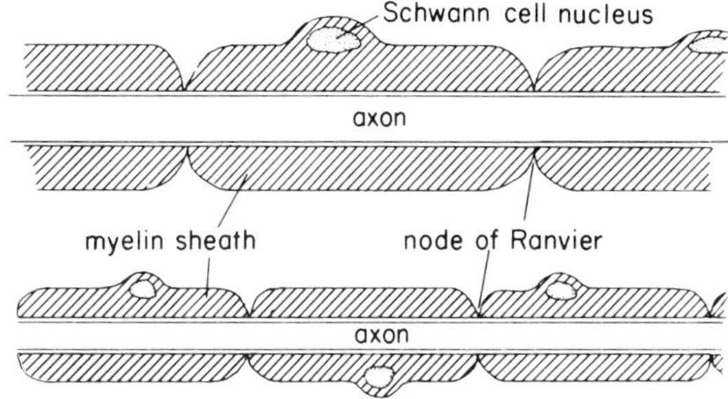

FIGURE 8-17 Diagrammatic representation of longitudinal sections of myelinated nerves of different diameter showing variations in the frequency of nodes of Ranvier in thick and thin nerve fibers.

fibers have been so designated because under LM they appear to lack a myelin sheath. However, electron microscopy has revealed that unmyelinated nerve fibers are also covered by Schwann cells; but their covering is only one layer thick and is visible only at the EM level. Several unmyelinated fibers may be invested in the cytoplasm of a single Schwann cell, as indicated in Fig. 8-13. Figure 8-18 is an EM in which unmyelinated fibers are invested in the cytoplasm of a single centrally located Schwann cell.

IMPULSE PROPAGATION ALONG NERVE FIBERS

As described earlier, nerve impulses are waves of action potentials propagated as a result of an influx of sodium ions. Since most of the unmyelinated nerve fibers are exposed to the tissue space, the propagation of an impulse along them is uninhibited and therefore proceeds at an equal speed within a given type of nerve.

In contrast to the impulse conduction in unmyelinated nerves, the propagation of action potential in myelinated nerves is dictated by the presence of the myelin sheath which insulates the nerve fiber, thereby preventing the influx of sodium ions between nodes. Therefore, the impulse jumps from node to

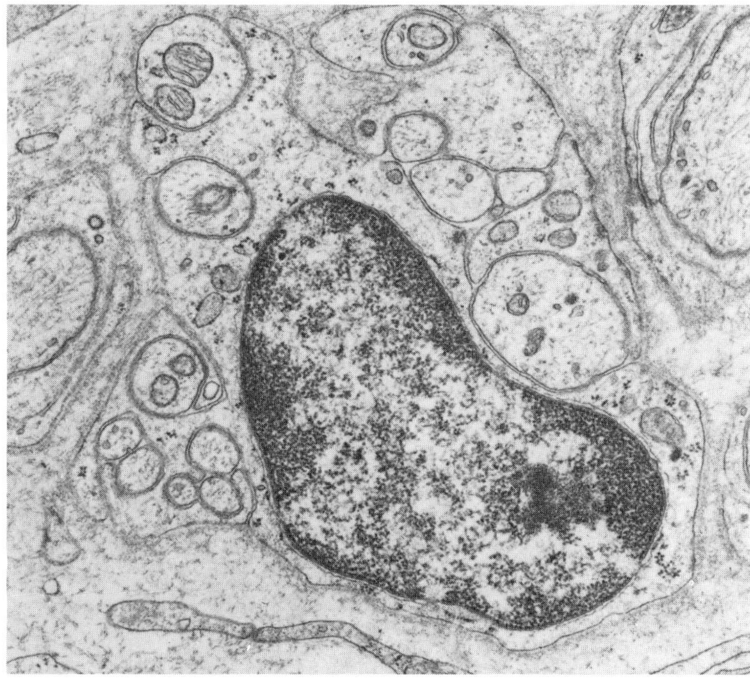

FIGURE 8-18 Electron micrograph of unmyelinated nerve fiber. Several nerve fibers are contained by the cytoplasm of a single Schwann cell.

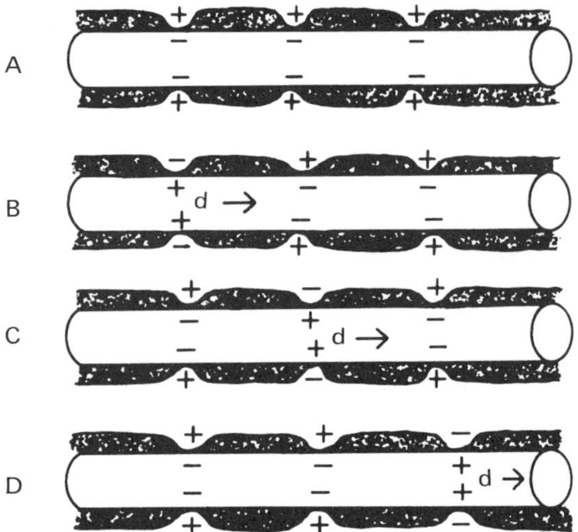

FIGURE 8-19 Diagram illustrating the process of saltatory conduction along a myelinated nerve (d = depolarization).

node in myelinated nerves, resulting in an increased rate of conduction (Fig. 8-19). This mode of impulse propagation between nodes of Ranvier is called *saltatory* conduction. (*Saltatory* means "leaping or jumping.") For this reason, the rate of impulse conduction in myelinated nerves is variable, depending on the internodal distance. The closer together the nodes are (as in a nerve with smaller diameter), the slower the rate of conduction, since there are more nodes of Ranvier per unit length. As the Schwann cells are smaller in nerve fibers of reduced diameter, the velocity of impulse conduction is slower in smaller myelinated nerves (Fig. 8-17). Thus the direct relationship between the conduction speed and the size of the nerve fiber is due primarily to differences in the internodal distance.

NEUROGLIA

Cells of the central nervous system not involved in impulse conduction are called *neuroglial cells,* of which the Schwann cells are one type. Another important group is the neuroglia proper, or *glia*. Figure 8-20 shows drawings of a few types of neuroglial cells.

One glial cell type is the *oligodendrocyte,* or oligodendroglia (*a* in Fig. 8-20). Oligodendroglia appear similar to astrocytes but are slightly smaller. Their cytoplasmic processes are short and have few branches (*oligo-* means "few"). They are found in close association with nerve fibers, largely in the white matter. All glial cells perform protective functions: some by becoming phagocytic, and others by being interposed between blood vessels and the neurons of the brain.

The *microglia* are the smallest of the glial elements (*b* in Fig. 8-20). They are generally associated with the blood vessels, and they may transform into phagocytes under certain conditions. Presumably they ingest foreign particles that may be toxic to the neurons. They have small dark-staining cell bodies with small nuclei.

The largest of the glial cells are the *astrocytes,* found in the brain. Astrocytes come in two forms: *protoplasmic* and *fibrous,* as diagramed in Fig. 8-20*d* and *c*, respectively.

Protoplasmic astrocytes are plump, large cells found only in the gray matter of the brain. Fibrous astrocytes are found in the white matter. Both types have long processes which extend between blood vessels and nerve fibers. Figure 8-21 is a region of the gray matter of the brain which shows a number of large neurons along with numerous glial cell nuclei that are scattered throughout the field. The large nuclei belong to protoplasmic astrocytes, while the small, dense nuclei are those of microglia.

In fact, there is no connective tissue space in the brain, and all nutrients that blood vessels supply reach the neurons via glial cell cytoplasm. Glial cells are therefore a system of effective barriers that exist between the neurons and blood vessels. This anatomical relationship between the two cell populations and the structure of blood vessels does not permit any bloodborne materials that may be unwanted to reach neurons of the brain. It is called the *blood-brain barrier,* a critical protective mechanism for the brain cells. Similar

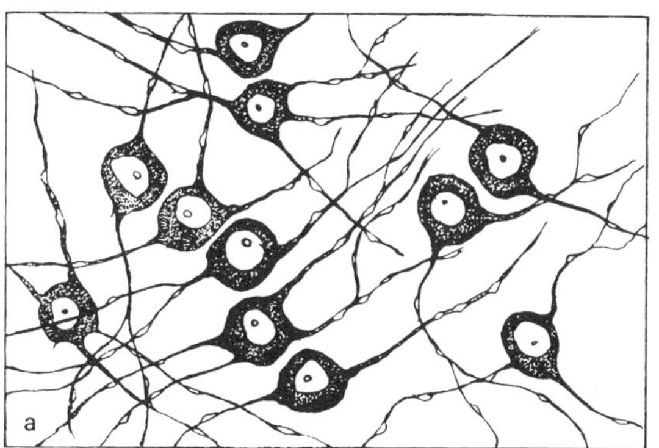

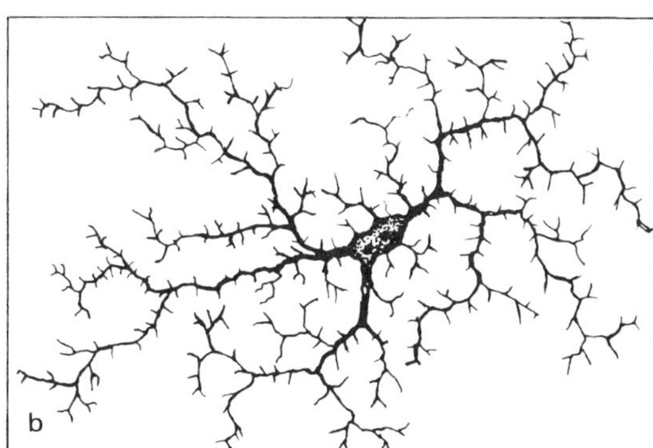

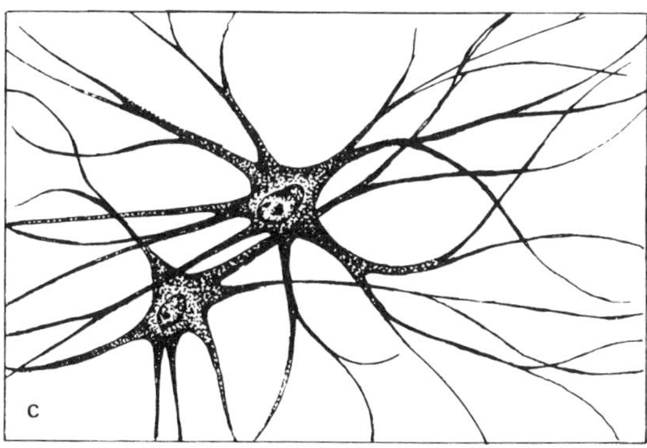

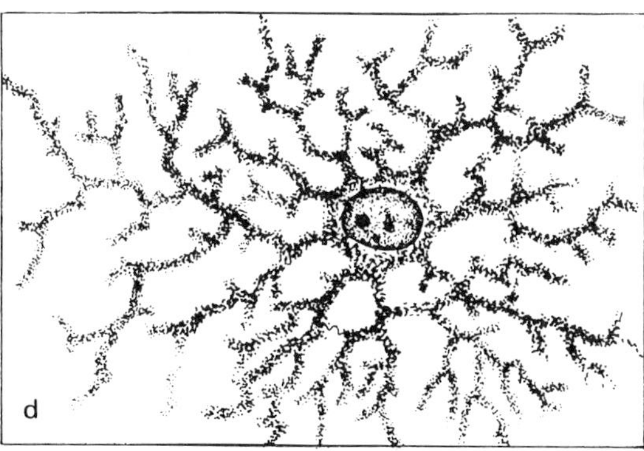

FIGURE 8-20 Schematic drawings of neuroglial cells of the central nervous system (*a*, oligodendroglia; *b*, microglia; *c*, fibrous astrocyte; *d*, protoplasmic astrocyte).

barriers between blood vessels and certain critical organs are present elsewhere in the body.

SENSORY ENDINGS

The afferent elements of the peripheral nervous system collect sensory information from various types of *sensory endings* and conduct this information to the CNS. Several types of sensory endings are each specialized for reception of a particular sensation.

Free Nerve Endings. These structures located throughout epithelia are unmyelinated and highly branching fibers. They are involved in touch and pain reception. The diagram in Fig. 8-5 illustrates such free nerve endings amongst the stratified squamous epithelium of the skin. In recent years electron microscopy has demonstrated the mode of penetration of free nerve fibers into stratified squamous epithelium where they terminate. Figure 8-22 shows an unmyelinated nerve as it approaches the basal surface of the epithelium where the basal lamina has fused with the glycoprotein coat of the investing Schwann cell surface (arrows, Fig. 8-22). Thus, the nerve fiber enters the *intraepithelial space* without being exposed to the connective tissue space. This arrangement suggests that the protection of nerve fibers reflected in the blood-brain barrier continues throughout the peripheral nervous system, even to the finest sensory fibers innervating stratified squamous epithelium. Figure 8-23 shows the profile of a nerve fiber close to its free ending in the granulosum layer of stratified squamous epithelium.

Meissner's Corpuscles. These structures lie just beneath the epidermis in the papillae of

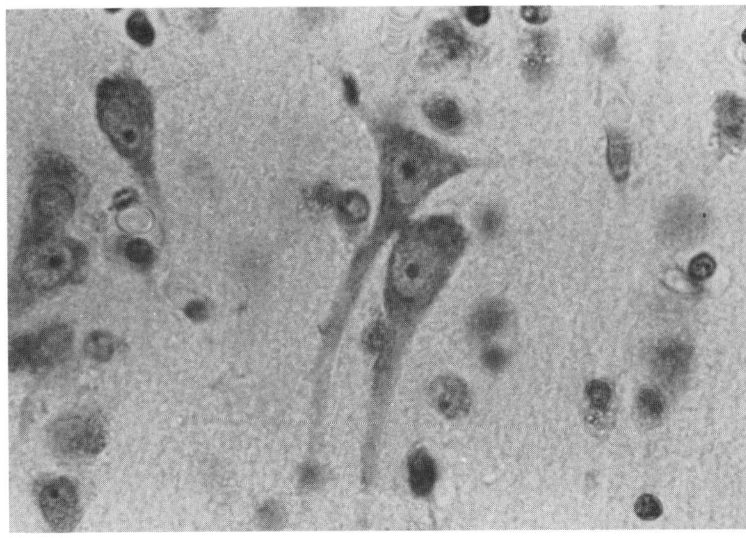

FIGURE 8-21 Light micrograph of motor neurons in the gray matter of the brain surrounded by glial cells.

dermis (Fig. 8-24). They are ovoid bodies composed of flattened cells. Myelinated nerves enter the Meissner's corpuscle, losing their myelinization as they enter and forming a complex spiral before they terminate. The Meissner's corpuscle responds to delicate tactile stimuli. Note the nerve fiber (arrows, Fig. 8-24) and nuclei of Schwann cells and fibroblasts as they approach the corpuscle.

Pacinian Corpuscles. These structures are found deep within the subcutaneous con-

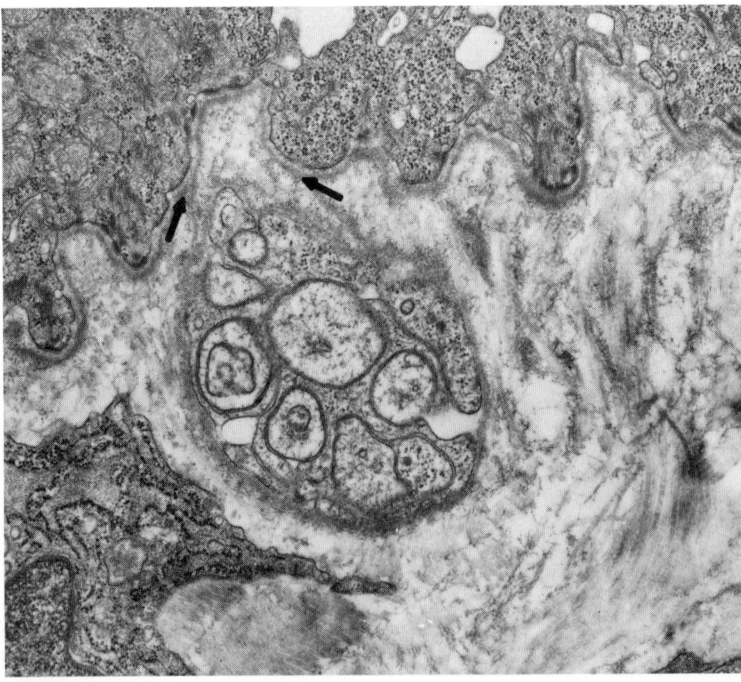

FIGURE 8-22 Electron micrograph of the basal cell layer of palatal epithelium. The unmyelinated nerve is approaching the basal layer. The basal lamina is fusing with the basal lamina of the epithelium (arrows).

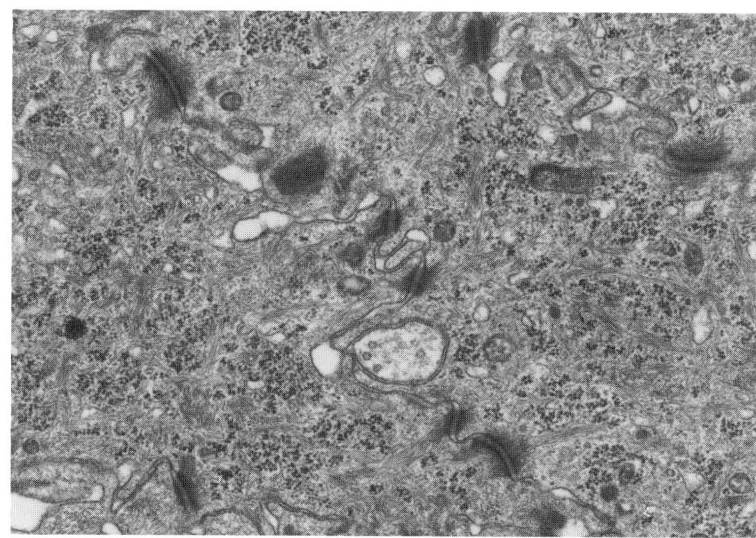

FIGURE 8-23 Electron micrograph of cells in stratum spinosum of palatal epithelium. The intraepithelial nerve ending is seen in the intercellular space. The nerve ending has less opaque cytoplasm than the epithelial cells and contains a few neurotubules.

nective tissue, in the mesentery, and in the vicinity of tendons. Figure 8-25 illustrates a pacinian corpuscle in LM. The onionlike lamellae of cells make up the bulbous sensory terminal. A single myelinated nerve loses its myelin as it enters and promptly terminates within the capsule. As indicated in Fig. 8-26, which shows a diagram of a pacinian corpuscle sectioned transversely and longitudinally, a nerve fiber terminates in the inner bulb. Pacinian corpuscles respond to pressure and gross tactile stimuli.

Muscle Spindles. As discussed in Chap. 6, these are elongated spindle-shaped structures in skeletal muscle containing several specialized fibers (intrafusal fibers), connective tissue, blood vessels, and unmyelinated nerve fibers. Muscle spindles are stretch receptors and hence are important in proprioception.

Golgi Tendon Organs. These structures are located within the dense regular connective tissue of tendons. Numerous branches of free

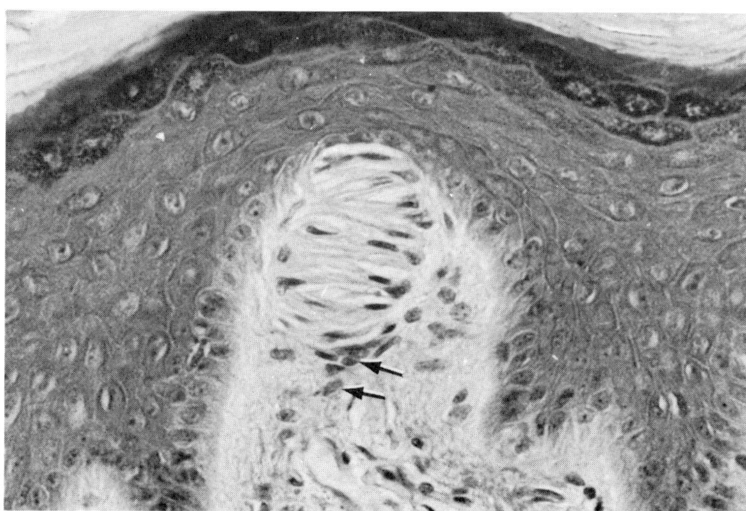

FIGURE 8-24 Light micrograph of skin. A Meissner's corpuscle is seen in the dermal papilla. Nerves enter the corpuscle (arrows).

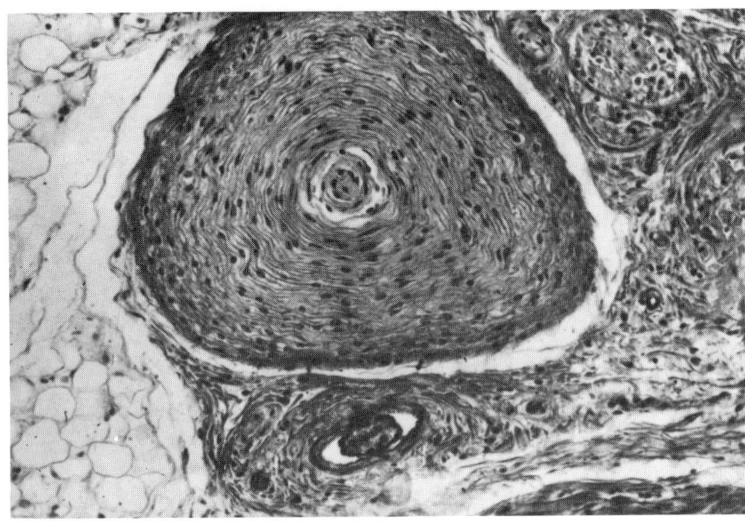

FIGURE 8-25 Light micrograph of cross section of pacinian corpuscle. This sensory ending is made up of onionskin–like lamellae.

nerve endings extend from their body, resulting in the treelike appearance. Goldi tendon organs are also stretch receptors and, like muscle spindles, are concerned with proprioception.

Kraus End Bulbs. These are similar in structure to pacinian corpuscles but are smaller and have fewer onionlike lamellae. The Kraus end bulbs function in temperature perception.

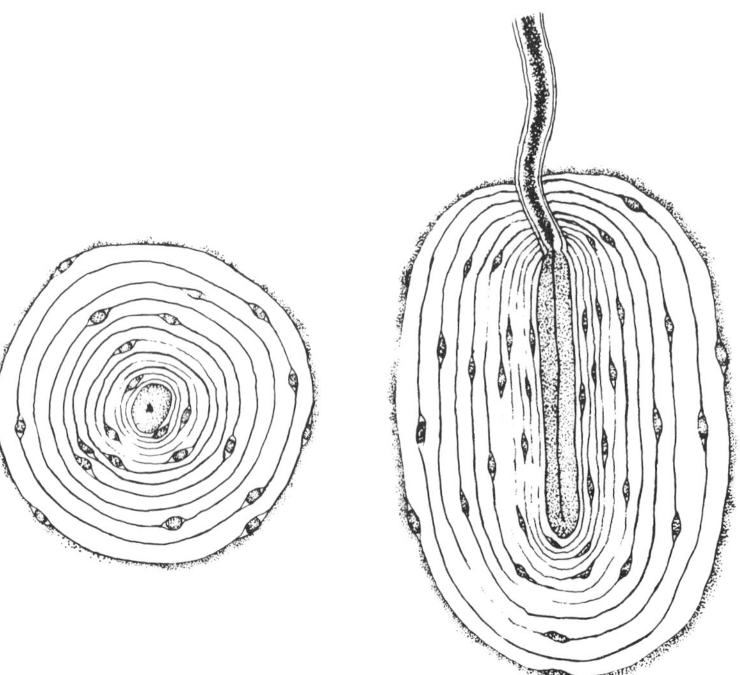

FIGURE 8-26 Diagram of cross-sectional and longitudinal views of a pacinian corpuscle. A single nerve terminates in each corpuscle.

REVIEW SECTION

1. Refer to Fig. 8-27. The perikaryon of a large neuron, as seen in the center of Fig. 8-27, contains many dark (basophilic) regions. These are called ________. The small dark-staining rounded body in the center of the cell is ________. Motor cell bodies such as this one are located in ________.

Some basophilic regions due to patches of RER and interspersed ribosomes are called *Nissl bodies*. Nucleoli in active neurons are quite large and dark-staining, while nuclei are light-staining. Many motor neuron cell bodies show features similar to the one above. Autonomic motor cell bodies are found in the brain and spinal cord and in various ganglia throughout the body, such as sympathetic ganglia. Motor cell bodies supplying the somatic system are located only in the brain and the spinal cord.

2. Refer to Fig. 8-28. The perikaryon of a sensory neuron cell is typically rounded in shape. Nissl bodies are much smaller than in motor neurons. Surrounding the sensory neurons are the numerous dark-staining nuclei of ________ cells. When numerous neuron cell bodies are found in a particular locale, performing a similar function, they are referred to as composing a structure known as a ________. Where would you expect to find sensory cell bodies?

Surrounding the sensory perikaryon are numerous satellite cells. They probably support neurons physically and sustain them metabolically. A group of neurons involved in a similar function are called a *ganglion* in the peripheral nervous system and a *nucleus* in the CNS.

3. Refer to Fig. 8-29. Sheaths of myelinated nerves often have a vacuolated or foamy appearance in a section, due to the dissolution of the lipid composing much of the sheath. What cell type makes up the myelin sheath? In this figure, several

The myelin sheath is made up of Schwann cells; the many nuclei throughout the field belong to Schwann cells, although some of the darker nuclei are those of fibroblasts. Indentations along the myelin sheath are called *nodes of Ranvier*.

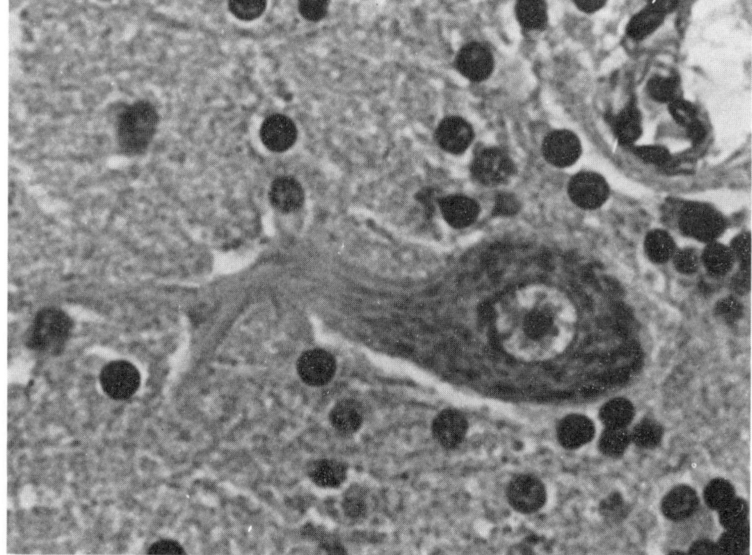
FIGURE 8-27

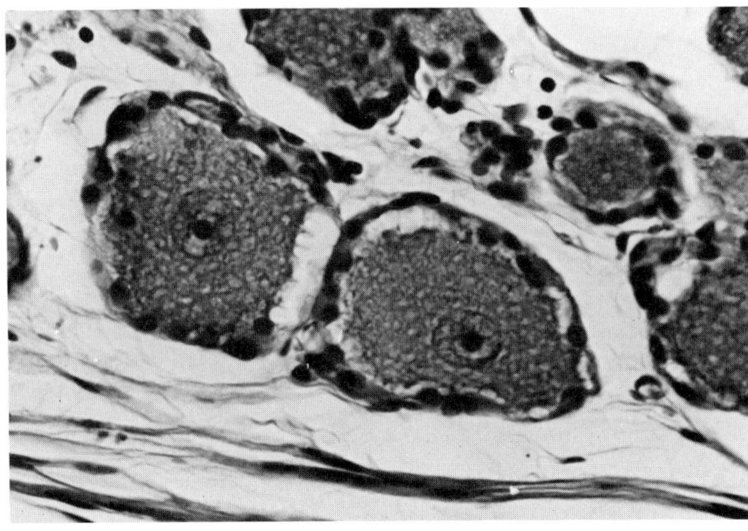

FIGURE 8-28

elongate nuclei are seen. These belong to __________ cells. Apparent indentations in the myelin sheath are called __________.

BIBLIOGRAPHY

Bloom, W., and D. W. Fawcett: *A Text Book of Histology,* 10th ed., Saunders, Philadelphia, 1975.

Davison, A. N., and A. Peters: *Myelination,* Charles C Thomas, Springfield, Ill., 1970.

Eccles, J. C.: *The Physiology of Nerve Cells,* Johns Hopkins, Baltimore, 1957.

———: *The Physiology of Synapses,* Academic Press, New York, 1964.

Peters, A., S. L. Palay, and H. F. de Webster: *The Fine Structure of the Nervous System: The Cells and Their Processes,* Harper & Row, New York, 1970.

Rhodin, J. A. G.: *Histology: A Text and Atlas,* Oxford University Press, New York, 1974.

FIGURE 8-29

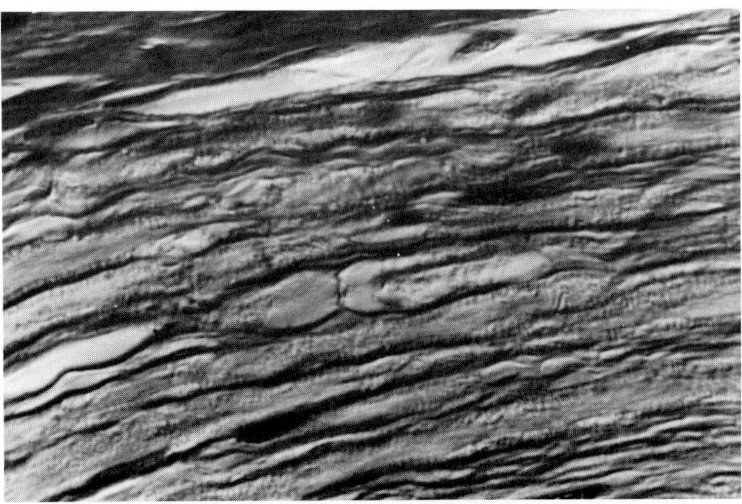

ADDITIONAL READINGS

Bodian, D.: "The Generalized Vertebrate Neuron," *Science* **137:**323–326 (1962).

Cauna, N.: "The Mode of Termination of the Sensory Nerves and Its Significance," *J. Comp. Neurol.* **113:**169–210 (1959).

——— and L. L. Ross: "The Fine Structure of Meissner's Touch Corpuscle, of Human Fingers," *J. Biophys. Biochem. Cytol.* **8:**467–482 (1960).

Davison, P. F., and A. Peters: *Myelination,* Charles C Thomas, Springfield, Ill., 1970.

de Robertis, E. D. P.: "Submicroscopic Morphology of the Synapse," *Int. Rev. Cytol.* **8:**61–96 (1959).

———: *Histophysiology of Synapses and Neurosecretion,* Pergamon, Oxford, 1964.

de Webster, H. F.: "The Geometry of Peripheral Myelin Sheaths during Their Formation and Growth in Rat Sciatic Nerves," *J. Cell Biol.* **48:**348–367 (1971).

Eames, R. A., and H. J. Gamble: "Schwann Cell Relationships in Normal Human Cutaneous Nerves," *J. Anat.* **106:**417–435 (1969).

Friede, R. L., and T. Samorajski: "The Clefts of Schmidt-Lantermann: A Quantitative Electron Microscopic Study of Their Structure in Developing and Adult Sciatic Nerve," *Anat. Rec.* **165:**89–102 (1969).

Glees, P.: *Neuroglia: Morphology and Function,* Charles C Thomas, Springfield, Ill., 1955.

Gray, E. G., and R. W. Gallery: "Synaptic Morphology in the Normal and Degenerating Nervous System," *Int. Rev. Cytol.* **19:**111–182 (1966).

Hyden, H. (ed.): *The Neuron,* Elsevier, Amsterdam, 1967.

Iggo, A., and A. R. Muir: "The Structure and Function of a Slowly Adapting Touch Corpuscle in Hairy Skin," *J. Physiol.* **200:**763–736 (1969).

Mentuzals, J.: "Ultrastructure of the Nodes of Ranvier and Their Surrounding Structures in the Central Nervous System," *Z. Zellforsch.* **65:**719–759 (1965).

Robertson, J. D.: "The Ultrastructure of Adult Vertebrate Peripheral Myelinated Nerve Fibers in Relation to Myelinogenesis," *J. Biophys. Biochem. Cytol.* **1:**271–278 (1955).

———: "The Ultrastructure of Schmidt-Lantermann Clefts and Related Shearing Defects of the Myelin Sheath," *J. Biophys. Biochem. Cytol.* **4:**39–46 (1958).

Uzman, B. G.: "The Spiral Configuration of Myelin Lamellae," *J. Ultrastruct. Res.* **2:**208–212 (1964).

Walker, R. B., and J. B. Kirkpatrick: "Neuronal Microtubular Neurofilaments and Microtubules," *Int. Rev. Cytol.* **33:**45–75 (1972).

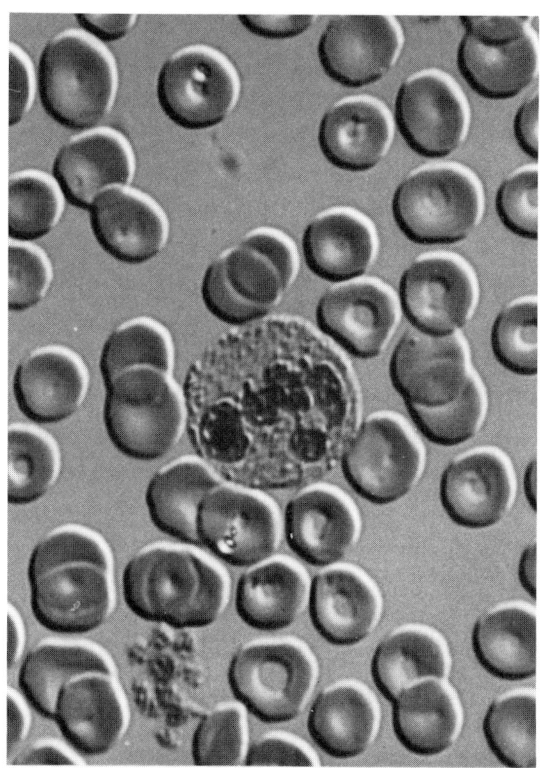

9 BLOOD

OBJECTIVES

Upon completion of this chapter, the student will be able to:

1 Define blood by identifying its major components.

2 Describe the following formed elements of the blood:
(a) erythrocytes (b) leukocytes (lymphocyte, monocyte, neutrophil, eosinophil, basophil) (c) platelets

3 Identify the following (when applicable) for each of the above:
(a) percent of total leukocytes (b) size (c) life span (d) morphology (e) functions

4 Differentiate between T cells and B cells by identifying the following for each:
(a) origin (b) life span (c) function

5 List four morphological characteristics used to differentiate between the various types of leukocytes.

6 Identify the staining reactions and contents of the granules found in granulocytes.

Blood is one chapter in which the use of your slides stained for different white blood cells will significantly add to what you will learn. For this reason you are encouraged to study the blood smear slides available in the student loan collection.

Blood consists of corpuscular elements suspended in a fluid called *plasma.* Derived from mesenchyme, this unique type of connective tissue circulates throughout the body within an enclosed system called the *cardiovascular system.* It transports oxygen and nutrients to and waste products from every cell, and provides corpuscular elements at the sites where they are needed. The blood moves mechanically through the vessels with the aid of the cardiovascular system, composed of a centrally located pump (heart) and a series of tubes (blood vessels). The blood maintains a dynamic equilibrium with other body fluids. Every time blood circulates, a good portion of the plasma leaves the vascular lumen to become tissue fluid and vice versa. This constant exchange is controlled by many physiologic factors.

CHEMICAL CONSTITUENTS OF PLASMA (SERUM) — TABLE 9-1

	CONSTITUENT	AMOUNT
Proteins	Albumin	4.5–5.5 g/100 mL
	Globulin	1.5–3.0 g/100 mL
	Fibrinogen	150–300 g/100 mL
Ions	Sodium	138–145 meq/L
	Potassium	4.0–5.0 meq/L
	Calcium	9.0–11.0 mg/100 mL
	Chloride	100–106 meq/L
	Bicarbonate	26–28 meq/L
	Phosphate	Small amount
Nutrients	Glucose	70–100 mg/100 mL
	Cholesterol	150–280 mg/100 mL
	Total lipids	470–750 mg/100 mL
Wastes	Urea nitrogen	8–20 mg/100 mL
	Uric acid	3.5–6.0 mg/100 mL
Gases	CO_2 content	26–28 meq/L
	O_2 content	15–23 vol %

PLASMA

Plasma, a homogenous straw-colored fluid, contains a variety of chemical substances whose concentrations remain remarkably constant, while they vary in tissue fluids. Basically, plasma is composed of water, with different ions including sodium, potassium, calcium, bicarbonate, chloride, and phosphate, and different types of proteins. These constituents maintain the constant, slightly alkaline pH (7.35 to 7.45) and the osmotic pressure of plasma. The proteins include albumin, globulins, and several clotting factors such as fibrinogen. When the clotting factors are activated, a fibrin mesh forms, entrapping the blood elements and leaving a clear yellowish fluid called *serum.* Since nutrition is one of its fundamental functions, plasma also carries food products absorbed from the gastrointestinal tract to the tissues, while it collects the cellular waste products. *Chylomicrons,* found in plasma, are especially evident after digestion of a fatty meal. Even oxygen, although it is primarily carried in the red blood corpuscles (RBCs), dissolves first in the plasma before reaching the RBCs. Secretory products of the various endocrine glands are also transported by plasma. Table 9-1 lists pertinent plasma constituents, along with quantitative data, found in an average healthy person.

ELEMENTS

The free-floating structural components of blood are not all true cells because some lack a nucleus. Since this is true, they are called *formed elements.* Their formation, covered in the next chapter, takes place in the reticular connective tissue of the blood-forming organs, and when they enter the blood, they are in a fully complete, functional form. The formed elements of the blood include the

platelets

FIGURE 9-1 Drawing depicting the formed elements of the blood.

erythrocytes (RBCs), white blood cells (WBCs or leukocytes), and platelets (thrombocytes) (Fig. 9-1). Most numerous of the elements, the RBCs, perform their oxygen-carrier function entirely within the bloodstream. The leukocytes, the only true cells in the blood, make up a much smaller proportion of the elements. Their chief function of defense is usually performed in the extravascular space of the loose connective tissue. The blood only provides for their transport. Blood platelets are tiny nonnucleated pieces of cytoplasm that, among other functions, contribute to blood clotting. The relative size and amount of the formed elements of the blood are summarized in Table 9-2.

Study of Blood. Blood elements may be studied by two methods: by counting and by observing their microscopic structure. The relative quantity of RBCs in plasma, called the *hematocrit*, is determined by centrifuging blood in a graduated tube. In the average person, RBCs make up almost half of the volume, as illustrated in Fig. 9-2. The total number of RBCs, WBCs, or platelets per cubic millimeter of blood can be counted under a microscope, using a hemocytometer or an electronic counter.

The structure of blood elements is commonly observed under a light microscope by staining a thin film of blood spread on a glass slide with a number of different dyes. *Wright's stain,* which contains eosin and methylene blue, is most often used. By combining microscopic observations with cell type count, the relative percentages of the several different kinds of WBCs can be determined. Such a determination is called *differential counting.*

Subcellular structures can be observed with greater detail in an electron microscope. This has allowed an improved understanding of the function of blood cells in physiology and pathology.

FORMED ELEMENTS OF BLOOD

TABLE **9-2**

ELEMENT	NUMBER, mm^3
RBCs	5.0×10^6 (♂) 4.5×10^6 (♀)
WBCs	$7.0 \sim 8.0 \times 10^3$
Platelets	3.0×10^5

RED BLOOD CORPUSCLES

Number. As mentioned earlier, RBCs are the most numerous formed elements in the blood; about 5 million are found in each cubic millimeter. Normal fluctuations in number are caused by such factors as sex, age, diet, exercise, barometric pressure, and variations in fluid intake. A delicately balanced correlation exists between their formation and destruction; their life span in humans is only 120 days. Destruction is accomplished chiefly in the spleen, by reticuloendothelial cells, which engulf the whole RBC or fragments of it. After digestion by the phagocytes, the hydrolyzed products are then reused by newly forming RBCs in the

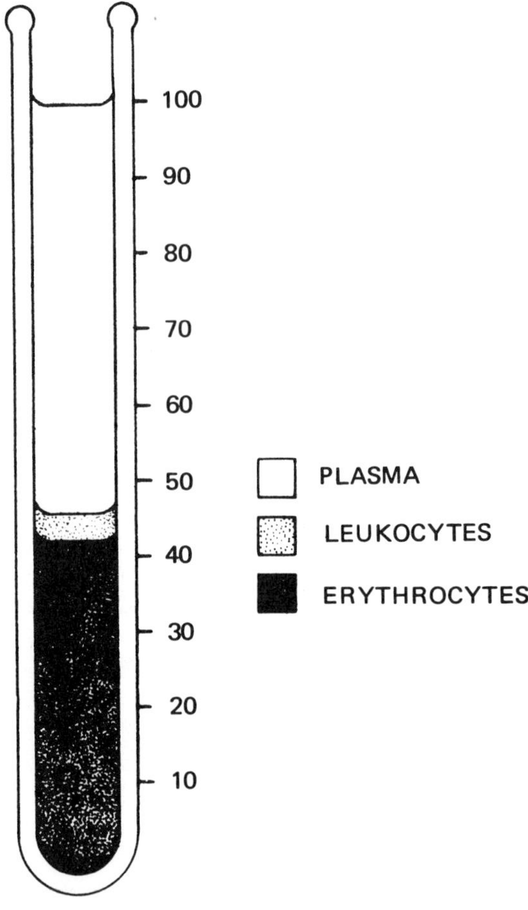

FIGURE 9-2 Schematic drawing of the components of blood as seen in a test tube after centrifugation.

bone marrow. Thus, much of the proteinaceous hydrolysate and, more importantly, iron are preserved within the body despite regular replacement of RBCs. Mature RBCs formed in the bone marrow are continuously being released, maintaining a constant number in the circulating blood.

Morphology. The RBCs are nonnucleated, membrane-bound structures about 7 to 8 μm in diameter. They have a biconcave discoidal shape which allows for great flexibility, enabling them to squeeze through the narrowest of capillaries. Their biconcave disk shape (see Fig. 9-3) provides a large surface area efficient for gaseous exchange. In an average man the total surface area of erythrocytes is about 3820 m^2, or 2000 times greater than the total body surface.

ERYTHROCYTES

Under the light microscope (Fig. 9-4), RBCs appear round and have a reddish, homogenous cytoplasm with a slight central pallor due to the biconcave shape. The biconcave nature of RBCs is well shown in this LM, which was taken with an interference contrast microscope.

Figure 9-5 is a scanning EM of RBCs in which their surface morphology is clearly demonstrated. Notice the uniformity of the RBCs in size and shape. The size, which varies only slightly from cell to cell, can be used as an index to compare the size of the other formed elements of blood and other histologic structures throughout the body. Variations in size, color, shape, and number of red corpuscles may indicate abnormal processes such as anemias occurring in the body.

In the EM (shown in Fig. 9-6), the RBC cytoplasm appears as an amorphous, dense material devoid of organelles, surrounded by a plasma membrane similar in structure to that of other cells. The flexion of the RBC indicates its elastic nature. Recent studies have shown that the RBCs have microtubule skeletons present around the periphery of

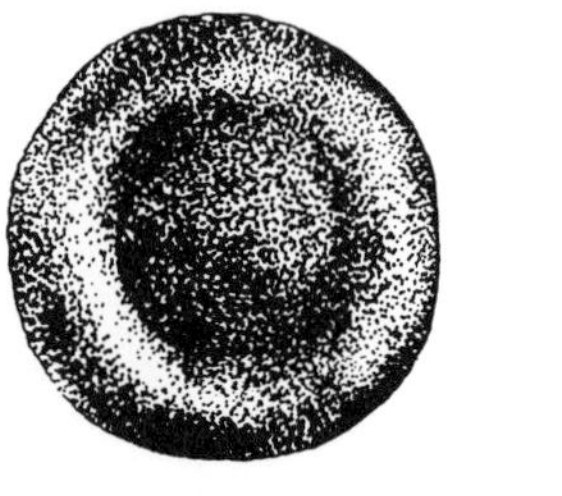

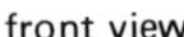

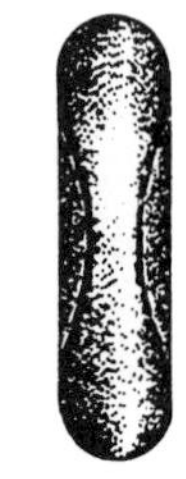

FIGURE 9-3 Drawing of RBC in *en face* and side views.

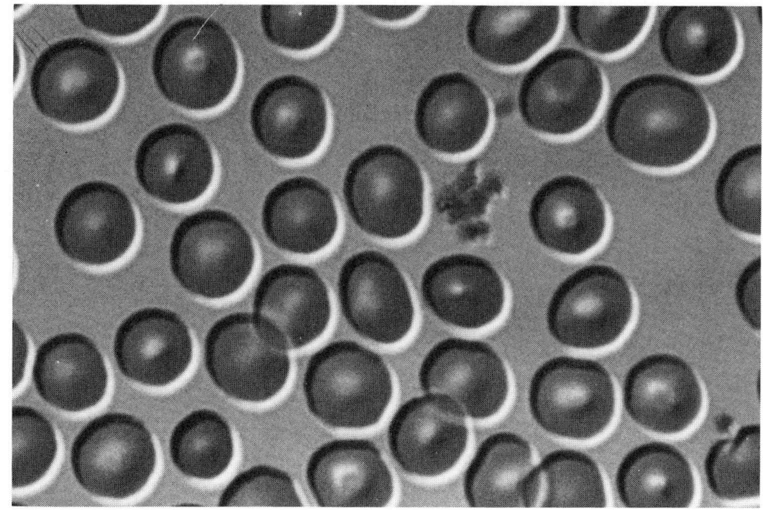

FIGURE 9-4 Light micrograph of peripheral blood smear. Numerous RBCs are seen. Among them is a cluster of platelets.

the doughnut-shaped disk. As was mentioned earlier, microtubules function as a cytoskeleton as well as contributing to the motility of certain cytoplasmic elements.

Function. Body cells require a continuous supply of oxygen which the lungs deliver by means of the circulatory system. Since oxygen is not dissolved to any great extent in plasma, an oxygen-binding protein called *hemoglobin* has evolved in higher forms of life. Hemoglobin, confined to the RBCs in higher vertebrates, consists of a protein moiety, *globin*, joined to an iron-containing pigment, *heme*. The red color in blood is caused by hemoglobin's chemical structure, which is shown in Fig. 9-7.

Hemoglobin has the unique characteristic of binding oxygen loosely, forming *oxyhemoglobin*. In areas like the lungs where the oxygen concentration is high, hemoglobin readily combines with large quantities of oxygen dissolved in the plasma. Oxyhemoglobin travels to the peripheral tissues

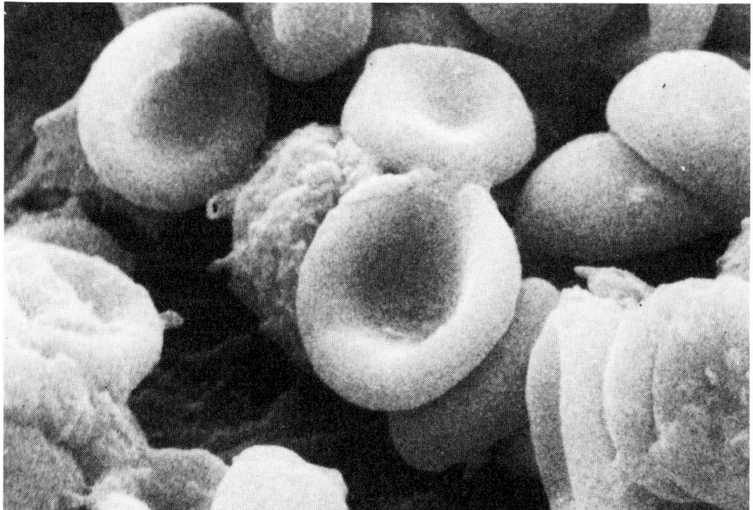

FIGURE 9-5 Scanning electron micrograph of RBCs. Their biconcave shape is clearly visible.

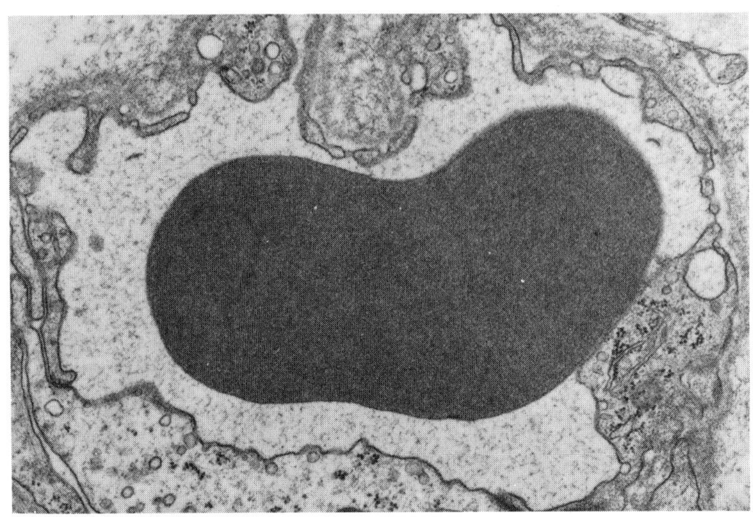

FIGURE 9-6 Transmission electron micrograph of an RBC. The RBC has a dense amorphous cytoplasm which is devoid of organelles.

where the oxygen is used for metabolic processes. With the resultant lowering of the oxygen tension, oxygen is released from the hemoglobin. The deoxygenated hemoglobin is called *reduced hemoglobin*. Hemoglobin can retrieve some of the carbon dioxide released by the cells. However, CO_2 is carried in plasma in the form of bicarbonate. Although hemoglobin makes up only about one-third of the chemical composition of the RBCs, it transports 100 times as much oxygen as the plasma alone.

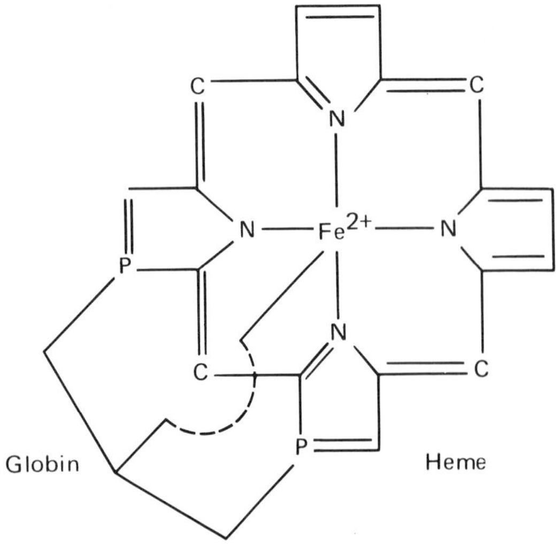

FIGURE 9-7 The structural formula of hemoglobin.

RETICULOCYTES

The reticulocytes (1 percent) are larger than the average RBC and have a bluish hue to their cytoplasm. When stained with special vital dyes, a basophilic structure with a reticulated appearance is visible. Their color is due to the clumping of the basophilic material of the cell. Known to be actively engaged in hemoglobin synthesis, the basophilic materials of the cytoplasm represent free ribosomes responsible for such synthetic activities. Figure 9-8 shows a portion of a reticulocyte seen in an electron microscope. Note the presence of ribosomes, as well as some vesicles and a few mitochondria (arrows). Therefore, reticulocytes are the youngest RBCs in the circulating blood. Because reticulocytes have just been released by the bone marrow, their number in peripheral blood is an important indicator of abnormal erythropoiesis, or RBC production.

WHITE BLOOD CELLS

Number. White blood cells are much less numerous in the circulating blood than RBCs, about 7000 to 8000 per cubic millimeter. Al-

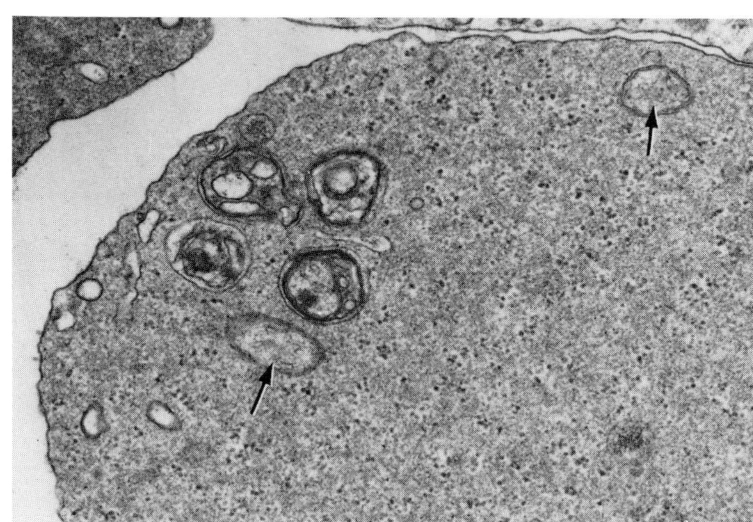

FIGURE 9-8 Electron micrograph of a portion of a reticulocyte. The cytoplasm contains ribosomes, vesicles and some mitochondria (arrows).

though certain of the white blood cells have a long life span, most live only several days to a few weeks. They must be constantly replenished by new cells from the blood-forming organs. White blood cells enter and leave the bloodstream, where they may discharge some of their functions. Those that leave the blood are found in large numbers in connective tissues; there they perform a variety of defensive functions. For instance, large numbers of WBCs mobilize to loci when a foreign matter enters into the body. The number of WBCs in local tissues as well as in peripheral blood provides an important clue to the pathologic conditions in many diseases.

General Morphology and Function. Colorless until stained, leukocytes form the white or "buffy" layer on top of the RBCs when placed in a test tube. Possessing a nucleus, they are true cells. They can perform an ameboid movement in passing through capillary walls and traveling within the connective tissues. Based on the presence or absence of granules in their cytoplasm, the WBCs can be divided into two groups: granular and agranular. A total of five different white blood cell types in circulating blood can be distinguished microscopically. *Agranular, mononuclear* cells include lymphocytes and monocytes. These cells actually do contain small irregular granules that stain blue with aniline dyes and are therefore called *azurophilic granules.* These are considered nonspecific granules because they are present in several unrelated cell types. The *granular* leukocytes of blood are neutrophils, eosinophils, and basophils, so named because of the presence of granules with particular staining characteristics.

The relative percentages of WBCs, as shown in Fig. 9-9, can be determined by counting the number of different cell types on a stained blood film based on the following morphologic characteristics: (1) cell size; (2) the size, shape, density, and pattern of chromatin in the nucleus; (3) the color and relative amount of cytoplasm; and (4) the size, number, and staining of granules. Variations in number or morphology of WBCs can be used in differential diagnoses of several categories of disease.

AGRANULOCYTES

The agranular WBCs, often called *agranulocytes,* are comparatively undifferentiated, since they are able to reproduce by mitosis under certain conditions. They are capable of transforming into different cell types. They

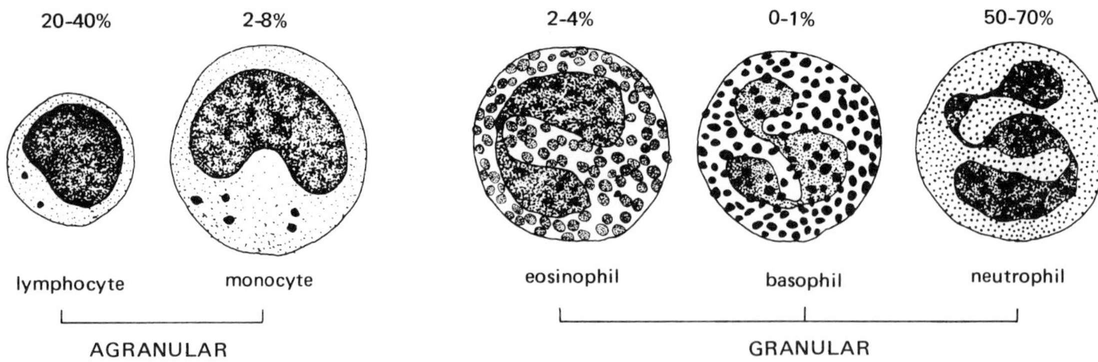

FIGURE 9-9 Diagram illustrating the relative percentages of the various white blood cells in peripheral blood.

are also called *mononuclear leukocytes* because they have a single nucleus in contrast to the segmentation or lobation of nuclei seen in most granular leukocytes.

LYMPHOCYTES

About 20 to 40 percent of the WBCs in blood are lymphocytes. They are small cells, usually slightly larger than RBCs. Most are about 9 to 10 μm in diameter, although some may be as large as 20 μm. In the typical small lymphocyte (Fig. 9-10), the round nucleus almost fills the entire cytoplasm, thus having a high nucleocytoplasmic ratio. One side of the nucleus has an indentation. The nucleoplasm is dense and patchy in appearance, indicating that the nucleus must be fairly quiescent. There is a scant amount of clear to bluish cytoplasm which shows a number of small azurophilic granules. Figure 9-11 is an EM of a small lymphocyte. The cytoplasm contains a few mitochondria, a small number of free ribosomes, vesicles, and a small Golgi complex. In the EM in Fig. 9-12, a portion of a small lymphocyte is shown at a higher magnification. Note the presence of a few tubular profiles of RER in addition to the structures mentioned above.

The condensed nuclear chromatin and the simple cytoplasmic structure at one time caused the lymphocyte to be considered a *trephocyte*, a cellular nutritional packet which

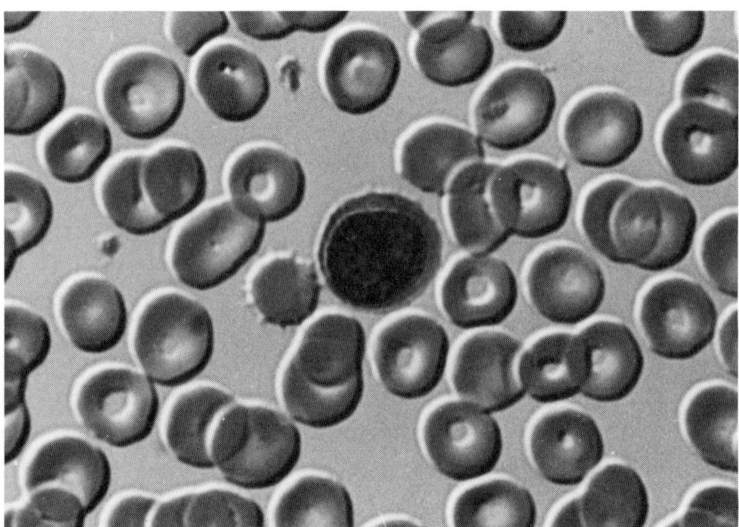

FIGURE 9-10 Light micrograph of peripheral blood smear. A small lymphocyte is present among the RBC. Its nucleoplasm is dense because of a large amount of heterochromatin. The cytoplasm surrounding the nucleus is sparse.

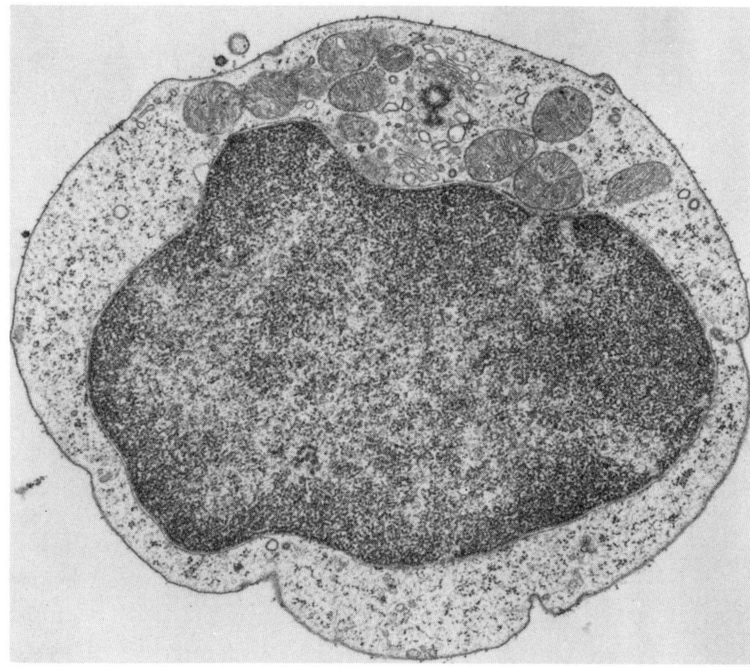

FIGURE 9-11 Electron micrograph of small lymphocyte. The nucleus has a condensed chromatin pattern. The cytoplasm contains a small Golgi complex, centrioles, and a few mitochondria in the region of the nuclear indentation. Sparse tubular profiles of RER are scattered throughout the cytoplasm.

carried nucleic acids to needed areas, such as in wound repair or inflammation. Because of a great deal of information recently gained, lymphocyte function is now considered to be responsible for most if not all types of immune phenomena. A particular lymphocyte varies in its contribution to immune phenomena according to its origin and previous life history, so that a large number of subpopulations are being characterized with respect to these different functions. Two major subpopulations are recognized: T-cell precursors move from bone marrow to the thymus early in development, differentiate, and

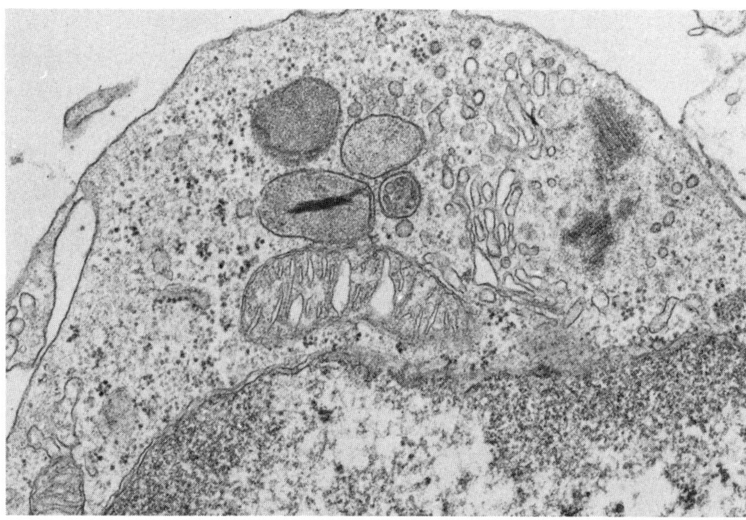

FIGURE 9-12 Electron micrograph of a portion of a small lymphocyte. A pair of centrioles, a small Golgi complex, a few mitochondria and a few azurophilic granules are present in the cytoplasm. Sparse tubular profiles of RER are also apparent.

start producing T lymphocytes that are long-lived; they may survive for years and possibly for the entire lifetime of the individual. Thus, they are thought to be carriers of immunologic memory. Capable of registering an initial contact with foreign matter (antigen), they can facilitate a quicker response when the same antigen is introduced for a second time. Therefore, the T cells are primarily responsible for cell-mediated immunity, in which they recognize and attack the foreign substance in a localized reaction. One of the best examples is the phenomenon of a transplantation rejection.

B cells, which also originate from the bone marrow, are short-lived and survive only a few weeks. With proper helper functions provided by other lymphocytes and macrophages and on contact with antigen, they are capable of transforming into plasmablasts and plasma cells, which produce circulating antibodies. Thus, B cells are responsible for humoral immunity. Recent evidence indicates that the interaction between T- and B-cell populations and their subpopulations may be much more complex than previously thought.

MONOCYTES

Measuring 12 to 20 μm in diameter, monocytes are larger than lymphocytes but fewer in number. They compose 2 to 8 percent of the total WBCs. The cytoplasm is more voluminous than that of lymphocytes and the nucleus is large and usually horseshoe-shaped (Fig. 9-13). The fine delicate pattern of the chromatin gives it a lacy appearance. The abundant grayish blue cytoplasm contains occasional vacuoles and a number of small azurophilic granules, which are better visualized in this figure taken under phase optics. As in lymphocytes, a Golgi complex is seen near the nuclear indentation in EM surrounded by azurophilic granules and mitochondria. A small Golgi complex of a monocyte is shown at high magnification in Fig. 9-14. The electron-dense granules seen in this EM (arrows) are the azurophilic granules presumed to be formed by the Golgi complex. A pair of centrioles are frequently seen in association with the Golgi complex. The cell has a number of small vesicles and free ribosomes but little RER and other organelles.

The EM in Fig. 9-15 compares the differences between lymphocytes (L) and a monocyte (M), in terms of both the volume and the organelle content of their cytoplasm. Table 9-3 lists the comparative feature mentioned

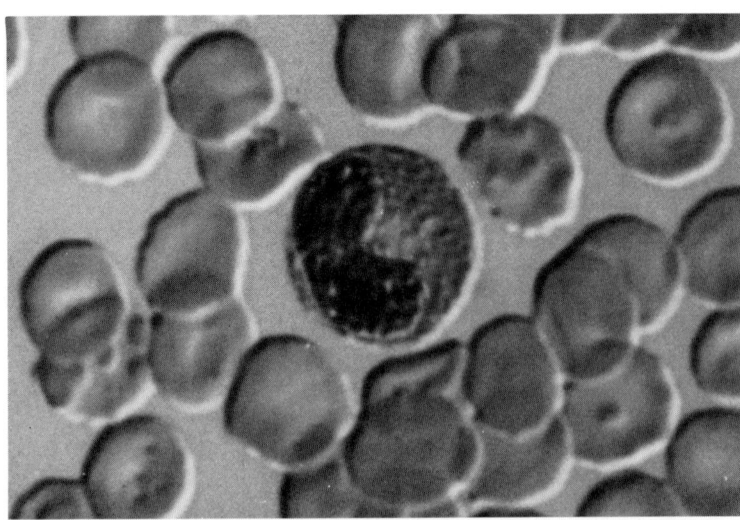

FIGURE 9-13 Light micrograph of peripheral blood smear. A monocyte is seen among the RBCs. It has a horseshoe-shaped nucleus and abundant cytoplasm.

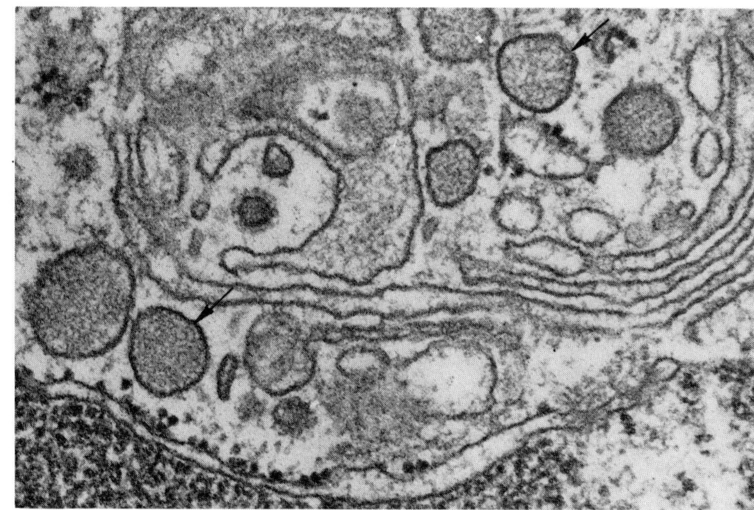

FIGURE 9-14 Electron micrograph of a portion of a monocyte near the nuclear indentation. A small Golgi complex with associated azurophilic granules (arrows) are seen.

above. Active monocytes continuously send out pseudopodia, enabling them to pass through the capillary walls by ameboid movement and enter the connective tissue space. Here they have been shown to transform into macrophages, which display phagocytic functions under certain conditions.

GRANULOCYTES

The granular leukocytes are also known as myeloid elements, since they develop in the bone marrow. They are characterized by the presence of typical granules in their cytoplasm. Through the staining reaction of the granules, the granular leukocytes are classified as neutrophils, eosinophils, and basophils on the basis of the granules. Granulocytes are further characterized by their many-lobed nuclei, consisting of segments connected by thin strands. In neutrophils, this is most pronounced, and they are often called *polymorphonuclear leukocytes* (PMN).

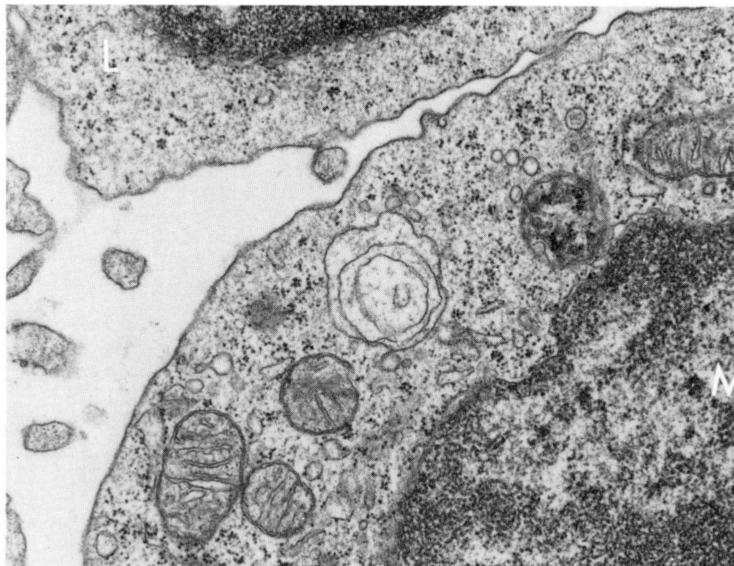

FIGURE 9-15 Electron micrograph illustrating differences in the cytoplasm between a monocyte (bottom) and a lymphocyte (top).

AGRANULOCYTES

TABLE 9-3

CHARACTERISTICS	LYMPHOCYTE	MONOCYTE
Size	8–10 μm	12–20 μm
Nucleus	Round	Indented
Chromatin	Dense	Lacy
Cytoplasm	Clear blue	Bluish gray
Granules	Azurophilic Few, distinct	Azurophilic Many, indistinct
Quantity	20–40%	2–8%

In contrast to the agranulocytes, the mature granulocytes are more highly differentiated and are unable to divide. The distinguishing features of the three classes of granular leukocytes are listed in Table 9-4.

NEUTROPHILS

Neutrophils are the most numerous of the leukocytes, making up 50 to 70 percent of the total WBCs. They are intermediate in size between the lymphocyte and the monocyte, being about 10 to 12 μm in diameter. Neutrophils have a multilobated nucleus with three or four nuclear segments connected by thin chromatin threads (Figs. 9-16 and 9-17). Those nuclear lobes increase in number with age. The pink cytoplasm is filled with fine granules (Fig. 9-16). However, they are more easily visualized under the phase microscope (Fig. 9-17) than in an H and E preparation. Neutrophils contain two types of granules: granular (specific) and azurophilic (nonspecific), (Fig. 9-18). They also have a small Golgi apparatus and a few profiles of RER (Fig. 9-18). A higher view of a portion of a neutrophil (Fig. 9-19) depicts details of the specific granule's structure. Nonspecific or azurophilic granules appear early in development and therefore are few in number in the mature neutrophil. The specific neutrophilic granules are formed later in development and are more numerous, particularly in fully mature neutrophils. In LM these granules are small and indistinct with a pinkish blue hue which gives the cell the name neutrophil. Most mature neutrophils are characteristically dense in routine preparations for EM, indicating that they contain a greater amount of proteins in their cytoplasm than other granulocytes. This is illustrated in Fig. 9-20, which shows a mature neutrophil next to a mature eosinophil, contrasting the difference in the density of the cytoplasm between these two cells. The difference in

GRANULOCYTES

TABLE 9-4

CHARACTERISTICS	NEUTROPHIL	EOSINOPHIL	BASOPHIL
Size	9–12 μm	10–14 μm	8–10 μm
Nucleus	Lobed	Lobed	Lobed
Chromatin	Condensed	Condensed	Condensed
Cytoplasm	Pink	Pink	Pink
Granules	Pinkish blue	Bright orange	Dark blue
Granules content	Hydrolases	Peroxidase, hydrolases	Histamine, heparin, serotonin
Quantity	50–70%	2–4%	0–1%

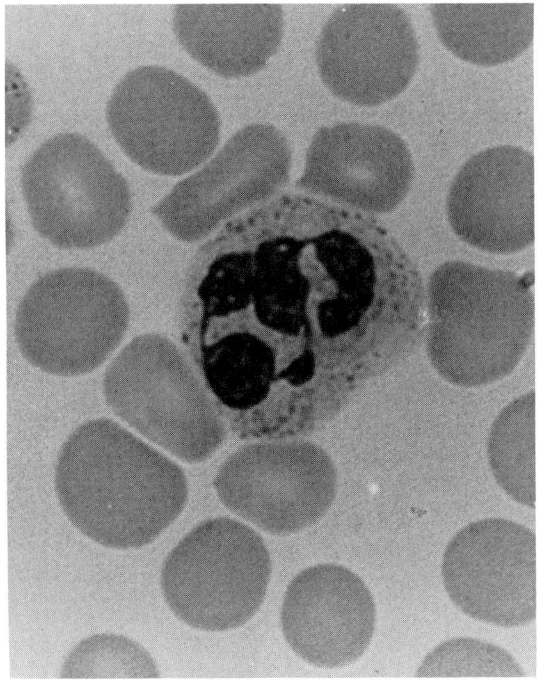

FIGURE 9-16 Light micrograph of peripheral blood smear. A neutrophil is present among the RBCs. Note its multilobated nucleus.

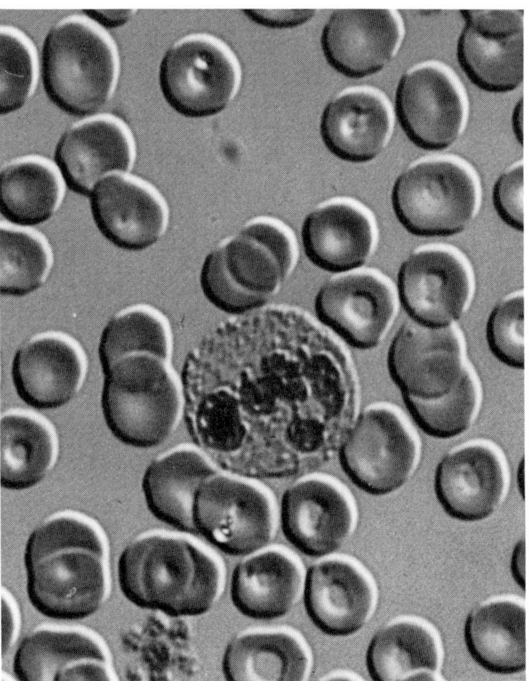

FIGURE 9-17 Phase contrast micrograph of peripheral blood smear. The neutrophilic leukocyte is more easily recognized.

electron density of the cytoplasm is also marked among cells of different maturity. In general, darker cells are smaller and contain fewer cytoplasmic organelles (Fig. 9-21).

Neutrophils, capable of recognizing and engulfing foreign materials, serve as the first line of defense against invading organisms. They participate in the acute inflammatory

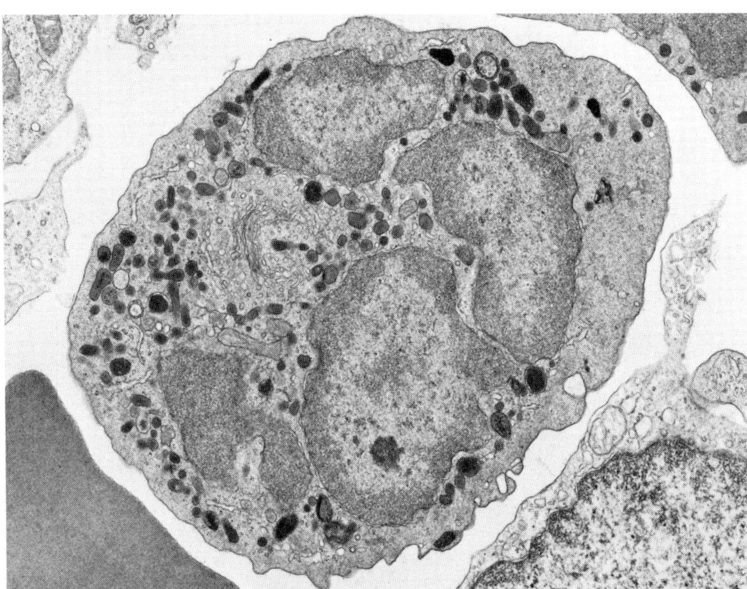

FIGURE 9-18 Electron micrograph of neutrophil. The nucleus appears multilobated. A neutrophil has a small Golgi apparatus and a few profiles of RER. The cytoplasm contains two types of granules, specific and azurophilic (nonspecific).

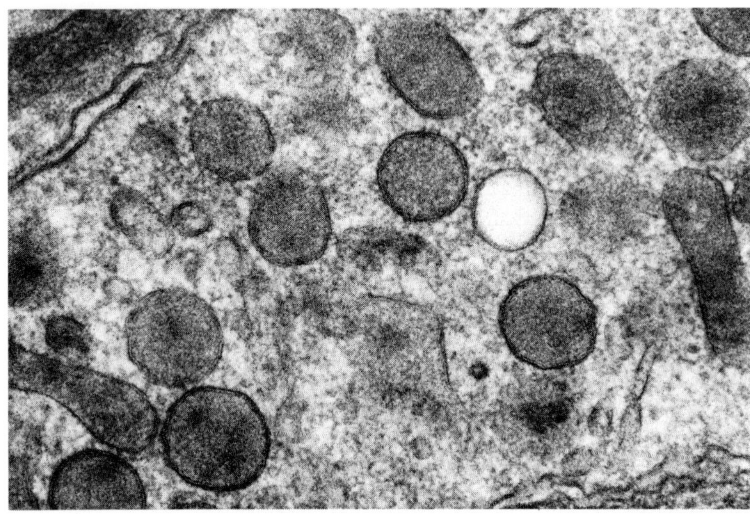

FIGURE 9-19 Electron micrograph of a portion of a neutrophil illustrating the structure of specific granules.

response and other infectious processes that require fast action. Neutrophils are the first cells to appear at the site of infection, migrating from blood vessels and through the connective tissue space by ameboid movement. They are attracted to the site by chemotaxis caused by the presence of bacteria or substances produced by the injured tissues.

The mechanism causing the removal of noxious material is attributed to two properties of the neutrophil. Its phagocytic ability enables the cell to engulf bacteria and other small foreign bodies. The cell is also able to digest the engulfed material. The cytoplasmic granules contain lysosomal hydrolases which, when joined with the phagocytic vacuoles, cause lysis of the ingested material. Neutrophils may also release their lysosomal enzymes into the extracellular environment, thereby degranulating themselves and initiating lysis of tissue components in the region. The cell itself may die during either of these processes, forming pus, which is largely a

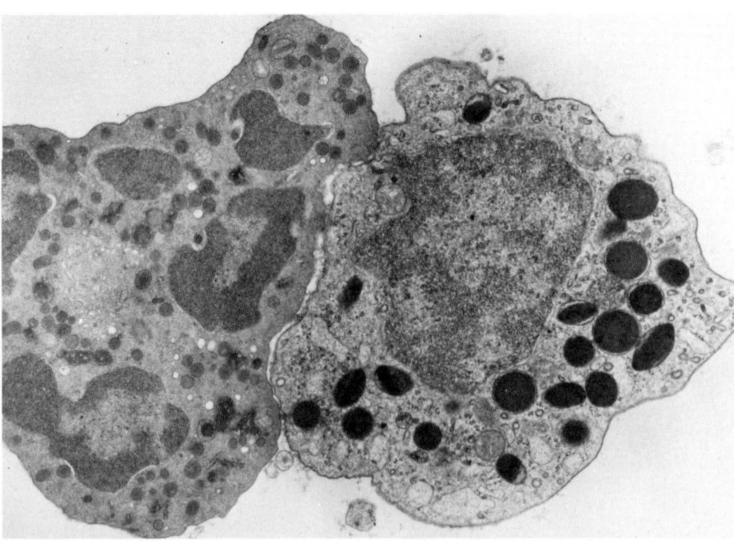

FIGURE 9-20 Electron micrograph illustrating differences in cytoplasmic density among mature leukocytes. A mature neutrophil (left) has a denser cytoplasm than a mature eosinophil (right).

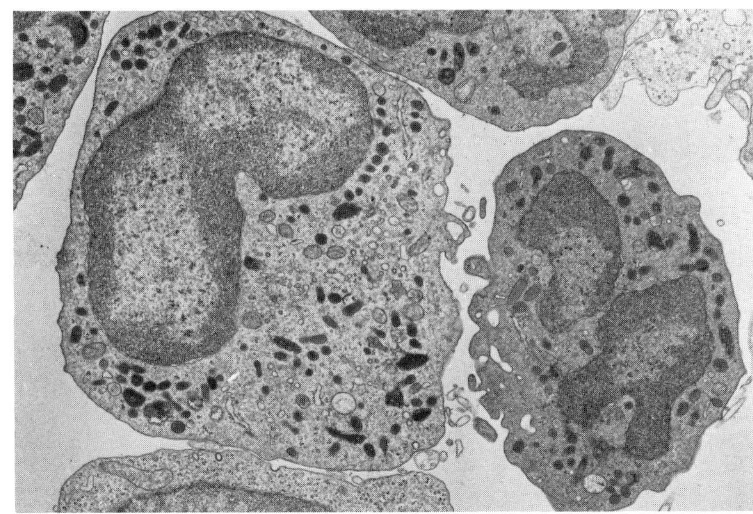

FIGURE 9-21 Electron micrograph illustrating variations in cytoplasmic density during differentiation. In general more mature cells are smaller and darker than less mature cells.

collection of dead neutrophils and bacteria. In summary, the neutrophil is capable of ameboid movement, phagocytosis, lysis of ingested material, and degranulation.

EOSINOPHILS

Approximately 24 percent of the WBCs are made up of eosinophilic leukocytes (eosinophils). Slightly larger than neutrophils, they usually contain a bilobed nucleus with condensed chromatin (Fig. 9-22). These cells are characterized by an abundance of large (0.51 μm in diameter), round cytoplasmic granules that are intensely eosinophilic. Biochemical studies show that the granules contain histamine and such enzymes as peroxidase and hydrolases. Eosinophils are also capable of ameboid movement, phagocytosis, and degranulation. They are thought to have a role in mediating certain cellular immune responses, as they are greatly increased under allergic conditions.

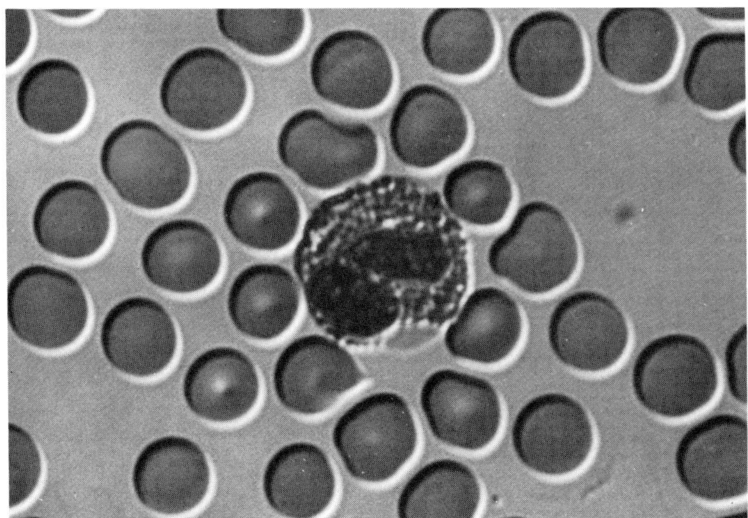

FIGURE 9-22 Light micrograph of peripheral blood smear. An eosinophil is present among the numerous RBCs.

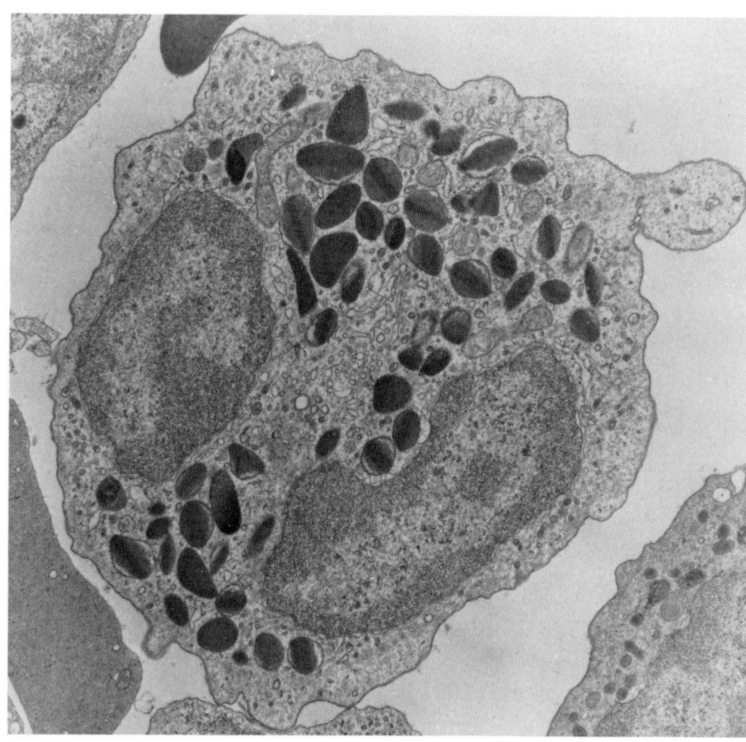

FIGURE 9-23 Electron micrograph of an eosinophil. The eosinophil has a bilobed nucleus, and the cytoplasm contains numerous granules, each having a single discoid crystal within. In addition, they have a small Golgi complex and sparse profiles of RER in their cytoplasm.

In rodent cells these granules have a single discoid crystal in an equatorial plane (Fig. 9-23). Human granules contain single or multiple crystals which are variable in shape. However, in all species, the crystals are embedded in a finely granular or amorphous matrix with relatively low electron density (Fig. 9-24). This characteristic makes it very easy to identify an eosinophil in an EM.

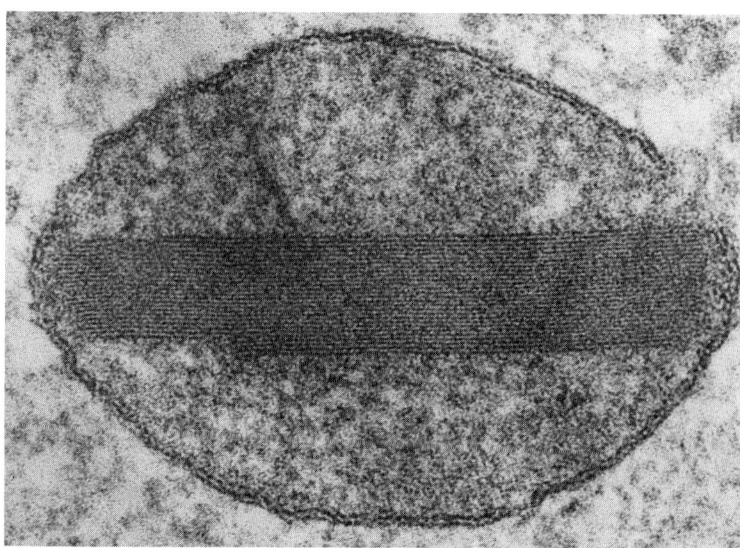

FIGURE 9-24 Electron micrograph of an eosinophilic granule. The granule is membrane bound and has an equatorial crystal which is embedded in a finely granular matrix.

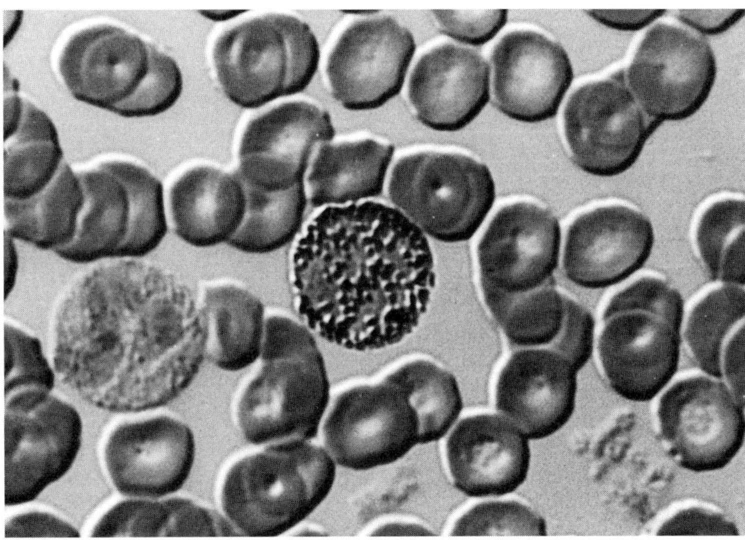

FIGURE 9-25 Light micrograph of peripheral blood smear. A basophil is seen among the numerous RBCs. It is filled with basophilic granules.

BASOPHILS

The basophilic leukocytes (basophils) make up less than 1 percent of the total WBCs. They are the smallest of the granulocytes, measuring about 8 to 10 μm in diameter. Their relatively large nucleus is not as obviously located as in the other granulocytes. It often presents an S shape and takes a pale stain (Fig. 9-25). The nucleus of the basophil is usually difficult to distinguish because it is covered by large, dark blue–staining granules. Basophilic granules are somewhat irregular and variable in shape and size. Although they take up the basic dye, the granules are water-soluble and may appear as empty vacuoles. In electron micrographs, the basophil is full of large, rounded granules with few other intracellular organelles (Fig. 9-26). As in the eosinophil, the ground cytoplasm of the basophil is less electron-dense

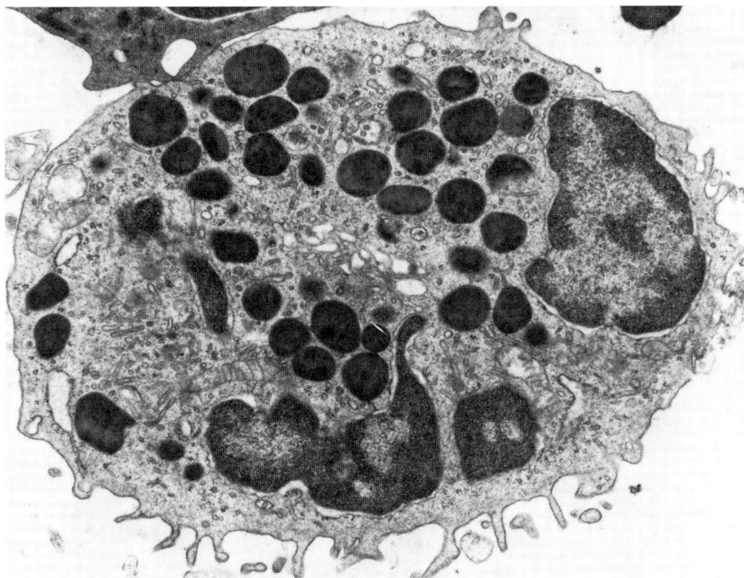

FIGURE 9-26 Electron micrograph of basophil. The cytoplasm contains numerous, round, membrane-bound granules. In general, cytoplasmic organelles are sparse.

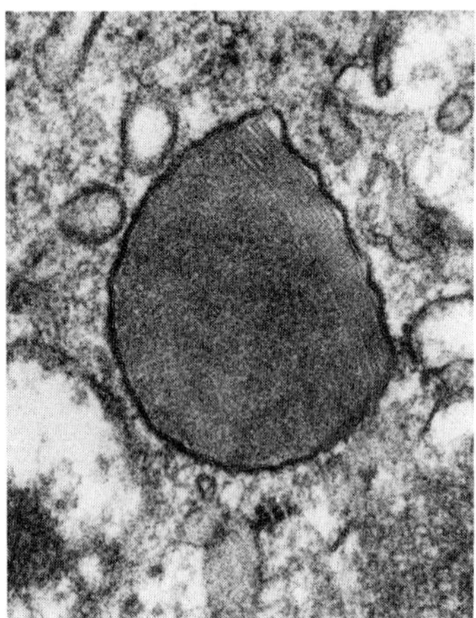

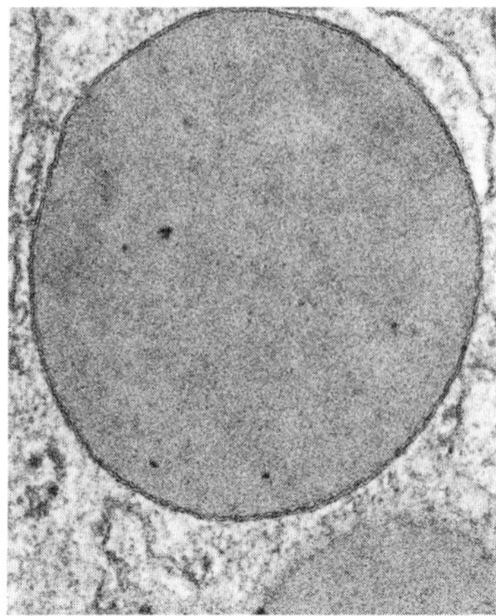

FIGURE 9-27 (*far left*) Electron micrograph of a basophilic granule. It contains a lamellar and crystalloid structure in addition to a more homogeneous substance.

FIGURE 9-28 (*near left*) Electron micrograph of a basophilic granule. This granule has a homogeneous structure.

than that of the neutrophil. The internal structure of a basophilic granule may contain a lamellar and crystalloid structure or homogeneously dense interior (Figs. 9-27 and 9-28). Basophilic granules contain histamine, a vasodilator, and heparin, an anticoagulant. Along with serotonin, these substances contribute to changes in blood vessel permeability as well as the physical nature of the connective tissue ground substance. Thus, basophils closely resemble mast cells of connective tissue. For this reason, mast cells, located along the capillary walls, are sometimes called *tissue basophils.*

PLATELETS

Platelets, also called *thrombocytes,* are numerous in blood; there are about 300,000 per cubic millimeter. As seen in the center of Fig. 9-29, they are small irregular bodies about 2 to 4 μm in diameter. Platelets contain no nuclei; they are made up of cytoplasm with a number of azurophilic granules. In a smear preparation, the granules are concentrated in the central area called the *granulomere,* leaving a pale-blue periphery called the *hyalomere.*

In EMs of platelets, dark, rounded to club-shaped granules, small mitochondria,

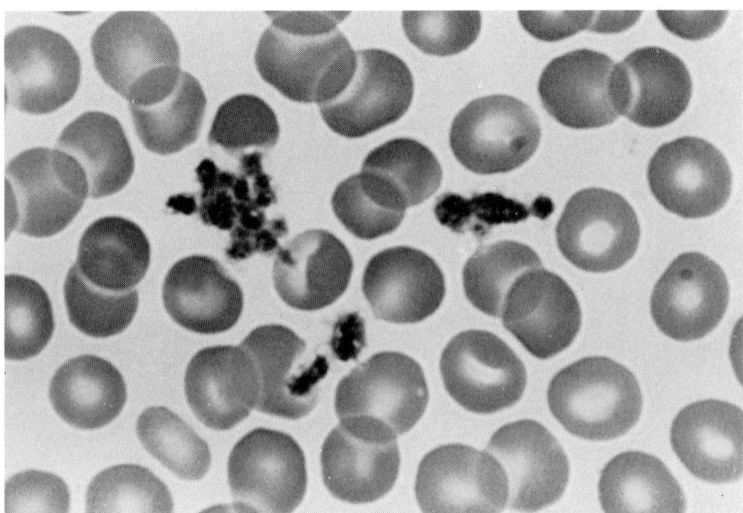

FIGURE 9-29 Light micrograph of peripheral blood smear. Several clusters of platelets are seen among the RBCs.

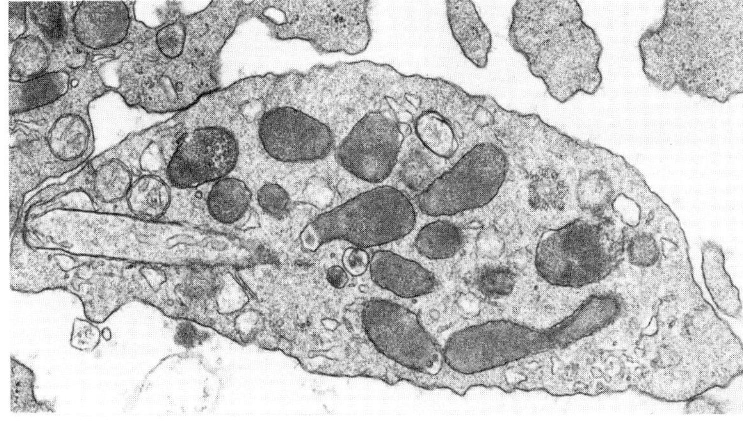

FIGURE 9-30 Electron micrograph of platelet. Platelets have granules which are rounded to club-shaped. They also contain mitochondria and a few vacuoles in the granulomere.

and occasional vacuoles make up the granulomere (Fig. 9-30). Vesicular elements of RER and free ribosomes may also be found in the granulomere, since platelets are small pieces of cytoplasm detached from the megakaryocyte, which is the largest cell in the bone marrow. The hyalomere is simply the exaggerated peripheral cytoplasm caused by stretching during smear preparation. Average survival time of platelets is about 5 to 9 days, so they are continually produced to maintain their remarkably constant number in the peripheral blood.

Platelet granules contain a number of biogenic amines and thromboplastin, which can initiate the coagulation of blood when released. Platelets are inherently adhesive and often aggregate spontaneously. They therefore agglutinate under certain physiologic conditions. Within a few seconds after injury of a small vessel, platelets start to adhere to and aggregate along damaged endothelial cells, tissue fibers, and each other. Their loose aggregates eventually fuse into a structureless mass that effectively stops loss of blood from the vessel. Fibrin strands, formed from plasma by a series of reactions initiated by thromboplastin, reinforce this hemostatic plug. Simultaneously, blood cells become caught in this agglutinated mass, resulting in the red color of a blood clot. This is the beginning of the repair process. When the aggregating platelets and fibrous material from the plasma contract, they close the ruptured vessel and eliminate the bleeding.

REVIEW SECTION

1. Refer back to the text as necessary and fill in the table below. Include the characteristic appearance of the nucleus, cytoplasm, and granules in the table.

	CELL SIZE	NUCLEUS	CYTOPLASM	GRANULES
Lymphocyte				
Monocyte				
Neutrophil				
Eosinophil				
Basophil				

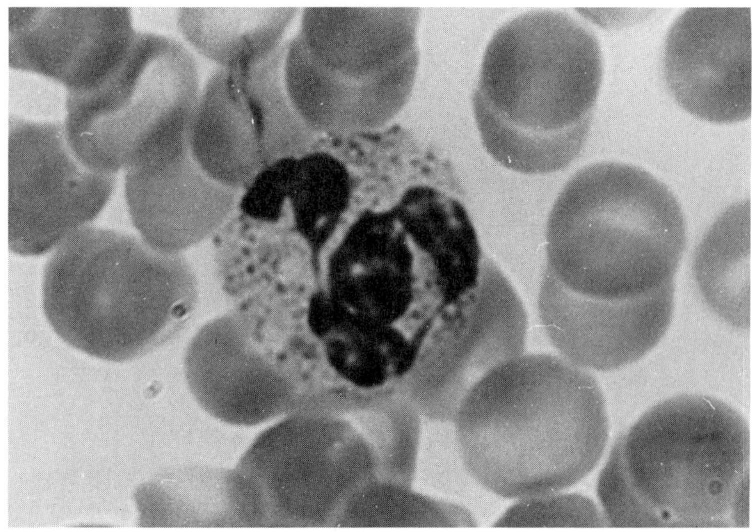

FIGURE 9-31

2. Using information from your table above (check charts in text to verify yours), you should be able to identify as they appear microscopically, various WBCs. Figure 9-31 shows a WBC and some RBCs. Which is larger, the RBC or WBC? How would you describe the shape of the leukocyte nucleus and its chromatin pattern? How would you describe the granules? What color would the cytoplasm be in your own slides? Now identify the cell (Fig. 9-31).

As you noted, the WBC is larger than the uniformly sized RBCs. The appearance of the lobated, condensed nucleus, and light cytoplasm may have helped you in identifying the WBCs. The many granules are also a distinguishing characteristic of the neutrophil.

3. A in Fig. 9-32 shows another leukocyte. What is the most obvious feature? Describe the granules. Do you recall what these granules contain? What is the distinctive feature of these granules under EM? Identify the cell.

You probably recognized this leukocyte by its very large granules, which contain peroxidase. An EM of these eosinophilic granules shows a dense crystalloid core.

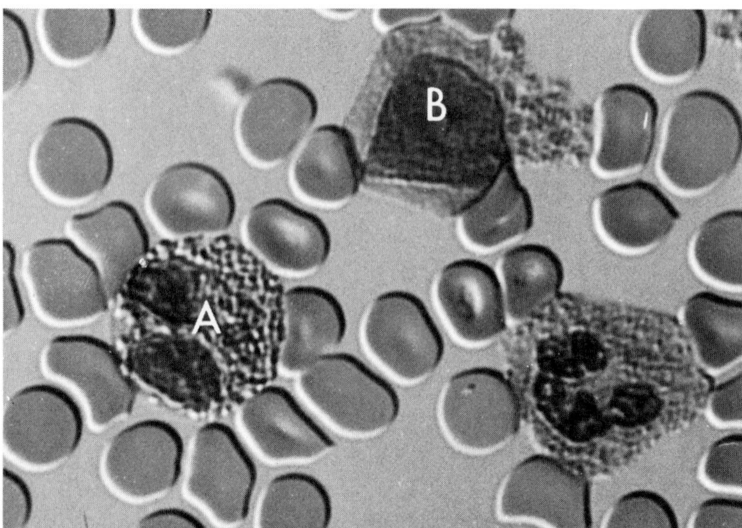

FIGURE 9-32

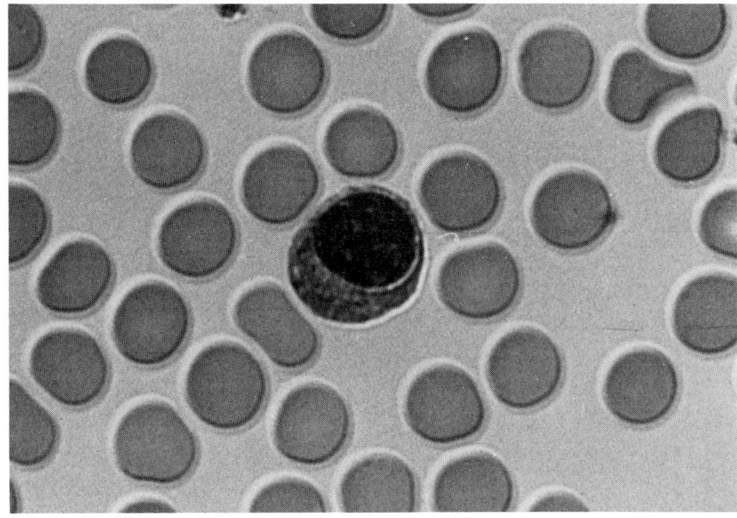

FIGURE 9-33

4. B in Fig. 9-32 shows a leukocyte surrounded by RBCs and platelets. Describe the nucleus in terms of its size, shape, and chromatin pattern. Are there any granules present? If so, what are they called? What type of leukocyte is this? Its function is __________.

This large leukocyte usually has a deeply indented, lacy-appearing nucleus. The azurophilic granules are quite indistinct and seem to blend with the cytoplasm, giving it a bluish gray appearance. The monocyte functions in phagocytosis, both by its inherent capability and by its transformation into a macrophage. In this way, it contributes to the defense mechanism of the body.

5. Figure 9-33 shows one more leukocyte with RBCs. How does this leukocyte compare in size with the one in the previous figure? How does the nucleus here compare to the one in Fig. 9-32 in terms of its size, shape, and chromatin pattern? What type of granules do you see? How would you describe the cytoplasm? Identify the cell.

This lymphocyte is smaller than the monocyte, but larger than the RBCs. The nucleus is smaller, rounder, and denser than that of the monocyte. Only a few azurophilic granules may be seen within the cytoplasm of this lymphocyte.

6. Figure 9-34 is an EM of a leukocyte. The nuclear chromatin in the two lobes might be described as __________. Which are larger, the specific granules (SG) or the azurophilic granules (AG)? How else do the granules differ? Is the RER abundant or sparse? Can you specify a reason for this amount? What type of cell is this?

The fairly dense chromatin and the sparse RER indicate the cell is not very active in protein synthesis. The larger, denser azurophilic granules and more numerous light-specific granules help identify this cell as a neutrophil.

7. Refer to Fig. 9-35, which contains a few profiles of a blood element called __________. Why does this element have this particular shape? What organelles, if any, can you expect to find in the cytoplasm? What does the cytoplasm contain? What is its function?

The biconcave disk shape of the RBC gives it an increased surface area for oxygen exchange. No organelles are visible in this particular EM, although microtubules are present along the periphery of the discoidal structure. This is because microtubules are preserved only with a special fixation procedure. The cytoplasm does contain a great deal of hemoglobin, used in its oxygen-carrying function.

8. Identify the structure shown in Fig. 9-36. What do you see in this small structure? Why does it not have a nucleus? This smallest formed element of the blood functions in __________.

The platelet contains azurophilic granules in the central granulomere. The cytoplasm may also contain small vacuoles, mitochondria, vesicular profiles of RER, and free ribosomes, since it is a piece

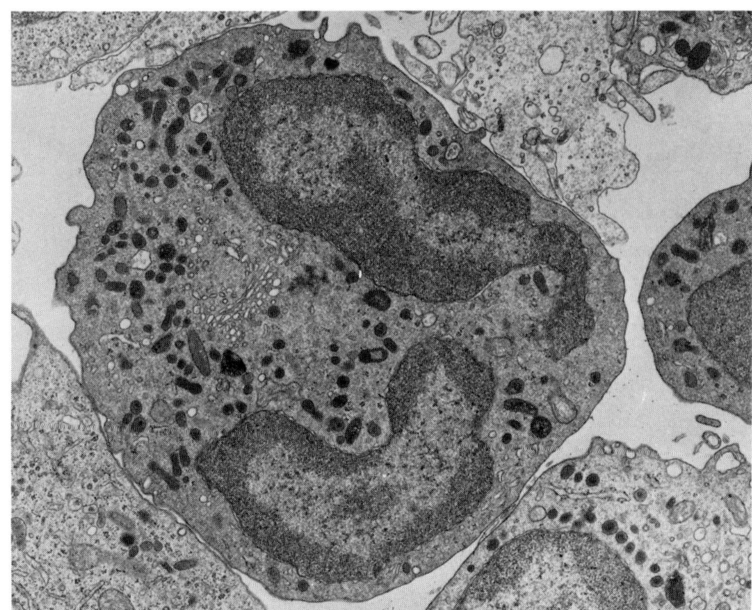

FIGURE 9-34

of cytoplasm detached from a megakaryocyte. The granules contain various chemical substances that aid in the blood-clotting mechanism. In addition, free ribosomes may be observed in a small percentage of circulating platelets.

9. Figure 9-37 shows an EM of a leukocyte. Would you classify this cell as a granulocyte or an agranulocyte? Describe the typical cytoplasmic granules. How do these compare with the one in Fig. 9-34? Identify this cell. What might its granules contain?

The granules here are larger than those of the neutrophil in Fig. 9-34. They have a dark, crystalloid core, giving it a "hamburger-on-a-bun" appearance. These eosinophilic granules usually contain peroxidase along with several other hydrolases.

FIGURE 9-35

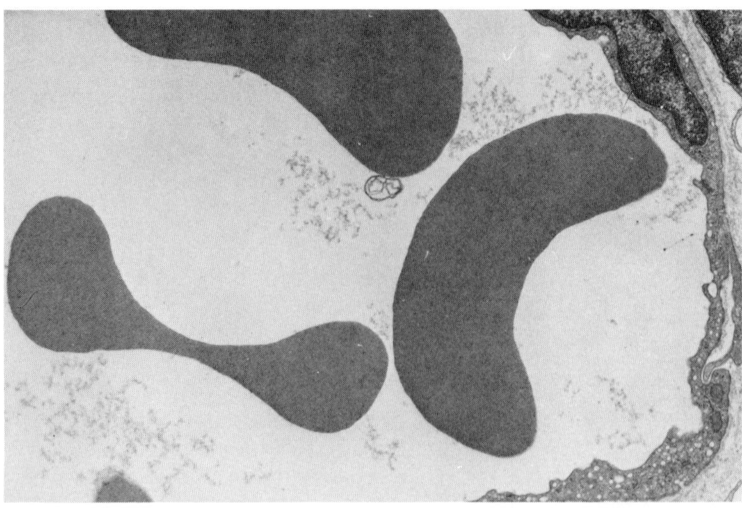

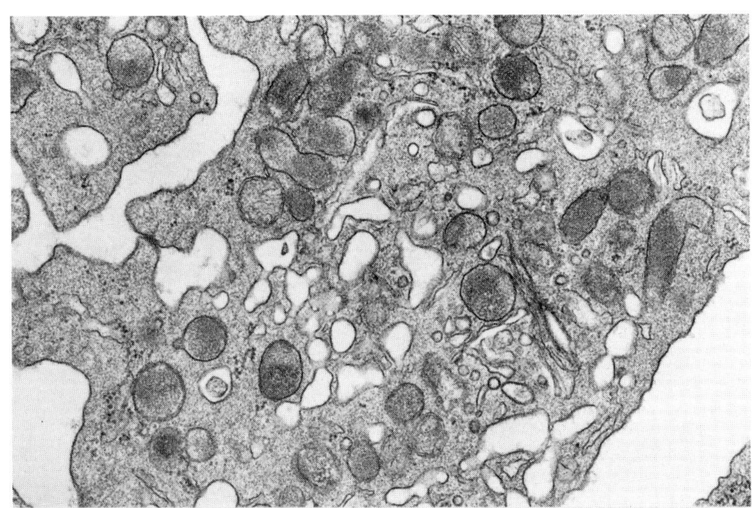

FIGURE 9-36

BIBLIOGRAPHY

Bishop, C., and D. M. Surgenor (eds): *The Red Blood Cells,* Academic Press, New York, 1964.

Bloom, W., and D. W. Fawcett: *A Text Book of Histology,* 10th ed., Saunders, Philadelphia, 1975.

Elves, M. W.: *The Lymphocytes,* Lloyd Luke, London, 1966.

Jerne, N. K.: "The Immune System," *Sci. Am.* **229:** 52–60 (1973).

Marcus, A. J., and M. B. Zucker: *The Physiology of Blood Platelets,* Grune & Stratton, New York, 1965.

Rebuck, J. W. (ed.): *The Lymphocyte and Lymphocytic Tissue,* Hoeber-Harper, New York, 1960.

FIGURE 9-37

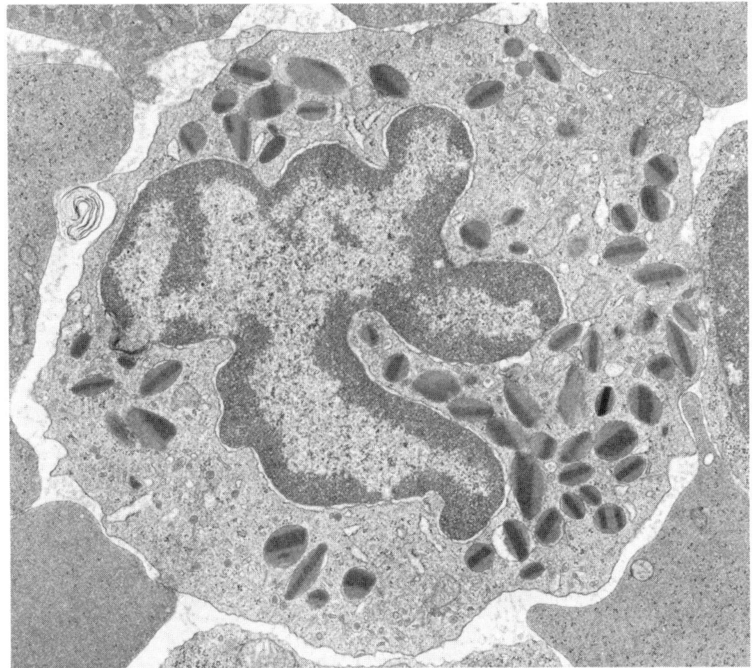

Rhodin, J. A. G.: *Histology: A Text and Atlas,* Oxford University Press, New York, 1974.

Weiss, L., and R. O. Greep: *Histology,* 4th ed., McGraw-Hill, New York, 1977.

ADDITIONAL READINGS

Anderson, D. R.: "Ultrastructure of Normal and Leukemic Leukocytes in Human Peripheral Blood," *J. Ultrastruct. Res.* (suppl.) **9**:5–42 (1966).

Archer, R. K.: "On the Functions of Eosinophils in the Antigen-Antibody Reaction," *Brit. J. Haemat.* **11:**123–129 (1965).

Athens, J. W.: *Granulocyte Kinetics in Health and Disease,* National Cancer Institute Monograph, no. 30, 1969, pp. 135–155.

Bainton, D. F., and M. G. Farquhar: "Segregation and Packaging of Granule Enzymes in Eosinophil Leukocytes," *J. Cell Biol.* **45:**54–73 (1970).

Behnke, O.: "Electron Microscopical Observations on the Surface Coating of Human Blood Platelets," *J. Ultrastruct. Res.* **24:**51–69 (1968).

Cohn, Z. A.: "The Structure and Function of Monocytes and Macrophages," *Adv. Immunity* **9:** 163–214 (1968).

Daems, W. T.: "On the Fine Structure of Human Neutrophilic Leucocyte Granules," *J. Ultrastruct. Res.* **24:**343–348 (1968).

David-Ferreira, J. F.: "The Blood Platelets: Electron Microscopic Studies," *Int. Rev. Cytol.* **17:**99–148 (1964).

Lowenstein, L. M.: "The Mammalian Reticulocyte," *Int. Rev. Cytol.* **8:**136–174 (1959).

Miller, F., E. DeHarven, and G. E. Palade: "The Structure of Eosinophil Leukocyte Granules in Rodents and in Man," *J. Cell Biol.* **31:**349–362 (1966).

Spicer, S. S., and J. H. Hardin: "Ultrastructure, Cytochemistry and Function of Neutrophil Leukocyte Granules: A Review," *Lab. Invest.* **20:**488–497 (1969).

Terry, R. W., D. F. Bainton, and M. S. Farquhar: "Formation and Structure of Specific Granules in Basophilic Leukocytes," *Lab. Invest.* **21:**65–76 (1969).

Wivel, N. A., M. A. Mandel, and R. M. Asofsky: "Ultrastructural Study of Thoracic Duct Lymphocytes of Mice," *Am. J. Anat.* **128:**57–72 (1970).

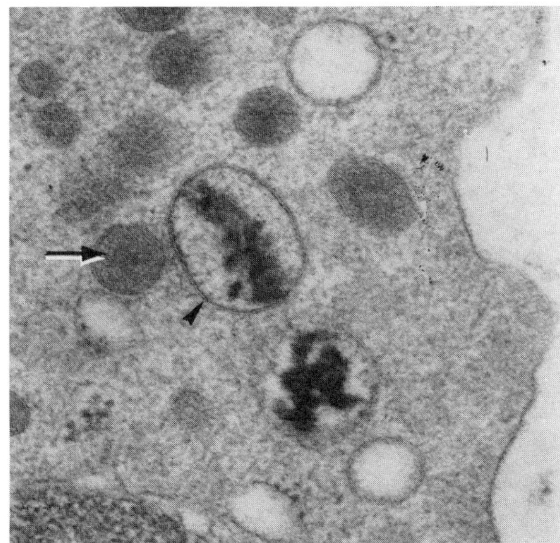

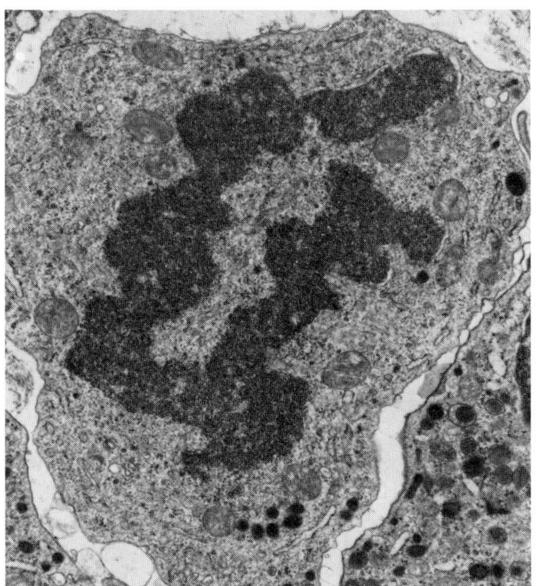

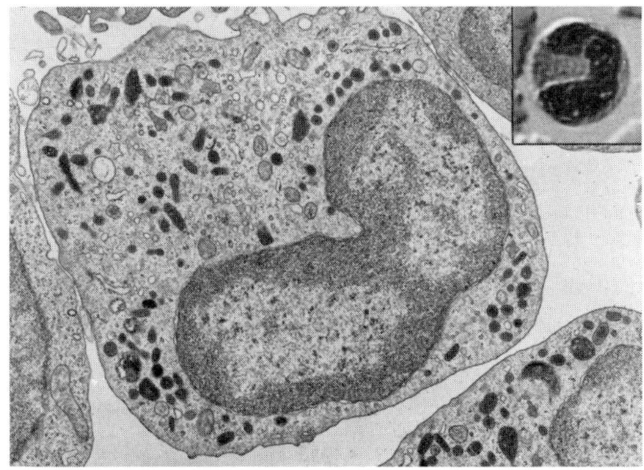

10 BLOOD FORMATION

OBJECTIVES

Upon completion of this chapter, the student will be able to:

1 Identify the embryonic origin, function, and structure of a stem cell.

2 Identify the general changes during maturation and differentiation of each of the formed elements of the blood with regard to (when applicable):
(a) size (b) nucleocytoplasmic ratio (c) cytoplasmic contents (d) nuclear shape and appearance (e) granules

3 Define rhopheocytosis.

4 Name in order the stages of the maturation of both granulocytes and red blood corpuscles from the primitive stem cell to the mature circulating element.

The number of formed blood elements in a unit volume of blood remains remarkably constant throughout life, even though these elements are continuously being formed, dying, and leaving and reentering the circulation. The blood-forming organs must be linked to a feedback mechanism which monitors these elements. They must also be able to inform the hemopoietic cells, so that the proper number of new cells may be produced from different stem cells. Replacing worn-out cells while maintaining the dynamic balance is a major function of blood-forming organs. Because they are linked to the blood circulatory system, lymphatics, and lymphoid organs, the tissue space through which the blood elements traverse must be included in our consideration of the dynamic balance.

Blood elements are lost for various reasons. Lymphocytes emigrate constantly from the tissue spaces through the epithelia of the tonsils and from the entire gastrointestinal tract, as well as from other mucous membranes. RBCs must be replaced after menstruation hemorrhage, or even after blood donations. Various feedback mechanisms ensure the responsiveness of the many organs concerned with maintenance of hemodynamics. A feedback message must be delivered to the cells that are capable of recognizing the message and differentiating to form the correct number of the needed elements. Because the phenomena involved are complex, there has been, and still is, intensive research on the details of the feedback system. These systems are known to involve a number of cellular, endocrine, and neural factors; and since each formed element has a specific life span and function, it is probable that independent mechanisms are responsible for its regulation, proliferation, and differentiation. For instance, *erythropoiesis* is stimulated by a hormonelike factor, *erythropoietin*, which responds to shifts in the balance between tissue oxygen supply and demand. In the hypoxic state which occurs in individuals in the process of adapting to life at a high altitude, the erythropoietin level may increase with subsequent increasing numbers of RBCs. More complicated factors govern *leukopoiesis* and *thrombopoiesis* (formation of blood platelets). While a compensatory increase in the number of circulating elements operates in every blood cell line, specific conditions cause certain types of WBCs to increase above the usual numbers. For example, during acute inflammation, several times the normal amount of neutrophils circulate. Similarly, an abnormally large number of eosinophils may be present in certain types of allergy.

INTERRELATIONSHIP OF BLOOD CELLS

Traditionally, there are two explanations for the production of the various formed elements of the blood. One theory, the *polyphyletic* theory, considered that mature blood elements came from different types of *stem cells*. Another, the *monophyletic* theory, advocated a basic stem cell population that was pluripotential so that it could differentiate into any of the WBCs, RBCs, or platelet precursors. Recent investigations favor the latter approach. Therefore the stem cell, or *hemocytoblast*, can be regarded as the progenitor of these blood elements.

Stem cells are derived from embryonic connective tissue, the mesenchyme. The continuous process of hemopoiesis requires a constant supply of stem cells through the entire life span. They must be able to maintain their own population by mitosis, as well as differentiate into the various mature blood elements. The stem cells are normally located in the blood-producing organs, which include the bone marrow and lymphoid organs. The lymphoid organs, such as lymph nodes, tonsils, spleen, and thymus, are capable of producing various types of lymphocytes. The

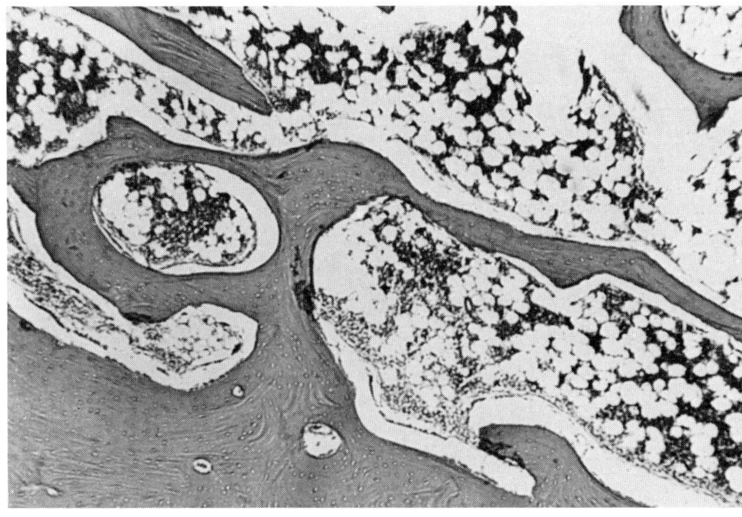

FIGURE 10-1 Light micrograph of spongy bone. The marrow spaces are filled with bone marrow tissue.

bone marrow is responsible for the production of granulocytes, RBCs, and platelets, as well as some of the lymphocytes and monocytes. In the bone marrow, granulocyte precursors outnumber the RBC precursors about three to one, although there is a greater proportion of RBCs in peripheral blood. This occurs partly because granulocytes have a shorter life span than RBCs.

Bone Marrow. The gelatinous bone marrow tissue shown in Fig. 10-1 is encased within bone. Its stroma is connected to the endosteum of the bone and consists of reticular cells and fibers which form a reticular framework. Through the space created by the stroma runs a vast capillary and sinusoidal network which not only provides nutrition to the forming elements but also serves as the portal for release of mature elements into the circulating blood. The bone marrow contains a complex mixture of blood cells in various stages of differentiation, as well as stem cells. In Fig. 10-2, a smear preparation of bone marrow, cells with many different morphological characteristics can be visual-

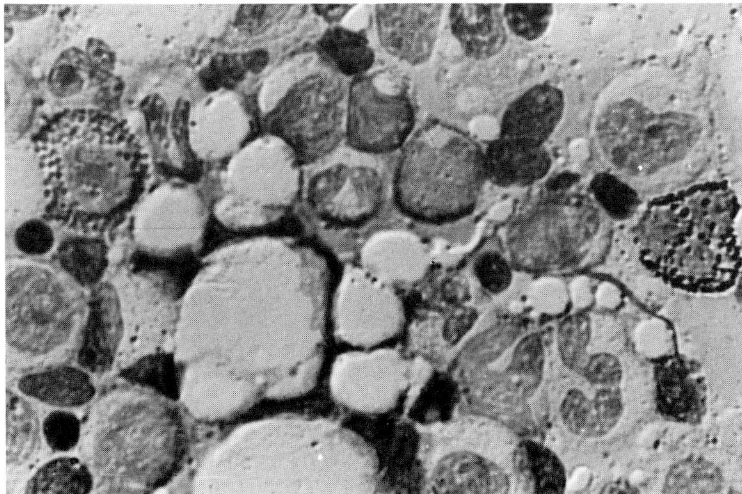

FIGURE 10-2 Phase contrast micrograph of a smear preparation of bone marrow. A mixture of blood cells in different stages of development are seen. A few fat cells are also present.

ized. Depending largely on the individual's age and metabolic state, varying numbers of fat cells will also be present. In general, fat cells increase with age; in older bone marrow they render it a yellowish color. The bone marrow with many fat cells is called *yellow marrow,* while in younger individuals the marrow appears more reddish, so it is called *red marrow.*

As primitive stem cells differentiate to fully mature cells, certain characteristic changes occur in all cell lines. With increasing age the cell generally becomes smaller. The nuclear chromatin becomes condensed. Concomitantly, the nucleus becomes smaller and finally becomes indented and lobated, as exemplified by the granulocytes. The cytoplasm loses its basophilia with maturation. In the granulocytic series, production of specific granules occurs at certain stages and continues until the cells become fully mature.

Viewed by electron microscopy, the cellular machinery necessary for protein synthesis during cellular growth and differentiation is complex. Therefore, the fine structure of the stem cell's cytoplasm, since it is first directed towards growth, typifies a cell undergoing active protein synthesis. Later the cytoplasm becomes more specialized according to the ultimate function of the type of cell produced. Thus, morphological studies of maturing blood elements are important in understanding the nature of hemopoiesis.

STRUCTURE OF STEM CELLS

Stem cells resemble immature blood cells with basophilic cytoplasm and an undifferentiated nucleus (Fig. 10-3). They are large cells, 15 μm or more in diameter. The nucleus is large and round, with a light chromatin pattern. It contains at least one eccentric nucleolus. In LM, some azurophilic granules may be observed in the cytoplasm. These primitive cells have a small number of RER profiles and a poorly developed Golgi complex, but a large number of free monosomes and polysomes in varying proportions. The

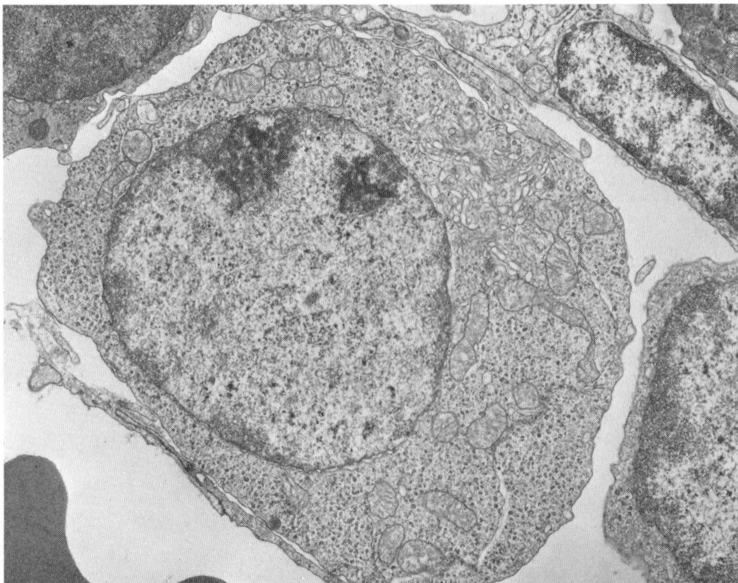

FIGURE 10-3 Electron micrograph of stem cell. The nucleus is large with a light chromatin pattern. The cytoplasm contains a large number of free ribosomes and polysomes, a few profiles of RER, a poorly developed Golgi apparatus, and a few mitochondria with poorly developed cristae and an electron-lucent matrix.

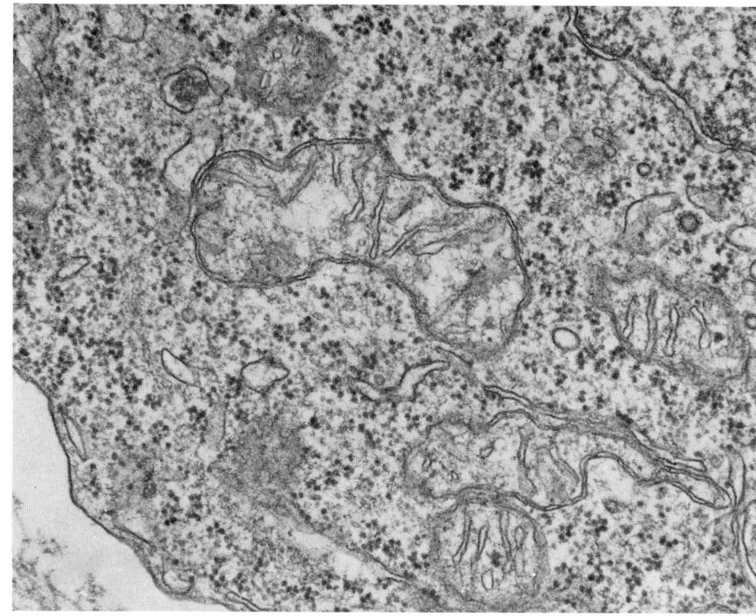

FIGURE 10-4 Portion of a stem cell. The cytoplasm contains numerous free polysomes and monosomes. Mitochondria are large and are irregular in shape. They have an electron-lucent matrix.

polysome/monosome ratio generally reflects the level of protein synthesis (translation) of the developing cell. A few large mitochondria with irregular shapes and poorly differentiated cristae are observed (Fig. 10-4). The lucent mitochondrial matrix rarely contains intramitochondrial granules.

As pointed out earlier, the stem cell divides to preserve its own population, while some daughter cells may become committed to differentiate toward specific cell lines. In this first state, a committed stem cell is called a *blast*. For example a *myeloblast* represents a committed precursor to a *granulocytic* line

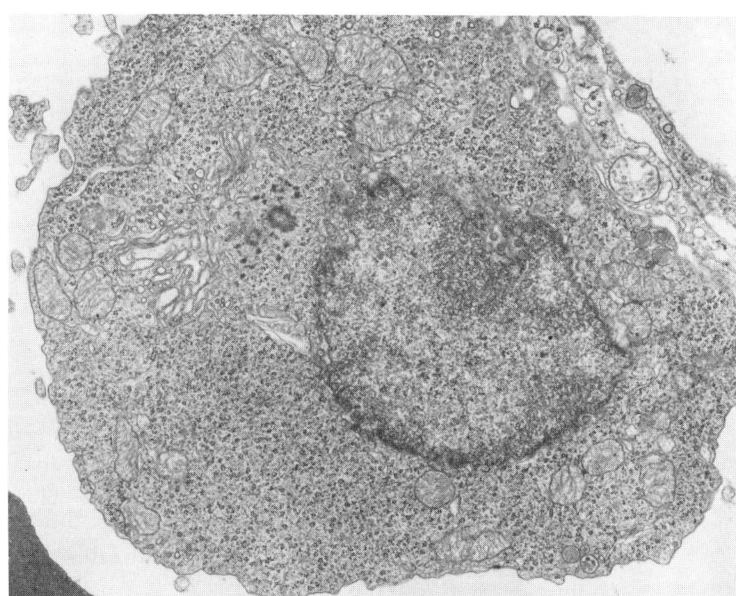

FIGURE 10-5 Electron micrograph of stem cell. It shows signs of some differentiation. Numerous polysomes are present in the cytoplasm. Profiles of RER are developing. The cell contains a few dense granules and has a poorly developed Golgi apparatus.

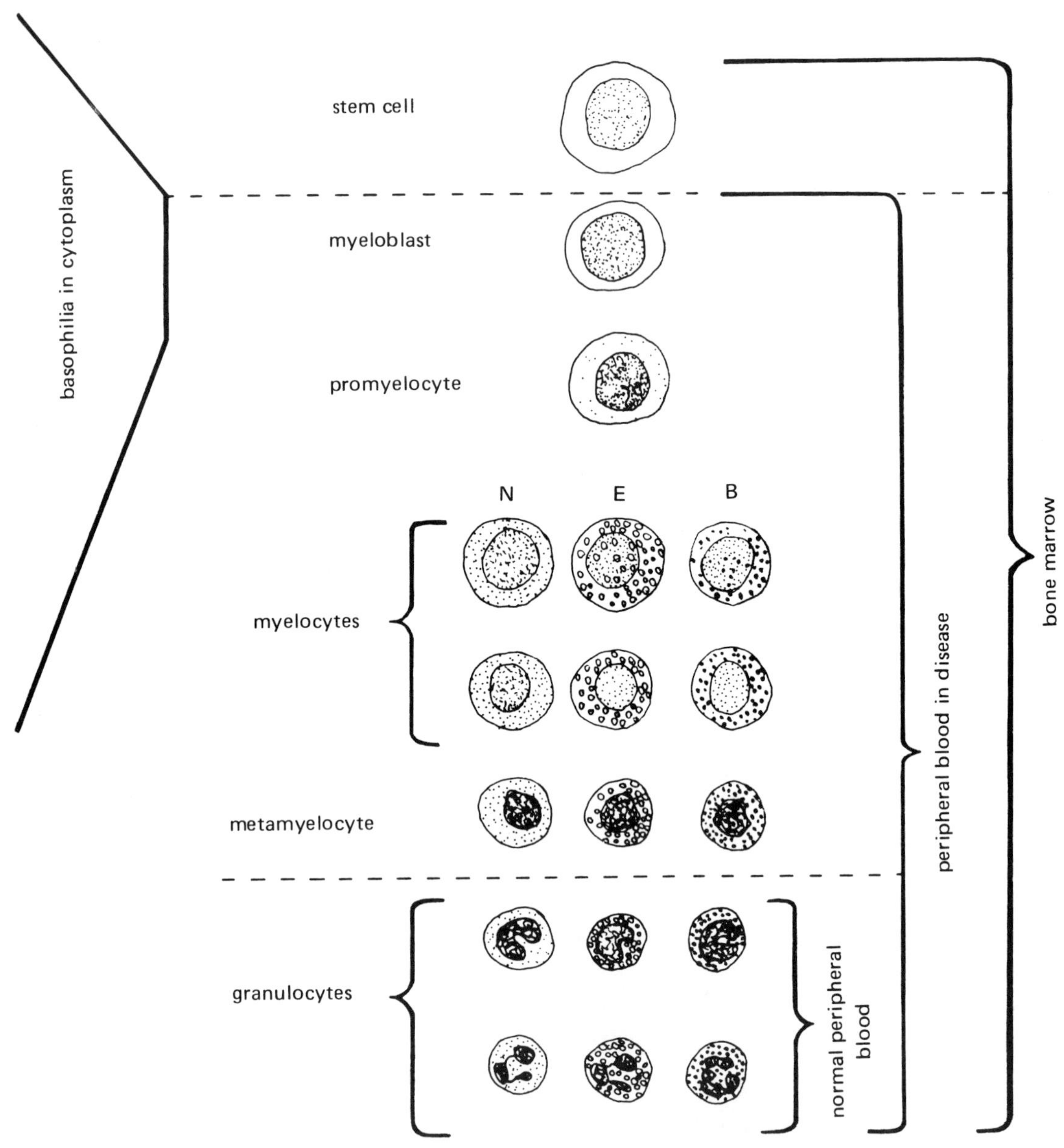

FIGURE 10-6 Diagram illustrating the development of granulocytes. (1) The cell size decreases as the cell matures. (2) The cytoplasm becomes less basophilic. (3) The nucleus becomes smaller. (4) Specific granules appear.

and a *lymphoblast* will differentiate to become a *lymphocyte*. Since the different blast cells are in the earliest stage of differentiation, it is difficult to distinguish them morphologically at the LM level.

The stem cell in Fig. 10-5 has a great number of polysomes, suggesting that it might be in the process of undergoing rapid differentiation. It contains several profiles of developing RER, a number of dense granules, and a large but poorly organized Golgi apparatus.

DEVELOPMENT OF GRANULOCYTES

Similar in developmental features, the three types of granulocytes (neutrophils, eosinophils, and basophils) differ only by specific granules and relative cell numbers. Before becoming mature granulocytes, they must progress through three recognizable stages of differentiation: *promyelocyte, myelocyte,* and *metamyelocyte* stages. These stages are diagramed in Fig. 10-6 as they are seen with the light microscope.

Promyelocytes

The transition from myeloblast to promyelocyte is characterized by an increase in cell size and by an increase in the amount of cytoplasm. The nucleus is still rounded but may be ovoid; the chromatin is denser and the nucleoli are still large. The cytoplasm remains highly basophilic, and the quantity of azurophilic granules is variable. All azurophilic granules are formed before the end of the promyelocyte state, so the relative number reflects the maturity of the cell.

Cytoplasmic organelles appear more numerous and somewhat more developed than in the myeloblast or stem cell stage (Fig. 10-7). Note the increase in RER and better differentiated mitochondria (arrows). A number of azurophilic granules are also visible (arrowheads).

Myelocytes

During this stage, specific granules for each of the granulocytes appear, making it possible to identify neutrophilic, basophilic, or eosinophilic myelocytes at the LM level. The cell tends to become smaller, the nuclear chromatin is somewhat denser and more compact, and there is a more prominent indentation of the nucleus (inset in Fig. 10-8). Nucleoli are still fairly prominent. Electron microscopy reveals that the myelocyte has well-developed cytoplasm. The Golgi complex is extensively developed, with many small transport vesicles. Unlike other granulocytes, the neutrophils have two cytoplasmic granules, namely, azurophilic and specific. The neutrophilic myelocyte seen in Fig. 10-8 has azurophilic granules and smaller specific granules. The

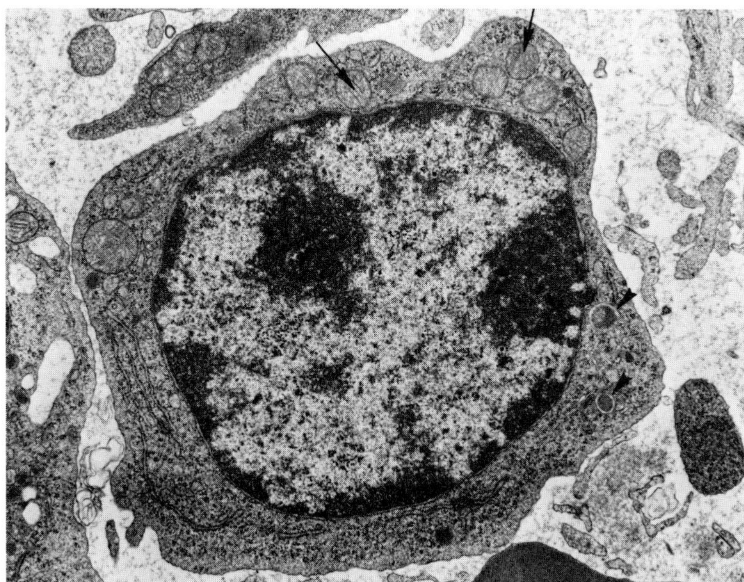

FIGURE 10-7 Electron micrograph of promyelocyte. The cytoplasmic organelles are more numerous and better developed than in the stem cell. Mitochondria (arrow) and RER are better developed. Some azurophilic granules (arrowheads) are also visible. Nucleoli are prominent in the nucleus.

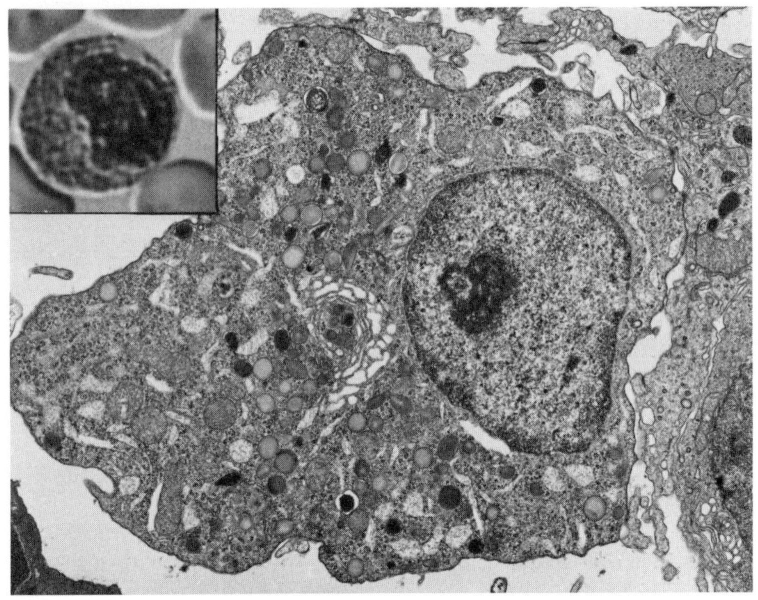

FIGURE 10-8 Electron micrograph of myelocyte. The nuclear chromatin is denser than in the promyelocyte, and a prominent nucleolus is present. The Golgi complex is well developed, and the cytoplasm is filled with RER. In addition to azurophilic granules, specific granules are being formed. Note the numerous small mitochondria. Inset, light micrograph of myelocyte.

cytoplasm is packed with RER, with relatively fewer free ribosomes compared with the promyelocyte. Note the numerous small and ovoid mitochondria. They have only a few but well-differentiated cristae and dense matrix. A portion of a neutrophilic myelocyte demonstrates the appearance of specific (arrows) and azurophilic (arrowheads) granules (Fig. 10-9). Figure 10-10 is an EM of a late neutrophilic myelocyte. The myelocytes are actively involved in the elaboration of various specific granules which contain neurohumoral substances or lysosomal enzymes. The organization of the RER and Golgi complex, along with the number of specific granules, indicate the degree of differentiation of a given cell type.

It should be emphasized that the struc-

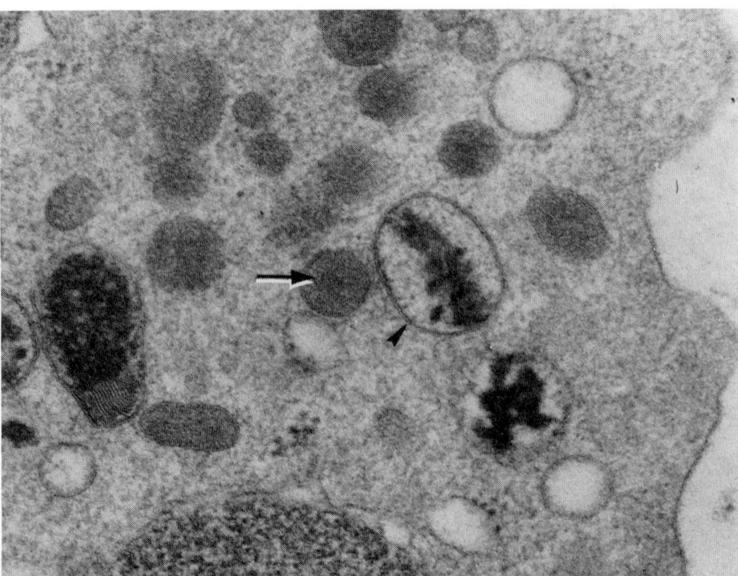

FIGURE 10-9 Portions of neutrophilic myelocyte. The appearance of the azurophilic (arrowheads) and specific (arrows) granules is demonstrated.

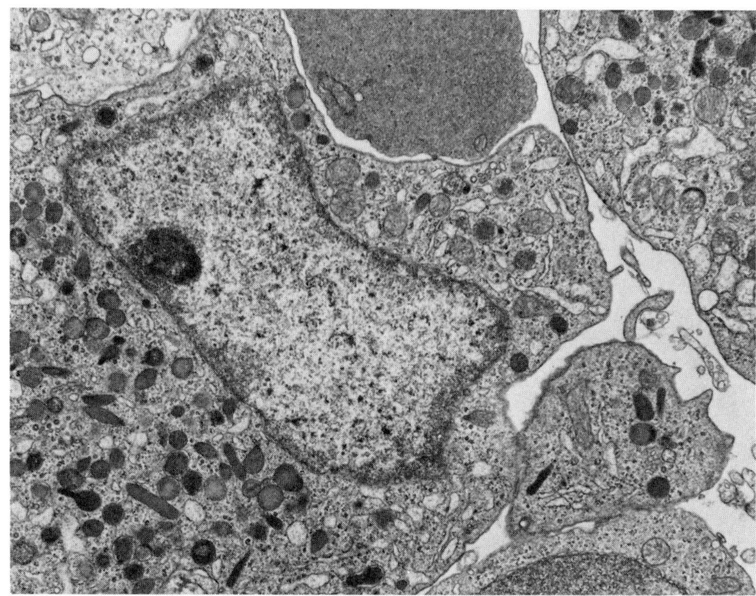

FIGURE 10-10 Electron micrograph of late neutrophilic myelocyte. This cell has the structure of a cell actively involved in the elaboration of specific granules; i.e., a well-developed Golgi apparatus, abundant RER, transport vesicles, and numerous mitochondria. A prominent nucleolus is present in the nucleus.

tural changes just described represent a continuous process. A given cell under a microscope would exhibit a majority of features belonging to a stage of differentiation for that line. Through the myelocyte stage, the granulocytic precursors are able to divide (Fig. 10-11). Further differentiation of neutrophils is characterized by a decrease in RER and lobation of the nucleus. The EM in Fig. 10-12 is a neutrophil in "band cell" stage, which is so called because of the bent appearance of the nucleus.

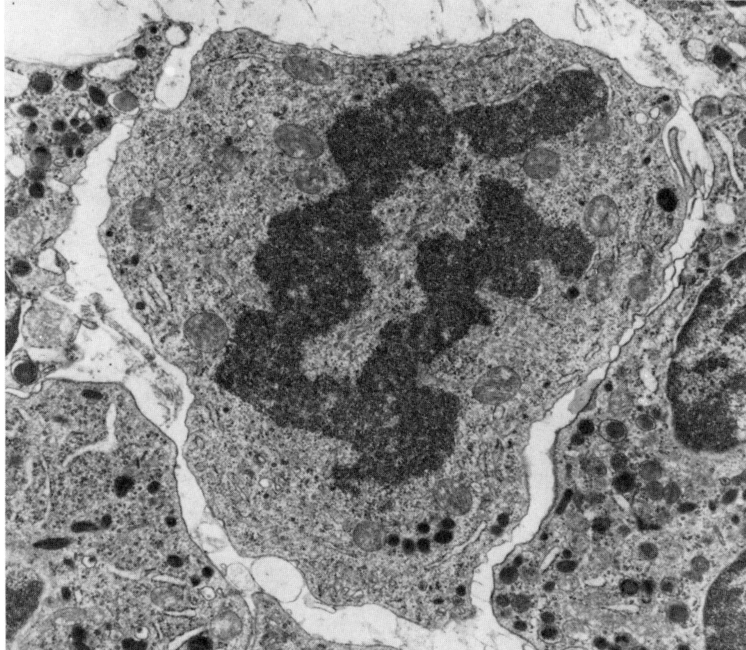

FIGURE 10-11 Electron micrograph of myelocyte in mitosis.

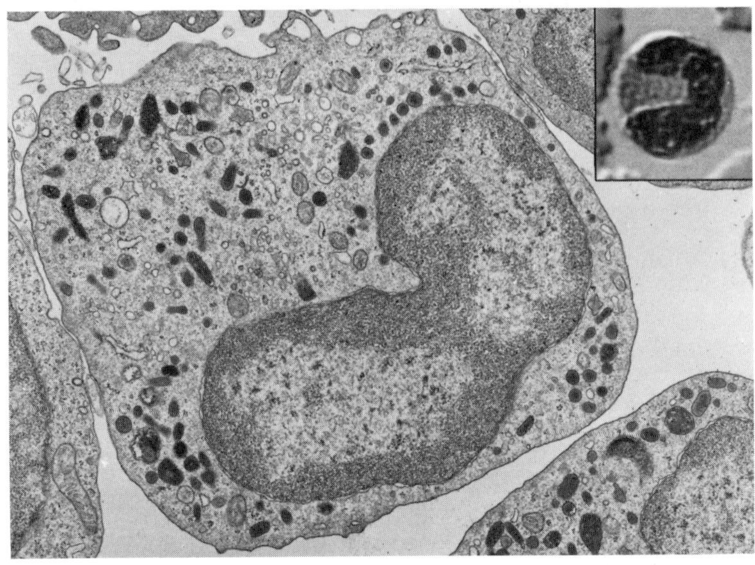

FIGURE 10-12 Electron micrograph of neutrophil in band stage. There is a beginning of the lobation of the nucleus. The nucleus appears bent. There is a decrease in the amount of RER, and the Golgi complex is reduced in size. Inset, light micrograph of neutrophilic band cell.

Mature Granulocytes

With increasing maturity, there is a gradual diminution of cytoplasmic organelles as described previously. The cytoplasm loses its basophilia, but becomes loaded with the various specific granules (Fig. 10-13). The nucleus has a dense chromatin and develops into a lobated form, which is seen most

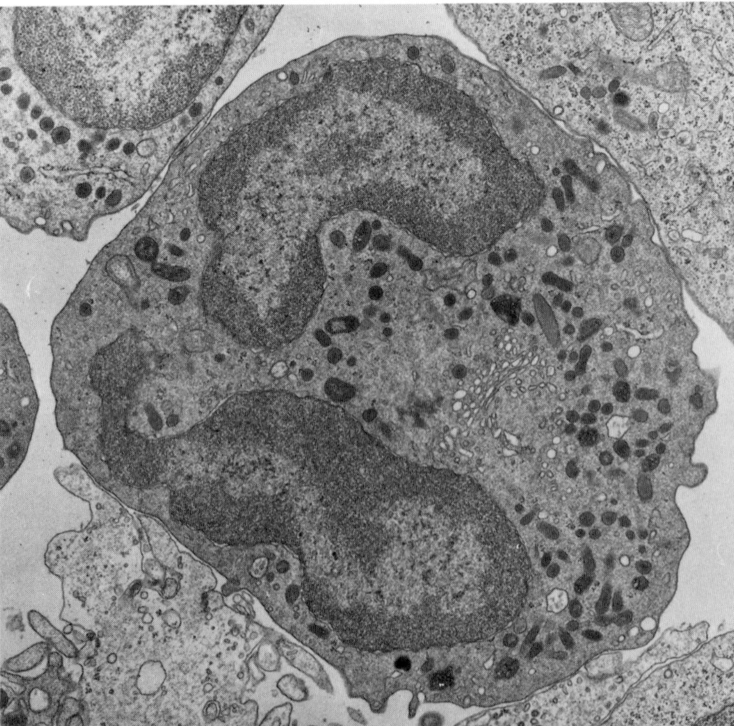

FIGURE 10-13 Electron micrograph of mature neutrophil. The nucleus appears lobated with a more condensed chromatin pattern than in the immature stages. Ribosomes are reduced in number, and the mitochondria and RER profiles appear smaller. The Golgi complex is inconspicuous. Specific granules are abundant in the cytoplasm.

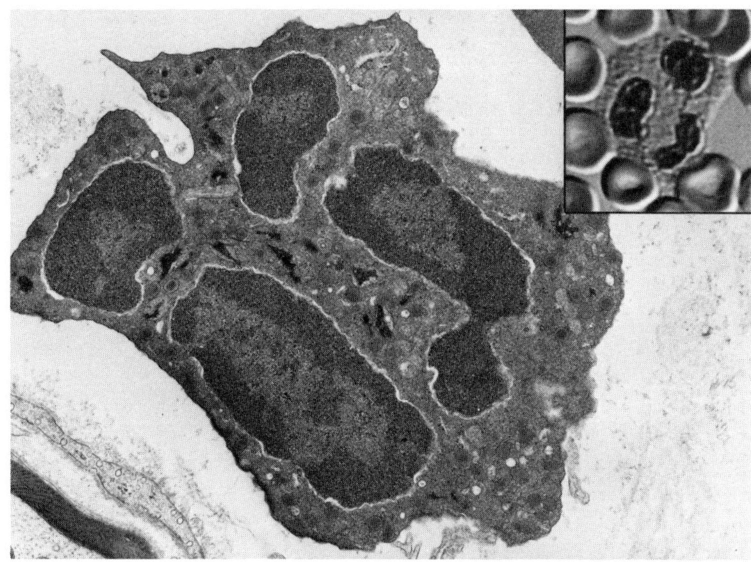

FIGURE 10-14 Electron micrograph of mature neutrophil. The nucleus is multilobated, and the cytoplasm appears dense.

extensively in fully mature neutrophils. A marked reduction in the number of ribosomes is seen, along with a diminution of RER profiles and the size of mitochondria. The density of the ground cytoplasm varies with the type of granulocyte. The eosinophils and basophils have a rather lucent cytoplasm, whereas neutrophils have a moderate to extremely dense ground cytoplasm, shown by the neutrophil in Fig. 10-14. At this point the

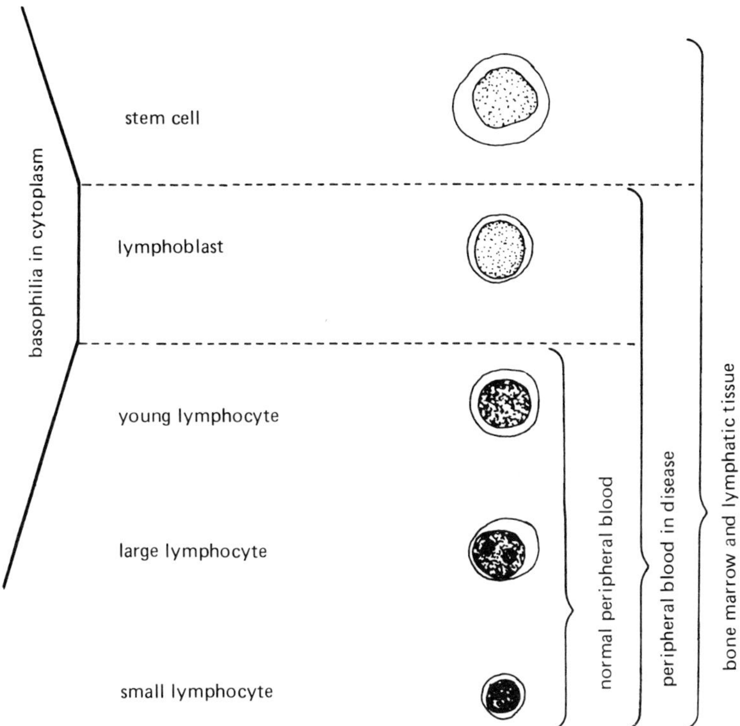

FIGURE 10-15 Diagram illustrating the development of agranular leukocytes. (1) The cell size decreases as the cell matures. (2) The cytoplasm becomes less basophilic. (3) The nucleus becomes smaller and denser.

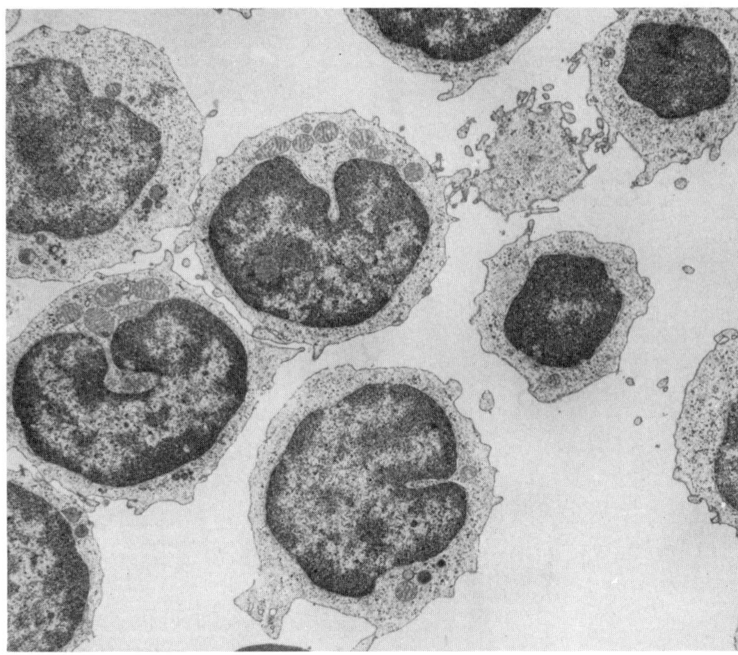

FIGURE 10-16 Electron micrograph of lymphocytes of varying sizes. Note the dense nuclei and the sparse amount of cytoplasm.

mature cells are ready to be released into the circulating blood.

DEVELOPMENT OF AGRANULAR LEUKOCYTES (LYMPHOCYTES)

Lymphocytes

The most primitive cell in the lymphocyte line of development, the lymphoblast, resembles a myeloblast. In appearance, it has an undifferentiated nucleus, which becomes condensed with maturity. One or more nucleoli may be found in the nucleus of a lymphoblast. Figure 10-15 depicts the maturation in the lymphoid series.

These cells proliferate rapidly by mitosis, undergo differentiation, particularly in the thymus, and become the small lymphocytes that enter the bloodstream via the lymphatics. Lymphocytes develop in lymphoid organs as well as in the bone marrow. Lymphocytes do not produce complicated cytoplasmic structures in the course of their differentiation. In contrast, a gradual diminution of free ribosomes, mitochondria, and Golgi complex leads to the typical appearance of mature lymphocytes with a high nucleo-cytoplasmic ratio. Their Golgi complex and mitochondria, though smaller than those in less-differentiated forms, have a more discretely differentiated appearance. The Golgi complex is composed of small but neatly stacked lamellar components, while mitochondria have a few distinct cristae and a dense matrix. In Fig. 10-16, lymphocytes of different sizes are shown. A typical small lymphocyte is seen in the EM of Fig. 10-17. The small amount of its cytoplasm contains only a few mitochondria and ribosomes. When one considers the complex immunologic and defense phenomena in which lymphocytes are engaged, the simplicity of their cytoplasmic structures is somewhat deceiving. It is pertinent to keep in mind the heterogenous populations of adult lymphocytes that are different in function and ontogeny.

Monocytes

Little is known about the development of

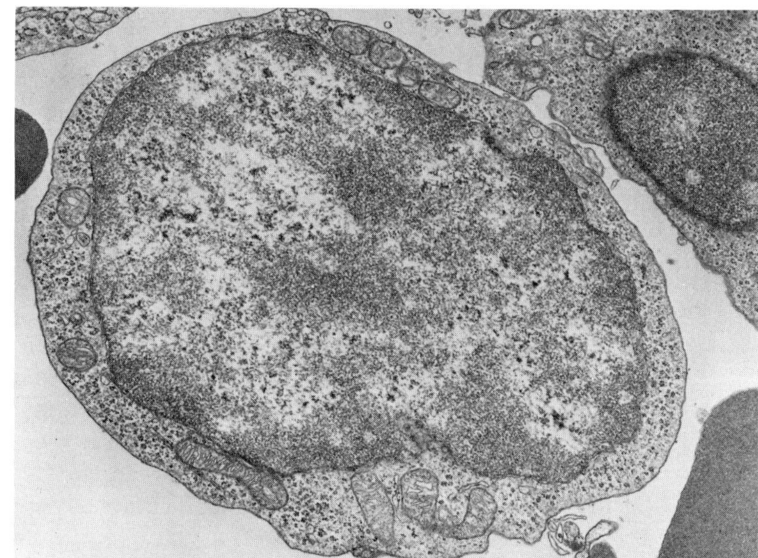

FIGURE 10-17 Electron micrograph of typical small lymphocyte. The nucleus is large with a condensed chromatin pattern. The small amount of cytoplasm contains a few mitochondria and free ribosomes.

monocytes. However, they are thought to develop primarily in the bone marrow. Monocytic differentiation results in the formation of a horseshoe-shaped nucleus with a moderately dense chromatin pattern. Azurophilic granules are produced and appear around the Golgi complex located against the nuclear indentation. Figure 10-18 shows the schematic development of monocytes. The fine structural aspects of changes occurring in monocyte differentiation resemble to some degree those described for lymphocytes.

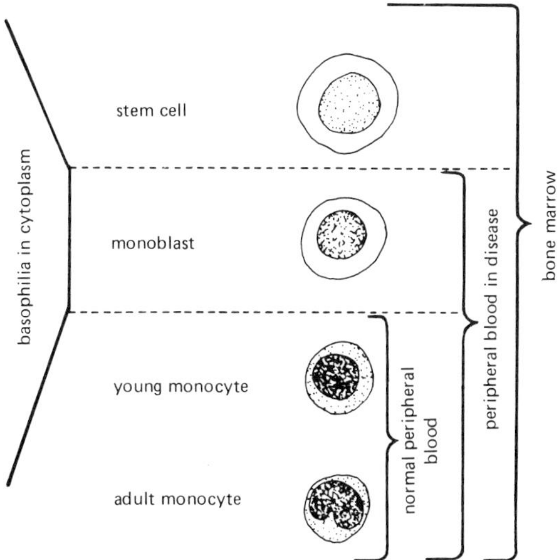

FIGURE 10-18 Diagram illustrating monocyte development. (1) The cell size decreases as the cell matures. (2) The cytoplasm becomes less basophilic. (3) The nucleus becomes smaller, denser, and horseshoe-shaped. (4) Azurophilic granules are produced.

DEVELOPMENT OF RBCs

Red blood corpuscles differentiate from stem cells, passing through stages of basophilic erythroblast, polychromatic erythroblast, and normoblast before becoming mature RBCs. This series is diagramed in Fig. 10-19.

Basophilic Erythroblast

In the early stages of RBC differentiation, the cell decreases in size from the stem cell. The cytoplasm is intensely basophilic (inset in Fig. 10-20), possessing a notably circular contour with a round nucleus. The nuclear pattern is somewhat patchy as the chromatin

basophilia in cytoplasm
stem cell
erythroblast
basophilic erythroblast
polychromatic erythroblast
normoblast
reticulocyte
red blood corpuscle
normal peripheral blood
peripheral blood in disease
bone marrow

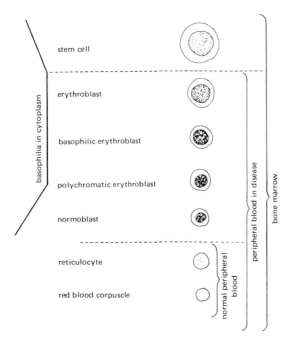

FIGURE 10-19 Diagram illustrating the development of RBCs. (1) The cell size decreases as the cell matures. (2) The cytoplasm becomes less basophilic. (3) The nucleus becomes smaller and denser. (4) The nucleus is extruded.

becomes denser. Fair-sized nucleoli are frequently found.

The cytoplasm contains many polysomes that are made up of only several ribosomes per polysome (Fig. 10-20). The uniformity in polysome size reflects the fact that they are primarily involved in the production of a single type of protein, hemoglobin. Elsewhere in the cytoplasm sparse RER is present, a Golgi apparatus is rare, and mitochondria are small but present in moderate numbers. The nucleus in EM often reveals clear nuclear channels leading to the nuclear pores. Occasional signs of pinocytosis may appear with invaginations of the cell membrane (Figs. 10-21 and 10-22). These invaginations have been shown to recover ferritin-like granules from worn-out RBCs (arrows in Figs. 10-21 and 10-22) that are stored in macrophages, a process called *rhopheocytosis.* Studies have shown that RBCs at the end of their life span are phagocytized and degraded by macrophages, which in turn hydrolyze hemoglobin into ferritin-like form. Accordingly, erythroblasts develop around such macrophages that have ingested expired RBCs; this presumably

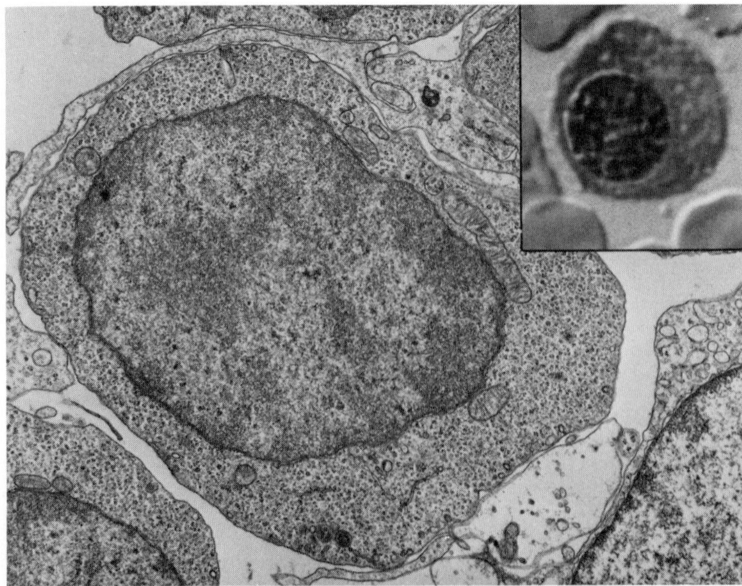

FIGURE 10-20 Electron micrograph of basophilic normoblast. The nucleus is rounded and has a patchy chromatin pattern. The cytoplasm contains numerous polysomes. In the cytoplasm, RER is sparse, and Golgi complex is rarely present. Inset, light micrograph of basophilic normoblast.

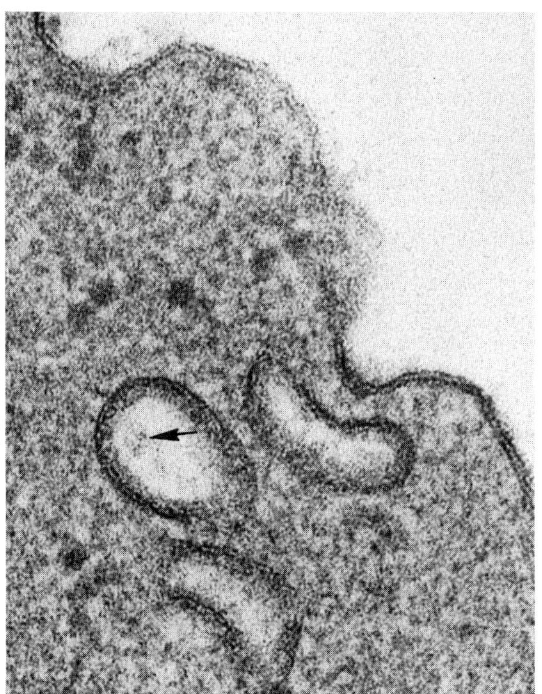

FIGURE 10-21 Portion of basophilic normoblast. Signs of pinocytosis—invaginations of the cell membrane—is apparent. Vesicles formed by these invaginations contain ferritin-like granules (arrow).

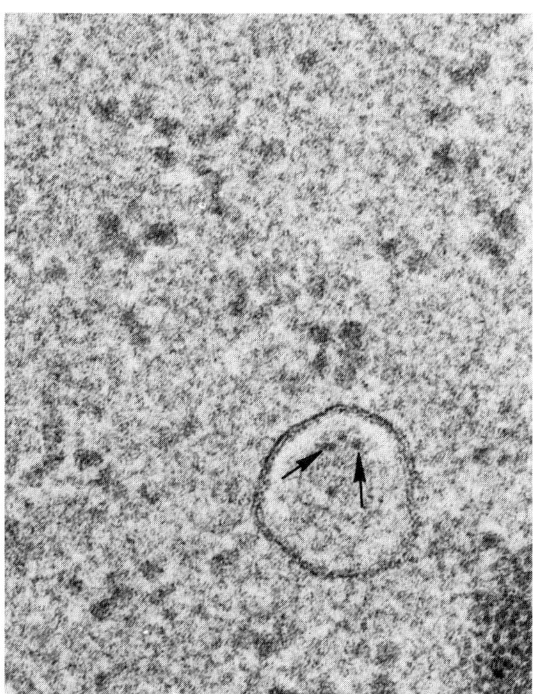

FIGURE 10-22 Electron micrograph of pinocytic vesicle at higher magnification. The vesicle contains ferritin-like granules. Note the polysomes in the surrounding cytoplasm.

signifies a functional interrelationship between macrophages and erythroblasts, and therefore has been named an *erythroblastic islet.* Figure 10-23 is an EM which depicts an erythroblastic islet where a number of developing erythroid cells (*E*) are surrounding a centrally located macrophage (*M*).

In erythroblasts, an aggregation of ferritin-like granules are frequently found in the cytoplasm. These are called *siderosomes.*

Since the hemoglobin synthesis appears to start very early in differentiation of an erythroblast, the increasing concentration of hemoglobin results in a denser cytoplasm. A polychromatic erythroblast (Fig. 10-24) can be compared with the basophilic erythroblast (Fig. 10-20). Note the increasingly irregular condensation of nucleoplasm, the uniform-sized polysomes, and the increasing density of the cytoplasm. As hemoglobin production continues, it accumulates in the cytoplasm of the developing erythroblasts. Because hemoglobin is acidophilic and polyribosomes are basophilic, a characteristic polychromatic staining reaction can be seen in LM. As the cell becomes differentiated further, polyribosomes become fewer in number, but the density of the cytoplasm becomes even greater than before. The polysome to monosome ratio becomes reduced as well, indicating the probable slowdown in the rate of hemoglobin synthesis. The nucleus decreases in size and the heterochromatin becomes more condensed, producing a checkerboard pattern of the nucleoplasm.

Normoblast

During this stage of RBC development, hemoglobin pervades the entire cytoplasm and gives it an orange-pink appearance which approximates the color of a circulating RBC (inset in Fig. 10-25). At this point the

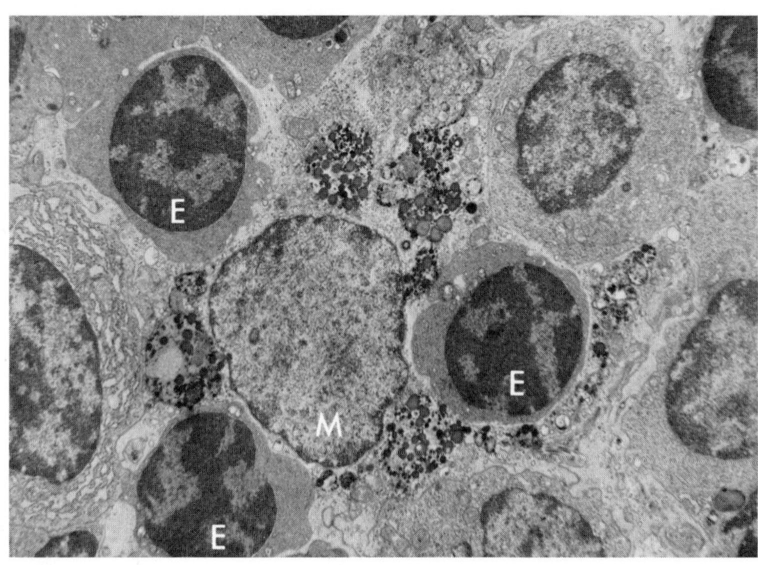

FIGURE 10-23 Electron micrograph of erythroblastic islet. Developing erythroid cells (E) are surrounding a macrophage (M).

cellular machinery has become quiescent, so the nucleus is very small and dense, or pyknotic. No further cell division is possible. The cell's nucleus is lost before it enters the circulating blood. It was once thought that the nucleus was lost by a process of karyorrhexis (nuclear fragmentation) or karyolysis. In recent years, it has been shown to be extruded intact by constriction of the cytoplasm, squeezing it toward one side, as in Fig. 10-25. The cytoplasm then pinches off, forming an RBC and leaving the denuded nucleus behind. Although the exact mechanism for the nuclear extrusion is not understood, microtubules present in the RBC may play a role.

DEVELOPMENT OF PLATELETS

Like all other formed elements of the blood, platelets originate from stem cells. The stem

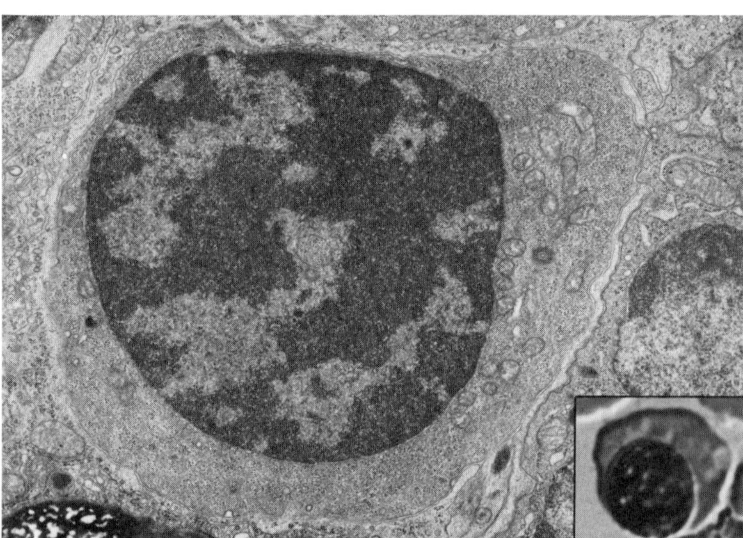

FIGURE 10-24 Electron micrograph of polychromatic erythroblast. The nucleus has an irregular condensation of the nucleoplasm. There is an increase in the number of uniform-sized polysomes and an increasing density of the cytoplasm because of the accumulation of hemoglobin. Inset, light micrograph of polychromatic erythroblast.

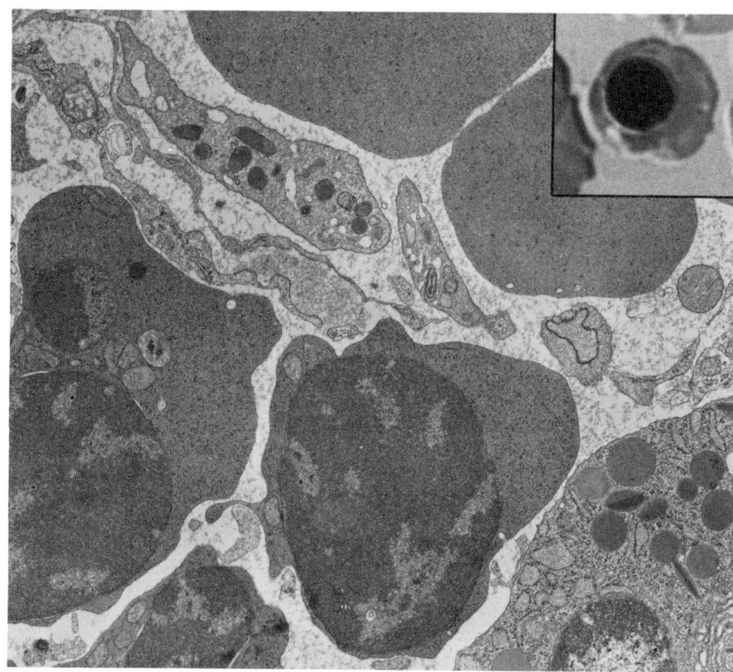

FIGURE 10-25 Electron micrograph of normoblast. The nucleus is small and pyknotic. There is an increased accumulation of hemoglobin giving the cytoplasm still more electron density. Inset, light micrograph of normoblast.

cell first differentiates into a megakaryoblast, which matures into a multinucleated giant cell called the *megakaryocyte*. These large cells, found only in the bone marrow of adults, measure at least 30 μm in diameter (Fig. 10-26). Repeated nuclear divisions without corresponding cytokinesis cause both the complex nuclear structure and the large size of the cell.

The megakaryocyte's cytoplasm is generally eosinophilic. However, EMs show numerous free ribosomes close to the nucleus as well as small stacks of Golgi lamellae, where granules are packaged. Figures 10-27

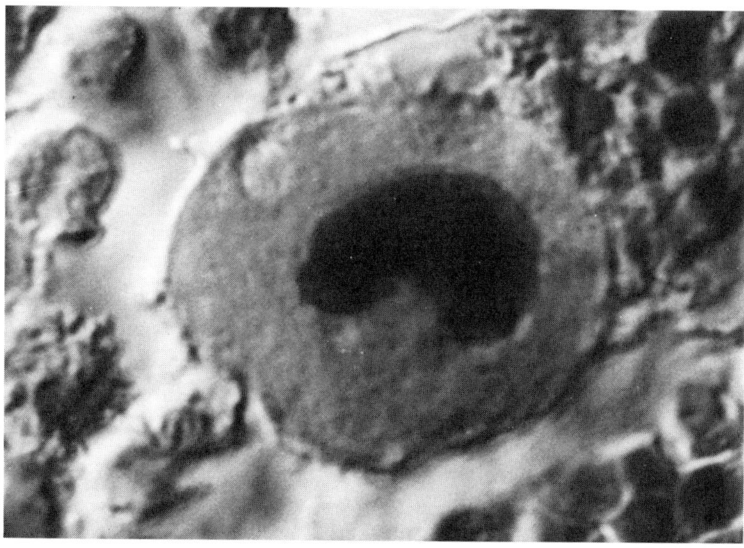

FIGURE 10-26 Light micrograph of megakaryocyte. Note the large size of the cell and the complex nuclear structure.

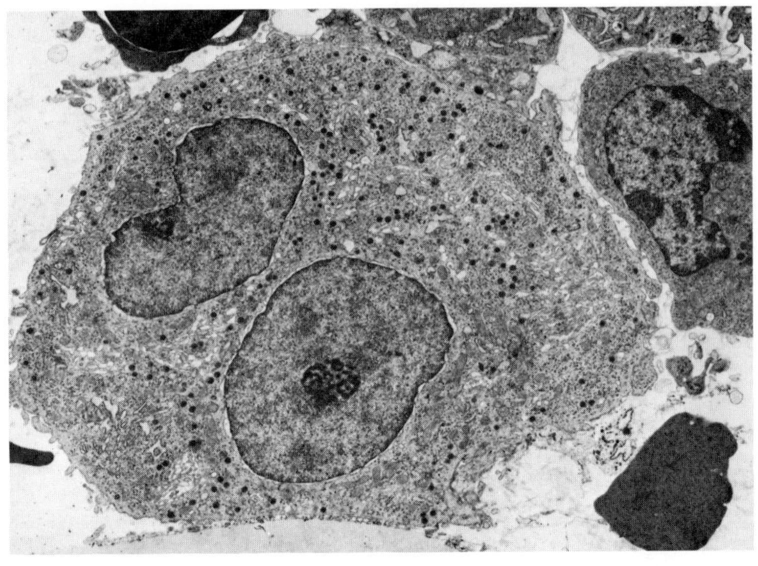

FIGURE 10-27 Electron micrograph of megakaryocyte. Two portions of the nucleus are present, one of which contains a prominent nucleolus. The cytoplasm is filled with numerous granules, ribosomes, polyribosomes, Golgi stacks, RER, and mitochondria.

to 10-29 are EMs of megakaryocytes in which numerous granules (azurophilic in LM), polyribosomes, Golgi stacks, and mitochondria are seen. Once differentiation is complete in the megakaryocyte, the "pinching off" of the platelets in the peripheral cytoplasm occurs, as shown in Fig. 10-30. This is accomplished by the development of *demarcation lines* along the periphery, which can be seen even with the light microscope. The EM shown in Fig. 10-30 reveals that these lines are made up of many small vesicles (dotted line), which eventually fuse and release a piece of cytoplasm as a platelet. This piece may contain a few ribosomes, small mitochondria, RER, vesicles, and vacuoles, as well as platelet granules. As mentioned earlier, these platelet granules are often club-shaped and correspond to the granulomere portion. Nuclear materials are never seen in platelets.

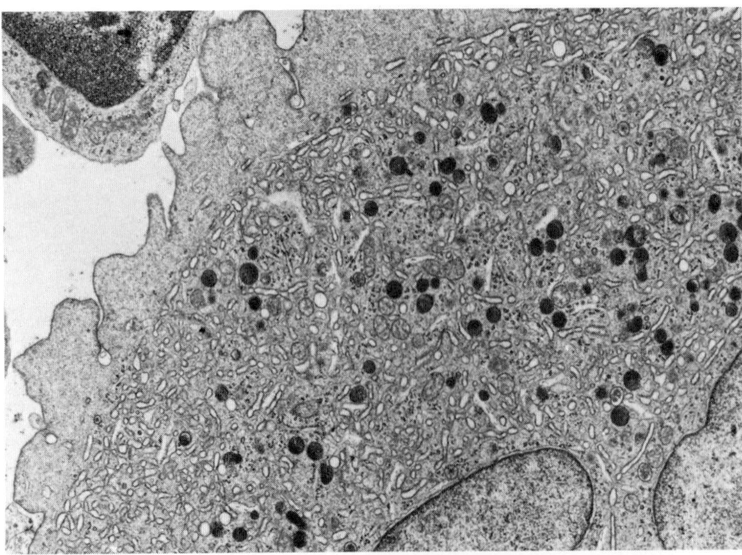

FIGURE 10-28 Portion of megakaryocyte. The cytoplasm is filled with granules. Demarcation lines are being formed by the alignment of vesicles. Future platelets are made by pinching off of the peripheral cytoplasm by fusion of these vesicles.

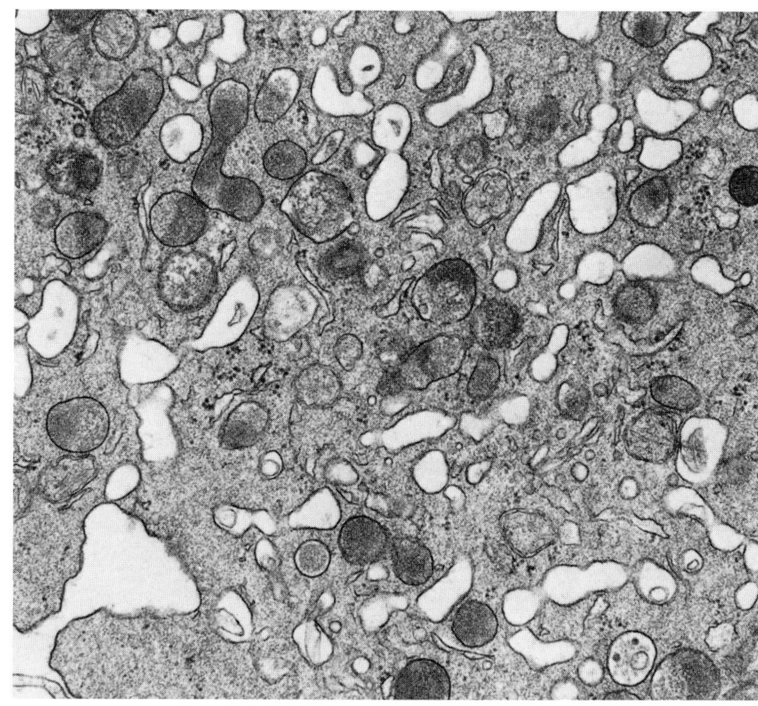

FIGURE 10-29 Portion of the peripheral cytoplasm of megakaryocyte viewed at higher magnification. Vesicles outline the periphery of future platelets.

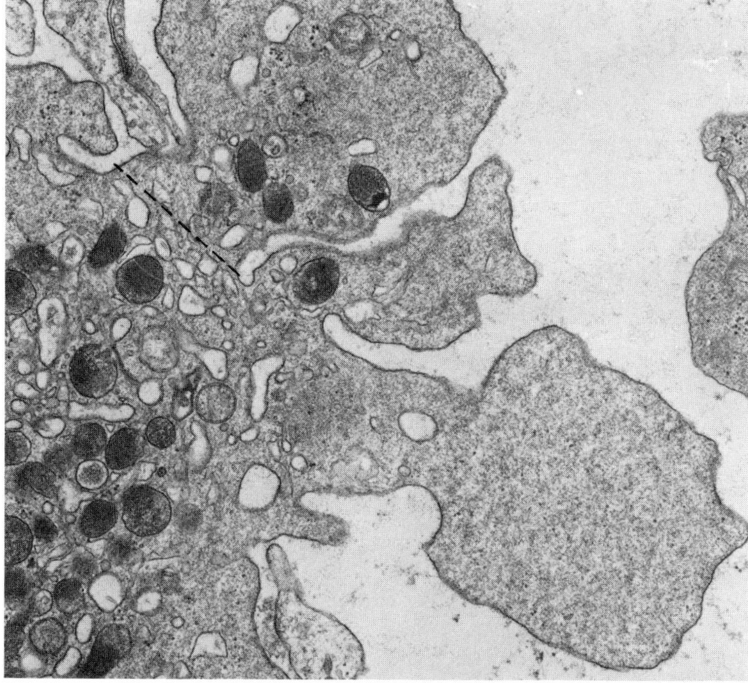

FIGURE 10-30 Electron micrograph of the peripheral cytoplasm of megakaryocyte. The demarcation lines are apparent, and mature platelets are pinching off.

SUMMARY OF DIFFERENTIATION

In summary, similar changes usually occur in the maturation of the specific blood elements.

1. The cell size becomes smaller as the cell matures.

2. The cytoplasm becomes less basophilic as ribosomes and/or RER diminish, reflecting a reduction in protein synthesis. Other cytoplasmic structures such as mitochondria become fewer in number but more differentiated in appearance.

3. The nucleus becomes smaller (and is even lost in the RBC) and the chromatin becomes denser.

4. Specific granules appear during the myelocyte stage of granulocyte formation.

REVIEW SECTION

1. Figure 10-31 is a cell from bone marrow seen in an EM. Would you consider this a young or mature cell? Why? What is its name?

This stem cell, or hemocytoblast, is fairly young because it has few organelles, a scarcity of RER, and poorly developed mitochondria. These cells may be found primarily in the bone marrow and lymphoid organs.

2. Refer to Fig. 10-32. Is this cell more or less advanced than the one in the preceding figure? What cellular components indicate advancement? What cytoplasmic organelles can you identify?

This promyelocyte represents a later stage in blood formation than did the stem cell. The more condensed peripheral nucleoplasm and abundant organelles, such as RER (Golgi complex) and mi-

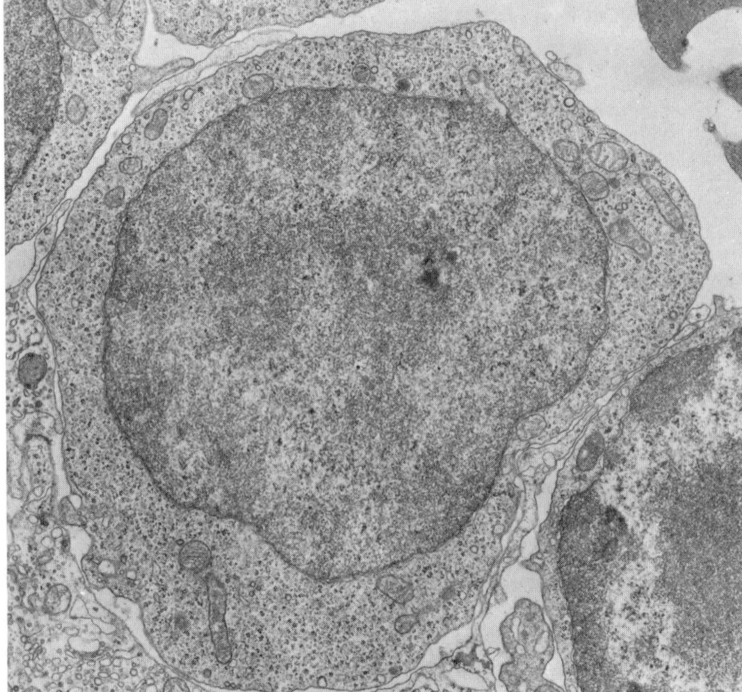

FIGURE 10-31

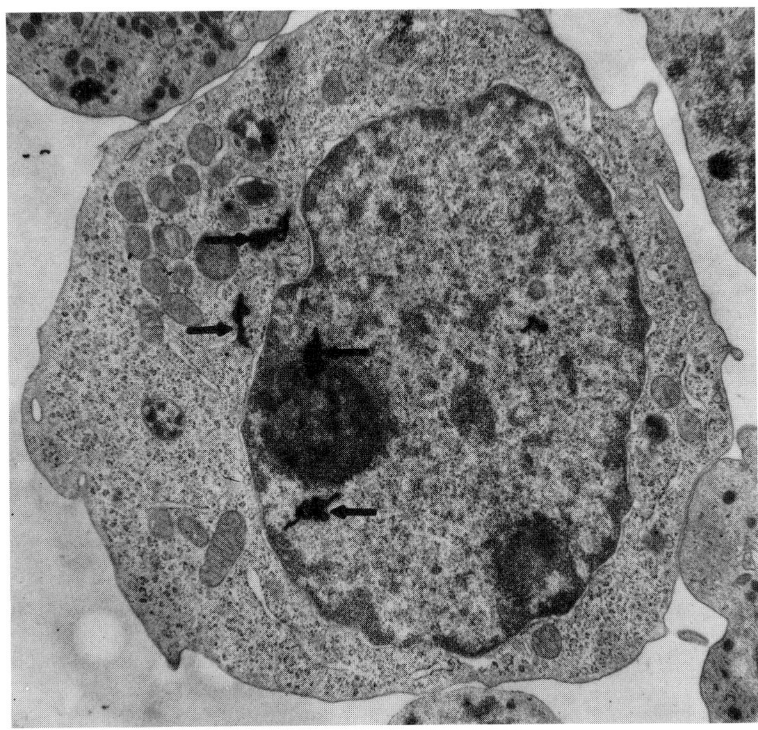

FIGURE 10-32

What would the cytoplasm appear in an LM? Name the type of cell.

tochondria, indicate a more developed cell. Azurophilic granules can also be seen. The cytoplasm would be basophilic under LM. This EM is

FIGURE 10-33

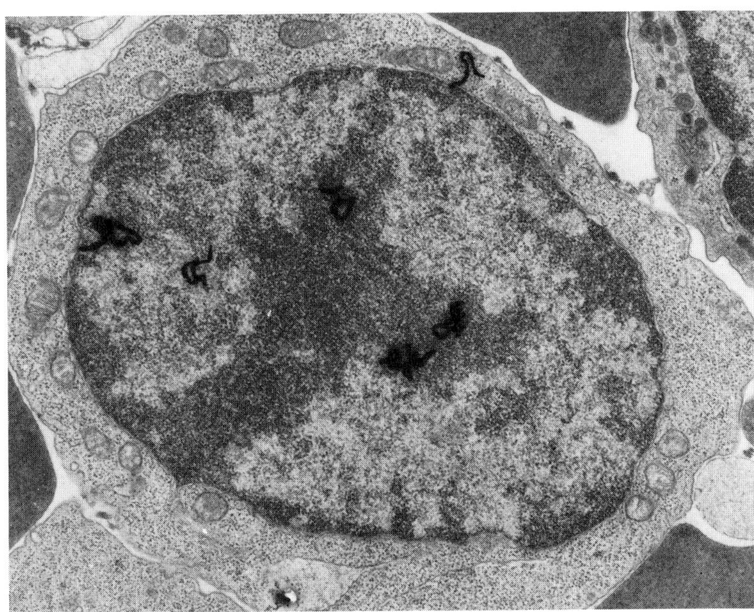

a radioautograph in which a radioactive amino acid was localized over the active portions of the nucleus and cytoplasm (arrows).

3. Refer to Fig. 10-33. This represents two stages in the development of what element? Which of the two cells is younger? What structural features indicate that?

The developing RBC initially has extensively basophilic cytoplasm and is called a *basophilic erythroblast.* Its large size also indicates a blast stage. In the more mature, smaller cell, the cytoplasm has a slight pinkish hue reflecting the accumulation of hemoglobin molecules in it. At this stage the cell may be called a polychromatic erythroblast. As in Fig. 10-32, the site of amino acid incorporation has been identified.

BIBLIOGRAPHY

Bessis, M.: *Living Blood Cells and Their Ultrastructure,* translated by R. I. Wood, Springer-Verlag, Berlin, 1973.

Bloom, W., and G. W. Bartelmez: "Hematopoiesis in Young Human Embryos," *Am. J. Anat.* **67:**21–54 (1940).

Gordon, A. S.: *Regulation of Hematopoiesis,* 2 vols., Appleton Century Crofts, New York, 1970.

Rhodin, J. A. G.: *Histology: A Text and Atlas,* Oxford University Press, New York, 1974.

Stohlman, F.: *Symposium on Hemopoietic Cellular Proliferation,* Grune & Stratton, New York, 1970.

Weiss, L.: *The Cells and Tissues of the Immune System,* Prentice-Hall, Englewood Cliffs, N.J., 1972.

——— and R. O. Greep: *Histology,* 4th ed., McGraw-Hill, New York, 1977.

Williams, W. J., E. Bentler, A. J. Erslav, and D. W. Dundles: *Hematology,* McGraw-Hill, New York, 1972.

ADDITIONAL READINGS

Ackerman, G. A.: "Ultrastructure and Cytochemistry of the Developing Neutrophil," *Lab. Invest.* **19:** 290–302 (1968).

———: "The Human Neutrophilic Promyelocyte," *Z. Zellforsch.* **118:**467–481 (1971).

———: "The Human Neutrophilic Myelocyte," *Z. Zellforsch.* **121:**153–170 (1971).

Bainton, D. F., J. L. Ulzot, and M. G. Farquhar: "The Development of Neutrophilic Polymorphonuclear Leukocytes in Human Bone Marrow," *J. Exp. Med.,* **134:**907–934 (1971).

Berman, I.: "The Ultrastructure of Erythroblastic Islands and Reticular Cells in Mouse Bone Marrows," *J. Ultrastruct. Res.,* **17:**291–313 (1967).

Bessis, M., and J. Thiery: "Electron Microscopy of Human White Blood Cells and Their Stem Cells," *Int. Rev. Cytol.* **12:**199–214 (1961).

Caffrey, R. W., N. B. Everett, and W. O. Rieke: "Radioautographic Studies of Reticular and Blast Cells in the Hemopoietic Tissues of the Rat," *Anat. Rec.* **155:**41–58 (1966).

Campbell, F.: "Ultrastructural Studies of Transmural Migration of Blood Cells in Bone Marrow of Rats, Mice and Guinea Pigs," *Am. J. Anat.* **135:** 521–536 (1972).

Capone, R. J., E. I. Weinreb, and G. B. Chapman: "Electron Microscopic Studies on Normal Human Myeloid Elements," *Blood* **23:**300–320 (1964).

Hardin, J. H., and S. S. Spicer: "An Ultrastructural Study of Human Eosinophil Granules: Maturation Stages and Pyroantimonate Reactive Cation," *Am. J. Anat.* **128:**283–310 (1970).

Marks, P. A., and R. A. Rifkind: "Protein Synthesis: Its Control in Erythropoiesis," *Science* **175:**955–961 (1972).

Murphy, M. J., J. R. Bertles, and A. S. Gordon: "Identifying Characteristics of the Hemopoietic Precursor Cells," *J. Cell Sci.* **9:**23–47 (1971).

Nichols, B. A., D. F. Bainton, and M. G. Farquhar: "Differentiation of Monocytes: Origin, Nature and Fate of Azurophilic Granules," *J. Cell Biol.* **50:** 498–515 (1971).

Nowell, P. C., and D. B. Wilson: "Lymphocytes and Hemic Stem Cells," *Am. J. Pathol.* **65:**641–652 (1971).

Yamada, E.: "The Structure of the Megakaryocyte in Mouse Spleen," *Acta Anat.* **29:**267–290 (1957).

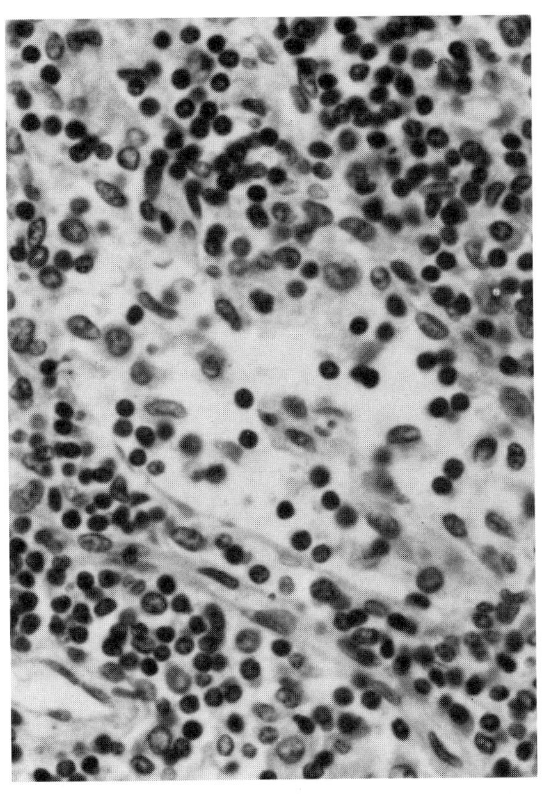

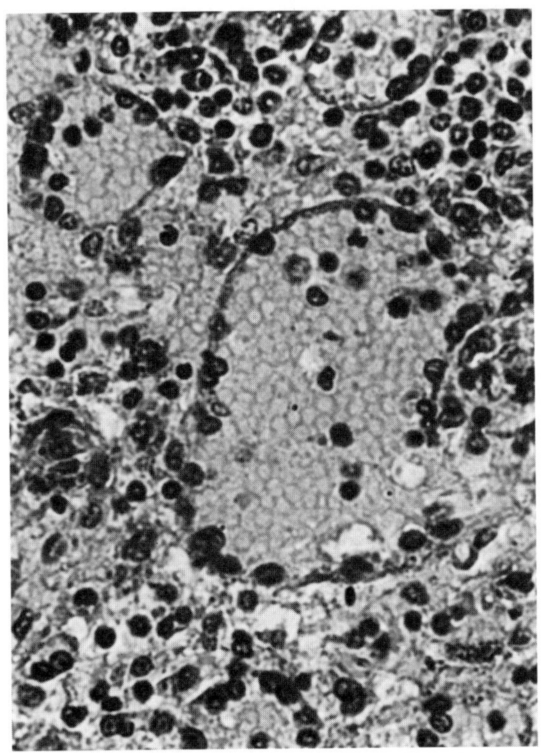

11 LYMPHOID SYSTEM

OBJECTIVES

Upon completion of this chapter, the student will be able to:

1 Identify:
(a) how lymph originates (b) morphological features that characterize lymphoid tissues (c) three different major lymphoid organs (d) the function of the lymphoid system

2 Identify on a slide or diagram of a section of lymph node the location and function (if applicable) of the following structures:
(a) afferent and efferent lymphatics and valves (b) capsule (c) septa (d) hilum (e) medulla (f) cortex (g) sinuses (subcapsular, perinodular, and medullary) (h) primary nodule (i) germinal centers or secondary nodule (j) medullary cord

3 Identify the following cellular elements:
(a) macrophages (b) plasmablasts (c) reticular cells

4 Describe the response of lymphoid cells to various antigens in cell-mediated immunity and in humoral immunity by identifying the:
(a) function of T cells (b) function of B cells (c) process of transplant rejection

5 Identify the location of the spleen in the body.

6 Identify the location and structure of the following:
(a) trabecular artery (b) trabecular vein (c)

periarteriolar sheath (d) splenic nodule (e) central arteriole (f) red pulp (g) splenic sinuses (h) cord of Billroth (or splenic cord) (i) white pulp (j) pulp artery (k) sheathed artery (l) terminal arteriole (m) pulp vein (n) germinal center

7 Describe the functions of the spleen by identifying the:
(a) mechanism of immune response occurring in the spleen (b) means by which "worn-out" blood elements are destroyed (c) means by which blood is stored and mobilized when needed

8 Trace the blood through the 10 major arteries and veins of the splenic blood vascular system from the splenic artery to the splenic vein.

9 Identify the morphology and function of the splenic sinuses that are interposed between arteries and veins in the spleen.

10 Identify three ways in which splenic endothelial lining cells differ from other endothelial cells within the body.

11 Identify the structures that account for the appearance of the following:
(a) white pulp (b) red pulp (c) cords of Billroth

12 Describe the thymus by identifying its:
(a) origin (b) location in the body (c) stroma (d) organization into lobules, medulla, and cortex (e) function

13 Identify the three changes unique to the thymus with aging past puberty with regard to the following:
(a) number of Hassall's corpuscles (b) presence of adipose tissue (c) level of function

14 Identify the location and structure of Hassall's corpuscle.

15 Identify the embryonic origin of thymic reticular cells.

The biological significance of the lymphoid system has received increasing attention during the last 15 years or so. This is largely due to an increase in our understanding of the cellular mechanism of antibody production, the nature of antigen processing, and different populations of lymphoid cells that have been known to play highly differentiated roles in different aspects of immunity. The key structure that is central to all lymphoid functions is the *lymphocyte*, which used to be considered as a "terminal" cell that might be a package of reserve nucleic acids for reutilization in augmenting nutritional needs of other cells of the tissues. This is an understandable concept and one that prevailed until the late forties, since much of earlier observation revealed no apparent dramatic changes in appearance of lymphocytes throughout the life of an organism. The structure of a lymphocyte, as you have seen in the chapter on blood, is extremely simple. It is made up of a small, dense nucleus surrounded by a thin rim of cytoplasm. Even in an EM (Fig. 11-1) the cytoplasm shows only a moderate number of ribosomes, a few small mitochondria, and occasional vesicles (or azurophilic granules). The Golgi complex, when present, is made up of a small stack of lamellar elements and might contain a small centriole in its vicinity.

It was during the early fifties that a large number of investigations came up with more careful observations of the subtle changes in histologic appearance of lymphocytes during antibody formation and a host of other immune responses which definitely pointed to the lymphocytes as being the central figure in our body defense mechanisms. Thus it has been made clear that lymphocytes not only are capable of being activated by a variety of challenges but also are composed of at least two major subpopulations that differ in function, origin, and life span.

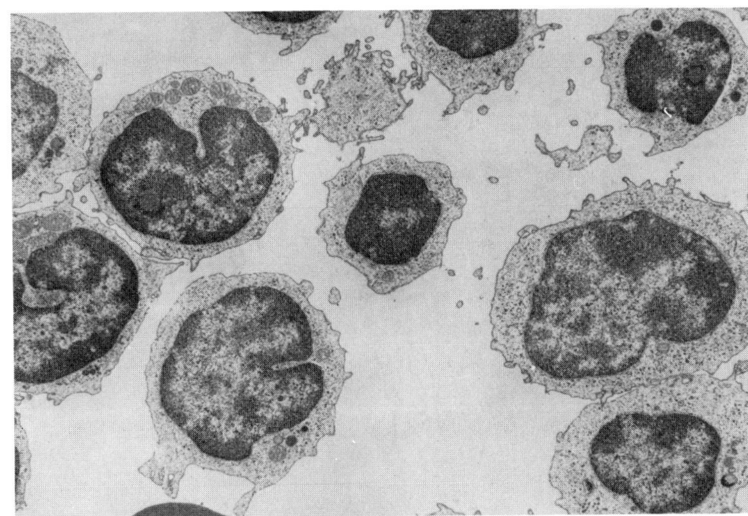

FIGURE 11-1 Electron micrograph of lymphocytes. The nuclei are dense and show large amounts of heterochromatin. Their thin rim of cytoplasm contains a moderate number of ribosomes, a few mitochondria, and occasional vesicles. Notice the large nucleocytoplasmic ratio.

Briefly, the lymphocytes can be regarded to consist of at least two different cell types: One which originates in bone marrow, called *B cell,* is short-lived and responsible for production of humoral antibodies; the other, which arises in the thymus, called *T cell,* lives for years and serves as an immunologic memory cell responsible for cell-mediated immunity. Cell-mediated immunity is expressed in such vital reactions as immune removal of tumor cells, host-versus-graft reaction, and rejection of transplanted organs. The interactions between T and B cells are under intensive study. Helping these cells function are a number of other cell types such as macrophages and various reticular cells that make up the stroma of all lymphoid organs.

The basic and collective functioning of the lymphoid organs consists primarily of elimination of noxious and foreign substances that enter the body. The lymph node is concerned with the elimination of foreign materials that may invade the body through tissue spaces, while the spleen is responsible for elimination of like substances that come via blood. The thymus, along with bone marrow, is the organ where precursors of lymphocytes are continuously produced. In addition, connective tissues underlying thin epithelia, such as respiratory and digestive tract linings, contain isolated aggregates of lymphocytes that counter transepithelial invasions along the body surface.

The lymphatic system begins throughout the body as a series of blind ends that gather tissue fluids. The ducts with blind ends join similar lymphatics to produce larger ducts which eventually lead to the central collecting ducts, namely, thoracic and right lymphatic ducts. These ducts pour the lymph into the venous blood near the heart. Along the system of lymphatic ducts are located regional lymph nodes which may carry names identifiable with the particular locale. The main lymphatic channels and nodes of the human body are shown in Fig. 11-2.

LYMPH NODES

STRUCTURE

The lymph node is a small, usually bean-shaped body which is encapsulated by fibroelastic connective tissue. It may be barely visible or may measure up to an inch in length, depending on the functional status. It is indented on one side, where the blood vessels enter and leave and where the efferent

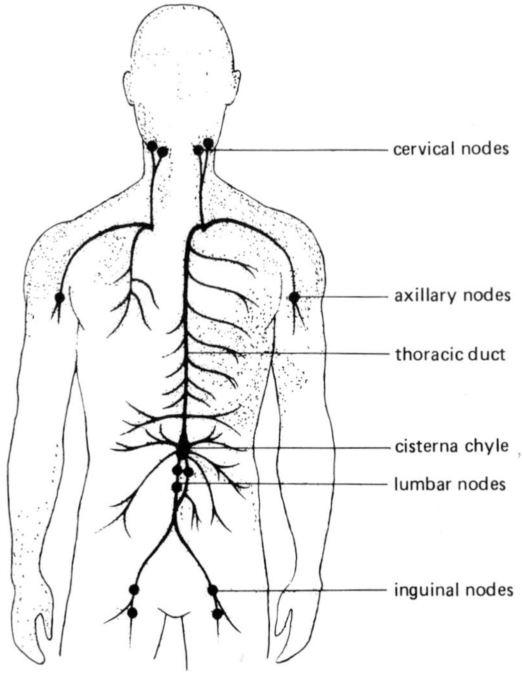

FIGURE 11-2 Drawing illustrating the main lymphatic channels and nodes of the human body.

lymphatics exit. This region is known as the *hilum*. The afferent lymphatics bring the lymph to the node and enter at various points along its capsule. Since the lymphatic system does not have a pumping mechanism, such as the heart in the vascular system, many one-way valves are necessary in both afferent and efferent lymphatics to prevent backflow of lymph. Figure 11-3 is a diagrammatic representation of a lymph node. Only two efferent lymphatics can be seen exiting at the hilum, while numerous afferent lymphatics empty into the node along the capsule. Figure 11-4 is a low-power LM of a lymph node in which the capsule (c) and a couple of cortical nodules (n) are found.

The lymphocytes and stroma of the lymph node are organized so that the flow of lymph from afferent lymphatics passes through a series of sinusoidal spaces lined with phagocytic and endothelial elements. The space immediately under the capsule is the *subcapsular sinus,* which is continuous with *perinodular sinuses* around aggregates of lymphocytes called *lymphoid nodules.* From the perinodular sinus, the lymph flows into a system of *medullary sinuses,* which are bordered by cords of cells that contain plasma cells and macrophages. Figure 11-5 depicts

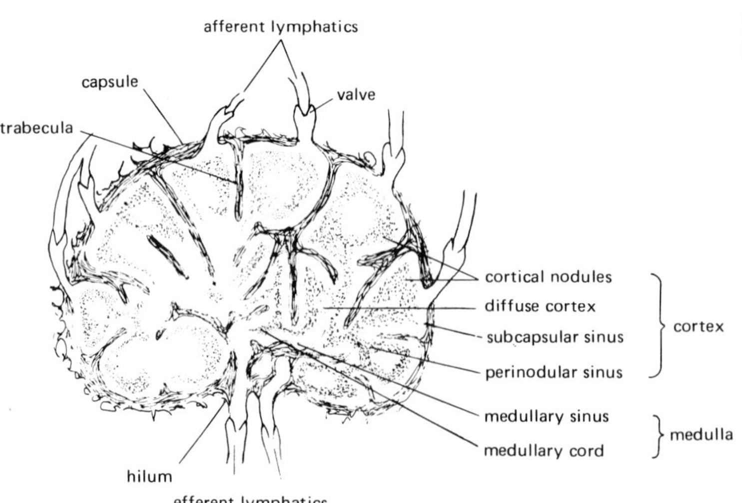

FIGURE 11-3 Diagram of a lymph node.

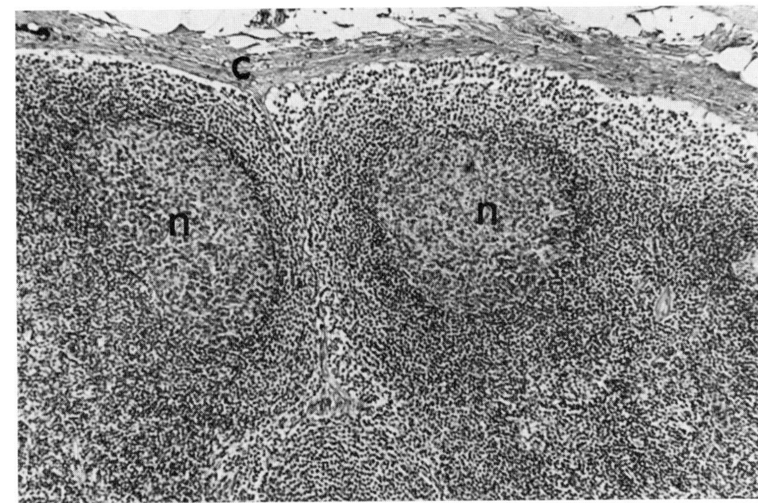

FIGURE 11-4 Light micrograph of the cortex of a lymph node. The capsule (c) and two cortical nodules (n) are seen in the upper half and the diffuse cortex occupies the lower half.

the general pattern of orientation of lymphoid nodules along the more superficial region called *cortex* (c), as well as the irregularly arranged, deeply situated *medulla* (m). The cortex may be subdivided into (1) the superficial region, where nodules of lymphocytes (n) are present, and (2) an inner, *diffuse cortex,* indicated by double-ended arrows.

The capsule of a lymph node merges with the surrounding connective tissue, as shown in Fig. 11-6. It is penetrated by afferent lymphatics which are filled with lymphocytes (arrowheads in Fig. 11-6). Immediately subjacent to the capsule is the subcapsular sinus (arrows). The sinus as well as the lymphoid stroma has a loose network of reticular cells (Fig. 11-7). Note the relative absence of other cellular elements in the sinus areas. The appearance of reticular cells traversing through the medullary sinuses is clearly depicted in light micrographs taken at higher magnification (Fig. 11-8).

Traditionally, the reticular cells of the lymphoid organs were thought to be of two kinds, primitive and phagocytic, and the reticular fibers were thought to be the product

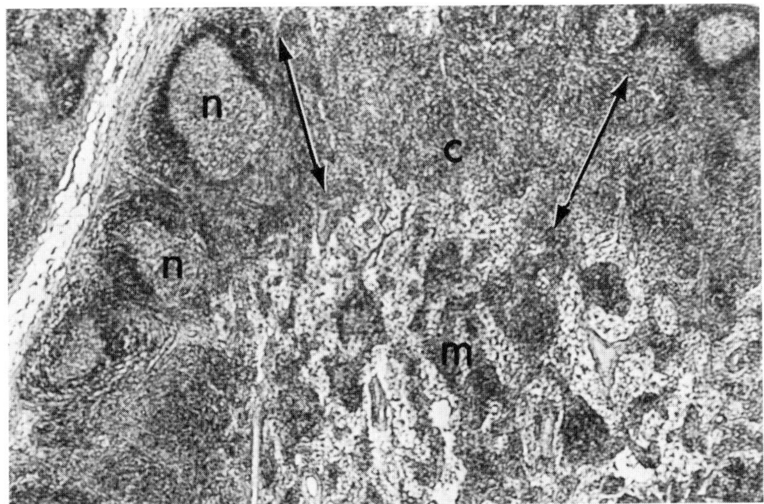

FIGURE 11-5 Light micrograph illustrating the general pattern of organization of lymph nodes. Lymphoid nodules are located in the superficial region called the *cortex* (c). The medulla (m) is irregularly arranged and situated deep in the center of the lymph node. The cortex may be divided into an outer region which contains the nodules and an inner region, the diffuse cortex (double-ended arrows). The nodules seen in this micrograph are secondary nodules, since they have a lightly stained center. Such nodules, called *germinal centers,* may contain many lymphoblasts.

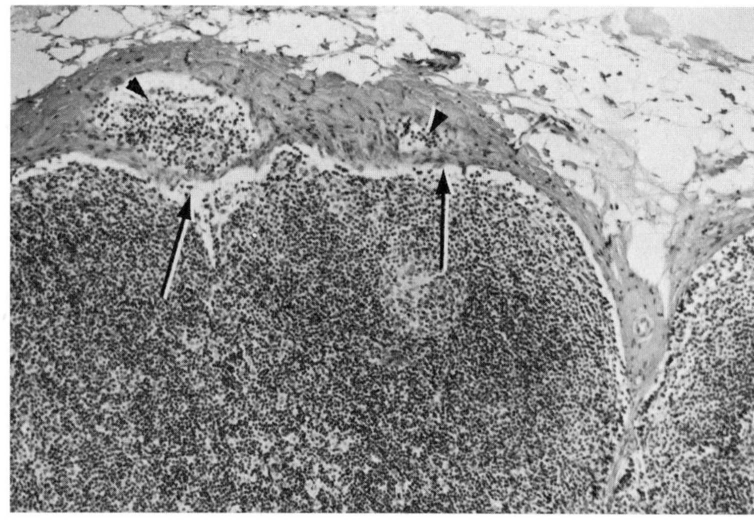

FIGURE 11-6 Light micrograph of a portion of a lymph node. The capsule which consists of dense FECT merges with the loose FECT surrounding the lymph node. It is penetrated by two afferent lymphatic vessels which contain lymphocytes (arrowheads). The subcapsular sinus (arrows) separates the capsule from the cortex.

of the primitive variety. Electron microscopy has confirmed that there are phagocytic reticular cells in the lymph node which are in close proximity to the reticular fibers (f in Fig. 11-9). These phagocytic reticular cells show all the structural characteristics of a macrophage, namely, primary lysosomes, phagosomes, vesicles, vacuoles, and ruffling of plasma membranes.

With respect to the so-called primitive reticular cell, there appears to be no clear structural definition acceptable to current workers. However, recent investigations have revealed that the fine structure of certain reticular cells associated with reticular fibers is not unlike that of fibroblasts. Such a reticular cell has a cytoplasm which is filled with RER, mitochondria, and a juxtanuclear Golgi complex (g) and therefore resembles a fibroblast (Fig. 11-10). Usually, the *fiber-associated reticular cell* has a cytoplasm that wraps around a reticular fiber which in an EM appears identical with a collagenous fiber (f in Fig. 11-10). Since the electron microscopic

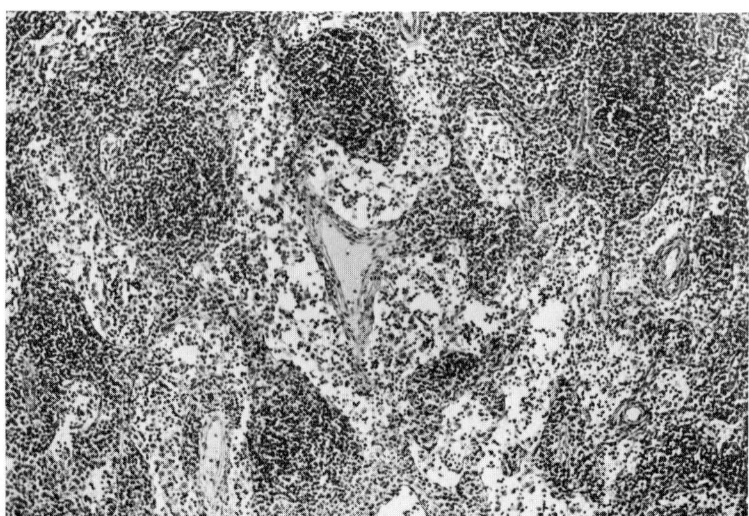

FIGURE 11-7 Light micrograph of a portion of the medulla of a lymph node. It shows the appearance of the stroma and the medullary sinuses. They contain a loose network of reticular cells. Note the relative absence of other cellular elements in the sinuses.

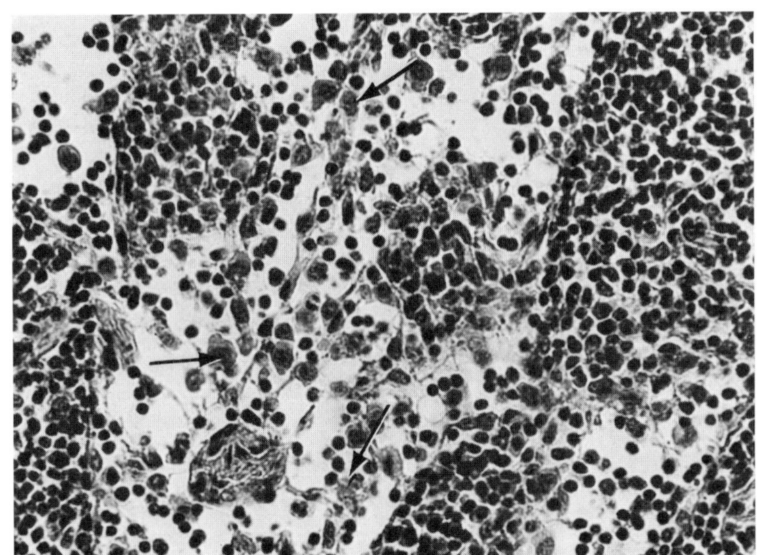

FIGURE 11-8 Light micrograph of reticular cells (arrows) in the medullary sinuses. Note the relative absence of lymphocytes and other cell types within the sinuses.

appearance of the reticular fibers is the same as that of collagenous fibers, and since the lymphoid reticulum is continuously turning over, the close positional relationship between the reticular cell and fibers and the structural resemblance between fibroblasts and fiber-associated reticular cells are teleologically understandable from an evolutionary standpoint.

Those reticular cells that are facing the sinusoidal spaces of the lymph node are somewhat different from other reticular cells in that they do have small micropinocytic vesicles along the surface (arrows, Fig. 11-11). Frequently they are phagocytic, and these cells also have cellular processes which wrap around small reticular fibers (f in Fig. 11-11) that are close to the sinusoidal lumen (l in

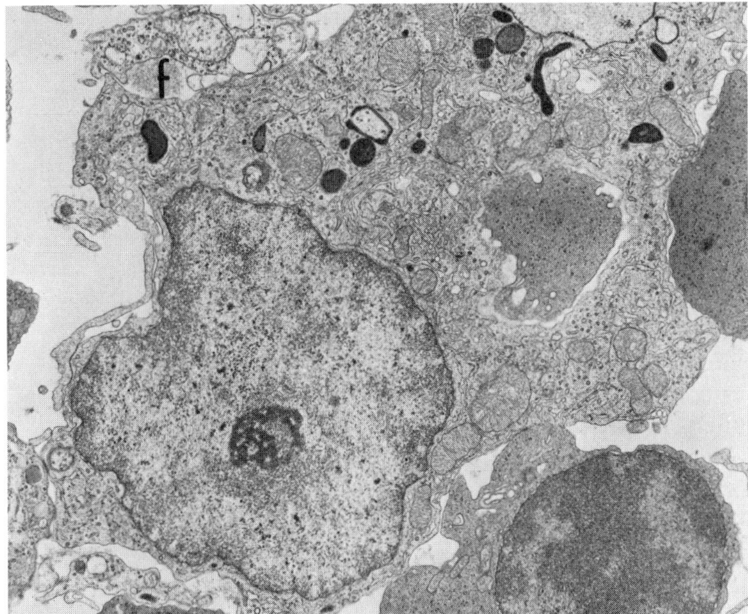

FIGURE 11-9 Electron micrograph of a phagocytic reticular cell which is in close proximity to reticular fibers (f). The somewhat irregularly shaped nucleus contains a nucleolus and some heterochromatin around the periphery. This cell has all the structural characteristics of a macrophage. Namely, the cytoplasm contains primary lysosomes, phagosomes, vesicles, vacuoles, RER ,and a well-developed Golgi apparatus. Its cell membrane also shows ruffling.

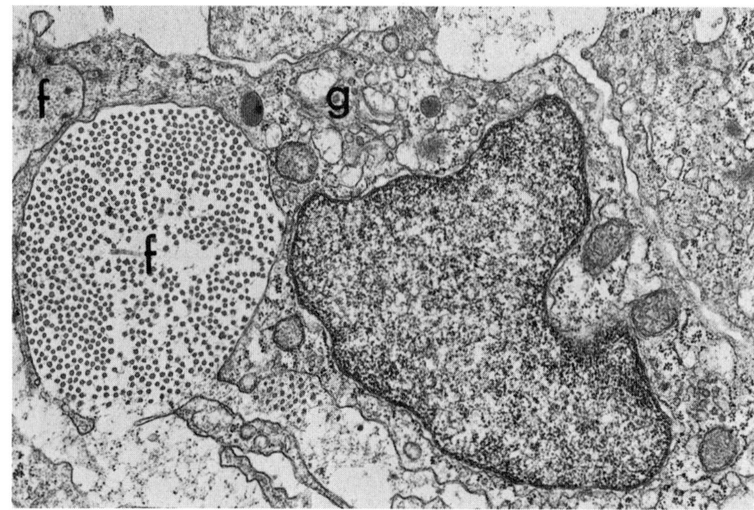

FIGURE 11-10 Electron micrograph of a fiber-associated reticular cell. It has an indented nucleus and a cytoplasm containing RER, mitochondria, and a Golgi complex (g). The cytoplasm wraps around reticular fibers (f), which in this micrograph appear similar to collagenous fibers. Therefore the structure of this cell seems to be similar to that of a fibroblast.

Fig. 11-11). In addition to the micropinocytic vesicles mentioned above, lining cells of the lymphoid sinusoids have been found to be phagocytic. For these reasons the cells covering the sinusoids can be regarded as a special type of *endothelial reticular cell* that shares structural and functional characteristics of the vascular endothelium and fiber-associated reticular cells. This type of phagocytic endothelial reticular cell is also found in endocrine organs and the liver. The EM in Fig. 11-12 is another example of the sinusoidal endothelial reticular cell from the subcapsular sinus. They are often active in pinocytosis of small particulate matters, as shown by arrows in Fig. 11-12. The barely visible small particles are ferritin molecules introduced to the subcapsular sinus by injection.

The cortex of the lymph node is divided into compartments by connective tissue *septa* or *trabeculae,* which are extensions of the capsule (Fig. 11-3). These compartments contain the lymphoid nodules, which are frequently called *primary nodules* and usually appear dark because of the large nucleocytoplasmic ratio of lymphocytes in them. Primary nodules are continuous with the diffuse cortex and medullary cords (Fig. 11-5). The cortex, seen in Fig. 11-13, contains a few primary nodules of dense lymphoid cells (circled).

Frequently the cells in the center of the primary nodule are active in lymphocytopoiesis. Since such an area contains large lymphoblasts, it appears light in LM and therefore has been called a *germinal center* or *secondary nodule* (n in Fig. 11-4). Since germinal centers appear during activation of lymphoid tissues by foreign matters, the term *reaction centers* has often been used to designate the large germinal centers present in lymph nodes subjected to challenge (Fig. 11-5).

As mentioned earlier, the medullary cords are made up of many plasma cells. These cells of the medullary cord function in the production of antibodies. Figure 11-14 shows a portion of the medulla in which medullary cords and sinuses are clearly defined. Note the sinusoidal reticular cells.

FUNCTIONAL CONSIDERATIONS

Within the lymph node, the first line of defense against bacteria is performed by macrophages and fixed phagocytic reticular cells. Free macrophages can move about in the sinuses, as they do in other tissue spaces, and engulf foreign material. Phagocytic re-

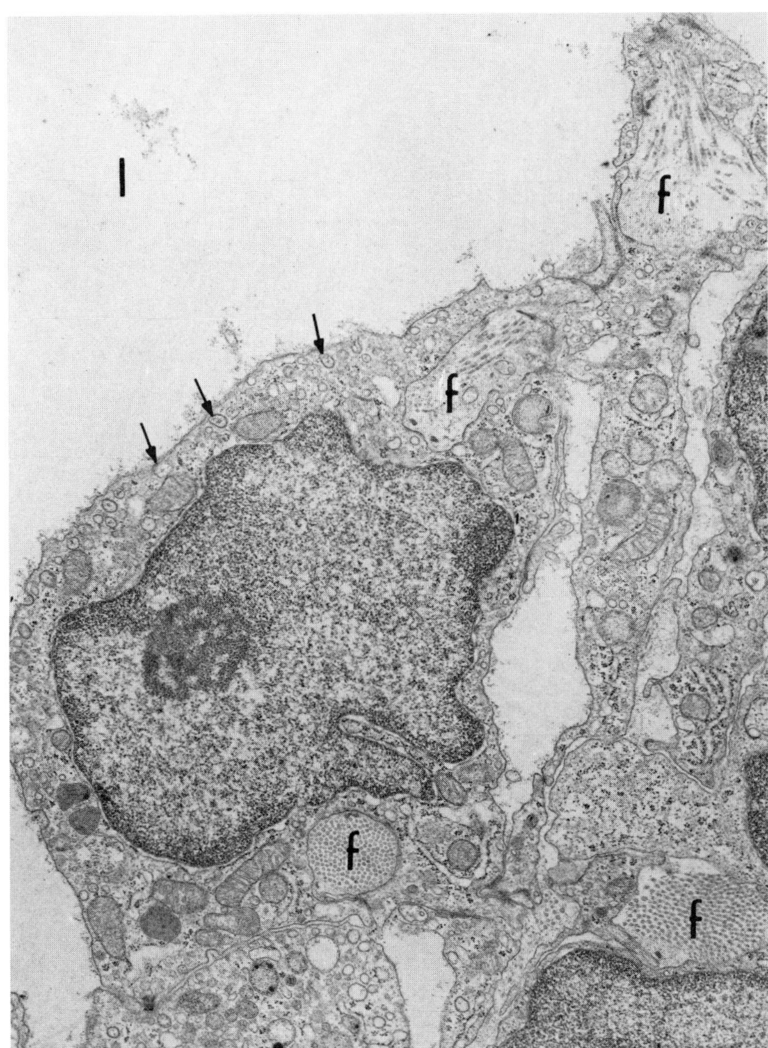

FIGURE 11-11 Electron micrograph of reticular cells lining a sinusoidal space of lymph nodes. These cells have small micropinocytic vesicles along their luminal surface (arrows). They also have cellular processes which wrap around small reticular fibers (f) that are close to the lumen (l). In addition, these lining cells are phagocytic. Therefore, they can be regarded as a special type of endothelial cells, i.e., they share structural and functional characteristics of vascular endothelial cells and fiber-associated reticular cells.

ticular cells (fixed macrophages) lining the sinuses are kept in place by the stroma of reticular fibers. As mentioned previously, littoral cells that are nonphagocytic may be activated when challenged. The precursor of macrophages is thought to be the monocyte, which becomes an active macrophage upon leaving the blood and entering lymphoid organs. The macrophages digest most foreign materials, including bacteria and other antigenic substances. Since plasma cells and plasmablasts, which produce antibody proteins, are not capable of phagocytosis, the transfer of antigenic information has to take place between the phagocytic elements and antibody-producing cells. The functional relationship between the two cell types is often illustrated in microscopic preparations which show a number of developing plasma cells surrounding a macrophage. This has been called a *plasmacytic islet* (Fig. 11-15). The relationship between a macrophage and developing plasma cells in a plasmacytic islet is not unlike that found in erythropoiesis. Thus the macrophage plays an important role by providing appropriate stimuli in the form

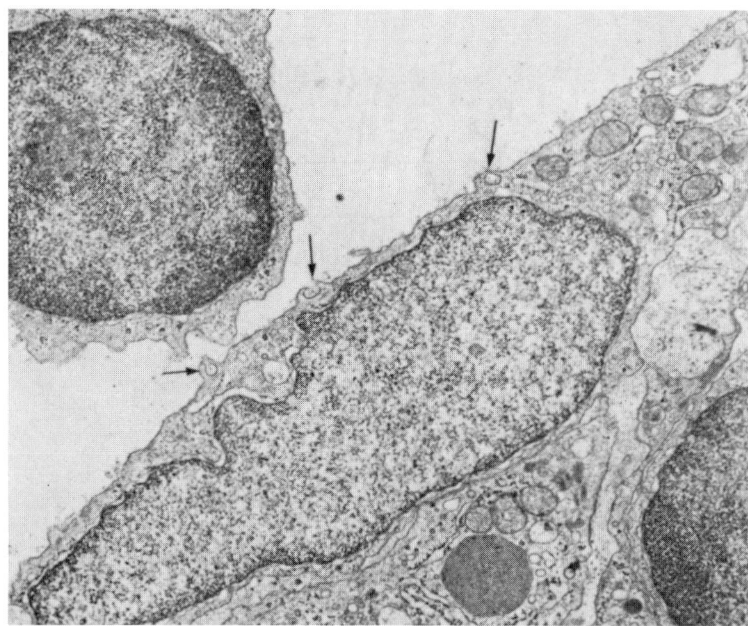

FIGURE 11-12 Portion of a sinusoidal lining cell. The luminal surface shows forming micropinocytic vesicles (arrows).

of recycling tissue components or partially digested antigens.

Recent experiments have shown that the two major subpopulations of lymphocytes (viz T and B cells) occupy distinct histologic regions. For example, much of the B cells are situated in the cortical nodules, whereas the diffuse cortex contains most of the T cells. The life span of B cells is short. Once they leave the site of origin, the B cells die within a matter of weeks. On the other hand, the long-living T cells have been shown to persist and recirculate for years.

SPLEEN

The filtration of tissue fluids is carried on by lymph nodes; the spleen, on the other hand,

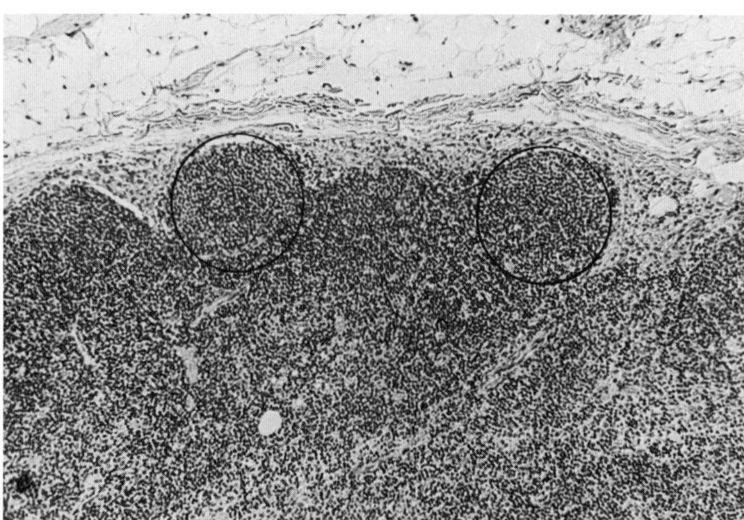

FIGURE 11-13 Portion of a lymph node. Primary nodules (circles) which are continuous with the diffuse cortex and the medullary cords are seen in the outer part of the cortex next to the subcapsular sinus. They are dark because of the dense aggregation of lymphoid cells.

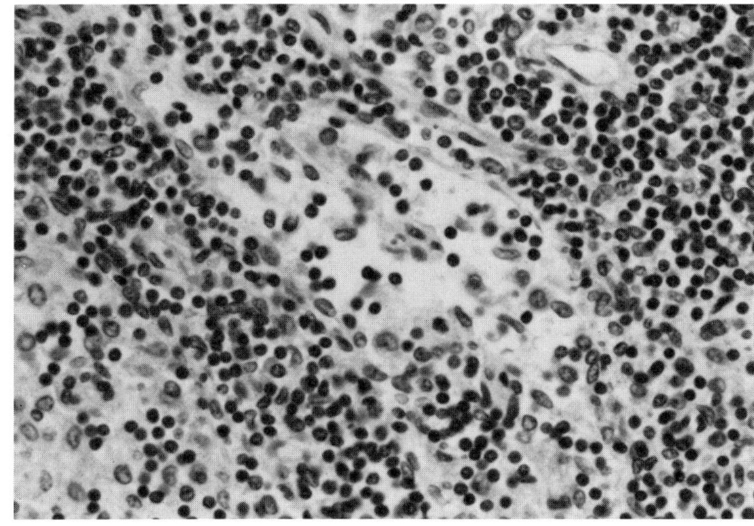

FIGURE 11-14 Portion of the medulla. It consists of medullary cords and medullary sinuses. Many of the cells in the medullary cords are plasma cells which are involved in the production of antibodies. Note the sinusoidal reticular cells.

is concerned with the filtering of blood, which may carry bloodborne foreign matter. Foreign substances may enter blood directly, or they may enter it via the lymphatics if the regional lymph nodes cannot adequately handle the foreign materials in the lymph. In its fucntion, the spleen is aided by the liver, which depends on its population of phagocytes, called *von Kupffer cells,* to engulf and digest the foreign substances. The spleen is

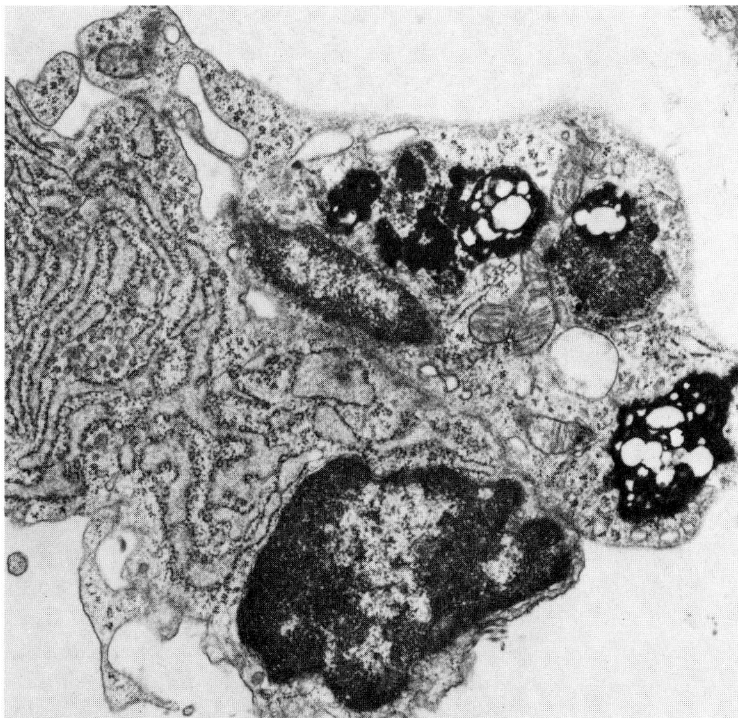

FIGURE 11-15 Electron micrograph of a plasmacytic islet. A plasma cell is in intimate contact with a macrophage.

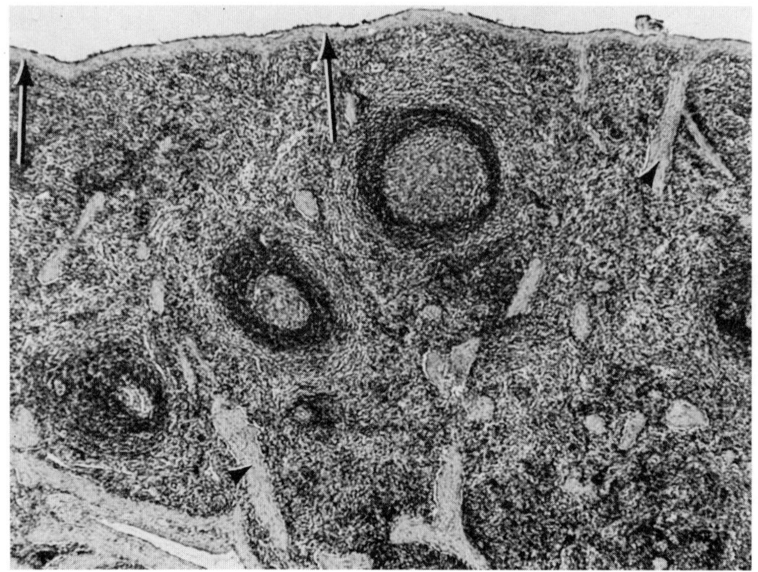

FIGURE 11-16 Light micrograph of a portion of the spleen. It is invested in a capsule (arrows) of FECT which contains smooth muscle. Trabeculae (arrowheads) of dense FECT extend from the capsule into the parenchyma and partially subdivide it. Several large, splenic nodules which contain germinal centers are seen.

different from the liver in that it processes the foreign materials in such a way that an immune response results.

In this section the structure of the spleen will be considered in light of its basic functioning as a blood filtration system and antibody production site. The role of the liver in the disposal of foreign materials will be discussed in a later chapter.

STRUCTURE

The spleen is located on the left side of the abdominal cavity posterior to the stomach. Since the spleen filters the blood and, in the process, responds immunologically, an understanding of the structure of the spleen, particularly with respect to the organization of its vessels, is essential in comprehending the biologic roles of the spleen. A capsule of connective tissue covers the spleen. Unlike other lymphoid organs, the capsule of the spleen contains numerous smooth muscle cells invested in dense FECT. Trabeculae extending from the capsule are made of similar tissue components and partially divide the parenchyma of the spleen. A low-power view of the spleen showing the dense lymphoid tissue, thick capsule (arrows), and the trabeculae (arrowheads) is seen in Fig. 11-16. It may be noted that there are several lymphoid nodules that are stained dense because of the concentration of lymphocytes. Note also the two large lymphoid nodules in which germinal centers are present.

When a fresh spleen is sectioned, it discloses distinct red and white areas. For this reason the parenchyma of the spleen has been divided into *white pulp* and *red pulp*. The white pulp is composed of the splenic nodules, trabeculae, and connective tissue. The red pulp of the spleen consists of the splenic sinuses and the cells that fill the remaining tissue space. Since these cells are present in the form of cords, these arrangements are called *cords of Billroth*, or splenic cords. The red pulp appears red because of the high content of RBCs. The cords of Billroth (Fig. 11-17) contain macrophages that break down the worn-out RBCs, granulocytes, and platelets. Distended venous sinuses (S in Fig. 11-17) are filled with RBCs. The regions of the red pulp other than the sinuses are the cords of Billroth.

Figure 11-18 is a diagrammatic presentation of the vasculature of the spleen. The *splenic artery* penetrates the capsule at the hilum and branches into a number of *trabecular arteries*, which become smaller as the

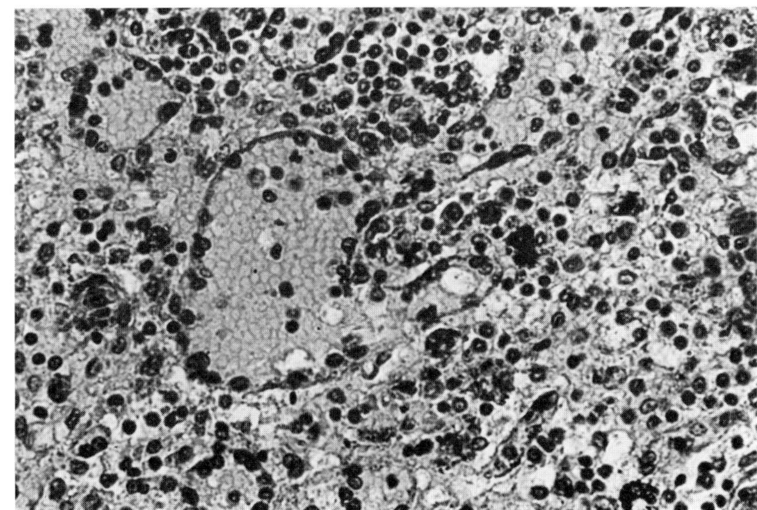

FIGURE 11-17 Red pulp of the spleen. Distended venous sinuses (S) are filled with RBCs. Macrophages in the cords of Billroth, which occupy the regions between the sinuses, are actively breaking down worn-out RBCs.

arterial tree continues to branch. When the trabecular artery leaves a trabecula, the artery is covered by a periarteriolar sheath of lymphocytes. Along this segment of artery are many isolated lymphoid nodules, here called *splenic nodules.* In histologic sections, the artery that passes through the splenic nodule is usually eccentrically located but is called a *central artery.* The central artery eventually becomes a *pulp artery* because it emerges from the lymphocyte sheath where cords of splenic cells and sinusoids are present. The terminal end of the pulp artery is surrounded by a few layers of epithelioid cells and is called a *sheathed artery.* The sheathed artery branches into smaller *terminal arterioles,* each of which empties into the space of a splenic sinus. Blood from the splenic sinuses is collected into a *pulp vein,* which by joining with similar veins and entering a trabecula be-

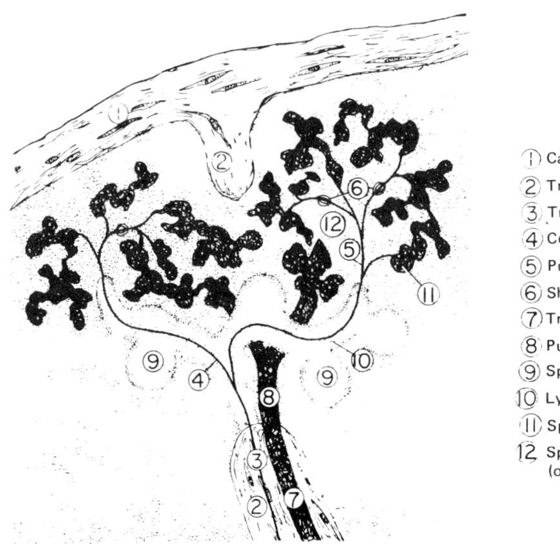

FIGURE 11-18 Diagram illustrating the vasculature of the spleen.

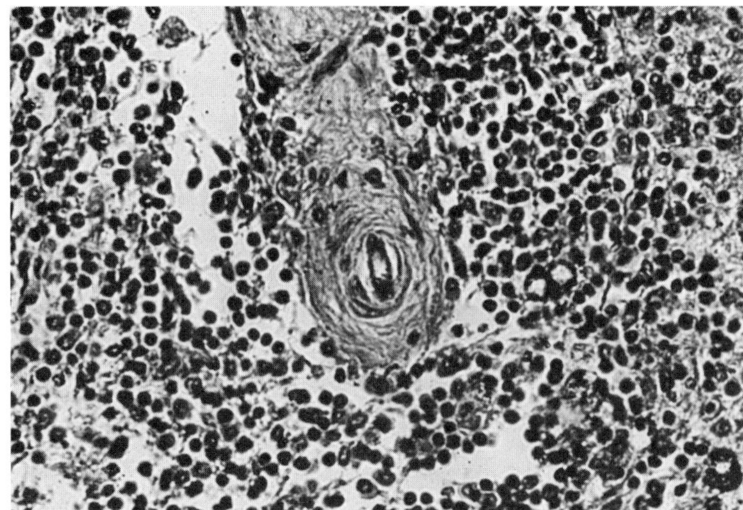

FIGURE 11-19 Trabecular artery.

comes a *trabecular vein*. Trabecular veins merge and form the *splenic vein*.

In the following series of photomicrographs, the various vascular segments will be illustrated. A few trabecular vessels are present. In Fig. 11-19, a cross section of a trabecular artery is located near the center of the photomicrograph. A splenic nodule at a lower magnification is seen in Fig. 11-20. A central artery is seen in the upper periphery of the nodule, surrounded by a periarteriolar sheath of lymphocytes. Under certain conditions, the splenic nodule (primary nodule) attached to the sheath may enlarge and show a light-colored germinal center as seen here. In the upper right quadrangle of Fig. 11-21, two pulp arteries (arrows) are present. They continue to become sheath arteries. In humans the sheathed artery is rather short and may be difficult to find in random histologic sections.

In Fig. 11-22, several terminal arterioles (arrows) are seen. The close physical relationship of these arterioles (arrowheads) in-

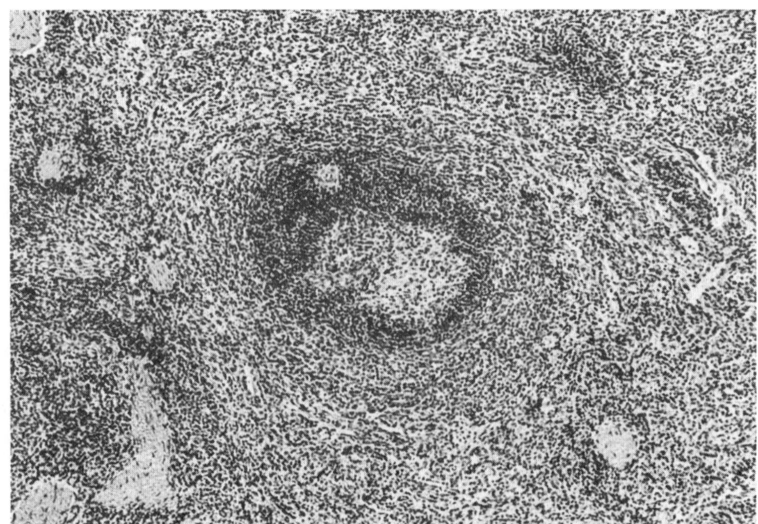

FIGURE 11-20 Low-power micrograph of splenic nodule. The central artery (arrow) is located in the upper periphery of the nodule surrounded by a periarteriolar sheath.

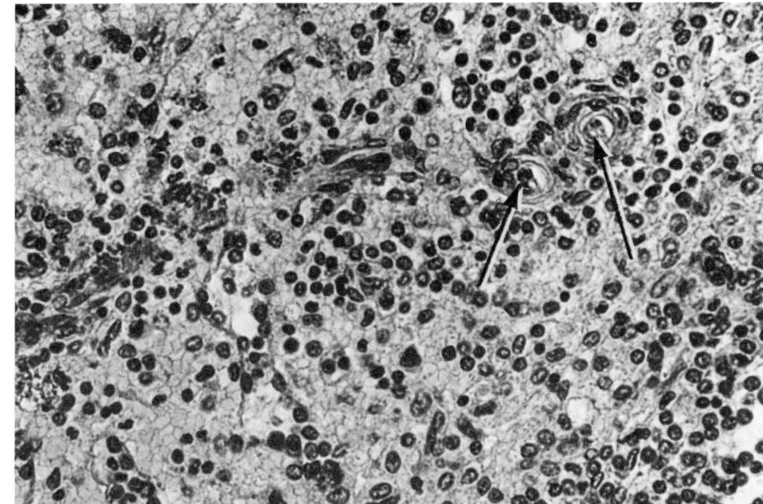

FIGURE 11-21 Pulp arteries (arrows). They are covered by a sheath of thin cells.

dicates that they might have just branched off from a sheathed artery.

The spleen processes the blood much like the lymph nodes process the lymph. We have noted the small diameter of terminal arterioles, each of which pours the blood into a large sinusoidal space in the red pulp. This abrupt change in size of the vessel results in a drastic slowing of blood flowing through the sinus. The relatively long stay of blood within the sinusoid and the structural differentiation of the sinusoidal wall allow the interaction between bloodborne elements and macrophages and other cells of the cord of Billroth.

Along the walls of splenic sinusoids is a broken or incomplete basement membrane which is covered by endothelial cells. They are spindle-shaped and phagocytic. Since they are spindle-shaped, endothelial cells of splenic sinusoids may be in contact with each other at their ends, leaving patent spaces between them. Seen in cross sections, the endothelial cells protrude into the lumen of the sinuses (Fig. 11-23).

Electron microscopy has revealed the

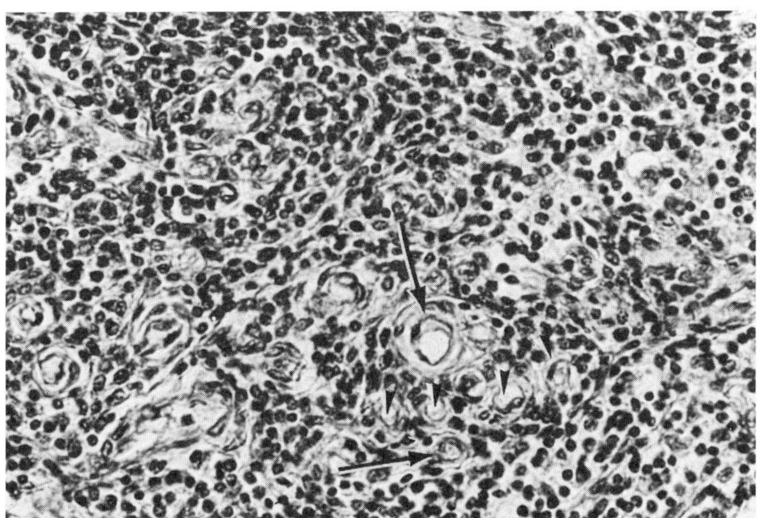

FIGURE 11-22 Light micrograph of terminal arterioles (arrows) in the red pulp. The close physical relationship between some of them (arrowheads) indicate that they might have just branched off a pulp artery.

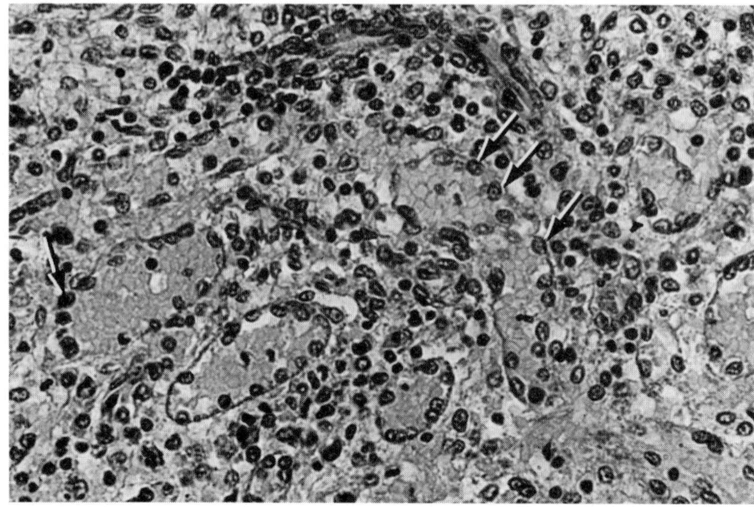

FIGURE 11-23 Light micrograph of several sinuses seen in cross section. They are lined by endothelial cells (arrows) which protrude into the lumen. The endothelial lining appears to be discontinuous.

presence of the gaps in the basement membrane (bm in Fig. 11-24). The basement membrane separating the splenic sinus (s in Fig. 11-24) from the cord of Billroth (B in Fig. 11-24) is interrupted. A loose network of reticular fibers holds the lining cells together and surrounds the entire sinusoidal space. It is through this space between endothelial cells that RBCs migrate into the splenic cords where the tired out ones are phagocytized. During contraction of smooth muscle cells in the capsule and trabeculae, RBCs and other elements of the cord may be mobilized from the spaces to join the blood. In this way, the spleen serves as a reservoir of blood which can be used by the body when needed. The schematic drawing in Fig. 11-25 demonstrates various sections of splenic sinuses. It shows a cutaway view of a sinus, demonstrating spindle-shaped lining cells and the reticular network of fibers produced by the fiber-associated reticular cells.

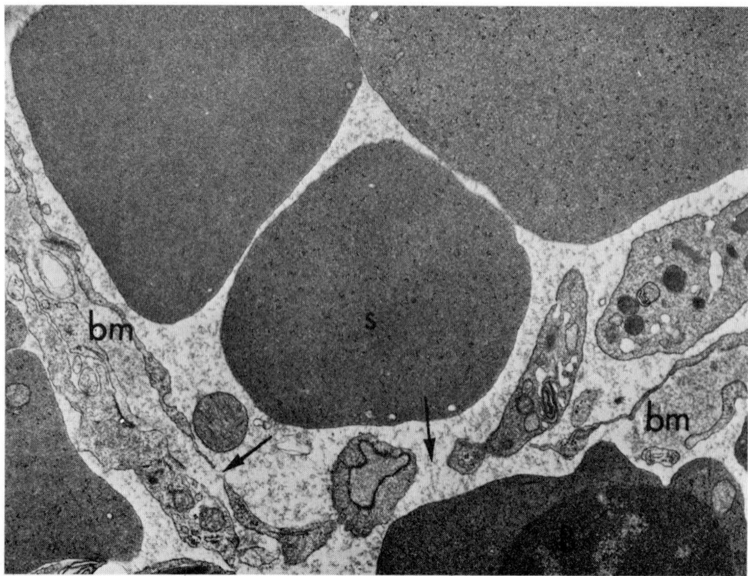

FIGURE 11-24 Electron micrograph of a portion of a splenic sinus. Gaps (arrows) are apparent between the endothelial cells. The basement membrane (bm) separating the sinus (s) from the cord of Billroth (B) is discontinuous. A loose network of reticular fibers holds the endothelial cells together and surrounds the entire sinus.

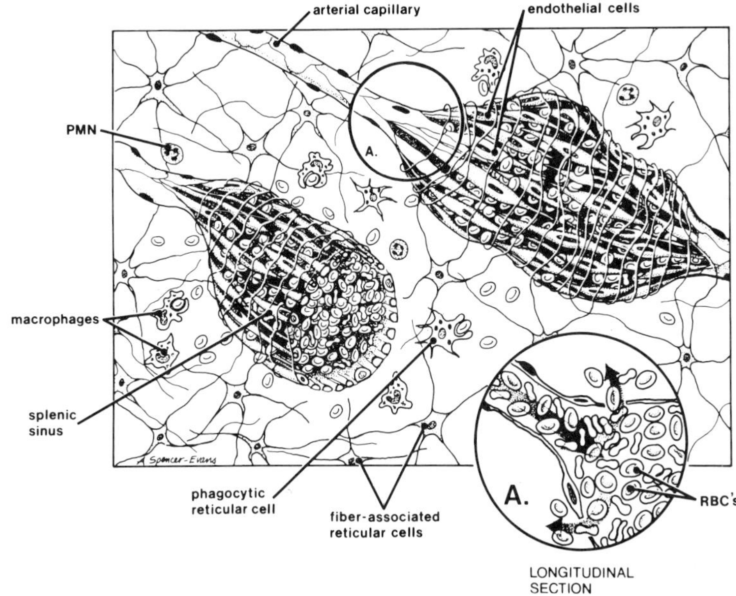

FIGURE 11-25 Schematic drawing of various sections of splenic sinuses.

THYMUS

The thymus is a major lymphoid organ located in the anterior mediastinum. As pointed out earlier, the central role thymic lymphocytes play in various defense functions has only recently been acknowledged.

STRUCTURE

The thymus develops as a bilateral structure which arises embryologically from the third and fourth branchial arches. The thymus grows rapidly during the postnatal period. It continues to develop and becomes most active at puberty; thereafter degeneration takes place. In old age, the thymus becomes smaller and is mainly composed of a fatty tissue with few lymphoid cells. This is referred to as involution of the thymus caused by senescence.

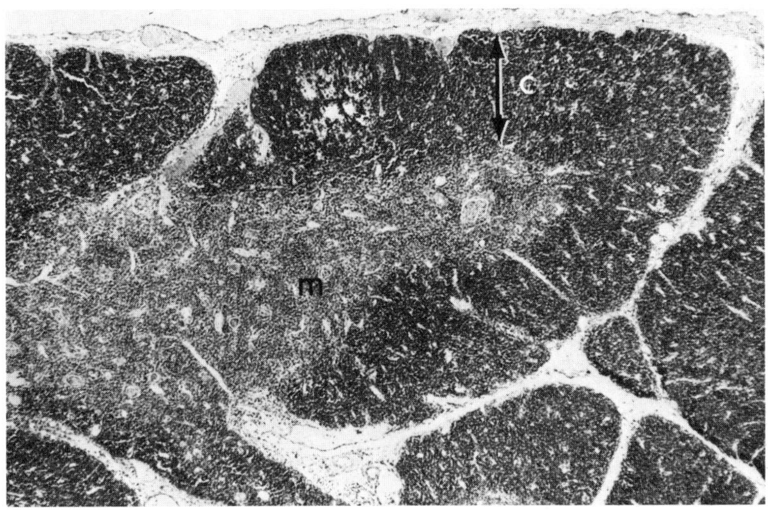

FIGURE 11-26 Light micrograph of a lobule from a child's thymus. A dense collection of thymic lymphocytes fill the periphery. This region constitutes the cortex (c) of the thymus. The lighter-stained region in the center of the lobule is the medulla (m). It is occupied mainly by thymic reticular cells.

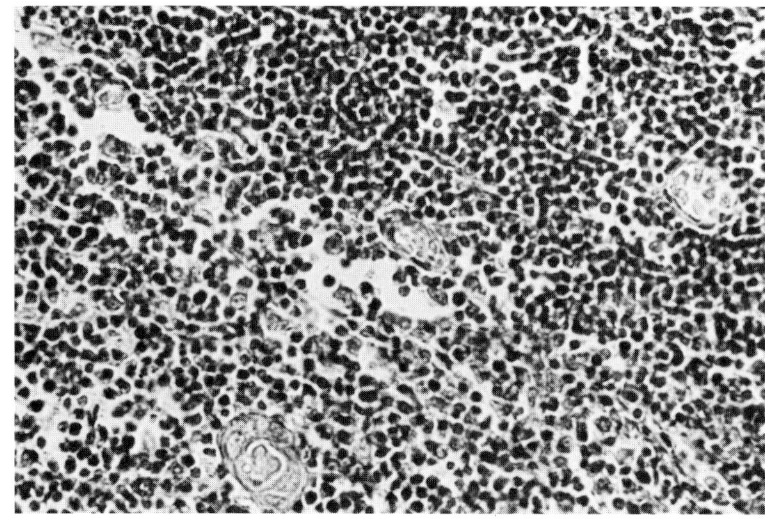

FIGURE 11-27 Light micrograph of a portion of thymic cortex. It has a dense accumulation of thymic lymphocytes (T cells) into spaces that are supported by a loose stroma of reticular cells which are of epithelial origin.

The thymus is composed of two lobes separated by a connective tissue capsule. The capsule sends trabeculae, or septa, into deeper portions of the thymus, subdividing the organ into small units called *thymic lobules*. The interior of each lobule is supplied by blood vessels that travel through the connective tissue septa. A photomicrograph of a lobule of a child's thymus is shown in Fig. 11-26. It shows the lightly stained trabeculae that break thymic tissues into many lobules. Within each lobule, dense collections of thymic lymphocytes fill the periphery. This distinguishes the lymphocytic cortex (c) from the less dense medulla (m), which is occupied by thymic reticular cells. Under higher magnification the thymic cortex shows numerous small lymphocytes that are crowded in the space supported by a loose stroma of reticular cells (Fig. 11-27). The medulla, on the other hand, has fewer lymphocytes and is composed of many reticular cells (arrows in Fig. 11-28). The thymic reticular cells are epithelial in origin. That the thymic reticular cells are

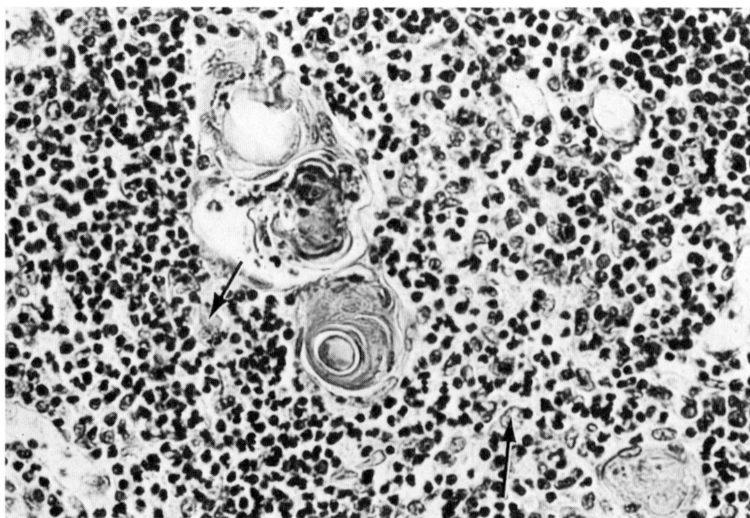

FIGURE 11-28 Light micrograph of thymic medulla. The medulla contains many reticular cells (arrows) which also are of epithelial origin. Lymphocytes are relatively sparser in the medulla than in the cortex. Notice the structure and appearance of Hassall's corpuscles.

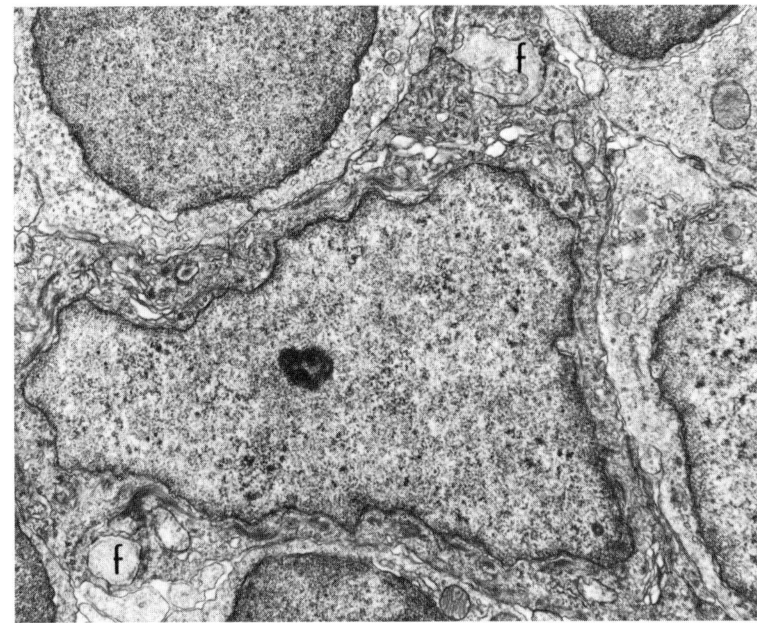

FIGURE 11-29 Electron micrograph of a thymic reticular cell. Cytoplasmic processes wrap around reticular fibers (f). The cytoplasm contains bundles of tonofilaments.

epithelial in nature has been confirmed by electron microscopic study of the organ, which has revealed bundles of tonofilaments, desmosomes, and other cytoplasmic features common to epithelial cells in general (Figs. 11-29 and 11-30). The appearance of the epithelial reticular cell of the thymus and its relationship to fibers is well depicted in Fig. 11-29 (f indicates fibers). Thymic reticular cells form desmosomes and hemidesmosomes and have a basal lamina (arrowheads in Fig. 11-30) between them and the fibers

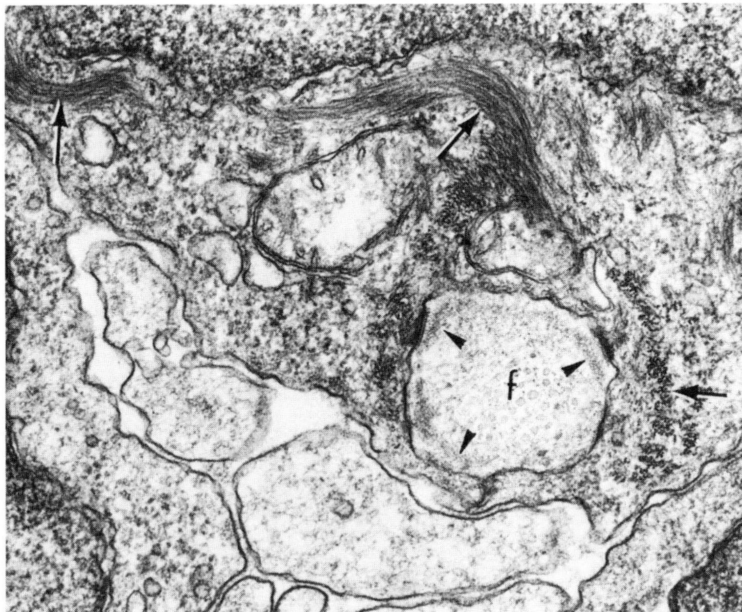

FIGURE 11-30 Electron micrograph of a portion of a reticular cell. Cytoplasmic processes wrap around reticular fibers (f). A basal lamina is seen between the reticular cell and the fibers. It has hemidesmosomes along this junction (arrowheads). The cytoplasm contains thick bundles of tonofilaments (arrows) seen in longitudinal and cross section.

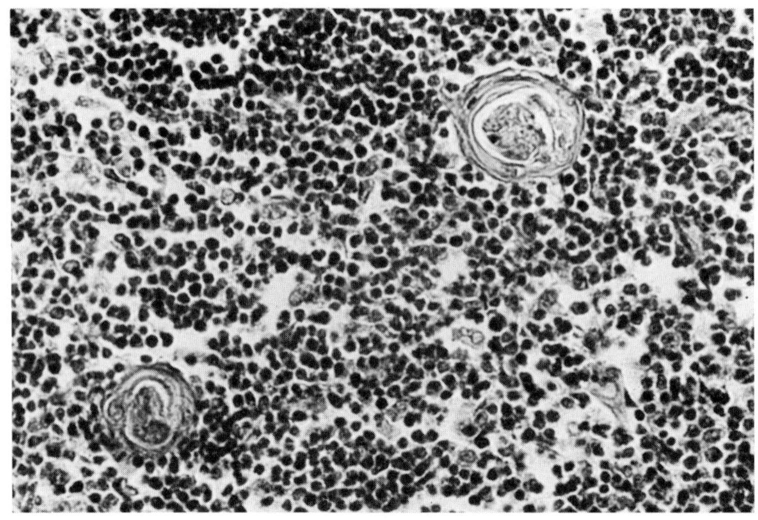

FIGURE 11-31 Light micrograph of a thymic medulla in a thymus obtained at birth. In addition to lymphocytes and reticular cells, a couple of Hassall's corpuscles are seen.

(f). The cytoplasm of the reticular cells contains bundles of tonofilaments (arrows in Fig. 11-30).

The thymic medulla contains only a small number of lymphocytes, but often shows a unique structural feature called *Hassall's corpuscle* (Fig. 11-31 and also Fig. 11-28). These concentric rings of epithelioid cells are present exclusively in the thymic medulla and increase in number until puberty. The function of Hassall's corpuscle has not been clarified, although a number of hypotheses, including production of *lymphocytopoietic* factors, have been proposed. Hassall's corpuscles are present in the medulla of the thymic lobules at birth. These are more numerous in the adult thymic lobule.

As described previously, much of the cortical lymphoid cells and some of the medullary elements are replaced by adipose cells during the involution of the thymus. The thymic involution begins in the early teens and continues until most thymic parenchyma is lost. This results in an extensive replace-

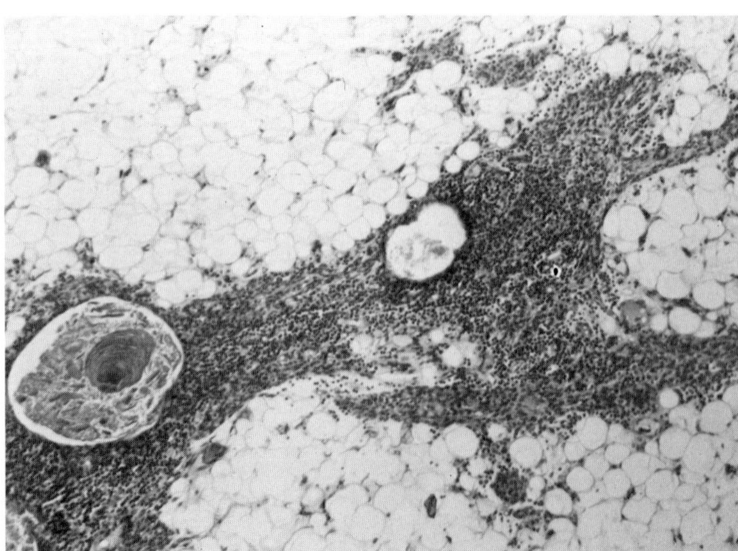

FIGURE 11-32 Light micrograph of a portion of an involuted thymus. The parenchyma is almost completely replaced by adipose cells. Some Hassall's corpuscles (arrows) are present.

ment of lymphocytes with adipose tissue in older persons (Fig. 11-32), as mentioned above. However, Hassall's corpuscles do not seem to disappear.

REVIEW SECTION

1. The irregularly shaped organ in Figure 11-33 which is dense in its periphery is called __________. What are the dense, rounded areas with a light central portion? Identify the structure.

The dark-staining periphery of a lymph node is called the *cortex*. It contains dense lymphoid tissue and germinal centers. The medulla, or central region, contains medullary sinuses and cords. Afferent lymphatics enter at the capsule, and lymph travels through the cortex, where lymphocytes are exposed to it. It exits the organ at the hilar region via an efferent lymphatic.

2. In Fig. 11-34 a clear space is seen underlying the connective tissue capsule (arrows). The clear-appearing area is termed the __________. What is the light-stained region in the center? Identify the organ.

The sinus underlying the capsule of lymph node is called the *subcapsular sinus*. Many lymphocytes are found in the parenchyma of lymph nodes. Germinal centers stain lighter than surrounding lymphoid tissue.

3. Figure 11-35 shows a number of lymphoid cells arranged in irregular cords. Identify the organ and region.

A portion of the medulla from a lymph node demonstrates aggregations of cells that make up the medullary cords and delineate its sinuses.

4. Figure 11-36 shows areas in which there is a light, rounded eosinophilic structure (arrows). What are they called? Identify the organ. Is this organ from a young or old individual?

The eosinophilic structure is called a *Hassall's corpuscle*. The thymus in this figure is from an older individual. It has undergone thymic involution, as evidenced by the presence of adipose cells.

5. Identify the organ in Fig. 11-37.

This is the thymus, indicated by the presence of the dark thymic cortex and light medulla in the lobule. At a higher magnification one should be able to recognize Hassall's corpuscles in the medulla.

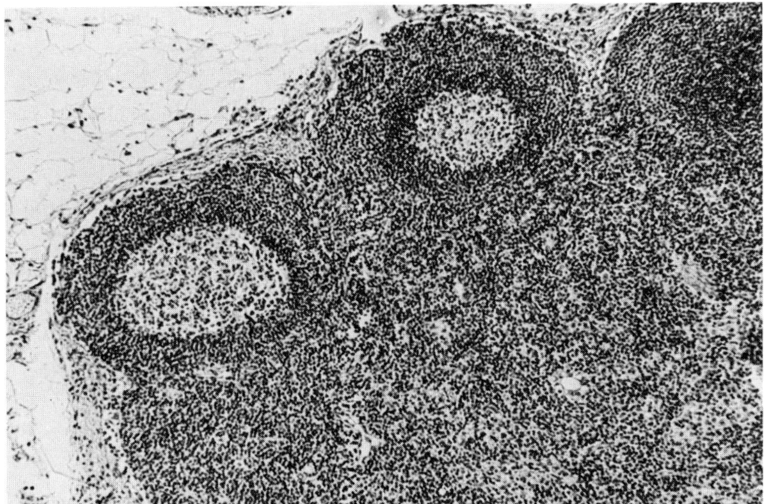

FIGURE 11-33

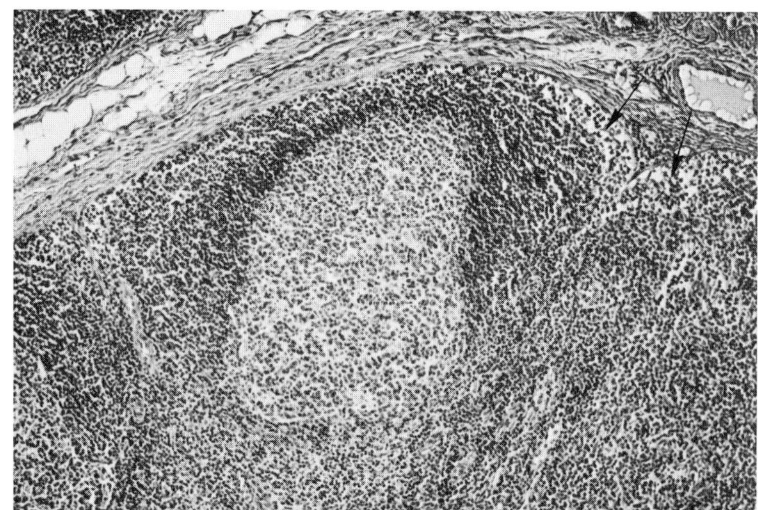

FIGURE 11-34

6. Would you describe the lymphoid organ in Fig. 11-38 as being lobated? Identify the organ.

This figure shows one of the randomly located lymphoid nodules in the spleen. There is no lobulation as in the thymus, and the trabeculae appear as patchy areas. There is no differentiated cortex or medulla, as in the lymph node and thymus.

7. Figure 11-39 is a cell from a lymphoid organ. Based on your understanding of cells that make up the lymphoid organs, would you call this a lymphocyte, reticular cell, monocyte, or lymphoblast?

The cell appearing here is a lymphoblast. Note the large nucleus with prominent nucleoli and a voluminous cytoplasm in which membranes of ER, polysomes, mitochondria, and an immature Golgi complex are seen.

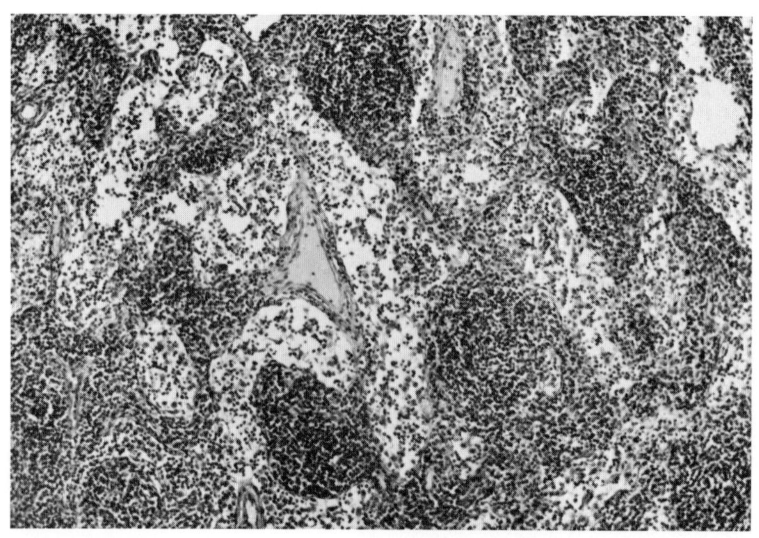

FIGURE 11-35

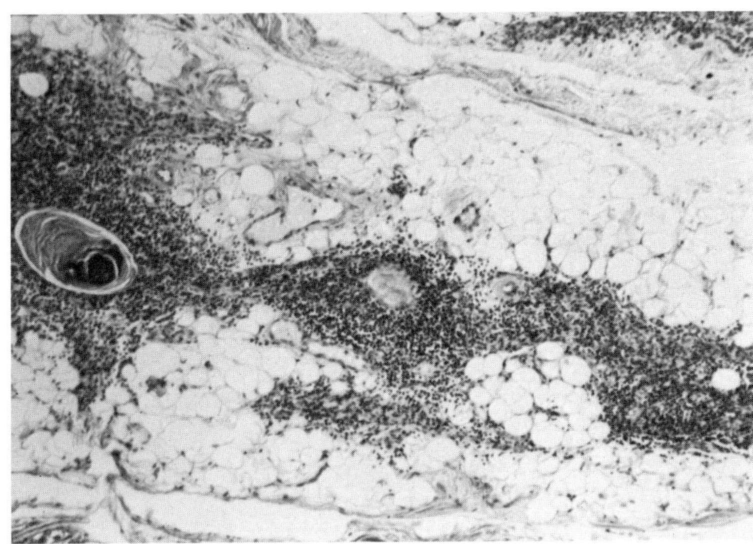

FIGURE 11-36

8. A germinal center appears in Fig. 11-40. Are germinal centers present in the thymus? An arteriole is present in the vicinity of the germinal center (arrow). What is the arteriole called?

Germinal centers are not found in the thymus. The arteriole associated with the lymphoid nodule in this figure is called a *central artery*.

9. Identify the regions indicated by a in Fig. 11-41. What name is given to the cells on either side of this structure?

The regions indicated are splenic sinusoids, and the cords of cells that separate the sinuses are called *cords of Billroth*.

FIGURE 11-37

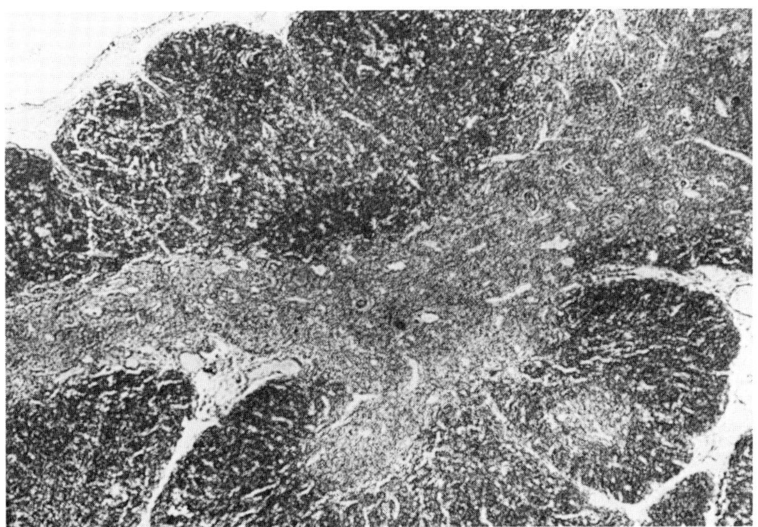

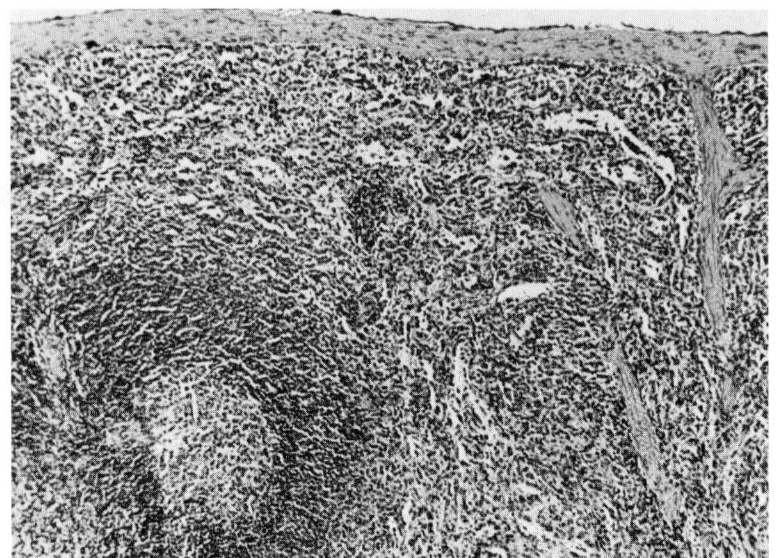

FIGURE 11-38

BIBLIOGRAPHY

Goldstein, G., and I. R. Mackay: *The Thymus,* W. H. Green, St. Louis, 1969.

Good, R. A., and A. E. Gabrielson: *The Thymus in Immunobiology,* Hoeber-Harper, New York, 1964.

——— and D. W. Fisher (eds.): *Immunobiology,* Simaner Associates, Inc., Stanford, Conn., 1971.

Weiss, L.: *The Cells and Tissues of the Immune System,* Prentice-Hall, Englewood Cliffs, N.J., 1972.

——— and R. O. Greep: *Histology,* 4th ed., McGraw-Hill, New York, 1977.

Yaffey, J., and F. Gourtice: *Lymphatics, Lymph, and Lymphoid Tissue,* Harvard University Press, Cambridge, Mass., 1956.

ADDITIONAL READINGS

Anderson, A. O., and N. D. Anderson: "Studies on the Structure and Permeability of the Microvasculature in Normal Rat Lymph Nodes," *Am. J. Pathol.* **80:**387–418 (1975).

FIGURE 11-39

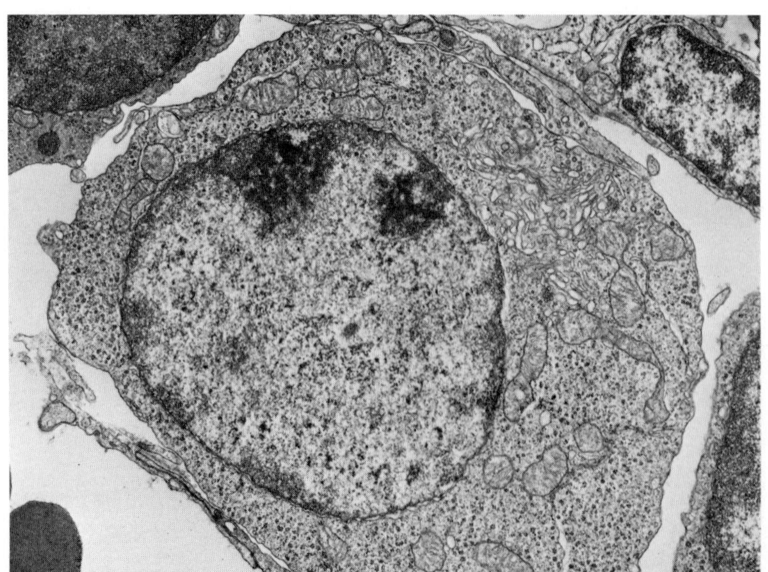

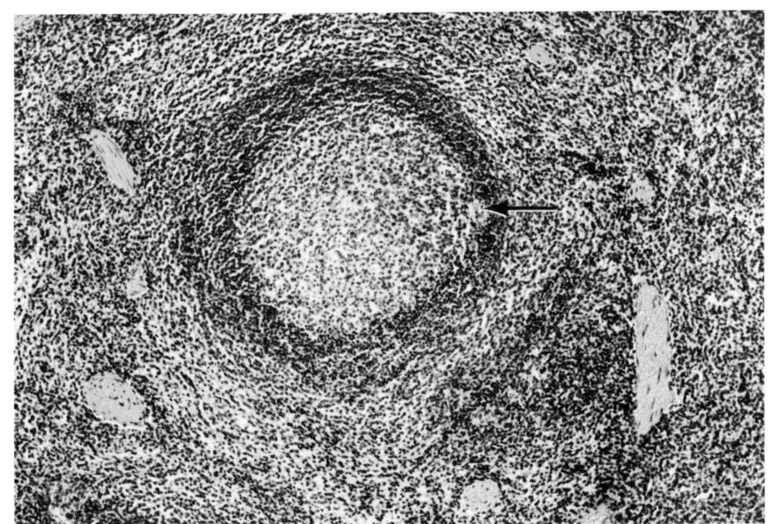

FIGURE 11-40

Bjorkman, S. E.: "The Splenic Circulation. With Special Reference to the Function of the Spleen Sinus Wall," *Acta. Med. Scand.* (*Suppl. 191*) **128:** 1–89 (1947).

Chapman, W. L., and J. R. Atlas: "The Fine Structure of the Thymus of the Fetal and Neonatal Monkey (*Macaca mulatta*)," *Z. Zellforsch.* **114:**220–233 (1971).

Chen, L. T., and L. Weiss: "The Role of the Sinus Wall in the Passage of Erythrocytes through the Spleen," *Blood* **41:**529–537 (1973).

Clark, S.: "The Reticulum of Lymph Nodes in Mice Studied with the Electron Microscope," *Am. J. Anat.* **110:**217 (1962).

Drinker, C. K., M. E. Field, and H. K. Ward: "The Filtering Capacity of Lymph Nodes," *J. Exp. Med.* **59:**393–405 (1934).

Edwards, V. C., and G. T. Simon: "Ultrastructural Aspects of Red Cell Destruction in the Normal Rat Spleen," *J. Ultrastruct. Res.* **33:**187–301 (1970).

Everett, N. B., and R. W. (Caffrey) Tyler: "Lymphopoiesis in the Thymus and Other Tissues; Functional Implications," *Int. Rev. Cytol.* **22:**205–286 (1967).

Han, S. S.: "The Ultrastructure of the Mesenteric Lymph Node of the Rat," *Am. J. Anat.* **109:**183–225 (1961).

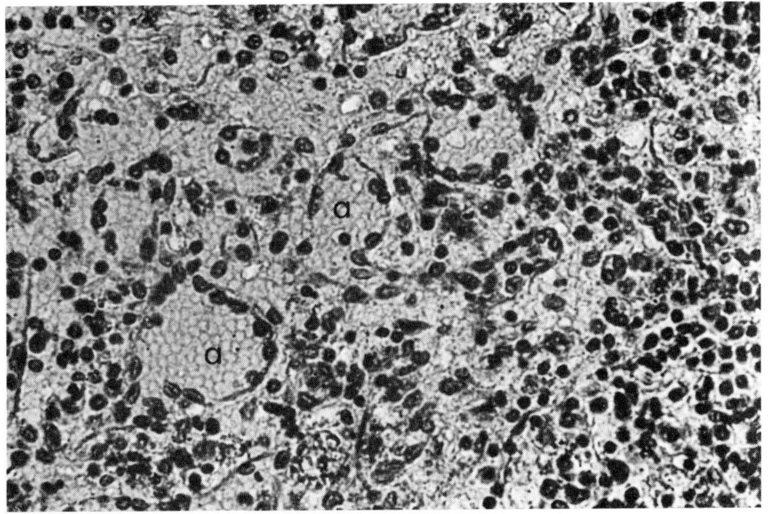

FIGURE 11-41

Hashimoto, T.: "Electron Microscopic Studies of the Epithelial Reticular Cells of the Mouse Thymus," *Z. Zellforsch.* **59:**513–529 (1963).

Kohnen, P., and L. Weiss: "Electron Microscopic Study of Thymic Corpuscles in the Guinea Pig and the Mouse," *Anat. Rec.* **148:**29–57 (1964).

Leak, L. V., and J. F. Burke: "Fine Structure of the Lymphatic Capillary and the Adjoining Connective Tissue Area," *Am. J. Anat.* **118:**785–809 (1966).

Lewis, O. J.: "The Blood Vessels of the Adult Mammalian Spleen," *J. Anat.* **91:**245–250 (1957).

Millikin, D. D.: "Anatomy of Germinal Centers in Human Lymphoid Tissue," *Arch. Pathol.* **82:** 499–505 (1966).

Moe, R. E.: "The Fine Structure of the Reticulum and Sinuses of Lymph Nodes," *Am. J. Anat.* **112:** 311–335 (1963).

———: "Electron Microscopic Appearance of the Parenchyma of Lymph Nodes," *Am. J. Anat.* **114:** 341–369 (1964).

Mori, Y., and K. Lennert: *Electron Microscopic Atlas of Lymph Node Cytology and Pathology,* Springer-Verlag, Berlin, 1969.

Pictet, R., L. Orci, W. G. Forsman, and L. Girardier: "An Electron Microscope Study of the Perfusion-fixed Spleen. I. The Splenic Circulation and the RES Concept," *Z. Zellforsch.* **96:**372–399 (1969).

Simon, G. T., and J. S. Burke: "Electron Microscopy of the Spleen. III. Erythro-Leukophagocytosis," *Am. J. Pathol.* **58:**451–469 (1970).

Snook, T.: "A Comprehensive Study of the Vascular Arrangements in Mammalian Spleens," *Am. J. Anat.,* **87:**31–77 (1950).

Sprent, J.: "Circulating T and B Lymphocytes of the Mouse. I. Migratory Properties," *Cell Immunol.* **7:**10–39 (1973).

van Haelst, U. G. T.: "Light and Electron Microscopic Study of the Normal and Pathological Thymus of the Rat. I. The Normal Thymus," *Z. Zellforsch.* **77:**534–553 (1967).

12 HEART

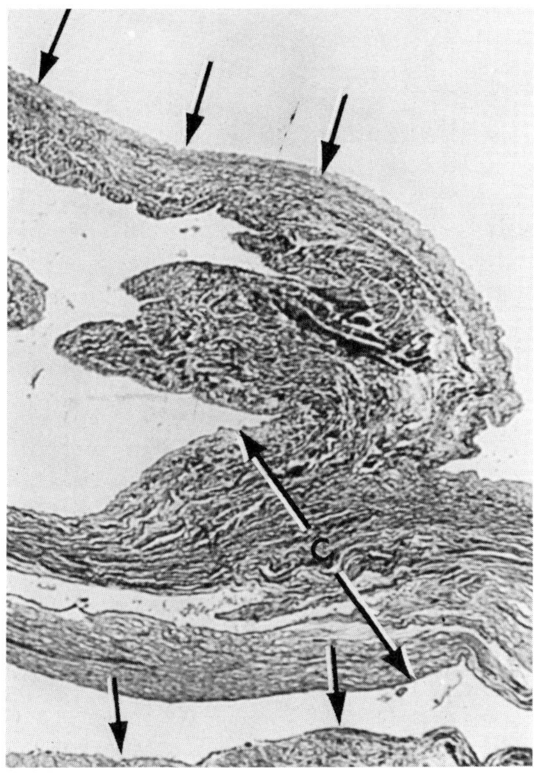

OBJECTIVES

Upon completion of this chapter, the student will be able to:

1 List the three layers of the heart wall.

2 Identify the location, structure, and function of the following:
(a) endocardium (b) myocardium (c) epicardium (d) cardiac muscle (e) Purkinje fibers (f) chordae tendinae

3 Describe the impulse conduction system of the heart by identifying:
(a) the origin of the impulse (b) the pathway followed by the impulse from its origin to the periphery fo the heart

4 Identify the method by which each of the following systems supplies the heart:
(a) vascular (b) nervous

The cardiovascular system, which includes the heart and blood vessels of the body, is responsible for the supply of nutritive substances to and removal of waste products from every cell of the body. This includes the gaseous exchanges. The circulation of blood, which is the transport vehicle of all metabolites, is initiated by the heart in association with the vascular distribution system. For this reason the heart can be regarded as a central pump for blood circulation and therefore is equipped with facilities to receive venous blood, oxygenate it, and redistribute it through the vessels.

STRUCTURE OF THE HEART

In the heart there are four chambers to which the aorta, the vena cavae, the pulmonary trunk, and the pulmonary veins are attached. The chambers of the heart are a right and a left atrium and a right and a left ventricle. The separation of atria and ventricles is marked on the surface by the coronary (atrioventricular) sulcus. The ventricles are separated from each other by an interventricular septum, the position of which is indicated by the anterior interventricular and posterior interventricular sulci of the heart. These are indicated in Fig. 12-1 which diagrams the anterior and posterior surface views of the heart.

The atria are separated by an interatrial septum. The right atrium receives all the systemic venous blood by way of the superior vena cava, the inferior vena cava, the coronary sinus, and the anterior cardiac veins. It passes this blood through the right atrioventricular or the tricuspid valve into the right ventricle (Fig. 12-2). The right ventricle, by its contraction, forces the blood out through the semilunar valves of the pulmonary trunk to the lungs via the pulmonary arteries; the right atrioventricular valve closes at the same time to prevent reflux of blood into the right atrium. The oxygenated blood from the lungs is returned to the left atrium of the heart by the pulmonary veins. From here it enters the left ventricle through the left atrioventricular or bicuspid (mitral) valve. Contraction of the musculature of the left ventricle, together with concomitant closure of the left atrioventricular valve, drives the aerated blood into the arterial system of the body via the aorta through the semilunar valves. When the heart relaxes after contraction, the back flow of blood into the ventricles is prevented by closure of the semilunar valves of the aorta and the pulmonary trunk.

At the junction between the atria and

FIGURE 12-1 Diagram of the human heart illustrating the anterior view to the right and the posterior view to the left.

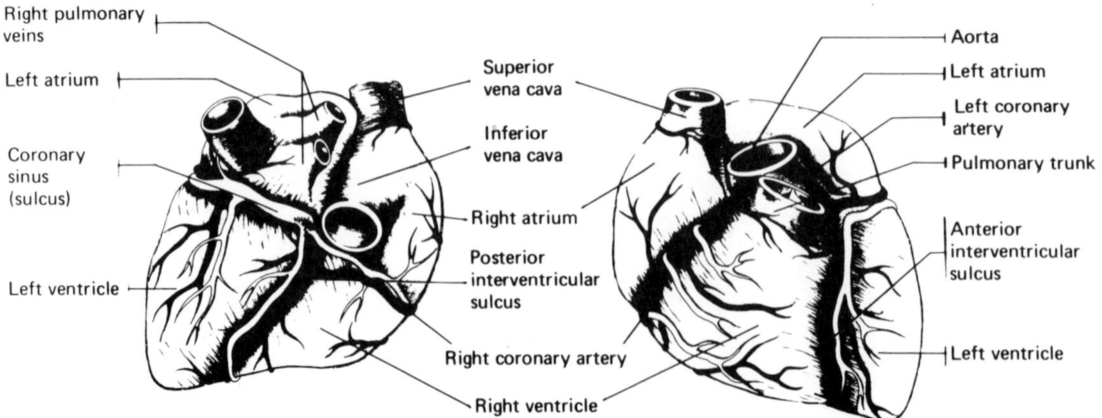

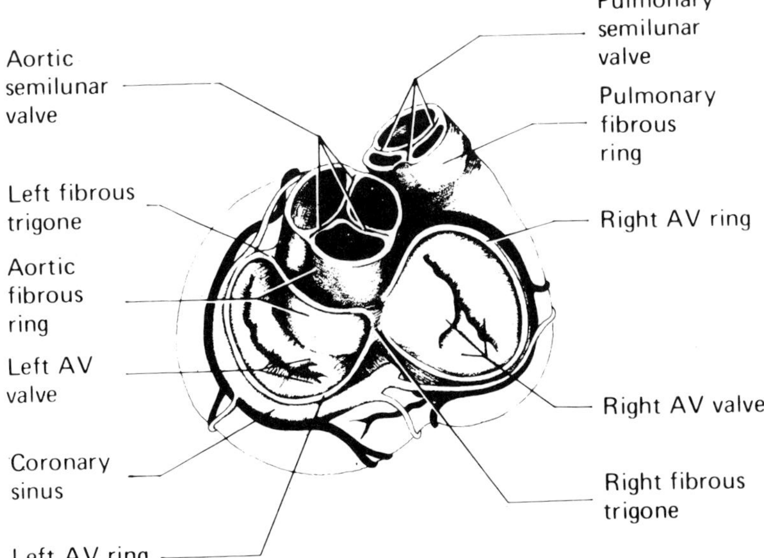

FIGURE 12-2 A schematic section of the heart through the coronary sulcus.

ventricles where the coronary sulcus is present is a collection of dense connective tissue. This serves as the "skeleton" of the heart and therefore is called the *cardiac skeleton*. This fibrous connective tissue exists as fibrous rings, giving circular form and rigidity to the atrioventricular apertures and to the roots of the aorta and pulmonary trunk. The fibrous rings send thin, sheetlike extensions into the valves. The aortic fibrous ring is massive and sleevelike (Fig. 12-2). It furnishes semicircular attachments for the cusps of the aortic *semilunar valves* and has a deep extension into the left atrioventricular junction region. Here the atrial musculature arises. The *septum membranaceum* is a continuation of the aortic sleeve or cuff into the septum, and right and left fibrous trigones are small offshoots of the cuff which provide for muscular origins. The pulmonary fibrous ring is smaller than the aortic fibrous ring, but it is peaked like the latter for the attachment of the cusps of its valves.

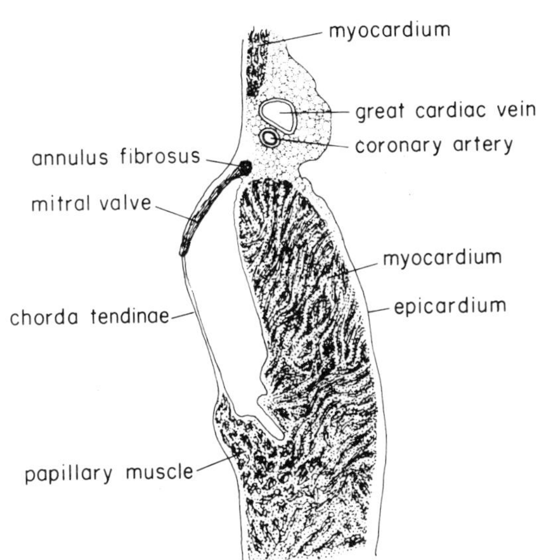

FIGURE 12-3 Schematic drawing of a longitudinal section through the heart wall.

HISTOLOGY OF HEART WALLS

The walls of the heart are made of three basic layers (Fig. 12-3): endocardium (inner lining layer), myocardium (muscle layer), and epicardium (outer covering layer).

ENDOCARDIUM

The endocardium, the inner layer of the heart, is homologous and continuous with the intima, or inner layer, of the major blood

FIGURE 12-4 Light micrograph of the heart wall showing the endocardium (e) and the myocardium (m).

vessels. It is thickest in the atria (particularly the left atrium) and relatively thin in the ventricle.

The endocardium contains several distinct layers (e, endocardium in Fig. 12-4). The innermost layer toward the chambers of the heart is simple squamous epithelium, or *endothelium* (arrows, Fig. 12-5). These flat cells have oval to rounded nuclei and, like other epithelial cells, rest on a thin basement membrane. However, they are of mesodermal origin in contrast to other epithelial cells, which are of ecto- or endodermal origin. The *subendothelial layer* of the endocardium consists of fine collagen fibrils, elastic fibers, and smooth muscle cells (S-1 in Fig. 12-5). Most areas have another layer of loose FECT known as the *subendocardium* (S-2 in Fig. 12-5). It is subjacent to the subendothelial layer and contains Purkinje cells of the impulse-conducting system and various blood vessels. A few smooth muscle cells are found in the subendothelial layer (arrows, Fig. 12-6). The subendocardial connective tissue is often quite thick (double-ended arrow in Fig. 12-6).

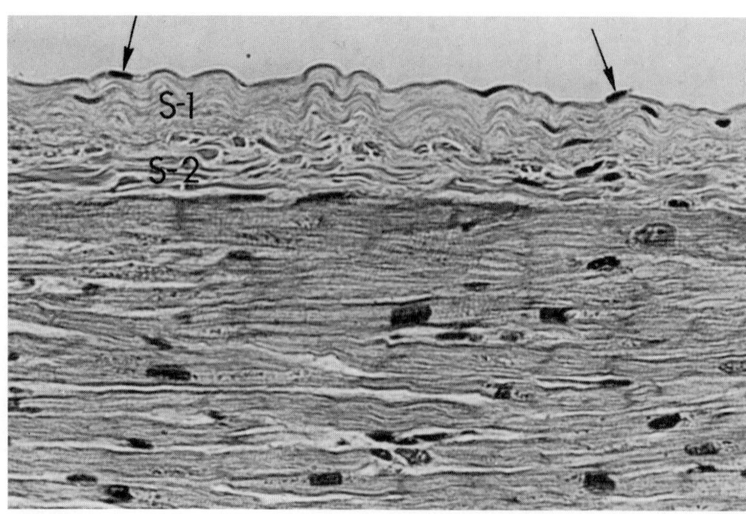

FIGURE 12-5 Light micrograph of the endocardium and the outermost portion of the myocardium. The endocardium consists of endothelium (arrows) and subendothelial layer (S-1). In most areas, it is held to the myocardium by a subendocardial layer (S-2).

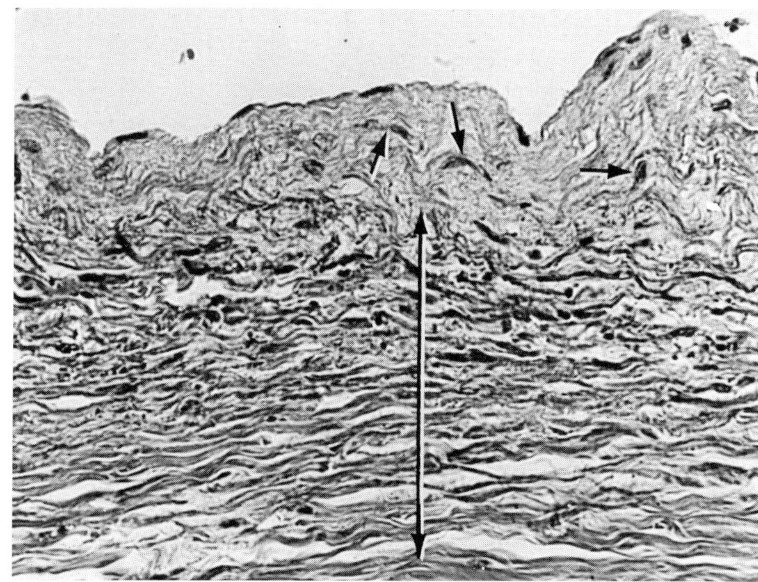

FIGURE 12-6 Light micrograph of the endocardium. Flattened endothelial cells constitute the lining toward the lumen. A few smooth muscle cells (arrows) are seen in the subendothelial layer in addition to fibroblasts and collagenous fibers. The subendocardial layer is quite thick (double-ended arrow).

MYOCARDIUM

The myocardium makes up the bulk of heart tissue. It is composed of cardiac muscle classified as striated and involuntary. This layer varies in thickness, depending on the relative work load of the heart chambers. Because they must pump the blood through the circulatory system, the ventricles have the greatest development. Of the two ventricles, the left has a myocardium three times the thickness of the right, since it provides blood for the circulation of the entire body. In contrast, the right ventricle provides only blood for the pulmonary circulation. The atria have a relatively thin myocardium, since they pass their blood directly to the ventricles. Cardiac muscle is arranged in an interconnecting latticework of bundles (m in Fig. 12-4), with spaces where the connective tissue of the subendocardium may join with the subepicardium (to be discussed later). In the atrial walls the external bundle layers tend to run circumferentially around both atria. The inner bundles are oriented at right angles to outer bundles and tend to be independently associated with each atrium.

The ventricular musculature arises from the fibrous trigones and the fibrous rings at the base of the heart. The spiral loops of which this musculature is composed embrace the cavities of either one or both ventricles, one limb of a loop lying on the outer surface of the heart and one in the interior. The superficial fibers of the sternocostal surface pass downward and toward the left; those of the diaphragmatic surface pass toward the right. The principal anterior superficial bundle arises from the posterior aspect of the left atrioventricular and right atrioventricular fibrous rings and passes apicalward across the right ventricle (Fig. 12-7). The bundle twists upon itself in a tight vortex at the apex and turns inward to spread out on the inner surface of the left ventricle, inserting into papillary muscles, into the septum, and into the opposite fibrous rings (Fig. 12-7). Another prominent superficial bundle arises from the root of the aorta and from the left atrioventricular fibrous ring. It spirals toward the apex, making a double circle around the heart, somewhat like a figure eight that is wider open at the top. The fibers of this bundle turn into the vortex of the left ventricle and insert into the opposite side of the tendinous structures from which the bundle arose. The deeper bundles also arise from the

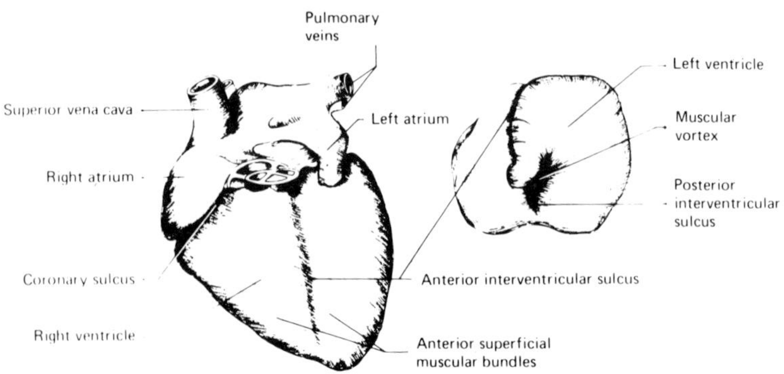

FIGURE 12-7 Drawing illustrating the superficial musculature of the heart; anterior and apical views.

fibrous rings and form spirals and figure eights, but they do not reach the apex of the heart. One deep bundle makes a double circle of the cavity of the left ventricle, whereas others encircle only the right ventricle or loop around both ventricles.

The configuration of these intermingling muscle bundles of the ventricular part of the heart (as seen in Fig. 12-7) is such as to compress the cavities by both a direct squeezing action (the deeper bundles) and a wringing motion (the more superficial spiral bundles). The combination of these actions powerfully empties the ventricles, and the pull of the spiral muscles causes the apex of the heart to impinge on the anterior chest wall, resulting in the characteristic apex beat.

Histologically, the myocardial fibers are arranged in discrete bundles separated by connective tissues (Fig. 12-8). Blood vessels and lymphatics are found in this connective tissue (arrows in Fig. 12-8). It should be remembered that every heart muscle cell is in continuous rhythmic contraction throughout life.

The connective tissue and the rich blood vessels invested within it are in continuity with the epicardial connective tissue and the coronary vessels.

EPICARDIUM

The epicardium is the outer layer of the heart. The surface cells are simple, flat to cuboidal mesothelium (mesodermally derived epithelium, Figs. 12-9 and 12-10). The mesothelium

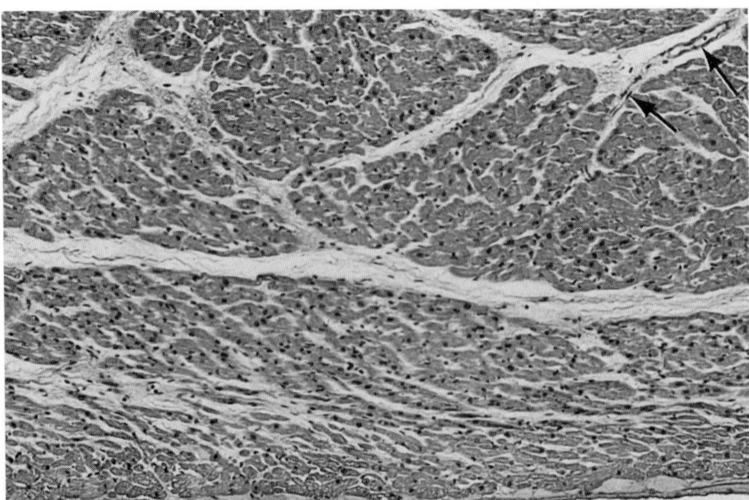

FIGURE 12-8 Light micrograph of the myocardium. The cardiac muscle fibers are arranged in discrete bundles which run in various directions within the different layers of the myocardium. These bundles are separated by loose FECT. Blood vessels and lymphatics are found in this connective tissue (arrows).

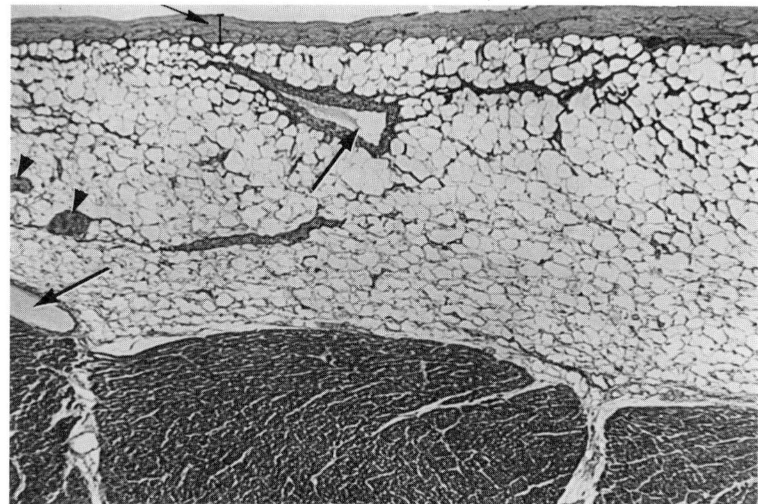

FIGURE 12-9 Light micrograph of the epicardium. It is covered by a single layer of flat to cuboidal mesothelial cells. Underlying the mesothelium is a layer of connective tissue which contains elastic fibers (bracket). The epicardium is attached to the myocardium by a layer of loose connective tissues. This layer is called the subepicardial layer and contains blood vessels (arrows), nerves (arrowheads), and adipose cells.

is the layer of cells composing the infolded visceral layer of the pericardial sac in which the heart is found. Underlying the mesothelium is a layer of connective tissue containing abundant elastic fibers, present in its deeper portions (bracket in Figs. 12-9 and 12-10). The epicardium is attached to the myocardium by a layer of loose FECT. As in the subendocardium, this layer is called a *subepicardial layer.* This connective tissue layer contains blood vessels (arrows in Fig. 12-9) to nourish the heart, adipose tissue, and nerve elements (arrowheads in Fig. 12-9). The epicardium is continuous with the adventitia of the blood vessels that enter and leave the heart.

VALVES

The endocardial layer folds inward to form valves that prevent backflow of blood from the ventricles to the atria (Fig. 12-11). The valves are formed from flaps which can open into the ventricles to allow the blood to pass in that direction, but are prevented from

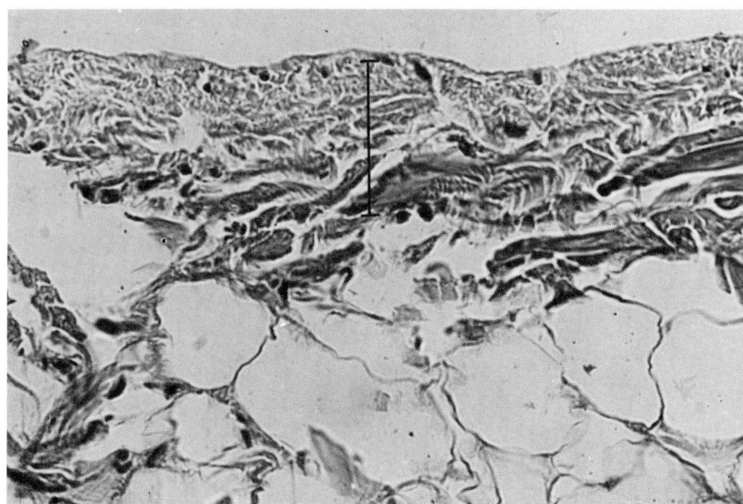

FIGURE 12-10 Epicardium viewed at higher magnification. Note the mesothelium and the relatively dense connective tissue underlying it. Immediately below, the portion of the subepicardial layer is seen (bracket). It contains an abundance of adipose cells.

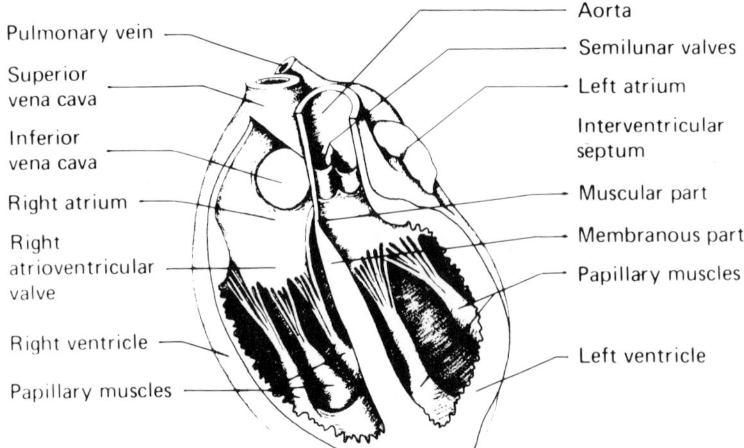

FIGURE 12-11 A schematic drawing illustrating the valves of the heart.

opening into the atria (Fig. 12-12) by the *chordae tendinae,* which are kept tight by the papillary muscles during contraction. The endocardial flap has a thick, elastic subendocardium on the atrial side to enhance its flexibility, but it is tough and fibrous on the ventricular side. The fibers of the valve are continuous with the fibrous tissue of the anulus fibrosus and the chordae tendinae of the cardiac skeleton. The mitral valve (arrows) is attached to the annulus fibrosus (af) between the left atrium (a) and ventricle (v) (Fig. 12-13). Elastic fibers are abundant on the atrial surface (Fig. 12-14), which represents a valve stained with Weigert's method for elastic fibers. They appear dense (ef) and dark when stained with Weigert's method. Heavy collagenous fibers (cf) can be seen in bundles at the ventricular surface.

As mentioned previously, the papillary muscles originating from the apex of the ventricles are connected with the tips of the valves by chordae tendinae (tendinous cords). This attachment to the chamber wall prevents inversion of the flaps. The papillary muscle portion of this attachment compensates for lost dimension of the ventricle during contraction (Fig. 12-12).

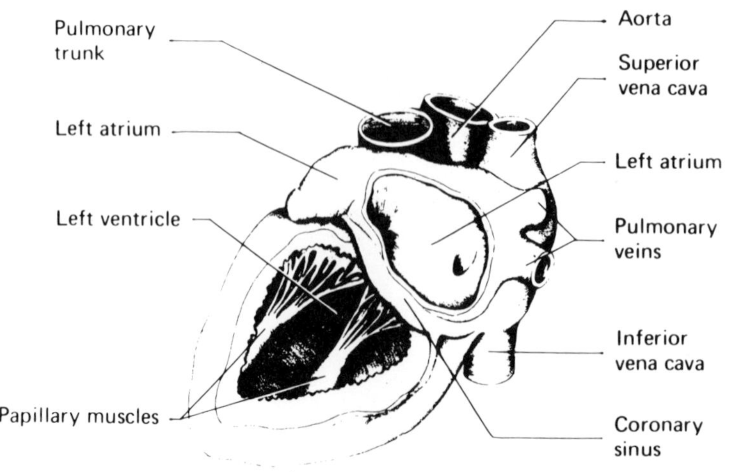

FIGURE 12-12 The interior of the left atrium and left ventricle as seen through openings in the posterior wall of the heart.

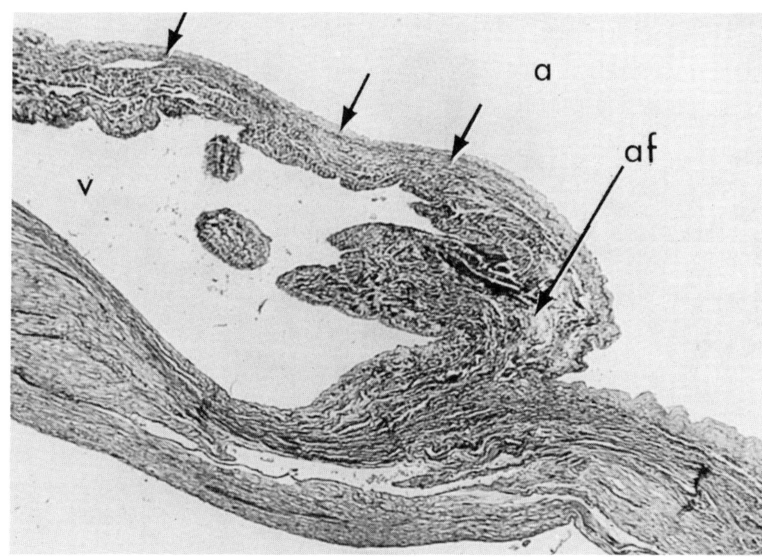

FIGURE 12-13 Light micrograph of the mitral valve (arrows). It is situated between the left atrium (a) and left ventricle (v) where it prevents backflow of the blood from the ventricle to the atrium during the contraction of the heart. It orginates from and is attached to the annulus fibrosus (af) at the base of the heart.

CONTRACTION

The efficiency of the heart as a pump depends on the coordinated contraction of muscle fibers. This is in turn dependent on an elaborate conduction system as diagramed in Fig. 12-15. The depolarization which initiates contraction of the specialized muscle cells begins at the *sinoatrial (SA) node,* a group of cells near the entrance of the superior vena cava in the right atrium. Depolarization passes from one atrial muscle cell to another, stimulating them to contract in a wavelike order. This causes contraction of the atrial chambers. You may recall that cardiac muscle cells have regions of nexuses between adjacent cells along intercalated disks.

Purkinje fibers are special fibers which carry the depolarization impulse directly to all parts of the ventricles via the interventricular septum. As such, they can be best located in the subendocardial region along the septum. The Purkinje fibers pick up the impulse from a group of cells in the right

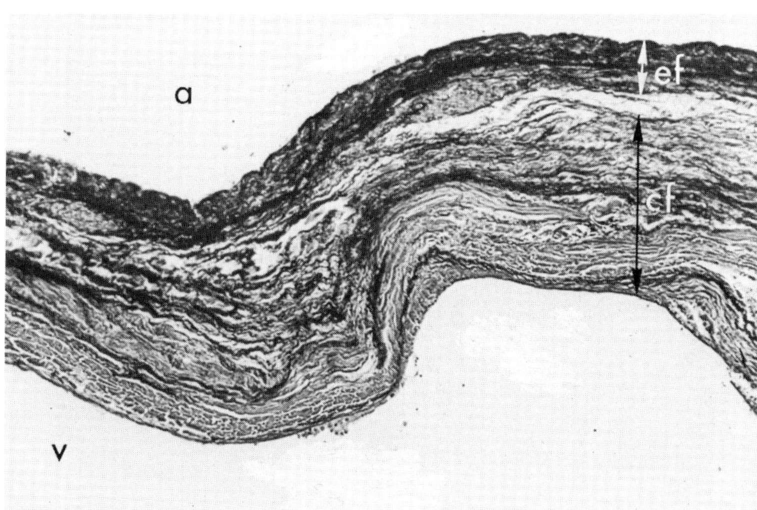

FIGURE 12-14 Light micrograph of the mitral valve. The endocardium of the valve is thick. Elastic fibers (ef) are abundant in the endocardium on the atrial side (a), while heavy collagenous fibers (cf) can be seen on the ventricular side (v).

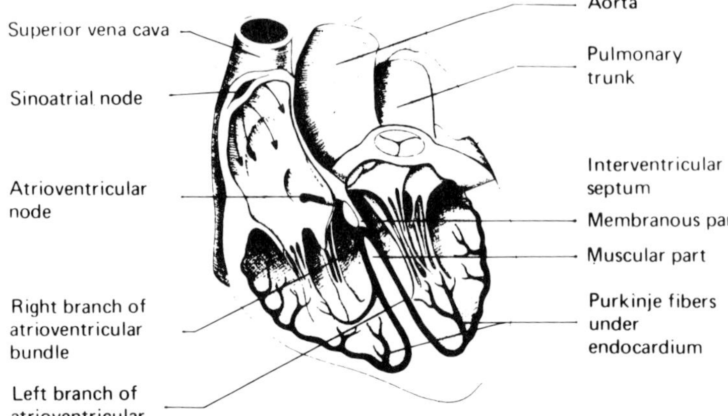

FIGURE 12-15 The conduction system of the heart.

atrium called the *atrioventricular (AV) node*. The impulse conduction by Purkinje fibers is rapid and therefore allows the ventricular muscles to contract simultaneously.

The Purkinje fibers are best developed toward the terminal portions of the conduction system and are larger than normal cardiac muscle fibers, with fewer myofibrils (Fig. 12-17). Compare Fig. 12-16 and Fig. 12-17, which represent cross sections of myocardium and Purkinje fibers at the same magnification. Purkinje fibers have a clear perinuclear region where large glycogen deposits are frequently found (Fig. 12-17 and arrows in Fig. 12-18). They lie in the subendocardium and characteristically bulge between the intercalated

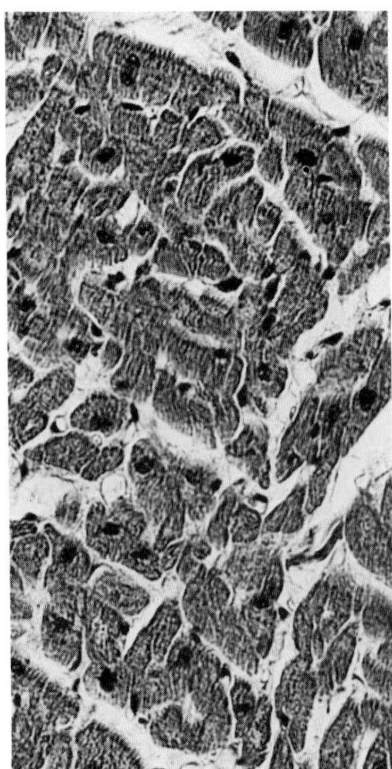

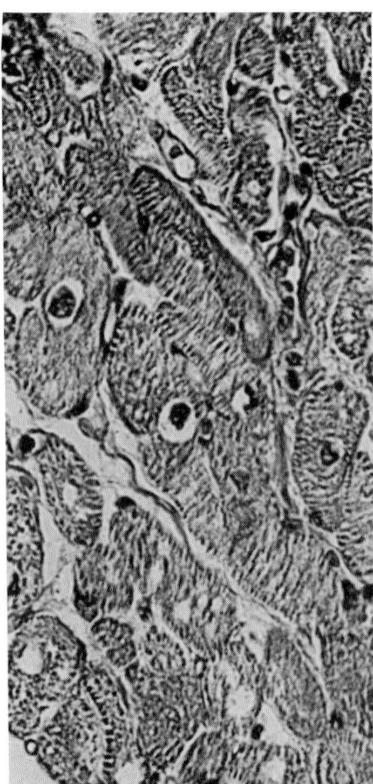

FIGURE 12-16 Light micrograph of the myocardium seen in cross section.

FIGURE 12-17 Light micrograph of Purkinje fibers seen in cross section.

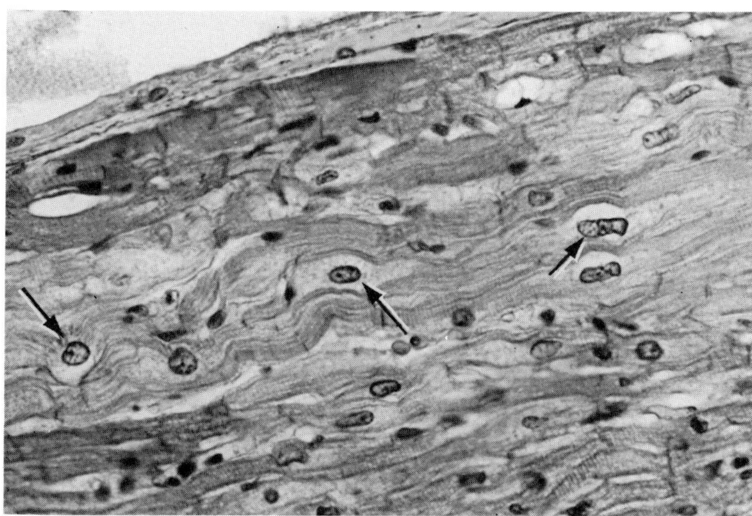

FIGURE 12-18 Light micrograph showing Purkinje fibers (arrows). They lie in the subendocardium. The endocardium is at the top and the myocardium at the bottom.

disks joining them together. Again compare the ordinary cardiac muscle cells occupying the lower portion in Fig. 12-18 with the Purkinje cells. The nuclei of Purkinje cells are larger and lighter than those of ordinary myocardial cells.

Blood and Nerve Supplies to Heart

Blood is supplied to the heart by way of the coronary arteries, as shown in Fig. 12-1. Each artery supplies one-half of the heart. They give off one major branch to the respective atrium and continue on between the ventricles. Capillary flow is picked up by cardiac veins (shown in Fig. 12-1) leading to the coronary sinus, which in turn empties into the right atrium. A number of small veins empty directly into the heart.

The heart is innervated by the cardiac nerves which modify the intrinsic rhythmicity of heart action. These include both *sympathetic* and *parasympathetic* innervations. Stimulation through the sympathetic nerves increases the rate and force of the heartbeat; slowing and reduction in force are the results of parasympathetic stimulation. The parasympathetic cardiac nerve is the *vagus,* which supplies three branches on each side. Three sympathetic branches on each side also reach the heart from the cervical sympathetic ganglia or the associated trunk, and in addition, direct branches from the upper four or five thoracic ganglia join the cardiac plexus.

REVIEW SECTION

1. Does the *a* or *b* side of the section in Fig. 12-19 have more developed muscle? Which side would represent the ventricle? The area indicated by bracket and *c* is called ________.

The *a* side of the figure is the ventricle, as indicated by its greater degree of muscular development. The atrial section (*b*) of the heart only receives blood and passes it to the ventricle, whereas the ventricular portion (*a*) must force the blood through the pulmonary circulation or the entire systemic vasculature. The area in brackets (*c*) is the membraneous portion of interventricular septum. Portions of atrioventricular valves of both sides are also visible (arrows).

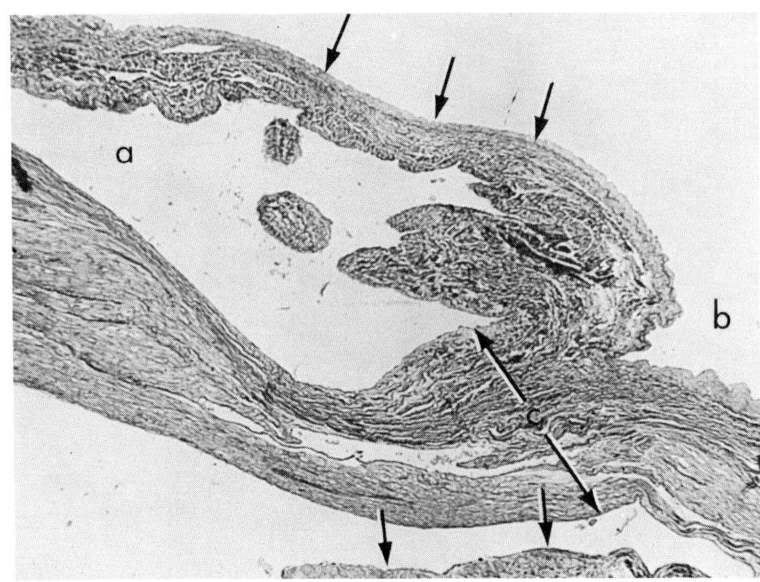

FIGURE 12-19

2. Figure 12-20 is a section through the heart septum. The light-stained tissue (*a*) is __________. The light large cells indicated by arrows are __________. The cells indicated by arrowheads are __________.

The section of the septum shows the thin endocardium (*a*), Purkinje fibers (arrows), and smaller myocardial cells. The muscular portion of the interventricular septum has the same structure on both sides as the specialized Purkinje fibers run on both sides. Nuclei of the regular myocardial cells are indicated by arrowheads.

3. Figure 12-21 is another section of the heart. What is the predominant type of tissue in this section? How does the orientation of the tissue in

The myocardium makes up the major bulk of the heart, with its different bundles of myocardial cells running in distinct directions. The upper half of the

FIGURE 12-20

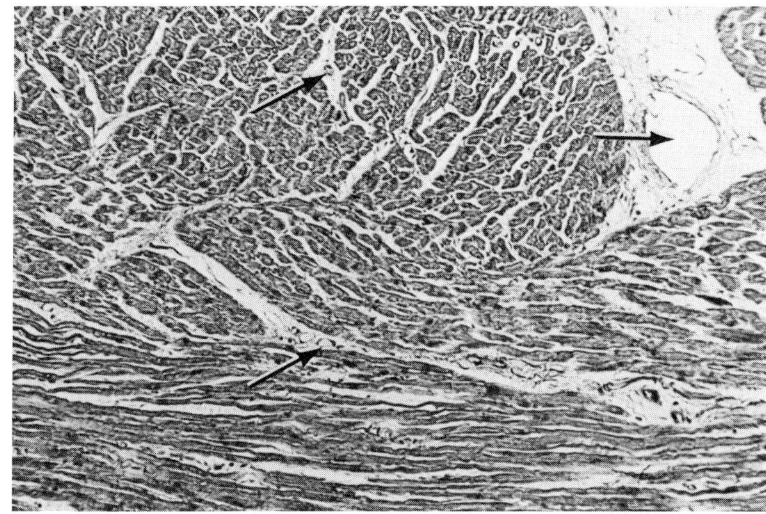
FIGURE 12-21

the upper half differ from that in the lower half? What do the spaces (arrows) contain?

field shows them in cross section, while in the lower half the muscle fibers are longitudinally oriented. Spaces between bundles contain connective tissues in which blood vessels nourishing the cardiac tissue may be found (arrows).

4. Figure 12-22 is a cross section though an AV valve. Which layer of the heart wall extends to form the valve? What kind of tissue is indicated by the arrows? How did you reach this conclusion? This layer, and hence the valve, consists of__________. Is the top of the tissue the ventricular or atrial side of the valve?

The valves of the heart are extensions of the endocardium and thus include mostly endothelium, connective tissue, and occasional muscle fibers. The side toward the ventricle is a wide, tough fibrous layer. The layer of elastic tissue (arrows) is the atrial side of the flap, serving as a cushion and allowing the flap to swing away into the ventricle.

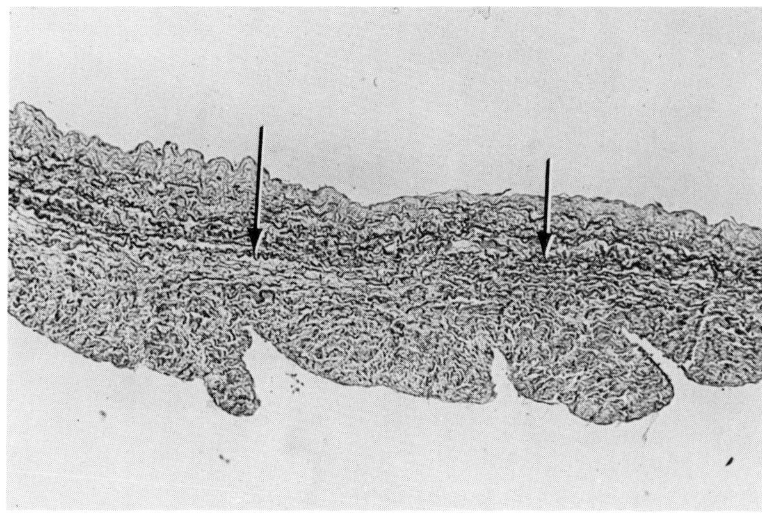
FIGURE 12-22

BIBLIOGRAPHY

Abraham, A.: *Microscopic Innervation of the Heart and Blood Vessels in Vertebrates Including Man,* Pergamon Press, New York, 1961.

Bloom, W., and D. W. Fawcett: *A Textbook of Histology,* 10th ed., Saunders, Philadelphia, 1975.

Davies, F.: "The Conducting System of the Vertebrate Heart," *Br. Heart J.* **4:**66–76 (1942).

Luisada, A. A. (ed.): *Development and Structure of the Cardiovascular System,* McGraw-Hill, New York, 1961.

McNutt, N. S., and D. W. Fawcett: "Myocardial Ultrastructure," in G. A. Langer and A. J. Brady (eds.), *The Mammalian Myocardium,* Wiley, New York, 1974.

Rhodin, J. A. G.: *Histology. A Text and Atlas,* Oxford University Press, New York, 1974.

Statler, W. A., and R. A. McMahon: "The Innervation and Structure of the Conductive System of the Human Heart," *J. Comp. Neurol.* **87:**57–83 (1947).

ADDITIONAL READINGS

Chiba, T., and A. Yamauchi: "On the Fine Structure of Nerve Terminals in the Human Myocardium," *Z. Zellforsch.* **108:**324–338 (1970).

Copenhaver, W. M., and R. C. Truex: "Histology of the Atrial Portion of the Cardiac Conduction System in Man and Other Mammals," *Anat. Rec.* **114:** 601–626 (1952).

Davies, F., and E. T. B. Francis: "The Conduction of the Impulse for Cardiac Contraction," *J. Anat.* **86:**302–309 (1952).

Forsmann, W. G., and L. A. Girardier: "A Study of the T-System in Rat Heart," *J. Cell Biol.* **44:**1–19 (1970).

Hoffman, B. F.: "Physiology of Atrioventricular Transmission," *Circulation* **24:**506–517 (1961).

James, T. N.: "Anatomy of the Cardiac Conducting System in the Rabbit," *Circ. Res.* **20:**638–648 (1967).

———, L. Sherf, and F. Urthaler: "Fine Structure of the Bundle-Branches," *Br. Heart J.* **36:**1–18 (1974).

Kolb, R., A. Pischinger, and L. Stockinger: "Ultrastructur der Pulmonalisklappe des Meerschwinchens," *Z. Miker. Anat. Forsch.* **76:**184–211 (1967).

Lannigan, R. A.: "Ultrastructure of the Normal Atrial Myocardium," *Br. Heart J.* **28:**785–795 (1966).

Mitomo, Y., K. Nakao, and A. Angrist: "The Fine Structure of the Heart Valves in the Chicken. I. Mitral Valve," *Am. J. Anat.* **125:**147–167 (1967).

Muir, A. R.: "Observations on the Fine Structure of the Purkinje Fibers in the Ventricles of the Sheep's Heart," *J. Anat.* **91:**251–258 (1957).

Novi, A. M.: "An Electron Microscopic Study of the Innervation of Papillary Muscles in the Rat," *Anat. Rec.* **160:**123–142 (1968).

Rhodin, J. A. G., P. Del Missier, and L. C. Reid: "The Structure of the Specialized Impulse-Conducting System of the Steer Heart," *Circulation* **24:** 349–367 (1969).

Sommer, J. R., and E. A. Johnson: "Cardiac Muscle. A Comparative Study of Purkinje Fibers and Ventricular Fibers," *J. Cell Biol.* **36:**497–526 (1968).

Thaemert, J. C.: "The Fine Structure of Neuromuscular Relationships in Mouse Heart," *Anat. Rec.* **163:**575–586 (1969).

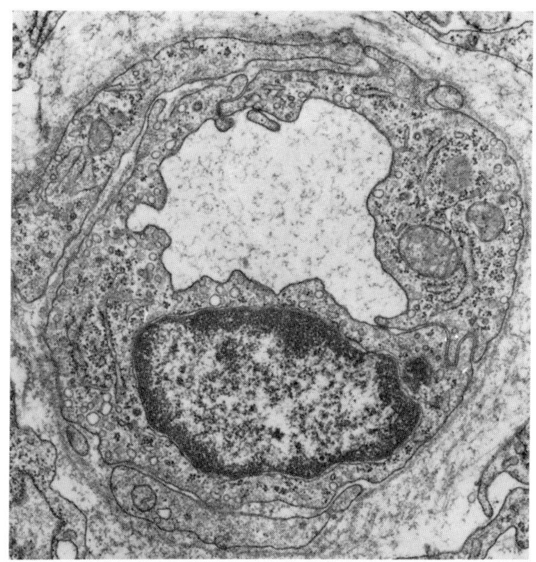

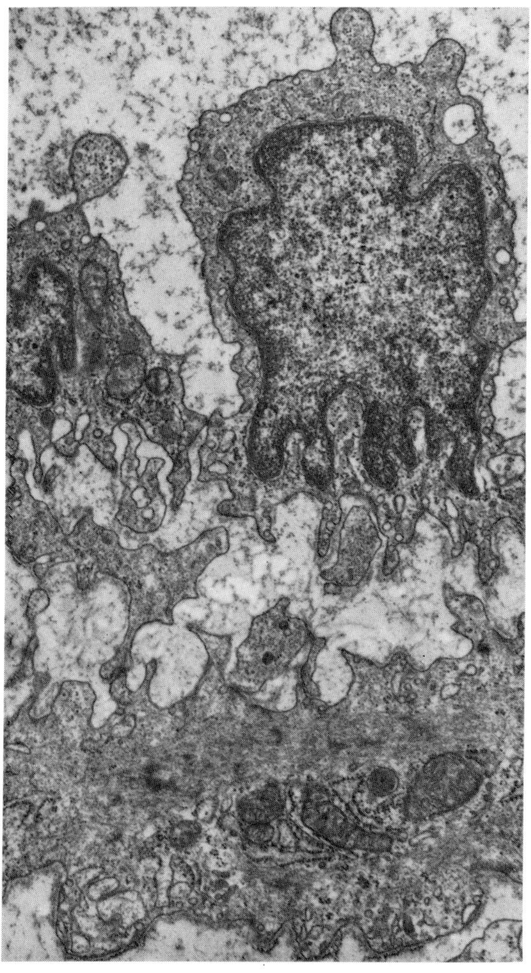

13 BLOOD VESSELS

OBJECTIVES

Upon completion of this chapter, the student will be able to:

1 Describe the endothelial layer of the capillary wall by identifying the location, structure, and function of the following: (a) small and large functional pores (b) basal lamina (c) cytoplasmic components (d) intracellular filaments (e) intercellular connections

2 Identify the location, structure, and function of the pericytes (cells of Rouget).

3 Describe the differences between small, medium, and large arteries and veins by identifying the location, structure, and function of the following three layers of each: (a) intima (b) media (c) adventitia

4 Identify small, medium, and large arteries and veins in slides.

Major blood vessels joining the heart distribute and collect the blood to and from every region of the body. They show certain specializations that reflect different functional needs at different levels of the distributing and collecting system of the blood. The ultimate functional area of the circulatory system is the series of capillary beds which show wide variations in structure, depending on the organs and tissues of the body where they are located. In the spleen and liver they are of sinusoidal structure and are lined with specialized endothelial cells that are capable of phagocytosis.

Similar but less diffusely organized sinusoids with continuous basement membranes exist in many of the endocrine tissues, while in the glomerulus of the kidney, endothelial cells have highly fenestrated cytoplasm, allowing for rapid and massive movement of fluids across the capillary. Elsewhere throughout the body, the general structure of the capillary beds has certain common features, since these capillaries perform similar functions of supplying nutrients and gases to the peripheral tissues and retrieving waste substances from them. For this reason the general structure of capillaries as they exist in most tissues of the body will be considered first and in some detail.

STRUCTURE OF CAPILLARIES

The capillary, usually 7 to 9 μm in diameter, is the smallest vessel of the blood-vascular system. It is so small that formed elements of blood must squeeze through in single file. Basically, a capillary is lined with flat endothelial cells with attenuated cytoplasm resting on a thin basement membrane. External to the basement membrane are undifferentiated connective tissue cells which have been variously referred to as pericytes, perivascular cells of Rouget, or reserve mesenchymal cells. The flat endothelial cell lining the capillary has a bulging nucleus with a dense chromatin pattern when seen under the light microscope (Fig. 13-1). The cytoplasm is barely visible only where it runs around the nucleus (arrow in Fig. 13-1). The light microscope does not readily reveal the thin basement membrane beneath the endothelial cell or the attenuated cytoplasmic border. Cell boundaries may be delineated with the use of a silver impregnation technique.

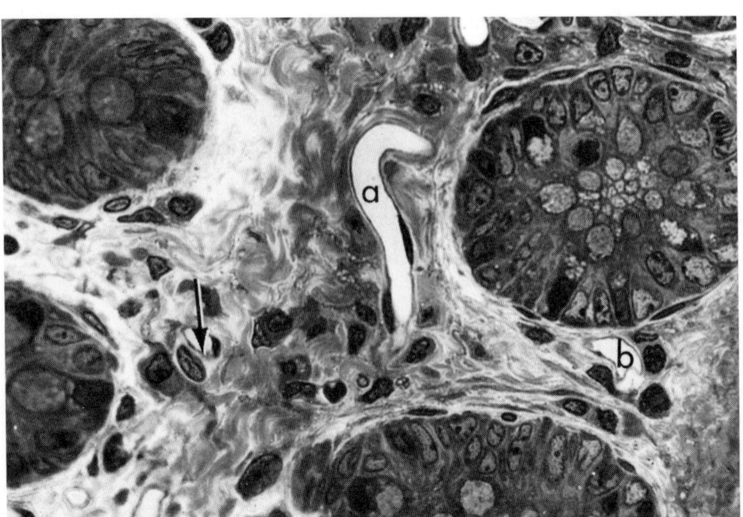

FIGURE 13-1 Light micrograph of the wall of the small intestine illustrating longitudinal (a) and cross-sectional (b) views of capillaries. The nuclei of the endothelial cell bulge into the lumen and the cytoplasm is visible only in the nuclear portion (arrow).

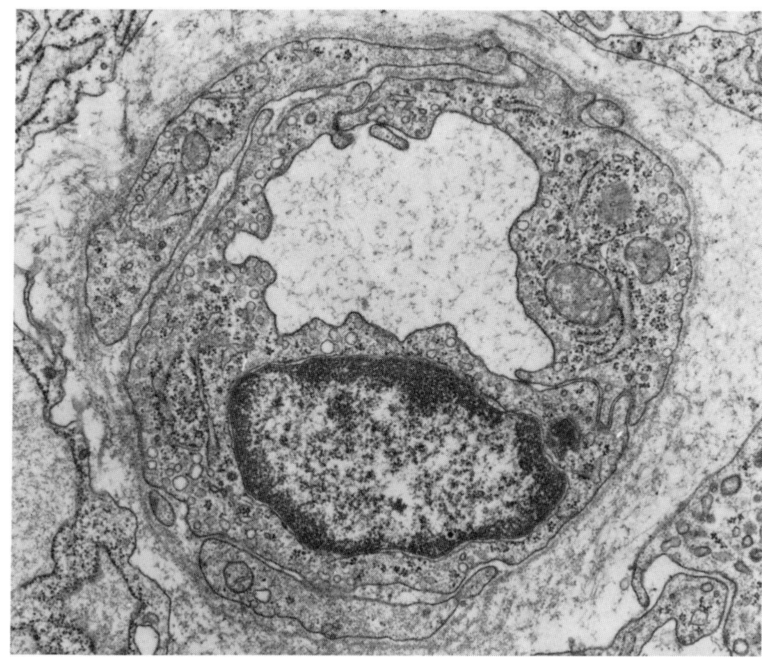

FIGURE 13-2 Electron micrograph of a capillary. The oval nucleus has a dense chromatin pattern at the periphery. Mitochondria, RER, ribosomes, and micropinocytic vesicles are apparent in the cytoplasm. Intermediate junctions hold the epithelial cell together. A thin basal lamina separates the endothelial lining from the connective tissue.

ELECTRON MICROSCOPIC APPEARANCE

The endothelial cell rests on a thin basal lamina (Figs. 13-2 and 13-3) which separates it from the adjacent connective tissue. The cytoplasm around the nucleus contains a small number of RER profiles, ribosomes, vesicles, and an occasional Golgi complex (circles in Fig. 13-3). Throughout the peripheral cytoplasm, where it becomes more attenuated, small vesicles of about 70 to 80 nm in diameter may be seen. These vesicles, called *micropinocytic vesicles* (arrows in Fig. 13-4), appear to originate by invaginations from either the luminal or basal surfaces of the attenuated endothelial cytoplasm. They are thought to be responsible for "in quantum" transport of materials across the capil-

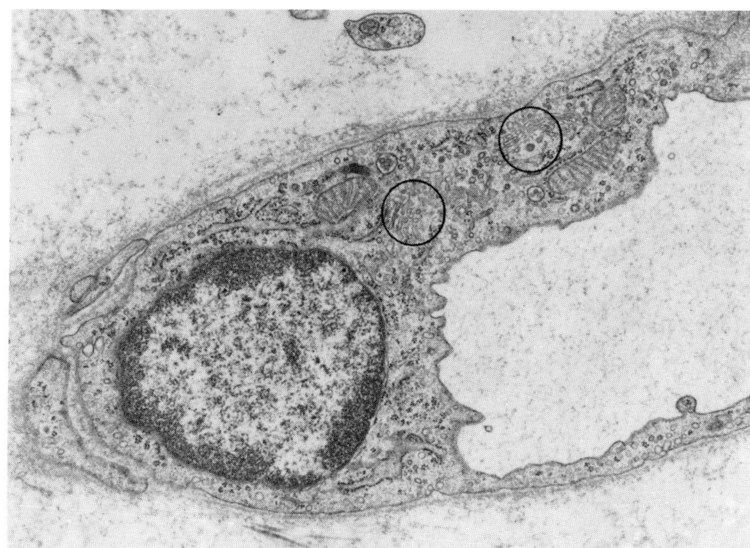

FIGURE 13-3 Electron micrograph of a capillary. The chromatin is condensed along the periphery of the nucleus. Mitochondria, RER, ribosomes, and a couple of small Golgi complexes (circles) are seen in the cytoplasm. Micropinocytic vesicles are abundant, particularly in the attenuated portions of the cytoplasm. A thin basal lamina separates the endothelium from the connective tissue.

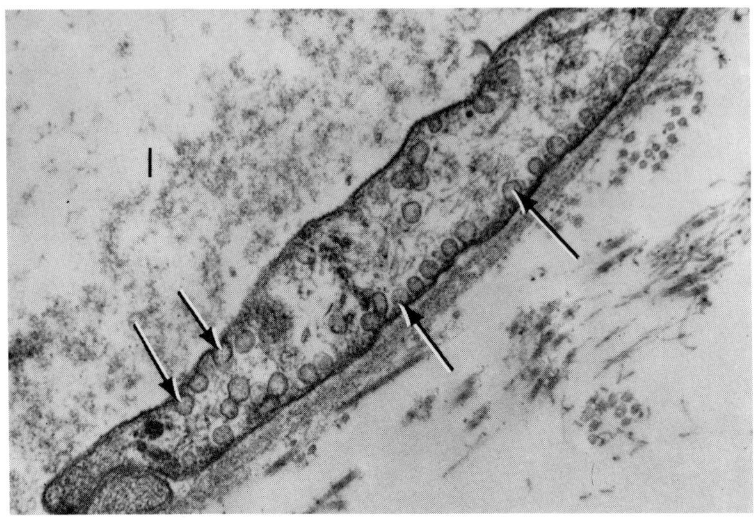

FIGURE 13-4 Attenuated portion in the periphery of the cytoplasm of an endothelial cell. Numerous micropinocytic vesicles seem to originate by invaginations either from the luminal (l) or basal surfaces of the attenuated cytoplasm.

lary wall. In certain areas of attenuated endothelial cytoplasm such as a capillary from a metabolically active organ, one finds fenestrated regions (see both Figs. 13-5 and 13-6). Figures 13-5 and 13-6 are low- and high-power EMs of a capillary from the exocrine pancreas, which produces much protein for secretion. Such fenestrations are covered by a thin diaphragm and therefore resemble those fenestrated capillaries found in the renal glomerulus and certain endocrine organs.

Frequently intracellular filaments run randomly throughout the cell and often form bundles or tracts (t in Fig. 13-7). Near the intercellular boundaries, they converge into specialized regions of attachment of the plasma membrane. Some scientists feel that

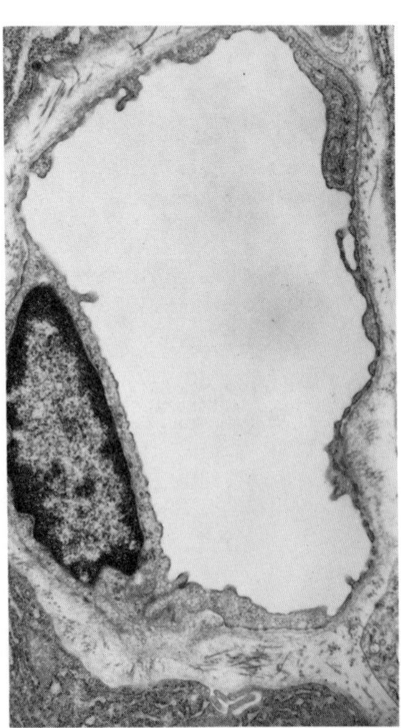

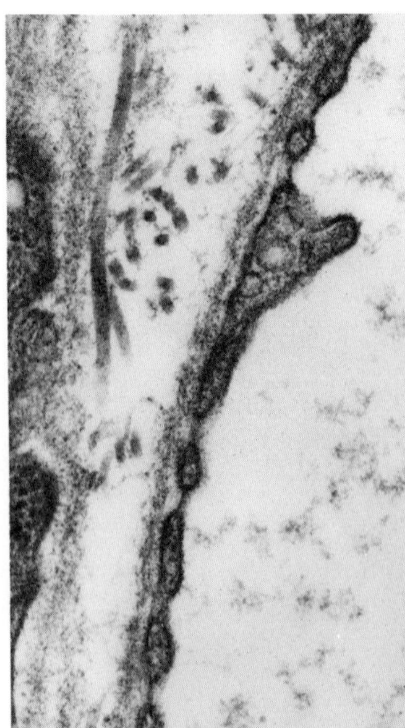

FIGURE 13-5 Electron micrograph of a portion of an endothelial cell. In the attenuated part, micropinocytic vesicles and fenestrations are seen. A diaphragm covers each fenestration.

FIGURE 13-6 Portion of another endothelial cell having fenestrations. A diaphragm also covers these fenestrations. Also note the forming micropinocytic vesicles.

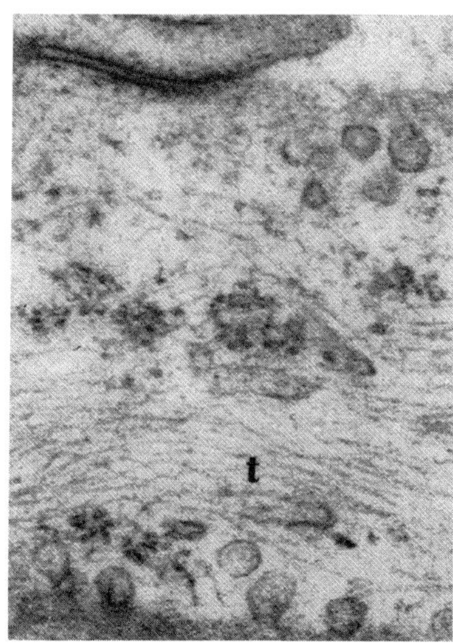

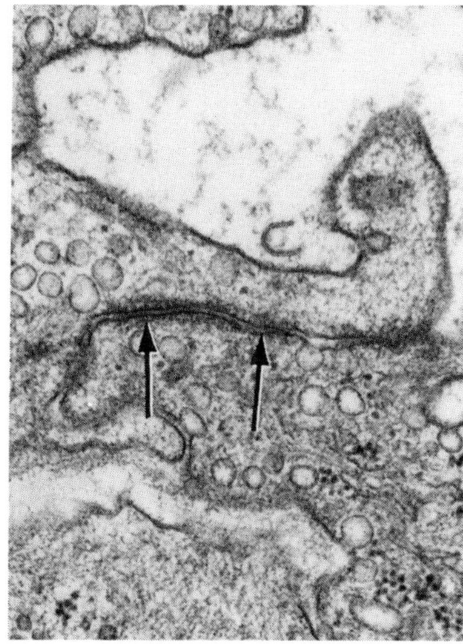

FIGURE 13-7 Portion of the cytoplasm of an endothelial cell. Microfilaments run randomly throughout the cytoplasm and in some parts may form tracts (t).

FIGURE 13-8 Electron micrograph of the intercellular boundary of two endothelial cells. Poorly developed intermediate junctions (arrows) and zonula occludens–like junctions are the types of junctional complexes that connect endothelial cells.

endothelial cells are capable of contracting and that the filaments represent contractile elements, while others view these filaments as being analogous to the tonofilaments of epithelium. Indeed structures resembling poorly developed "minidesmosomes" or intermediate junctions can be seen between endothelial cells (arrows in Fig. 13-8).

At the junction between endothelial cells, a small flap that extends into the lumen is often observed in one or both cells (f in Fig. 13-9). Note also the presence of intermediate junctions (arrows). Such flaps of endothelial junctions could be simple cytoplasmic tongues (Fig. 13-10) or may form elaborate interdigitations (Fig. 13-11). Under normal con-

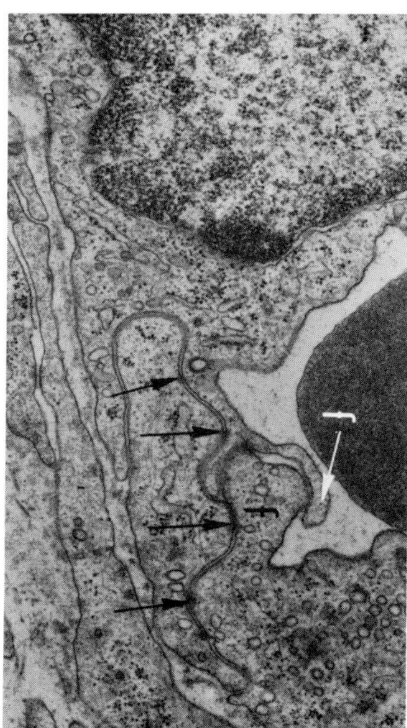

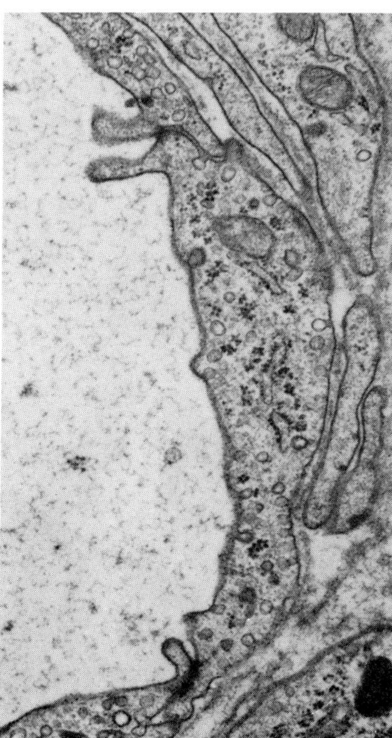

FIGURE 13-9 Electron micrograph of the junction between two endothelial cells. A cytoplasmic flap (f) extends from each cell into the lumen. A zonula occludens–like junction and intermediate junctions (arrows) connect the cells.

FIGURE 13-10 Intercellular junctions between a small portion of three endothelial cells. Small cytoplasmic tongues extend into the lumen at the junction.

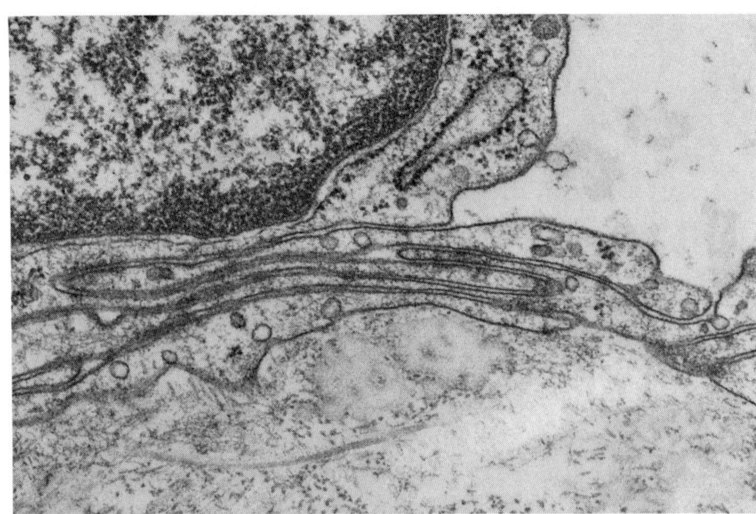

FIGURE 13-11 Elaborate interdigitations of cytoplasmic processes on the luminal side of the intercellular junctions of two endothelial cells in a capillary.

ditions, the intercellular junction is closed and does not allow passage of cells nor influx or efflux of larger molecules, but may open when irritations or inflammatory conditions prevail. This has been shown experimentally through introduction of such agents as histamine or serotonin followed by india ink or other electron markers, which demonstrated a leakage along the capillary wall at the intercellular junctions. Thus, the leakage of the capillary wall (including hemorrhage) during inflammation appears to be due to separation of intercellular junctions. It has been shown that the desmosome-like regions between endothelial cells tend to stay intact, even during the separation of the junctions induced by inflammatory processes. Under physiologic conditions the intercellular junctions also provide a passage for the ameboid white blood cells without allowing nonameboid RBCs to leave the blood vessel.

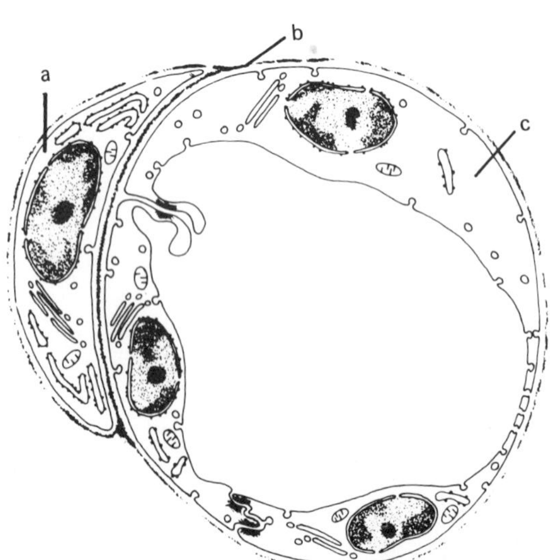

FIGURE 13-12 Diagram illustrating the parts of a capillary. (a) pericyte; (b) basal lamina; (c) endothelial cell.

Physiological studies in the past have indicated that the transport of materials across the capillary may require two functional pores: one that is very small and another large enough to allow the transport of macromolecules. While there are some conflicting views that have yet to be resolved, recent experiments using various electron markers have suggested that the micropinocytosis mentioned previously may account for the large pores advocated by many physiologists, while the small pore system may be represented by the intercellular junctions. As shown by the lack of accumulation of

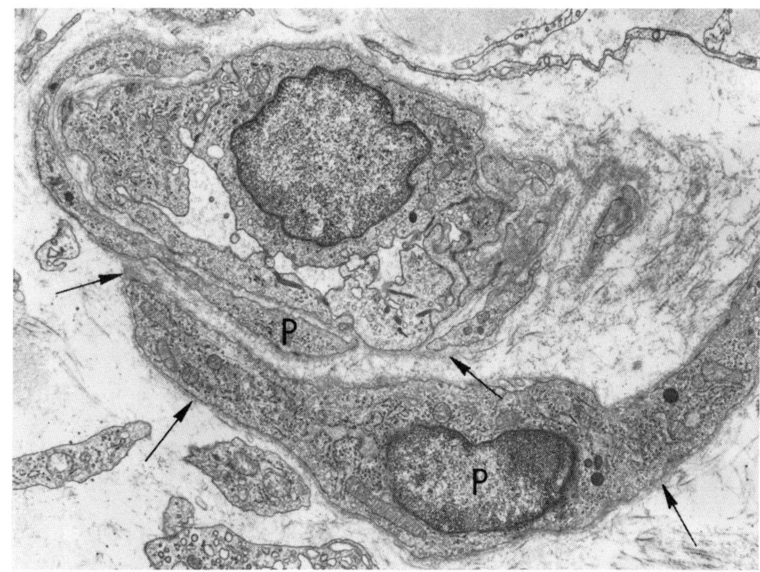

FIGURE 13-13 Electron micrograph of a capillary. The bulging nucleus of the endothelial cell partially occludes the lumen. Two pericytes (p) are localized within a basal lamina (arrow) of the capillary. One of them is sectioned through the nucleus.

electron markers between endothelial cells and basal lamina, the latter does not seem to play the role of a functional barrier.

Figure 13-12 diagrams the essential features described so far and in addition indicates one of the periovascular cells. These undifferentiated connective tissue cells associated with the capillary are called *pericytes* or *Rouget cells* (Figs. 13-10 and 13-11). One or more (Fig. 13-13) pericytes, which may or may not include the nucleus, are seen along the outside of the capillary. Often long cytoplasmic processes extend from the perinuclear region over large areas of the capillary. The pericytes are located within a basal lamina (arrows in Figs. 13-13 and 13-14), which is often contiguous with that of the endothelium.

Pericytes are undifferentiated cells that have the potential to become fibroblasts or macrophages when conditions are appropriate. The quiescent cell has a modest amount

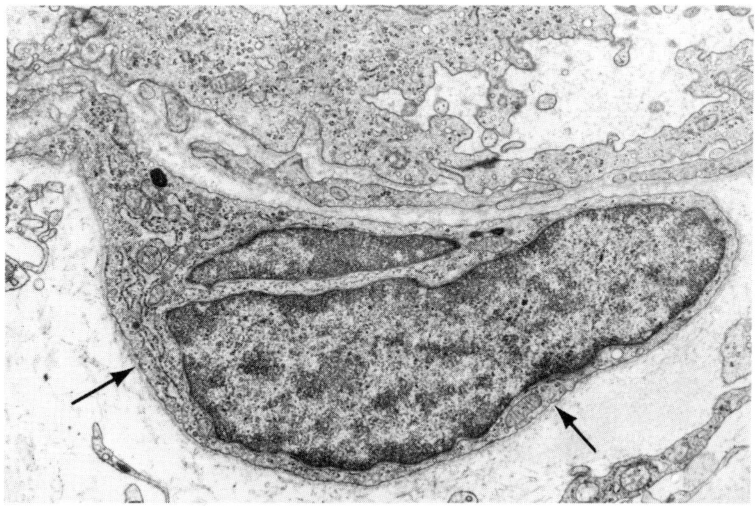

FIGURE 13-14 Electron micrograph of a pericyte. It is surrounded by a basal lamina (arrows).

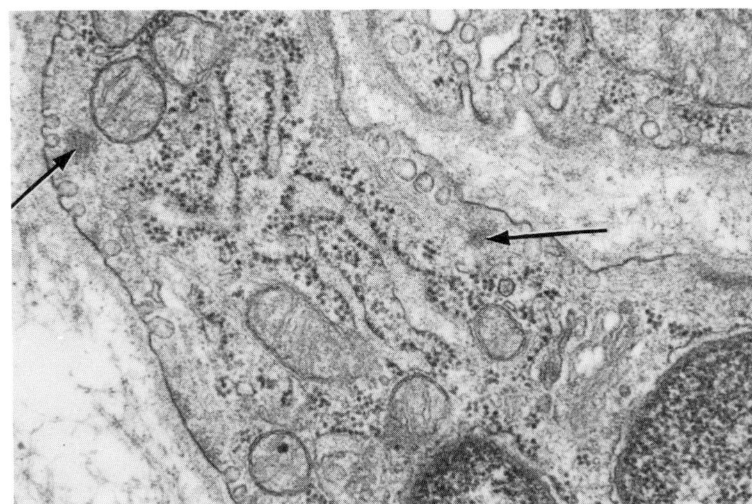

FIGURE 13-15 Portion of a pericyte. Cytoplasmic organelles seen in this cell are RER, polysomes, mitochondria, and a Golgi complex. Bundles of fibrils (arrows) are present in the cortical cytoplasm. Micropinocytic vesicles are forming along the surface of the cell. Notice the continuous basal lamina.

of RER, a Golgi apparatus, and a few small mitochondria (Fig. 13-15). There may be occasional intracellular fibrils which are bundled along the cortical cytoplasm (arrows in Fig. 13-15), as in some of the fibroblasts. Thus the pericyte, even when quiescent, resembles an immature fibroblast. Pericytes become rich in cytoplasmic components during wound healing or inflammatory response. They may divide in the regenerating areas, differentiating into fibroblasts or macrophages. Under such conditions, the cytoplasm of the endothelial cells also becomes rich in polysomes and RER, since they, too, divide to produce new vessels.

STRUCTURE OF LARGE VESSELS

Arteries and veins make up the remainder of the circulatory system and link capillary beds with the heart. They are classified into three major divisions based on size: small, medium, and large. The small arteries and veins (40 to 400 μm in diameter) generally do not have specific names. Medium-sized arteries and veins usually do have individual names, such as the brachial artery or the femoral vein. The large or conducting vessels include such structures as the aorta and the venae cavae.

Throughout the vasculature, the following three organizational layers of the vessel walls can be recognized. The first and innermost layer is continuous with the endocardium and is called *tunica intima*. It is composed of the endothelium, basement membrane, and a small amount of connective tissue. The middle layer, *tunica media*, is generally the thickest and shows wide variations in cells and fibers of which it is composed. The outermost layer, the *adventitia*, contains a large amount of connective tissue cells and fibers that are continuous with like structures that invest the vessel. In the following discussion of blood vessels, structures that are specifically differentiated to meet the unique functional needs of different classes of vessels will be emphasized.

ARTERIES

Since all arteries are concerned with the distribution of the blood under high pulsating pressure, the walls must be resilient. This is accomplished by elastic fibers and smooth muscle cells that are abundant in the tunica media.

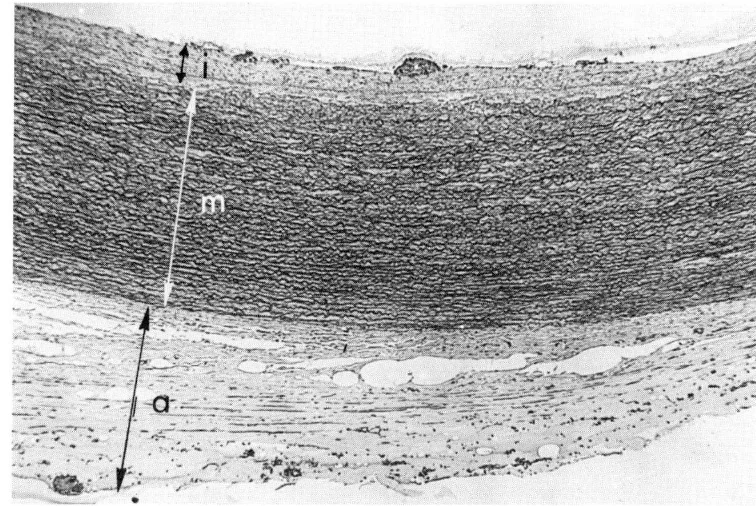

FIGURE 13-16 Light micrograph of a cross section of a large artery (aorta). The intima (i) is thick and contains thin elastic fiber. A wavy internal elastic membrane separates the intima from the media (m) which is a heavy layer. The latter has many elastic membranes that are fenestrated. A thick adventitia (a) surrounds large arteries.

Large Arteries

The intima of the large elastic or conducting arteries is thick and, in portions, blends with tunica media (i in Fig. 13-16). Thin elastic fibers are found throughout the intima. The internal elastic membrane is wavy and refractile, has less affinity for the stains than the other elastic fibers, and marks the beginning of the tunica media. The media (m in Fig. 13-16) is a relatively heavy layer of elastic membranes that are fenestrated in the largest vessel. In H- and E-stained preparations, the elastic elements of the media predominate the field as refractile lines, and the boundary between the intima and media is often indistinct (Fig. 13-17). Figure 13-18 illustrates the appearance of fenestrated elastic membranes (arrows) of the media stained dark with Weigert's stain for elastic fibers. Collagenous fibers and smooth muscle cells run through the fenestrations as well as between the layers of elastic membranes. There are relatively fewer smooth muscle cells in larger arteries, as these vessels function primarily

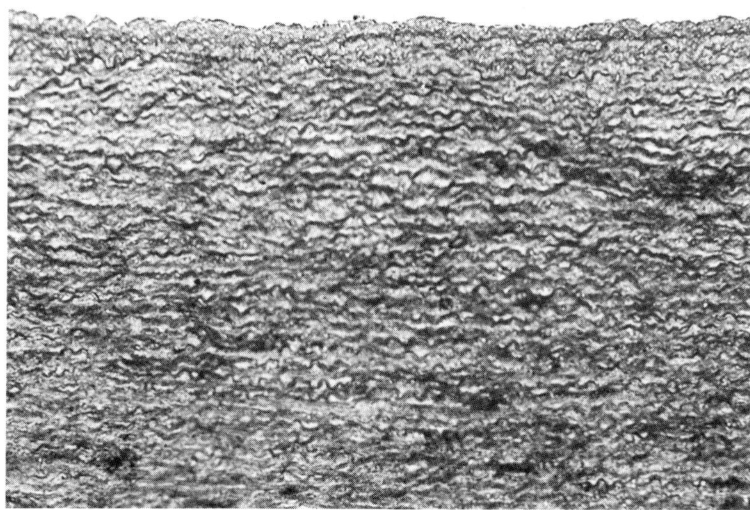

FIGURE 13-17 Light micrograph of aorta, in this section stained with H and E. The elastic membranes appear as refractile lines, and the boundary between the intima and the media is indistinct.

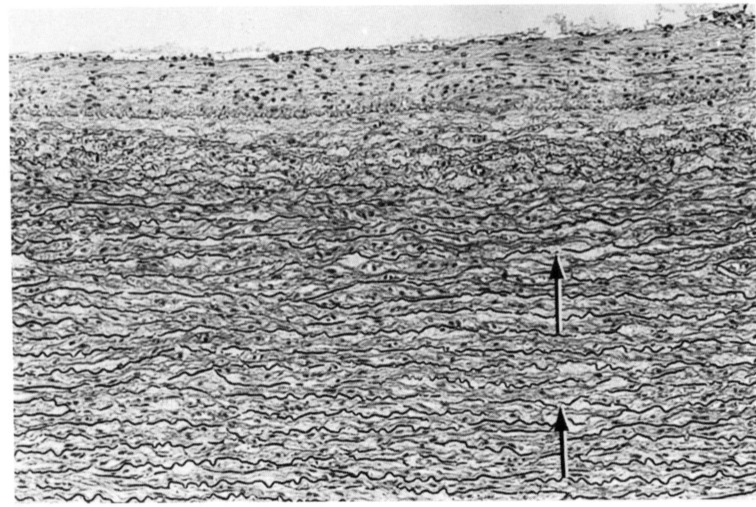

FIGURE 13-18 Light micrograph of aorta stained with Weigert's method. The fenestrated membranes of the media appear dark, and the boundary between the intima and the media is clearly indicated by the internal elastic membrane (arrows indicate fenestrations).

to conduct blood under high pressure and thus require more elastic fibers. A thick adventitia surrounds the large arteries (a in Fig. 13-16).

Medium-sized Arteries

Medium-sized arteries contain a thick intima (Fig. 13-19). In contrast to the large artery, the *internal elastic membrane,* or *elastica interna,* is prominent in medium-sized arteries (iem in Fig. 13-19). Note the appearance of thick internal elastic membranes seen at successively higher magnifications (arrows in Fig. 13-20; iem in Fig. 13-21). The smooth muscle in the tunica media becomes more prominent, interspersed with many collagenous and reticular fibers and fibroblasts (arrowheads in Fig. 13-20). Elastic fibers are also present in the tunica media, which forms a clear *external elastic membrane,* or *elastica externa,* outside the muscular middle layer (eem in Fig. 13-20). The adventitia is thick, showing many fibroblasts, occasional smooth muscle cells, and elastic fibers within the connective tissue (Fig. 13-19).

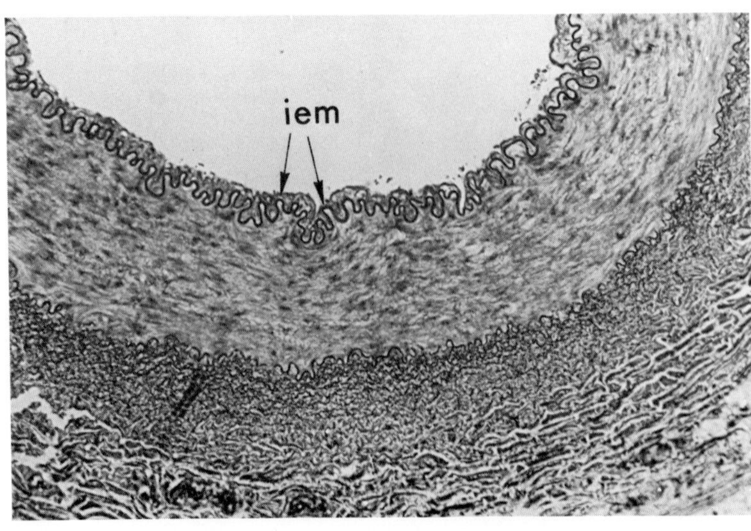

FIGURE 13-19 Light micrograph of a medium-sized artery. The intima is thick and is separated from the media by a distinct internal elastic membrane. Smooth muscle cells are prominent in the media, interspersed with many collagenous and reticular fibers and fibroblasts. The external elastic membrane constitutes the border between the media and the adventitia which is thick. Because of the construction of the vessel wall, the internal and external elastic membranes appear wavy in histologic sections (iem = internal elastic membrane).

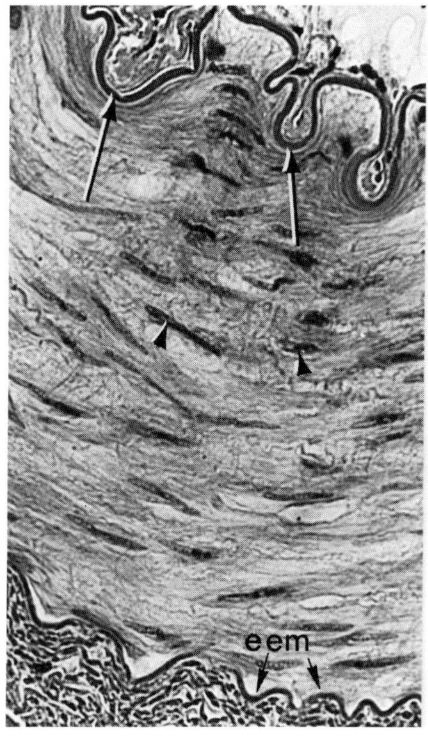

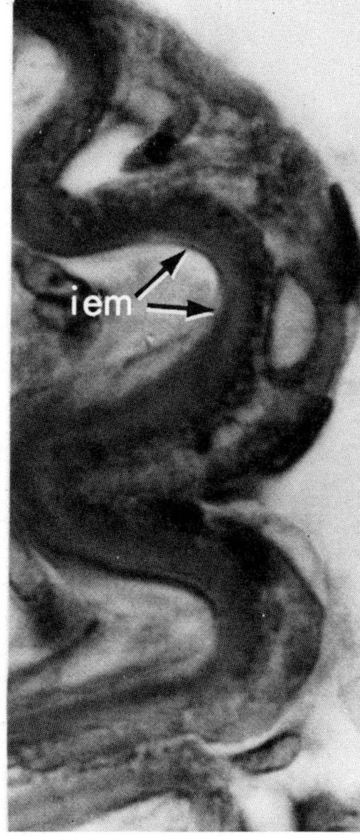

FIGURE 13-20 Light micrograph of medium-sized artery. A distinct internal elastic membrane (arrows) is seen at the boundary between the intima and the media. Smooth muscle cells (arrowheads) are prominent in the media and circularly arranged. Elastic membranes are also present in the media and form an external elastic membrane outside the media. The adventitia is thick (eem = external elastic membrane).

FIGURE 13-21 The internal elastic membrane (iem) of a medium-sized artery viewed at higher magnification.

Small Arteries

The intima is composed of endothelium and its basement membrane with connective tissue visible only under the electron microscope (Fig. 13-22). The elastic membrane is thinner but visible throughout the arterioles. In tunica media the smooth muscle cells are more tightly arranged than in larger arteries. This is because there are fewer connective tissue fibers. The adventitia is composed of

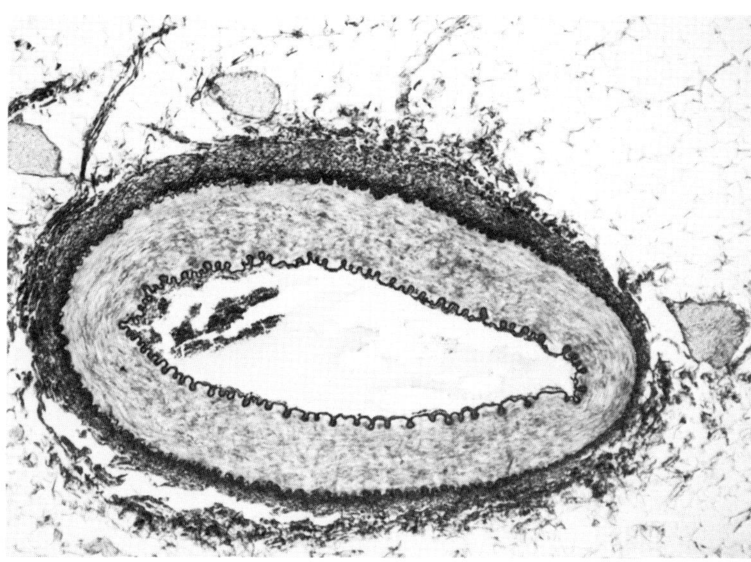

FIGURE 13-22 Cross section of a small artery. The intima is very thin. Note the wavy irregular contour of the internal elastic membrane. The media is made up of smooth muscle cells that are more tightly arranged than in larger arteries. It is the thickest layer in the wall of small arteries. The external elastic membrane is much thinner than the internal elastic membrane and it is barely visible. A relatively thick adventitia forms the outer layer.

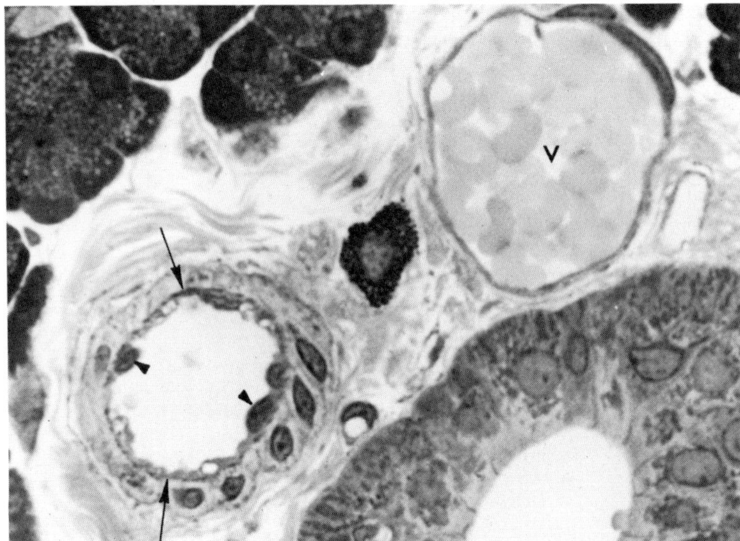

FIGURE 13-23 Light micrograph of arteriole. The intima is thin and consists of a single layer of small endothelial cells. It is separated from the media which is composed of a single layer of smooth muscle cells by the internal elastic membrane (arrows). Contraction of the media causes the endothelial cell to protrude into the lumen (arrowheads). A few connective tissue cells make up the adventitia. A venule (v) accompanies the arteriole.

a fair amount of connective tissue cells and fibers that run more or less circularly. As the small arteries become even smaller, the smooth muscle cells of the tunica media become fewer in number and smaller in size. In the smallest arterioles, the intima is thin and is composed of corresponding small endothelial cells and a nearly invisible basement membrane. It is separated from the media, which is composed of a single layer of smooth muscle cells, by a thin elastic membrane (arrows in Fig. 13-23). Contraction of the media of small arteries and arterioles often causes their endothelial cells to protrude into the lumen (arrowheads in Fig. 13-23 and e in Fig. 13-24). The adventitia at this level is

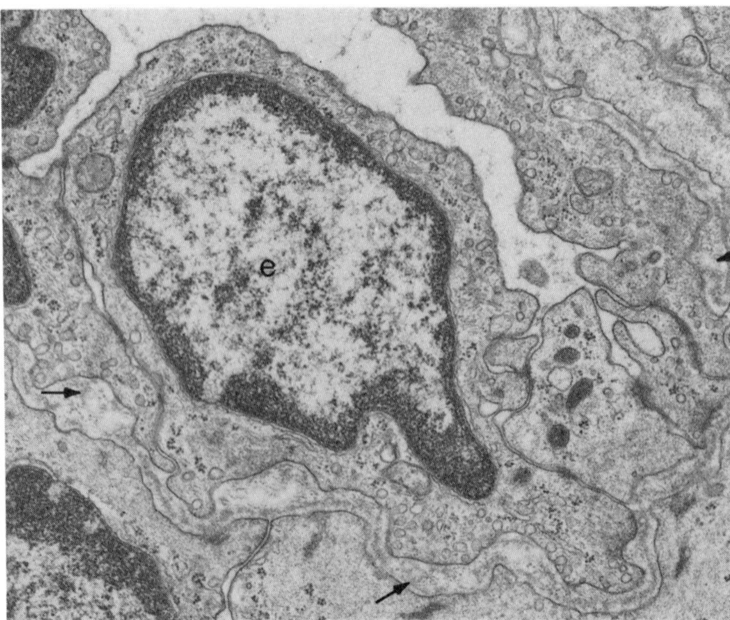

FIGURE 13-24 Electron micrograph of an arteriole. One of the endothelial cells (e) of the intima practically obliterates the lumen. The elastic membrane is nonexistent except for a few elastic fibers (arrows) embedded in the basal lamina.

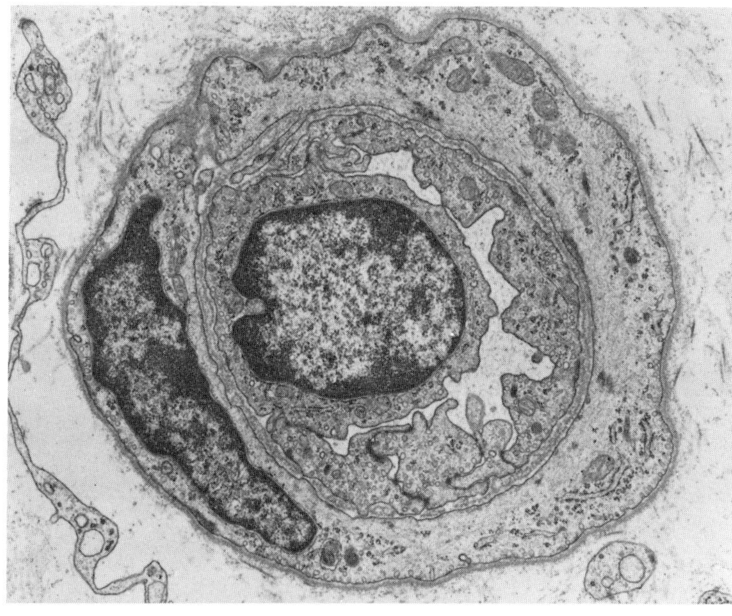

FIGURE 13-25 Electron micrograph of an arteriole showing the structure of intima, media, and the adventitia. The lumen is almost obliterated by a protruding endothelial cell. The elastic membrane appears scanty, and the media is made up of a single smooth muscle cell layer.

composed of a few connective tissue cells, some of which are difficult to distinguish from pericytes.

Electron microscopy reveals additional details of the structure of the three layers of the wall of arterioles. The elastic membrane is largely gone except for a few elastic fibers invested in the basal lamina (arrows in Fig. 13-24). Figure 13-25 is an EM of another arteriole showing the structure of the intima, media, and adventitia. Note the scanty appearance of the elastic membrane and the

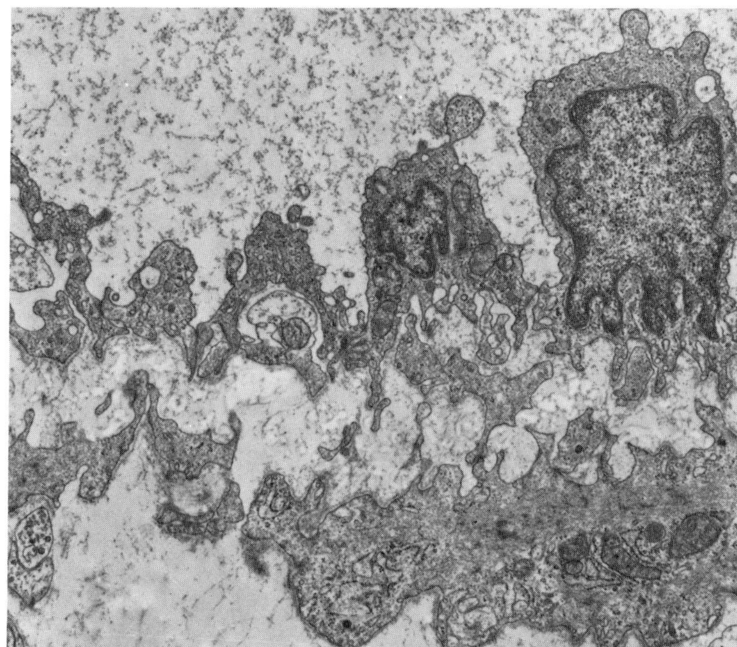

FIGURE 13-26 Portion of an arteriolar wall. Both endothelial (e) and smooth muscle cells (m) have cytoplasmic processes that penetrate the connective tissue of intima, and contact is made between some of them.

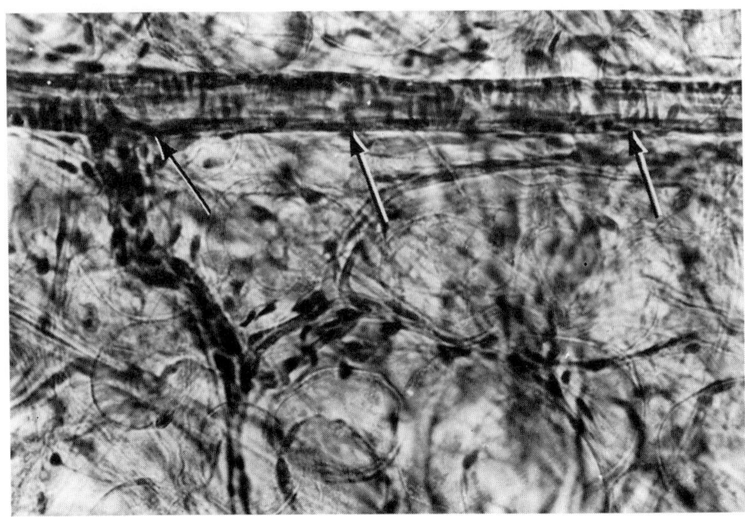

FIGURE 13-27 Phase contrast micrograph of a spread of adipose tissue. The terminal portion of an arteriole (arrows) is seen before branching into capillaries. Note the appearance of the smooth muscle and endothelial cells seen in this longitudinal view of an arteriole.

diminished thickness of the media. The adventitia is practically absent. The arterioles are the prime means of arterial blood pressure regulation. This is accomplished by the constriction and dilatation of their vessel wall, which responds in a most delicate manner because of the thin musculature. There appears to be a wide variation in the structure of arteriolar walls. One such example is seen in Fig. 13-26, in which both the endothelial (e) and smooth muscle (m) cells send out cytoplasmic processes that penetrate the intimal connective tissue and make contacts between them.

The junction between the final segment of the arteriole and the capillary network is characterized by the presence of a discontinuous muscle layer. This segment is often referred to as *metarteriole*. A metarteriole can best be seen in a spread preparation rather than a sectioned tissue. Figure 13-27 is a phase contrast micrograph of adipose tissue which shows the terminal region of the arteriole (arrows) in longitudinal section prior

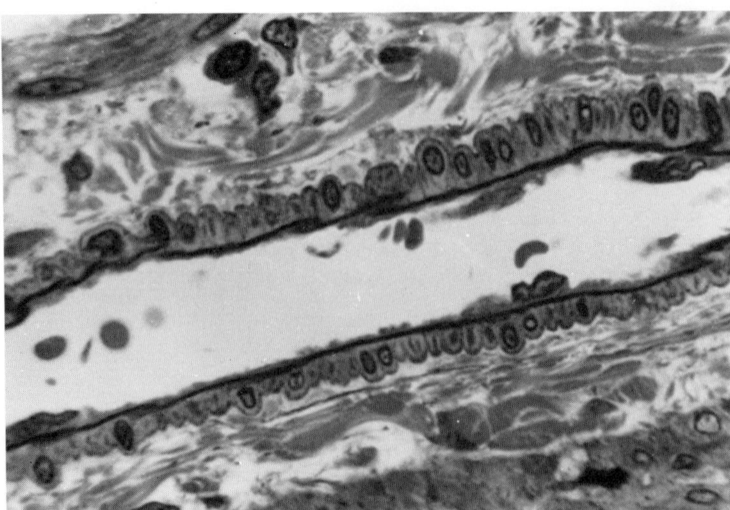

FIGURE 13-28 Light micrograph of a 1-μm-thick plastic embedded tissue. The arteriole is seen in longitudinal section. Note the appearance of the endothelial cells seen in cross section. The adventitia merges with the surrounding connective tissue.

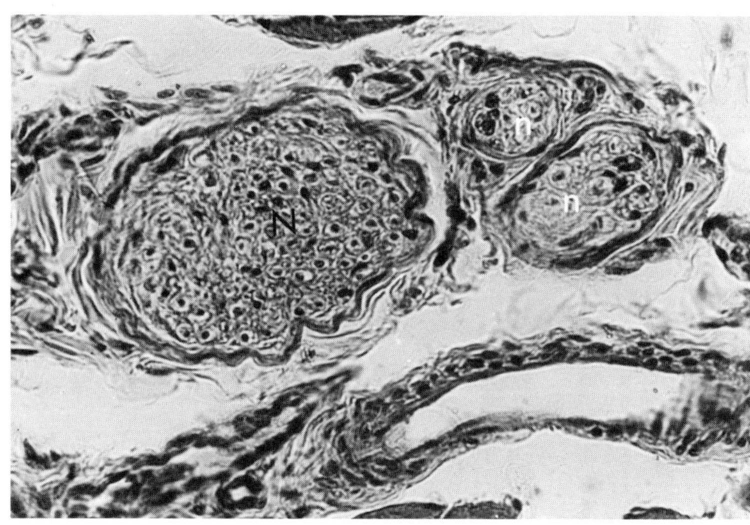

FIGURE 13-29 Arteriole seen in longitudinal view in a section of paraffin-embedded tissue. Note the three small nerves (N) seen in cross section.

to branching into the capillary bed. The discrete appearances of smooth muscle cells, basement membrane, and a few endothelial cells are well depicted in 1-μm-thick longitudinal sections of plastic embedded tissues (Fig. 13-28). Several RBCs are also visible.

In paraffin-embedded tissues sectioned at 5 μm, the small arteriole (Fig. 13-29) can also be identified. Note the presence of three small nerve bundles seen in cross section (N). The terminal segment of a metarteriole branches into a number of capillaries (a and c in Fig. 13-30).

VEINS

The veins passively conduct the blood from capillaries back to the heart. They are usually present next to an artery, since veins generally accompany the branching arterial tree. The lumen of veins is often irregularly shaped in sections and larger than in arteries (a and v in Fig. 13-31). Venous blood pressure is correspondingly lower. Because of the continuous postmortem movement of blood, veins are often filled with formed elements of blood (Fig. 13-31).

The venule has a thin intima with an attenuated endothelium, a very thin media, and some collagenous fibers in the adventitia, as shown in the EM in Fig. 13-32. As in arterioles, small veins or venules demonstrate a range of structural heterogeneity in terms of their size and wall structure. In longitudinal sections the distended venule contains

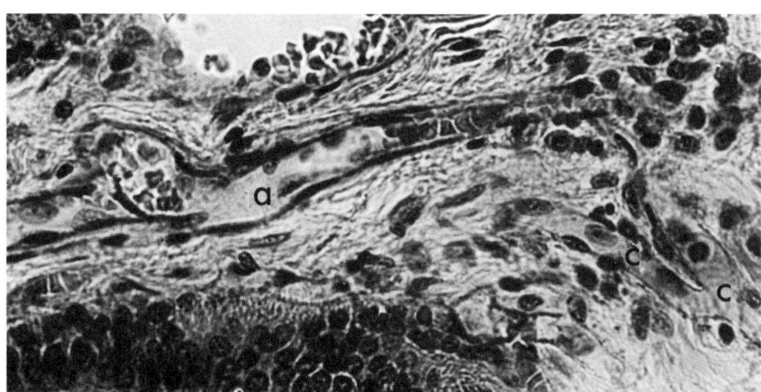

FIGURE 13-30 Light micrograph of the terminal portion of a metarteriole (a). It appears to branch into a number of capillaries (c).

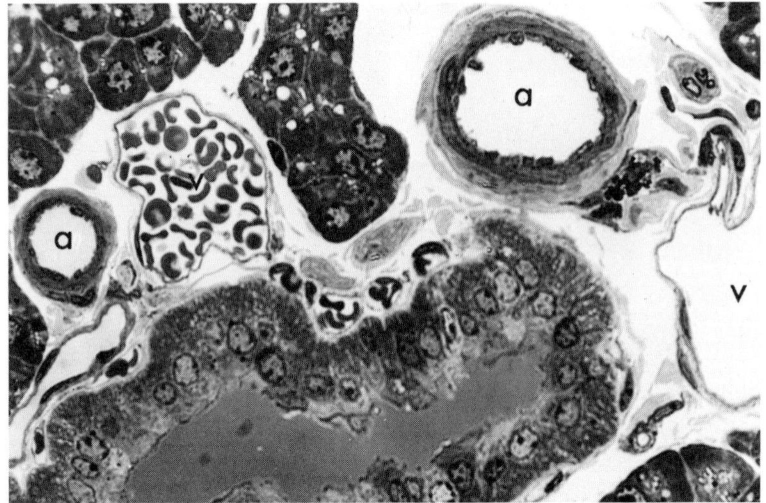

FIGURE 13-31 Light micrograph of a 1-μm-thick plastic embedded tissue. Note the clear definition of the cells in glands, ducts, and blood vessels. Two small arteries (a) with their accompanying veins (v) are seen. The veins are irregularly shaped and have a thinner wall than the arteries. Often the veins are filled with formed blood elements.

a large number of RBCs and has a thin wall made up of an endothelium and an indistinct basement membrane. Such distended veins are frequently found where interstitial tissues are loose and subjected to a greater motility (Fig. 13-33).

The EM in Fig. 13-34 is an oblique section of a venule in the uvula. Note the irregularity along the basal surface of endothelial cells and sparse connective tissue elements. Seen at higher magnification, these vessels have a nominal basal lamina (arrows in Fig. 13-35) and a few collagen fibrils in the underlying connective tissue space, which is filled with microfibrils (arrowheads) running in random directions. Very little smooth muscle is found in the wall structure of the smallest veins (Fig. 13-36). Collagenous fiber bundles that

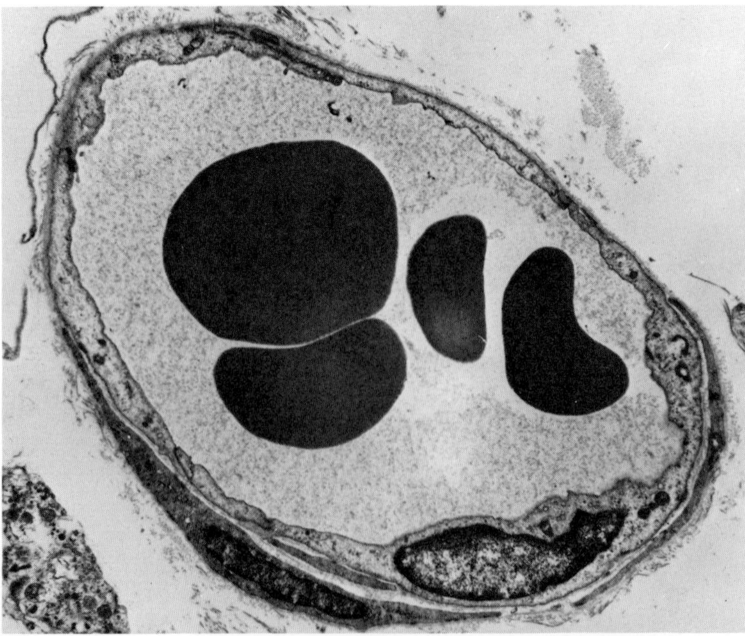

FIGURE 13-32 Cross section of a venule. The intima is composed of endothelial cells and the basal lamina. The media is thin and some collagenous fibers contribute to the adventitia.

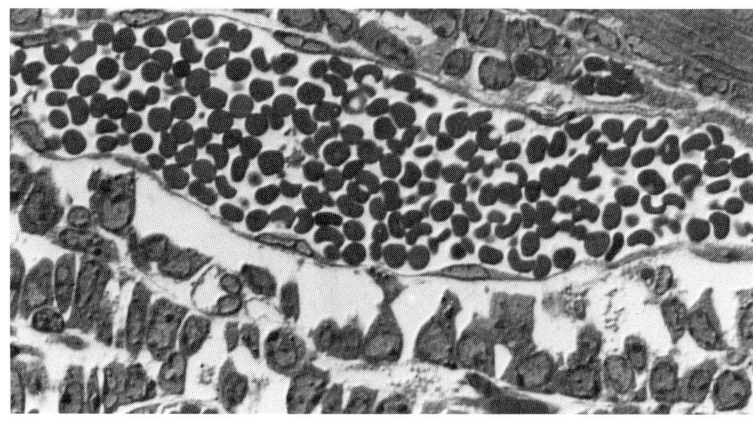

FIGURE 13-33 Longitudinal section of a venule. The wall consists of endothelial cells and an indistinct basal lamina. The distended lumen contains numerous RBCs.

run longitudinally provide necessary structural reinforcement (Fig. 13-35).

As the vein becomes larger, the media becomes thicker; elastic elements and a number of smooth muscle cells appear in the media (Figs. 13-37 and 13-38). The internal and external elastic membranes are lacking, and there are fewer smooth muscle cells in the media than in the wall of arteries of similar size. This is clearly illustrated in Fig. 13-39. In medium-sized veins, the intimal layer becomes modified to produce valves and prevent backflow of blood. They are especially numerous in the veins of the trunk and limbs, which must conduct the blood against gravity. The smooth muscle cells in the tunica media of veins are generally longitudinal in orientation. In the largest veins, such as the vena cava, the intima is thin and the media is little developed. In contrast, the adventitia is thick and well-developed and contains bundles of longitudinally oriented smooth muscle cells. This difference in relative thickness of the two layers is compared

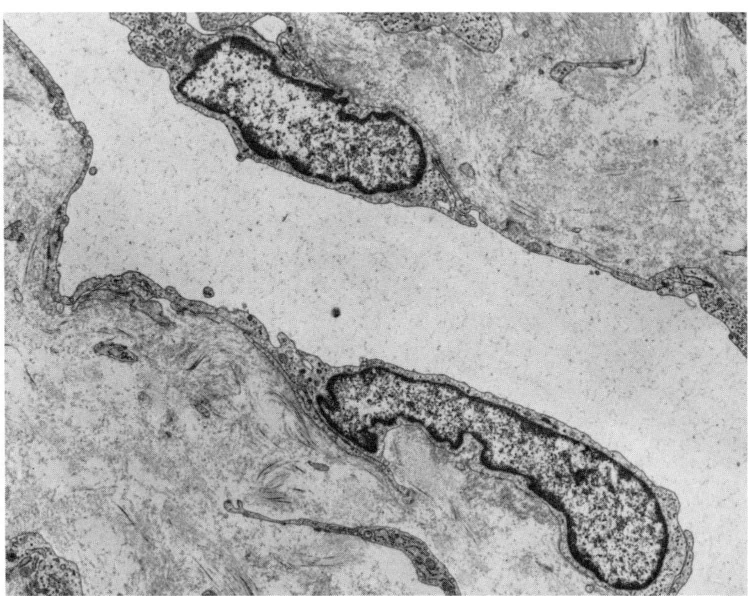

FIGURE 13-34 Electron micrograph of a venule. The junction between the basal surface of the endothelial cells and the surrounding connective tissue is irregular. A delicate basal lamina follows this irregular outline of the endothelial lining. Sparse connective elements are present in surrounding FECT.

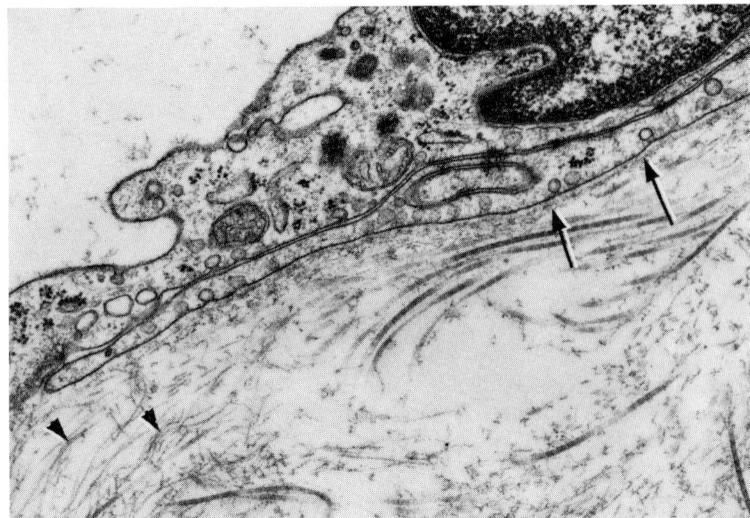

FIGURE 13-35 Electron micrograph of a portion of the wall of a venule. A delicate basal lamina separates the endothelial cells which contain numerous micropinocytic vesicles from the adjacent connective tissue. Sparse collagenous fibers (arrows) are present in underlying tissue space which is filled with randomly running microfibrils (arrowheads).

in Figs. 13-39 and 13-40, which show cross sections of a medium-sized vein and vena cava, respectively.

DISTINCTIONS BETWEEN ARTERIES AND VEINS

A comparison between an artery and vein of equal size shows some of the physical differences that are useful in differential diagnosis of vessels. The diagrams in Figs. 13-41 to 13-43 attempt to demonstrate them. Figure 13-41 compares a cross section of a small artery with that of a small vein. In general, the artery has a definite circular structure reinforced by the circularly arranged smooth muscle cells, collagen, and elastic fibers, while the vein is less regular in shape, lacking the degree of support found in the artery. Note that the artery has a thicker intima, media, and elastic membranes than the vein. Both vessels have similar adventitia. Similarly, a comparison of medium-sized artery and vein is shown in Fig. 13-42. Figure 13-43 depicts the increased thickness of media and

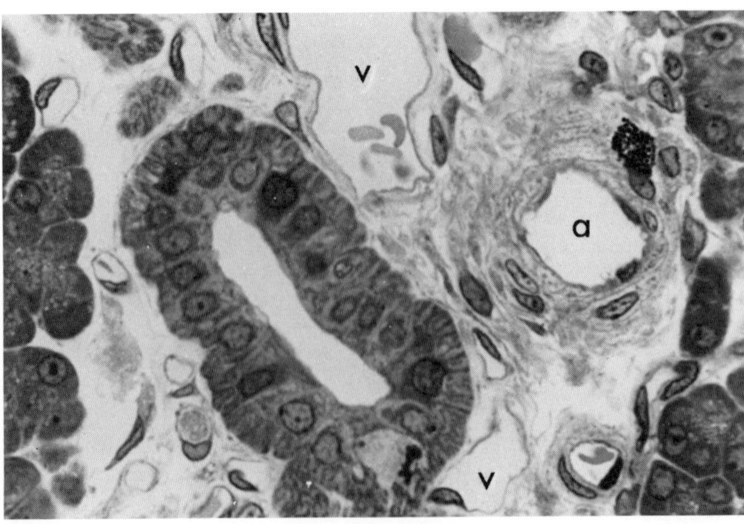

FIGURE 13-36 This section shows a pair of small veins (v) which have a thin wall and very little smooth muscle. The surrounding connective tissue provides reinforcement for the wall. A small artery (a) accompanies the small veins.

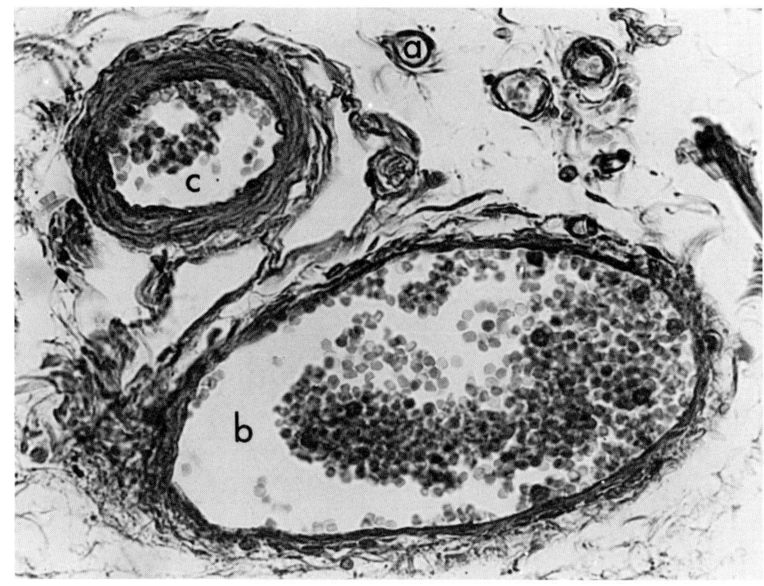

FIGURE 13-37 Cross-sectional view of capillary (a), small artery (c), and small vein (b). Compare the layers of the walls of these three vessels. The media is thicker in this vein than in the one shown in previous figures. Elastic fibers are present in the media, but internal and external elastic membranes are lacking.

the rich elastic membrane that accompanies it in the media of the aorta. In contrast, a large vein shows a fair number of cross-sectioned smooth muscles in the media which is thin.

REVIEW SECTION

1. Name the two circular structures with lumen appearing in Fig. 13-44. On the right is a __________. On the left is a __________. The cells of the vessel which stain dark (pink or red) are

The first clue to the identification of an artery in cross section is its circular shape, which is maintained by the contractility of the smooth muscle cells. In H- and E-stained material the muscle of

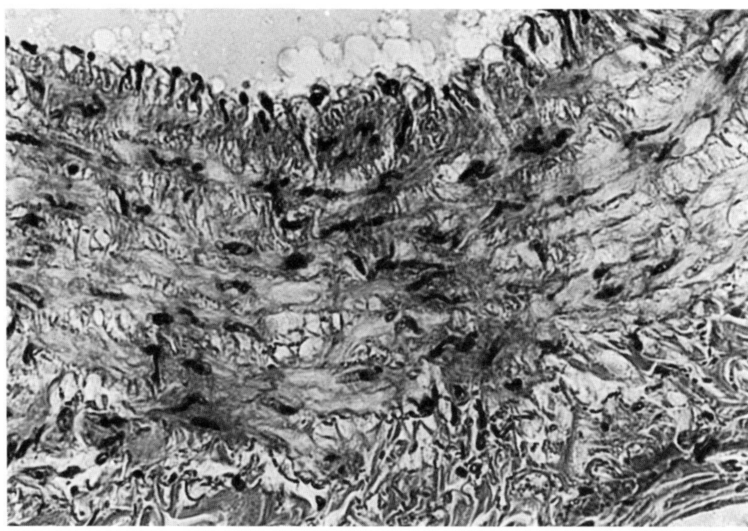

FIGURE 13-38 Light micrograph of medium-sized vein. All layers have increased in thickness. The media contains elastic membranes and an increased number of smooth muscle cells. The adventitia is better developed.

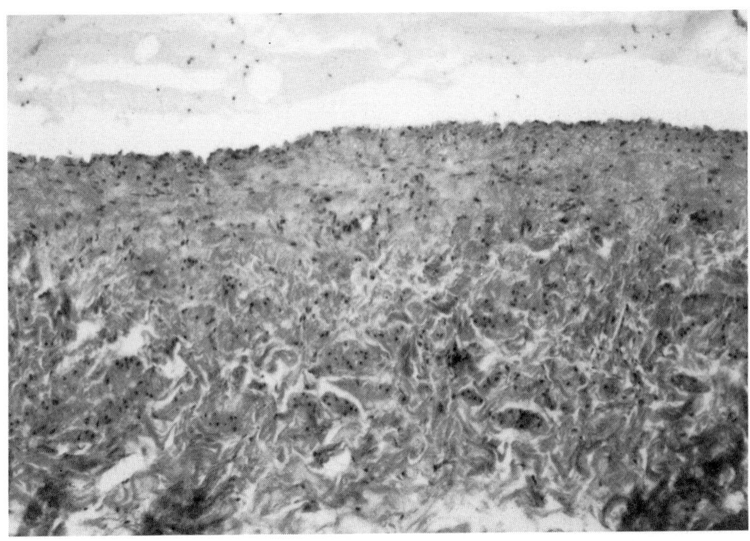

FIGURE 13-39 Cross-sectional view of medium-sized vein. Note the elastic membranes and the increased amount of smooth muscle in the media.

__________. What maintains the shape of the small, rounded vessel? Which tissue do both vessels share?

both arteries and veins will usually stain pink or dark pink, rather than the red or orange-colored staining of collagenous fibers. The vein to the right of the artery has far less muscle in its tunica media. The connective tissue of tunica adventitia of juxtaposed vessels often merges, as suggested in this figure.

2. What type of vessel is shown in Fig. 13-45? What size is it?

This micrograph shows a cross section through a small artery. There is a dark (pink) muscle layer and a thin tunica intima lining the lumen with endothelial cells and basal lamina. The adventitia blends into adipose tissue in this section. Note the difficulty in discerning an internal elastic mem-

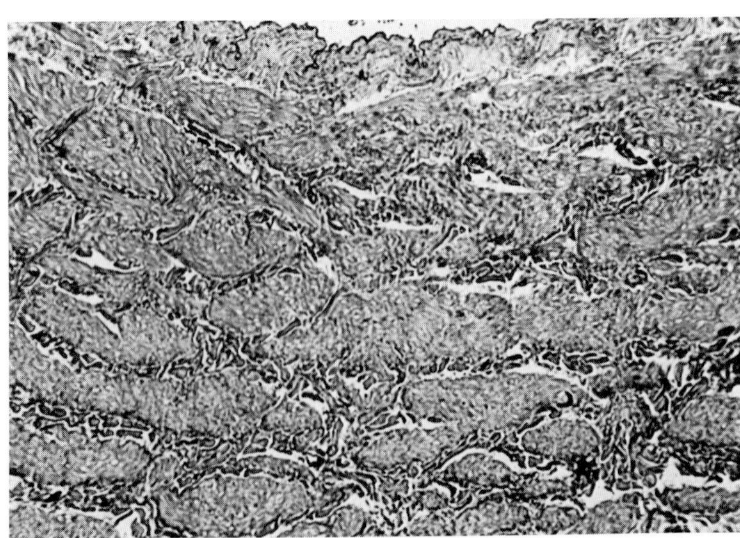

FIGURE 13-40 Light micrograph of the wall of vena cava. The intima is thin and the media is little developed. In contrast, the adventitia is well-developed and contains bundles of longitudinally oriented smooth muscle cells.

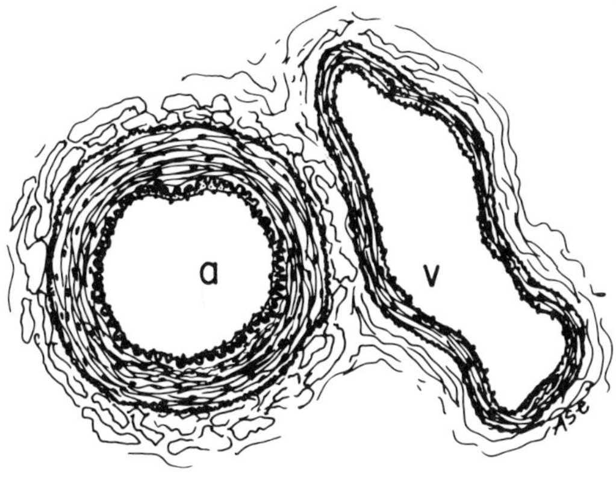

FIGURE 13-41 Diagram of a small artery (a) and a small vein (v). Compare the structure of their walls.

3. What is the circular structure identified by a in Fig. 13-46? What is the structure indicated by b? What is the dense round object in b? The structure labeled "c" is the lumen of a __________. Can you identify the cells indicated by arrows?

4. What is the predominant tissue occupying Fig. 13-47? The narrow strip of tissue at the surface delineated by a dense horizontal line would be made of __________. This inner layer would be termed the __________. The wavy linear structure present throughout the field is __________. What type and size is this vessel?

brane. Note also the presence of many RBCs in the lumen, a feature that is not frequently seen in an artery.

In this photomicrograph, cross sections of arterioles (a) and venules (b) are seen. Several capillaries (c) are also apparent. Endothelial cells lining the capillaries are indicated by arrows.

This section of the aorta shows several features typical of a large artery. The intima is quite thick, since connective tissue and elastic fibers increase in this area. The internal elastic membrane marks the beginning of the tunica media. The tunica media in the large arteries contains a large number of fenestrated elastic membranes between and through which the muscle fibers run. The thick connective tissue adventitia is not included in this photomicrograph.

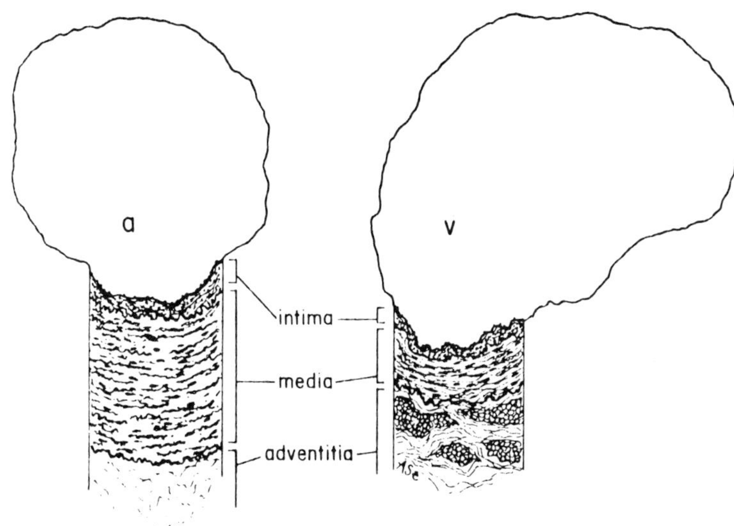

FIGURE 13-42 Diagram of the walls of medium-sized artery (a) and medium-sized vein (v). Compare the structure of their walls.

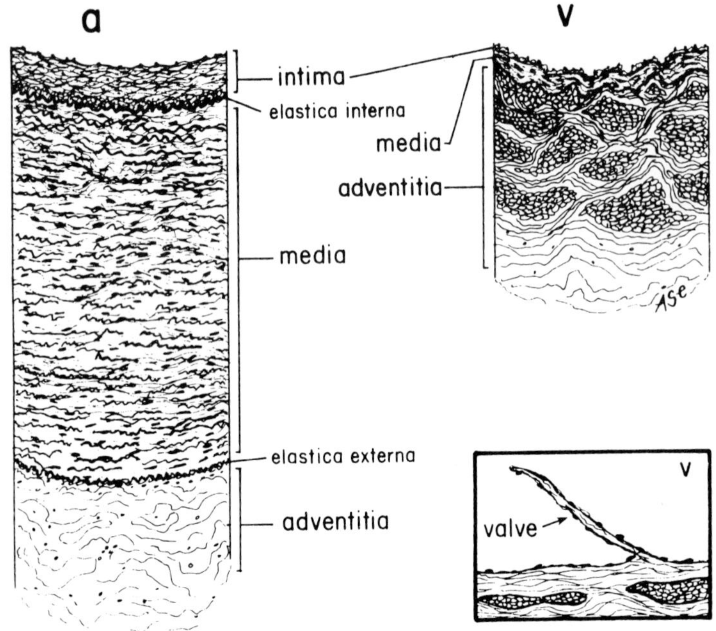

FIGURE 13-43 Diagram illustrating the wall of aorta (a) and large vein (v). Compare their structure. The structure of a venous valve is illustrated in the rectangle.

5. The small longitudinally sectioned tubular structure in Fig. 13-48 is _________. What are the cells indicated by the arrows? How could you distinguish it from a smooth muscle cell? What type of tissue does this field represent?

This figure shows a longitudinal section of a capillary in the middle running through adipose tissue. Note the thin wall. The pericytes (arrows) in this figure can be distinguished from a smooth muscle cell because of their parallel orientation with respect to the vessel.

6. Figure 13-49 is a cross-sectioned EM view of a vessel. Identify the structures indicated by a, b, c, and d. What kind of vessel is shown?

This EM shows a portion of a vein in which RBCs (a) in the lumen, endothelium (b), fibroblasts (c) in the media, and cross-sectioned smooth muscle

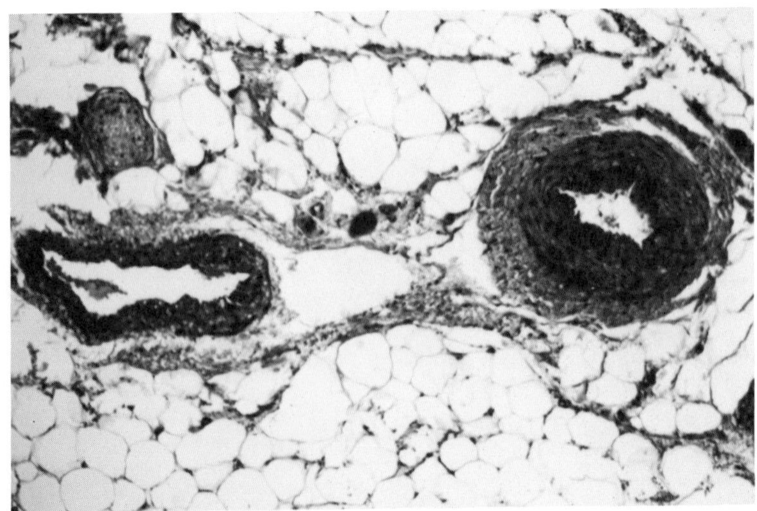

FIGURE 13-44

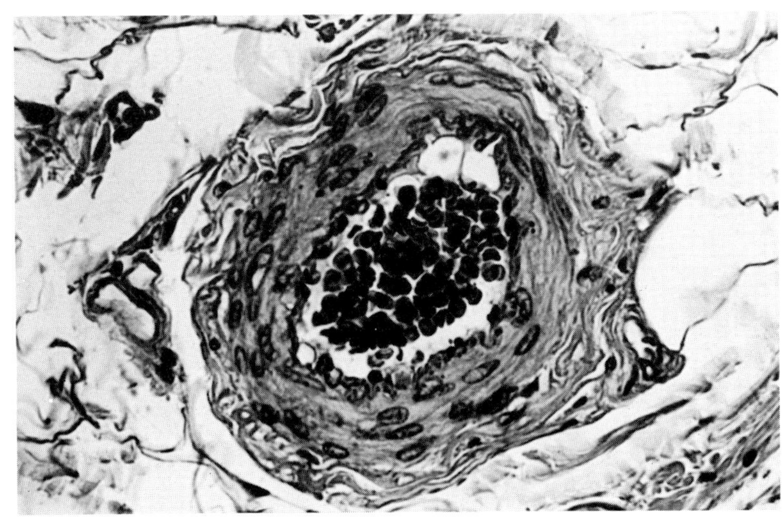

FIGURE 13-45

cells (d) are seen. The longitudinal smooth muscle of the adventitia increases as the caliber of the vein increases.

BIBLIOGRAPHY

Abramson, D. I. (ed.): *Blood Vessels and Lymphatics,* Academic Press, New York, 1962.

Bennett, H. S., J. H. Luft, and J. C. Hampton: "Morphological Classification of Vertebrate Blood Capillaries," *Am. J. Physiol.* **196:**381–390 (1959).

Karnovsky, M. J.: "The Ultrastructural Basis of Capillary Permeability Studied with Peroxidase as a Tracer," *J. Cell Biol.* **35:**213–236 (1967).

Landis, E. M., and J. R. Pappenheimer: "Exchange of Substances through the Capillary Walls," in W. F. Hamilton, and P. Dow (eds.), *Handbook of Physiology,* sec. 2, vol. 2, American Physiological Society, Washington, D.C., 1963, pp. 961–1034.

FIGURE 13-46

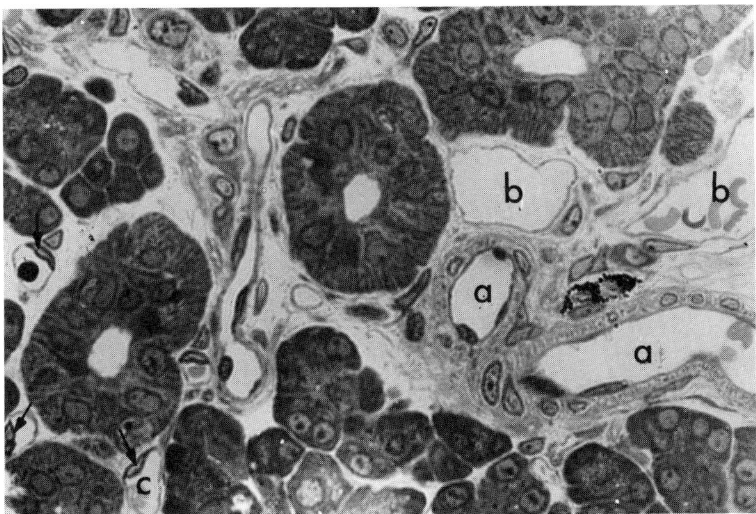

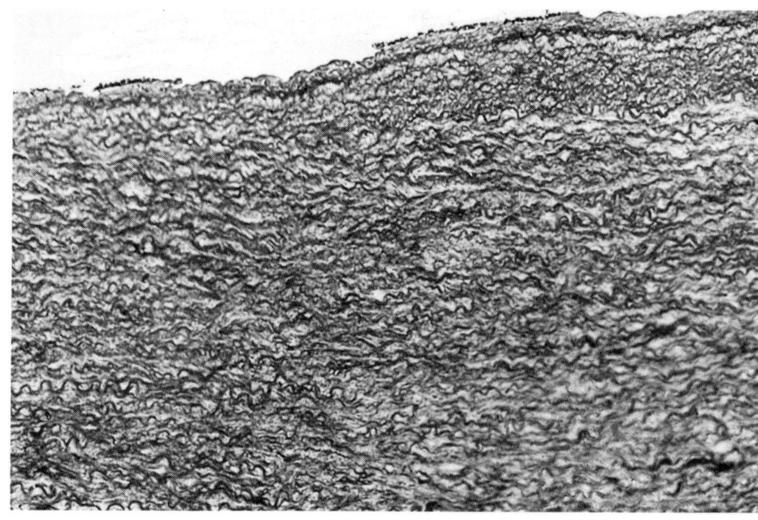

FIGURE 13-47

Palade, G. E.: "Blood Capillaries of the Heart and Other Organs," *Circulation* **24:**368–384 (1961).

Rhodin, J. A. G.: *Histology. A Text and Atlas,* Oxford University Press, New York, 1974.

Weiss, L., and R. O. Greep: *Histology,* 4th ed., McGraw-Hill, New York, 1977.

ADDITIONAL READINGS

Brightman, M. W., and T. S. Reese: "Functions between Intimately Opposed Cell Membranes in the Vertebrate Brain," *J. Cell Biol.* **40:**648–677 (1969).

Bruns, R. R., and G. E. Palade: "Studies on Blood Capillaries. I. General Organization of Blood Capillaries in Muscle," *J. Cell Biol.* **37:**244–276 (1968).

——— and ———: "Studies on Blood Capillaries. II. Transport of Ferritin Molecules across the Wall of Muscle Capillaries," *J. Cell Biol.* **37:**277–299 (1968).

Cecio, A.: "Ultrastructural Features of Cytofilaments within Mammalian Endothelial Cells," *Z. Zellforsch.* **83:**277–299 (1968).

Crone, C., and N. A. Lassen (eds.): *Capillary Permeability,* Academic Press, New York, 1970.

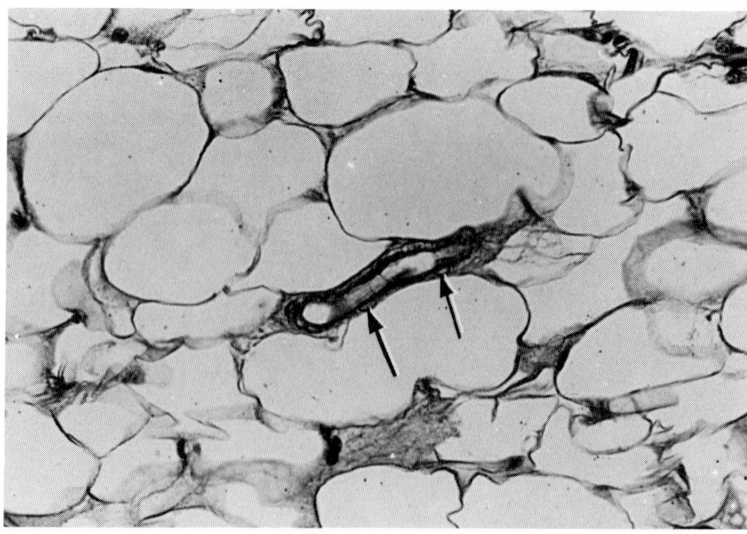

FIGURE 13-48

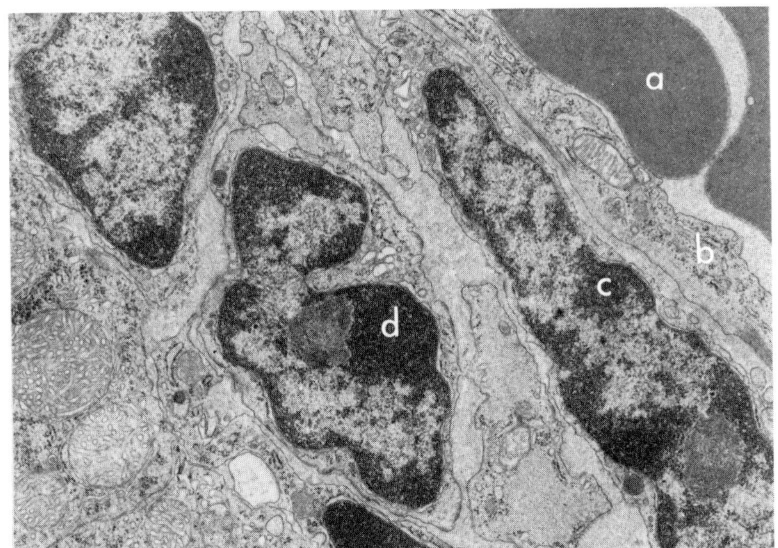

FIGURE 13-49

Fernando, N. V. P., and H. Z. Movat: "Fine Structure of the Terminal Vascular Bed. II. The Smallest Arterial Vessels: Arterioles and Metarterioles," *Exp. Mol. Pathol.* **3:**1–9 (1964).

——— and ———: "Fine Structure of the Terminal Vascular Bed. III. Capillaries," *Exp. Mol. Pathol.* **3:** 87–97 (1964).

Florey, H.: "The Endothelial Cell," *Biol. Med. J.* **2:** 487 (1966).

Maul, G. G.: "Structure and Formation of Pores in Fenestrated Capillaries," *J. Ultrastruct. Res.* **36:** 768–782 (1971).

Movat, H. Z., and N. V. P. Fernando: "The Fine Structure of the Terminal Vascular Bed. II. Small Arteries with an Internal Elastic Lamina," *Exp. Mol. Pathol.* **2:**549–563 (1963).

——— and ———: "The Fine Structure of the Terminal Vascular Bed. IV. The Venules and their Perivascular Cells," *Exp. Mol. Pathol.* **3:**98–114 (1964).

Palade, G. E.: "Transport in Quanta across the Endothelium of Blood Capillaries," *Anat. Rec.* **136:** 254 (1960).

Pease, D.C., and S. Molinari: "Electron Microscopy of Muscular Arteries: Pial Vessels of the Cat and Monkey," *J. Ultrastruct. Res.* **3:**447–468 (1960).

——— and W. J. Pauli: "Electron Microscopy of Elastic Arteries: The Thoracic Aorta of the Rat," *J. Ultrastruct. Res.* **3:**469–483 (1960).

Rhodin, J. A. G.: "The Ultrastructure of Mammalian Arterioles and Precapillary Sphincters," *J. Ultrastruct. Res.* **18:**181–223 (1967).

———: "The Diaphragm of Capillary Endothelial Fenestrations," *J. Ultrastruct. Res.* **18:**181–223 (1967).

———: "Ultrastructure of Mammalian Venous Capillaries, Venules and Small Collecting Veins," *J. Ultrastruct. Res.* **25:**462–500 (1968).

Simionescu, N., M. Simionescu, and G. E. Palade: "Segmental Differentiation of Cell Junctions in the Vascular Endothelium. The Microvasculature," *J. Cell Biol.* **67:**863–885 (1965).

———, ———, and ———: "Segmental Differentiation of Cell Junctions in the Vascular Endothelium. Arteries and Veins," *J. Cell Biol.* **68:**705–723 (1976).

———, ———, and ———: "Characteristic Endothelial Junctions in Different Segments of the Vascular System," *Thromb. Res.* (Suppl. II) **8:** 247–256 (1976).

Smith, O., J. W. Ryan, D. D. Michie, and D. D. Smith: "Endothelial Projections as Revealed by Scanning Electron Microscopy," *Science (Wash., D.C.)* **173:**925–927 (1971).

Somlyo, A. P., and A. V. Somlyo: "Vascular Smooth Muscle. I. Normal Structure, Physiology, Biochemistry and Biophysics," *Pharmacol. Rev.* **20:**197–272 (1968).

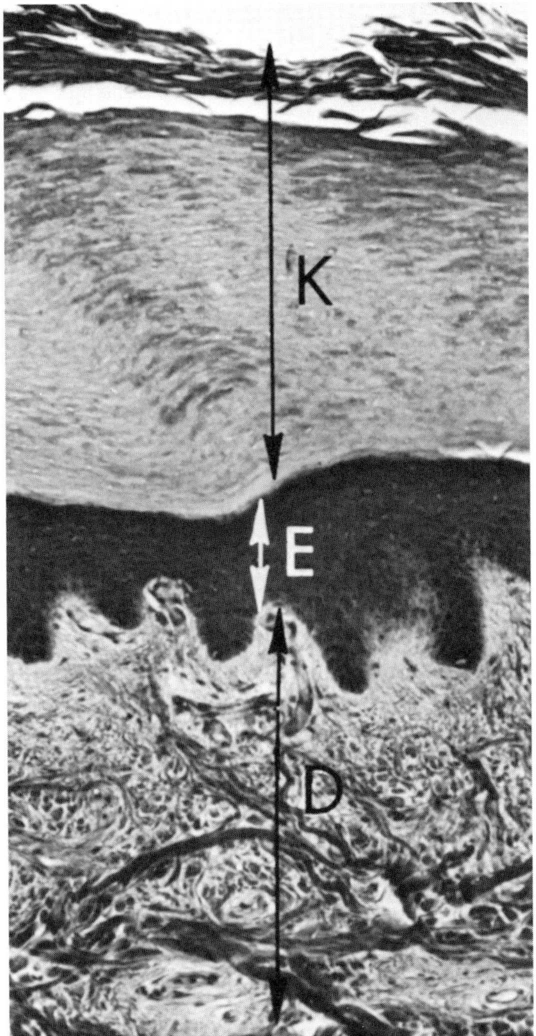

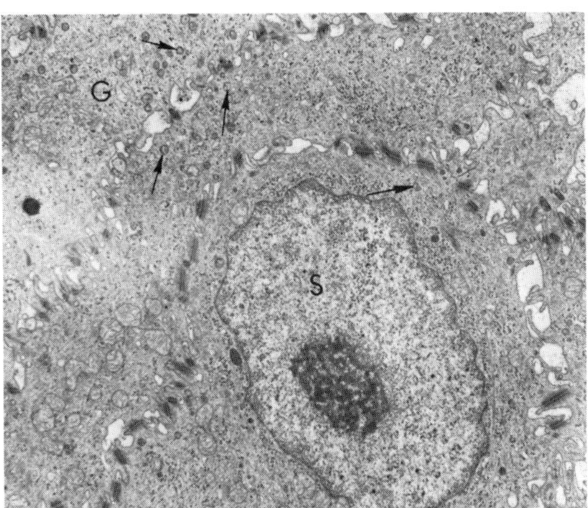

14 INTEGUMENT

OBJECTIVES

Upon completion of this chapter, the student will be able to:

1 Identify the major functions served by the skin.

2 List the five layers of the epidermis and describe the location, morphology, and function of each, when applicable.

3 Name the two layers of the dermis and identify the location and structure of each.

4 Identify the distribution of nerves and blood vessels in the integument.

5 Identify the location and morphology of the following:
(a) hypodermis or subcutis (b) sebaceous glands (c) sweat glands (d) papillae

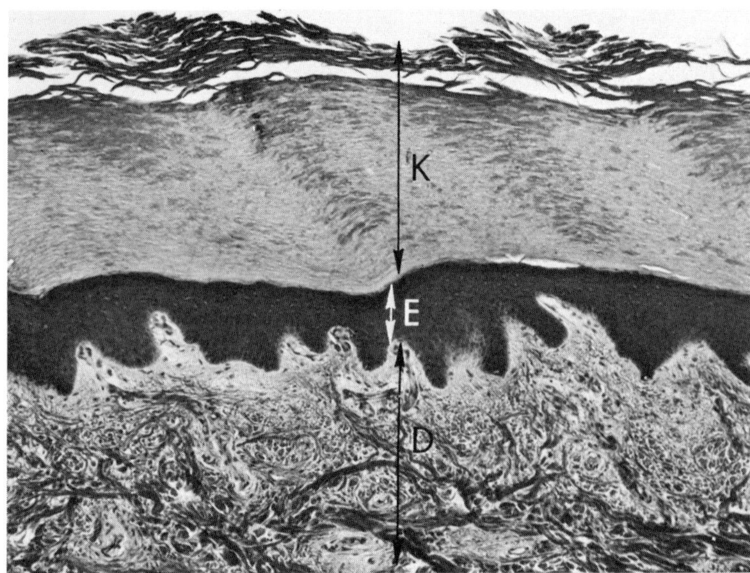

FIGURE 14-1 Light micrograph of skin. Dark-stained stratified squamous cells (E) run horizontally through the center of the field. Above there is a lighter-stained keratinized layer (K). The interface between epidermis and dermis (D) is irregular because of the papillation of the latter.

The integument is an impressive organ in terms of the variety of functions it performs for the body. This organ comprises the skin covering the entire body surface, the glands of the skin, and the nail, hair, and associated structures. Functionally, skin serves to *protect* underlying tissues from abrasion and other physical abuses, as well as *contain tissue fluids* of the body. Depending on the site and amounts of physical forces acting upon it, the skin shows a wide range of variations with respect to thickness and degree of keratinization of epidermis.

Just as the subjacent connective tissue shows different thicknesses as a result of functional adaptation, so does the integument. Compare the thickness of your eyelid with that of the sole of your foot. In addition to physical protection, the skin is important in *conserving body heat*. Finally, it serves in the *regulation* of fluid metabolism and disposal of waste substances, since it *excretes* varying amounts of sweat in the course of a day.

SKIN

Basic Structure

The skin consists of an outer layer of epithelial cells, the *epidermis*, and the underlying connective tissue, or *dermis* (Fig. 14-1). The epidermis is composed of stratified squamous epithelium with keratinization. The different layers vary in cell size, shape, structure, and degree of intracellular specializations, as discussed in Chap. 2.

The irregular interface between the epithelium and the dermis is due to the high papillation of the latter and may be quite variable depending on the region. The dense connective tissue of the dermis (D in Figs. 14-1 and 14-2) is supported by a loose connective tissue which contains much adipose tissue, and thereby allows the movement of the skin. This layer is called *hypodermis* or *subcutis* (H in Fig. 14-2). In places where the skin is subject to few abrasive forces, the epidermis is thin. Compare the thickness of epidermis in Fig. 14-3 (E), which is from eyelid, with that shown in Fig. 14-2. Both micrographs are taken at the identical magnification. However, the epithelial layers are not clearly visible in thin skin as they are in thick skin (Fig. 14-4).

EPIDERMIS

The primary layers of the epidermis are stratum basale (B), stratum spinosum (S), stratum granulosum (G), stratum lucidum (L), and

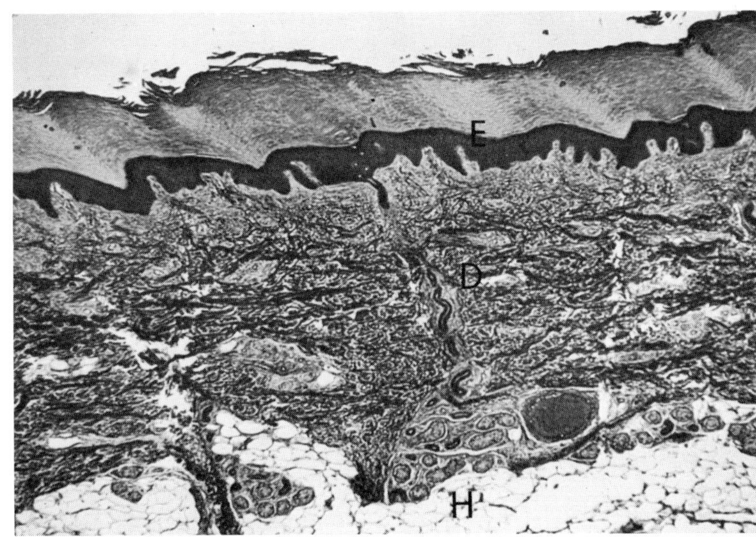

FIGURE 14-2 Light micrograph of skin. Epidermis (E) and dermis (D) is supported by the hypodermis (H), which contains much adipose tissue. This allows movement of the skin.

stratum corneum (K), in ascending order from the basement membrane (Fig. 14-4).

Stratum Basale

The cells of the stratum basale, or basal layer, are of tall cuboidal or columnar shape. They are closely packed together, giving an impression of a darker line because of a greater density and crowding of nuclei (Fig. 14-5). Mitotic figures (arrows in Figs. 14-5 and 14-6) are frequently present among basal cells, since they are responsible for the rapid replacement of all epidermal cells in more superficial strata.

The fine structure of the stratified squamous epithelium has been dealt with in some detail in Chap. 2. The diagrams included here will help the recall and reinforcement of the structural characteristics of different epithelial layers of the skin mentioned previously. Because of the rapid and continual generation of new cells, the cytoplasm of the basal cells contains abundant ribosomes (Fig. 14-7). The localization of mitochondria along

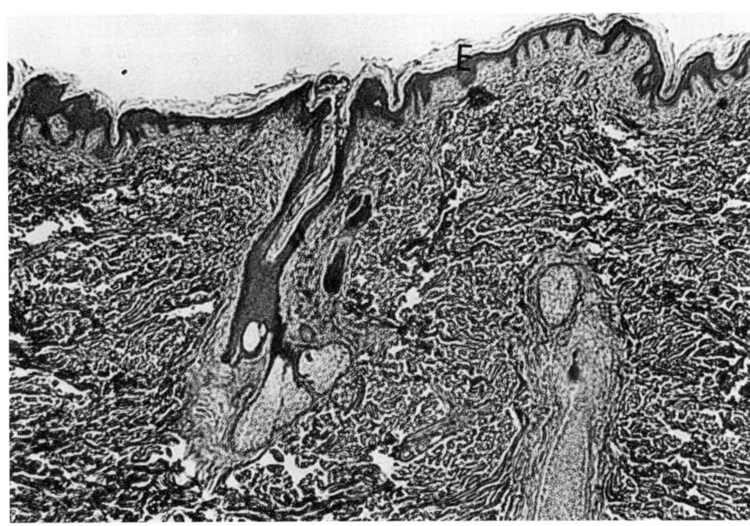

FIGURE 14-3 Light micrograph of thin skin from the eyelid. This part of the body is subjected to little abrasion. Therefore the epidermis is thin.

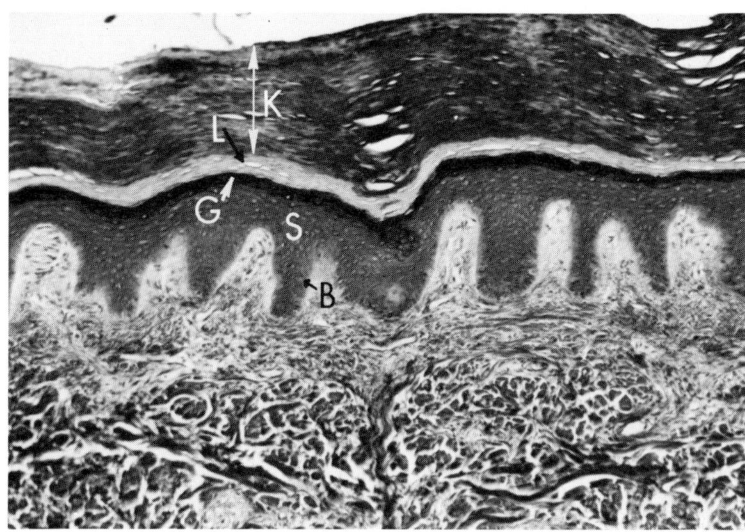

FIGURE 14-4 Thick skin, consisting of the following layers in ascending order from the basement membrane: stratum basale (B), stratum spinosum (S), stratum granulosum (G), stratum lucidum (L), and stratum corneum (K).

the base of the cells indicates that the basal cell cytoplasm is polarized. Desmosomes are numerous between basal cells; they are small, however, and intracellular tonofilaments show only modest differentiation. Numerous hemidesmosomes are present along the basal lamina, which connects the epidermis with the underlying tissue. The underlying connective tissue is highly vascularized, as evidenced by the many capillaries (c in Fig. 14-6), reflecting the need for nourishment of epidermis, which is avascular.

Among the cells of stratum basale are present special pigment-containing cells, *melanocytes*. The dark-stained aggregations in the basal cells in Fig. 14-8 represent melanin pigments. Melanocytes are derived from the ectoderm of the neural crest. They have a small central body with many numerous long, thin, and branching processes that extend up in between the cells of stratum spinosum. Melanocytes lack desmosomes. They have a large Golgi apparatus and rounded mitochondria (Fig. 14-9). The cyto-

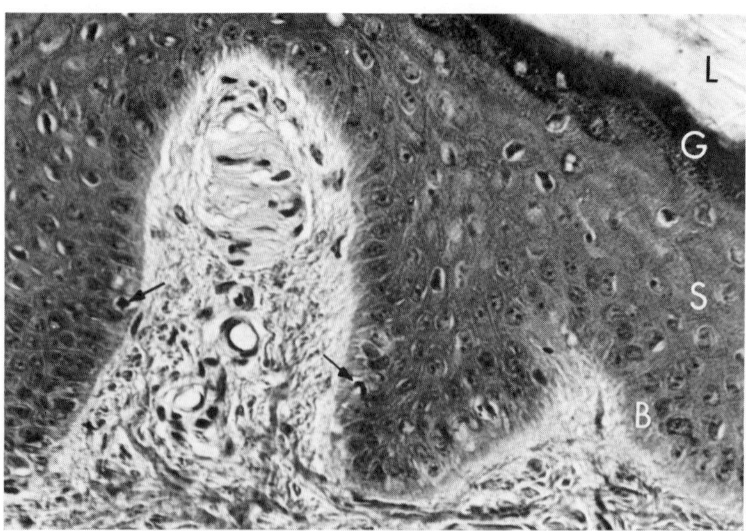

FIGURE 14-5 Epidermis viewed at higher magnification. The various layers are depicted in order from the irregular junction with the dermis seen at the bottom. They are stratum basale (B), stratum spinosum (S), stratum granulosum (G), and stratum lucidum (L). Stratum corneum is barely visible in the upper right corner. Mitotic figures (arrows) are present among the basal cells.

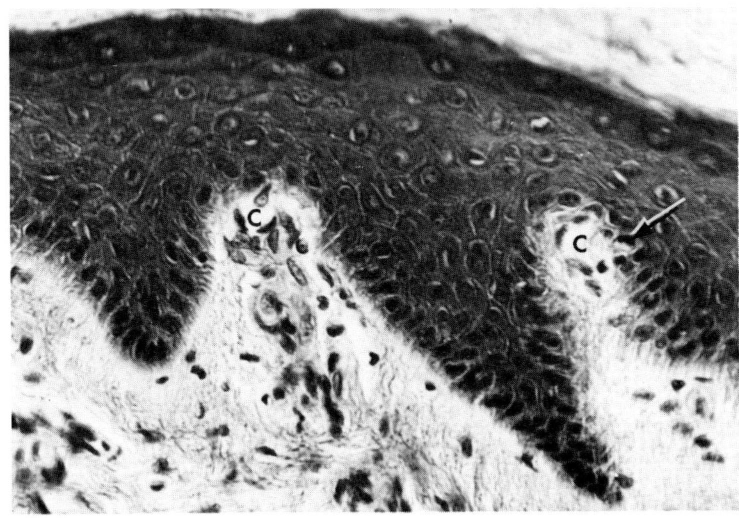

FIGURE 14-6 Light micrograph of skin. The border between the epidermis and the dermis is irregular because of the papillation of the latter. Epidermis is avascular and receives its nourishment by diffusion from capillaries (c) present in the papillae. Mitotic figure is indicated by arrow.

plasm also contains specific pigment granules, melanosomes, which are derived from the Golgi apparatus (Fig. 14-9).

Melanocytes are present among the basal cells of all individuals in all races. They are also present in albinos, individuals with light skin and white hair, but there the melanosomes are lacking electron-dense material. The basal cells of the epidermis also contain melanosomes which are thought to have been transferred from the processes of the melanocytes to the epidermal cells. In blacks, melanosomes are also found in other strata of epidermis.

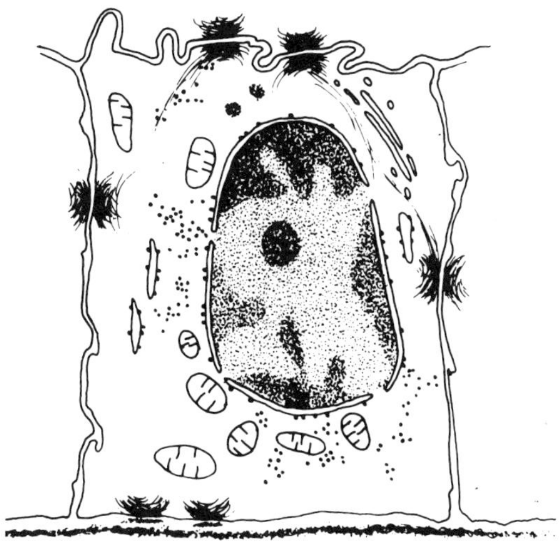

FIGURE 14-7 Diagram of basal cell. See text for explanation, and refer to Chap. 2 for details.

Stratum Spinosum

The spinosum cells are highly developed to withstand stress from the surface. The cells immediately above the basal layer are larger than the basal cells and polygonal in shape (Fig. 14-6). Prior to the advance of electron microscopy, these cells were thought to have a "prickly" appearance along their periphery, hence the term *stratum spinosum,* or spinous cell layer, was introduced. This prickly appearance is due to the highly differentiated desmosomes (maculae adherens) and tonofilaments (Fig. 14-10) that the cells in this layer contain.

These desmosomes between cells once were thought to represent intercellular bridges. Desmosomes of this layer are not only larger and better differentiated than those of the basal layer, but they are also more numerous. Accordingly, heavier bundles of tonofilaments can be seen even at the light microscopic level under suitable conditions (arrows in Fig. 14-11).

The desmosomes or maculae adherens can be seen as "prickles" or lines that appear to "connect" adjacent cells via tonofilaments.

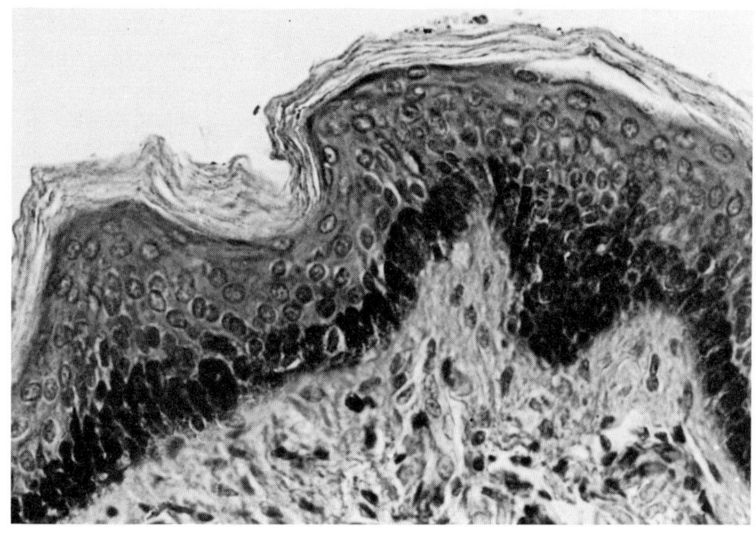

FIGURE 14-8 Light micrograph of skin. Dark-stained aggregation in the basal cells are melanin pigments. Among the basal cells are melanocytes, but they cannot be differentiated from the basal cells in H- and E-stained sections.

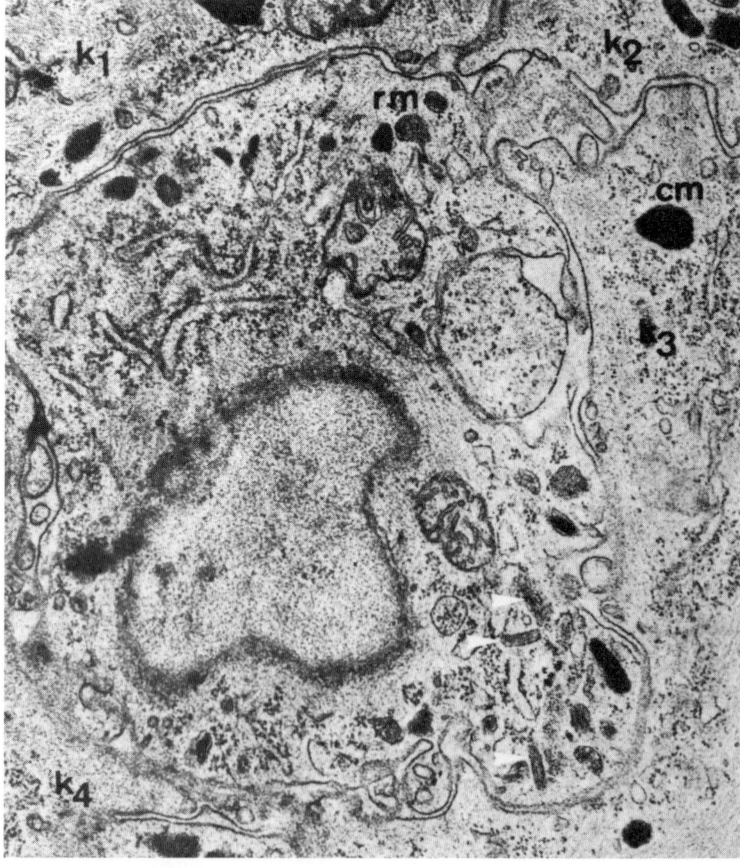

FIGURE 14-9 Electron micrograph of active melanocyte. It is located within the intercullular space between the four keratinocytes (k_1–k_4) of the lower stratum spinosum. Delicate, slightly electron-dense filaments (cf) appear in the cytoplasm of the melanocyte. Premelanosomes, either with round shapes and granular structure (rm) or ovoid shape and striated bars are present. The stratum spinosum cells contain single or composite membranebound melanosomes (cm). (*Courtesy of H. E. Schroeder.*)

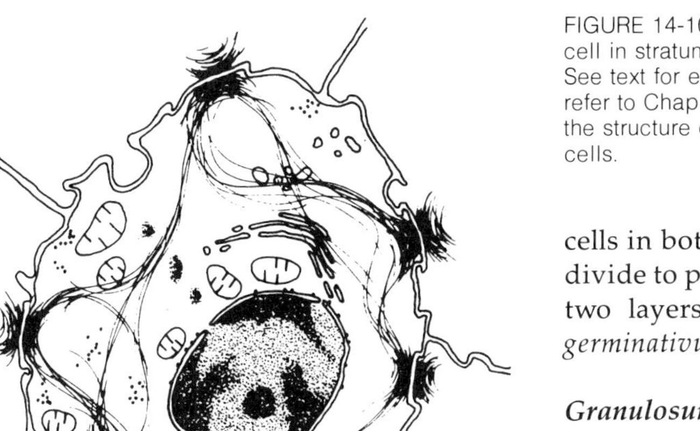

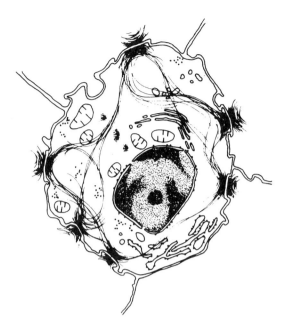

FIGURE 14-10 Diagram of cell in stratum spinosum. See text for explanation and refer to Chap. 2 for details on the structure of epithelial cells.

The mitochondria are spread throughout the cell, having moved from their previous basal position. On the basis of these structural differentiations, it might be considered that the spinous cell layer is best fitted to withstand physical stresses. The spinous cell layer contains occasional mitotic figures (Fig. 14-5) and therefore also contributes to replacement of cells for more superficial layers. Since cells in both the basal and spinous cell layers divide to provide more superficial cells, these two layers together are called the *stratum germinativum*, or *germinal layer*.

Granulosum Layer

As the cells move away from the spinosum cell layer, they become more flattened and keratohyalin granules start to appear rather abruptly (Fig. 14-5). It may be noted that in cells that are loaded with keratohyalin granules, the nucleus appears as a clear negative image in contrast to the dark, cytoplasmic granules. This layer may be one or more cells thick (Figs. 14-5 and 14-12). When the granulosum layer is seen at a lower level of magnification, it appears as a dark linear array of cells (Fig. 14-4), but at a higher level of magnification (Fig. 14-12) individual cells are readily identified and the granular appearance of their cytoplasm is revealed as well. In the granulosum layer, there is a reversal of developmental trends between the

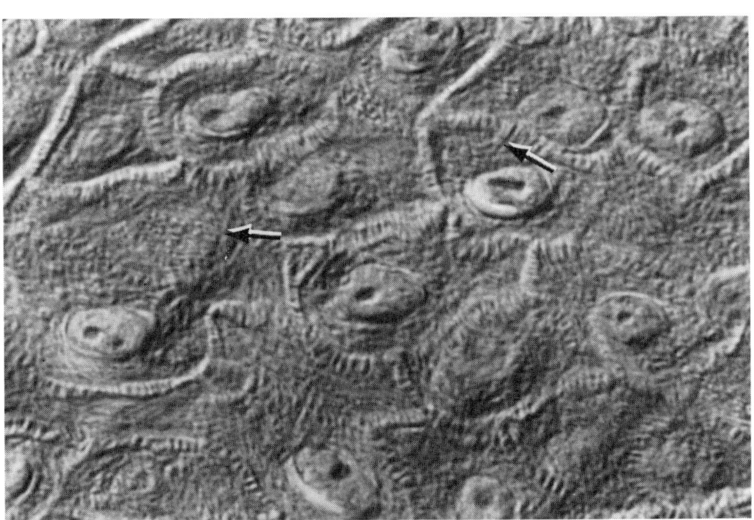

FIGURE 14-11 Light micrograph of the spinous layer in skin. Desmosomes are seen as "prickles" that appear to connect adjacent cells. Thick bundles of tonofilaments (arrows) extend into the "spines" of the cells and attach to the desmosomes.

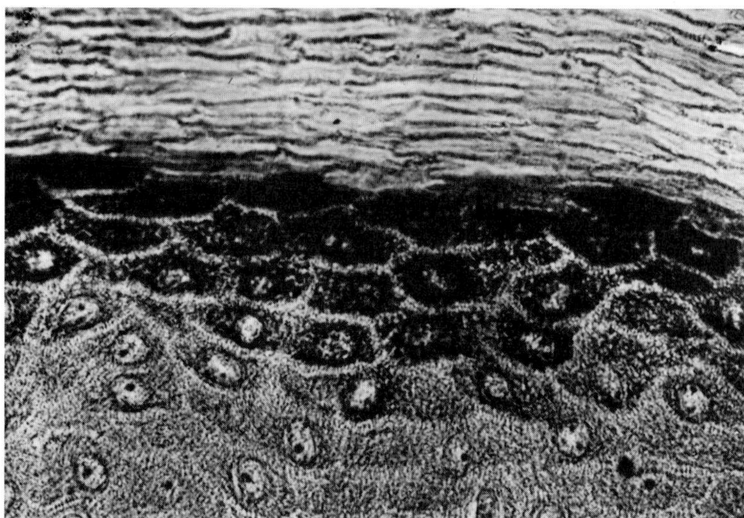

FIGURE 14-12 Light micrograph of skin. Stratum corneum is seen at the top, stratum granulosum is at the middle, and stratum spinosum at the bottom of the micrograph. Stratum granulosum is made up of several cell layers. Individual cells with keratohyalin granules in their cytoplasm can be distinguished. In some cells the granules are so abundant that the nucleus is obscured.

basal and spinous cell layers. The desmosomes and tonofilaments are reduced in size and number (Fig. 14-13).

The keratohyalin granules are not bordered by a membrane but are surrounded by a rim of ribosomes (Fig. 14-16). The more superficial cells of this layer are filled with keratohyalin granules, to the exclusion of other cell organelles. The mechanism by which the keratohyalin granule–containing cells produce the stratum corneum is not clearly understood. However, a rather abrupt transition between this layer and the next layer, which could be either stratum corneum or lucidum, occurs in all keratinized epithelium (Fig. 14-12).

The cells of stratum granulosum as well as stratum spinosum contain another type of granules, membrane-coating granules, which are believed to originate in the Golgi apparatus. Sparse membrane-coating granules begin to appear in the upper cell layers of stratum spinosum (arrows in Fig. 14-14). They become more numerous in stratum granulosum where migration of the granules to the peripheral cytoplasm occurs in successively higher layers (Fig. 14-15). Eventually this migration results in fusion of the gran-

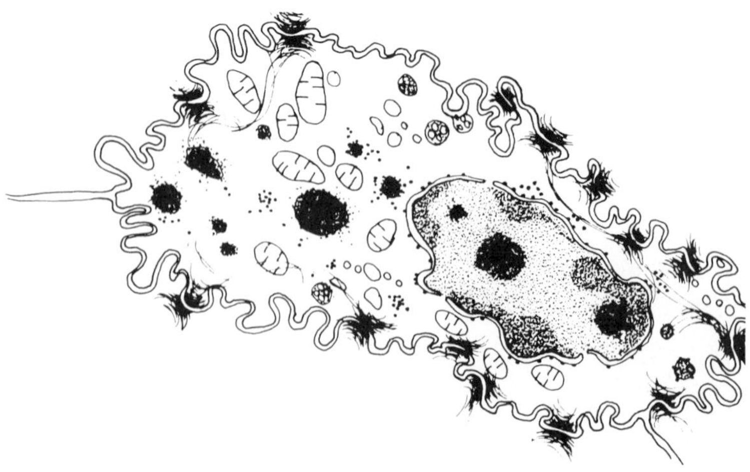

FIGURE 14-13 Diagram of granulosum cell.

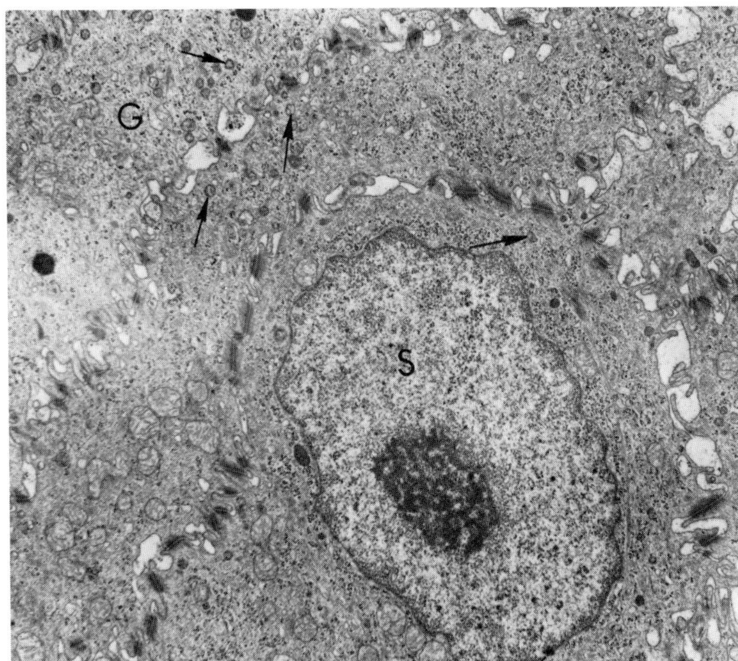

FIGURE 14-14 Electron micrograph of spinosum (S) and granulosum (G) cells. Membrane-coating granules (arrows) first appear in the cytoplasm in the cells of the upper layers of stratum spinosum and become more numerous in stratum granulosum where they are randomly distributed in the cytoplasm of the cells in the lower layers.

ules with the plasma membrane, as in the case with other secretory granules during release (arrow in Fig. 14-16). In this way, the content of the membrane-coating granules is exteriorized, occupies the space between cells, and covers the surface of keratinizing cells (Fig. 14-17). This coating process is believed to contribute to subsequent modification and disintegration of the intercellular relationship.

Stratum Corneum

The stratum corneum is the most superficial layer of keratinized epithelia and represents scaly, degenerated masses of cells which have lost their nuclei. The membrane-coating

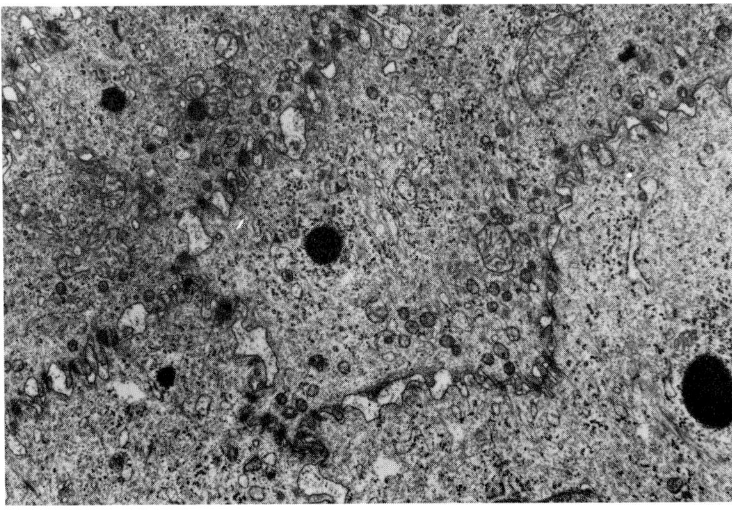

FIGURE 14-15 Electron micrograph of stratum granulosum. Membrane-coating granules are numerous in the higher layers and migrate to the periphery of the cells.

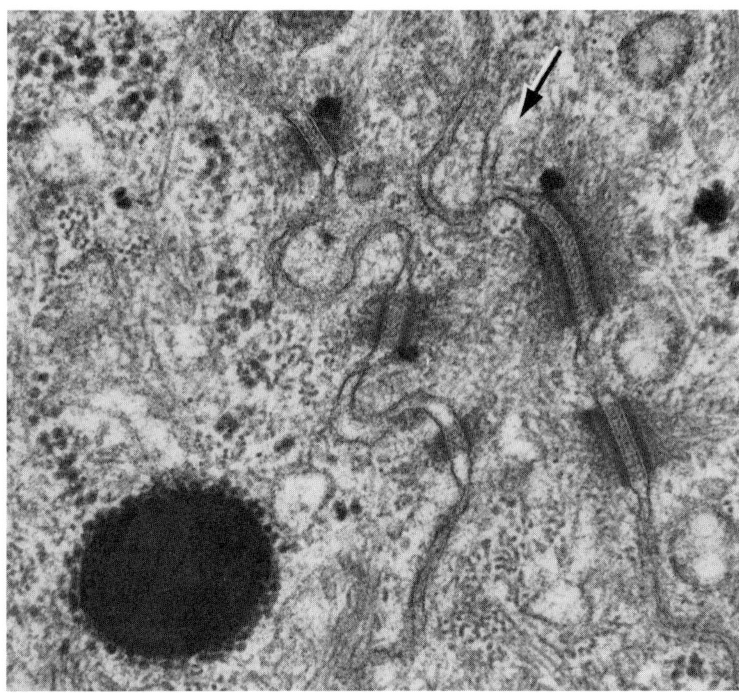

FIGURE 14-16 Portions of cells in stratum granulosum. Membrane-coating granules are seen in the peripheral cytoplasm where they fuse with the cell membrane (arrow) and the content of the granule is released into the intercellular space. Note the keratohyalin granule which is surrounded by a rim of ribosomes.

granules and keratohyalin granules also disappear in this stratum, but the mechanism that accomplishes this is not known. The cytoplasm becomes homogenous with fewer fibril masses, as disorganization of the tonofilaments and desmosomes occurs. Cross sections of tonofibril bundles are easily identified in cells of the spinous layer (Fig. 14-

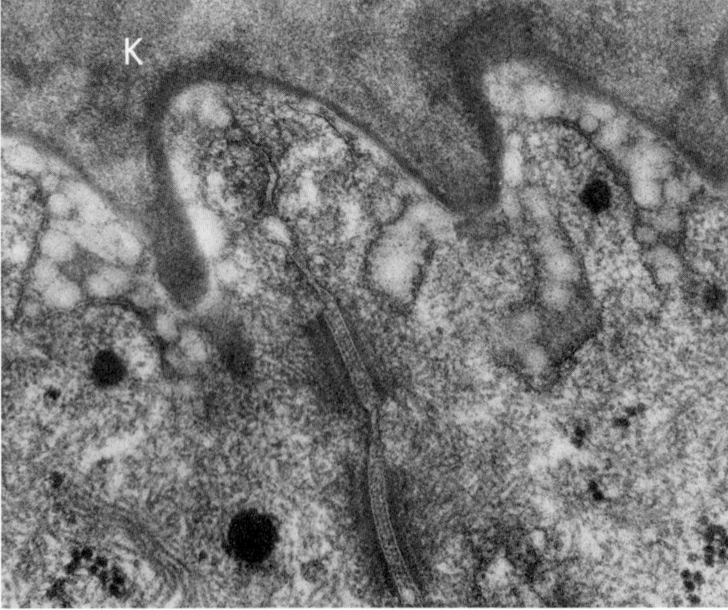

FIGURE 14-17 Portions of border cells of stratum granulosum and stratum corneum (K). Several membrane-coating granules are exteriorizing their content into the intercellular space.

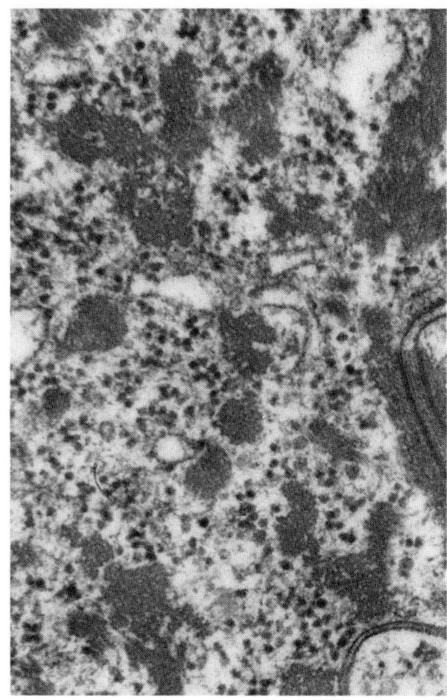

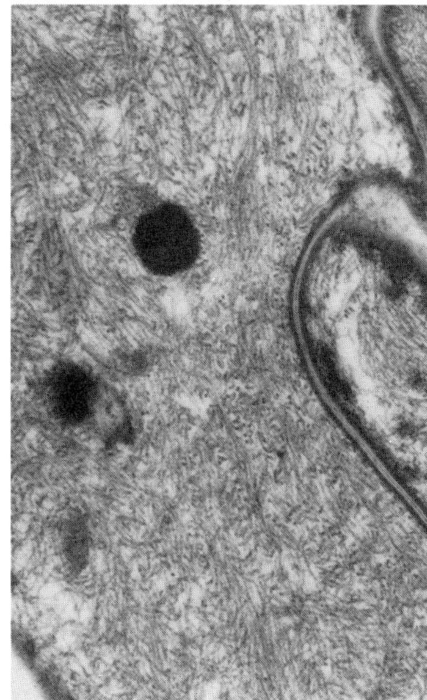

FIGURE 14-18 Portion of a stratum spinosum cell. Cross sections of tonofilament bundles are easily identified.

FIGURE 14-19 Portion of a stratum corneum cell. Cell organelles have disappeared and tonofilament bundles are disorganized and individual filaments have a random orientation.

18). In the cells of the keratinized layer, the ribosomes have disappeared and fibril bundles are disorganized, showing random orientation of individual fibrils (Fig. 14-19). For further details of this region, refer to Figs. 2-28 to 2-31 in Chap. 2.

Stratum Lucidum

In areas of the body where physical stresses are maximal, such as the soles of the feet, the keratinization of epidermis is also maximal, producing a stratum corneum that is several millimeters thick. In such regions, a thin layer called the *stratum lucidum* is often present between the stratum granulosum and the keratinized layer. This layer is called stratum lucidum because it is a lucid, or translucent, layer of yellowish, unstainable material (L in Fig. 14-4).

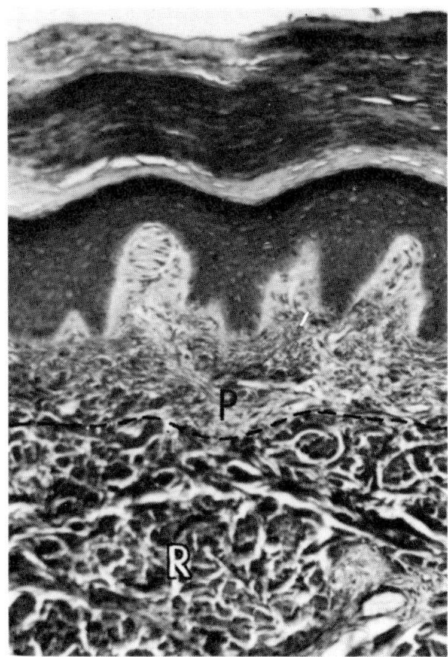

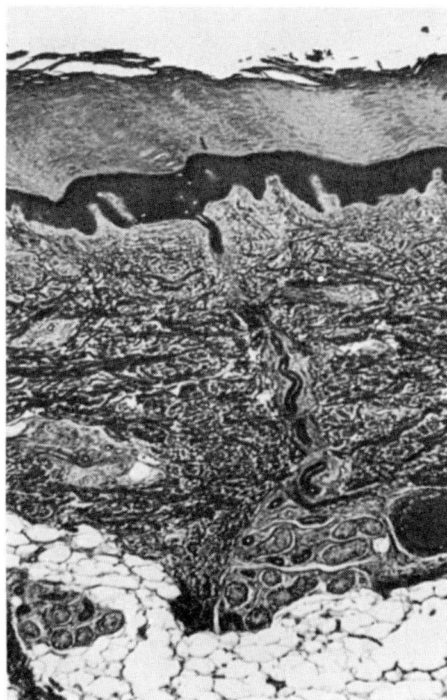

FIGURE 14-20 Light micrograph of skin. Epidermis is seen at the top and dermis at the bottom. The latter consists of a papillary (P) and a reticular (R) layer. As the name indicates, the papillary layer has numerous papillae which extend into the epidermis and give the epidermodermal junction an irregular outline. This irregular boundary aids in the attachment of epidermis to dermis.

FIGURE 14-21 Light micrograph of skin. Note that the two layers of dermis are easily identified at this low magnification.

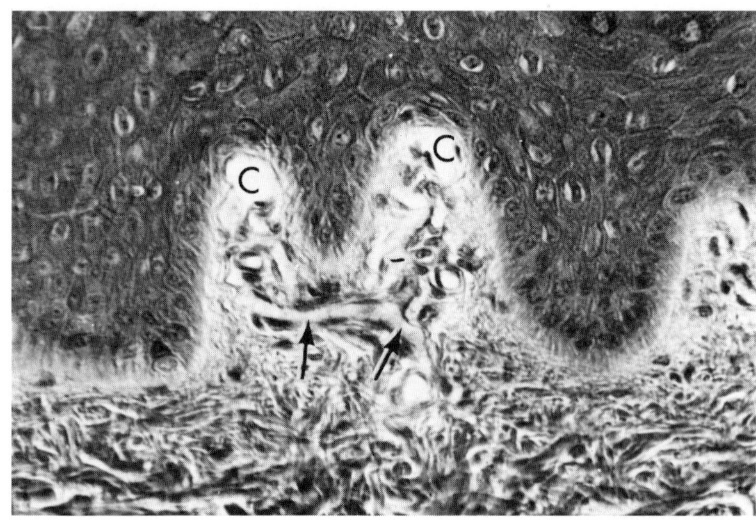

FIGURE 14-22 Light micrograph of the germinal layer of epidermis and the papillary layer of dermis. Capillaries (C) are seen in the papillae closely associated with the basal layer of epidermis. They supply nourishment to epidermis, which is avascular. Numerous small vessels (arrows) are seen deeper in the papillae.

DERMIS

Basic Structure

The dermis consists of dense, irregular connective tissue (dense FECT), which in turn lies on the hypodermis, a subcutaneous loose FECT. While the major vessels and nerves run through the subcutaneous connective tissue, the dense FECT of the dermis is specialized to provide physical support for the epidermis. It also maintains a fine network of blood vessels, nerves, and delicate sensory endings that have to be in an integral structural relationship with the epithelial cells.

The dermis is made up of two distinguishable layers which differ primarily in the thickness of connective tissue fibers of which each layer is made up (Fig. 14-20). The more superficial layer is called the *papillary layer,* because it closely follows the irregular basal surface of the epidermis, resulting in a papillated appearance. In contrast, the deeper *reticular layer* is made up of heavy collagenous

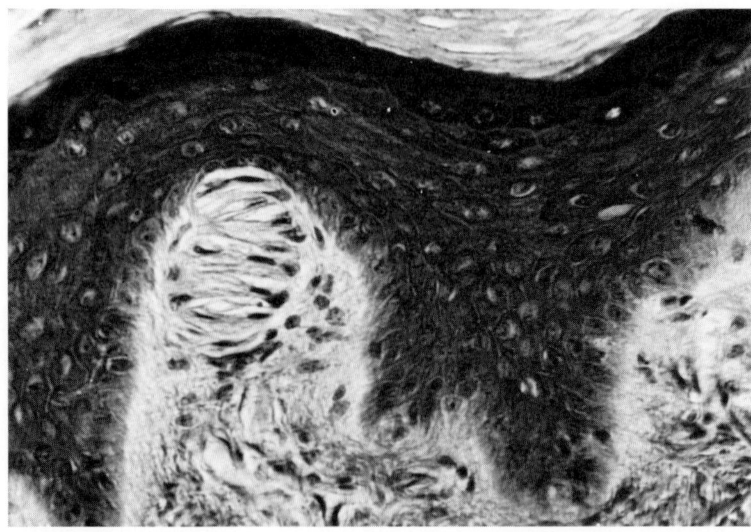

FIGURE 14-23 Light micrograph of Meissner's corpuscle. Note the lamellar appearance of the capsule.

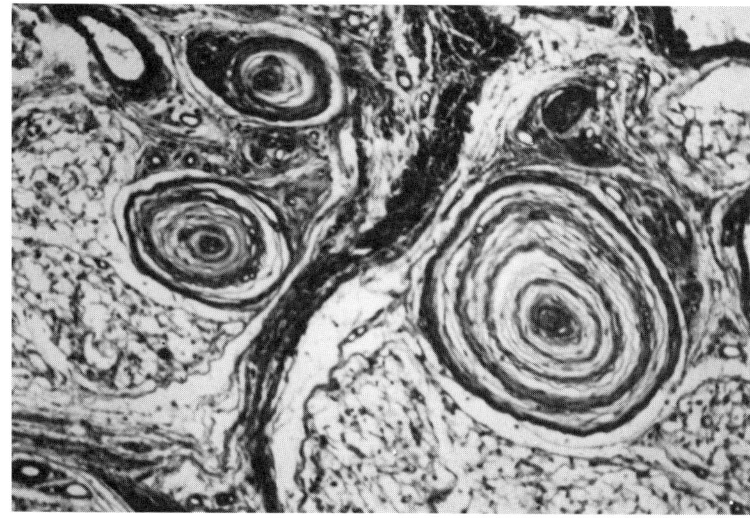

FIGURE 14-24 Pacinian corpuscles seen in cross section. Note the onion ring configuration of their capsule.

fibers and is the thicker of the two. Because of the differences in fiber texture, the two layers of the dermis can be readily identified even at an extremely low power (Fig. 14-21).

The papillary layer, which is composed of delicate connective tissue fibers filling the space immediately under the irregular basal layer of the epidermis, contains many capillary networks (C) in each of the papillae (Fig. 14-22). Note the numerous small vessels that are sectioned in different planes throughout the layer (arrows) and form loops of various shapes within the papillae. This reflects the significance of this layer with respect to nourishment of the avascular epidermis.

As mentioned in Chap. 8 on nervous tissues, the fine free sensory nerve endings that could not be distinguished without special stains are present in the papillary layer where they penetrate the basal lamina. In addition, the papillary layer also contains numerous, more differentiated nerve endings for tactile sensation, such as Meissner's cor-

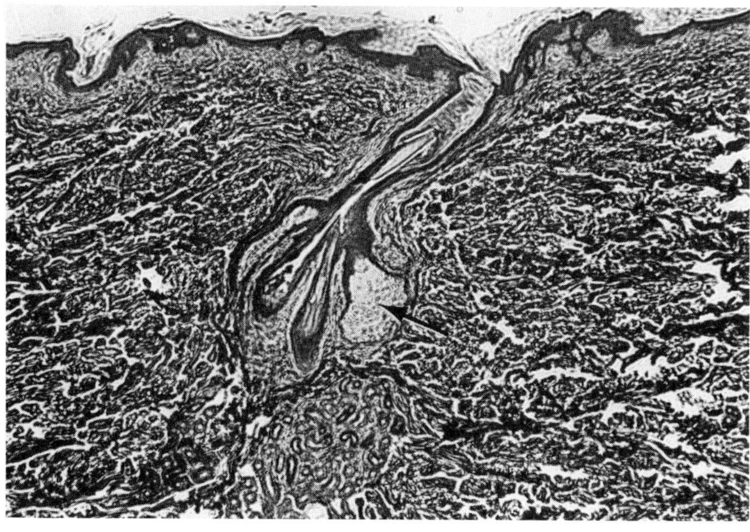

FIGURE 14-25 Light micrograph of skin. In the center, a hair shaft with its follicle is seen. A sebaceous gland is associated with the hair (arrow).

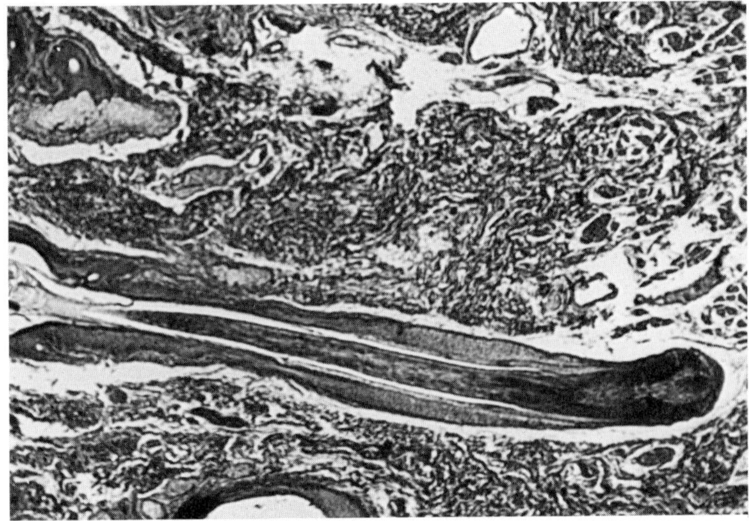

FIGURE 14-26 Light micrograph of a hair. The shaft extends from the bulbous follicle which is the basal end of the hair. Cells in the follicle undergo mitosis and this cell proliferation causes the hair to grow. The basal cell layer in the follicle corresponds to the basal layer of epidermis and the cornified layer to the hair shaft.

puscles (Fig. 14-23). These endings are specialized for fine touch discrimination. They are located in the apex of the papillae and fit snugly against the basal cell layer. Other sensory organs, such as *pacinian corpuscles* for deep pressure sensation, are located under the dermis in the loose FECT of the subcutis (Fig. 14-24). Consult Chap. 8 for further details.

In addition, the papillary layer of the dermis in the skin of blacks as well as the skin of nipples and perineal regions of whites contains dermal chromatophores which darken the basal layer (Fig. 14-10).

SPECIALIZATIONS OF INTEGUMENT

HAIR AND SEBACEOUS GLANDS

An obvious specialization of the integument is the hair shaft and its follicle (Figs. 14-25 and 14-26). The hair follicle is the bulged

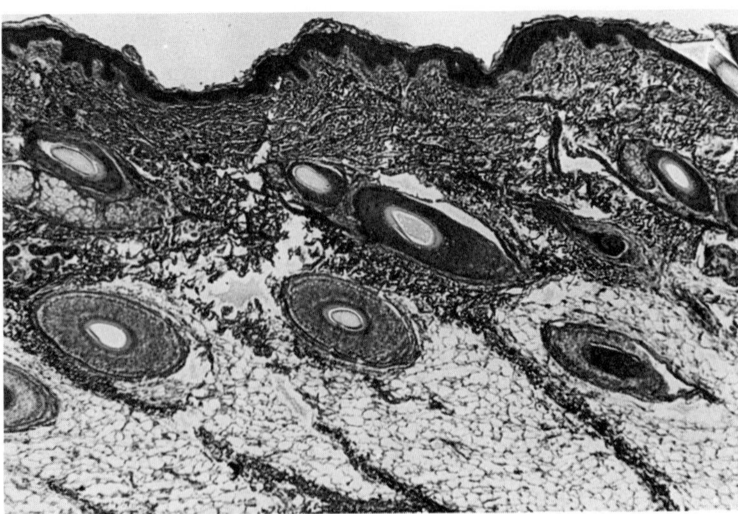

FIGURE 14-27 Cross section of hair seen at different levels in the dermis.

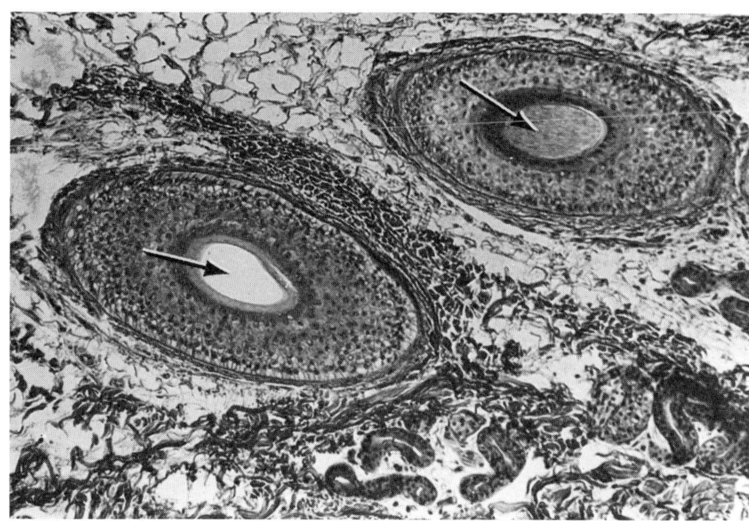

FIGURE 14-28 Cross section of hair seen at higher magnification. The shaft (arrows) in the center of each hair corresponds to stratum corneum and is thus keratinized.

basal end of the hair which is composed of several cell layers. Each layer of the follicle is a derivative of one of the epidermal layers. The shaft and follicle penetrate from the surface through the dermis and into the subcutaneous connective tissue (Figs. 14-26 and 14-27). At the growing end of the hair the cells undergo mitosis. The central portion of the hair shaft corresponds to the cornified layer and often appears yellow or translucent (arrows in Fig. 14-28).

A number of sebaceous glands are present along the hair shaft. Notice the bulbous gland associated with the hair shaft (arrow in Fig. 14-25). Its duct is somewhat obliquely sectioned. A thin strand of muscle called *arrector pili muscle* is seen running obliquely closely associated with the deep portion of the sebaceous gland (arrows in Fig. 14-29). This muscle originates from the hair follicle and extends into the connective tissue of the dermis, the reticular layer. Through periodic contraction, the arrector pili muscle serves to squeeze the content of the sebaceous gland

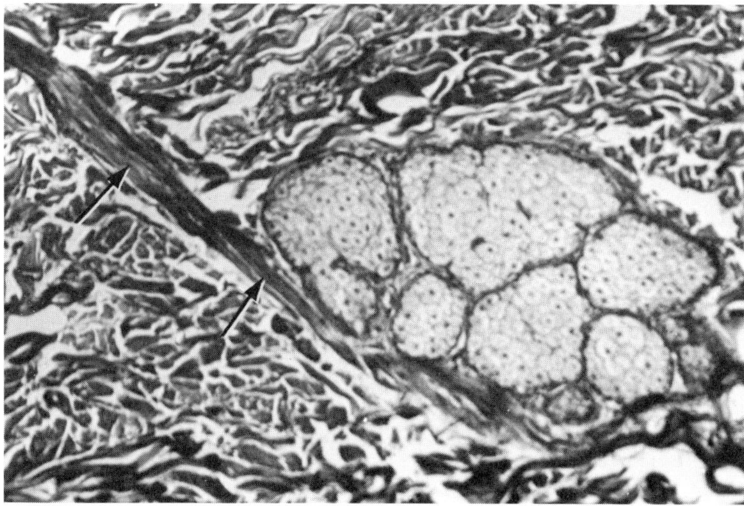

FIGURE 14-29 Light micrograph of sebaceous gland. The arrector pili muscle (arrows) runs obliquely through the field.

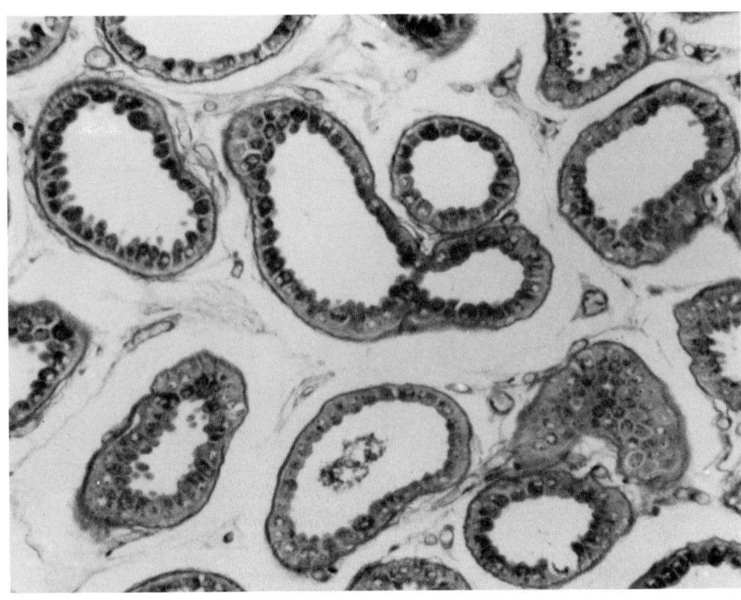

FIGURE 14-30 Light micrograph of apocrine sweat glands. Note the large lumen of the secretory end pieces.

out onto the lateral aspect of the hair shaft. In so doing, the entire cells or debris are secreted; this mode of secretion is called *holocrine*. In addition, contraction of the muscle causes the hair to stand up and creates "goose bumps."

The sebaceous gland, as the name implies, produces an oily substance called *sebum*, which aids in the lubrication of the hair. The sebaceous gland is a holocrine gland, excreting entire cells after their cytoplasm becomes loaded with the product. In the production of sebum, the peripheral cells of the gland reproduce and push previous progeny away from the basement membrane, closer to the duct. During this time, the cells continue to produce lipid material which overwhelms the cytoplasm and thus causes a degeneration of intracellular organelles. The cells are then detached and moved into the

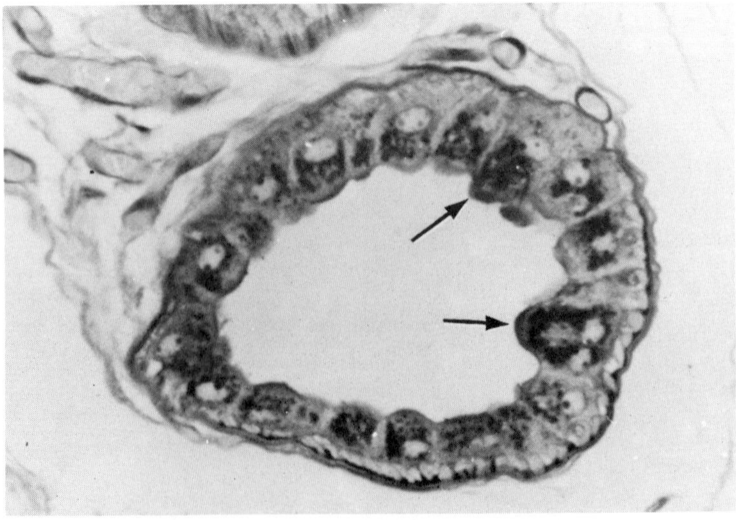

FIGURE 14-31 Light micrograph of a secretory end piece of an apocrine sweat gland. The apical portion of some of the cells is beginning to pinch off (arrows).

space along the hair shaft and eventually work their way up to the skin surface. The peripheral cells near the basement membrane are small in size and have larger nuclei than the cells located towards the center of the gland where the cells are larger, the nuclei become pyknotic, and the cytoplasm is highly vacuolated. If the narrow space between the opening of the sebaceous gland and hair shaft becomes clogged or invaded by bacteria, an infection occurs, resulting in acne.

SWEAT GLANDS

Another important set of glands of the skin are the sweat glands. Like other glands, they represent ingrowth of the epidermal cells. The sweat glands can be grouped into two categories: one which is large and present in certain regions, such as axillae and perineal regions; and the other, which is small and present throughout the entire skin surface. The large sweat glands are often called *apocrine sweat glands* because they were once thought to be a typical apocrine gland in which portions of the apical cytoplasm were lost during secretion (Fig. 14-30). Note the appearance of the large secretory portions (Fig. 14-30) and the apical cytoplasm that is about to pinch off (arrows in Fig. 14-31).

The smaller, more numerous glands that are scattered throughout the skin or body surface are caled *eccrine sweat glands*. In historical preparations, the small sweat glands are made up of a coiled secretory portion (arrows in Fig. 14-32) and a less coiled duct portion (arrowheads in Fig. 14-32). The duct of the sweat gland is lined by two layers of cuboidal cells or stratified cuboidal epithelium. It passes through the two layers of dermis into the subcutaneous connective tissue. At the base of the duct, the tubular secretory portion appears irregularly coiled. The secretory ends are lined by a single layer of cuboidal cells (Fig. 14-33). Notice the large nuclei (arrows) that are somewhat basally located in the secretory cells. Close to the basement membrane are several myoepithelial cells whose nuclei often appear as dark triangular areas between and at the base of sweat gland cells (arrowheads). The ducts of sweat glands have a larger lumen than the secretory ends, and their lining epithelial cells stain darker than those of the secretory ends (compare the appearance of the secretory ends with that of the duct in Fig. 14-33).

The duct of the sweat gland is tortuous as it penetrates through the dermis (arrows in Fig. 14-34). Note the light-stained secretion

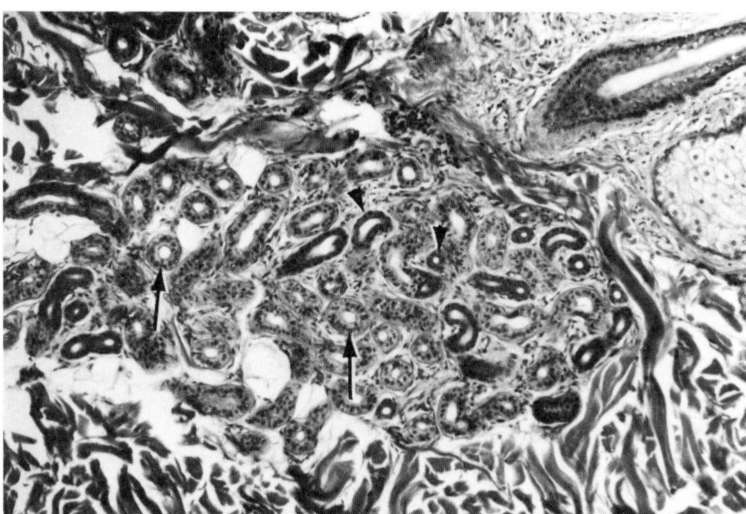

FIGURE 14-32 Light micrograph of sweat gland. The secretory end pieces (arrow) appear light, whereas the ducts (arrowheads) appear dark. The lumen of the end pieces is smaller than that of the ducts.

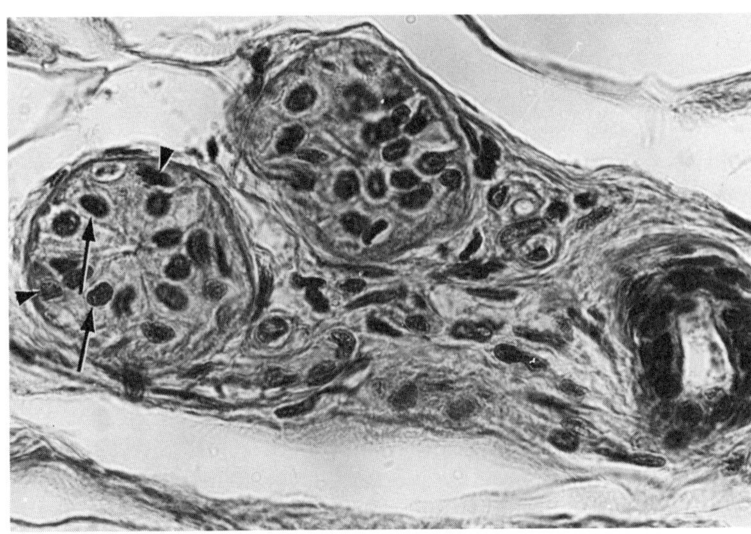

FIGURE 14-33 Eccrine sweat gland. The secretory end pieces are lined by a single layer of cuboidal cells that have a basally located nucleus (arrows). At the base of these cells, the dark nuclei of myoepithelial cells are seen (arrowheads).

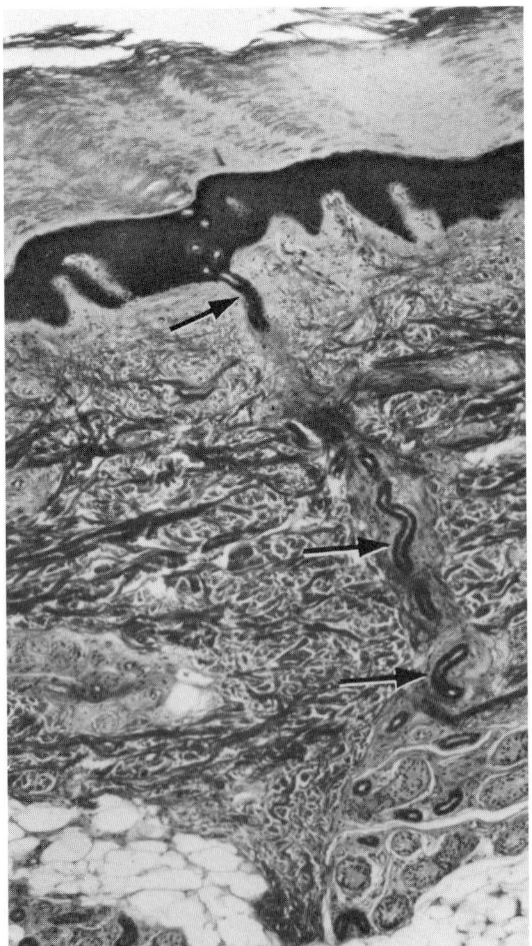

FIGURE 14-34 Light micrograph of skin. An eccrine sweat gland is seen in the lower portion of the micrograph. The tortuous duct (arrows) winds its way through the dermis and epidermis to reach the surface.

portions scattered throughout the lower third of the micrograph (Fig. 14-34). These ducts are very tortuous as they pass through the epidermis (D in Fig. 14-35), and they are lined by a stratified epithelium.

As mentioned previously, myoepithelial cells are located around the larger secretory cells along the base of the sweat gland. The cytoplasm of a myoepithelial cells looks like that of a pericyte, with a three-dimensional structure akin to an octopus (Fig. 14-36). The contraction of the many armlike processes of myoepithelial cells forces the gland to squeeze out its contents to the skin surface. Note that the myoepithelial cell is present within the basement membrane.

THE NAILS

Like the hair shaft, the nails are derived from layers of the epidermis. A nail is a cornified extension of the epidermis, providing protection for the ends of the fingers and toes (Fig. 14-37).

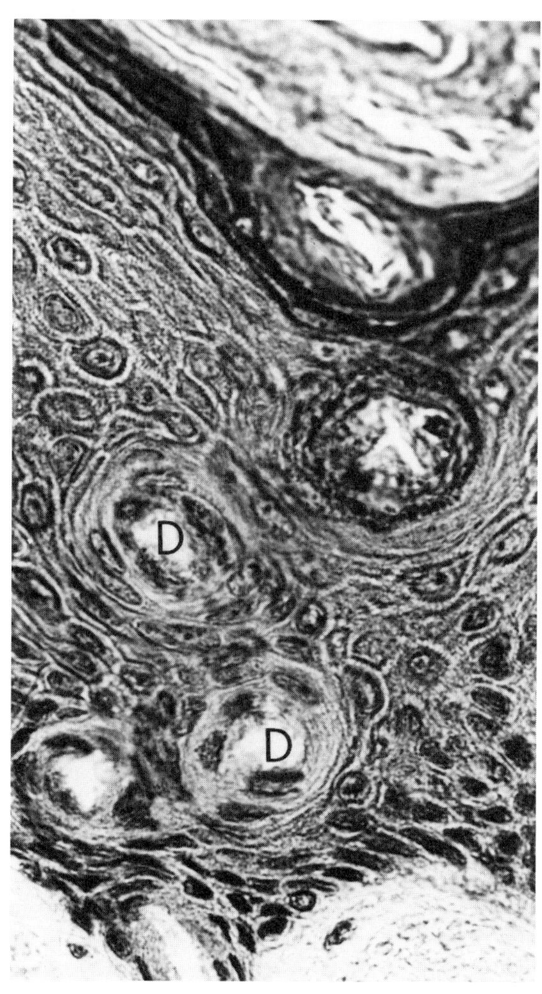

FIGURE 14-35 Light micrograph of a duct (D) passing through the epidermis. It is very tortuous and is lined by stratified epithelium.

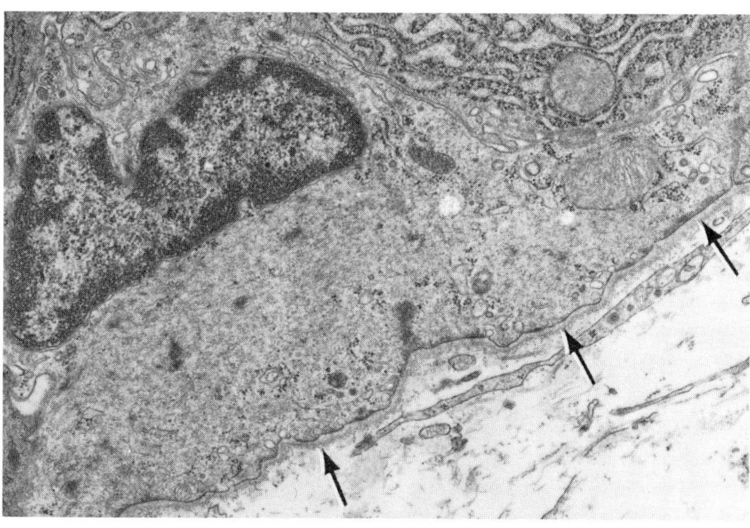

FIGURE 14-36 Electron micrograph of a myoepithelial cell. It is located in a triangular area close to the basal lamina (arrows) between the bases of the secretory cells.

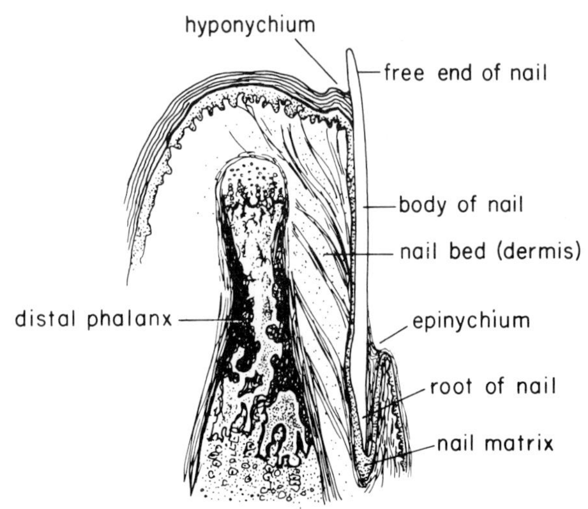

FIGURE 14-37 Drawing of a longitudinal section through a nail. The nail matrix corresponds to stratum germinativum of epidermis and the body of the nail which is cornified to stratum corneum.

REVIEW SECTION

1. Figure 14-38 is a section through the skin of the thumb. Is the cornified layer thick or thin? The spiraling structure passing vertically through this field (arrows) is part of ________. The dark-staining band traversing horizontally through the upper one of the field (arrowheads) is undergoing what process? The light-staining extensions of connective tissue into the epidermis are called ________.

The surface layer, the stratum corneum, is very thick in areas that withstand a lot of abrasion, such as the palm of the hand. Sweat gland ducts pass through the layer, as well as hair shafts. The cells which rise through the layers to be sloughed off at the surface originate in the basal layer and the stratum spinosum. Together they are referred to as the stratum germinativum. Cornification of the upper layers is first initiated at the stratum granulosum, a narrow layer of dark-staining cells between the stratum corneum and the stratum spinosum. The stratum lucidum is a clear to yellow layer above the stratum granulosum, as visible in this section.

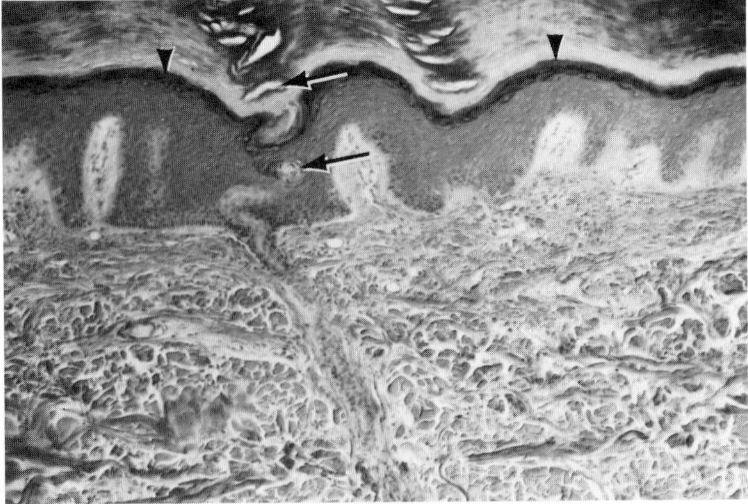

FIGURE 14-38

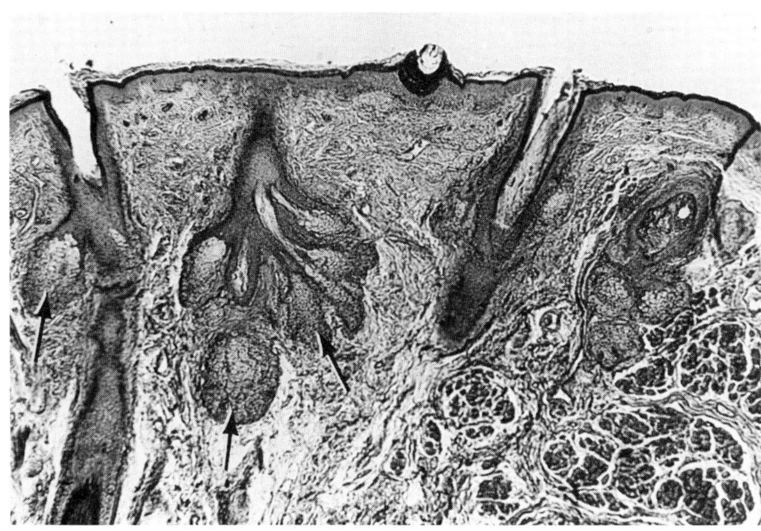

FIGURE 14-39

2. The section in Fig. 14-39 is a section through a portion of skin. A few irregular structures invaginating from the surface represent __________. What is the name given to the lighter-staining groups of cells indicated by arrows? What is the name given to the lighter-staining structure after leaving this region? What is its function?

Sebaceous glands are found in association with hair shafts and follicles. The cells near the duct are secreted as sebum. Sebum aids in lubrication of the hair. The darker-staining cells at the peripheries of the gland reproduce to provide replacements for those cells secreted. Because whole cells are secreted, this gland is a holocrine gland. Hair shafts are shown in cross section.

3. What is the distinctive configuration of structures indicated by arrows in Fig. 14-40? The tissue surrounding this structure is __________. What portion of the skin contains this type of tissue? Its function is __________.

The pacinian corpuscle is found in the deeper hypodermis or subcutaneous connective tissue. It responds to deep or heavy pressure.

FIGURE 14-40

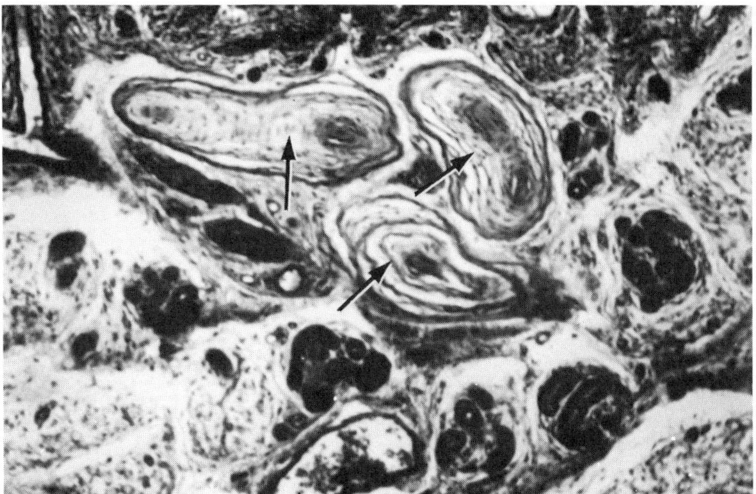

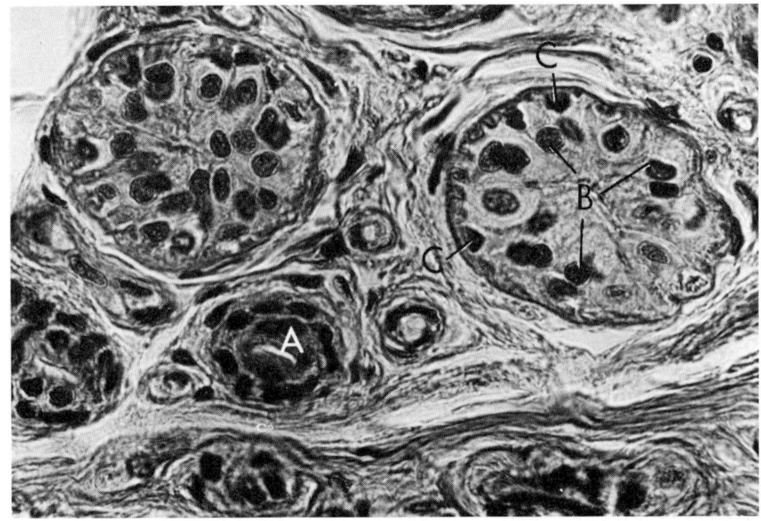

FIGURE 14-41

4. Refer to Fig. 14-41. Name the structures indicated by A, B, and C. What do the structures produce?

Figure 14-41 is a portion of eccrine gland in which the duct (A), secretory cells (B), and myoepithelial cells (C) are clearly visible. They produce sweat.

5. In what region of the skin is the spindle-shaped structure in Fig. 14-42 found? What is its name? The function of this structure is ________ .

The Meissner's corpuscle is found in the papillary layer of dermis where it responds to delicate tactile stimuli. It is much smaller than the pacinian corpuscle.

6. Match the following labels with the appropriate letter in Fig. 14-43.

_____ (1) stratum lucidum
_____ (2) stratum spinosum
_____ (3) stratum corneum
_____ (4) stratum granulosum
_____ (5) dermal papillae

1 (B), 2 (D), 3 (A), 4 (C), 5 (E).

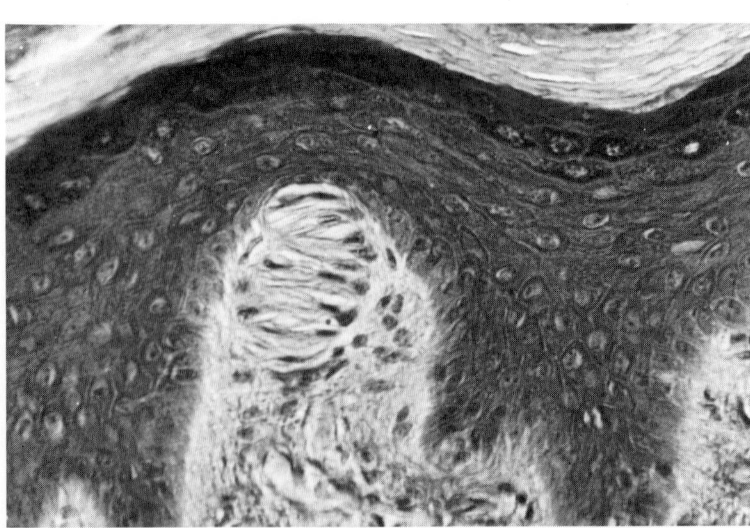

FIGURE 14-42

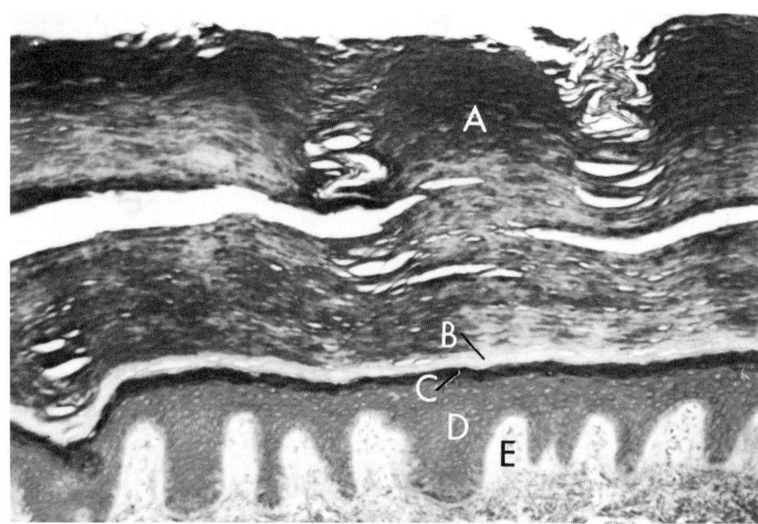

FIGURE 14-43

BIBLIOGRAPHY

Butcher, E. O., and R. F. Sognaes (eds.): *Fundamentals of Keratinization,* American Association for Advancement of Science, Washington, D.C., 1962.

Lyne, A. G., and B. Short (eds.): *Biology of the Skin and Hair Growth,* Angus and Robertson, Sydney, Australia, 1965.

Montagna, W., and W. C. Lobitz, Jr. (eds.): *The Epidermis,* Academic Press, New York, 1964.

——— and P. F. Parakkal: *The Structure and Function of Skin,* 3d ed., Academic Press, New York, 1974.

Rhodin, J. A. G.: *Histology. A Text and Atlas,* Oxford University Press, New York, 1974.

Weiss, L., and R. O. Greep: *Histology,* 4th ed. McGraw-Hill, New York, 1977.

Zelickson, A. S. (ed.): *Ultrastructure of Normal and Abnormal Skin,* Lea & Febiger, Philadelphia, 1967.

ADDITIONAL READINGS

Breathnach, A. S.: "Aspects of Epidermal Ultrastructure," *J. Invest. Dermatol.* **65:**2–15 (1965).

Brody, I.: "An Ultrastructural Study of the Role of Keratohyaline Granules in the Keratinization Process," *J. Ultrastruct. Res.* **3:**84–104 (1959).

———: "The Keratinization of Normal Guinea Pig Skin as Revealed by Electron Microscopy," *J. Ultrastruct. Res.* **2:**488–511 (1959).

———: "The Ultrastructure of the Tonofibrils in the Keratinization Process of Normal Human Epidermis," *J. Ultrastruct. Res.* **4:**264–297 (1960).

———: "The Modified Plasma Membranes of the Transition and Horny Cells of Normal Human Epidermis," *Acta Derm-Venereol.* **49:**128–138 (1969).

———: "The Electron Microscopic Study of the Fibrillar Density in the Normal Human Stratum Corneum," *J. Ultrastruct. Res.* **30:**209–217 (1970).

———: "Variations in the Differentiation of the Fibrils in the Normal Human Stratum Corneum," *J. Ultrastruct. Res.* **30:**601–614 (1970).

Charles, H., and J. T. Ingram: "Electron Microscope Observations of the Melanocyte of the Human Epidermis," *J. Biophys. Biochem. Cytol.* **6:**41–44 (1959).

Chase, H. B.: "Growth of the Hair," *Physiol. Rev.* **34:**113–126 (1954).

Farbman, A. I.: "Plasma Membrane Changes during Keratinization," *Anat. Rec.* **156:**269–282 (1966).

Geisenheimer, J., and S. S. Han: "A Quantitative Electron Microscopic Study of Desmosomes and Hemidesmosomes in Human Crevicular Epithelium," *J. Periodontol.* **42:**396–405 (1971).

Hashimoto, K.: "The Ultrastructure of the Skin of Human Embryos. VIII. Melanoblast and Intrafollicular Melanocyte," *J. Anat.* **108:**99–108 (1971).

———: "Ultrastructure of the Human Toe Nail," *J. Ultrastruct. Res.* **36:**391–410 (1971).

Huxley, H. J., and W. B. Shelley: *The Human Apocrine Sweat Gland in Health and Disease,* Charles C Thomas, Springfield, Ill., 1960.

Jessen, H.: "Two Types of Keratohyalin Granules," *J. Ultrastruct. Res.* **33:**95–115 (1970).

Lavker, R. M., and A. G. Matoltsy: "Formation of Horny Cells. The Fate of Organelles and Differentiation Products in Luminal Epithelium," *J. Cell Biol.* **44:**501–512 (1970).

———: "Membrane Coating Granules: The Fate of the Discharged Lamellae," *J. Ultrastruct. Res.* **55:**79–86 (1976).

Martinez, I. R., and A. Peters: "Membrane-coating Granules and Membrane Modifications in Keratinizing Epithelia," *Am. J. Anat.* **140:**93–120 (1971).

Menon, D. N., and A. Z. Eisen: "Structure and Organization of Mammalian Stratum Corneum," *J. Ultrastruct. Res.* **35:**247–264 (1971).

Montagna, W., R. A. Ellis, and A. F. Silver (eds.): *Advances in Biology of Skin,* vol. 3, *Eccrine Sweat Glands and Eccrine Secretion,* Pergamon Press, New York, 1962.

———, ———, and ——— (eds.): *Advances in Biology of Skin,* vol. 4, *The Sebaceous Glands,* Pergamon Press, Oxford, 1963.

——— and F. Hu (eds.): *Advances in Biology of Skin,* vol. 8, *The Pigmentary System,* Pergamon Press, Oxford, 1967.

———, J. P. Bently, and R. L. Dobson (eds.): *Advances in Biology of Skin,* vol. 10, *The Dermis,* Appleton Century Crofts, New York, 1970.

Snell, R. S.: "The Fate of Epidermal Desmosomes in Mammalian Skin," *Z. Zellforsch.* **66:**471–487 (1965).

Zaias, N., and J. Alvarez: "The Formation of the Primate Nail Plate," *J. Invest. Dermatol.* **51:**120–136 (1968).

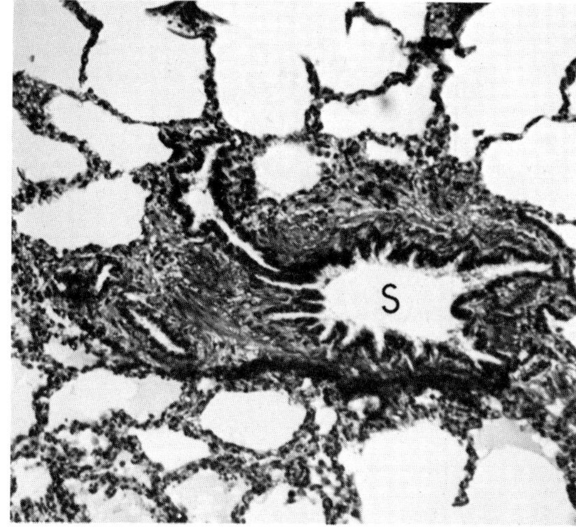

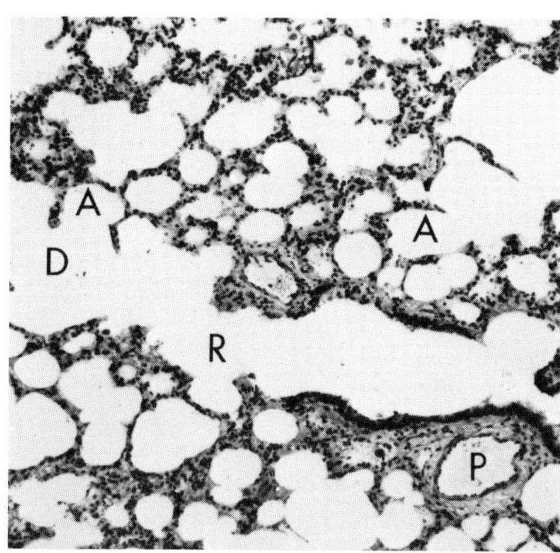

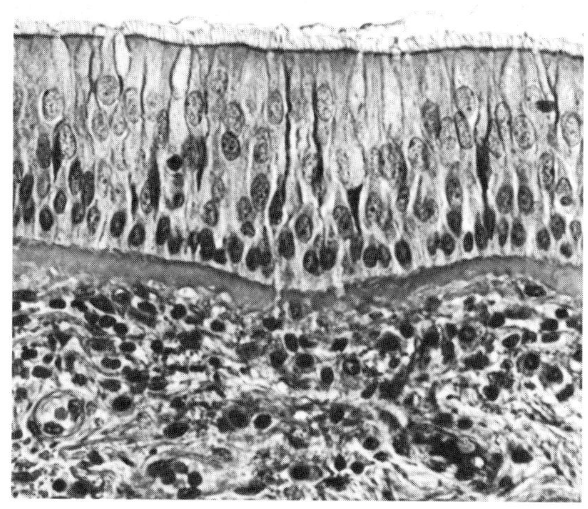

15 RESPIRATORY SYSTEM

OBJECTIVES

Upon completion of this chapter, the student will be able to:

1 Identify the following structures of the trachea:
(a) trachealis muscle (b) cartilage ring (c) epithelium (d) lamina propria (e) elastic membrane (f) submucosa (g) connective tissue

2 Identify the characteristics of the two different types of epithelium found in the trachea.

3 Identify the functions served by each of the following structures of the trachea:
(a) cartilage ring (b) heavy basement membrane

4 Identify for certain portions of the lower respiratory tract (trachea, bronchus and segmented bronchi, terminal bronchioles, respiratory bronchioles, and alveolar duct) the function and characteristics of the following:
(a) epithelial cells (b) lamina propria (c) muscularis mucosa (d) submucosa (e) cartilage plates (f) submucosal glands

5 Identify the three possible fates of a dust particle entering the respiratory tract as follows:
(a) captured by mucous lining and removed with sputum (b) captured by phagocytes and taken to lymph nodes (c) captured by phagocytes in alveoli and digested

6 Identify the structures of the alveolar sac

through which an oxygen molecule must pass on its way to the blood.

7 Identify the three main "conditioning" treatments given inspired air by the respiratory system.

All body cells need a constant supply of oxygen to function. The organs and passageways of the respiratory system allow the blood to exchange carbon dioxide for oxygen. These gases are transported out of and into the body tissues by the circulatory system. The respiratory system provides a series of air passages from outside of the body to the lungs, where the exchange of gases occurs. The conducting passages of the respiratory system begin with the nasal cavity and the nasopharynx. The air passes through the larynx into the trachea, another segment of the passageway. The trachea bifurcates into the main bronchi, which continue to become intrapulmonary bronchi. Within the lung the bronchi subdivide into bronchioles, and after a series of further branching, the distal ends of the passageway continue into the alveolar air sacs (Fig. 15-1).

Functions of conducting passages are to *warm, humidify,* and *clean* the air as it travels toward the alveoli. The conditioned air passes oxygen to and receives carbon dioxide from the blood through the thin integrated walls of the capillaries and alveoli. As the body cells respire in generating useful energy, oxygen is consumed and carbon dioxide released back into the venous blood.

CONDUCTING PASSAGES

Nasal Cavity

Air enters the respiratory system through the nasal cavity, which extends from each nostril to the opening of the nasopharynx. The dilated region extending from the nostril is called the *vestibule;* it is lined by stratified squamous epithelium which loses layering as it becomes gradually replaced by the respiratory epithelium. Hair and sebaceous glands are frequently found near the nostril. The respiratory portion of the nose is divided into two lateral halves by a cartilaginous and bony septum. The septal wall of the nasal cavity is smooth, but the lateral wall has an irregular contour due to the presence of turbinate protrusions called *conchae.*

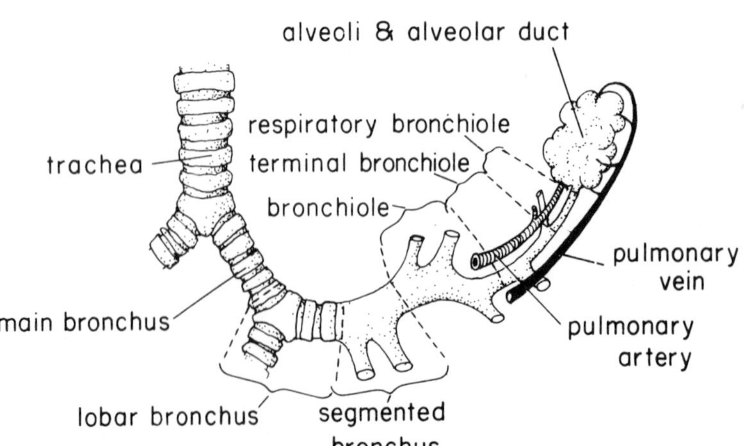

FIGURE 15-1 Diagram depicting the trachea and its extra- and intrapulmonary subdivisions.

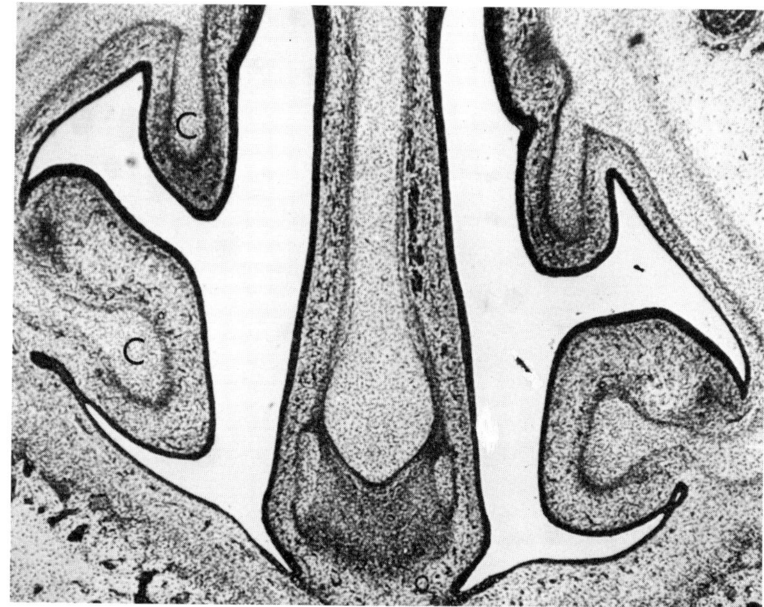

FIGURE 15-2 Frontal section through the nasal cavity from a fetus. The nasal septum, which runs vertically, bisects the nasal cavity. Irregular protrusions (C) from the lateral walls are the developing conchae. Both the nasal septum and the conchae have a cartilaginous skeleton at this stage of development. Respiratory epithelium that appears dark in the micrograph lines the entire nasal cavity.

Figure 15-2 is a frontal section of the nasal cavity from a developing fetus. Note the nasal septum, which runs vertically and bisects the nasal cavity into two cavities. Along the lateral walls are nasal conchae, which at this time contain a cartilaginous skeleton (C). The respiratory mucosa covers the entire surface of the nasal cavity. It is composed of respiratory epithelium and underlying connective tissue, the lamina propria (Fig. 15-3). At this prenatal stage, the respiratory epithelium is already made up of pseudostratified columnar ciliated cells. As fetal life progresses, there is further development of the nasal mucosa (Fig. 15-4). The epithelium has undergone further differentiation, and there is an increase in the vascularity of the lamina propria. This developing plexus of blood vessels can be identified even at low magnification (arrows in Fig. 15-4). Submucosa is absent in the nasal mucosa. Postnatally, the functioning

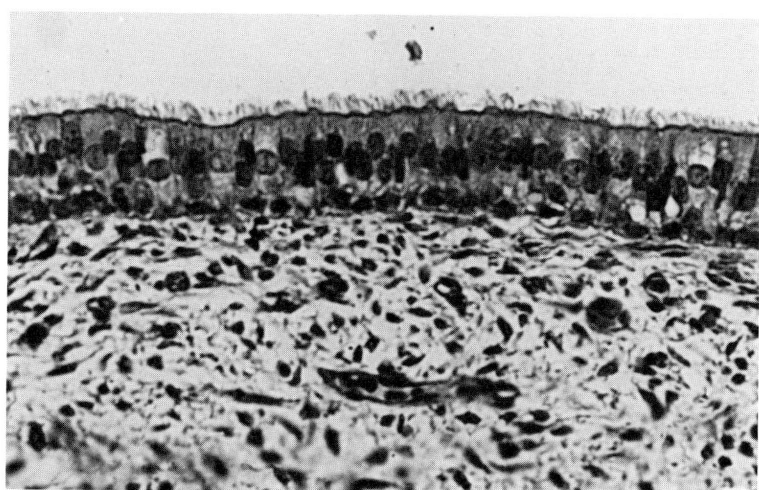

FIGURE 15-3 Nasal mucosa from a human fetus seen at higher magnification. At this stage, the epithelial lining already consists of pseudostratified columnar ciliated cells. The underlying connective tissue, the lamina propria, is separated from the epithelium by a basement membrane. It consists mainly of cells and ground substance. The connective tissue fibers are delicate.

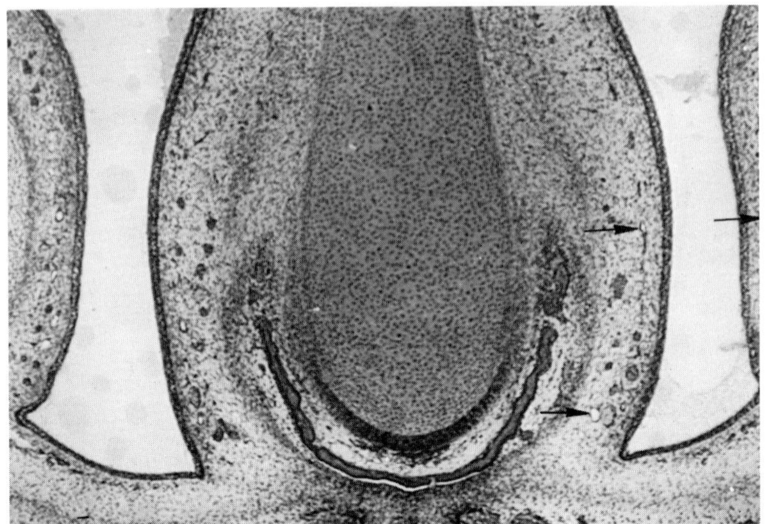

FIGURE 15-4 Frontal section through the lower portion of the nasal cavity from a fetus that is older than the one in Fig. 15-2. The cartilaginous nasal septum bisects the cavity that appears clear. At the lower border, the vomer bone is developing. The respiratory epithelium is more differentiated and a vascular plexus is developing in the lamina propria (arrows).

mucosa shows further development of a dilated vascular plexus, seromucous glands, and goblet cells (Fig. 15-5).

Many dilated veins which appear as clear spaces are found in the lamina propria. This network of venous plexuses allows the rapid heating of the air and begins the conditioning process. The seromucous nature of the glands is better depicted in Fig. 15-6. The large, clear cells are fat cells, and the smaller vacuolated cells are the mucous cells. Granulated cells within the acini are the serous cells. There are mixed cells as well. Often mononuclear cells, particularly lymphocytes, infiltrate the mucosa. Figure 15-7 shows a portion of the mucosa where the irregular duct of the mucosal gland (D) is approaching the respiratory epithelium. The dark, rounded nuclei scattered throughout the figure belong to lymphocytes that have invaded the connective tissue underlying the epithelium (arrows).

The olfactory region (epithelium that is specialized for the special sense of smell) is located in the superior portion of the nasal

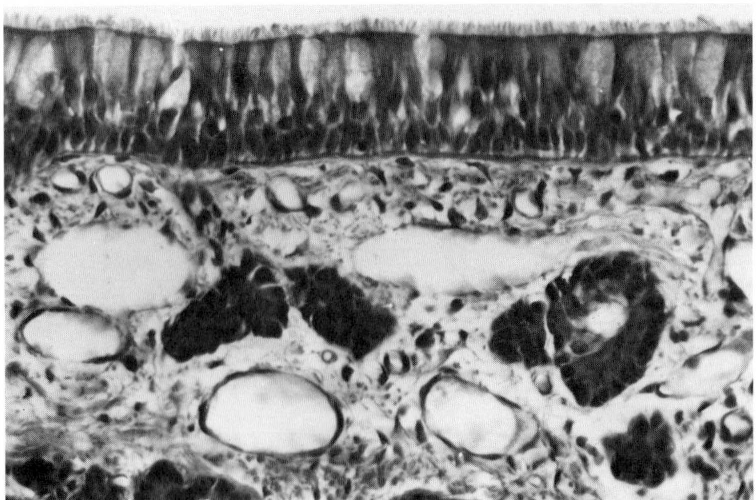

FIGURE 15-5 Light micrograph of postnatal functioning mucosa. A fully developed pseudostratified ciliated columnar epithelium with goblet cells rests on a distinct basement membrane. The lamina propria contains seromucous glands and a plexus of distended blood vessels. The latter appear as clear spaces.

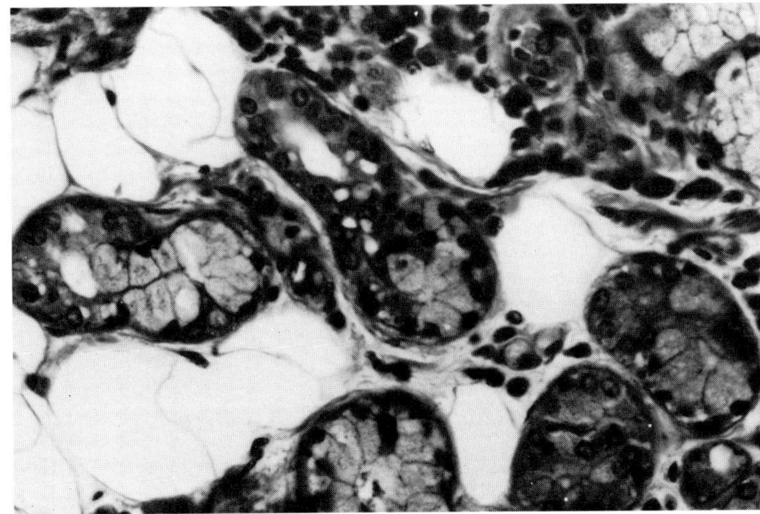

FIGURE 15-6 Light micrograph of mucosal glands. Mucous cells are vacuolated and appear empty, while serous cells are dark and contain granules. Some of the cells are mixed and are called *seromucous*. Large vacuolated fat cells are present in the surrounding connective tissue.

cavity. Placed among epithelial cells of this region are slender neural cells that are specialized olfactory receptors.

Nasopharynx

After passing through the nasal cavity, air enters the nasopharynx. Pseudostratified ciliated columnar epithelium with subjacent mixed serous and mucous glands lines the *nasopharynx,* except for those surfaces that are frequently brought in contact with each other (such as the area of contact between the soft palate and the posterior wall of the nasopharynx). These contacting surfaces have a high rate of cell attrition and are therefore lined with a stratified squamous epithelium. Under the mucosa is a layer of elastic tissue which is analogous to the muscularis mucosae to be described in later chapters on the gastrointestinal tract.

Larynx

The larynx is the next structure through which air passes after leaving the nasopharynx. A mucosa, a poorly defined submucosa, and a series of irregularly shaped cartilages

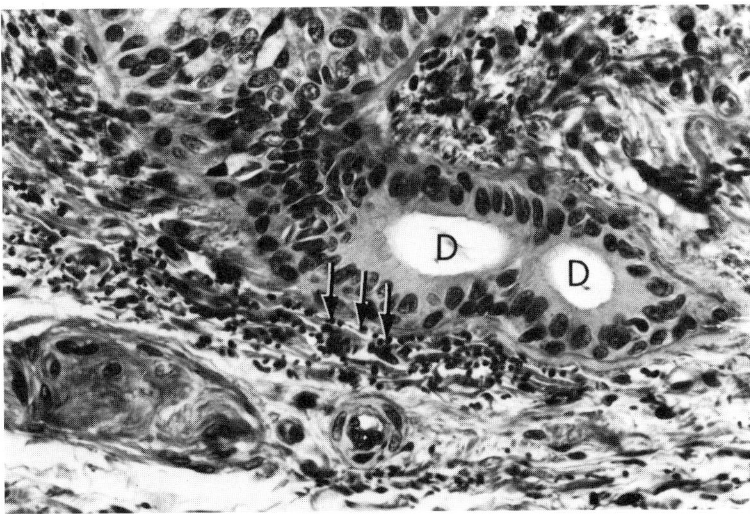

FIGURE 15-7 A duct (D) of a mucosal gland is approaching the lining epithelium. Numerous lymphocytes (arrows) have infiltrated the lamina propria. In addition, blood vessels and nerves are seen in the latter. Note the thick basement membrane that separates the respiratory epithelium from the underlying connective tissue.

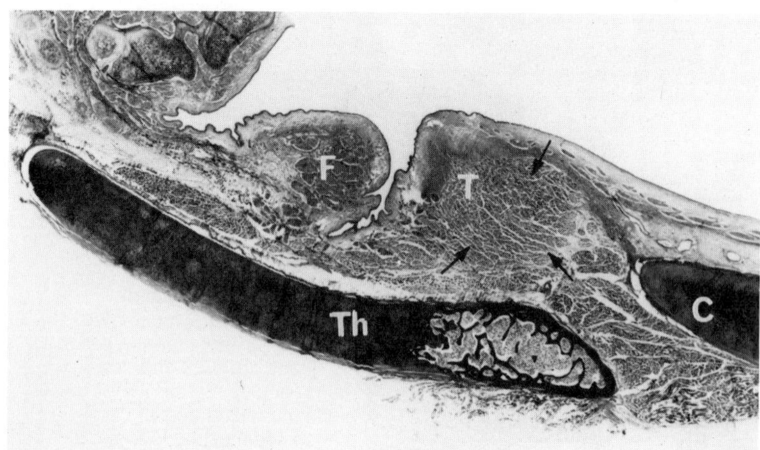

FIGURE 15-8 Frontal section through larynx. (C, cricoid cartilage; F, false vocal fold; T, true vocal fold; Th, thyroid cartilage; arrows outline the vocalis muscle.)

are present in the *larynx*. A group of intrinsic skeletal muscles which are attached to the cartilages are also present. Two prominent folds that run anteroposteriorly are present on each lateral wall of the larynx. The cephalic pair of folds, the *ventricular folds,* are also known as the *false vocal cords*. The caudal pair, the *vocal folds,* are called the true vocal cords, because they have the primary function in developing air vibrations necessary for sound production.

Two types of epithelium line the walls of the larynx. Stratified squamous epithelium lines such areas as the true vocal cords, where abrasion occurs; pseudostratified ciliated columnar epithelium is found lining all other areas. In regions where transition of the two types of epithelium occurs, ciliated stratified columnary epithelium is present. Mixed glands are present throughout, except in the vocal folds.

Figure 15-8 represents a frontal section of the larynx which shows the false vocal cord (F) on the left of the figure and the true vocal cord (T) on the right. Underlying the two vocal cord regions is a portion of the supporting thyroid cartilage (Th). Note also the laryngeal muscle, which attaches to the cartilage along the bottom of the field. The small cartilage at the right is the cricoid cartilage (C). Note that both cartilages are partially ossified. The muscle between the two cartilages is the cricothyroid muscle. Even at this low magnification the two vocal cords can be readily identified on the basis of the presence of *vocalis muscle,* which fills the true vocal cord as a mass of cross-sectioned skeletal muscle (arrows). In contrast, the false vocal cord has a more irregular epithelium and contains glands instead of muscular cords. The clear area extending deeply between the cords is the *ventricle* of the larynx, which functions in providing proper resonance for intonations.

Trachea

The *trachea* conducts air from the larynx to the bronchi. The basic shape of the trachea is maintained by the series of hyaline cartilage rings which are the struts, or supports, of the body of the tube (Fig. 15-9). These cartilaginous rings are C-shaped. There are 16 to 20 of them. The circular shape of the trachea is completed by the trachealis muscles, which run horizontally in the posterior part of the trachea (Fig. 15-10). Around the outside of the cartilaginous rings is the adventitia of the trachea.

Air introduced into the respiratory system often contains dust particles and other foreign materials. Pseudostratified ciliated columnar epithelium lines the tracheal lumen as a protective mechanism against these foreign particles. The thick basement membrane of the epithelium provides another layer of

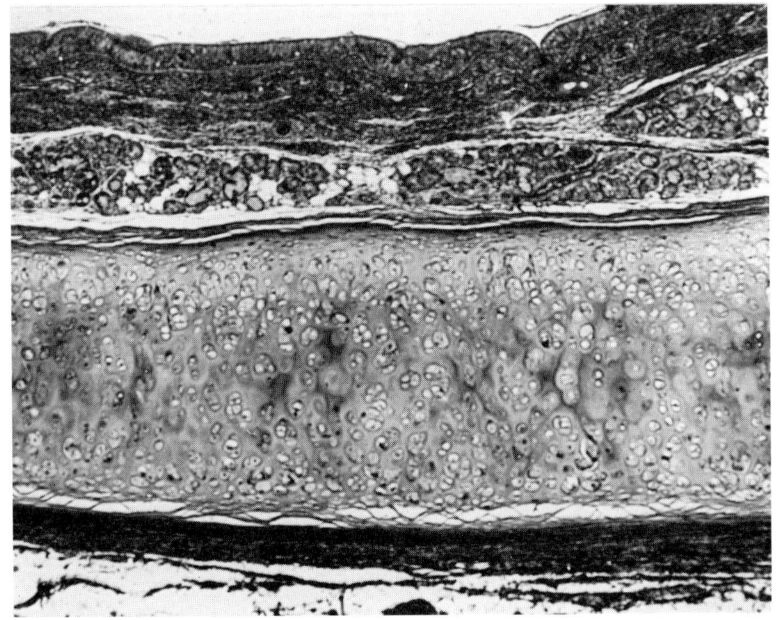

FIGURE 15-9 Cross section of trachea. A portion of a C-shaped ring is seen in the lower half of the micrograph. The mucosa, which is thick, is made up of respiratory epithelium and a lamina propria. It is attached to the perichondrium by the submucosa, which contains abundant mixed glands. The perichondrium surrounds the cartilages and becomes part of the investment of the trachea.

protection. This area also helps to warm the air before it reaches the lungs.

Goblet cells interspersed in the epithelium secrete mucus that captures the foreign particles and lubricates the tracheal surface. The epithelial cilia maintain a synchronized wavelike motion that moves the foreign particles trapped by mucus along the surface away from the lungs. A thin layer of connective tissue called *lamina propria* surrounds the epithelial lining of the trachea. The lamina propria is separated from the subjacent loose connective tissue, the submucosa, by an elastic membrane. Thus the mucosa consists of

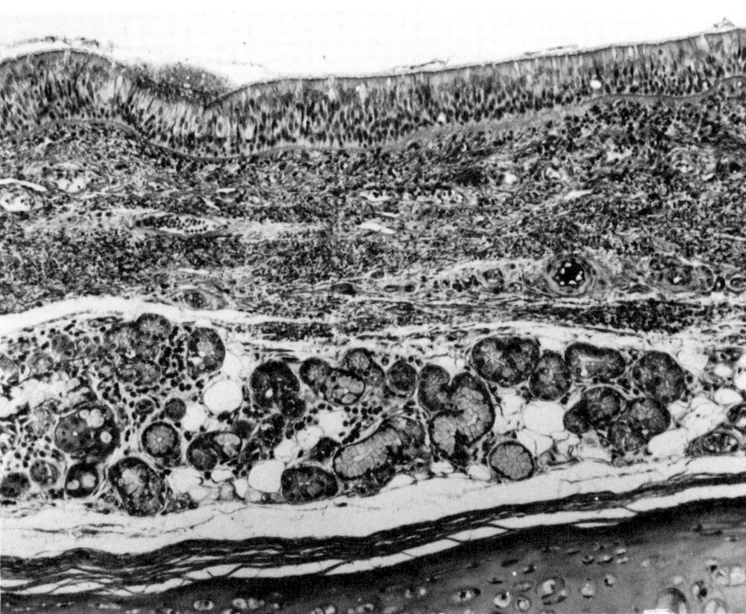

FIGURE 15-10 Light micrograph of the posterior part of the trachea. The mucosa is thick and is lined by respiratory epithelium. A thick basal lamina separates the epithelium from the lamina propria. The trachealis muscle is running horizontally deep to the mucosa.

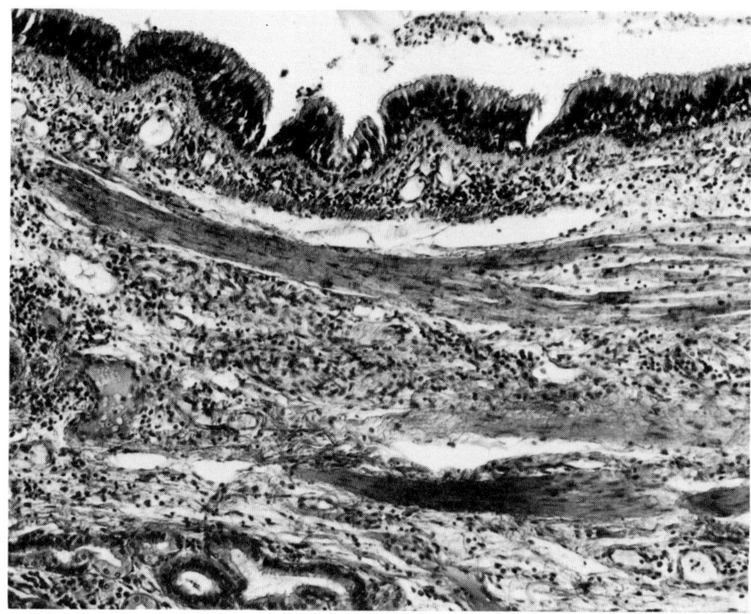

FIGURE 15-11 Light micrograph of tracheal mucosa. A pseudostratified ciliated epithelium lines the lumen. It contains numerous goblet cells and is separated from the connective tissue, the lamina propria, by a thick basement membrane. The lamina propria appears denser than the submucosa. The latter contains mixed tracheal glands.

the epithelium, the basement membrane, and the lamina propria, whereas the submucosa consists of loose FECT in which blood vessels, nerves, and mixed tracheal glands are present. The perichondrium of the trachea surrounds the cartilage and continues to become part of the connective tissue investment.

Many of the structures mentioned above are evident in Figs. 15-11 to 15-13. Along the surface of the photomicrographs is the respiratory epithelium lining the trachea. The thick basement membrane is well delineated. Dark nuclei of lymphocytes are present in the lamina propria. The submucosa contains the typical seromucous tracheal glands, which

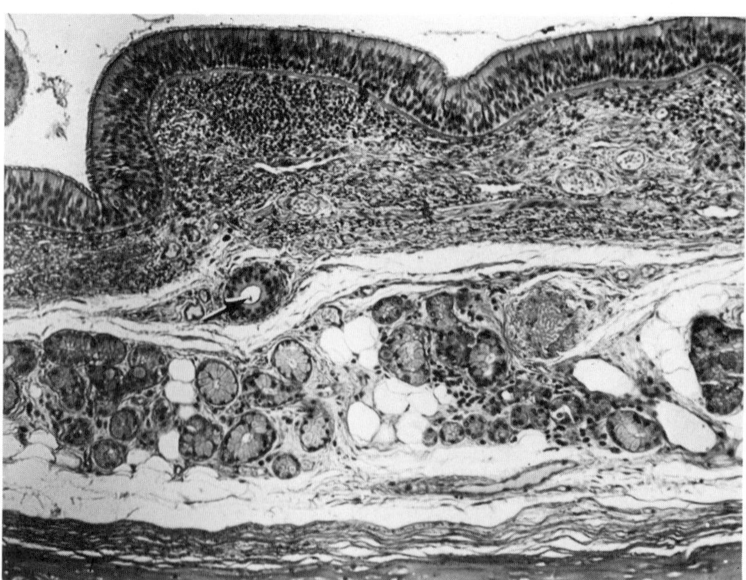

FIGURE 15-12 Light micrograph of tracheal mucosa. An elastic membrane separates the mucosa from the submucosa. The latter contains numerous mixed glands. Note the lymphocyte infiltration of lamina propria. Tortuous gland ducts (arrow) wind through the lamina propria to reach the surface. They penetrate the elastic membrane before reaching the glands in the submucosa.

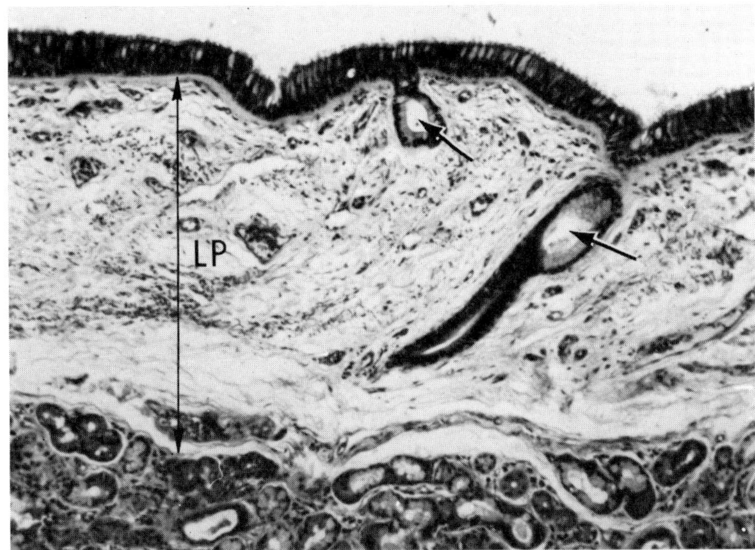

FIGURE 15-13 Light micrograph of tracheal mucosa. In this specimen the elastic membrane is thinner than in previous ones. The lamina propria (LP) is thicker than usual and contains only sparse mononuclear cells. Gland ducts (arrows) penetrate the lamina propria and open to the surface.

have a tortuous duct that opens to the tracheal lumen as indicated in Figs. 15-12 and 15-13. An unusually thick elastic membrane is broken where the duct penetrates into the submucosal gland. Figure 15-13 shows another example of the tracheal wall which illustrates certain variations in structure from the previous figures, as noted in the thin elastic membrane, more prominent submucosal glands, and a thick lamina propria (LP) which is not heavily infiltrated by mononuclear cells.

Figure 15-14 shows a higher power view of the tracheal mucosa. The ciliated border of the respiratory epithelium is well visualized. Under the thick basement membrane are nuclei of many lymphocytes, blood vessels, and some wisps of smooth muscle fibers that are sectioned obliquely. The wispy smooth muscle cells are generally in continuity with the elastic membrane, since the latter becomes thinner and is eventually completely replaced by smooth muscle cells. In areas of transition, such as in Fig. 15-15, the

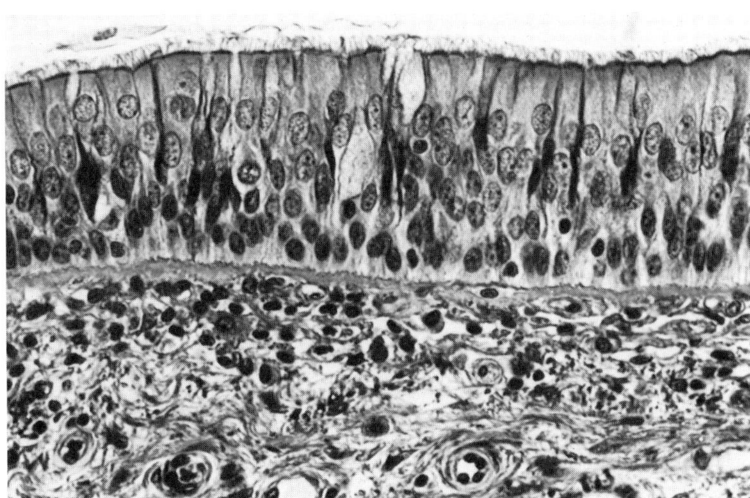

FIGURE 15-14 Tracheal mucosa seen at high magnification. The cilia along the apical border of the respiratory epithelium are readily visualized. Goblet cells are emptying their secretion to the surface. Numerous small blood vessels and mononuclear cells are present in the subjacent lamina propria. Note the thickness of the basement membrane.

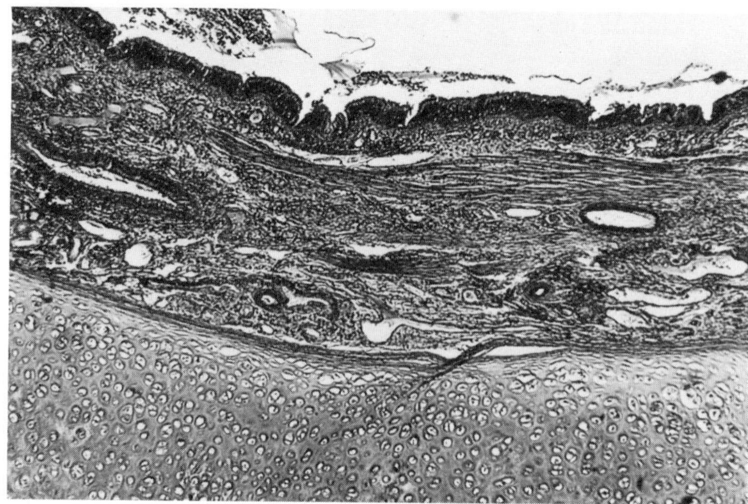

FIGURE 15-15 Cross section of trachea. Wisps of smooth muscle cells mingle with the elastic membrane. These wisps of smooth muscle will eventually completely replace the elastic membrane farther down in the respiratory passageways.

smooth muscle fibers often appear as muscular islets that may be intermingled with the thinning elastic membrane.

Bifurcation into Bronchi

As the air moves toward the lungs from the trachea, it passes into the *bronchus* where the air passage bifurcates. Here the epithelium is no longer pseudostratified, but is tall columnar. These cells are ciliated, and goblet cells remain abundant.

The lamina propria connects the epithelium to the muscularis mucosae which has replaced the elastic membrane of the trachea. Seromucous glands in the bifurcation area extend into the submucosal region. The cartilaginous rings of the trachea become broken into cartilage plates around the bronchi.

LUNGS

The lungs are the sum total of the smaller conducting passages, alveolar sacs, and associated blood vessels. They are lobated, following the pattern set up by the division

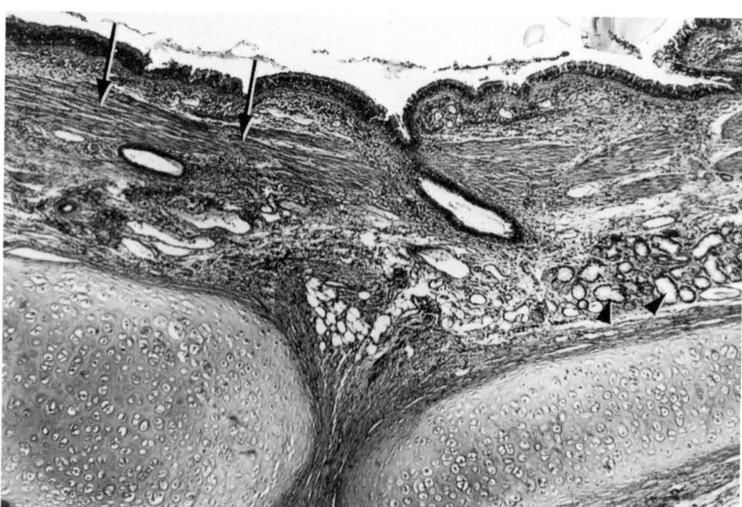

FIGURE 15-16 Portion of a lobar bronchus. The muscularis mucosae (arrows) is well developed. Contraction of the mucosa causes the irregular appearance of the mucosa. The submucosal glands (arrowheads) are reduced in number. The wall of the lobar bronchus is supported by cartilaginous plates.

of the bronchi. The right lung is divided into three lobes; the left lung is made up of two. The rib cage surrounding the lungs and the diaphragm supporting them expand and contract the thoracic cavity, aiding in the inflation and deflation of the lungs. Entering the hilum of each lobe are a bronchus, bronchial arteries and veins, lymphatics, and nerves. Connective tissues invest these structures and continue to provide appropriate matrix as these structures become modified to meet functional requirements at more distal regions. Each lung is composed of about 300 million alveolar sacs with associated capillaries of the pulmonary circulatory system.

The lung is covered by pleura, which is a serous membrane that dips into the interlobar fissures, covering the entire surface of the lung (visceral pleura). The pleura continues to line the walls of the thoracic cavity as it deflects along the periphery of pulmonary tissue (parietal pleura). The space between these layers of the pleura is termed the *pleural cavity*. The pressure in the pleural cavity, which is slightly less than that of the atmosphere, prevents the lung from collapsing, helps it maintain its shape, and aids in the inspiration of air.

Lobar Bronchi

Each bronchus divides as it enters the lungs to become intrapulmonary *lobar bronchi*. The epithelium at this level is simple columnar ciliated epithelium with goblet cells. One way in which lobar bronchi differ from the trachea and extrapulmonary bronchi is in the further development of the muscularis mucosae. The muscle cells become arranged in bundles traveling around the circumference of the bronchi. As mentioned previously, the muscularis mucosae in bronchi is not made up of a complete layer of muscles. This is in contrast to the distinct muscularis mucosae found along the gastrointestinal tract. Submucosal glands are reduced in number. Cartilaginous plates are smaller but continue to aid in the maintenance of the lumen of bronchi.

Many of the features of the lobar bronchi are found in Fig. 15-16, which shows a portion of a large lobar bronchus. Notice the two cartilage plates. The mucosa shows an epithelium which has a scalloped appear-

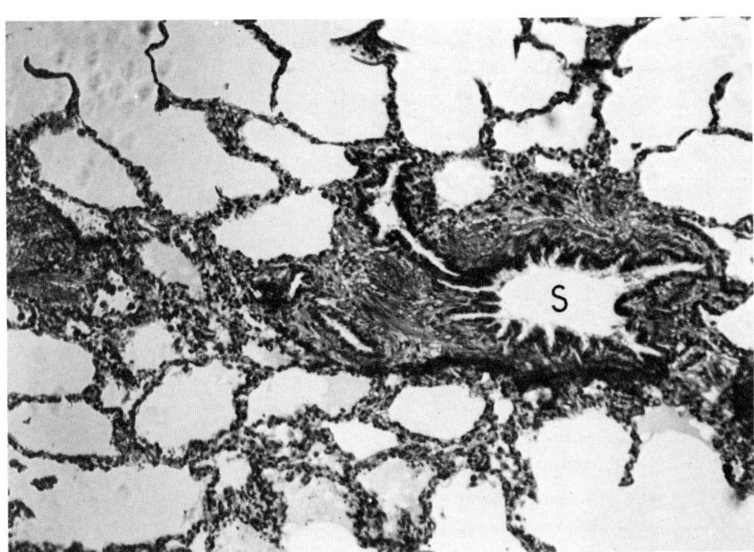

FIGURE 15-17 Segmental bronchiole (S) seen in cross sections. Bronchial blood vessels accompany the segmental bronchioles. Note the irregular appearance of the lumen caused by contraction of the smooth muscle in the bronchiolar wall.

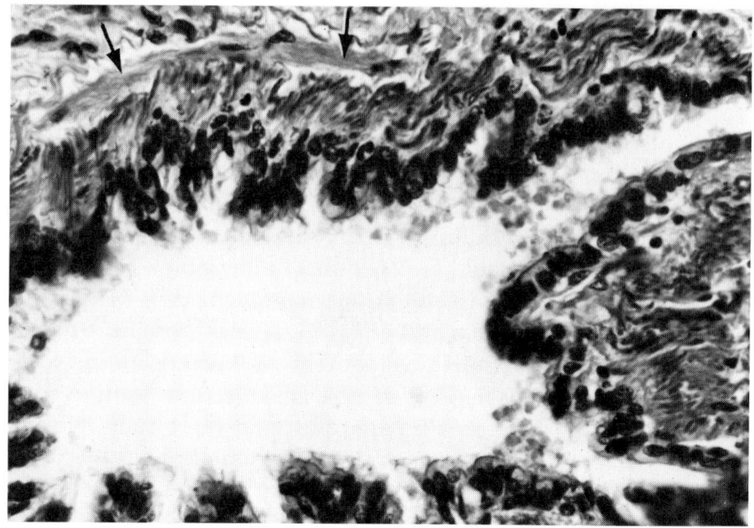

FIGURE 15-18 Cross section of segmental bronchiole viewed at higher magnification. Note the appearance of the smooth muscle cells (arrows). Segmental bronchioles lack submucosal glands.

ance. This reflects the heavier muscle development in the muscularis mucosae (arrows) which contracts and produces the squeezed appearance of the mucosa. Submucosal glands are also found but in a reduced number (arrowheads).

The lobar bronchi continue to divide into branches called *segmental bronchi*. These segmental bronchi repeatedly branch until they are called *bronchioles*. About 10 more divisions occur before the *terminal bronchioles* are reached. As the segmental bronchi become smaller and smaller through divisions, their lumen becomes more ruffled in appearance, as shown in Fig. 15-17. This indicates an even greater role that smooth muscles may play in the smaller bronchiole wall. A number of bronchial vessels may be seen along segmental bronchioles. Note the appearance of smooth muscle cells (arrows) at a higher magnification (Fig. 15-18). There are no submucosal glands in segmental bronchioles.

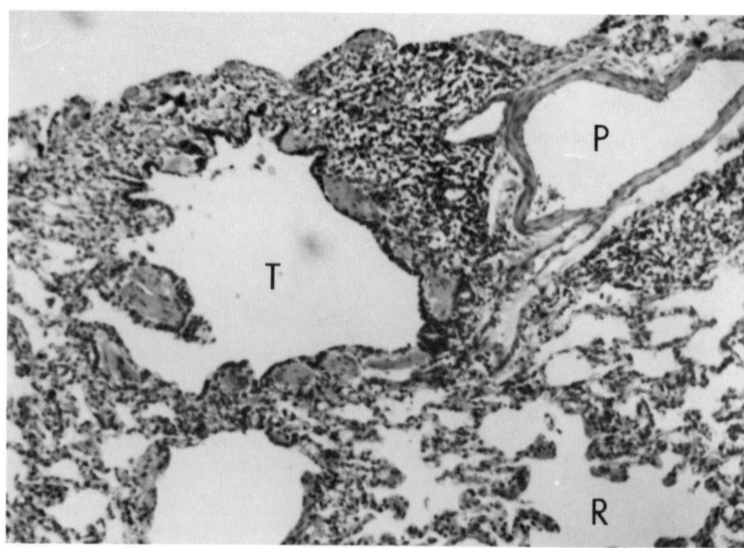

FIGURE 15-19 Portion of a pulmonary lobule. A terminal bronchiole (T), a respiratory bronchiole (R), and a branch of the pulmonary artery (P) are seen in this micrograph.

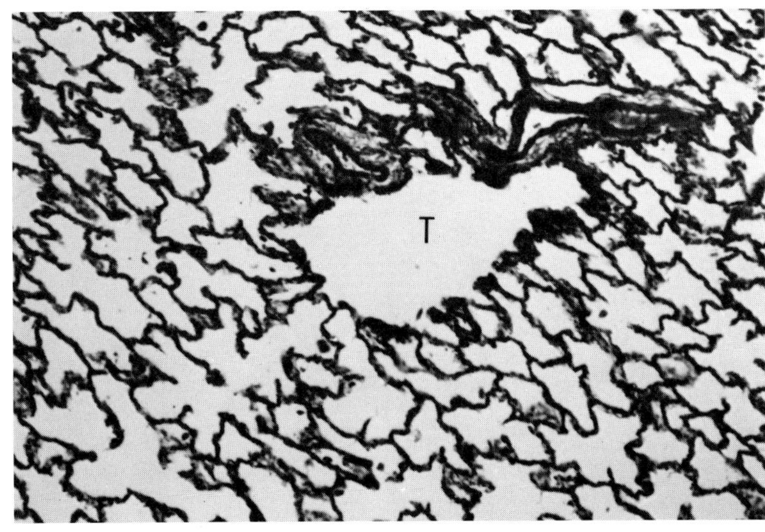

FIGURE 15-20 Longitudinal section of terminal bronchiole (T). It branches into a respiratory bronchiole, which has a thin wall.

Terminal Bronchioles

The terminal bronchiole is only 1 mm or less in diameter. There may be up to 80 terminal bronchioles to each pulmonary lobule. The histologic appearance of the terminal bronchiole is characterized by a reduction in the height of epithelial cells, which become a simple ciliated cuboidal epithelium containing no goblet cells. Having lost the cartilage plates and glands, terminal bronchioles no longer have a submucosa; hence the smooth muscle layer is invested by a FECT matrix which is shared by other structures such as blood vessels, lymphatics, and nerves (Fig. 15-19).

An example of a terminal bronchiole is shown in Fig. 15-19. The wall has a cuboidal epithelium and a discontinuous layer of smooth muscle cells which appear as wispy patches surrounding the mucosa. In a longitudinal view (Fig. 15-20), further branching of the terminal bronchiole into respiratory bronchioles may be seen. The latter has a very thin wall.

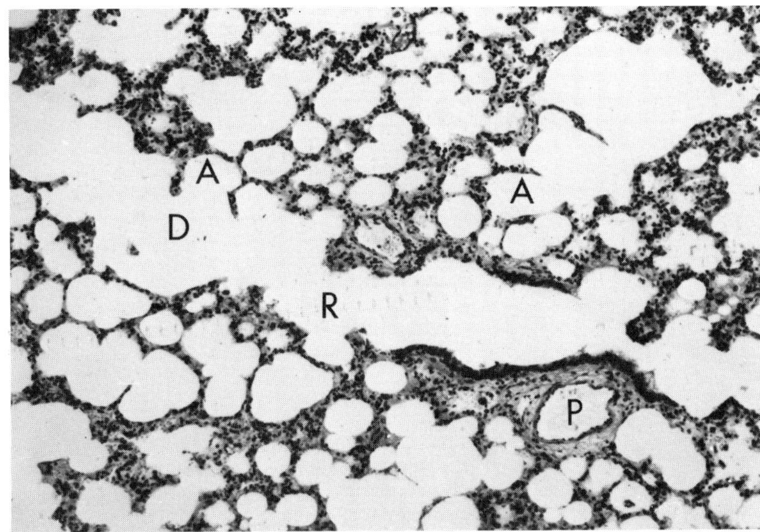

FIGURE 15-21 Portion of a lung lobule. A respiratory bronchiole (R) leads into an alveolar duct (D). In addition to respiratory epithelium it is lined by alveoli (A). The walls of the alveolar ducts and alveolar sacs are also lined by alveoli, the functional parts of the lung. A branch of the pulmonary artery (P) accompanies the respiratory bronchiole.

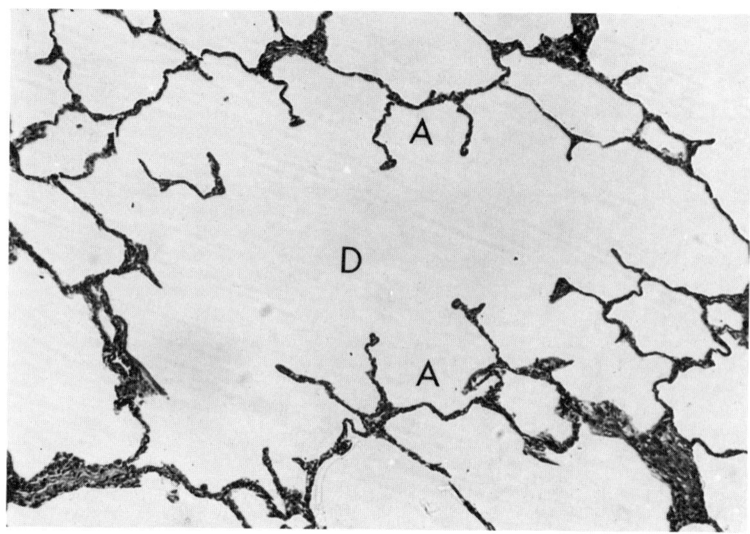

FIGURE 15-22 Light micrograph of an alveolar duct (D) opening into a number of alveoli (A).

Respiratory Bronchioles

The last segment of air passage is the *respiratory bronchiole* (Fig. 15-21). It is short and narrow (0.5 mm or less in diameter). The lining epithelium here is low cuboidal, which is continuous with the *alveolar duct*. Only wisps of smooth muscle cells and occasional blood vessels appear to support the cuboidal epithelium. A reduction in muscle occurs in the respiratory bronchiole. The epithelium changes from cuboidal (terminal bronchiole) to low cuboidal (respiratory bronchiole). This change is accompanied by a complete disappearance of cilia in the respiratory bronchiole. Respiratory bronchioles lead directly to alveoli, alveolar sacs, or alveolar ducts. Figure 15-22 is a view of alveolar ducts opening into a number of pulmonary alveoli and alveolar sacs. Smooth muscle bundles surround the opening of the respiratory bronchiole to an alveolus or an alveolar duct. The flattened epithelium and smooth muscle cells produce a characteristic bulge at the entrance of alveolar sacs and alveoli (Fig. 15-23).

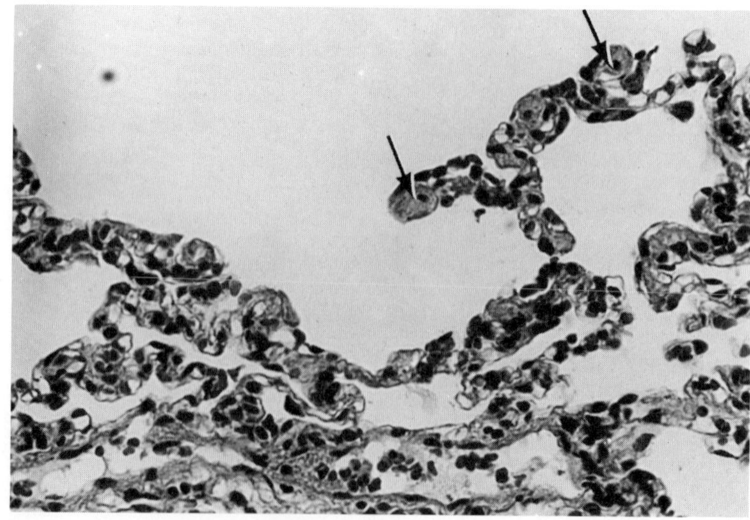

FIGURE 15-23 The wall of an alveolar duct seen at higher magnification. The flattened epithelium and the smooth muscle cells (arrows) produce a bulge at the openings of the alveoli.

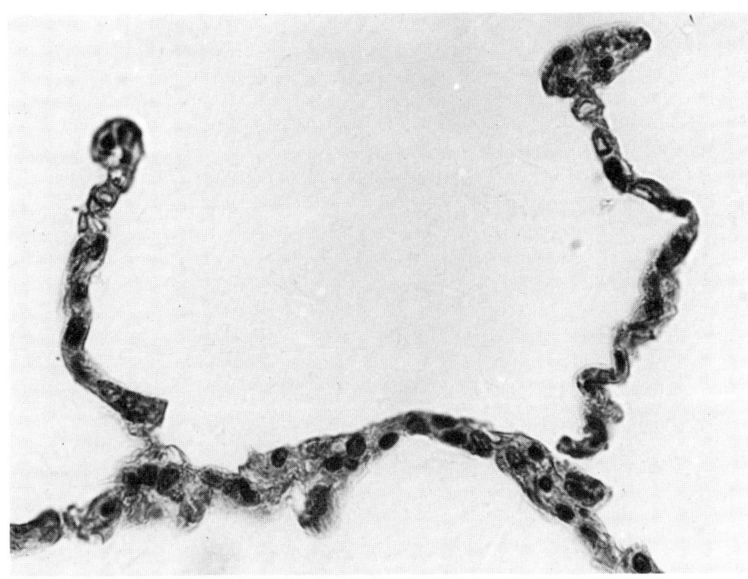

FIGURE 15-24 Light micrograph of a single alveolus. It demonstrates the thinness of the wall. Note the capillaries.

Alveoli

Most of the pulmonary parenchyma is composed of alveolar sacs which in turn are made up of individual alveoli. Alveoli and alveolar sacs are the final destination of inspired air where the exchange of oxygen and carbon dioxide between air and blood takes place. The total surface area of respiratory epithelia may approach 45 m^2. It is estimated that less than 10 percent of this is being used during rest.

A close-up of a single alveolus in Fig. 15-24 demonstrates the thin nature of the alveolar wall, also called *alveolar septum,* that facilitates the passage of gases. The clublike protrusions, a few capillaries, and other cells of the wall are visible. By making a thinner section as shown in Fig. 15-25, one can clearly observe the flat epithelial cells, septal connective tissue cells (arrows), and capillaries. The flat epithelial cells are made up of two different kinds: the type I pneumocyte, which

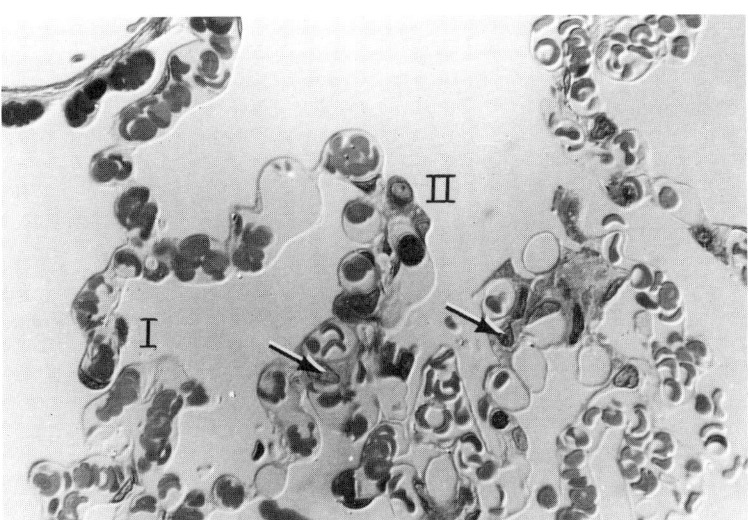

FIGURE 15-25 Phase photomicrograph of a section of a lobule in the lung showing several alveoli. The walls are made up of two types of cells: type I pneumocyte (I), concerned exclusively with gaseous exchanges; and type II (II) pneumocytes, which contain secretory granules that may be important in conditioning the surface. The alveolus also contains macrophages (arrows) that are called *dust cells.*

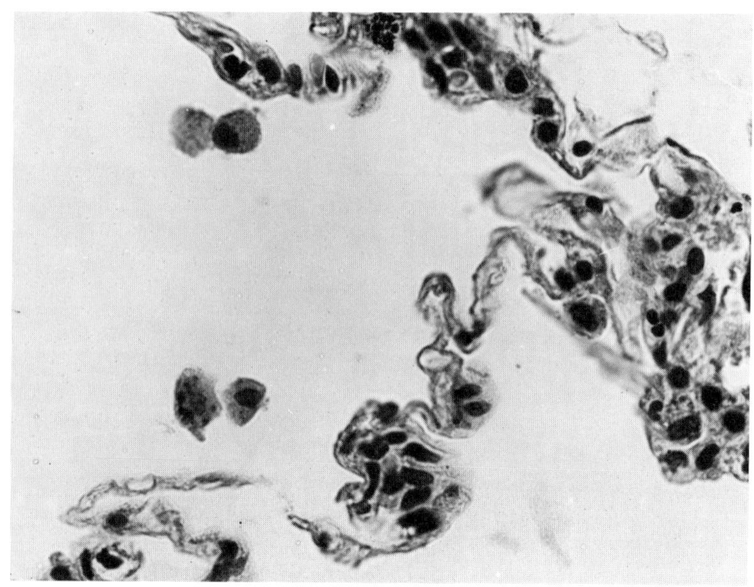

FIGURE 15-26 Light micrograph of an alveolus. Several dust cells are seen in the lumen in addition to those present in the alveolar septa.

is more numerous and is known to be concerned exclusively with gaseous exchanges; and the type II pneumocyte, which contains secretory granules that may be important in conditioning the surface. In addition to the lining cells of the alveoli, the alveolar septum contains macrophages, often called *dust cells.* Dust cells are more frequently found among people who have worked in such areas as coal mines, as well as those individuals who have smoked for an extended period of time. Since pulmonary alveoli do not have goblet cells or cilia, the macrophages and pneumocyte II represent vital elements in defense functions of the lung. It is not uncommon that the macrophages often leave the alveolar wall (Fig. 15-26). Additionally, the alveolar septum has fine collagenous and elastic fi-

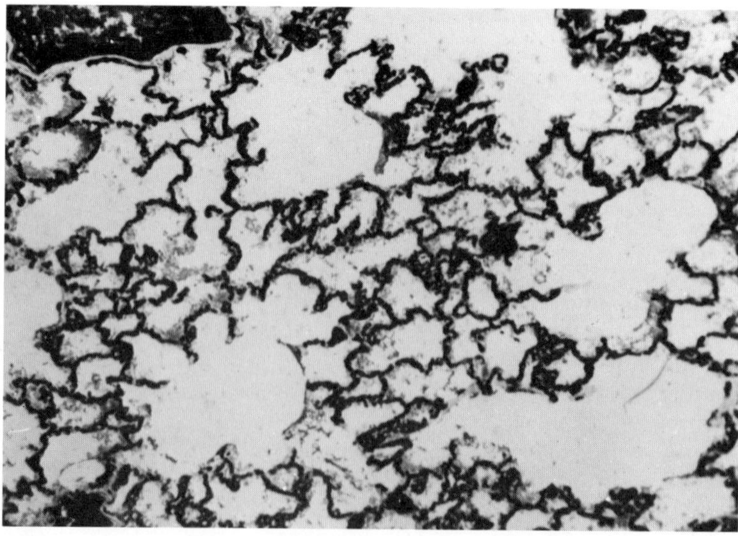

FIGURE 15-27 Light micrograph of a portion of a lung injected with india ink. It illustrates the extensive network of capillaries in the alveolar septa.

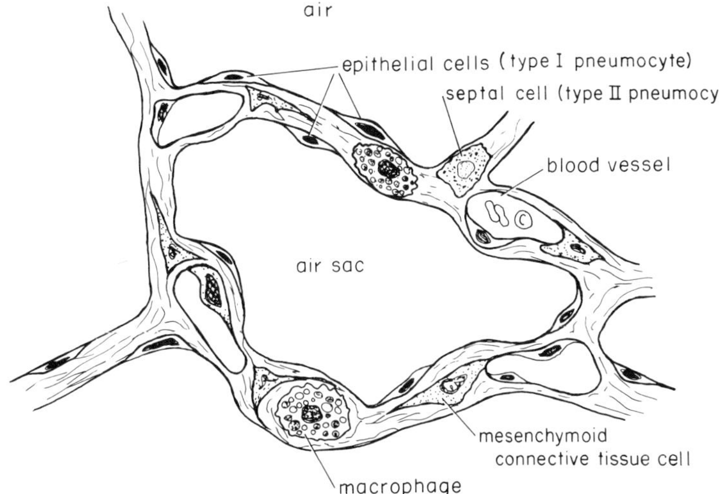

FIGURE 15-28 Diagram depicting the structure of the blood-air barrier.

bers, undifferentiated mesenchymal cells, and capillaries. A light micrograph of the lung taken after an injection of india ink, as shown in Fig. 15-27, depicts the capillary beds surrounding each of the alveolar sacs, forming an extensive network.

As mentioned previously, the extreme thinness of the alveolar wall and its close association with the capillary bed expedite gaseous exchange. A molecule of inhaled gas would have to pass through the luminal plasma membrane, cytoplasm, and basal plasma membrane of the squamous epithelium lining the alveoli, the basal lamina, connective tissue matrix present in some areas, and plasma membranes and cytoplasm of the endothelial cells of the capillary wall (Fig. 15-28). Together these structures constitute the alveolocapillary membrane, or the blood-air barrier.

The immature connective tissue cells that are present in the alveolar wall may demonstrate certain structural variations, since they serve the function of being reserve cells that can be mobilized to perform different functions. They are thought to differentiate into macrophages which are important for phagocytosis of any foreign material that might be carried there by the air. They ingest particles such as dust and carbon. As indicated earlier, lungs removed from coal miners have been shown to contain large quantities of dark carbon particles that can be diagnosed microscopically. This reserve population of immature cells can also differentiate to provide for tissue repairs in case of injury.

Blood vessels to the lung are of two sets. The *pulmonary arteries* (P in Figs. 15-19 and 15-21) and veins supply and drain blood for the purpose of respiration, i.e., exchange of gases. They follow the wall of bronchi and bronchioles, and eventually produce the extensive capillary networks around respiratory bronchioles, alveolar ducts, alveolar sacs, and alveoli proper. The *bronchial arteries* (P in Fig. 15-21) are much smaller and are often invested in the connective tissue of the bronchial tree. They serve to nourish the bronchial tissue and therefore terminate at the level of terminal bronchioles where the capillaries from the bronchial arteries anastomose with those from the pulmonary arteries.

REVIEW SECTION

1. Using the information provided in the text, complete the following chart.

	NASAL CONCHA	TRACHEA	LOBAR BRONCHI	SEGMENTED (INTRA-PULMONARY) BRONCHI	TERMINAL BRON-CHIOLE	RESPI-RATORY BRON-CHIOLE	ALVEOLI
Epithelium			Simple low pseudo-stratified with cilia	Same as lobar			
Basement membrane		Thicker than anywhere else in body		Present			Present with basal lamina
Lamina propria				Present			
Muscle				Increases, circular but wispy			
Elastic membrane	No			No			
Sub-mucosa (glands)	Mixed glands			No			
Cartilage	No						
Muscle	No						
Connec-tive tissue	No		Yes				Strands in matrix

2. Figure 15-29 shows several labeled structures. If the structure labeled "A" is typical respiratory epithelium, what type of epithelium is it? The structure labeled "B" is the underlying lamina propria and elastic lamina. What is C? What is D? The partially visible C-shaped cartilage ring is labeled "F." G labels the loose fibroelastic connective tissue and fat cells that cover the structure. What is the structure labeled "E?" What is this a section of?

This is, of course, a section taken from the trachea, which is lined with pseudostratified columnar epithelium with cilia and goblet cells (A). The submucosa (C) has many mixed glands (D). The fibrous perichondrium (E) surrounds the cartilage (F).

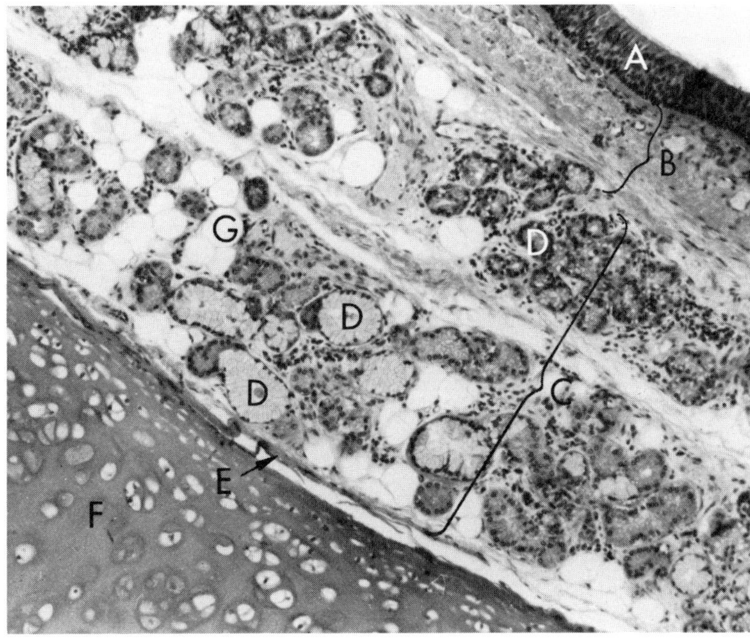

FIGURE 15-29

3. The thin-walled, irregularly shaped area indicated by A in Fig. 15-30 is called ________. It borders on a large clear area that is bounded by a continuous, dark-staining epithelium on the other side. The gray-staining structure between the irregularly shaped and large clear areas contains ________. The continuity of the border of this space to the region labeled "B" is called a ________. What is the structure labeled "C?" What is the structure labeled "D?"

The interruption of dark epithelia in this figure indicates that this is a section of a respiratory bronchiole (A). Note that the alveolar ducts (B) open directly into the lumen of this bronchiole. Alveolar sacs (C) are made up of a few to several individual alveoli (D).

FIGURE 15-30

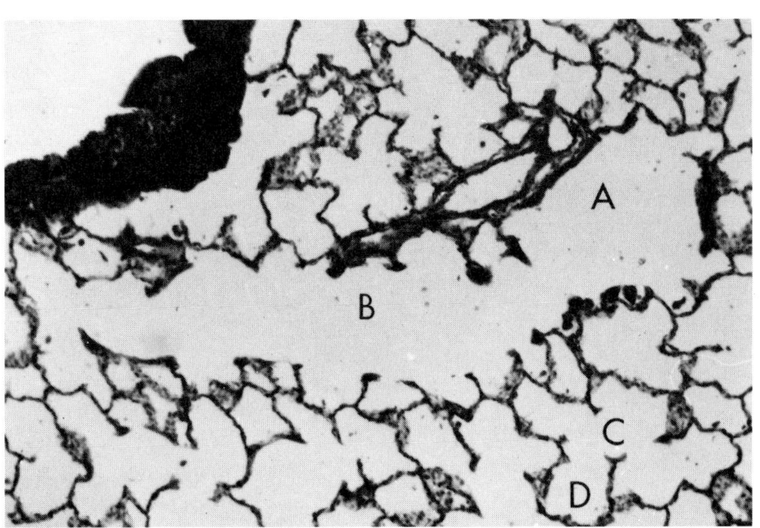

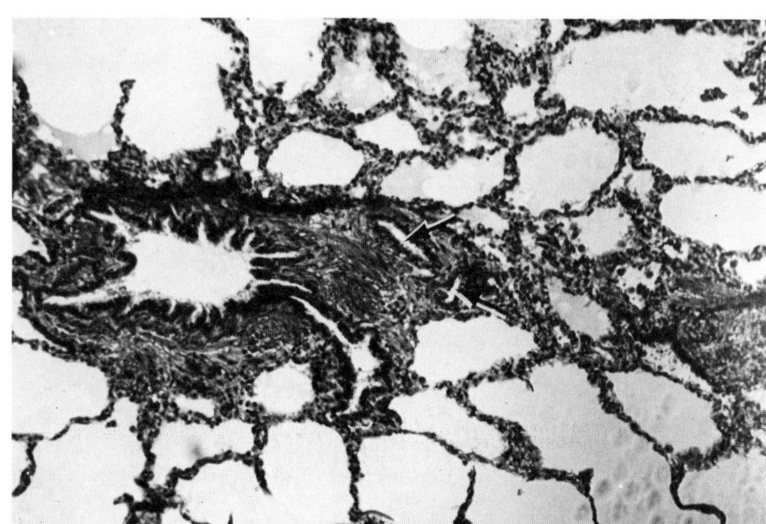

FIGURE 15-31

4. Name the tubular structure indicated by the arrows in Fig. 15-31. Name a structure that may resemble but is smaller than the above structure.

The arrows indicate a pulmonary artery that follows the bronchiolar tree. A similar but smaller branch of the bronchial artery may reach the terminal bronchiole, which is the structure seen to the right of the artery.

BIBLIOGRAPHY

Bertalanffy, F. D.: "Respiratory Tissue: Structure, Histophysiology, Cytodynamics. Part I. Review and Basic Cytomorphology," *Int. Rev. Cytol.* **16:** 233–328 (1964).

Engel, S.: *Lung Structure,* Charles C Thomas, Springfield, Ill., 1962.

Low, F. N.: "The Pulmonary Alveolar Epithelium of Laboratory Animals and Man," *Anat. Rec.* **117:** 241–263 (1953).

Rhodin, J. A. G.: *Histology. A Text and Atlas,* Oxford University Press, New York, 1974.

Weiss, L., and R. O. Greep: *Histology,* 4th ed., McGraw-Hill, New York, 1977.

ADDITIONAL READINGS

Adams, D. R.: "Olfactory and Non-olfactory Epithelia in the Nasal Cavity of the Mouse," *Am. J. Anat.* **133:**37–50 (1972).

Boyden, E. A.: "The Terminal Air Sacs and Their Blood Supply in a 37-day Infant," *Am. J. Anat.* **116:** 413–428 (1965).

Clements, J. H.: "Pulmonary Surfactant," *Am. Rev. Respir. Dis.* **101:**984–990 (1970).

Dawes, J. D. K., and M. M. L. Prichard: "Studies of the Vascular Arrangements of the Nose," *J. Anat.* **87:**311–322 (1953).

Divertie, M. B., and A. L. Brown: "The Fine Structure of the Normal Human Alveolocapillary Membrane," *J. Am. Med. Assoc.* **187:**938–941 (1964).

Frasca, J. M., O. Auerbach, V. R. Parks, and J. D. Jamieson: "Electron Microscopic Observations of the Bronchial Epithelium of Dogs," *Exp. Mol. Pathol.* **9:**363–379 (1968).

Hansell, M. M., and R. L. Morett: "Ultrastructure of the Mouse Tracheal Epithelium," *J. Morphol.* **128:** 159–170 (1969).

Hatasa, K., and T. Nakamura: "Electron Microscopic Observations of Lung Alveolar Epithelial Cells of Normal Young Mice with Special Reference to Formation and Secretion of Osmophilic Lamellar Bodies," *Z. Zellforsch.* **68:**266–277 (1965).

Karrer, H. E.: "The Ultrastructure of Mouse Lung: General Architecture of Capillary and Alveolar Walls," *J. Biophys. Biochem. Cytol.* **2:**241–252 (1956).

———: "The Ultrastructure of the Mouse Lung: The Alveolar Macrophage," *J. Biophys. Biochem. Cytol.* **4:**693–700 (1958).

Kikkawa, Y.: "Morphology of Alveolar Lining Layer," *Anat. Rec.* **167:**389–400 (1970).

Ladman, A. J., and T. N. Finley: "Electron Microscopic Observations of Pulmonary Surfactant and the Cells which Produce It," *Anat. Rec.* **154:**372 (1966).

Liebow, A. A., and D. E. Smith (eds.): *The Lung,* International Academy of Pathology Monograph, Williams & Wilkins, Baltimore. 1968.

Low, F. N.: "The Extracellular Portion of the Blood-Air Barrier and its Relation to Tissue Space," *Anat. Rec.* **139:**105–123 (1961).

Matulionis, D. H., and H. F. Parks: "Ultrastructural Morphology of the Normal Nasal Respiratory Epithelium in the Mouse," *Anat. Rec.* **178:**65–83 (1973).

Ryan, S. F.: "The Structure of the Interalveolar System of the Mammalian Lung," *Anat. Rec.* **165:** 467–484 (1965).

Smith, O., and J. Ryan: "Electron Microscopy of Endothelial and Epithelial Components of the Lungs: Correlations of Structure and Function," *Fed. Proc.* **32:**1957–1966 (1973).

Sorokin, S.: "A Morphologic and Cytochemical Study on the Great Alveolar Cell," *J. Histochem. Cytochem.* **14:**884–897 (1966).

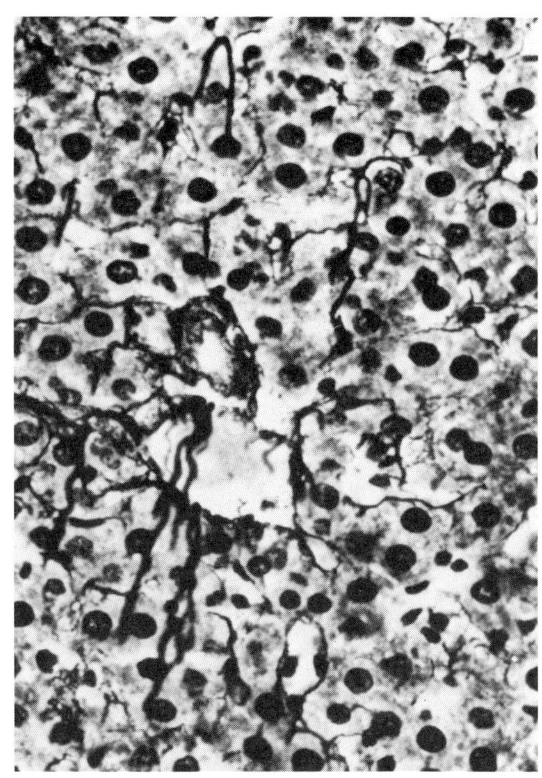

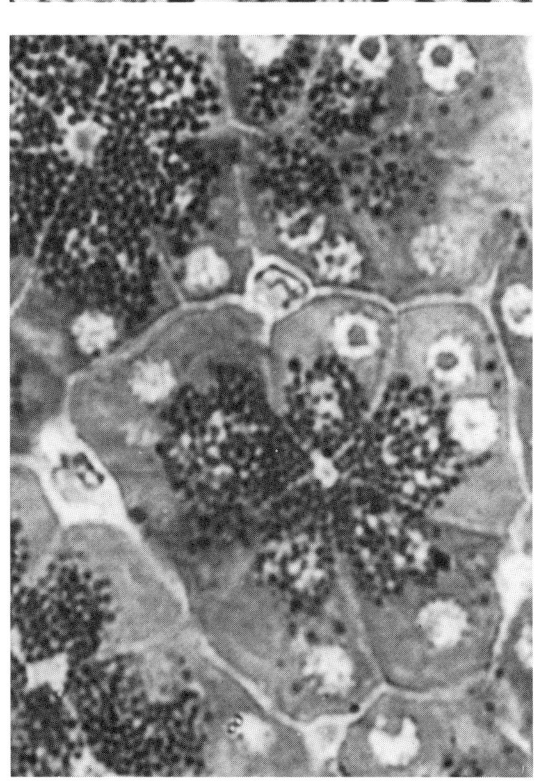

16 DEVELOPMENT OF OROFACIAL REGION

OBJECTIVES

After completing this chapter, the student will be able to:

1 Locate on illustration of an embryo of $5\frac{1}{2}$ weeks the following:
(a) oral opening (b) mandibular arch (c) frontal prominence (d) nasal (olfactory) pits (e) nasomedial and nasolateral processes

2 Indicate the migration of the structures listed in objective 1 during the development of the upper lip, jaws, nose, nasal septum, and nasolacrimal duct.

3 Specify in 4- to 8-week-old embryos the position and function of the following in the development of the palate:
(a) nasomedial processes (b) primary palate (c) secondary palate (d) maxillary processes

4 Identify common deviations from the normal processes of development resulting in anomalies of the face, jaws, and palate.

5 Locate in an illustration of a 4-week-old embryo the following:
(a) lateral lingual swellings (b) tuberculum impar (c) foramen cecum (d) copula

6 Identify the relations of the structures listed

in objective 5 with the various branchial arches.

7 Using their innervations as landmarks, locate the origins of the various portions of the tongue.

Because of the complexity of structures in orofacial regions, a brief account of their development is helpful in understanding the normal anatomy of the mouth and associated structures. During the fourth week of development all the primordial landmarks that are involved in the formation of the face and jaws become distinguishable. Among these primordial landmarks that can be seen from a frontal view are the following:

1. The frontal prominence and the paired olfactory (nasal) placodes cephalic to the oral opening
2. The paired maxillary processes lateral to the oral opening
3. The mandibular arch caudal to the oral opening

These are depicted in Fig. 16-1. During this period, each nasal placode becomes bordered by a quickly proliferating horseshoe-shaped elevation. The mesial and lateral borders of the tissue elevation are called *nasomedial* and *nasolateral processes,* respectively. Each nasal placode thus lies in the upper portion of the nasal cavity, sending nerve processes to the olfactory bulbs of the developing brain.

During the sixth to eighth weeks, the development of the upper face and jaws proceeds rapidly. The maxillary processes become more prominent and grow on a cephalolateral angle toward the midline. Concurrently, the nasomedial and nasolateral processes enlarge and overshadow the lower portions of the *frontal prominence.* The nasomedial processes then fuse along the midline, together forming the medial part of the nose and the philtrum. The growing *maxillary processes* form the remainder of the upper lip (sixth and eighth weeks in Fig. 16-1).

The nasolateral processes during this time become closely positioned to the cephalic borders of the maxillary processes. This forms the *nasooptic furrows* which are overshadowed during the eighth week by tissue from the maxillary processes, leaving

FIGURE 16-1 Drawing showing in frontal aspect some of the important steps in the formation of the face.

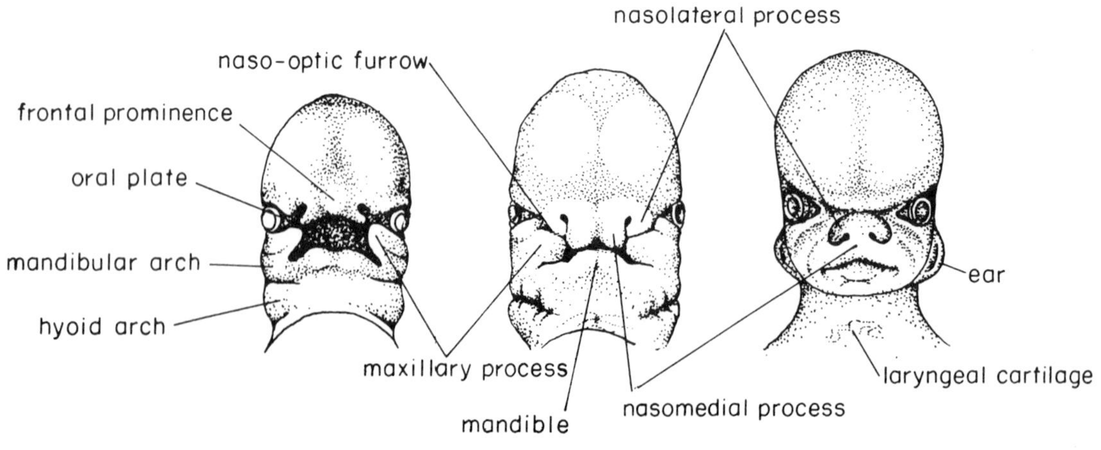

a tract that is destined to be the path of the *nasolacrimal duct*. Concurrently, the nasal pits are deepening, partly by the growth around them and partly by their own extension. They continue to deepen until the second month of development when they break through to the oral cavity to form the posterior nares (nostrils) of the nose.

Originally, an ectodermal plate called the *stomodeal* plate separates the cephalic end of the gut from the outside until the fourth week when it ruptures, leaving the oral opening. This stomodeal plate is positioned at the level of the future tonsillar region of the oral cavity. The deep oral cavity seen in the adult is the result of the forward growth by the developing structures of the face and jaws. The lower jaw forms during the fifth week by the merging of the first branchial arches. The branchial arches (and all other embryonic structures) are rapidly growing areas of mesenchymal tissue, which differentiate into connective tissue, cartilage, and bone. A medial notch in the mandible sometimes occurs with an incomplete merging of the branchial arches. This is readily recognized as a cleft chin in the adult.

THE PALATE

As mentioned earlier, the nasomedial processes fuse with each other and with the maxillary processes to form the upper lip. Deeper to the philtrum, the tissue from the nasomedial processes differentiates into two other structures:

1. The middle portion of the upper jaw and gingiva covering this portion of the jaw.

2. A wedge-shaped mass of mesoderm (cephalic) directly continuous with the nasal septum (Fig. 16-2). This wedge-shaped mass is called the *primary palate* (*medial palatal process*) and is closely applied to the arch of the upper jaw (Fig. 16-3).

The *secondary palate* (*lateral palatal shelves*) arises from the parts of the upper jaw that differentiate from the maxillary processes. These lateral palatal shelves during the seventh and eighth weeks of development grow toward the midline and fuse anteriorly with the primary palate (Figs. 16-2 and 16-3) and medially with each other. Since the process starts at the level of the tip of the tongue, the tongue must move caudally out of the way. The nasal septum then grows downward and posteriorly, fusing with the completed secondary palate. Thus, the nasal chambers are completely separated from the oral cavity. Toward the end of the second month of fetal life, formation of the deeper underlying bony components begins.

THE TONGUE

Development of the mucous covering of the tongue can first be seen in 5-week-old embryos as paired lateral thickenings on the internal face of the mandibular arch appear. These thickenings are known as the *lateral lingual swellings*. Between these swellings is a median elevation called the *tuberculum impar* (Fig. 16-4). Caudal to the tuberculum impar is another swelling, called the *copula*.

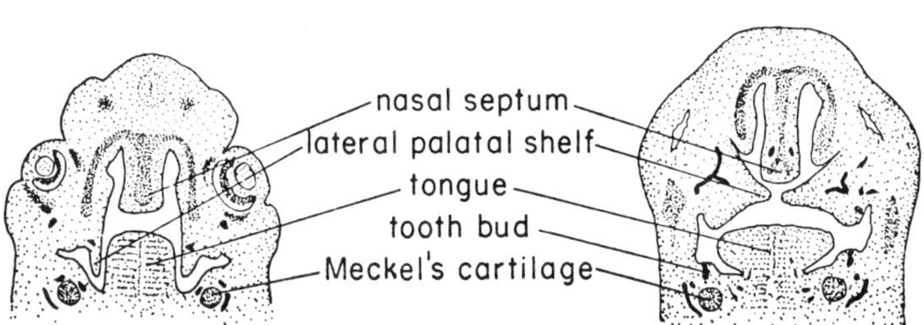

FIGURE 16-2 Diagram of frontal sections through the developing mouth of human embryos to show the relations before and after the retraction of the tongue from between the two lateral palatal shelves.

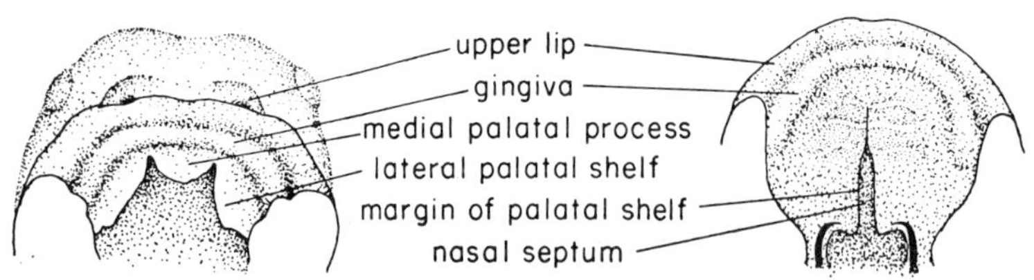

FIGURE 16-3 *En face* drawings illustrating the formation of the secondary palate.

Between the tuberculum impar, which becomes the body of the tongue, and the copula is the *foramen cecum* which is at the apex of the V-shaped *sulcus terminalis.* The sulcus terminalis separates the anterior two-thirds of the tongue from its posterior one-third and is the location where large circumvallate papillae are found. The foramen cecum thus represents a vestigial point from which the primordium of the thyroid gland has moved caudad.

The pattern of innervation of the adult tongue illustrates the developmental history of the lingual mucosa and muscular mass. Thus, the mandibular branch of the Vth nerve (tactile) and the chorda tympani branch of the VIIth nerve (gustatory) innervate the anterior two-thirds of the tongue, whereas the IXth nerve innervates the posterior third of the lingual mucosa exclusively (Figs. 16-5 and 16-6). The musculature of the tongue develops from a region further caudad than the arches mentioned above. As the developing musculature pushes forward, filling the expanding primordial mass of the tongue, it is accompanied by the XIIth nerve, which is the motor nerve to the tongue muscle in the adult.

FIGURE 16-4 Diagram depicting three stages in the development of the tongue.

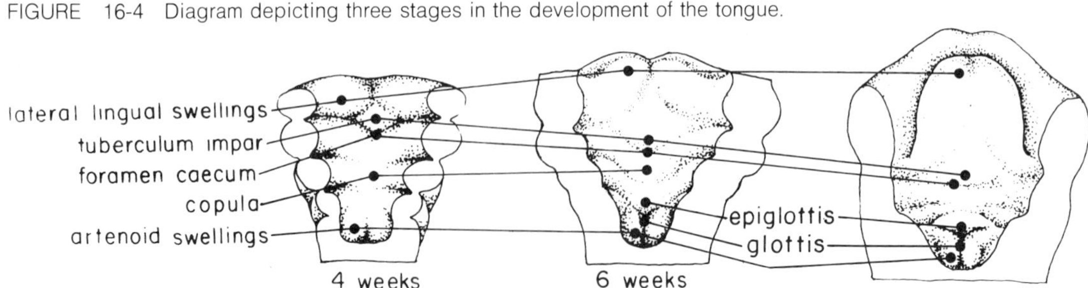

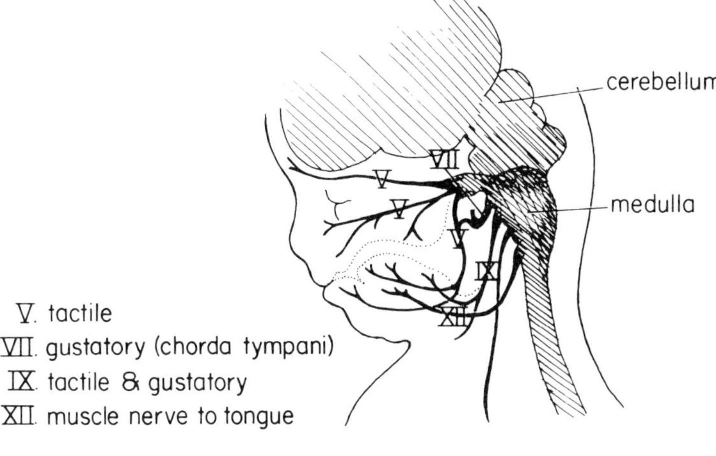

FIGURE 16-5 The arrangement of cranial nerves V, VII, IX, and XII in embryos of about 11 weeks.

REVIEW SECTION

1. Because of the path of the nerve of the tongue musculature (lingual muscles), it is assumed that the tongue musculature arises from that part of the wall of the pharynx that is in close apposition with the:
(a) mandibular nerve (b) VIIth nerve (c) IXth nerve (d) XIIth nerve

The XIIth nerve in an embryo of $5\frac{1}{2}$ weeks is positioned at the level of the caudal border of the fourth branchial arch. It is assumed that the tongue musculature arises at this level and then migrates cephalically to its adult position, carrying a branch of the XIIth nerve with it.

2. The upper lip and jaw are derived from which of the following structures present in an embryo of 6 weeks:
(a) nasolateral processes (b) nasomedial processes (c) maxillary processes (c) lateral palatal shelves

The upper lip and jaw are formed by the fusion of nasomedial processes with each other and with the medially growing maxillary processes.

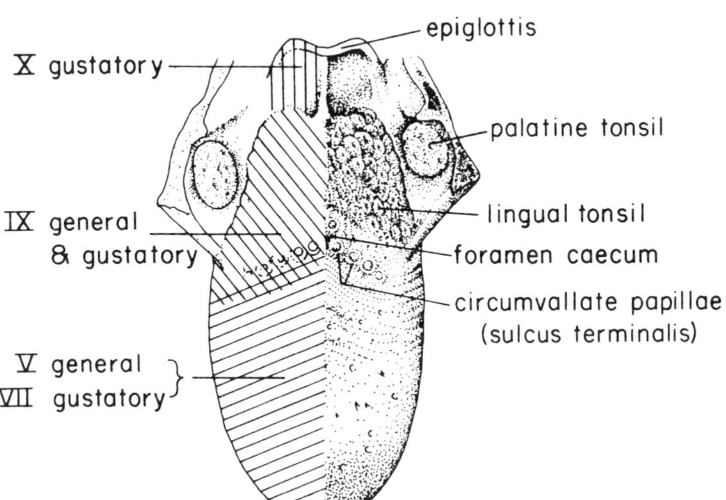

FIGURE 16-6 Diagram illustrating regions of the lingual mucosa innervated by cranial nerves V, VII, IX, and X, respectively.

3. A cleft chin results from the incomplete merging of:
(a) second branchial arches (b) lateral lingual swellings (c) first branchial arches (d) all of the above

The mandible results from the merging of the bilaterally symmetrical first branchial arches. A cleft chin results when this merger is incomplete.

4. The primary palate is a wedge-shaped mass of tissue that is:
(a) located on the floor of the oral cavity (b) closely applied to the medial portion of the upper jaw (c) a derivative of the nasomedial processes after their fusion (d) destined to be reached on its two remaining free lateral edges by the lateral palatal shelves

The primary palate results from differentiation of a wedge-shaped mass on the deeper aspect of the fused nasomedial processes. It is in close apposition to the upper jaw since the middle portion of the upper jaw is also derived from the fused nasomedial processes. It is destined to be reached by and fused with the two lateral palatal shelves to form the anterior aspect of the completed palate.

5. The second branchial arches which become part of the mucosa of the root of the tongue are innervated by:
(a) Vth nerve (b) IXth nerve (c) XIIth nerve (d) VIIth nerve

The root of the tongue is formed from the copula, a derivative of junction between the second and third branchial arches. Since the IXth nerve supplies this area, the posterior third of the lingual mucosa in the adult is innervated by a branch of the IXth nerve.

6. The foramen cecum is a vestigial region that one of the following structures has developed from:
(a) parotid gland (b) sublingual gland (c) thyroid gland (c) uvula

The primordium of the thyroid gland originates and descends from the region of foramen cecum to its adult location.

BIBLIOGRAPHY

Arey, L. B.: *Developmental Anatomy,* 7th ed., Saunders, Philadelphia, 1965.

Bhaskar, S. N. (ed.): *Orban's Oral Histology and Embryology,* 8th ed., Mosby, St. Louis, 1976.

Hamilton, W. J., and H. W. Mossman: *Human Embryology,* 4th ed., Williams & Wilkins, Baltimore, 1972.

Patten, B. M.: *Human Embryology,* 2d ed., McGraw-Hill, New York, 1953.

Pourtois, M.: "Morphogenesis of the Primary and Secondary Palate," in H. C. Slavkin and L. A. Bavetta (eds.), *Developmental Aspects of Oral Biology,* Academic, New York, 1972, pp. 81–108.

ADDITIONAL READINGS

Barry, A.: "Development of the Branchial Region of Human Embryos with Special Reference to the Fate of Epithelia," in S. Pruzansky (ed.), *Congenital Anomalies of the Face and Associated Structures,* Charles C Thomas, Springfield, Ill, 1961, pp. 46–62.

Holmstedt, J. O. V., and J. N. Bagwell: "Morphogenesis of the Secondary Palate in the Mongolian Gerbil (*Meriones unguiculatus*)," Acta. Anat. **97:** 443–449 (1977).

Kraus, B. S., H. Kitamura, and R. A. Latham: *Atlas of Developmental Anatomy of the Face: With Special Reference to Normal and Cleft Palate,* Hoeber-Harper, New York, 1966.

Walker, B. E., and F. C. Fraser: "Closure of the Secondary Palate in Three Strains of Mice," *J. Embryol. Exp. Morphol.* **4:**176–189 (1956).

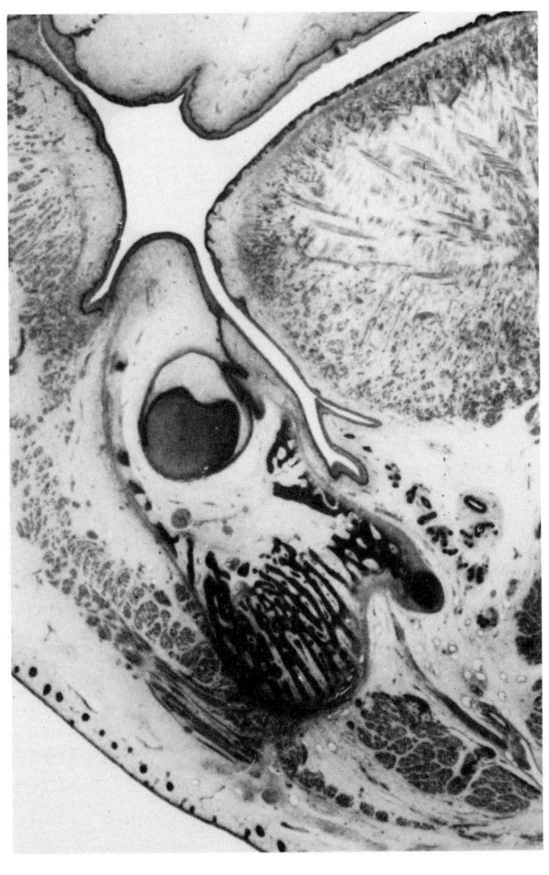

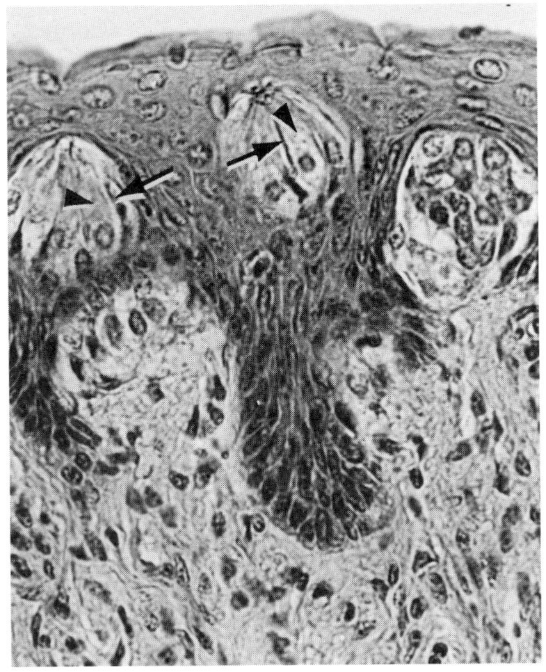

17

ORAL CAVITY I: ORAL MUCOSA, SALIVARY GLANDS, AND TONSILS

OBJECTIVES

Upon completion of this chapter, the student will be able to:

1 Describe the three general types of oral mucosa by identifying:
(a) type of epithelium (b) extent of keratinization (c) presence or absence of submucosa

2 Identify the characteristics of masticatory mucosa as compared to the general oral mucosa.

3 Identify the function of traction bands.

4 Identify two features of the vermilion border that cause its red color.

5 Name the four zones of the palate and identify the histologic characteristics of each.

6 Identify the location and structure of the palatal rugae.

7 Identify and describe the musculature and three types of papillae of the tongue.

8 Identify the location and function of each of the following taste bud elements:
(a) neuroepithelial cell (b) sustentacular (supporting) cell (c) taste hair and taste pore

9 Identify the locations, classifications, and secretions of the major and minor salivary glands.

10 Define the following terms:
(a) serous (b) mucous (c) demilune (serous) (d) myoepithelial cells (basket cells) (e) acinar cell

11 Indicate the structures and give histologic descriptions for the major salivary glands including:
(a) main excretory ducts (b) striated ducts (c) intercalated ducts (d) acini

12 Describe the basic secretory cycle and circadian effect on the flow of saliva.

13 Define immunoglobulin A (IgA) and identify the site of its production.

14 Identify the three types of tonsils:
(a) palatine (b) pharyngeal (c) lingual

15 Compare these types of tonsils with respect to the following:
(a) epithelium (b) glands (c) crypts (d) septa (e) size (f) location

16 Define the following terms:
(a) tonsillar ring (Waldeyer's) (b) salivary corpuscle (c) immunoglobulin A

The oral mucosa consists of oral epithelium and lamina propria. Oral epithelium is entirely stratified squamous in type. A submucosa and a muscle layer occur occasionally. Areas lacking the submucosa are the gingiva, the *median raphe* (located in the midline of the palate), and the dorsal surface of the tongue.

ORAL MUCOSA

Oral mucosa may be grouped into three types. The *lining mucosa* comprises the lips, the cheeks, the floor of the mouth, and the ventral (or under-) surface of the tongue. The lining mucosa is nonkeratinized because it is protected from masticatory stress. The *masticatory mucosa* is keratinized and present in the gingiva and hard palate. It is so called because of the presence of a keratinized layer, believed to be the evolutionary result of abrasion and stress in mastication. The dorsal surface of the tongue is often referred to as *specialized mucosa,* since it is made up of numerous papillae and such specializations as taste buds.

LINING MUCOSA

Lining mucosa is composed of a moist stratified squamous epithelium and an underlying lamina propria of loose fibrous elastic connective tissue (FECT). Portions of the muscles of facial expression are often found in association with lining mucosa. Connective tissue fiber bundles, called *traction bands,* extend from the various muscles through the submucosa to the lamina propria, connecting the muscle with the overlying epithelial tissue. This aids in coordinating the movements of muscles and the mucosa. Numerous small seromucous glands are present within the submucosa throughout the general lining mucosa and will be discussed in the appropriate sections.

The lip

Figure 17-1 shows a parasagittal view of the lip under low magnification. The bulk of the tissue is occupied by cross-sectioned masses of skeletal muscle representing the *orbicularis oris* muscle.

The skin side shows a number of hair shafts, whereas the mucosal side contains a

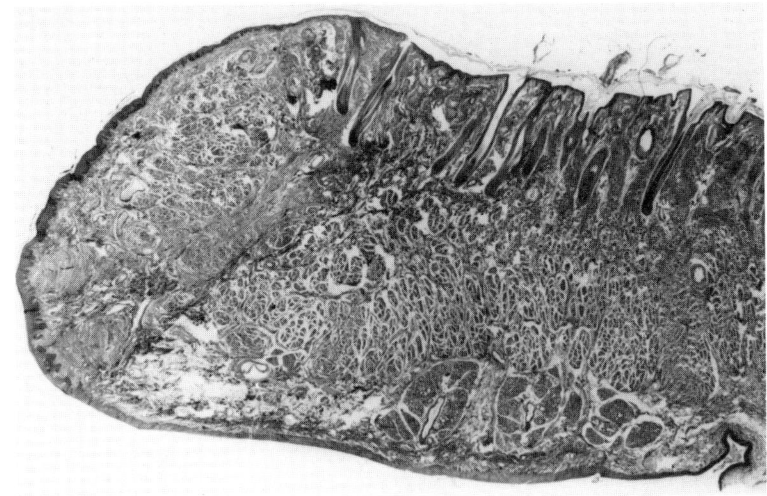

FIGURE 17-1 Light micrograph of a parasagittal section of the lip. The center of the lip is occupied by cross-sectioned skeletal muscle bundles (orbicularis oris muscle). The vermilion border is the transitional zone between the skin and the lining mucosa.

lightly stained band of tissue, which is due to the presence of labial glands. A transitional zone, which corresponds to the pink intermediate region between skin and mucosa, is present. The histologic details of this transitional zone, the *vermilion border,* are better presented in Fig. 17-2. Note the relatively thick epithelium in the vermilion border (red area) which has tall connective tissue papillae and the lack of glandular tissue in the underlying tissue. Although the vermilion border is slightly keratinized, high connective tissue papillae with blood vessels in this region give the unique red color to it.

The minor salivary glands are located just under the oral epithelium in the lip. Although there is no distinct border between the mucosa and submucosa, the deeper region

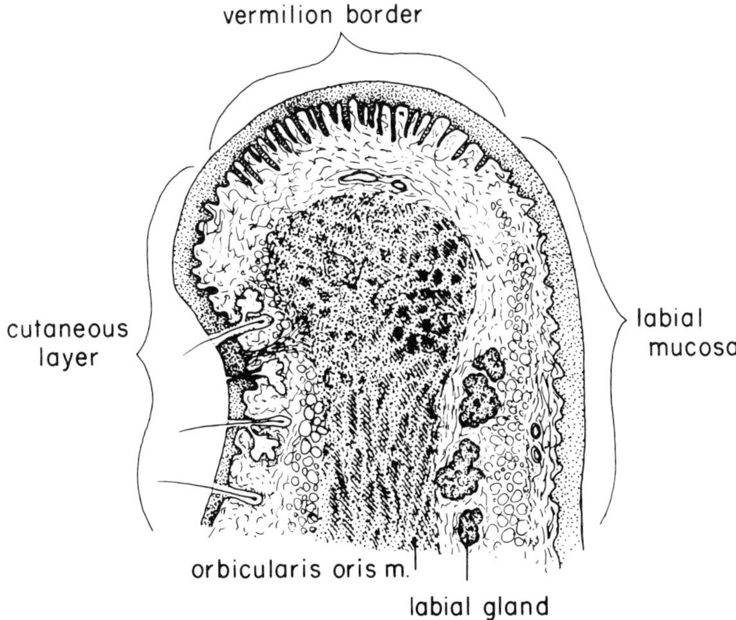

FIGURE 17-2 Drawing depicting the histologic details of the lip.

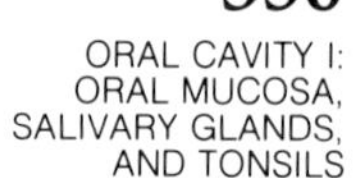

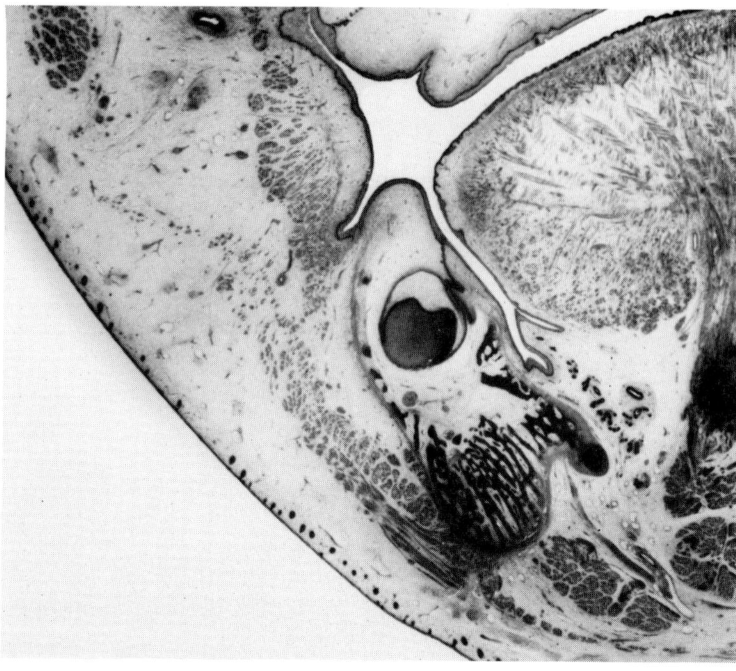

FIGURE 17-3 Frontal section through the oral cavity of a fetus. (b) Buccinator muscle. (f) Floor of the mouth.

where labial glands are located is analogous to the submucosa found elsewhere.

The cheek

The mucous membrane of the cheek is similar to that of the lip with the exception of the fatty areas. A low-power view of a frontal section of the oral cavity from a fetus is shown in Fig. 17-3. Note the various histologic landmarks as different structures are described in subsequent pages. The region through the cheek reveals portions of the buccinator muscle (b), in which all fibers have the same orientation. In the adult, numerous buccal glands are mixed with the fascicles of the buccinator muscle. The connective tissue fibers which originate from the buccinator muscle and insert in the lamina propria are called *traction bands* because they help to hold the basal fat-containing submucosa in place.

The ventral surface of tongue and floor of mouth

The ventral surface of the tongue and the floor of the mouth are also covered by the

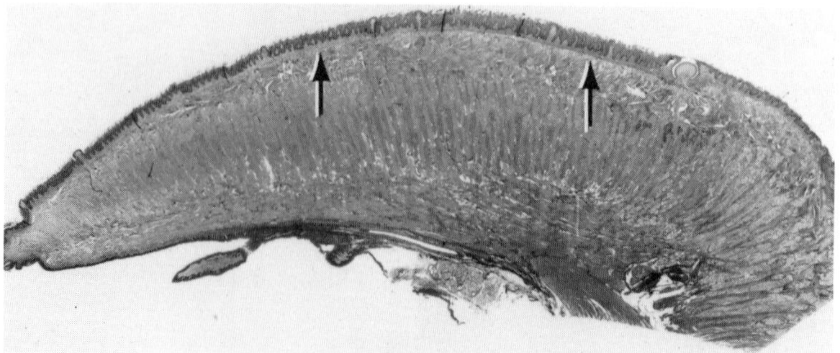

FIGURE 17-4 Longitudinal section through the tongue. A submucosa is lacking on the dorsum of the tongue (arrows).

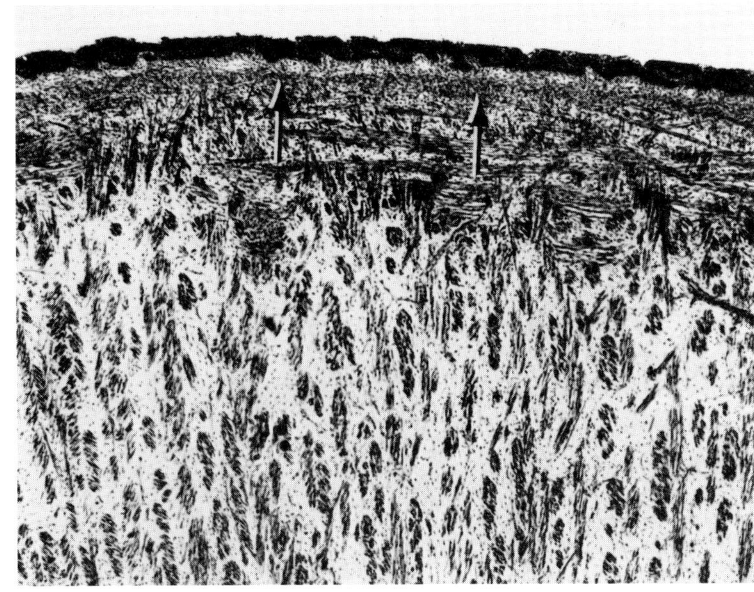

FIGURE 17-5 Light micrograph of the mucosa on the ventral surface of the tongue viewed at higher magnification. The submucosa is very thin (arrows).

ordinary lining mucosa and contain salivary glands. The floor of the mouth as shown in Fig. 17-3 (f) has a loose submucosa which contains the sublingual gland and the ducts of the sublingual and submandibular glands. In contrast to the relatively thick submucosa in the floor of the mouth, the ventral surface of the tongue has an extremely thin submucosa (Figs. 17-4 and 17-5, arrows).

MASTICATORY MUCOSA

As mentioned earlier, the masticatory mucosa is present in the hard palate and gingiva. It is so called because it is subject to abrasive forces related to mastication. It is usually keratinized to a minimum degree. Because of the thin keratinized layer, the masticatory mucosa has no stratum lucidum.

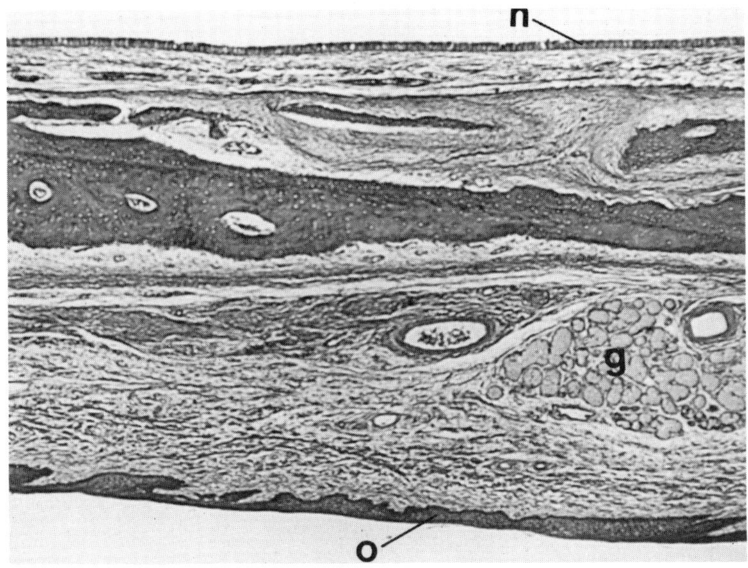

FIGURE 17-6 Section through the hard palate. It is covered by respiratory epithelium on the nasal side (n), while minimally keratinized epithelium lines the oral side (o). The submucosa contains mucous palatal glands (g).

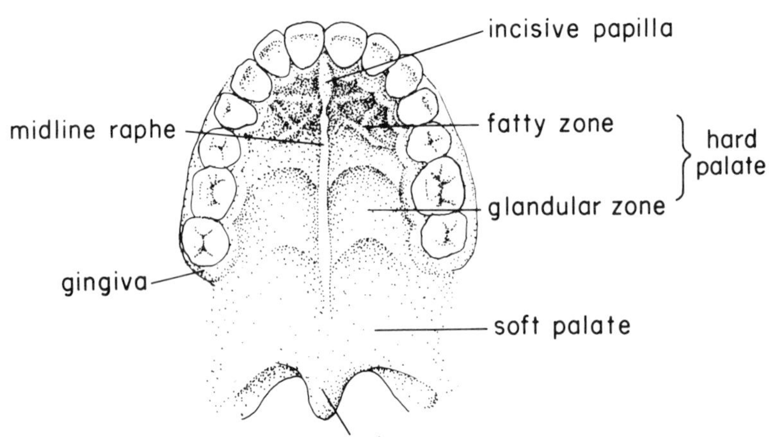

FIGURE 17-7 Diagram depicting the various zones of the hard palate.

Palate

The Hard Palate

As mentioned in the preceding chapter, the hard palate is formed when the two lateral palatine processes are fused during the seventh or eighth week of fetal life. The palate (Fig. 17-6) is covered by a respiratory epithelium on the nasal side (n), while the minimally keratinized oral epithelium covers the oral side (o).

The mucosa of the hard palate is commonly divided into four zones (Fig. 17-7). The *gingival zone* of the hard palate is composed of mucosa which is tightly attached to bone at the palatal surface of the teeth. The *raphe zone* or *median palatal raphe* is the midline area of a thin connective tissue septum which has resulted from the fusion of the palatal shelves. No glandular elements or submucosa are present in the raphe zone. The *fatty zone* of the palate contains adipose tissue which is retained in position by the connective tissue traction bands that attach the mucosa to the palatine bone. The *glandular zone* (g in Figs. 17-8 and 17-9) consists of numerous mucous glands and occasional traction bands (t in Fig. 17-8). Their terminal

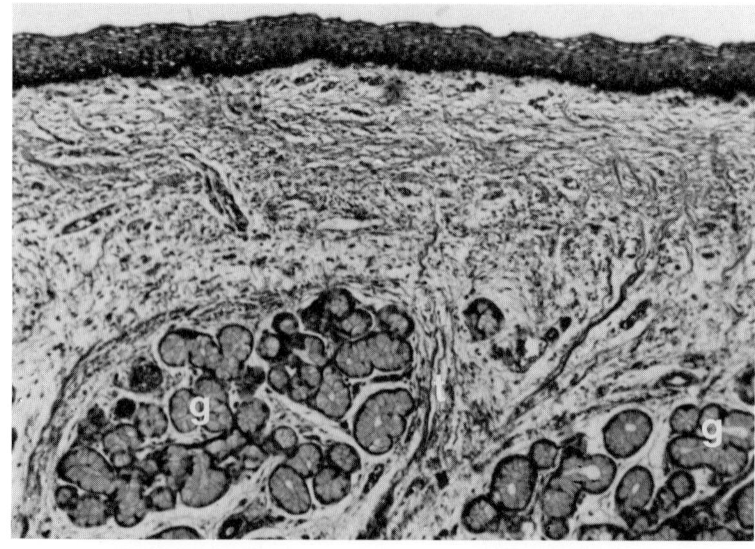

FIGURE 17-8 Light micrograph of palatal glands. They are located in the submucosa and are pure mucous glands (g). Their ducts open at multiple locations into the oral cavity.

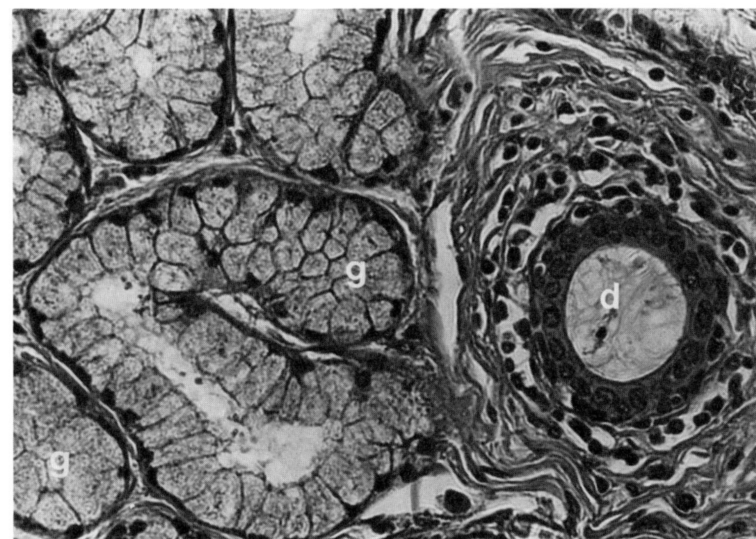

FIGURE 17-9 Terminal portion (g) of palatal glands viewed at higher magnification. They are tubular in shape and are comprised of secretory cells and myoepithelial cells. A duct (d) lined by a simple cuboidal epithelium is seen in the right portion of the micrograph.

portions are tubular in shape and the ducts (d) open into the oral cavity in multiple locations.

Palatal rugae are transverse ridges located in the anterior portion of the hard palate. Rugae provide a friction surface that plays a role in mastication. The *incisive papilla* is a raised midline area in the anterior portion of the hard palate. The incisive papilla is the site of emergence of vessels and nerves that supply the anterior palate.

The Soft Palate and Uvula

Like the hard palate, the soft palate is lined by minimum to nonkeratinized stratified squamous epithelium on its oral surface and respiratory epithelium on its nasal surface. The soft palate and the uvula, which comprise a muscular extension at the posterior end of the soft palate, are flexible because they have no bony structural reinforcement. The submucosa contains mucous glands (Figs. 17-8 and 17-9) on the oral side. A number of small skeletal muscles, such as the palatopharyngeus and palatoglossus, extend from the lateral aspects to insert in the soft palate and uvula (Fig. 17-10). These muscles constitute most of the soft palate and contribute to its movement. The soft palate closes off the nasal cavity during swallowing, sneezing, and vocalization by moving superiorly from its relaxed position.

Figure 17-10 is a photomicrograph of a parasagittal section of the palate and lip from a fetus at term. Note the appearance of the

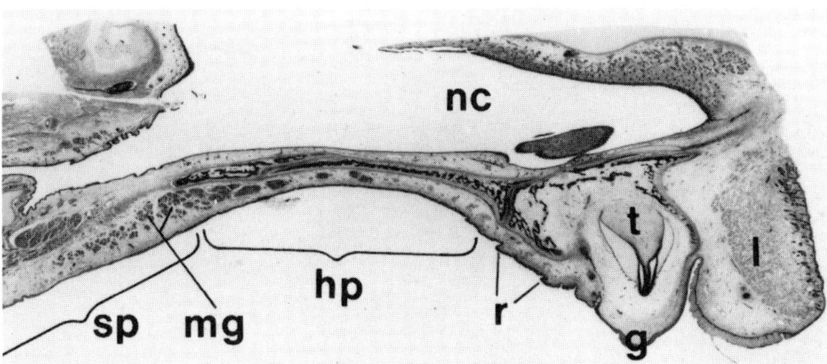

FIGURE 17-10 Parasagittal section through palate and lip of a fetus at term. (nc) Nasal cavity. (l) Lip. (t) Incisor tooth. (g) Gingiva. (r) Rugae. (hp) Hard palate. (sp) Soft palate. (mg) Mucous glands.

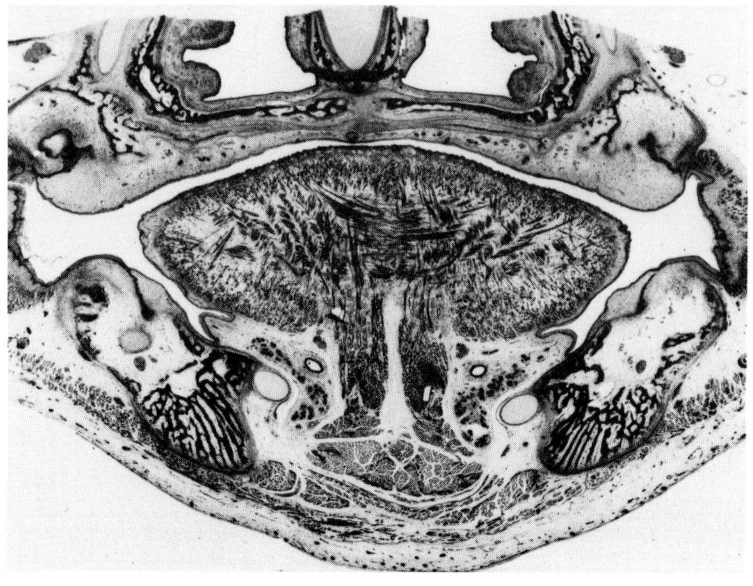

FIGURE 17-11 Micrograph of the lower half of the face of a fetus. Intramembranous bone formation is progressing in both the maxilla and mandible, and various stages of tooth development are seen. The flattened tongue occupies the oral cavity proper. Both intrinsic and extrinsic tongue muscles sectioned at various angles are apparent.

palatal mucosa from left to right as the photograph depicts the soft palate (sp), hard palate (hp), future gingiva (g), and lip (l). Mucous glands (mg) occupy the posterior third of the hard palate and approximately the anterior half of the soft palate. A couple of palatine rugae (r) are also visible. Under the presumptive gingiva is a developing incisor tooth (t) that is surrounded by the growing maxilla. In the lip the cutaneous surface, vermilion border, and ordinary labial mucosa can already be distinguished.

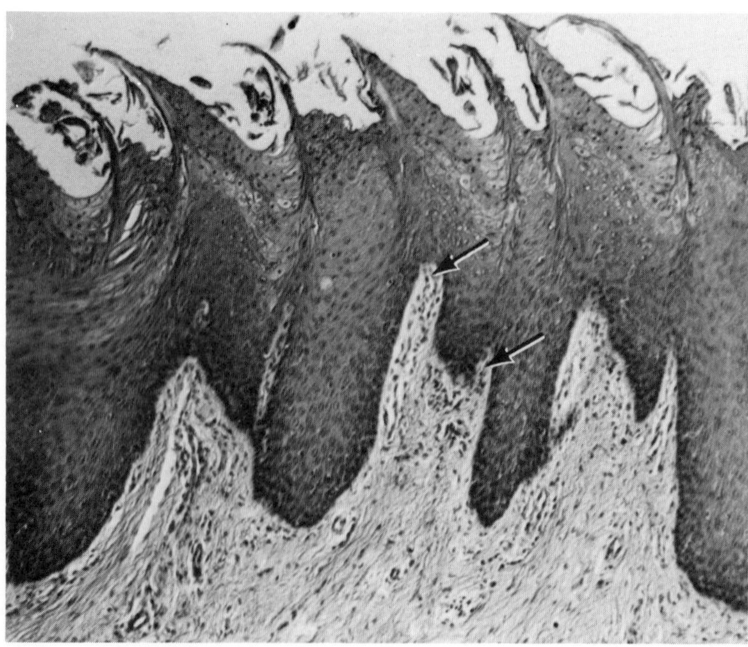

FIGURE 17-12 Light micrograph of the mucosa on the dorsum of the anterior two-thirds of the tongue. It has numerous filiform papillae with occasional fungiform papillae among them. The arrows indicate secondary connective tissue papillae within the filiform papillae.

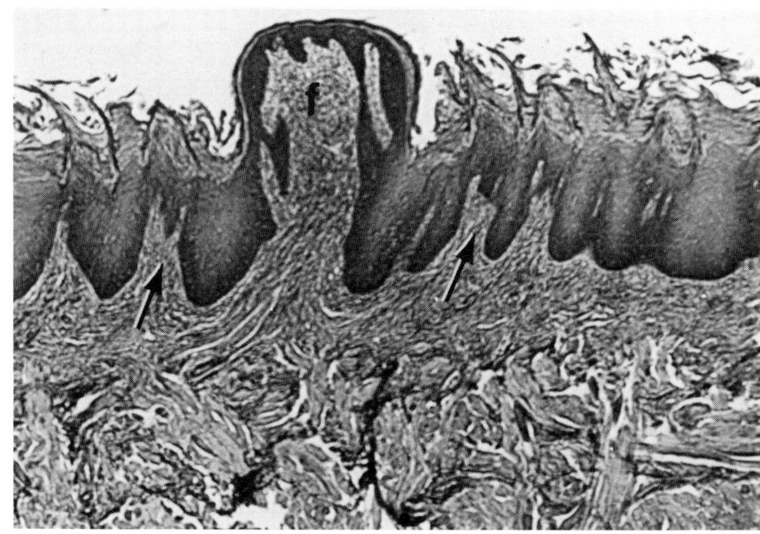

FIGURE 17-13 Filiform papillae viewed at higher magnification. They are pointed and are covered by a keratinized stratified squamous epithelium. Filiform papillae begin as protrusions of connective tissue (primary papillae) that may have several secondary connective tissue papillae within it (arrows).

SPECIALIZED MUCOSA

The dorsal surface and the sides of the tongue are covered by specialized mucosa in which different lingual papillae are situated. The functioning of these lingual papillae ranges from capturing of food in lower forms of life to the general gustatory sensation as discussed later. The substance of the tongue is formed by several different skeletal muscles that are necessary for phonation, mastication, and deglutition. These muscles can be grouped into two categories. One category represents *intrinsic muscles* of the tongue originating from one region of the connective tissue and inserting in another region. The other category is composed of several *extrinsic muscles* such as hyoglossus, styloglossus, palatoglossus, and genioglossus. Figure 17-11 is a low-power light micrograph of the frontal section of the tongue in which several intrinsic and extrinsic muscles may be seen. The mucosa of the dorsal lingual surface is thick because of the various papillae that are present.

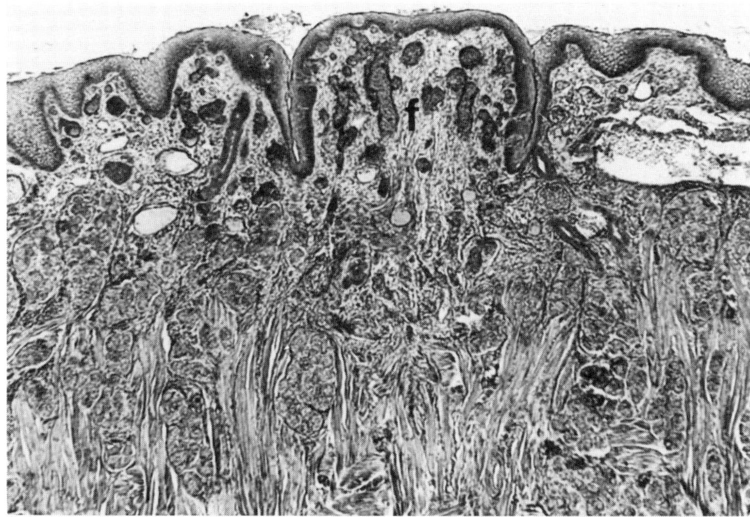

FIGURE 17-14 Fungiform papilla. It is covered by a nonkeratinized stratified squamous epithelium that has occasional taste buds embedded in it.

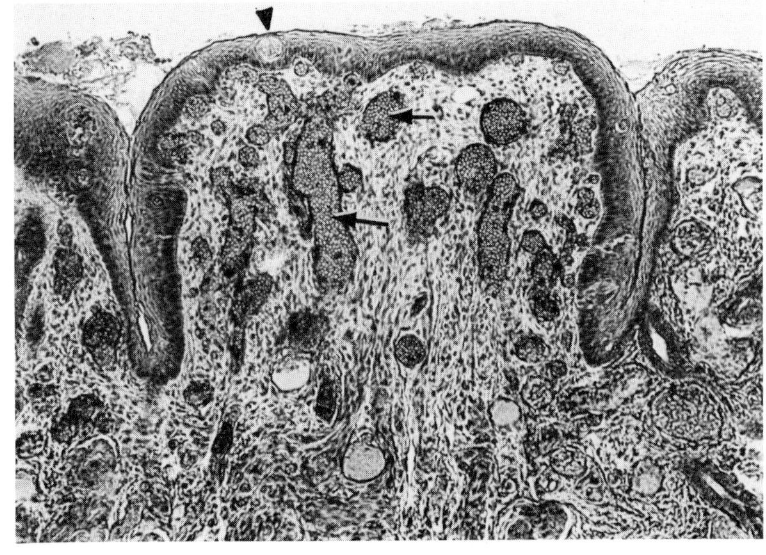

FIGURE 17-15 Fungiform papilla viewed at higher magnification. It is mushroom shaped and is covered by a nonkeratinized stratified squamous epithelium that contains occasional taste buds (arrowhead). The connective tissue core of fungiform papillae is highly vascularized (arrows). This gives the red appearance to these papillae in the living being.

Three types of papillae are seen. *Filiform papillae* are pointed and keratinized to provide a friction surface which aids in the tongue's role in mastication. They are the smallest, most numerous of the papillae and are found only in the anterior two-thirds of the tongue (Figs. 17-12 and 17-13). The filiform papillae begin as primary protrusions of the connective tissue, but may have several tall secondary connective tissue papillae within them (Figs. 17-12 and 17-13, arrows). These papillae may reach 3 mm in height.

Fungiform papillae are large and bulbous in shape and are covered by keratinized stratified squamous epithelium (f in Figs. 17-14 and 17-15), but the keratinized layer is thick. The red appearance of the fungiform papillae is due to a core of well-vascularized connective tissue which is readily visible at a higher magnification (Fig. 17-15, arrows). Fungiform papillae are sporadically present over the surface of the tongue. They appear mushroom-shaped and have occasional taste buds embedded within their epithelium.

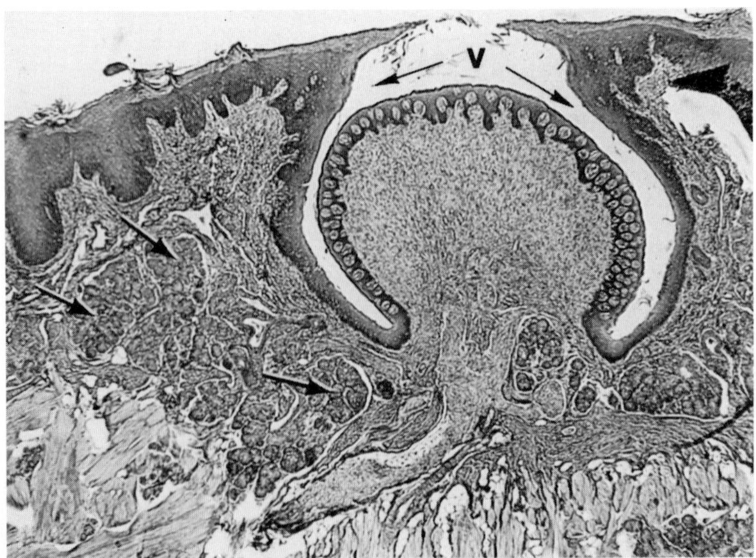

FIGURE 17-16 Circumvallate papilla. It is larger than a fungiform papilla and is surrounded by a trench or valley (v). The papilla is covered by a keratinized stratified squamous epithelium. In contrast to the thick keratinized layer on filiform papillae, it is rather thin on the vallate papilla. Taste buds are found in the epithelium, and pure serous glands, the von Ebner's glands (arrows), open into the valley. They aid in clearing the valley of tasted material.

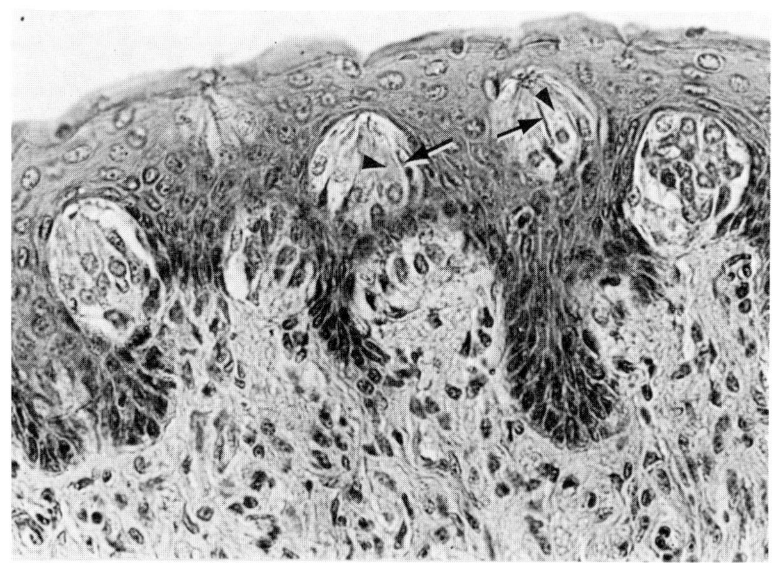

FIGURE 17-17 Taste buds are located within the epithelium of fungiform and vallate papillae. They are barrel-shaped and contain two types of cells: neuroepithelial cells (arrows) and sustentacular cells (arrowheads).

They are not as tall as filiform papillae, but their diameter is about 1 mm.

Circumvallate papillae (or the vallate papillae) are the largest as well as the least numerous of the three types of lingual papillae. On the average their diameter is approximately 3 mm. Along the *sulcus terminalis*, circumvallate papillae are arranged in a row, marking the boundary between the posterior third and anterior two-thirds of the tongue, as was shown in Fig. 16-6. Each circumvallate papilla is mound-shaped and sits in a depression, thereby producing a narrow valley around it (v in Fig. 17-16). At the base of this valley are openings of pure serous glands called *von Ebner's glands*. Von Ebner's glands (Fig. 17-16, arrows) function in clearing the valley of tasted materials. The posterior third, or root, of the tongue has no papillae. Instead, a series of elevations containing lingual tonsils are present. Numerous mucous glands are associated with the lingual tonsils.

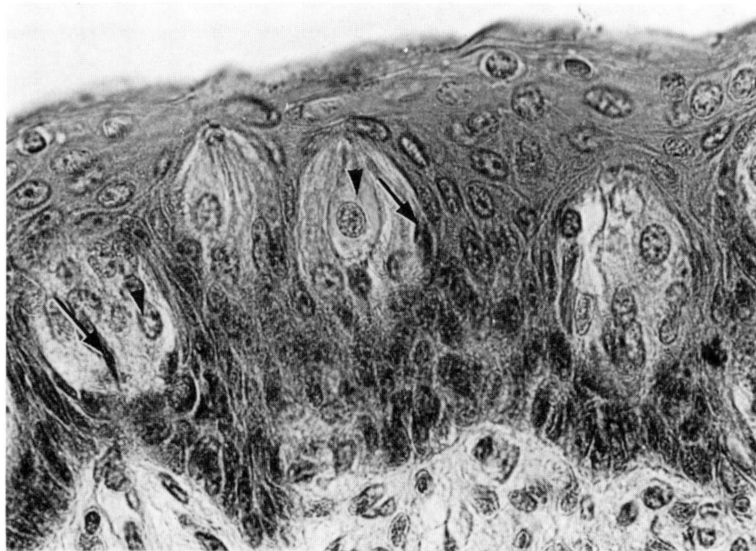

FIGURE 17-18 Taste buds viewed at higher magnification. Neuroepithelial cells (arrows) and sustentacular cells (arrowheads) are easily identified. The taste buds open into the oral cavity via taste pores. Cytoplasmic processes of the neuroepithelial cells, taste hairs extend into the pores.

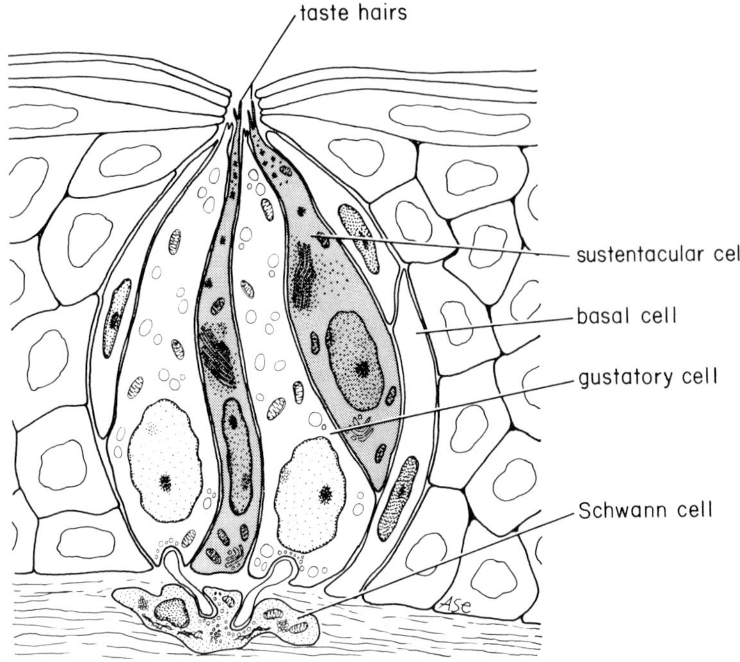

FIGURE 17-19 Schematic drawing of a taste bud. Each neuroepithelial cell has a single, stiff hairlike process that terminates in the taste canal. The nucleus of neuroepithelial cells is elongated with a clumped hyperchromatic chromatin pattern. The neuroepithelial cells rest upon a basal lamina. The sustentacular cells have a vesicular nucleus with a dispersed chromatin pattern. Gustatory nerve fibers penetrate the basal lamina and terminate between the neuroepithelial and sustentacular cells.

Taste buds

Taste buds are specialized receptor organs for gustatory sensation. They are barrel-shaped structures located within the lingual epithelium, principally along the sides of the circumvallate papillae (Figs. 17-17 and 17-18) and are made up of two types of cells. *Neuroepithelial cells* (arrows) are thin, dark-staining columnar epithelial cells specialized to serve as sensory receptors. *Sustentacular cells* (arrowheads) are larger and lighter-staining cells which supply and nourish the neuroepithelial cells.

Taste buds communicate with the oral cavity via small openings called taste pores (Figs. 17-18 and 17-19). Projections from the

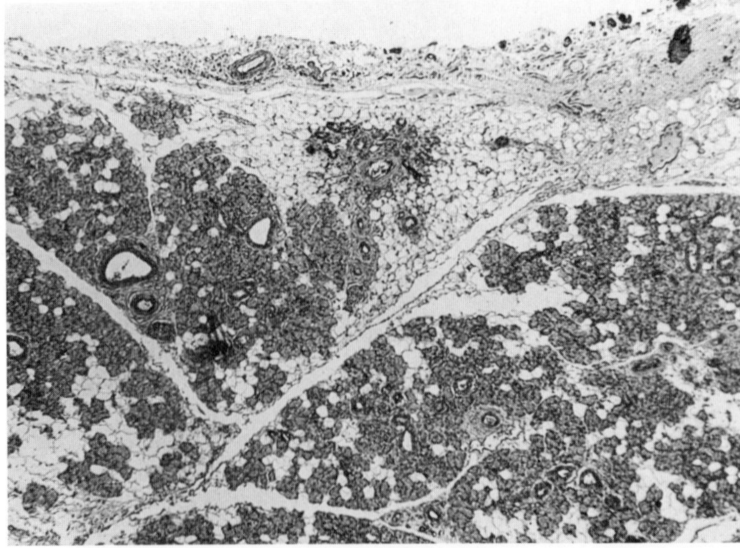

FIGURE 17-20 Light micrograph of the parotid gland. The parenchyma is divided into lobes and lobules by connective tissue septa. Larger ducts can be recognized within the parenchyma. A number of adipose cells are present.

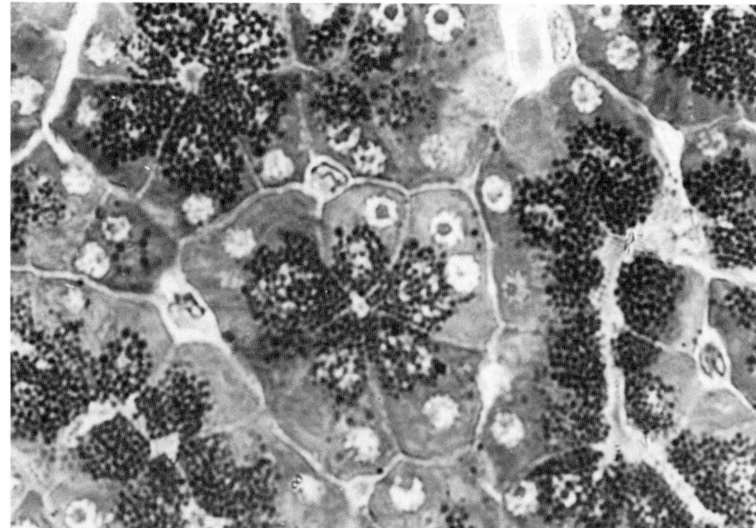

FIGURE 17-21 Light micrograph of a portion of the parotid gland. Several acini can be seen. They are lined by pyramidal cells. Their nuclei are vesicular with a diffuse chromatin pattern, and contain prominent nucleoli. The basal cytoplasm appears dark due to the basophilia of the RER. In some cells the Golgi apparatus appears as a light, irregular area between the nucleus and the numerous secretory granules that are present in the apical cytoplasm.

neuroepithelial cell extend into the taste pores. These projections are the *taste hairs*. A gustatory sensation is elicited when soluble materials enter taste pores and come into contact with taste hairs. Cranial nerves VII and IX convey impulses from the neuroepithelial cells to the brain. It is known that individual tastes such as sweetness and bitterness are regionalized by these nerves.

SALIVARY GLANDS

The salivary glands include a number of structures which secrete saliva. The smaller of these glands are situated in the oral mucous membrane and function mainly to lubricate and moisten the oral cavity. The larger of these glands (major or extrinsic) are situated outside the oral cavity, and their products are

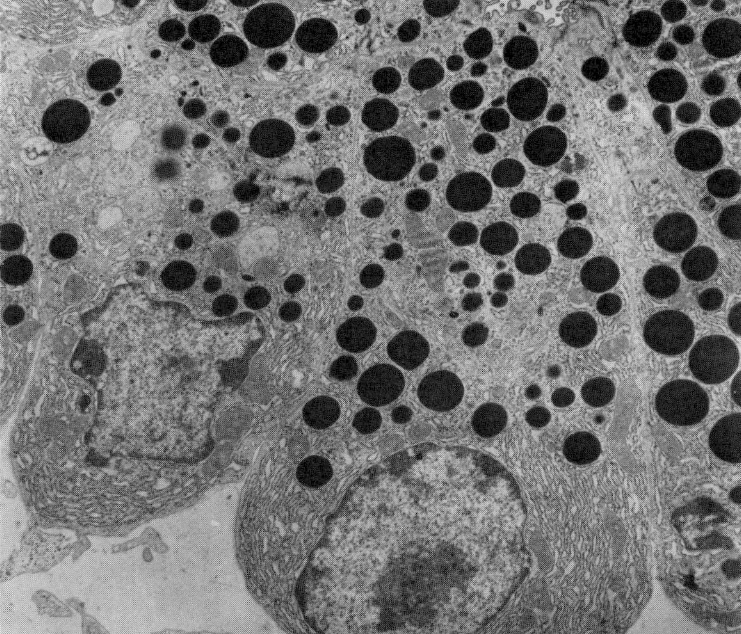

FIGURE 17-22 Electron micrograph of serous acinar cells. Note the abundant RER, the prominent nucleoli, and the varying morphology of the secretory granules.

transported through excretory ducts. Their products contain enzymes that digest food. The saliva, as observed in the oral cavity, includes all of the salivary secretions plus tissue exudates with cellular elements and bacteria. Therefore, it would be more appropriate to call this entity an *oral fluid.* Table 17-1 is a summary of location, size, and secretory products from these glands.

MAJOR SALIVARY GLANDS

The major salivary glands are the parotid, submandibular, and sublingual glands. All are paired glands that are encapsulated by connective tissue and are divided into lobes and lobules by septa composed of predominantly collagenous fibers with fine reticular fibers intermixed. These connective tissue septa contain the usual cell components including macrophages, fibroblasts, plasma cells, and adipose cells.

Parotid gland

The parotid in the adult is located lateral to the ramus of the mandible, around and inferior to the external auditory meatus, and is a compound, serous, tubuloalveolar gland. The serous alveoli (acini) are lined with pyramidal epithelial cells which rest upon a basement membrane and surround a narrow lumen. A number of adipose cells are usually present (Fig. 17-20). The details of individual alveoli are seen in Fig. 17-21, in which several acini are seen as sectioned at 1 μm and magnified at a higher level. The nuclei are vesicular with prominent nucleoli, and the dark cytoplasm at the base corresponds to the basal basophilia. Secretory granules are present in the apical portion of the cytoplasm. The Golgi apparatus is seen as a light, irregular area between the nucleus and the apical granules in some cells.

Ultrastructurally, serous acinar cells have the following features (Fig. 17-22): (1) abundant rough-surfaced endoplasmic reticulum (RER) at the base; (2) Golgi complexes of variable size, location, and orientation made

CLASSIFICATION OF SALIVARY GLANDS — TABLE **17-1**

LOCATION	NAME	TYPE OF SECRETION	SIZE
Lip	Superior labial	Mixed (predominantly mucous)	Minor
	Inferior labial	Mixed (predominantly mucous)	Minor
Cheek	Buccal glands	Mixed (predominantly mucous)	Minor
	Parotid	Pure serous	Major
Palate			
Hard	Posterolateral	Pure mucous	Minor
Soft	Palatine	Pure mucous	Minor
Tongue			
Corpus	Blandin-Nuhn (anterior lingual)	Mixed (predominantly mucous)	Minor
Root	von Ebner	Pure serous	Minor
Sublingual sulcus (floor of mouth)			
	Submandibular	Mixed (predominantly serous)	Major
	Sublingual (extrinsic)	Mixed (predominantly mucous)	Major
	Sublingual (intrinsic)	Mixed (predominantly mucous)	Minor
	Glossopalatine	Pure mucous	Minor

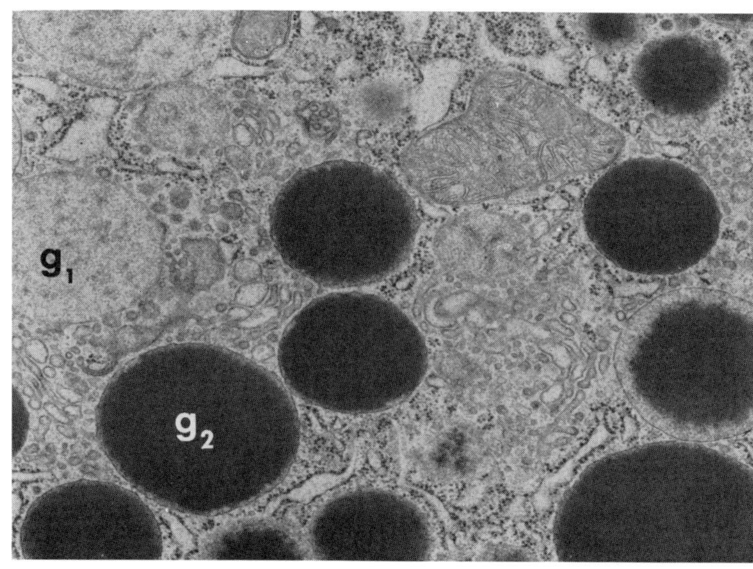

FIGURE 17-23 Electron micrograph of secretory granules. They show varying densities (g_1 and g_2) depending upon the degree of maturity.

of stacked saccules with numerous peripheral vesicles and condensing vacuoles; (3) secretory granules of a variable morphology, depending on their degree of maturity; and (4) usual assortment of lysosomes and mitochondria. Note the contrast in granule density between the immature (g_1) and the mature (g_2) granules depicted in Fig. 17-23.

There are delicate microvillous projections from the apical plasmalemma into the acinar lumen. Intercellular canaliculi may be present along the lateral border of acinar cells and appear more prominent during the postsecretory phase. A junctional complex made up of small desmosomes, zonula occludens, and zonula adherens is present along the luminal border between adjacent cells. The secretory product of the parotid acini is primarily composed of alpha amylase, but it also contains small quantities of ribonuclease (RNase), deoxyribonuclease (DNase), and proteases.

Each acinus is linked to a series of ducts which allow the flow of secretory products from the acinar lumen into the oral cavity. The first segment of this duct system is the *intercalated duct,* as diagramed in Fig. 17-24. It is lined with low cuboidal epithelial cells containing a centrally located nucleus surrounded by acidophilic cytoplasm. Figure 17-

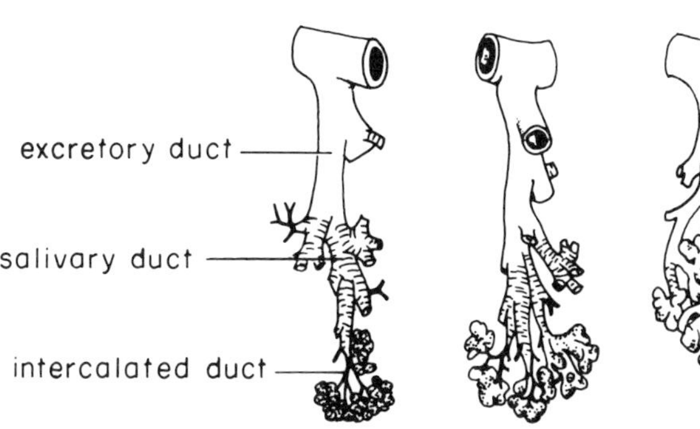

FIGURE 17-24 Schematic drawing to illustrate the structure of major salivary glands. Excretory ducts appear white and salivary ducts are cross-striped.

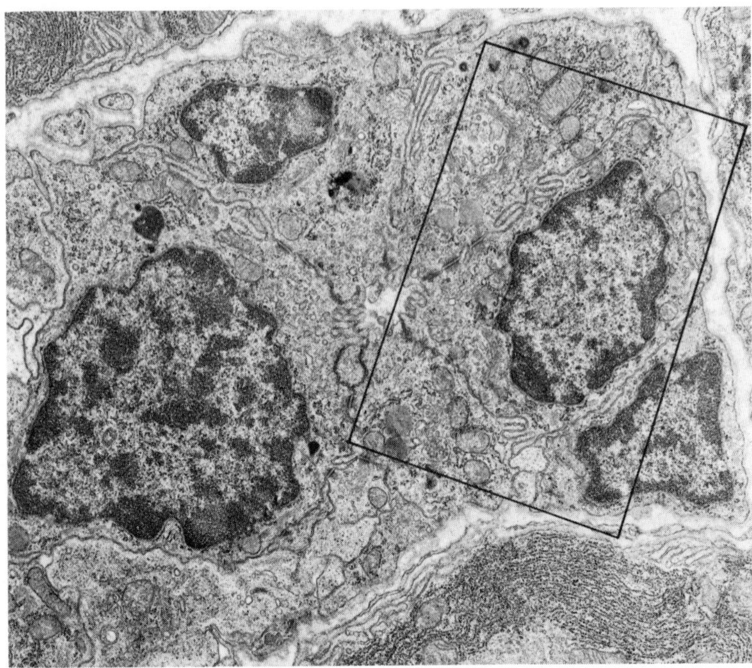

FIGURE 17-25 Electron micrograph of an intercalated duct. It is lined with low cuboidal cells that have a rounded nucleus and is separated from the connective tissue by a thin basal lamina. The cytoplasm contains the usual cell organelles in sparse numbers. There are small junctional complexes along the luminal border between adjacent cells.

25 is an electron micrograph of an intercalated duct. Junctional complexes are present along the luminal borders between cells (Fig. 17-26). Cells of the intercalated ducts may differentiate into acinar cells during recovery of the gland from injury.

Saliva passes through these ducts to reach the next segment, which is known as the *salivary, secretory,* or *striated duct*. The striated ducts are lined with simple columnar epithelial cells resting on a basement membrane (Fig. 17-27). They have a centrally located nucleus surrounded by finely granular acidophilic cytoplasm. They are called striated ducts because of the fine basal striations that are visible with appropriate staining. These basal striations consist of folded arrays of basal plasma membrane with rod-shaped mitochondria interspersed between them (Fig. 17-28). There is a close relationship between mitochondria (m) and the infolded plasma membrane. This configuration has been regularly observed in regions where active transport of electrolytes occurs. It has been shown that the striated duct cells actively resorb sodium ions from the saliva, thereby making it hypotonic. Since these striated ducts are located between lobules, they are often called *interlobular ducts*.

The last segment of ducts through which the saliva passes is the *excretory duct*. Excretory ducts arise by merger of a number of striated ducts and are lined with a simple columnar or stratified columnar epithelium which lacks basal striations. These ducts merge with each other and leave the parotid gland as the *main excretory duct* (*Stensen's duct*). The cells of excretory ducts have irregular microvillous projections at their apical surface and may be involved in the resorption of Na^+ and the excretion of K^+. Stensen's duct opens into the oral cavity in the buccal mucosa at the level of the upper second molar tooth.

Submandibular gland

The submandibular glands are located in the floor of the mouth and are mixed serous and mucous glands with serous cells predominating (Fig. 17-29). They are comprised of predominantly serous acini with occasional, lightly stained mucous elements scattered among them. There is a relative paucity of adipose cells in the submandibular gland in comparison with the parotid.

Figure 17-30 shows a higher magnification view of a submandibular gland in which serous acinar cells and several mucous acini are seen. The mucous acini create crescent-shaped areas of serous cells (arrows) that cover a portion of the mucous acini. These crescents are called *serous demilunes*. The electron micrograph in Fig. 17-31 depicts the appearance of serous cells forming a serous demilune (sd). The cells contain dense serous granules (arrows), which are smaller than light mucous granules (mg). Serous cells making up the serous demilune communicate with the acinar lumen through intercellular canaliculi between mucous cells. The mucous alveoli in hematoxylin and eosin (H and E) preparations present a light appearance. Its lumen may contain a large mass of mucin. The mucous cells rest on a basement mem-

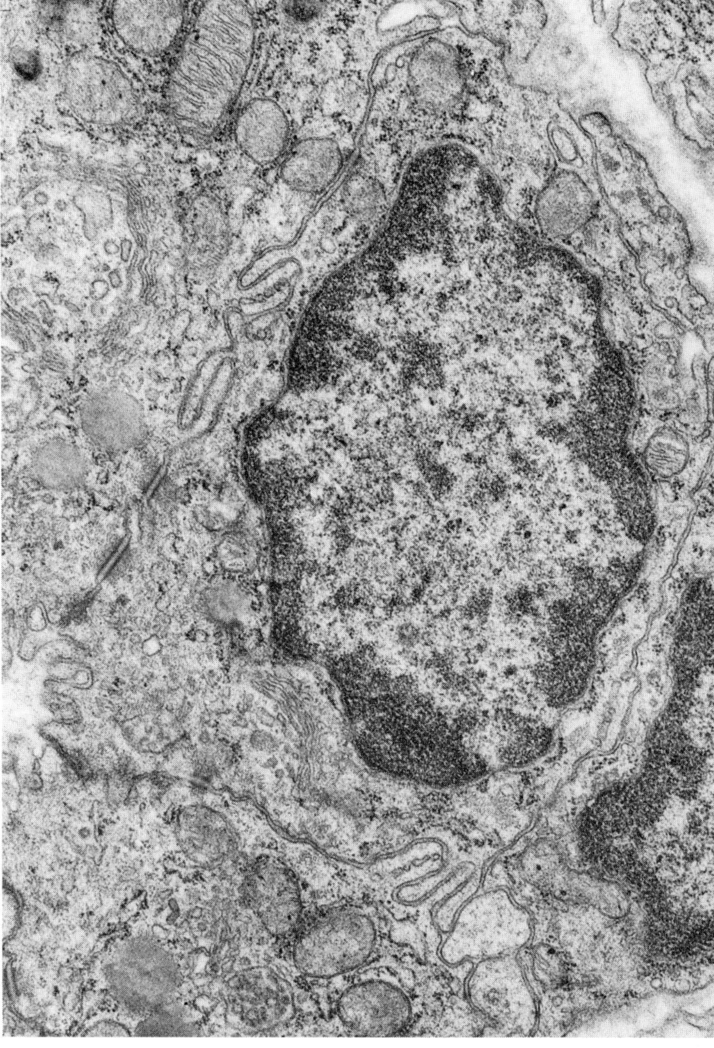

FIGURE 17-26 Intercalated duct cell viewed at higher magnification. It is a low cuboidal cell with a rounded nucleus that has a condensed chromatin pattern along its periphery. The cytoplasm contains the usual cell organelles in sparse numbers. A thin but distinct basal lamina separates it from the adjacent connective tissue. Note the junctional complex at the luminal border. It consists of a tight junction (zonula occludens), zonula adherens, and a desmosome.

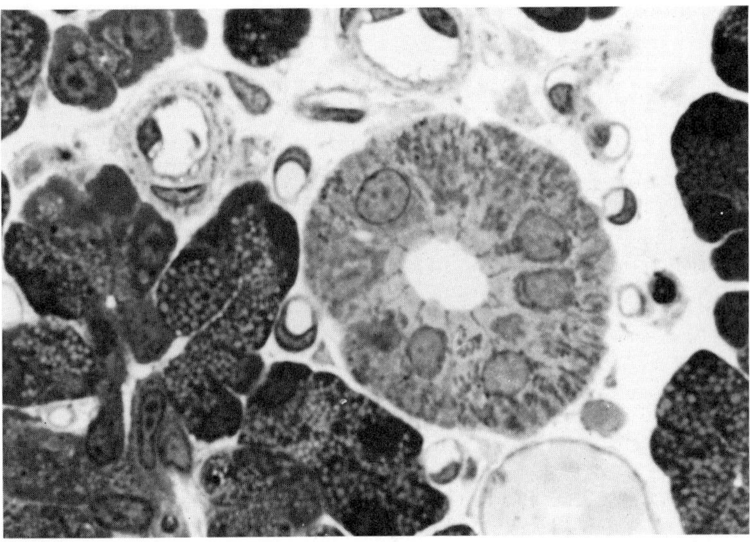

FIGURE 17-27 Light micrograph of a striated duct. The duct is lined with a simple columnar epithelium that rests upon a basement membrane. Duct cells have a centrally located nucleus surrounded by an eosinophilic cytoplasm. The basal portion of the cells have fine basal striations. Salivary ducts are also called striated ducts because of these striations.

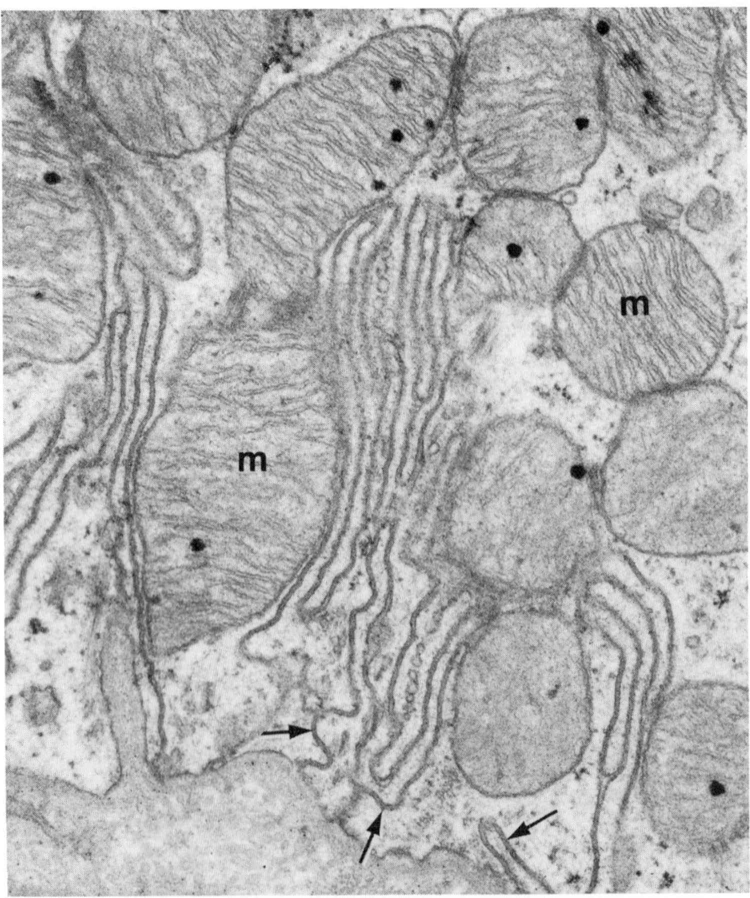

FIGURE 17-28 Basal portion of a striated duct cell. The basal plasma membrane has numerous infoldings. The cytoplasm between arrays of such infoldings contain abundant mitochondria (m) which provide energy for the active transport of electrolytes occurring in the striated ducts.

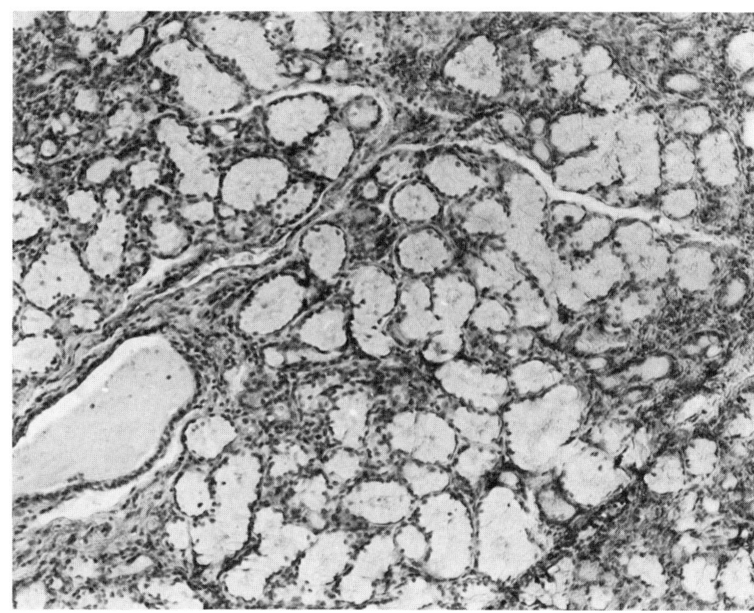

FIGURE 17-29 Light micrograph of a portion of the submandibular gland. It is comprised predominantly of serous acini with sparse mucous acini interspersed between them.

brane and have flattened nuclei along their basal surface. As the mucous product is removed during histologic preparation, the cytoplasm, where mucus was, often appears clear and lacy as in Fig. 17-30. In EM, the structure of the mucous granules shows a similarly empty appearance. Elsewhere in the cytoplasm, the RER is fairly abundant along the base, and the Golgi complex is present between the RER and the secretory granules. During the synthetic phase of the secretory cycle, the Golgi complex becomes more prominent (Fig. 17-32).

Occasionally, *myoepithelial cells* can be seen on the inner surface of the basement membrane (m in Fig. 17-33). The processes of these stellate cells girdle the acinus and, therefore, are often called *basket cells*. Ultra-

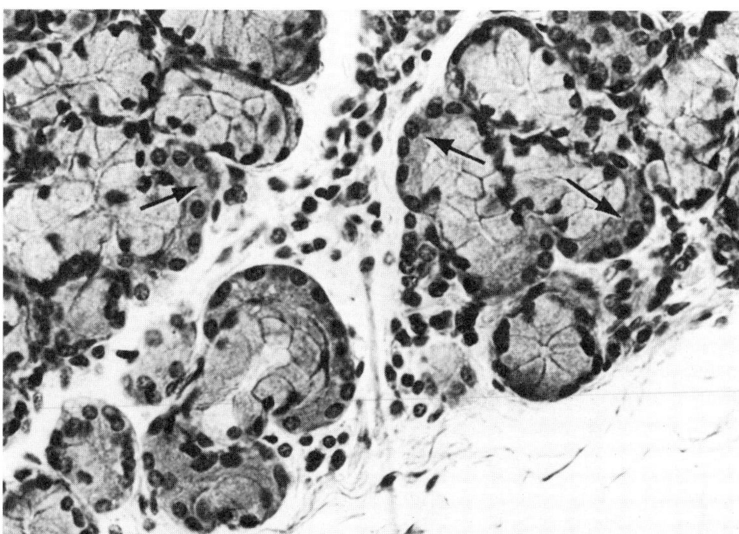

FIGURE 17-30 Light micrograph of a portion of the submandibular gland. Several mucous acini are present. Serous cells form crescent-shaped areas (arrows) that cover a portion of the mucous acini. These crescents are called serous demilunes.

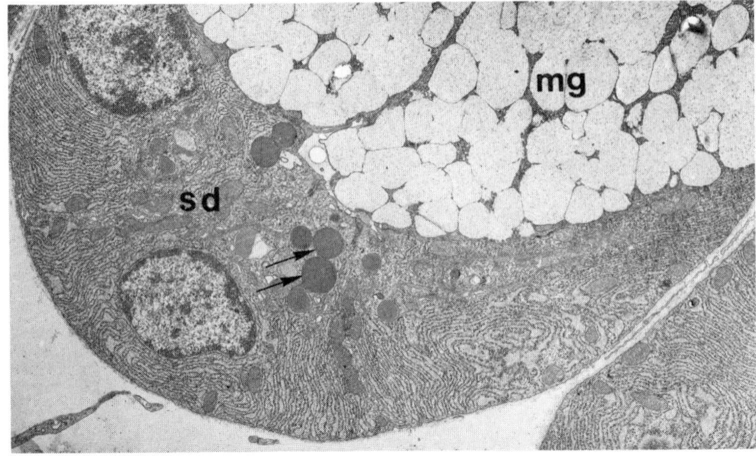

FIGURE 17-31 Electron micrograph of mucous acinar cells in the submandibular gland. A serous demilune cell (sd) covers the acinar cells. Mucous granules (mg) have an empty appearance, whereas the serous granules (arrows) are dark and dense.

structurally, the cytoplasm of myoepithelial cells resembles that of smooth muscle cells.

The duct system of the submandibular gland is similar to that of the parotid except that the striated ducts are generally longer and hence more conspicuous in sections. The main excretory duct of the submandibular gland is called *Wharton's duct* and opens at the caruncle of the sublingual frenulum.

Sublingual gland

The sublingual glands are situated in the floor of the mouth in association with the submandibular gland duct. The sublingual glands are largely mucous (Fig. 17-34). However they have a number of serous cells intermixed with the mucous acini. The mucous cells and serous cells are morphologically similar to those already described for the parotid and submandibular glands. Pure serous acini are rarely seen, and the majority of serous cells are involved in capping mucous acini as serous demilunes. Myoepithelial cells are also found in the acini present in these glands.

The duct system of the sublingual glands

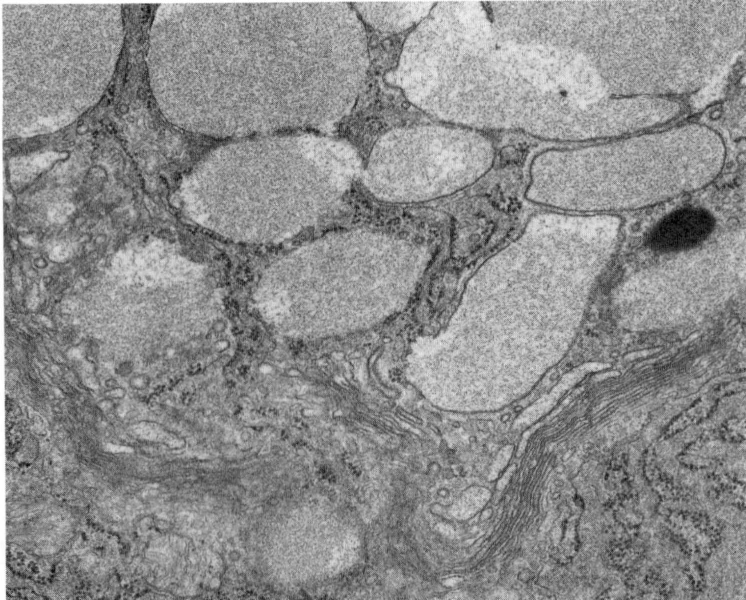

FIGURE 17-32 Portion of a mucous acinar cell that is in the secretory phase. A well-developed Golgi apparatus and abundant RER is seen. The mucous granules contain a flocculent material of low electron density. They appear to originate in the Golgi apparatus.

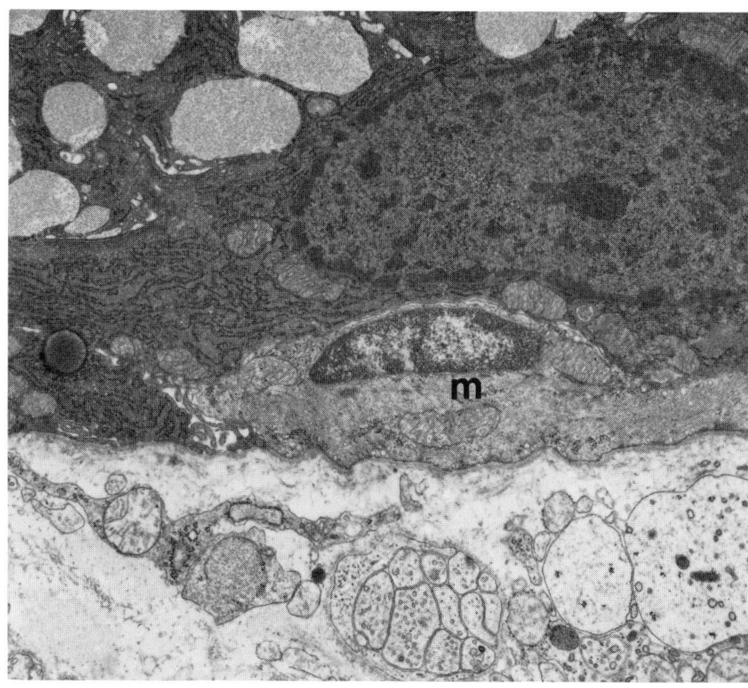

FIGURE 17-33 Electron micrograph of myoepithelial cells (m). The myoepithelial cell contains microfibrils and resembles a smooth muscle cell. It is contained within the basal lamina of the acinar cells and its processes girdle the acinus. Note the unmyelinated nerve fibers in the adjacent connective tissue.

differs from that of the parotid and submandibular glands in two respects. First, the intercalated ducts of the sublingual glands are lined with typical mucous cells instead of the low columnar to squamous epithelial cells found in the other major glands. Second, striated ducts, if present in the sublingual glands, are very short. Hence they generally escape detection in sections.

CYCLIC NATURE OF THE SECRETION OF SALIVARY ENZYMES

The amount of flow of saliva is dependent upon numerous variables. Experimentation

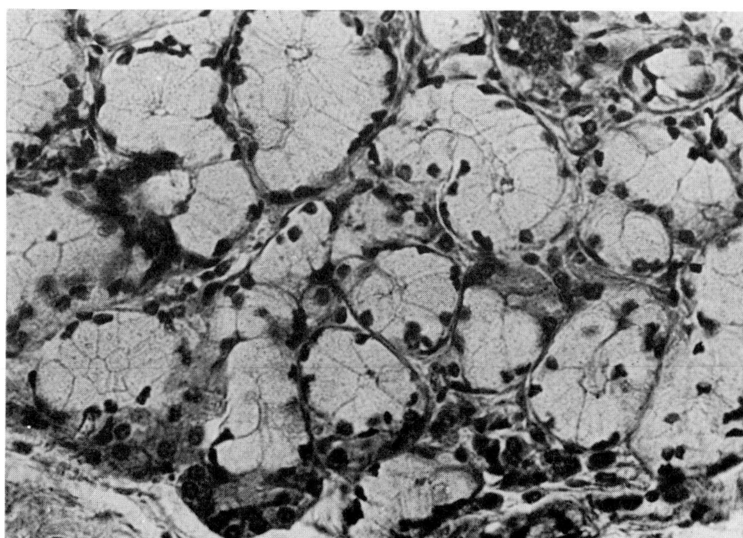

FIGURE 17-34 Light micrograph of a portion of the sublingual gland. The sublingual gland is composed largely of mucous acini that have serous demilunes. However, a few pure serous acini may be found among the mucous acini.

FIGURE 17-35 Diagram illustrating the structure of serum and secretory IgA.

has proved that the flow of saliva depends upon the season of the year and the time of day. In summer and during dark hours, the flow of saliva is less than in the winter months and during light hours. Saliva flows in larger quantities when an individual is standing as opposed to sitting or lying down. Smoking increases the flow of saliva, while sleeping decreases it. Certain flavored foods increase the flow of saliva to a greater extent than do others. These observations result from the intrinsic cyclic nature of secretory cells, circadian rhythm involving the thalamus of the brain, and other sensory inputs.

SALIVARY IMMUNOGLOBULIN A

Of the two types of immunoglobulin A (IgA), the larger one has been implicated as being an important component of the immune system of the body at the interfaces between the internal and external environments. This type of IgA is called *secretory IgA* and is found in most of the secretions of the body including the saliva. Because of its presence in these secretions (saliva, tears, breast milk, gastrointestinal secretions, and mucous lining of the respiratory and genitourinary tracts) and the unique locations of these secretions, IgA has been implicated as being the first line of immunologic defense against bacterial and viral invaders. Immunoglobulin G (IgG), the main immunoglobulin, outnumbers IgA by a ratio of 6:1 in the serum, while secretory IgA outnumbers IgG by a ratio of more than 20:1 in the secretions of the body. The structures of serum and secretory IgA are shown in Fig. 17-35.

Secretory IgA is larger than the IgA molecules in the serum, and it has a molecular weight of 400,000 as opposed to 170,000 for serum IgA. Secretory IgA is composed of two serum IgA molecules bound together by a nonimmunoglobulin component with a molecular weight of 60,000. This nonimmunoglobulin component bridges the two serum IgA molecules by disulphide bonds and has been labeled the *secretory piece*. The importance of this secretory piece stems from the fact that serum IgA molecules, when bound together by this secretory piece, are more resistant to degradation by proteolytic enzymes such as trypsin, chymotrypsin, and pepsin which are commonly found with secretory IgA in saliva. Evidence also indicates that this secretory piece may help to transport the protein IgA molecules across mucous membranes, but this evidence is far from conclusive.

Secretory IgA is produced locally at interfaces between the internal and external environments of the body. The salivary glands are no exception, and the following is the currently proposed scheme for secretory IgA production. Initially, plasma cells located in the interstitial spaces underlying the epithelial cells of the acini and ducts produce serum IgA. The serum IgA molecules then pass through these overlying epithelial cells where the secretory piece is produced and added to the serum IgA molecules. Thus, the molecule released into the saliva is a completed secretory IgA. The mechanism of transport of IgA is diagramed in Fig. 17-36.

The functions of secretory IgA can be assessed from individuals in whom secretory IgA is lacking, while the normal complement of serum immunoglobulins is present. Secretory IgA is thus implicated in (1) the destruction of certain inhaled pathogens which can cause respiratory infections, and (2) the control of bacteria constituting the gastrointestinal flora just after birth which is very important in the normal digestion of food. It is known that among the major salivary glands the parotid gland produces most of the secretory IgA. Indeed, a large number of plasma cells are found in the connective tissue supporting the glandular parenchyma of this gland.

MINOR SALIVARY GLANDS (INTRINSIC)

A number of small intrinsic salivary glands are present throughout the oral mucosa. They are generally named according to their location. Following are some of the commonly found minor salivary glands.

Sublingual glands

Eight to twenty of these glands are found in the sublingual sulcus near the major sublingual gland. The minor sublingual glands are largely mucous.

Labial glands

Several labial glands are present in the lamina propria of the internal surfaces of the upper and lower lips. They are mixed, seromucous in nature.

Glossopalatine glands

Several of these glands are found in the glossopalatine folds and the soft palate. They are mostly mucous glands.

Lingual glands

These glands are found in two areas: (1) the anterior two-thirds of the tongue on the ventral surface, where they are seromucous, and (2) the posterior third of the tongue behind the sulcus terminalis. Von Ebner's glands are pure serous, but other posterior lingual glands are pure mucous.

Palatine glands

In the lamina propria of the posterior lateral portion of the hard palate and in the submucosa of the soft palate and the uvula are located several hundred palatine glands. They are seromucous glands on the nasal side and pure mucous glands on the oral side.

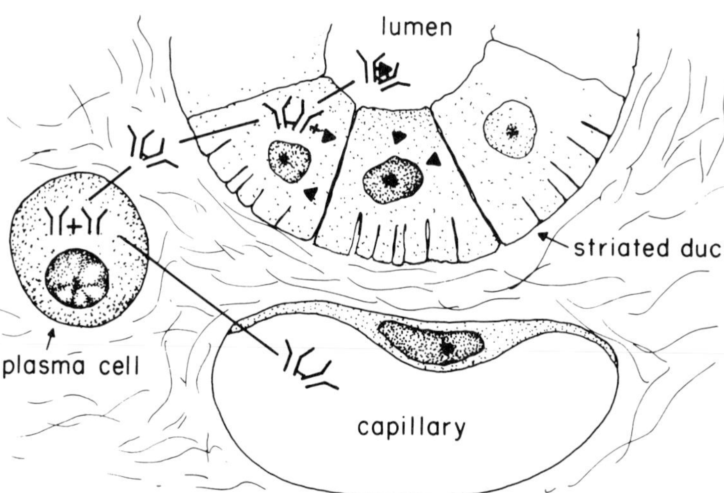

FIGURE 17-36 Diagram illustrating hypothetically the production and transport of secretory (salivary) IgA.

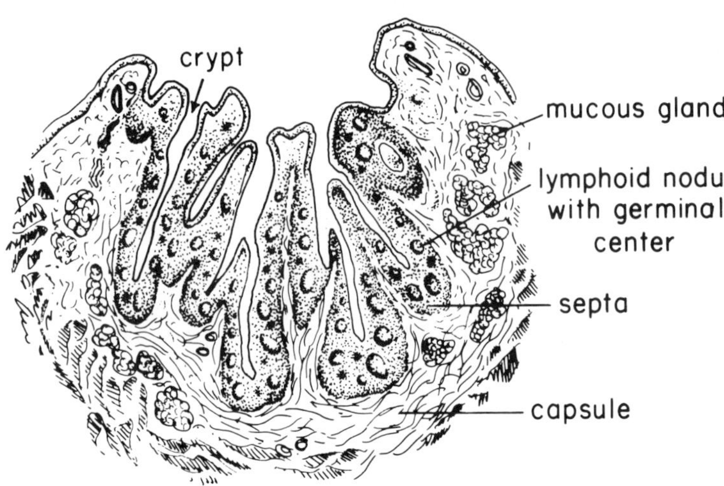

FIGURE 17-37 Diagram illustrating the structure of the palatine tonsils and their relationship with surrounding structures.

Buccal glands
Mostly mucous but some times mixed, buccal glands are present within the buccinator muscle.

Pharyngeal glands
These glands are small seromucous glands associated with regions covered by pseudostratified columnar epithelium.

TONSILS

The tonsils are aggregates of lymphoid nodules located in the oropharynx in a circular fashion. This circle of tonsillar tissue, previously called *Waldeyer's ring,* is formed by posterior lingual tonsils, the bilaterally located palatine tonsils, and pharyngeal tonsils that are invested in the pharyngeal mucosa of this region. The tonsils have traditionally been regarded as a ring of protective lymphoid elements against the entrance of foreign matter into the gastrointestinal tract and respiratory system.

Upon microscopic examination, all tonsils show dense aggregates of lymphocytes held by a delicate reticular stroma. Many germinal centers are present in active tonsils and are believed to produce and release lymphocytes into the pharynx and oral cavity. The lymphocytes appearing in the saliva are called *salivary corpuscles.* Some of them produce secretory antibodies (IgA) which function in the neutralization of foreign substances in the oral cavity. As mentioned earlier, IgA is produced by immunologically competent cells present along the entire gastrointestinal and respiratory mucosae including major salivary glands.

THE PALATINE TONSILS
The largest of the tonsils are the paired *palatine tonsils* or *faucial tonsils.* The palatine tonsils are ovoid masses of lymphoid tissue and are located within the lateral wall of the oropharynx near the posterior third of the tongue. Specifically, they are located between the palatoglossus and palatopharyngeus muscles, which are often called *anterior* and *posterior pillars* of the tonsil. Figure 17-37 shows the palatine tonsils in their gross relationship with surrounding structures. They measure about 2 cm along their long axis. The palatine tonsils are generally the most active of the three tonsils, and vary considerably in their structure, depending on the level of activity. A low-power microscopic section of the palatine tonsil is seen in Fig. 17-38.

The palatine tonsil is covered by the stratified squamous epithelium of the oral mucosa which makes deep branching crypts into the tonsillar substance. The tonsillar mass is surrrounded by a discrete but delicate

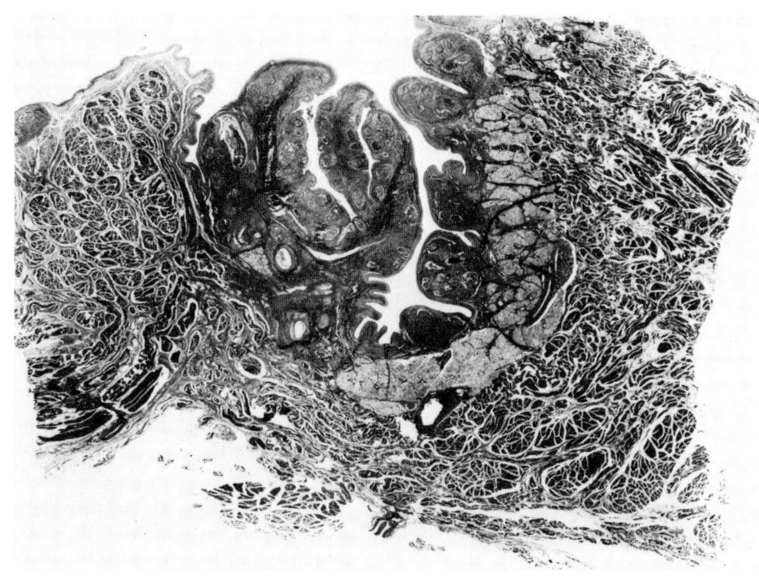

FIGURE 17-38 Low-power light micrograph of palatine tonsil.

connective tissue capsule which sends septa between the epithelial crypts. Mucous glands are often associated with the crypts of the palatine tonsil and function in flushing out cellular elements and infectious agents. These glands open onto the free surface of the tonsil. Within the mass of tonsillar tissue, lymphocytes and occasional plasma cells make discrete nodules including many germinal centers.

Figure 17-39 is an example of palatine tonsil, which clearly shows epithelial crypts (arrows) and connective tissue septa (t) as well as dense tonsillar lymphoid tissue. During inflammation, the tonsillar mass becomes notably enlarged, showing diffuse scattering of large lymphoid cells and numerous germinal centers (Fig. 17-40). Many lymphoid cells are in transit through the stratified epithelia (Fig. 17-41, arrows). Because of the

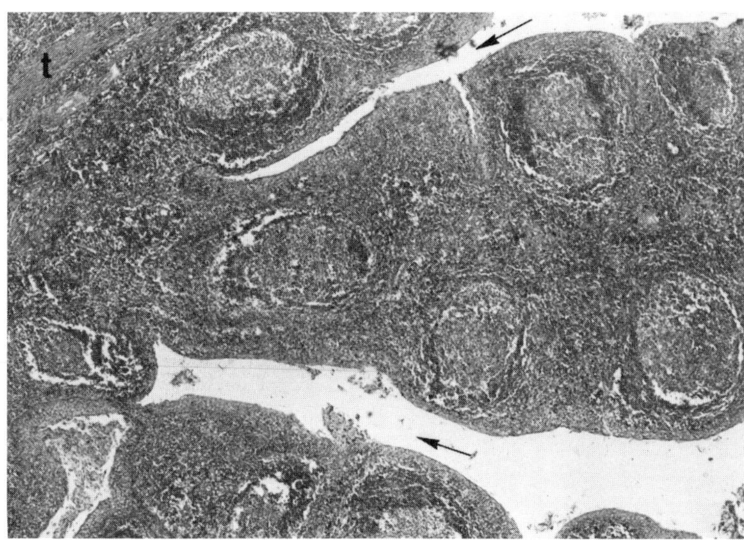

FIGURE 17-39 Portion of palatine tonsil. Note the deep crypts (arrows), the connective tissue septa (t), and the dense accumulation of lymphoid cells.

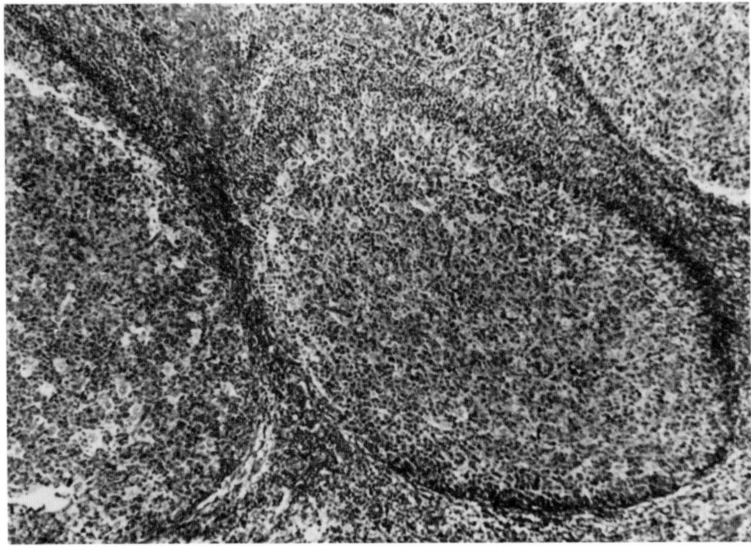

FIGURE 17-40 Light micrograph of germinal centers of the palatine tonsil.

possible involvement of palatine tonsils in immunologic defense mechanisms, the tonsillectomy is performed far less frequently as a therapeutic approach to tonsillitis today than in previous years.

THE PHARYNGEAL TONSIL

The *pharyngeal tonsil,* also known as the *adenoid,* occupies a broad area of the posterior wall of the nasopharynx. Since it is in the respiratory passage, it is covered by pseudostratified columnar epithelium (Fig. 17-42). The pharyngeal tonsil is much smaller than a palatine tonsil and appears less differentiated than the palatine tonsil in several respects. In general, the connective tissue capsule is thinner, with no septa. Although the pharyngeal tonsil is rather broad (3 cm

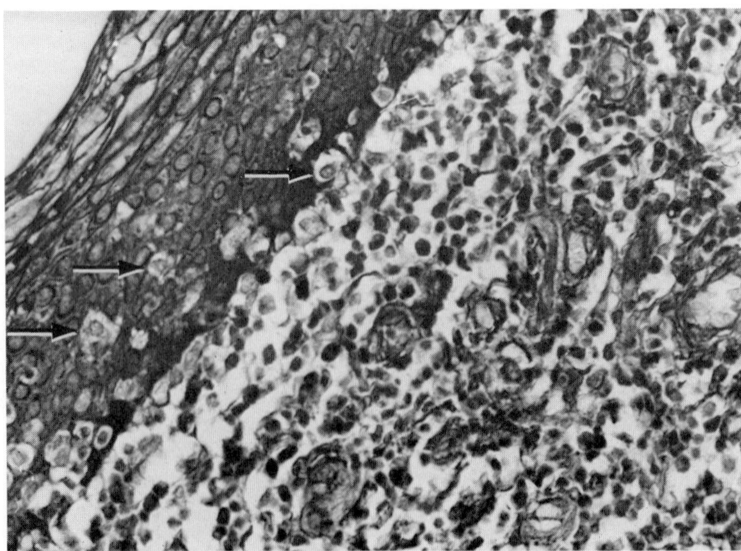

FIGURE 17-41 The stratified squamous epithelium covering the palatine tonsil viewed at higher magnification. Note the numerous lymphoid cells that are in transit through it from the connective tissue to the oral cavity (arrows).

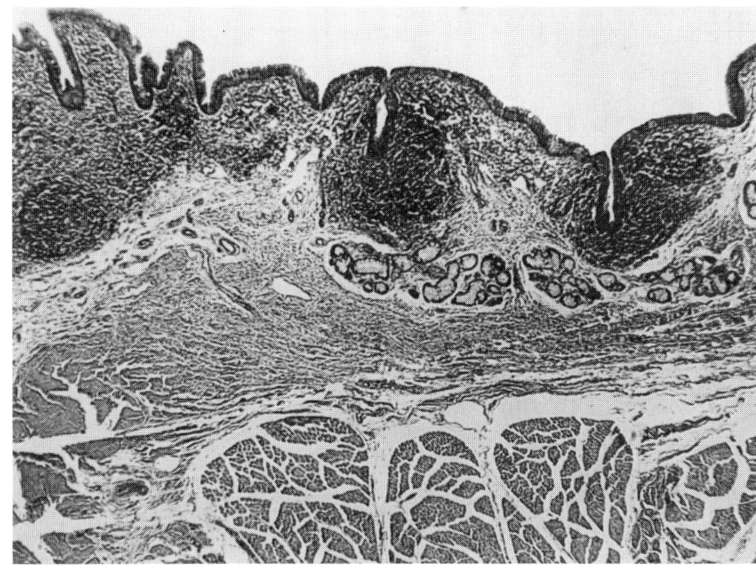

FIGURE 17-42 Low-power light micrograph of the pharyngeal tonsil. It is lined with pseudostratified columnar epithelium.

in horizontal dimension), it is shallow in depth (2 mm in thickness). No epithelial crypts are present, although its surface may show small indentations produced by aggregates of lymphoid nodules. Glands are seldom found in this area, but an occasional seromucous gland may be present. Figure 17-43 is a high-power view of the surface area of the pharyngeal tonsil which demonstrates the presence of respiratory epithelium and underlying lymphoid cells that are penetrating the region.

THE LINGUAL TONSIL

As mentioned in connection with the tongue, small mounds over the posterior third of the dorsal lingual surface are occupied by the lingual tonsils. As shown in Fig. 17-44, the bulbous projections are individual lymphoid nodules that are covered by lingual mucosa.

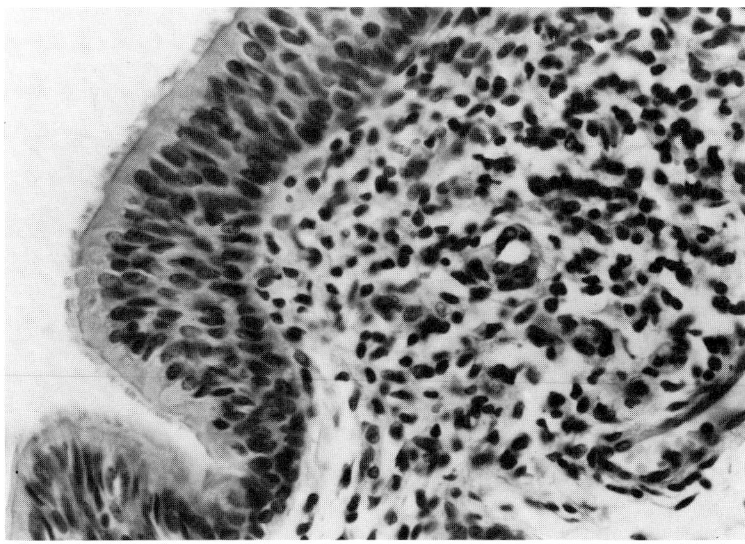

FIGURE 17-43 A portion of a pharyngeal tonsil. It is covered by respiratory epithelium. Numerous lymphoid cells are present in the underlying connective tissue.

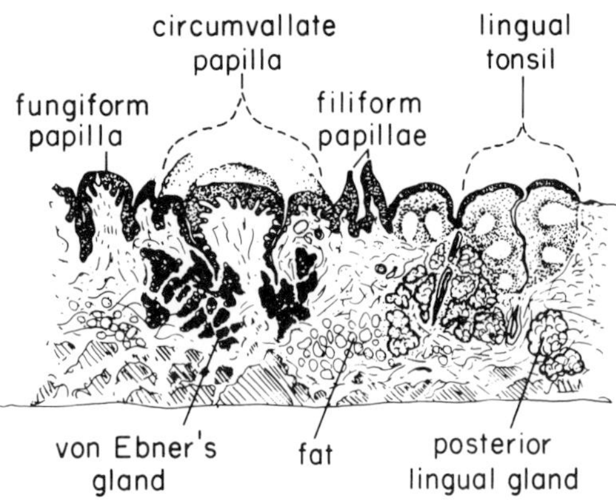

FIGURE 17-44 Diagram illustrating the structure of the lingual mucosa on the posterior third of the tongue.

Figure 17-45 shows the microscopic appearance of the lingual tonsil, which is seen in association with muscle, posterior lingual glands, and adipose tissue of the tongue. Single crypts are present in the lingual tonsil. A connective tissue capsule is present under each of the lingual tonsils and is frequently penetrated by the ducts of the posterior lingual glands which open into the epithelial crypts.

The lingual tonsil is intermediate between the palatine tonsil and the pharyngeal tonsil in terms of differentiation and histologic appearance. Germinal centers of the lingual tonsil are fewer and less active appearing than those of the palatine tonsil, but the germinal centers are more numerous and

FIGURE 17-45 Light micrograph of the lingual tonsil. It is covered with stratified squamous epithelium. A single crypt is present in each lingual tonsil. These tonsils are associated with tongue muscles, posterior lingual glands, and adipose tissue.

SUMMARY FOR DIFFERENTIAL IDENTIFICATION OF TONSILS — TABLE 17-2

LOCATION	PALATINE	LINGUAL	PHARYNGEAL
Epithelium	Stratified Squamous	Stratified Squamous	Pseudostratified Columnar ciliated
Glands	Mucous	Mucous	Seromucous (few)
Crypts	Deep, branching	No branching	None
Septa	Many long	None	None
Germinal centers	Numerous	Moderate	Few

appear more active than those of the pharyngeal tonsil.

The differential diagnostic features of tonsils are summarized in Table 17-2.

REVIEW SECTION

1. Figure 17-46 is a section of oral epithelium. The cells are arranged in many layers. Cells at the surface of the epithelium appear to be sloughing off. What type of epithelium appears here? Where might such epithelium be found?

2. Identify the structure appearing in Fig. 17-47.

A portion of the gingiva which is covered by the masticatory epithelium is shown here. Elsewhere, masticatory epithelia may be found in certain areas of the cheek and hard palate.

Minimally keratinized stratified squamous epithelium (masticatory epithelium) lines the hard palate and gingiva. These areas are subjected to considerable masticatory abrasion and stress.

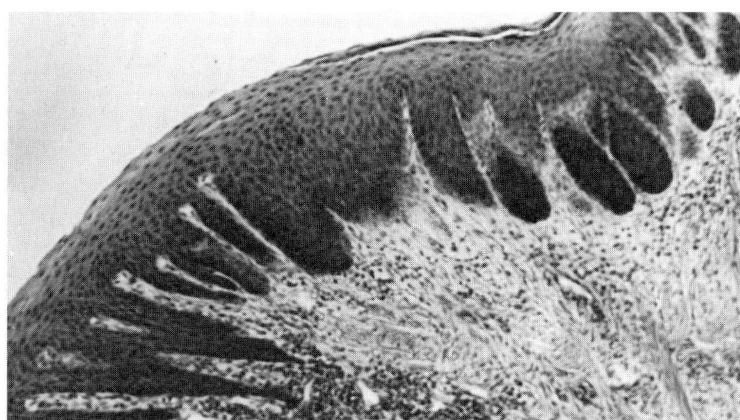

FIGURE 17-46

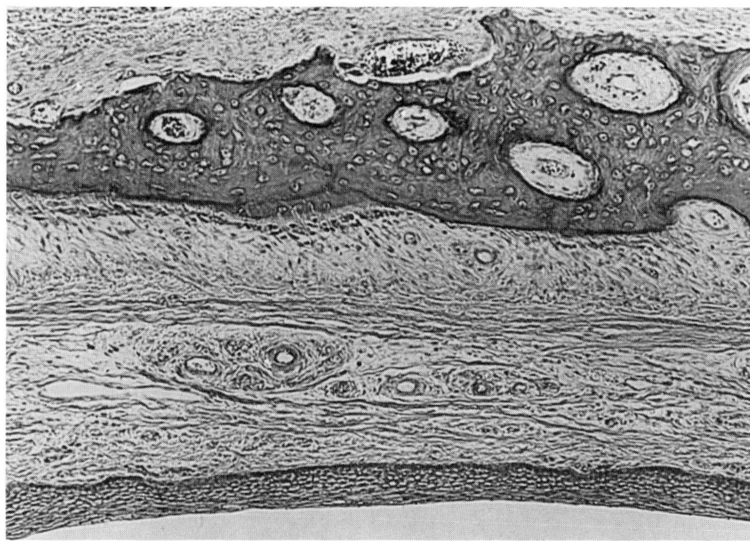

FIGURE 17-47

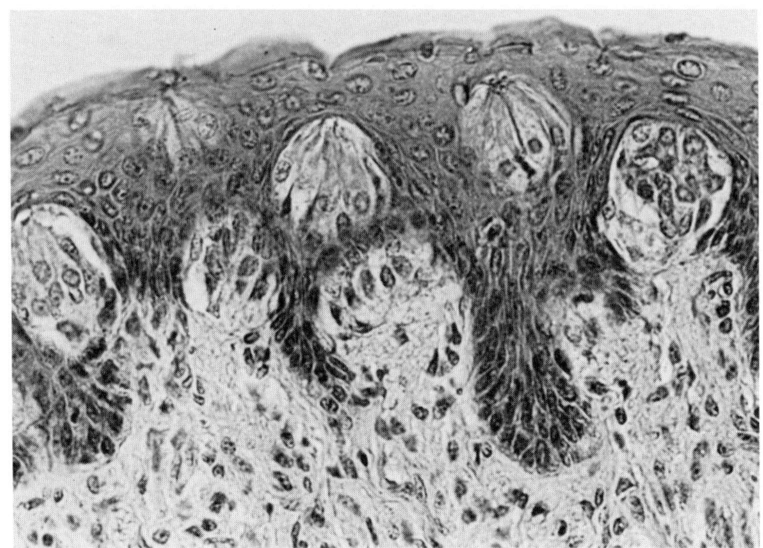

FIGURE 17-48

3. Figure 17-48 shows several barrel-shaped structures located in stratified squamous epithelium. What is the name of these structures? What function do they perform?

Taste buds appear in this Fig. 17-48. Taste buds are involved in gustatory perception. Materials in solution to be tasted enter the taste bud through the taste pore at the apex of the bud.

4. A section of an oral structure is seen in Fig. 17-49. What is the structure? What is its function? What tissue predominates? What composes the irregular border at the top of Fig. 17-49?

A longitudinal section through the tongue is seen in the Fig. 17-49. Skeletal muscle is predominant. The tongue is active in mastication, swallowing, and speech and is covered by stratified squamous epithelium.

5. What tissue is shown in Fig. 17-50?

Figure 17-50 represents a section of the submandibular gland. Notice the predominantly mucous cells with a serous demilune shown in the mixed

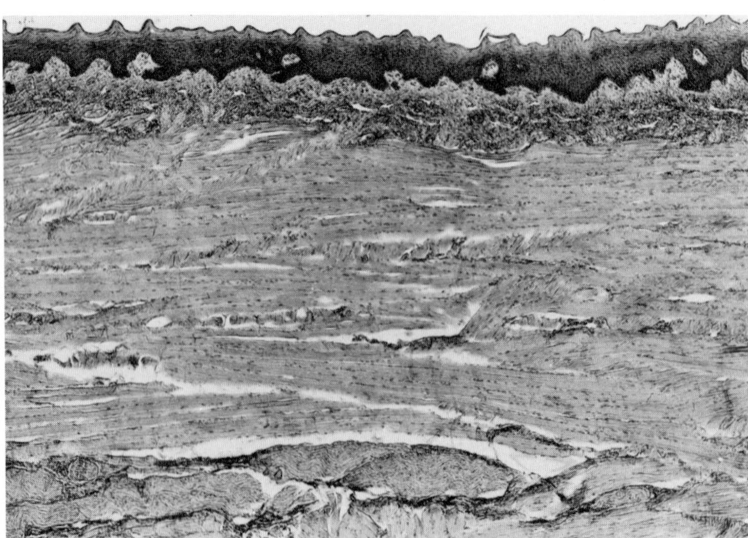

FIGURE 17-49

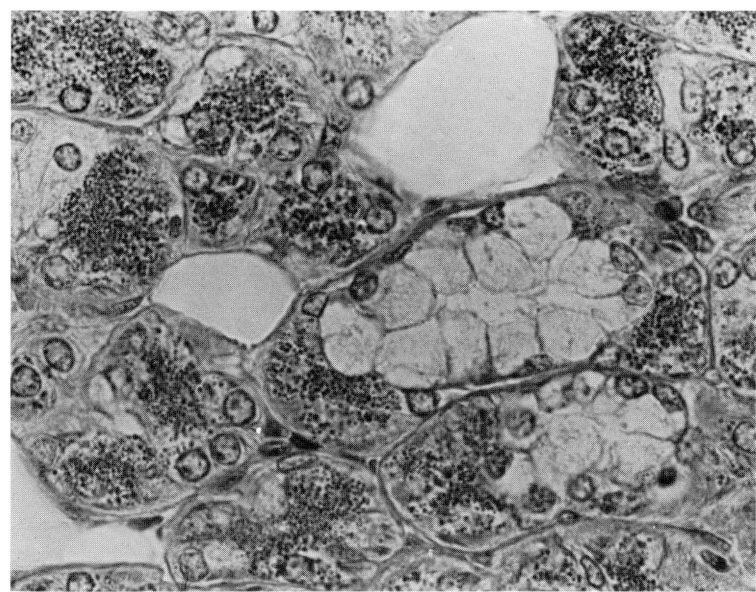

FIGURE 17-50

acini. In other regions of the submandibular gland you may observe more mucous cells. Mucous cells are not present in the parotid. On the other hand, the sublingual gland is largely mucous. Review other histologic differences between the major salivary glands such as the ducts and the amount of adipose cells.

6. How is the parotid gland classified?

The parotid is a predominantly serous tubuloalveolar gland.

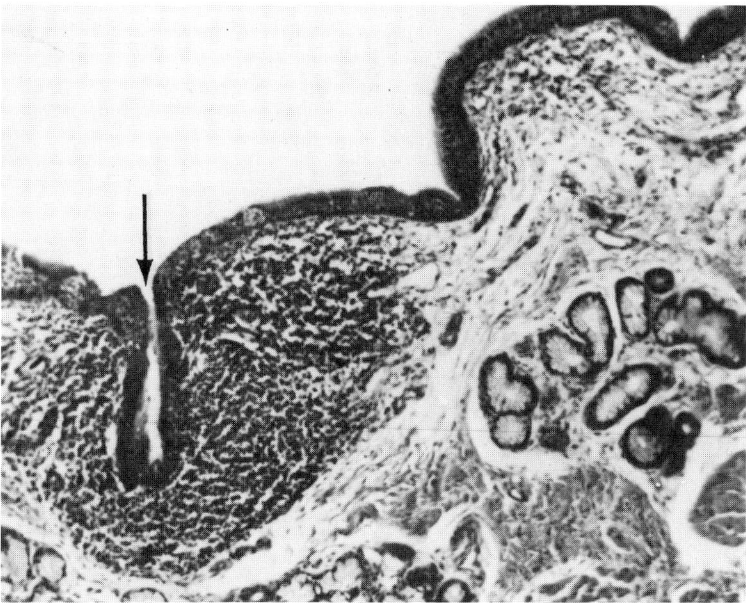

FIGURE 17-51

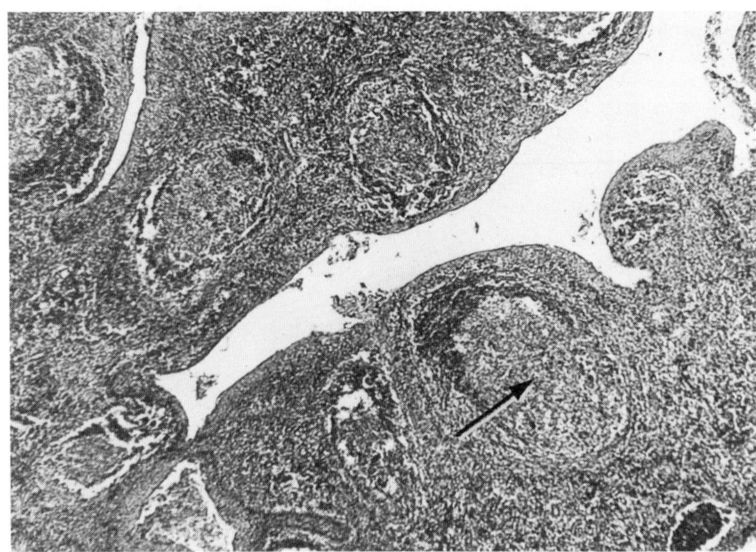

FIGURE 17-52

7. In which glands are the striated ducts most obvious?

Because the striated ducts are longest in the submandibular gland, a section through this gland will probably present more cross sections of striated ducts than one through the other salivary glands.

8. Why are striated ducts referred to as being "striated"?

Striated duct cells should not be confused with the striated border of intestinal cells. The striated ducts are so called because the basal portion of the cell appears striated due to the infolded plasmalemma with the interposed vertically oriented mitochondria, as seen in Fig. 17-28.

9. How is secretory IgA structurally adapted for functioning in an environment with proteolytic enzymes?

The secretory piece by connecting two IgA molecules makes the secretory IgA more resistant to degradation by proteolytic enzymes.

10. Figure 17-51 is a section through a lymphoid nodule. What is the clear, slitlike area in the nodule (arrow)? The tissue overlying the nodule is __________. What kind of tonsil is shown?

The lingual tonsil with its deep crypts occupies the posterior third of the tongue. The lingual tonsil is covered by a stratified squamous epithelium. In the vicinity of the lingual tonsil are several posterior lingual (pure mucous) glands.

11. Figure 17-52 shows another tonsil. The tissue indicated by the arrow is __________. Branching crypts are found in what type of tonsil?

The tonsil appearing in the field is a palatine tonsil. The palatine tonsils have deep branching crypts and large germinal centers, as indicated by the arrow.

BIBLIOGRAPHY

Bhaskar, S. N. (ed.): *Orban's Oral Histology and Embryology,* 8th ed., Mosby, St. Louis, 1976.

Bloom, W., and D. W. Fawcett: *A Textbook of Histology,* 10th ed., Saunders, Philadelphia, 1975.

Mason, D. K., and D. M. Chisholm: *Salivary Glands in Health and Disease,* Saunders, Philadelphia, 1975.

Provenza, D. V.: *Oral Histology,* Lippincott, Philadelphia, 1964.

Scheuer-Karpin, R., and A. Baudach: "Cytology of the Tonsils," *Lymphology* **3:**109–114 (1970).

Squier, C. A., and J. Meyer (eds.): *Current Concepts of the Histology of Oral Mucosa,* Charles C Thomas, Springfield, Ill., 1971.

ADDITIONAL READINGS

Amsterdam, A., I. Ohad, and M. Schramm: "Dynamic Changes in the Ultrastructure of the Acinar Cell of the Rat Parotid Gland during the Secretory Cycle," *J. Cell Biol.* **41:**735–773 (1969).

Brandtzaeg, P.: "Human Secretory Immunoglobulins. 4. Quantitation of Free Secretory Piece," *Acta Path. Microbiol. Scand.,* **79:**189–203 (1971).

———:"Mucosal and Glandular Distribution of Immunoglobulin Components," *Immunology* **26:** 1101–1114 (1974).

Castle, J. D., J. D. Jamieson, and G. E. Palade: "Radioautographic Analysis of the Secretory Process in the Parotid Acinar Cell of the Rabbit," *J. Cell Biol.* **53:**290–311 (1972).

Emmelin, N., and Y. Zotterman (eds.): *Oral Physiology,* Pergamon, Oxford, 1972.

Halpern, M. S., and M. E. Koshland: "Novel Subunit in Secretory IgA," *Nature* **228:**1276–1278 (1970).

Hand, A. R.: "Morphology and Cytochemistry of the Golgi Apparatus of Rat Salivary Gland Acinar Cells," *Am. J. Anat.* **130:**141–158 (1971).

Mandel, I. D., and H. Khurana: "Relation of Human Salivary IgA Globulin to Flow Rate," *Arch. Oral Biol.* **14:**1433–1435 (1969).

———, and S. Wotman: "Salivary Secretions in Health and Disease," *Oral Science Rev.* **8:**25–47 (1976).

Mestecky, J.: "Structure of Antibodies," *J. Oral Path.* **1:**288–300 (1972).

Munger, B. L.: "Histochemical Studies on Seromucous- and Mucous-secreting Cells of Human Salivary Glands," *Am. J. Anat.* **115:**411–429 (1964).

Ruthberg, U.: "Ultrastructure and Secretory Mechanisms of the Parotid Gland," *Acta Odont. Scand.,* vol. 19, suppl. 30, 1968, pp. 1–68.

Shannon, I. L., R. P. Suddick, and F. J. Dowd (eds.): *Saliva: Composition and Secretion,* Karger, New York, 1974.

Tamarin, A.: "Myoepithelium of the Rat Submaxillary Gland," *J. Ultrastruct. Res.* **16:**320 (1966).

Tandler, B.: "Ultrastructure of the Human Submaxillary Gland. I. Architecture and Histological Relationships of the Secretory Cells," *Am. J. Anat.* **111:**287–307 (1962).

———, C. R. Denning, I. D. Mandel, and A. H. Kutscher: "Ultrastructure of Human Labial Salivary Glands. I. Acinar Secretory Cells," *J. Morphol.* **127:** 382–408 (1969).

———, ———, ———, and ———: "Ultrastructure of Human Labial Salivary Glands. III. Myoepithelium and Ducts," *J. Morphol.* **130:**227–246 (1970).

Tomasi, T. B. and K. J. Biennenstock: "Secretory Immunoglobulins," *Adv. Immunol.* **9:**1–96 (1968).

Yohro, T.: "Nerve Terminals and Cellular Junctions in Young and Adult Mouse Submandibular Gland," *J. Anat.* **108:**409–417 (1971).

18
ORAL CAVITY II: TOOTH AND PERIODONTAL STRUCTURES

OBJECTIVES

After completing this chapter, the student will be able to:

1 Compare and contrast the characteristics and structure of enamel, dentin, cementum, and bone.

2 List the component parts of enamel and dentin.

3 Describe the interrelationship between enamel and dentin.

4 Name and describe the cause of the striations in enamel and dentin.

5 Define enamel tuft, spindle, and lamella.

6 Identify two types of cementum and compare its formation with bone formation.

7 List the components of pulp and give their functions.

8 List and define different pulpstones.

9 Identify the attachments, structure, and function of the three principal types of fiber bundles found in the periodontal ligament.

10 Describe the vasculature of the periodontal ligament.

11 Identify the characteristics of cementicles.

12 Define the three types of nerve endings found in the periodontal ligament and describe the stimuli to which these receptors respond.

13 Identify the point in time at which the alveolar bone becomes separate and distinct from the maxilla and mandible.

14 Identify the location, structure, and function of the alveolar bone proper and the cortical bone.

TOOTH STRUCTURES

The exterior of the adult tooth is covered by two types of tissue: enamel on the crown and cementum on the root (Fig. 18-1). Underlying this is the dentin which makes up the bulk of the tooth and gives it strength. The interior is occupied by the pulp, providing innervation and nutrition. These tissues differ in their origins, structures, properties, and functions.

ENAMEL

Characteristics

Enamel originates from ectodermally derived cells which produce a noncollagenous matrix of eukeratin. The formative cells of enamel, called *ameloblasts,* function as "glandular" or secretory cells in elaborating the matrix. Figure 18-2 shows a small portion of a developing tooth in which ameloblasts (a), the very first enamel (e), dentin (d), as well as the future dentinoenamel junction (dej), odontoblasts (ob), and the pulp (p) are seen. When enamel formation is complete, the ameloblasts become reduced in size and nonfunctional, making enamel the only tooth structure that cannot be regenerated or repaired.

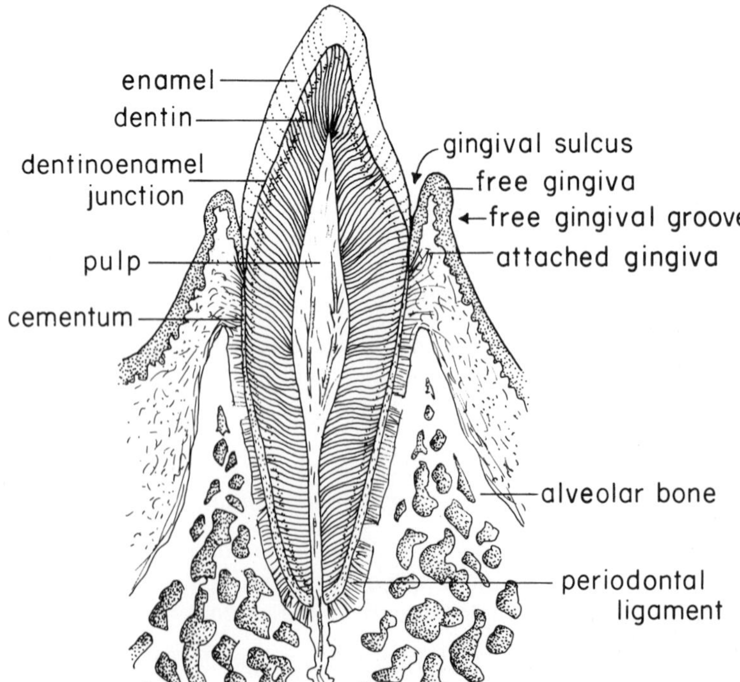

FIGURE 18-1 Diagram illustrating the structure of an adult tooth and its surrounding structures.

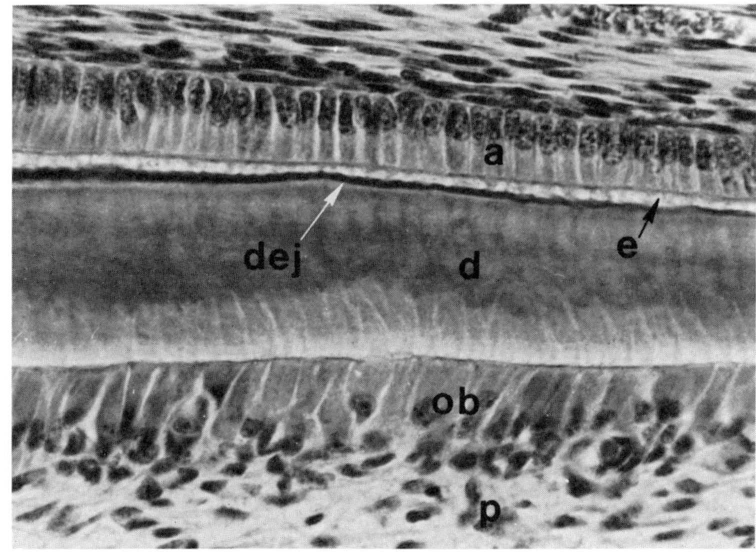

FIGURE 18-2 Portion of a developing tooth. Tall columnar ameloblasts (a) are secreting enamel matrix (e). The dentinoenamel junction (dej) separates the mineralized enamel from the dentin (d). Odontoblasts (ob) are producing the organic matrix of the dentin (the lighter staining area between the odontoblasts and the dentin). Odontoblasts send long processes that reach the dentinoenamel junction through narrow canals (dentinal tubules) in the dentin.

Enamel becomes highly calcified, enabling it to function as a protective coating for the underlying dentin that withstands masticatory forces (Fig. 18-3). Mature enamel consists mainly of inorganic material, 96 percent, with the remainder composed of water and a very small amount of the organic matrix. Due to its high mineral content and its crystalline arrangement, enamel is the hardest tissue in the body. Therefore, it is brittle and fractures very easily when it loses its dentin support. Enamel is fairly translucent, thereby reflecting to varying extents the yellowish color of the underlying dentin.

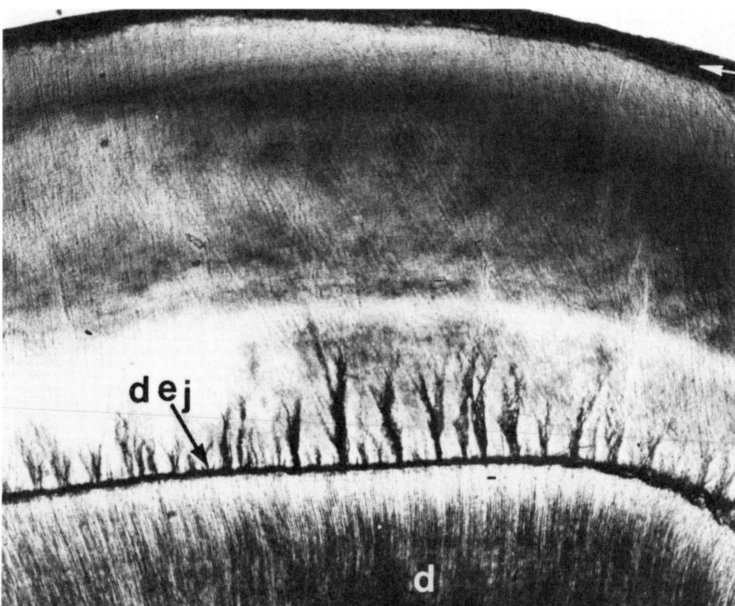

FIGURE 18-3 Ground section of a tooth. Enamel rods begin at the dentinoenamel junction (dej) and radiate outward to the surface (arrow). (d) Dentin with dentinal tubules.

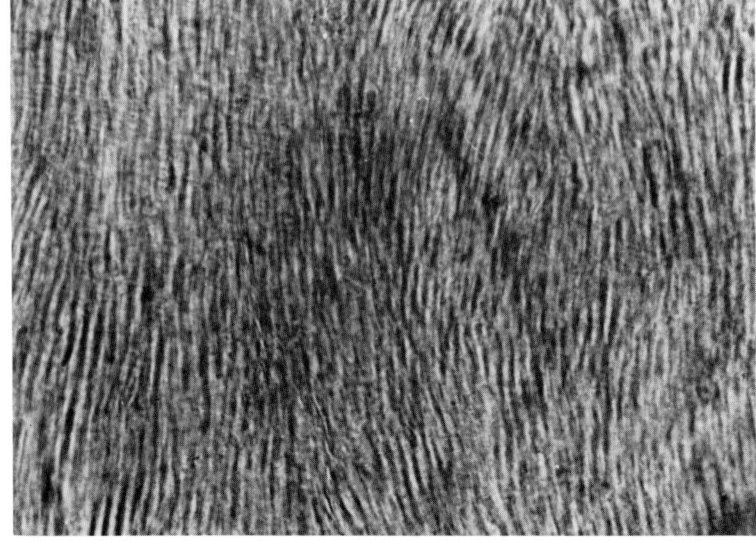

FIGURE 18-4 Enamel rods viewed at higher magnification. They appear light with a dark boundary, the interrod substance. Note the cross striations of the rods demarcating rod segments each 4 μm in length. The enamel rods are segmented because the enamel matrix is formed in rhythmic manner.

Structure

Since enamel is so highly mineralized, its structure is destroyed when decalcified for microscopic study because there is minimal organic matrix left to hold it together. For this reason, ground sections, which are dried and cut about 50 μm thick from nondecalcified enamel, are used to study the enamel structure. Structurally, enamel consists of two parts, rods (or prisms) and interrod (or interprismatic) cementing substance. The rods appear light, while the interrod substance is seen as dark lines. The rods are the calcified product of ameloblasts and originate at the dentinoenamel junction (DEJ) radiating outward to the enamel surface. The microscopic structure of the rods and interrod substances is better visualized at higher magnification in Fig. 18-4.

The course of the enamel rods from the dentin is initially straight, but then becomes wavy and twisted. This is especially evident in the regions of cusps and incisal edges, constituting *gnarled enamel* (Fig. 18-5). This interlocking of rods helps to provide strength to enamel under the forces of mastication. The rods average about 4 μm in diameter, being narrowest at the DEJ and doubling in

FIGURE 18-5 Ground section of enamel at the dentinoenamel junction. The enamel rods have a wavy and twisted course a small distance from the junction, producing the appearance of gnarled enamel.

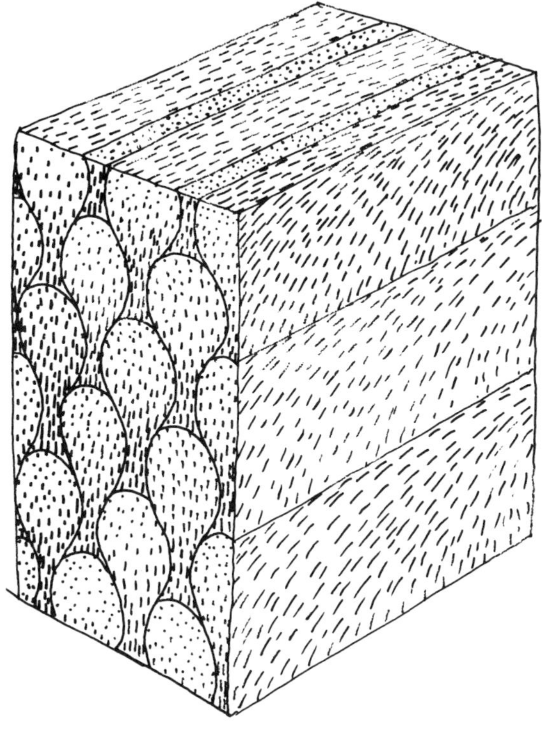

FIGURE 18-6 Model of enamel structure showing the outline of keyholes on the surface cut perpendicular to the direction of the prisms. The orientation of the apatite crystals within individual rods is also depicted.

diameter by the time they reach the surface. The rods are cemented together by interprismatic substance, less than 1 μm in thickness. This substance is similar in composition to the rods, being highly calcified, but is thought to be softer and more plastic than the rods.

With electron microscopy the common cross-sectional pattern of enamel consists of many keyhole or paddle-shaped prisms tightly packed together with the "tails" placed in between the "heads." Figure 18-6 is a drawing of the keyhole pattern of enamel indicating the orientation of the apatite crystals within individual rods. The apatite crystals have definite orientation in relation to each rod. They are usually parallel to the rod length in the head area and fan out in the tail region, thus producing a herringbone pattern with adjacent rods. When the crystals are first formed, they are needle-shaped. They then grow into hexagonal structures and form long ribbons stacked on top of each other with very little space in between. Figure 18-7 is a cross section of apatite crystals within an enamel rod. This gives enamel its strength.

The enamel rods are composed of seg-

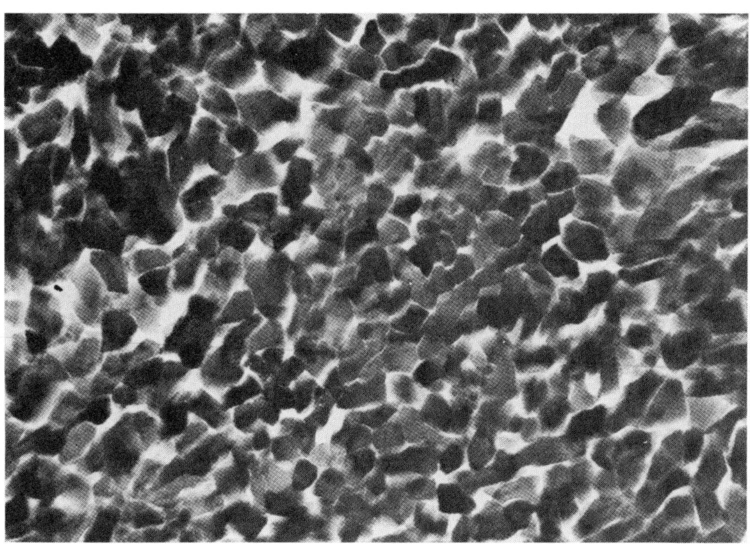

FIGURE 18-7 Electron micrograph of apatite crystals seen in cross section.

FIGURE 18-8 Ground section of tooth showing enamel and dentin. The striations present in the enamel are the lines of Retzius.

ments throughout their length. These segments are of uniform diameter, about 4 μm, and together with segments from adjacent rods produce a striated appearance, as seen in the ground section of enamel. These striations are perpendicular to the rod length and represent the resting stage in the rhythmic deposition of enamel matrix. Intercellular communication between ameloblasts is achieved via tight junctions, thus synchronizing their secretory function and producing incremental striations. Incremental *lines of Retzius* are brownish bands seen in ground sections of teeth (Fig. 18-8). They show the successive layers of enamel deposited during the formation of the crown and reflect the variations in structure and mineralization. The shock of birth is registered as an exaggerated line of Retzius, called the *neonatal line* (Fig. 18-9).

The neonatal line represents the abrupt change in the environment and nutrition of

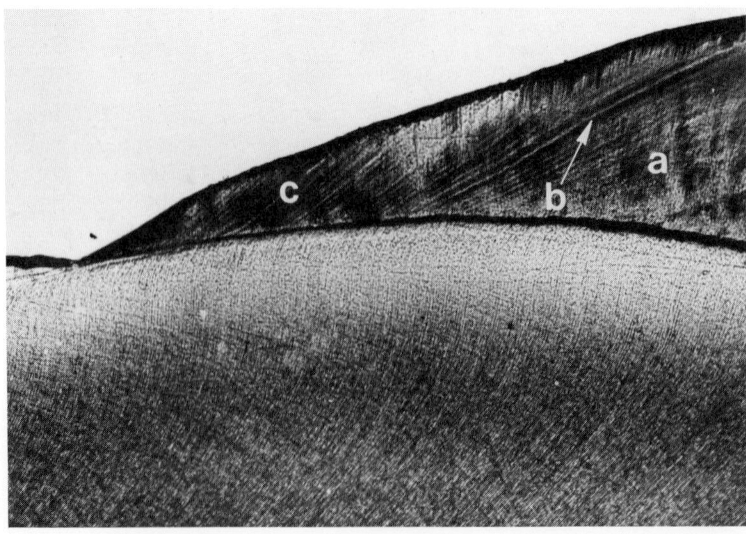

FIGURE 18-9 Ground section of tooth. The exaggerated incremental line at b is the neonatal line. It forms the border between enamel formed prenatally (a) and postnatally (c).

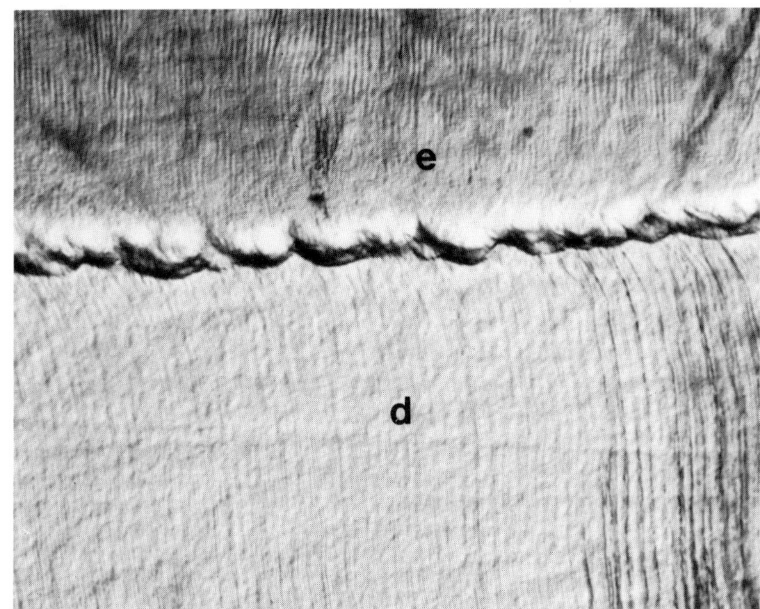

FIGURE 18-10 The dentino-enamel junction. Note the scalloped appearance. (d) Dentin. (e) Enamel.

the newborn infant. Several parallel lines of Retzius present in both prenatal and postnatal enamels similarly represent "growth arrest lines" of enamel due to various causes. It should be noted that such growth arrest lines are less prominent than the neonatal line. On the enamel surface, the Retzius lines are manifested as transverse wavelike grooves, called *perikymata*. They run parallel to each other and to the cementoenamel junction. The DEJ often has a scalloped appearance, with rounded projections of enamel fitting into shallow depressions in dentin (Fig. 18-10). This allows for a firmer connection between the enamel and dentin.

Thick ground sections of enamel show a number of specialized features. In enamel cut transversely, structures that resemble clumps of grass are frequently seen (Fig. 18-11). They are called *enamel tufts*. Enamel tufts originate at the DEJ and extend into the inner third of enamel. They represent an optical effect, being made up of enamel rods and interprismatic substance that are higher in organic content. The tuft is created by the multidirectional paths of the hypocalcified enamel rods as they leave the DEJ.

Other structures frequently occurring at the DEJ, especially in cervical areas, are *enamel spindles* (Fig. 18-12, arrows). They are spindlelike structures with blunt club-shaped ends in enamel (e). They appear to be continuous with dentinal tubules and are thus thought to be odontoblastic processes that pushed their way into the enamel epithelium before the hard substance of enamel was formed.

The orientation of the enamel spindles is perpendicular to the DEJ, therefore not conforming to the direction of enamel rods, but rather corresponding to the original direction of ameloblasts. Since enamel matrix is deposited at an angle to the long axis of ameloblasts, the direction of the two structures is different. This is diagramed in Fig. 18-13. The spindles appear dark when observed in dried ground sections due to the air (which appears dark due to its difference in refractive index from the calcified portion) replacing the disintegrated organic material.

Thin leaflike imperfections in enamel, resembling cracks, are called *enamel lamellae* (Fig. 18-14, arrows). In contrast to enamel tufts and spindles, the lamellae begin at the

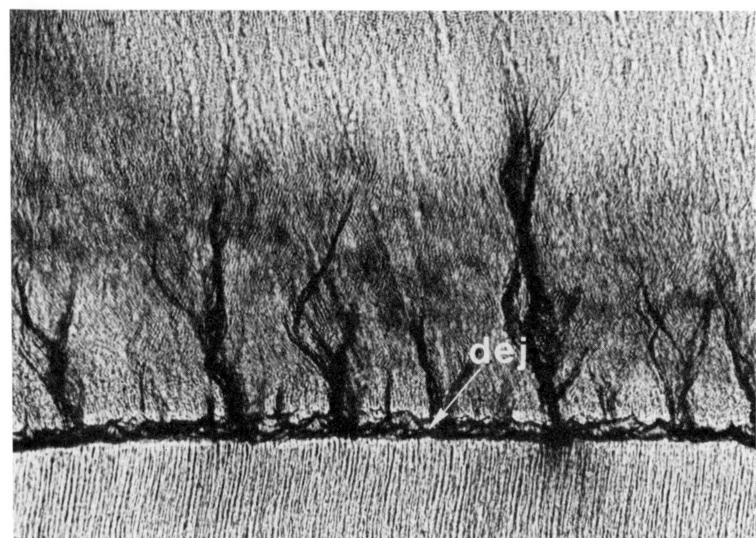

FIGURE 18-11 The dentino-enamel junction. Several enamel tufts extend from it. They consist of hypocalcified enamel rods and interprismatic substances and arise at the dej reaching into the enamel about one-fifth to one-third of its thickness.

enamel surface and proceed inward, sometimes reaching the DEJ or even possibly penetrating beyond it. Enamel lamellae run in a longitudinal direction from the cusp tip to the cervical area, while also proceeding in a spiral course. Thus, they can be best observed in horizontal sections of teeth. It is possible that enamel lamellae represent sites of bacterial entry contributing to the caries-producing process.

DENTIN

Characteristics

Dentin is a hard connective tissue encasing the pulp of the crown and root as well as supporting the enamel and cementum. It is somewhat elastic, thereby providing stability for the enamel which is much harder and unyielding. The mineral component of dentin is about 70 percent, making it harder than

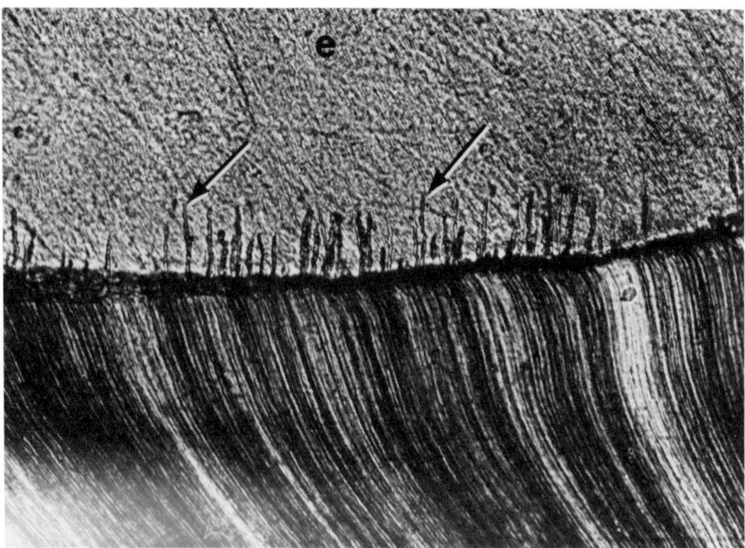

FIGURE 18-12 Enamel spindles (arrows). They occur at the dej and are spindlelike structures that appear to be continuations of dentinal tubules. (e) Enamel.

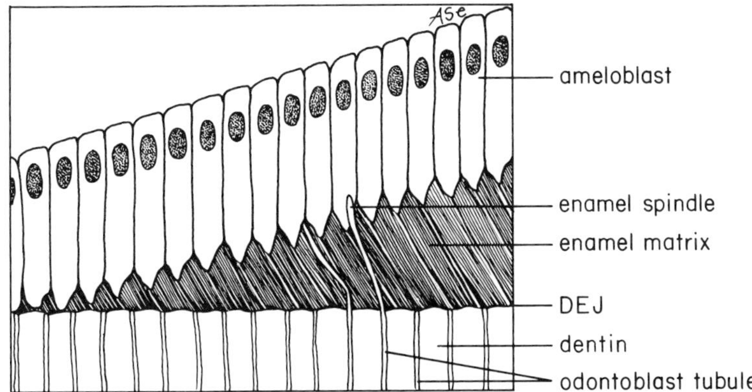

FIGURE 18-13 Diagram depicting the direction of enamel rods and enamel spindles.

bone, but softer than enamel. Because of its lesser mineral content, dentin is more radiolucent than enamel.

The formative cells of dentin, odontoblasts, form a circumpulpal layer with cell processes extending into dentin. Like ameloblasts, they function as "glandular" or secretory cells in elaborating matrix. In contrast to ameloblasts, which disappear once the tooth crown is completed, odontoblasts continue to lay down dentin matrix throughout life, especially after trauma to the tooth.

Dentin closely resembles bone in its physical and chemical properties. They are similar in matrix composition (collagen fibrils and glycoprotein), crystal type (apatite), germ layer origin (mesenchyme), and presence of entrapped cells or cell processes. They differ in that dentin is somewhat more highly mineralized than bone and dentin is completely avascular.

Structure

Dentin consists of two basic components: odontoblastic processes (op) and calcified matric (m) as seen in Fig. 18-15. The odontoblastic processes are cytoplasmic extensions of the odontoblasts, traversing the dentin within dentinal tubules to the DEJ or dentinocemental junction (DCJ) where they end in a branching network. Some terminal branches may extend across the DEJ into the

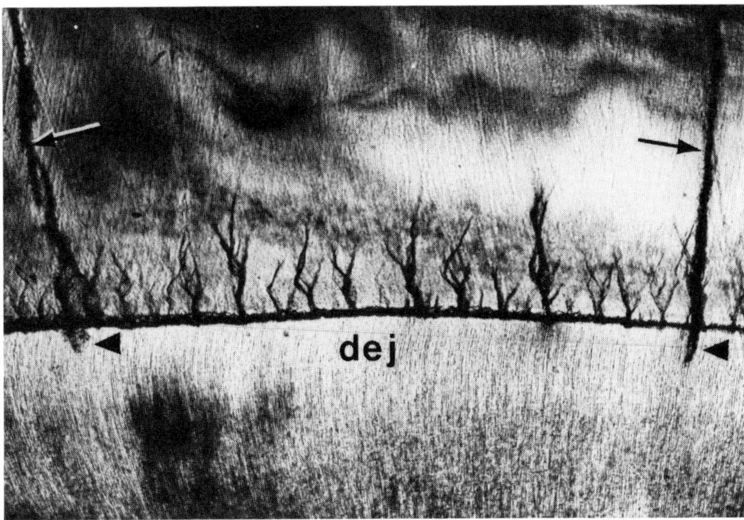

FIGURE 18-14 Ground section of tooth. Enamel lamellae (arrows) begin at the surface and extend inward into the enamel. Sometimes they reach the dej and even penetrate beyond it (arrowheads). Note the enamel tufts that begin at the dej and extend outward into the enamel.

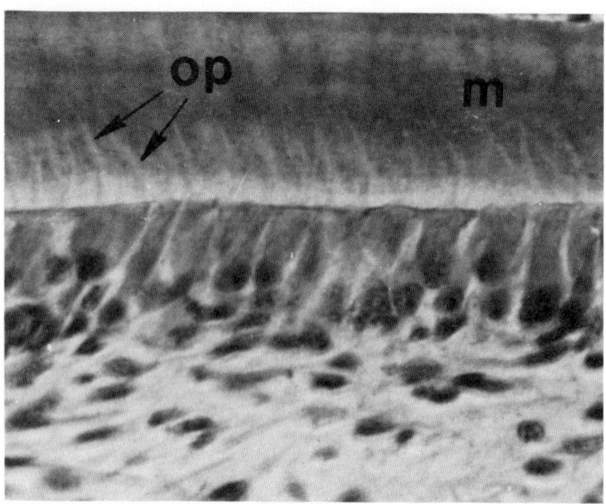

FIGURE 18-15 Light micrograph of dentin. Odontoblasts form a circumpulpal layer and have processes called odontoblastic processes (op) that extend into the calcified matrix (m). The odontoblastic processes have a diameter of 3 to 4 μm.

enamel, forming enamel spindles. Electron microscopy shows the cytoplasmic contents to be sparse, consisting of microtubules, mitochondria, and vesicles. The odontoblastic processes are thought to be highly irritable, perhaps resembling nerve fibers or stimulating nerve fibers which are in intimate contact with the odontoblastic cell body, thereby explaining the sensitivity of dentin. In this sense, odontoblasts with their processes may be regarded as "mechanoreceptors" in pain sensation.

Dentinal tubules radiate from pulp to DEJ and DCJ following a curved S-shaped course which is more prominent in the crown than in the root (Fig. 18-16). The diameter of the dentinal tubule varies throughout its length from 3 to 4 μm near the pulp to about 1 μm peripherally, resembling that of the odontoblastic process. This reflects a continued apposition of calcified materials along the inner aspect of dentinal tubules, which may be obliterated. The number of tubules per unit area is increased near the pulp, resulting in a greater quantity of intertubular matrix near the DEJ and DCJ. This results from the fact that the surface area of the pulp wall is much reduced from the total area of DEJ and DCJ.

As in enamel, a number of histologic features, which result from changes in the rate or pattern of matrix synthesis and mineralization of dentin during dentinogenesis, can be found in the adult dentin. The earliest formed matrix is almost entirely organic and later becomes mineralized. Because of the unique characteristics of this earliest formed matrix, *mantle dentin* (md in Fig. 18-16) is recognized as the first layer produced directly adjacent to enamel and cementum. It has a different microscopic appearance due to the

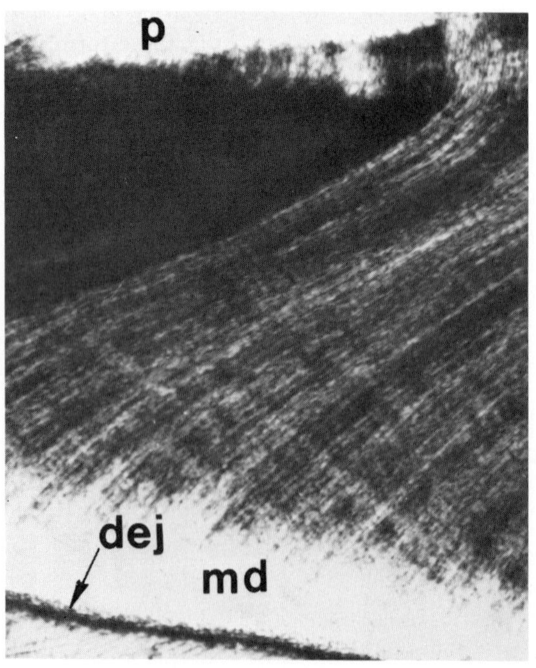

FIGURE 18-16 Dentinal tubules. They radiate from the dej following an S-shaped course which is most pronounced in the crown. The first layer of dentin formed during the development of a tooth is called mantle dentin (md). (p) Pulp.

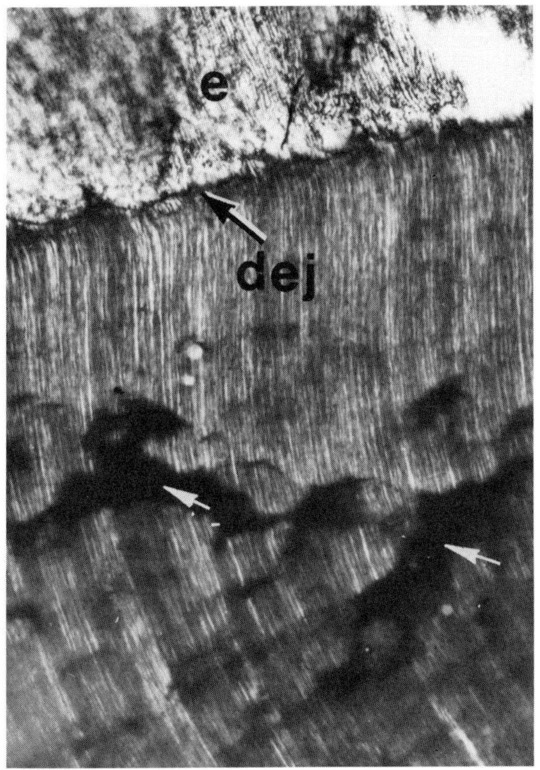

FIGURE 18-17 Interglobular dentin (arrows). It is a hypocalcified area of dentin that occurs adjacent to the dej. (e) Enamel.

perpendicular orientation of the coarse fiber bundles. The remaining dentin is called *circumpulpal dentin,* with a thin layer of predentin always present between calcified dentin and the odontoblasts.

The mineralization of dentin begins with the appearance of tiny initial calcification sites which expand peripherally. In a normal situation, these globules eventually merge to form a linear calcification front. If there is a delay in the rate of calcification, the fusion of the globules is prevented, thus forming spotty areas of highly calcified globules. The hypocalcified *interglobular dentin* is most often found in the crown just beneath mantle dentin and in the root near the DCJ. The globules are smaller and more closely packed together adjacent to the cementum, giving a granular appearance to the interglobular dentin in the root; therefore it is called *Tomes' granular layer.* Figure 18-17 is a ground section in which interglobular dentin appears dark with transmitted light (arrows). Figure 18-18 shows Tomes' granular layer (arrows) which is made up of fine granular-appearing linear aggregates of hypocalcified material immediately under the DCJ. The DCJ often is not clearly visible because of the similar refractive indices of dentin and cementum.

The structural pattern of dentin also reveals fine lines running at right angles to dentinal tubules called *incremental lines of von Ebner.* They represent the growth pattern of dentin, with the distance between lines of 4 to 8 μm, corresponding to the daily rate of apposition. These incremental lines are formed because of the synchrony of matrix elaboration by odontoblasts. Heavier lines, similar to the lines of Retzius in enamel, due to disturbances in the mineralization process are called the contour *lines of Owen* (Fig. 18-19). The most important line of Owen is the *neonatal line.*

As dentin is cellular, it is a vital tissue having the ability to react to physiologic and pathologic stimuli. The effects of aging or pathologic influences are characterized by the deposition of new layers of dentin or the alteration of original primary dentin. Throughout life, dentin formation continues on the entire pulpal surface, thereby decreasing the size of the pulp chamber. Dentin that is formed after eruption and therefore after the primary structure of the tooth is called *secondary dentin.* The secondary dentin shows irregularities or proceeds at an uneven rate, partly due to changes in the level of stimuli causing its formation. *Reparative* or irregular *dentin* is produced as the result of major trauma to the tooth, such as carious lesion. The tubules are often greatly reduced in number and extremely twisted in their course. Pathologic stimuli can also lead to changes in the dentin itself. *Transparent* or *sclerotic dentin* appears due to the increased

FIGURE 18-18 Tomes' granular layer (arrows). (c) Cementum.

mineralization around the degenerating odontoblastic processes. It can be observed only in ground sections. *Dead tracts* are empty dentinal tubules from which odontoblastic processes have been abruptly removed because of injuries and trauma.

CEMENTUM

Characteristics

Cementum is the hard connective tissue covering of the roots of teeth from the cementoenamel junction (CEJ) to the apex of the root. Not only does it function to protect the underlying dentin, but it also serves as an attachment structure for the periodontal fibers which anchor the tooth in the bony socket.

Cementum is generally less calcified than dentin, being about 45 to 50 percent inor-

FIGURE 18-19 Lines of Owen (arrows) seen in a ground section.

ganic. The remainder consists of organic matrix, collagen fibrils in a mucopolysaccharide ground substance, and water. Matrix is continually being formed by cementoblasts, as cementogenesis is a lifelong activity, especially in response to the forces on the tooth in occlusion. It is evident from the description of cementum that it closely resembles bone in function and physical and chemical properties, as well as in structure.

Structure

Two types of cementum are recognized on the basis of the presence or absence of cementocytes which represent trapped cementoblasts within the calcifying matrix of rapidly forming cellular cementum (Fig. 18-20). Acellular cementum is slower in formation, giving the cells a chance to retreat before calcification occurs. Locations of the two types are characteristic: the upper third to half of the root is covered by the acellular type, while the lower half is composed of cellular cementum, although the inner layer next to dentin may be acellular. For this reason, cellular cementum is usually present in the portion of the root where Tomes' granular layer is seen. Acellular cementum may even extend into the apical foramen. Since acellular cementum is formed first, it is sometimes called primary cementum, while cellular cementum is referred to as secondary cementum.

Growth patterns and markings, representing appositional and rest periods, seen in cementum are similar to those of other calcified tissues (bone, dentin, enamel). The cyclic incremental activity is seen as narrow bands following the root contour. These bands are called *imbrication lines.* The response of cementum to functional forces is further apposition. New cementum is deposited in successive layers called *lamellae* which vary in width, location, and composition, depending on the stimulus. If the stimulus is severe, the lamella is wide and contains many cementocytes.

PULP

Characteristics

The pulp is a somewhat primitive connective tissue encased by dentin, forming the core of the tooth. It develops from the dental

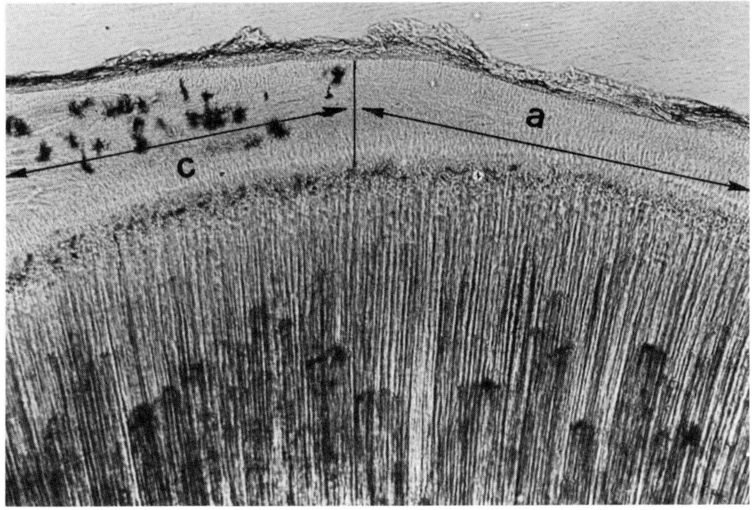

FIGURE 18-20 Ground section of tooth illustrating acellular (a) and cellular (c) cementum.

papilla, the mesenchymal portion of the forming tooth, which early becomes surrounded by and thus defined by the enamel organ. The pulp is considered to have *four functions:* formative, nutritive, sensory, and defensive. Since odontoblasts are an integral part of the pulp, the elaboration of dentin is a primary function. Dentinogenesis begins in the developing tooth and continues throughout life as long as the pulp remains vital. Since dentin has no vascular supply, it depends on the pulp for its nutrition, as evidenced by the capillary plexus near the odontoblasts that, in turn, supply the dentinal matrix via their processes. The nerves of the pulp consist of motor fibers to the smooth muscle cells in the arterial walls, as well as of sensory fibers which form a network under and around the odontoblasts, often sending fibers between them to follow the odontoblastic processes for varying distances into the dentinal tubules. All stimuli received by these nerve endings, such as heat, cold, or pressure, are interpreted in the same way as pain. The defensive reaction is expressed in two ways: by inflammation and formation of reparative dentin. Following insults from external injuries or irritants, such as caries, the pulp responds by forming a protective barrier of secondary or reparative dentin. With severe stimulation, inflammatory processes take place. The fluids leaving the blood vessels during the inflammatory process cause swelling in other tissues, but since the pulp is surrounded by an unyielding dentinal wall, the pressure from the accumulated fluids may cause the excruciating pain of pulpitis. The accumulation of fluids may also cause strangulation of the pulp, leading to necrosis due to lack of oxygen.

Morphology

The contour of the pulp generally follows the external outline of the tooth, consisting of coronal and radicular portions which are continuous with the periapical tissues via one or more apical foramina. In younger individuals the coronal pulp chamber is large, extending into the cuspal areas. With advancing age and continuing odontoblastic activity, the chamber decreases in size, sometimes irregularly. The developing radicular pulp is wide with a large open apical end which becomes considerably narrower as the tooth matures. Occasional small channels or accessory canals may run from the root canal out to the periodontal tissue, most commonly in the bifurcation area of multirooted teeth and in the apical one-third of the root. The constricted opening at the root tip, the apical foramen, is the passageway for the blood vessels, nerves, and lymph channels to and from the pulp. There may be multiple foramina, and they may vary in size and location, even changing with the deposition of hard tissues during tooth function.

The periphery of the pulp consists of the odontoblastic layer (Fig. 18-21). Other cellular elements consist primarily of fibroblasts, which are concerned with the production and maintenance of connective tissue matrix. In addition, there are those cells that are concerned with defense functions such as macrophages and wandering lymphoid cells.

The extracellular matrix of the pulp is composed of ground substance and fibers. The ground substance is gelatinous and quite viscous, reflecting the large amount of mucoproteins that embed the fine collagenous fibrils. The fibrous elements of the dental pulp are primarily collagenous in nature. Because of the fact that the pulp is encased in a limited space, the collagenous fibers are here rather delicate and oriented in random directions. Since these fine collagen fibrils are embedded in rich ground substance, many of them are argyrophilic, such as those found in the basement membrane. Hence they have been called *reticular fibers.* In fact, in young pulp most of these fibers cannot be

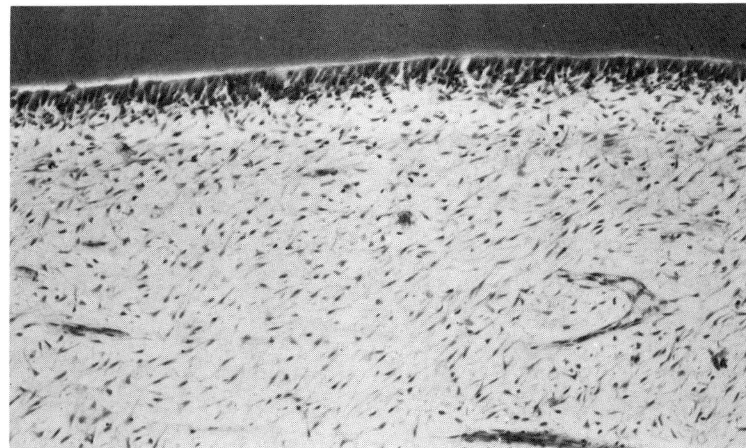

FIGURE 18-21 Light micrograph of tooth pulp. The peripheral layer next to the dentin consists of odontoblasts. The bulk of the pulp contains primarily fibroblasts and extracellular matrix. In addition, there are cells concerned with defense functions such as macrophages and lymphoid cells. Also note the presence of blood vessels and nerves.

stained with the usual dyes that demonstrate collagenous fibers, since the ground substance is richer in polysaccharides in younger pulp. In such young teeth, typical collagenous fibers can be found only around the blood vessels. As the pulp matures, reticular fibers decrease and collagenous fibers increase.

With normal aging, the cellular elements of the pulp decrease in number, while the fibers increase in both number and size. A number of calcified particles that are classified as free, attached, or embedded are found in the pulp and are called pulpstones or *denticles.* Figure 18-22 is a small portion of the pulp from an older individual in which several denticles are found. *Free denticles* (f) are located in the pulp tissue, separate from the dentinal wall (d), while *attached denticles* (a) are fused to it. *Embedded denticles* (e) are those that are completely surrounded by dentin. The number of denticles or pulpstones increases with age, as calcification in general

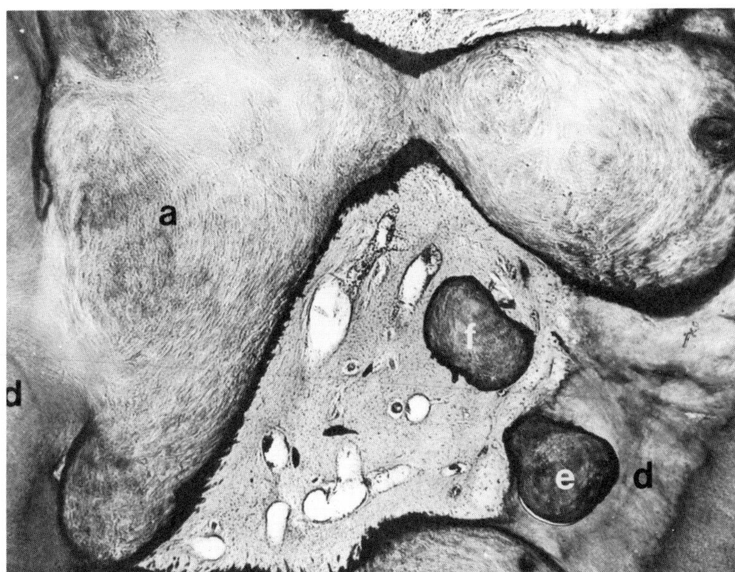

FIGURE 18-22 Light micrograph depicting the three types of denticles. (d) Dentin. (a) Attached denticle. (e) Embedded denticle. (f) Free denticle.

increases with age. Calcification in the pulp may occur in the form of a *diffuse calcification*. The diffuse calcification consists of irregular amorphous mineral deposits along the collagenous fiber bundles or blood vessels.

PERIODONTAL STRUCTURES

PERIODONTAL LIGAMENT

The *periodontal ligament* or membrane is defined as a series of fibrous connective tissue structures that surround and support teeth in their alveolar sockets (Figs. 18-23 and 18-24). It has many specialized bundles of fibers, all of which originate from the subgingival surface of the tooth and insert into the surrounding alveolar bone and the specialized connective tissues in the gingiva. Figure 18-23 illustrates the relationship of the tooth to the periodontal ligament and gingiva. The gingiva, as mentioned in Chap. 17, is covered by a minimally keratinized masticatory epithelium. Several regional specializations of the gingiva can be recognized. It has a flattened portion which is attached to the tooth called *epithelial attachment* (ea). Attached gingiva continues to become *crevicular gingiva* which borders the gingival sulcus (arrow). The connective tissue papillae under the attached and crevicular gingivae are short, sometimes being totally absent. The papillae become tall throughout the rest of the free gingiva, that is, the opposite surface from the tooth. Where the gingiva becomes attached to the alveolar bone there is a distinct groove called the *free gingival groove* (arrowhead). For this reason, the gingiva above the groove is called the *free gingiva* (fg), whereas the region below it is named the *attached gingiva* (ag).

The periodontal ligament carries blood vessels and nerves that must go through the periodontium to reach the tooth. The ligament is also a sensory organ, since it responds to the delicate pressure and tactile sensations originating in it as a result of mastication. The periodontal ligament contributes to the formation of oral fluid and prevents the entry of oral bacteria and other noxious agents into the periodontal region. A major problem confronting dentistry today is the question

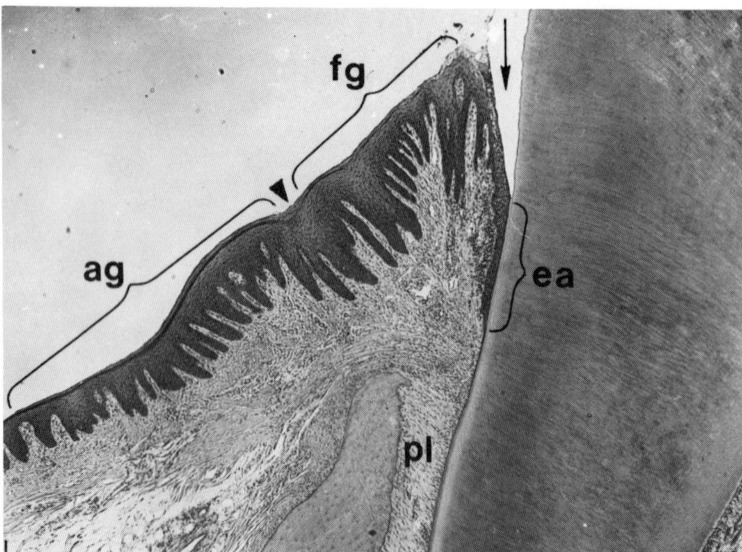

FIGURE 18-23 Light micrograph showing the relationship of the tooth to the gingiva and the periodontal ligament. (ag) Attached gingiva. (fg) Free gingiva. (ea) Epithelial attachment. (pl) Periodontal ligament. The crevicular gingiva borders the gingival sulcus (arrow). A groove, the free gingival groove (arrowhead), constitutes the border between free and attached gingiva.

of how to maintain the integrity of the periodontal ligament in its structure and function. Periodontal ligaments are made up of a number of principal fibers composed of collagen fiber bundles. These fiber bundles are named according to the region where they are present.

Principal fiber groups of the periodontal ligament

Free gingival fibers make up collagenous fiber bundles that originate from the cervical cementum. They travel outward and upward, spreading out at their distal end to intermesh with the collagenous fibers of the free and attached gingiva. *Transseptal fibers* are thick in nature, passing superior to the crest of the alveolar septum from one tooth to another. These bundles functionally unite adjacent teeth and support the proximal gingiva. *Alveolodental fibers* attach the tooth to the bone (b in Fig. 18-24). They can be further subdivided, depending on orientation and location of fibers.

Cellular components

Fibroblasts comprise the majority of cells in the periodontal ligament. They are active in establishing and maintaining the principal fibers and are thought to contribute to their reorganization during eruption and physiologic modulation of tooth positions. Osteoblasts deposit bone matrix around the ends of fiber bundles and thereby secure them to the alveolar wall. On the dental side, cementoblasts also lie between collagenous fibers near the surface of the cementum.

Epithelial structures

At the time of cementum formation, the epithelial root sheath of Hertwig which serves as the inductive structure in shaping the tooth contour breaks up, leaving small clusters and strands of epithelial cells near the cementum. These *epithelial rests* lie in the periodontal ligament adjacent to the cementum. These groups of cells may proliferate and give rise to pathological conditions such as cysts and other epithelial growths.

Calcifications called *cementicles* are occasionally found in the periodontal ligament of older individuals as free bodies or attached to or embedded in the cementum. They are

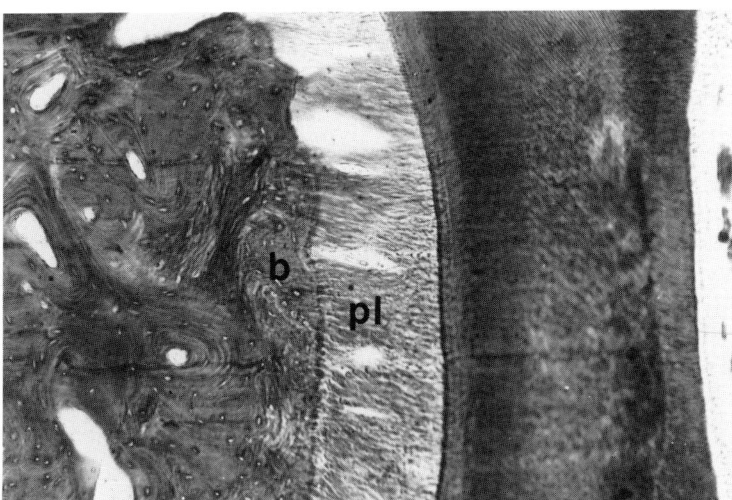

FIGURE 18-24 Light micrograph illustrating fibers of the periodontal ligament (pl). They run between the bone of the alveolar bone (b) and the cementum of the tooth.

presumed to be calcifications of degenerated epithelial cells.

Blood vessels and nerves

Unlike other dense fibrous tissue, the periodontal ligament has an abundant blood supply, which supplies nourishment to the dynamic bone and cementum. The artery which supplies the dental pulp sends branches into the apical portions of the periodontal ligament. Arteries supplying the alveolar bone also send branches through the alveolar wall into the periodontal ligament. These arteries also branch extensively as they emerge from the alveolar crest to supply the gingival tissue. Veins of the periodontal ligament form convolutions which empty and refill during mastication. In general, lymphatics follow the path of the blood vessels. Flow is directed from the ligament toward and into the adjacent alveolar bone.

A rich plexus of nerves is found in the periodontal ligament and is derived from nerves that follow the course of blood vessels. This plexus contains *three types of nerve endings:* free nerve endings (unmyelinated), knoblike swellings, and loops around principal fiber bundles. Pain, proprioception, and pressure are perceived by these receptors. All three play an important role in the protection of the masticatory system from sudden overload. In addition, sympathetic fibers of the autonomic nervous system innervate blood vessel walls which respond in accordance with the vascular response to autonomic input discussed elsewhere.

ALVEOLAR BONE

Strictly speaking, the alveolar process develops during tooth eruption only. However, preliminary organization and bone formation take place as early as the second month of fetal life. At this time, a bony groove, opening into the oral cavity and containing tooth germs, forms on both the maxilla and mandible (Fig. 18-25). Within these grooves and between tooth germs, bony septa begin to develop. During this period, the tooth germs (t) are contained in bony compartments, and they remain encased in bone despite their increasing size. This is due to the production of investing alveolar bone toward the oral cavity. Also during this period the alveolar bone increases in height and becomes part of the growing maxilla or mandible. However, this does not produce the alveolar processes, which are laid down during the eruption of the tooth. The process of tooth eruption begins when tooth development proceeds at a faster rate than alveolar bone formation.

The alveolar process is that portion of the mandible or maxilla which forms the socket and supports the tooth. There is no histologic distinction between alveolar bone and bone making up the remainder of the maxilla or mandible. The distinction is based only on development and function. On the basis of function, alveolar bone can be separated into two types. The first type is alveolar bone proper. The alveolar bone proper surrounds the tooth, forming the socket, and provides for the attachment of the periodontal fibers. The second type of bone represents the rest of the alveolar bone as it surrounds and supports the alveolar bone proper. It has been termed supporting alveolar bone.

The *alveolar bone proper,* forming the wall of the tooth socket, is traversed by many small openings which carry branches of the interalveolar blood vessels and nerves to the periodontal ligament. This bone is made up of both a laminated, compact type of bone and bundle bone, a form of bone anchoring periodontal fibers. Figure 18-24 shows a section of alveolar bone proper (b) anchoring periodontal fibers (pl). The cortical bone of the jaws is generally thicker in the mandible than in the maxilla, being thickest in mandibular molar regions. In both jaws, cortical

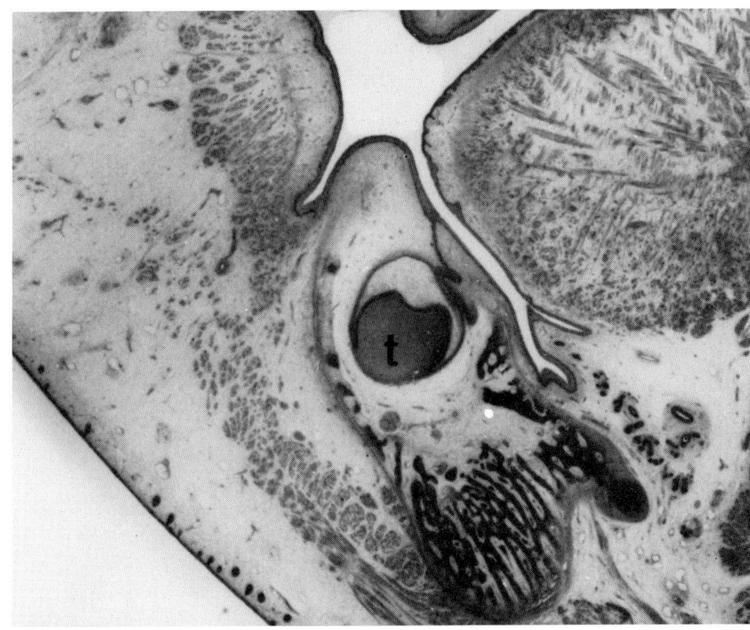

FIGURE 18-25 Portion of the developing jaw of the human fetus. The forming tooth (t) lies in a bony groove on the growing mandible.

bone is thinner in the anterior region. The cortical bone of the maxilla is perforated by many openings, allowing the passage of blood and lymph vessels. This is significant in considering the mechanism of action of local anesthetics which penetrate the openings in the cortical bone, allowing fast local anesthesia.

REVIEW SECTION

1. Figure 18-26 is a section through an adult tooth. The thin, dark line (arrow) in the enamel represents ________. Are these structures hypocalcified or hypercalcified?

These enamel lamellae appear dark because they are hypocalcified regions in the tooth. They result from developmental clefts or actual cracks after the formation of enamel.

2. Figure 18-27 is another section of tooth. What does the dark horizontal line (arrows) represent? What are the structures extending upward from this line? Why do these areas appear irregular?

In this section you see enamel tufts extending into the enamel from the DEJ. These tufts follow the course of enamel rods, thus reflecting their irregular growth path.

3. Refer to Fig. 18-28. Name the small, club-shaped objects indicated by the arrows. What do they represent?

In this particular section a large number of enamel spindles are seen as they extend into enamel from the DEJ. Enamel spindles represent the remnants of odontoblastic processes embedded in the enamel matrix during the early odontogenic period.

4. The three major groups of periodontal fibers originate in cementum. Where do they insert?

All three major groups of periodontal fibers originate in the cementum. They cross the periodontal ligament to embed in bone of the alveolar process, cross above the alveolar septum, and embed in cementum of the adjacent tooth or extend into the gingival tissues.

FIGURE 18-26

FIGURE 18-27

5. What function do the transseptal group of fibers perform?

By virtue of the cementum-to-cementum attachment, the transseptal fibers unite adjacent teeth and support interproximal gingiva.

6. Alveolar bone becomes apparent and distinct from maxillary and mandibular bone early in fetal life. (True or False?)

Although alveolar bone begins forming around the tooth germs early in fetal life, it does not become apparent and distinct from the mandible and maxilla until tooth eruption.

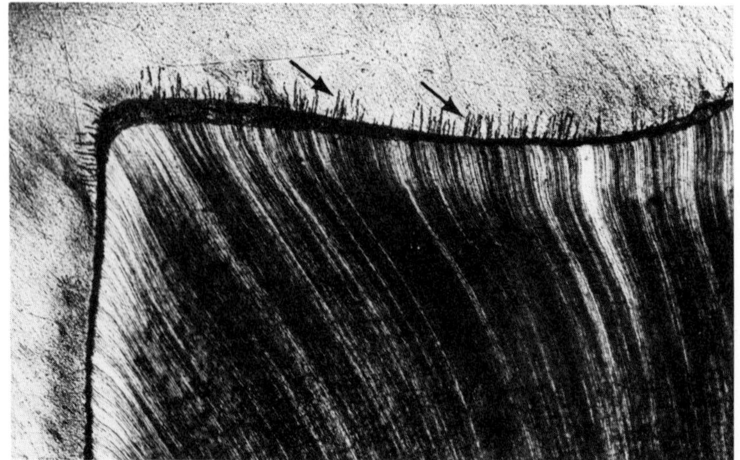
FIGURE 18-28

7. What purpose do the small holes in the alveolar bone proper serve?

The periodontal membrane derives its vascular and neural supply from vessels and nerves of the mandible and maxilla. The small openings in the alveolar bone proper allow passage of nerves and vessels to the periodontal ligament.

8. Cortical bone is (thicker/thinner) on the mandible when compared to cortical bone of the maxilla.

Cortical bone of the mandible is thicker and reaches its thickest state in the molar region.

BIBLIOGRAPHY

Bhaskar, S. N. (ed.): *Orban's Oral Histology and Embryology,* 8th ed., Mosby, St. Louis, 1976.

Melcher, A. H., and W. H. Bowen (eds.): *Biology of the Periodontium,* Academic, New York, 1969.

Mjor, I. A., and J. J. Pindborg (eds.): *Histology of the Human Tooth,* Munksgaard, Copenhagen, 1973.

Provenza, D. V.: *Oral Histology,* Lippincott, Philadelphia, 1964.

Siskin, M. (ed.): *The Biology of the Human Dental Pulp,* Mosby, St. Louis, 1973.

ADDITIONAL READINGS

Beust, T.: "Morphology and Biology of the Enamel Tufts with Remarks on Their Relation to Caries," *J. Amer. Dent. Assoc.* **19:**488 (1932).

Brabant, H., and L. Klees: "Histological Contribution to the Study of Lamellae in Human Dental Enamel," *Int. Dent. J.* **8:**539 (1958).

Gaunt, W. A., J. W. Osborn, and A. R. Ten Cate: *Advanced Dental Histology,* 2d ed., Wright, Bristol, 1971.

Hodson, J. J.: "An Investigation into the Microscopic Structure of the Common Forms of Enamel Lamellae with Special Reference to Their Origin and Contents," *Oral Surg.,* **6:**305–383, 495 (1953).

Meckel, A. H., W. J. Griebstein, and R. J. Neal: "Structure of Mature Human Dental Enamel as Observed by Electron Microscopy," *Arch. Oral Biol.* **10:**775–783 (1965).

———, ———, and ———: *Tooth Enamel,* Wright, Bristol, 1965.

Poole, D. F. G.: "The Use of the Microscope in Dental Research," *Brit. Dent. J.* **121:**71–79 (1966).

Schour, F.: "The Neonatal Line in the Animal and Dentin of the Human Deciduous Teeth and First Permanent Molars," *J. Amer. Dent. Assoc.* **23:**1946 (1936).

Shannon, I. L., R. P. Suddick, and F. J. Dowd (eds.): *Saliva: Composition and Secretion,* Karger, New York, 1974.

Tamarin, A.: "Myoepithelium of the Rat Submaxillary Gland," *J. Ultrastruct. Res.* **16:**320–338 (1966).

Tandler, B.: "Ultrastructure of the Human Submaxillary Gland. I. Architecture and Histological Relationships of the Secretory Cell," *Am. J. Anat.* **111:** 287–307 (1962).

———, C. R. Denning, I. D. Mandel, and A. H. Kutscher: "Ultrastructure of Human Labial Salivary Glands. I. Acinar Secretory Cells," *J. Morphol.* **127:** 382–408 (1969).

———, ———, ———, and ———: "Ultrastructure of Human Labial Salivary Glands. III. Myoepithelium and Ducts," *J. Morphol.* **130:**227–246 (1970).

Tomasi, T. B., and K. J. Biennenstock: "Secretory Immunoglobulins," *Adv. Immunol.* **9:**1–96 (1968).

Yohro, T.: "Nerve Terminals and Cellular Junctions in Young and Adult Mouse Submandibular Gland," *J. Anat.* **108:**409–417 (1971).

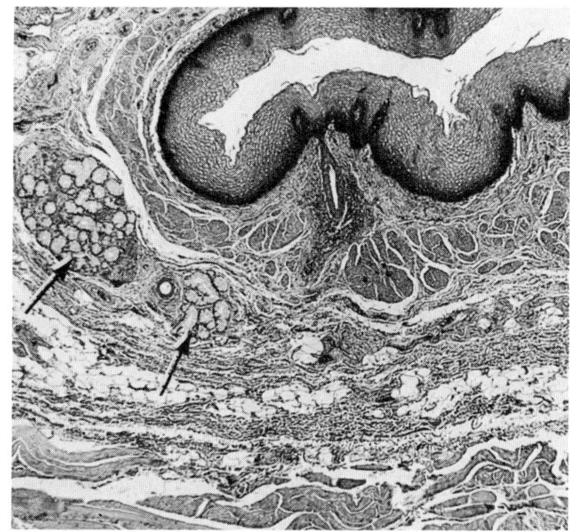

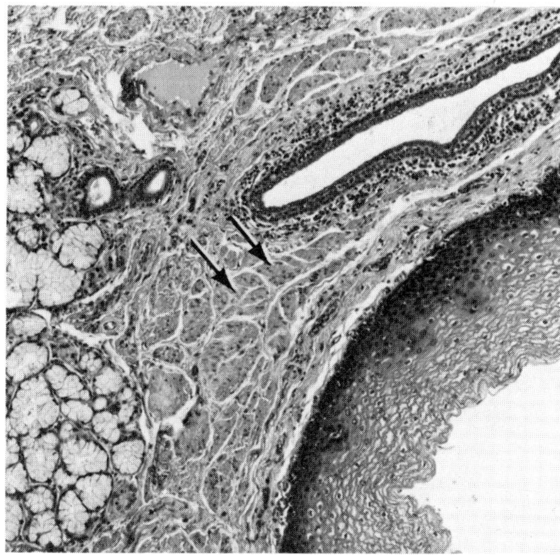

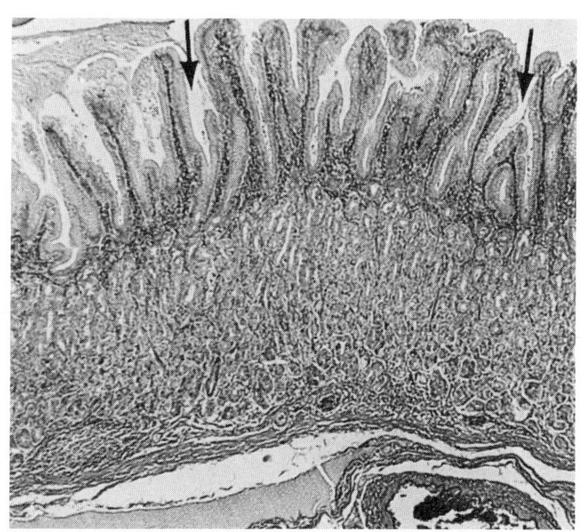

19

ESOPHAGUS AND STOMACH

OBJECTIVES

Upon completion of this chapter, the student will be able to:

1 Identify histologic sections of the following regions of the gastrointestinal (GI) tract:
(a) esophagus (upper, middle, and lower thirds) (b) cardioesophageal junction (c) cardia of the stomach (d) fundus of the stomach (e) pylorus of the stomach

2 Identify the changes in the following (where applicable) for the regions listed in objective 1:
(a) epithelium (b) glands (c) lamina propria (d) lamina muscularis mucosae (e) submucosa (f) tunica muscularis externa

3 When possible, correlate the structural modifications in each of the above with changes in function in the regions of the upper GI tract.

4 Identify the location, structure, and function of each of the following cells of the fundic gland:
(a) mucous neck cells (b) parietal cells (c) chief cells (d) argentaffin cells

The digestive system can be looked upon as a long tube passing through the body and associated glands. The digestive tube performs varying functions pertinent to food intake, its digestion and absorption, and excretion of waste products. Therefore, the digestive tube or gastrointestinal (GI) tract consists of specialized regions, namely, the mouth, pharynx, esophagus, stomach, small intestine, large intestine, and rectum.

The first step in the functioning of the GI system occurs in the oral cavity where foods are masticated and mixed with enzymes of the saliva. The mouth and paraoral structures allow the deglutition of the food through the esophagus into the stomach where it is mixed further with various enzymes and gastric acid. The structure of the stomach permits the temporary storage and thorough mixing of the food and gastric juice prior to passing the mixture into the small intestine. The small intestine receives the semidigested food in small quantities at a time and presents the bile and pancreatic enzymes which render changes in pH and further hydrolysis of food substances.

The major function of the small intestine is in the absorption of digested foods. The large intestine or colon absorbs water and contributes to the formation of fecal masses prior to their elimination. The histology of the mouth and pharynx has already been discussed. This chapter and Chap. 20 will cover the histology of the rest of the GI tract.

GENERAL STRUCTURE PLAN OF THE GI TRACT

Despite the functional heterogeneities that characterize the different segments noted above, the GI tract has a basic structure that is common throughout. An understanding of this basic general structure is helpful in considering the various structural specializations to be discussed later. Figure 19-1, a diagrammatic section of the GI tube, depicts these common features. The surface epithelium is moist throughout the GI tract. It is stratified squamous where abrasion and other physical forces are imposed. These areas include regions of the mouth as well as of the esophagus and the anal region. Glands of different types are present along the entire length of the GI tract. They are unicellular glands of the epithelium, such as goblet cells and certain mucosal and submucosal glands, as well as massive outgrowths including the major salivary glands, liver, and pancreas.

Subjacent to the epithelium is a layer of connective tissue called the *lamina propria.* The lamina propria is attached to an underlying muscle layer, the *lamina muscularis mucosae.* The lamina muscularis mucosae consists of two smooth muscle cell layers: a poorly developed inner circular layer and a well-developed outer longitudinal layer. These three structural elements—namely, the epithelium, lamina propria, and lamina muscularis mucosae—are collectively referred to as the *mucosa* of the GI tract.

Surrounding the mucosa is a layer of loose fibrous elastic connective tissue (FECT), the *submucosa.* This layer contains large blood vessels, lymphatics, and nerves which radiate into the lamina propria and muscularis mucosae. Many lymphocytes, plasma cells, and macrophages are also seen in the lamina propria and submucosa. A network of nervous elements that are part of the parasympathetic system, called *Meissner's plexus,* is found in the submucosa in the intestines.

The outer muscle layer, *tunica muscularis externa,* consists of two layers of smooth muscle cells external to the submucosa. The inner layer is primarily circular in orientation, while the outer layer is longitudinal. This pattern or organization, inner circular and outer longitudinal, aids in distinguishing between longitudinal and cross-sectional views of the GI tract.

The *myenteric plexus,* also called *Auer-*

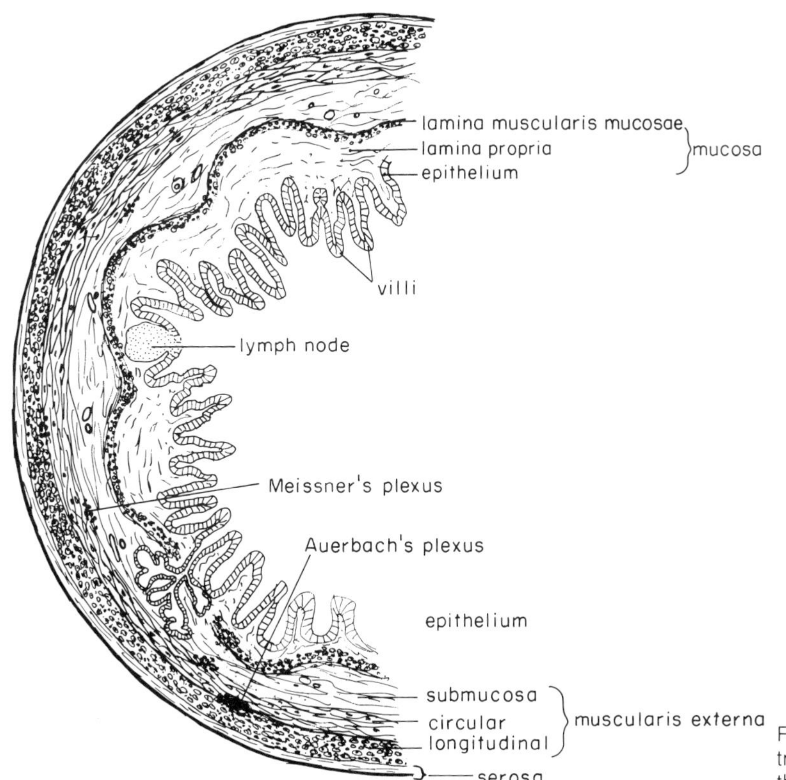

FIGURE 19-1 Diagram illustrating common features of the segments of the GI tract.

bach's plexus, is another parasympathetic plexus located between the two muscle layers of the muscularis externa. Impulses from these nerves initiate coordinated peristaltic contractions. Figure 19-2 is a section through the two layers of the lamina muscularis externa and shows a number of nerve fibers along with several parasympathetic ganglion cells of Auerbach's plexus (arrows). The large ganglion cells are surrounded by the more numerous, smaller nuclei of satellite cells in the plexus. Both the nuclei and nucleoli of the ganglion cells are prominent. The nuclei of satellite cells are smaller than those of ganglion cells and are oval in shape. The dense and spindle-shaped nuclei that are away from the ganglion cells belong to fibroblasts. Find the location of Auerbach's plexus on the diagram in Fig. 19-1. The muscle layers of the GI tract promote the mixing of the ingested material (chyme) with digestive enzymes and also cause digested food to come in contact with absorptive cells. The muscle layers also maintain the motion of the food through the tube via wavelike peristaltic contractions.

At certain junctions, the muscular layers are specialized to hold food in particular regions of the digestive system. For example, the cardiac (upper end) and the pyloric (lower end) sphincter muscles of the stomach prevent food from escaping during processing in the stomach. Anal sphincters retain the feces until voluntary elimination.

External to the muscularis externa is a layer of loose FECT, which is in turn covered by a mesothelium. Although the mesothelium is not present in areas such as the esophagus and portions of the intestines, much of the GI tube in the abdomen is covered by the visceral peritoneum which is the mesothelial covering of the tract. The mesothelium and the connective tissue layer outside the muscularis externa constitute the

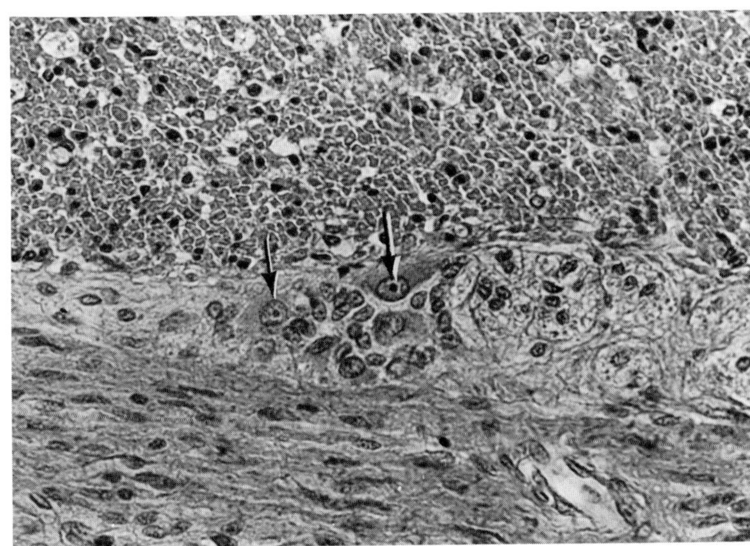

FIGURE 19-2 Light micrograph of tunica muscularis externa. The myenteric plexus (arrows) is located between the two layers of smooth muscle.

serosa. The term *adventitia* is used to designate the outermost connective tissue covering in regions where the mesothelium is absent.

ESOPHAGUS

Since the primary function of the esophagus is in aiding in deglutition and conducting the food into the stomach, the wall of the esophagus reflects structural modifications of the basic plan necessary for performing these functions (Fig. 19-3). The wall has longitudinal folds caused by the constriction of the esophagus which must remain empty between the passage of food. As indicated earlier, the esophageal epithelium is a moist stratified squamous type which provides protection against abrasion (Fig. 19-4).

Since such an abrasive force imposed upon the esophageal wall is only a moderate one, the stratified squamous epithelium of the esophagus is nonkeratinized. This epithelium ends abruptly at the cardioesophageal junction, that is, the junction between

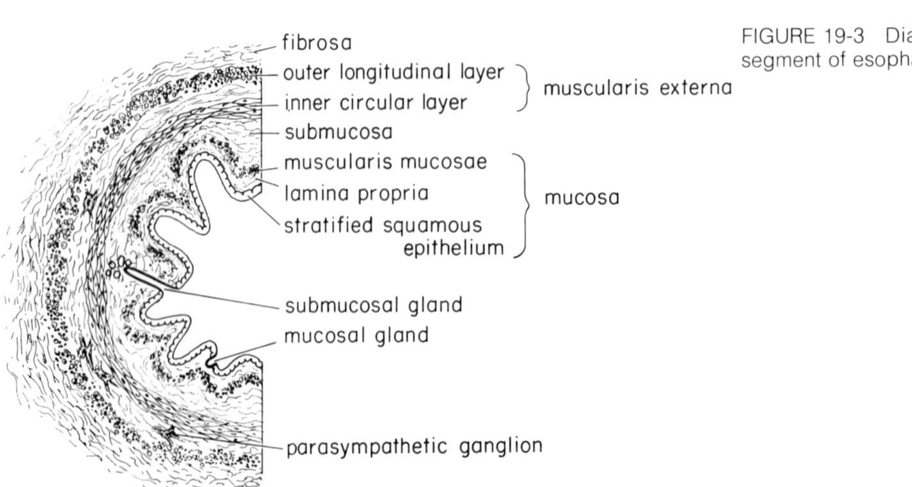

FIGURE 19-3 Diagram of a segment of esophagus.

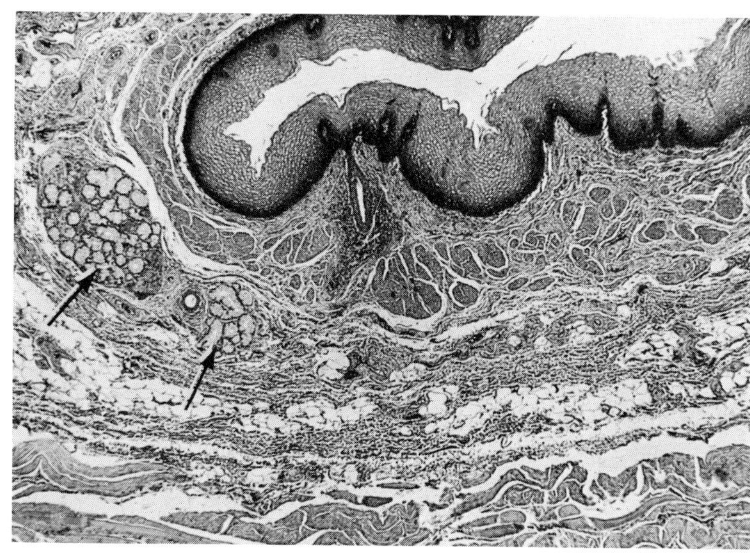

FIGURE 19-4 Light micrograph of a cross section of esophageal mucosa. It is lined with stratified squamous epithelium. Mucous glands (arrows) are embedded in the submucosa. Longitudinal muscles of lamina muscularis mucosae (m) are prominent.

the esophagus and the cardiac region of the stomach. Even though the chewed food is moistened by saliva in the mouth, the esophagus needs additional lubrication. Mucus-secreting *esophageal glands* (Fig. 19-4, arrows) embedded in the submucosa throughout the esophagus help provide this additional lubrication. Mucus-secreting glands are also present in the lamina propria of the upper and lower ends of the esophagus, although the upper glands are sometimes absent. These glands are called *cardiac glands* since they are structurally similar to the glands of the cardiac region of the stomach.

The cross section of the esophagus in Fig. 19-5 shows clearly the inner circular and the outer longitudinal layers of the muscularis externa, circular in brackets (c) and longitudinal in brackets (l).

The muscularis mucosae is rich in longitudinal muscles that are interrupted (Fig. 19-4). The stratified squamous epithelium

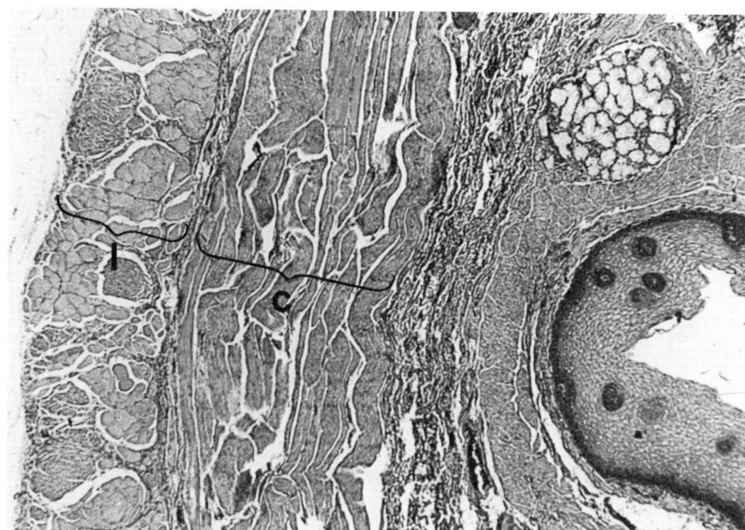

FIGURE 19-5 Cross section of the esophagus illustrating the arrangement of the layers of muscularis externa in the upper third. (c) Circular layers. (l) Longitudinal layers.

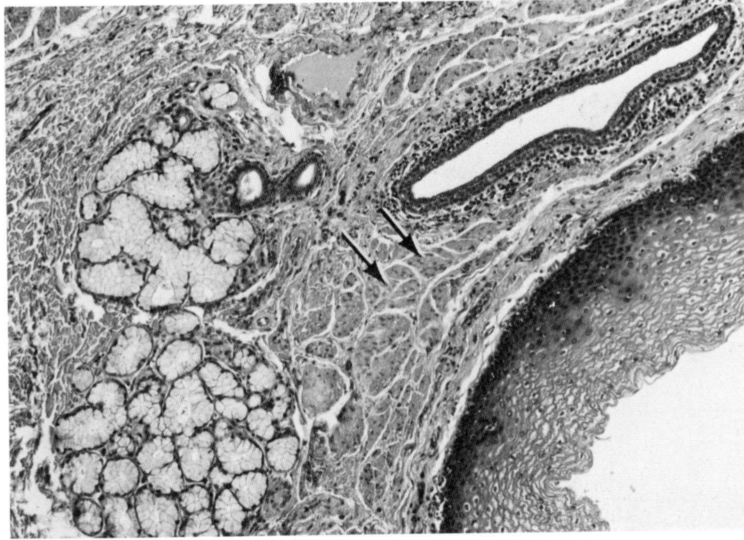

FIGURE 19-6 Cross section of esophagus viewed at higher magnification. Muscularis mucosae is made up of bundles of smooth muscle cells (arrows). A duct (d) leads from the esophageal gland to the lumen.

bordering the lumen shows a dark linear arrangement of basal cells. Esophageal glands are present in the submucosa. Figure 19-6, a cross section of the esophagus, shows at a higher magnification the epithelium on the right and the lamina propria and submucosa to the left. The lamina muscularis mucosae is made up of bundles of cross-sectioned smooth muscle cells (arrows). The mucus-secreting acini of the esophageal glands are clearly demonstrated in the left half of the field. Note also a duct of an esophageal gland which in the upper right-hand corner appears as a slit (d). It is lined by stratified cuboidal epithelium with accumulation of lymphocytes in the FECT surrounding it. The inner circular layer of muscle in the lamina muscularis mucosae is not developed in the esophagus. The longitudinal layer is quite prominent and tends to produce a discontinuous layer (m in Fig. 19-4). In contrast, both the inner and outer layers of muscle are well developed in the muscularis externa. Since the upper half or so of the esophagus is involved in the voluntary swallowing motion, the muscularis externa in the upper region is composed of skeletal muscles, as shown in Fig. 19-7, which depicts a cross-sectioned muscularis externa. Note the heavy longitudinal fibers present in the right half of the field.

In fact, many of the longitudinal fibers represent continuing muscular fascicles of pharyngeal constrictor muscles. The muscularis externa in the middle third of the esophagus is made up of a mixture of skeletal and smooth muscle cells, as it represents a transition of the muscularis externa into an involuntary region. Figure 19-8 is a cross section of the muscularis externa showing the inner circular (left half) and outer longitudinal (right half) layers of muscle. The layers show both large skeletal muscle fibers (k) and smaller smooth muscle cells (m). The amount of smooth muscle fibers increases as they approach the lower third of the esophagus where the muscularis externa is totally involuntary and is comprised entirely of smooth muscle. Neural elements of parasympathetic ganglia are also present in the lower third of the esophagus between the two layers of the muscularis externa.

STOMACH

The esophagus passes the food into the stomach, which represents a unique enlargement of the GI tract with regional specializations.

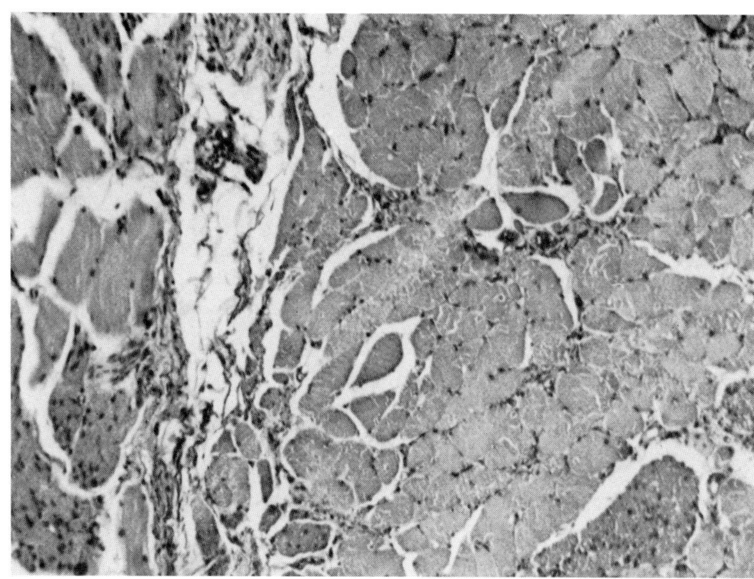

FIGURE 19-7 Light micrograph of muscularis externa in the upper part of the esophagus.

The stomach is divided into three histologic regions according to their function and structure. The region nearest the esophageal opening is termed the cardiac region, or *cardia*. The term cardia refers to the proximity of this area to the heart. The body of the stomach is called the fundic region, or *fundus*. The pyloric region, or *pylorus*, is that region of the stomach which tapers into the duodenum, or the beginning of the small intestine.

The surface of the gastric mucosa has numerous folds, called *gastric rugae*. These are more prominent in the fundic region where the rugae run more or less longitudinally. In addition, the mucosa of the stomach has small irregular mounds called *gastric areas*. Both gastric rugae and areas are gross anatomical features. Histologically, the entire gastric mucosa shows a third irregularity, namely, *gastric pits*. Gastric pits, are short invaginations of the mucosal surface (Fig. 19-

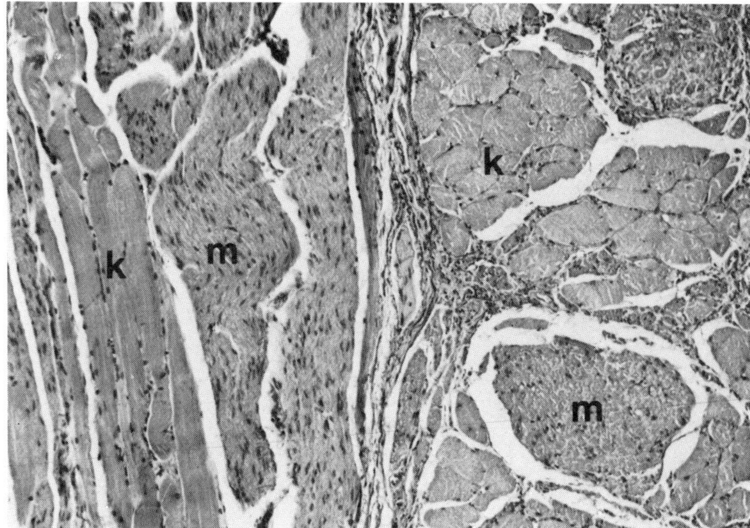

FIGURE 19-8 Light micrograph of a cross section of muscularis externa of the esophagus in the middle third. It is made up of a mixture of skeletal muscle (k) and smooth muscle (m) and comes from a region of esophagus where there is a transition from voluntary to involuntary action.

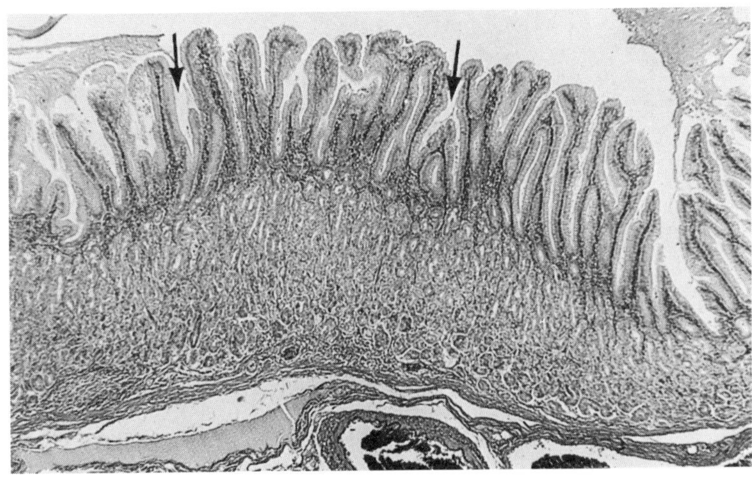

FIGURE 19-9 Light micrograph of gastric mucosa. Gastric pits (arrows) are short invaginations of the mucosal surface.

9, arrows). They differ in depth, being deepest in the pyloric region. All mucosal glands of the stomach open at the bases of gastric pits.

CARDIAC REGION OF THE STOMACH

The epithelial lining of the GI tract changes abruptly at the cardioesophageal region from stratified squamous epithelium to simple columnar epithelium. This junction, along with other regions where the type of covering epithelia changes abruptly, has been noted for its predilection for cancer development. Although not fully understood, the merging of two different types of epithelial tissue is believed to be conducive to malignant transformation.

Many mucous glands lubricate the cardiac region. Straight tubular gastric pits extend about halfway into the lamina propria. The muscularis mucosae is evident as the longitudinally running muscle bordering the lamina propria. Somewhat coiled, tubular *cardiac glands* that produce mucus open at the base of the gastric pits. The cardiac glands occupy the entire thickness of the lamina propria, making smooth muscle cells of the lamina muscularis mucosae barely visible. Throughout the lamina propria are scattered lymphocytes.

At either end of the stomach are found sphincters that hold the food while it is being churned in the stomach. As mentioned previously, the sphincter of the cardiac region is termed the *cardiac sphincter.* The marked development of the tunica muscularis externa forms the sphincter, which is normally in a contracted state. The cardiac sphincter is not well developed in newborn infants, which accounts for the familiar vomiting shortly after feeding. At the cardiac sphincter, a third, obliquely oriented layer of muscle appears internal to two layers of the muscularis externa. This third layer of muscle is present throughout the stomach and helps in developing the churning action by the stomach.

FUNDIC REGION OF THE STOMACH

Histologic modifications in the fundic region allow the production of gastric enzymes and acids here. As mentioned earlier, the mucosal surface forms a series of gastric folds called *rugae* which run more or less longitudinally. These folds allow volume changes of the stomach and increase its surface area, thus allowing the stomach to become distended and providing more area for glands which produce digestive juice.

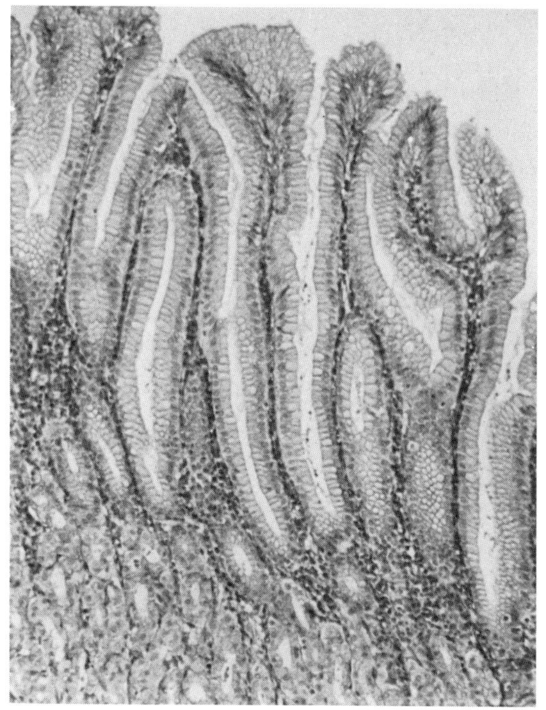

FIGURE 19-10 Light micrograph of gastric mucosa showing gastric pits.

FUNDIC GLANDS

Numerous gastric pits are found in the fundus, as noted by deep invaginations of the mucous surface (Fig. 19-10). All of the lining cells are mucous secretory cells which are stained with the periodic acid Schiff (PAS) procedure for demonstration of mucin (Fig. 19-11). Like cardiac glands, *fundic glands* (or *gastric glands*) extend from the bottom of each gastric pit (Fig. 19-11). Fundic glands are tubular in shape and branch occasionally. They are larger than cardiac glands and show regional specialization which may be identified as (1) the *mouth,* or opening from the pit into the gland, (2) the *neck,* or constricted portion, (3) the *body,* or main portion of the gland, and (4) the *fundus,* or base of the gland. The gastric pits extend only about a third of the way into the mucosa, and along the base of the glands is the muscularis mucosae. The structure of the fundic gland is presented in Fig. 19-12.

In longitudinal sections of gastric pits the appearance of the mucous epithelium is clearly seen (Fig. 19-13). Since all these cells are mucus secreting, the physical consistency of their cytoplasm is the same. This accounts for the lack of goblet cell–like appearance of

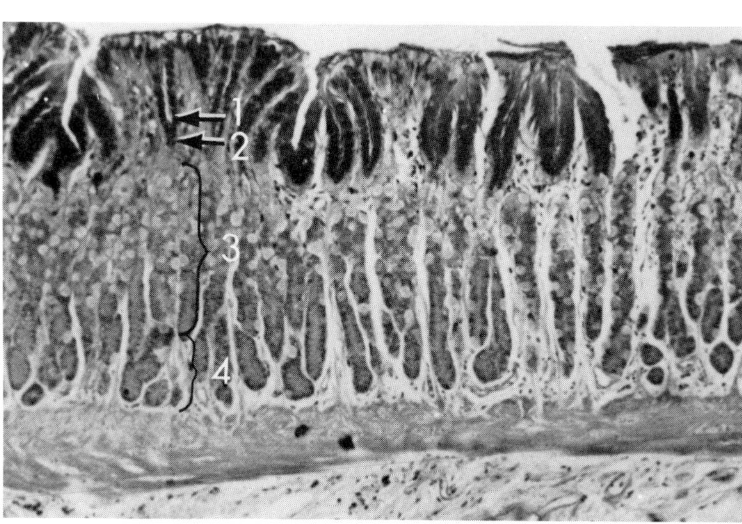

FIGURE 19-11 Light micrograph of gastric mucosa (fundic) stained with PAS. All surface cells are mucus-secreting cells. Fundic glands extend from the bottom of each gastric pit. They have regional specializations that are identified as (1) the mouth, (2) the neck, (3) the body, and (4) the fundus.

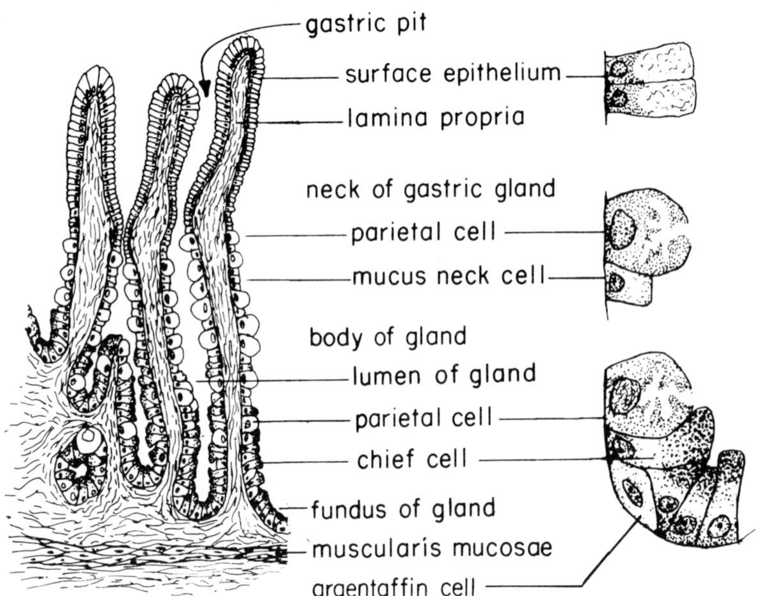

FIGURE 19-12 Diagram of the structure of the fundic gland.

mucous cells, as in the respiratory mucosa. Figures 19-14 to 19-17 are higher-power views of fundic glands. Note the appearance of PAS-stained mucous lining cells of the collapsed pits on the right in Fig. 19-14. In the neck region of the gland are *mucous neck cells.* They are an undifferentiated variety that is thought to be the progenitor cells (arrows). They are small and contain only a limited quantity of mucus. It has been shown that mucous neck cells divide, migrate from this area, and eventually replace both the surface epithelium and the glandular cells of the stomach.

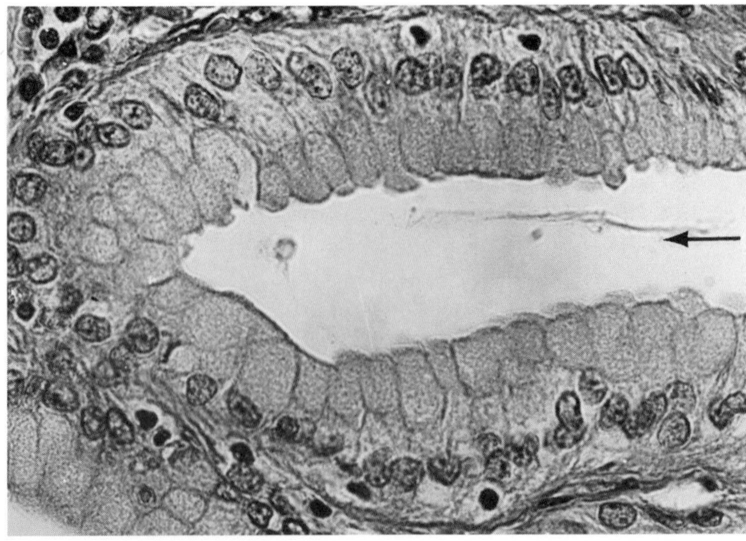

FIGURE 19-13 Longitudinal section of a gastric pit (arrow). It is lined by mucous cells that all have the same appearance.

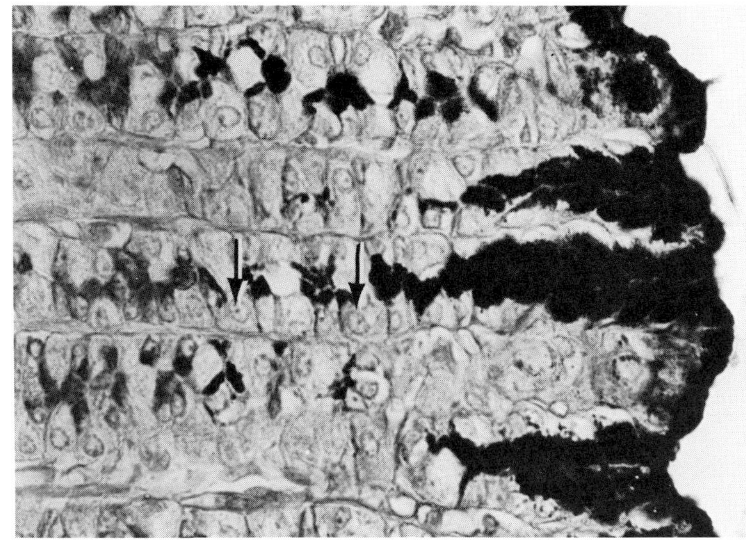

FIGURE 19-14 Light micrograph of fundic glands stained with PAS. The mucus in the mucous lining cells of the collapsed pit stains positively. In the neck region, mucous neck cells (arrows) are present. These cells contain only a small amount of mucus.

In addition to the mucous neck cells, three additional types of cells can be identified in the fundic gland proper. *Parietal cells* are numerous in the neck and body of the fundic gland. Parietal cells are large, are oval or polygonal in shape, and contain a cytoplasm that is eosinophilic (Figs. 19-15 and 19-16, arrows). Figure 19-15 is a view of the upper half of fundic glands that run vertically through the field. Note the large, lightly stained parietal cells. Cross sections of several fundic glands are shown in Fig. 19-16. This particular photomicrograph is close to the base of the gland where only occasional parietal cells are scattered among the more basophilic chief cells. The eosinophilia of the parietal cell reflects the lack of RER in these cells. The acidophilic region has been shown to contain many mitochondria (Fig. 19-17). Parietal cells secrete hydrochloric acid and

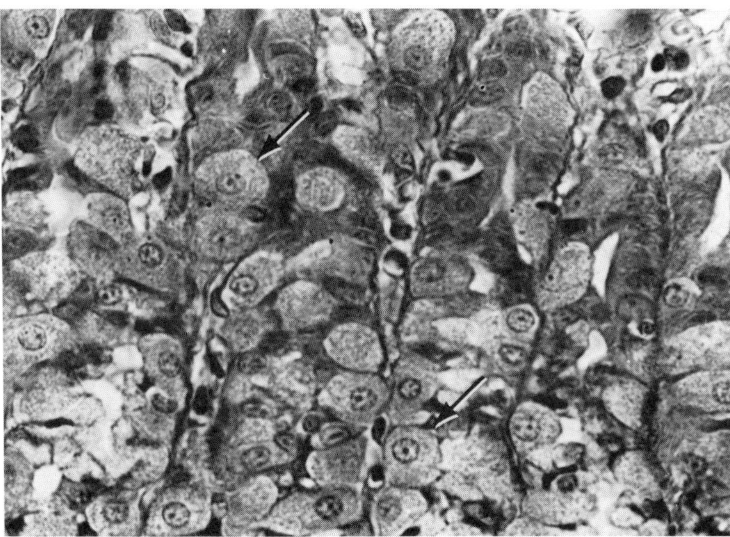

FIGURE 19-15 Light micrograph of longitudinal section of fundic glands. Note the lightly stained parietal cells (arrows).

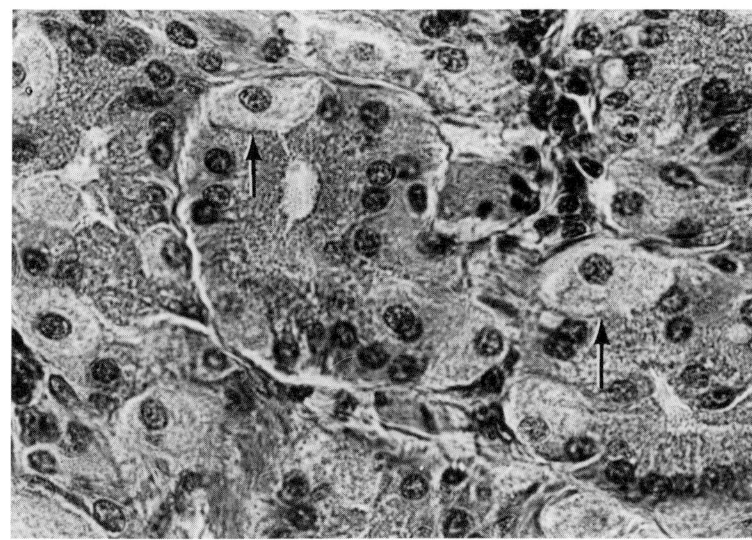

FIGURE 19-16 Cross section of fundic glands. Parietal cells (arrows).

create an acidic environment for the action of the enzymes secreted by the chief cells, to be described below.

In addition to the numerous rounded mitochondria with well-developed cristae, the parietal cell contains narrow channels in the cytoplasm formed by irregular invaginations of the apical plasma membrane. These channels are called *intracellular canaliculi.* Microvilli-like structures extend into the intracellular canaliculi. Intracellular canaliculi can be seen at the light microscopic level under appropriate conditions. Figure 19-18 is an LM of several fundic glands in which intracellular canaliculi (arrows) are visualized as lightly stained channels in parietal cells. The moderately dark granular cytoplasm reflects the presence of well-developed mitochondria as diagramed in Fig. 19-17.

Chief cells are the most numerous cells of the fundic gland. Chief cells secrete serous

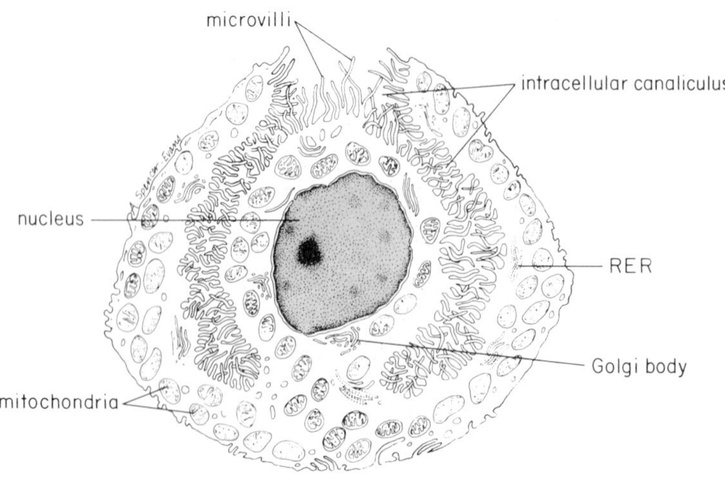

FIGURE 19-17 Drawing of the ultrastructure of the gastric parietal cell. It is a large cell with abundant large mitochondria and intracellular canaliculi lined with microvilli. (*Redrawn after S. Ito.*)

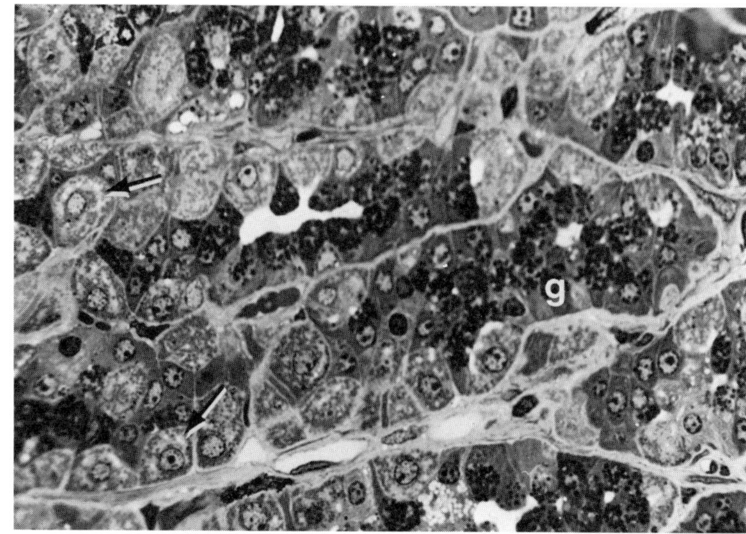

FIGURE 19-18 Light micrograph of fundic glands. Intracellular canaliculi (arrows) are visualized as lightly stained areas in the parietal cells. Chief cells are basophilic and contain zymogen granules (g) in the apical cytoplasm.

enzymes, such as pepsinogen and are large pyramidal-shaped cells (Fig. 19-12). Figure 19-18 demonstrates the zymogen granules (g) of chief cells prepared in a manner that preserves the granules. The chief cells increase in number toward the base of the gland (Fig. 19-19). The cytoplasm in the paranuclear region is dark (basophilic). Zymogen granules are present in the apical cytoplasm. In an EM these cells show cytoplasmic differentiation that is typical of most serous or protein-secreting cells. These include abundant RER, well-developed Golgi complex, and apically located zymogen granules, as well as a nucleus with prominent nucleoli.

Argentaffin cells, sometimes called *enterochromaffin cells,* are not visible in routine

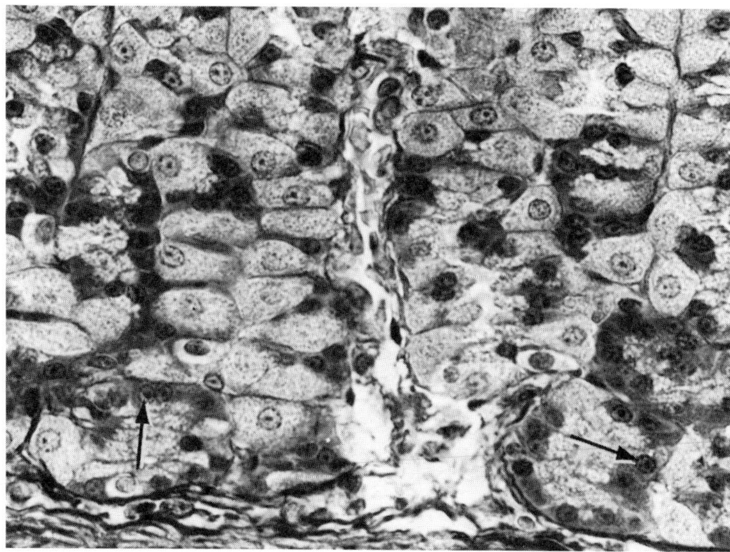

FIGURE 19-19 Light micrograph of the lower half of fundic glands. Chief cells (arrows) appear dark, while parietal cells are lightly stained.

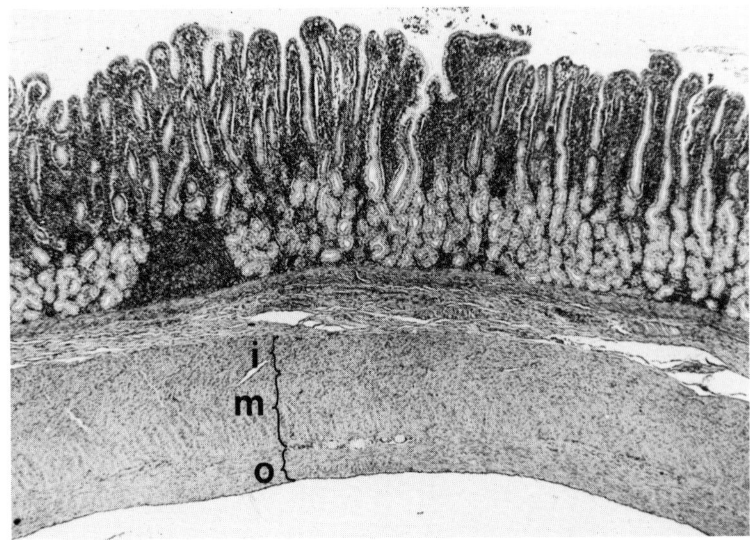

FIGURE 19-20 Light micrograph of the wall of the stomach in the pyloric region. The mucosa, submucosa, and muscularis externa can readily be identified. Muscularis externa consists of three layers: an outer longitudinal layer (o), a middle circular layer (m), and inner oblique layer (i). Note that the direction of the muscles are not as clearly defined as in the rest of the GI tract.

histologic preparations, but can be identified when stained black with silver ammonium oxide or brownish yellow with chromates. More recent immunocytochemical techniques have allowed localization of a group of such cells in the stomach and elsewhere in the gastrointestinal tract. Six different types of argentaffin cells in the stomach have been identified by electron microscopic and immunocytochemical studies. These cells are much smaller than the other fundic gland cells and are located primarily in the depths of the gland. Argentaffin cells produce and secrete *hydroxytryptamine,* or *serotonin,* a vasoconstrictor. Unlike the other cells of the gastric gland, argentaffin cells secrete their product into the underlying connective tissue and are therefore considered to be endocrine in nature. Serotonin induces the contraction of smooth muscle cells. Other products of argentaffin cells include gastrin which stimulates the secretion of hydrochloric acid by parietal cells.

MUSCULARIS EXTERNA

The tunica muscularis externa of the stomach is well developed, an adaptation for the churning function of the stomach. As pointed out previously, a layer of oblique muscle is positioned internal to the inner circular and outer longitudinal layers. Therefore the muscularis externa of the stomach can be said to be comprised of the *inner oblique, middle circular,* and *outer longitudinal* layers (Fig. 19-20). The inner oblique layer reinforces the muscular contractions necessary to mix the large amounts of food stored in the distended stomach with the digestive juice produced by it.

PYLORIC REGION OF THE STOMACH

The pyloric region conducts the processed food material to the duodenal portion of the small intestine. As mentioned, the *pyloric sphincter* regulates the release of small quantities of chyme into the duodenum. As in the cardiac region, there are large numbers of gastric pits and associated mucous cells to provide lubrication. Figure 19-20 is a light micrograph taken from the pyloric region and shows glands of the pylorus. The gastric pits are deepest in this region, and they extend up to three-quarters of the way into the mucosa. The difference in the ratio of the

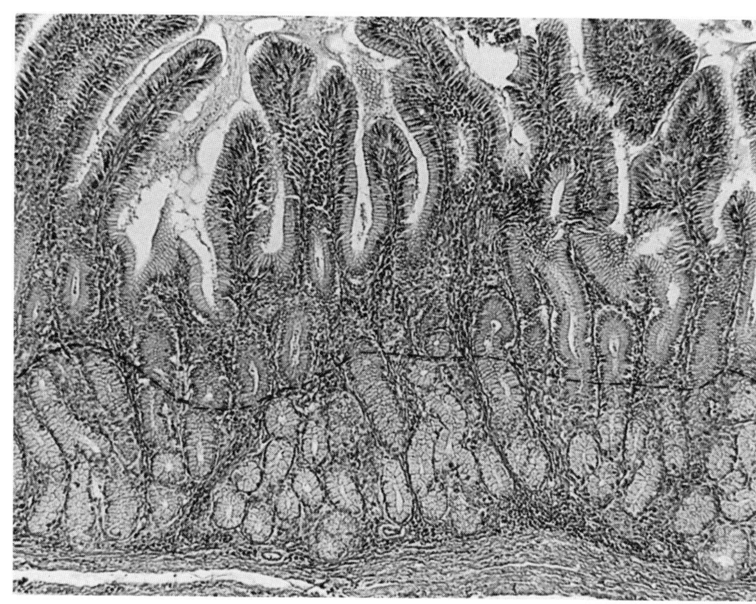

FIGURE 19-21 Light micrograph of pyloric glands. The bottom of the gastric pits are delineated by the broken line.

depth of the pits to the thickness of the mucosa permits the identification of different regions of the stomach. Recall that the pits of the cardia extend only one-quarter of the way into the mucosa. Other than the depth of the gastric pits, the pyloric glands are entirely mucous (Fig. 19-21) and have a histologic appearance that closely resembles the cardiac glands.

The circular layer of the muscularis externa becomes highly developed at the junction between the pylorus and duodenum. Figure 19-22 is the macroscopic appearance of the pyloroduodenal junction in which the muscularis externa forming the sphincter appears as a moderate thickening (arrow) between the pylorus (p) and the duodenum (d). The microscopic view of the sphincter is seen in Fig. 19-23, in which the cross-sectioned sphincter muscle (s) occupies the lower half of the field.

REVIEW SECTION

1. Figure 19-24 shows a cross section of the esophagus. The picture shows a longitudinal fold as an outfolding of lamina propria and epithelium. What type of epithelium lines the esophagus? Where would the glands, if any, be located in this section? The ill-defined two layers of muscle along the bottom of the micrograph make up the __________ and __________.

Moist stratified squamous epithelium lines the esophagus, and it provides a surface to protect against abrasion. Mucus-secreting glands are found in the submucosa. The muscularis external of the esophagus has an inner circular layer and an outer longitudinal layer.

2. Name the three types of glandular cells that are present in the section shown in Fig. 19-25.

The cells found in these fundic glands are mucous neck cells, parietal cells, and chief cells. In addition, argentaffin cells can be found if prepared for immunocytochemical demonstration or traditional silver staining of such cells.

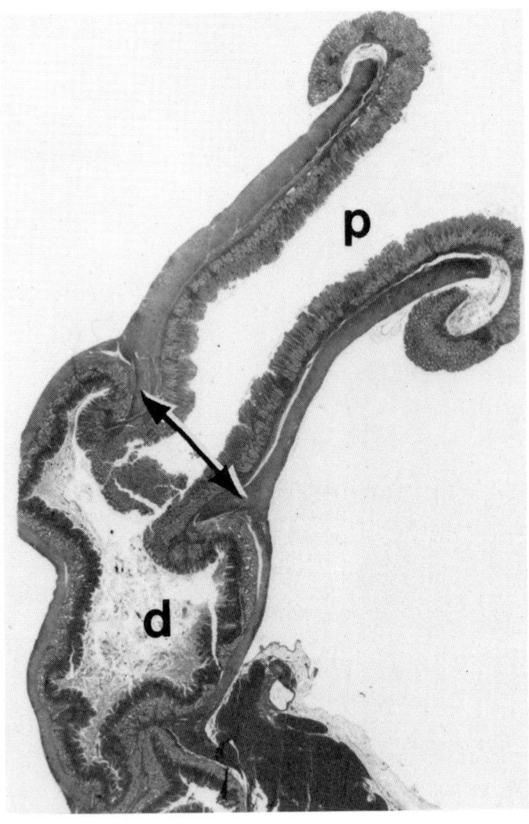

FIGURE 19-22 Photograph of the pyloroduodenal junction. The sphincter (arrow) appears as a slight thickening of muscularis externa between pyloris (p) and duodenum (d).

3. Name the type of epithelium that lines the gastric pits shown in Fig. 19-26. Note the distance the gastric pits extend into the mucosa. From what portion of the stomach was this section taken?

This is a section taken from the fundus of the stomach. The epithelium lining the gastric pits is made up of mucus-secreting columnar epithelial cells. The pit extends downward only a third or so of the thickness of the mucosa.

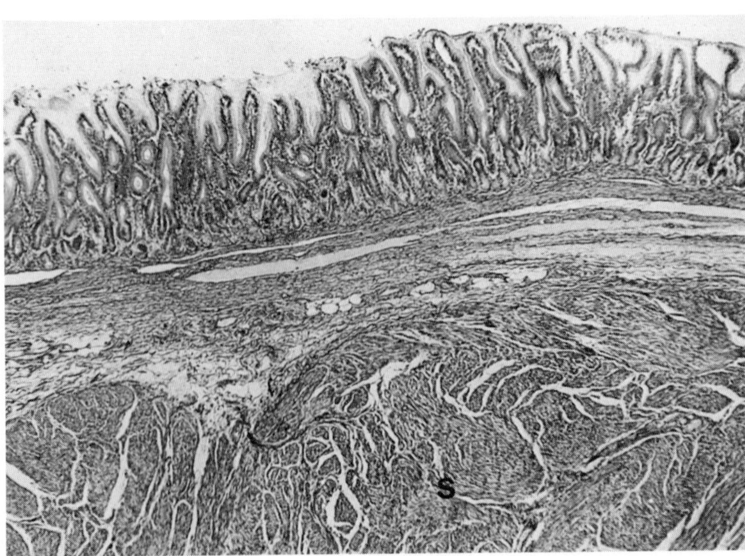

FIGURE 19-23 Light micrograph of the pyloroduodenal sphincter (s).

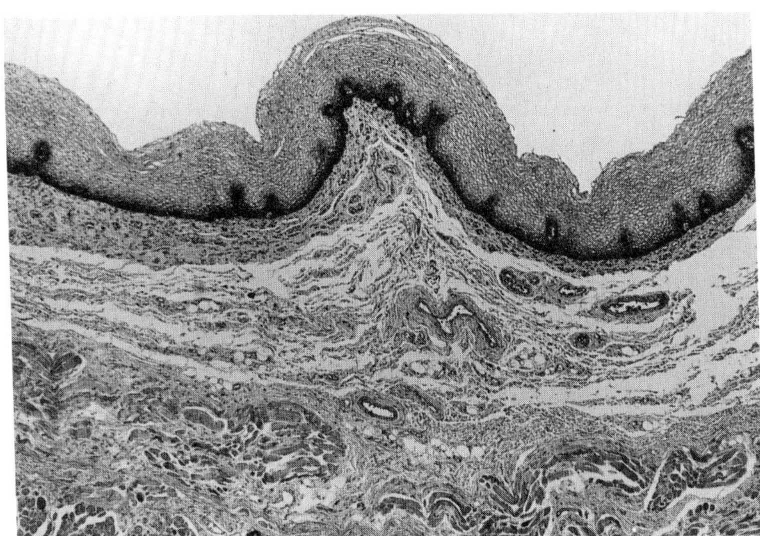
FIGURE 19-24

4. Note the length of the gastric pits in relation to the length of the glands in Fig. 19-27. This is a section of which portion of the stomach? What is b? What is c? What is a? What is d?

The deep gastric pits and short glands delineated (broken line) in this micrograph represent the pylorus of the stomach. The layers of the stomach are very evident: the lamina propria (a), muscularis mucosae (b), submucosa (c), and muscularis externa (d) are labeled.

BIBLIOGRAPHY

Bertlanffy, F. D.: "Cell Renewal in the Gastrointestinal Tract of Man," *Gastroenterology* **43:**472–475 (1963).

Edwards, D. A. W.: "The Esophagus," *Gut* **12:**948 (1971).

Goetsch, E.: "The Structure of the Mammalian Esophagus," *Am. J. Anat.* **10:**1–40 (1970).

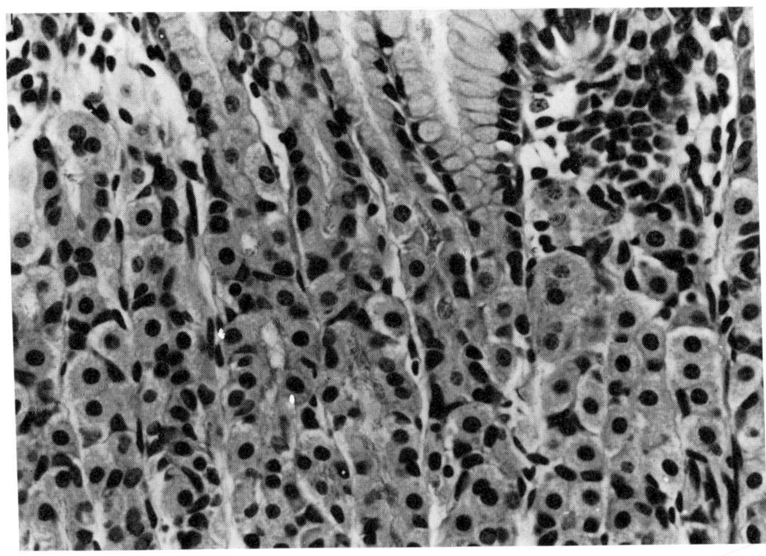
FIGURE 19-25

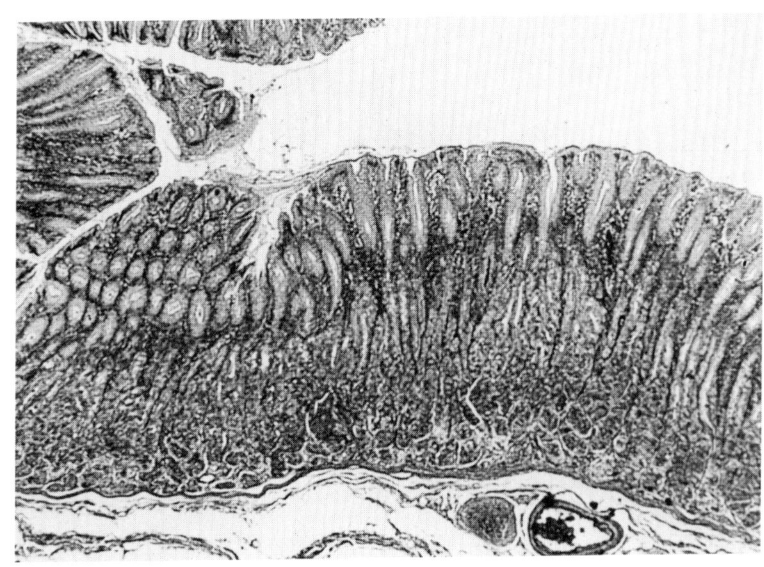

FIGURE 19-26

Ito, S.: "Anatomic Structure of the Gastric Mucosa," in C. F. Code and M. I. Grossman (eds.), *Handbook in Physiology,* sect. 6, vol. 2, American Physiological Society, Washington, 1967–1968, pp. 705–741.

Rhodin, J. A. G.: *Histology: A Text and Atlas,* Oxford University Press, New York, 1974.

Weiss, L., and R. O. Greep: *Histology,* 4th ed., McGraw-Hill, New York, 1977.

Wolf., S.: *The Stomach,* Oxford University Press, New York, 1965.

ADDITIONAL READINGS

Forssmann, W. G., and L. Orci: "Ultrastructure and Secretory Cycle of the Gastrin-producing Cells," *Z. Zellforsch.* **101:**419–432 (1969).

———, ———, R. Pictet, E. Renold, and C. Rouiller: "The Endocrine Cells in the Epithelium of the Gastrointestinal Mucosa of the Rat: An Electron Microscope Study," *J. Cell Biol.* **40:**692–715 (1969).

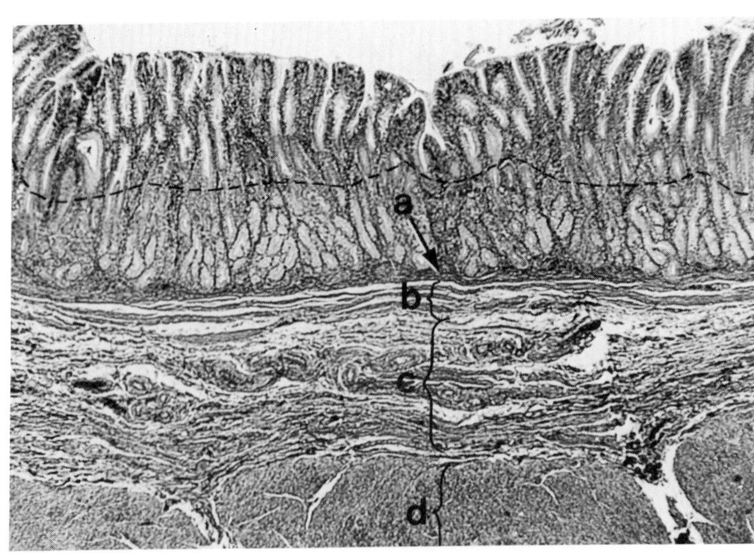

FIGURE 19-27

Hayward, A. F.: "The Fine Structure of Gastric Epithelial Cells in the Suckling Rabbit with Particular Reference to the Parietal Cell," *Z. Zellforsch.* **78:**474–483 (1967).

Helander, H. F.: "Ultrastructure of Fundus Glands of the Mouse Gastric Mucosa," *J. Ultrastruct. Res.*, (suppl.) **4:**1–123 (1962).

———: "Ultrastructure of Gastric Fundus Glands of Refed Mice," *J. Ultrastruct. Res.* **10:**160–175 (1964).

Ito, S., and G. D. Schofield: "Studies on the Depletion and Accumulation of Microvilli and Changes in the Tubuvesicular Compartment of Mouse Parietal Cells in Relation to Gastric Acid Secretion," *J. Cell Biol.* **63:**364–382 (1974).

Johns, B. A. E.: "Developmental Changes in the Esophageal Epithelium in Man," *J. Anat.* **86:** 431–442 (1952).

Johnson, F. R., and R. M. H. McMinn: "Microscopic Structure of the Pyloric Epithelium of the Cat," *J. Anat.* **107:**67–86 (1970).

Kaye, M. D., and V. Showalter: "Normal Deglutive Responses of the Human Lower Esophageal Sphincter," *Gut* **13:**352–360 (1972).

Lillibridge, C. B.: "The Fine Structure of Normal Human Gastric Mucosa," *Gastroenterology* **47:** 269–290 (1964).

Parakkal, P. F.: "An Electron Microscopic Study of Esophageal Epithelium in the Newborn and Adult Mouse," *Am. J. Anat.* **121:**175–196 (1967).

Pfeiffer, C. J.: "Surface Topology of the Stomach in Man and the Laboratory Ferret," *J. Ultrastruct. Res.* **33:**252–262 (1970).

Rohrer, G. V., J. R. Scott, W. Joel, and S. Wolf: "The Fine Structure of Human Gastric Parietal Cells," *Am. J. Dig. Dis.* **10:**13–21 (1965).

Rubin, W., L. L. Ross, M. H. Sleisenger, and G. H. Jeffries: "The Normal Human Gastric Epithelia: A Fine Structural Study," *Lab. Invest.* **19:**598–626 (1968).

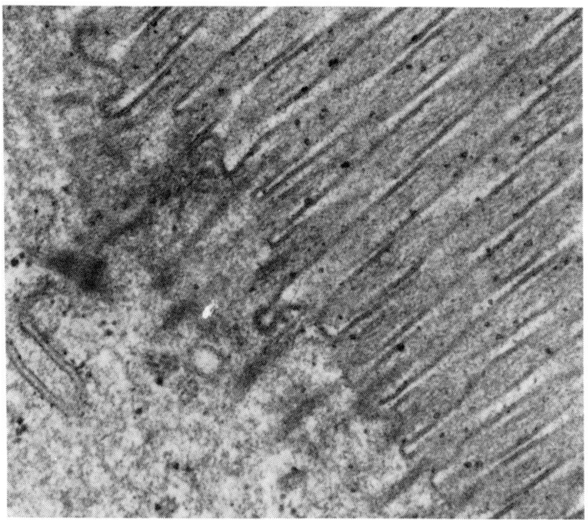

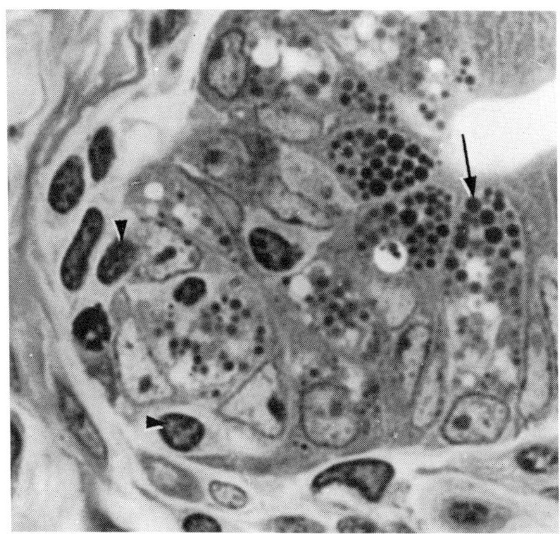

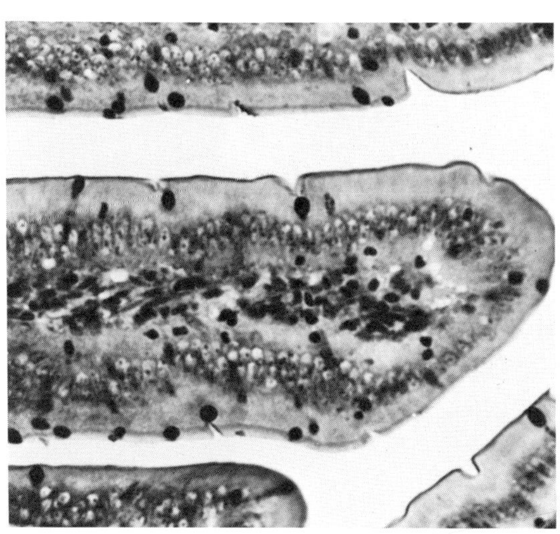

20 LOWER GASTRO-INTESTINAL TRACT

OBJECTIVES

Upon completion of this chapter, the student will be able to:

1 Identify three modifications of the intestinal wall that increase its absorptive surface areas.

2 Identify the functions of the cellular components of the crypts of Lieberkühn (intestinal glands).

3 Identify the role played by the following in absorption:
(a) microvilli (b) villi (c) central lacteal (d) blood vessels

4 Compare the small and large intestines in regards to the following:
(a) microvilli (b) villi (c) goblet cells (d) crypts of Lieberkühn (e) defense-oriented cells

5 Identify the structure, function, and location of each of the following:
(a) Peyer's patches (b) teniae coli (c) appendix (d) Brunner's gland (e) Meissner's plexus (f) Auerbach's plexus

SMALL INTESTINE

The small intestine is an organ of continued absorption and digestion. It is about 20 ft in length extending from the pyloroduodenal junction to the colon. The major segments of the small intestine are the duodenum, jejunum, and ileum. Since the basic microscopic structures of different segments have many common features and since the transition between the segments is gradual, the entire small intestine will be discussed as a single entity. As this is done, differences that are unique to the various segments will be pointed out.

MUCOSA

Three modifications of the mucosa of the small intestine increase the absorptive surface areas. Along the inner surface of the intestinal wall are circularly arranged infoldings called *plicae circulares* which are best developed at the beginning of the small intestine and become less prominent toward the ileocecal junction. The secondary projections which are leaflike or fingerlike in shape occur throughout the mucosa and are called *intestinal villi.* Figure 20-1 is a macrophotograph of a longitudinal section through the small intestine showing the circular infoldings (p) of the mucosa (plicae circulares) and the many fine intestinal villi that project from its inner surface. On the left is the pyloric sphincter (s) shown as a heavy mass of cross-sectioned muscles.

In Fig. 20-2 a pair of plicae circulares (p) are depicted as visualized in a segment of the small intestine sectioned longitudinally. The leaf-shaped intestinal villi (v) are also visible. Note the characteristic inner circular and outer longitudinal layers of the muscularis externa (me). The plica contains the muscularis mucosae (mm) and a portion of the submucosa accompanying it. Figure 20-3 is an example of an intestinal wall which is made up of the villi lined by the characteristic simple columnar epithelium with goblet cells which are stained dark (arrows). Along the apical surface of the epithelium is a dense linear border called *striated* or *striped border.*

EPITHELIUM

As mentioned previously in Chap. 2, in an electron micrograph the striated border has been shown to consist of microvilli of uniform size which constitute the third level of mucosal projections to increase the absorptive surface area. Each microvillus has supporting axial filaments which fan out at the base of the microvillus to form the terminal web (Fig. 20-4). The junctional complex that is present at the apical end of the cells is made up of a tight junction, zonula adherens and desmosomes, corresponding to the *terminal bar* of light microscopy. It serves to prevent the

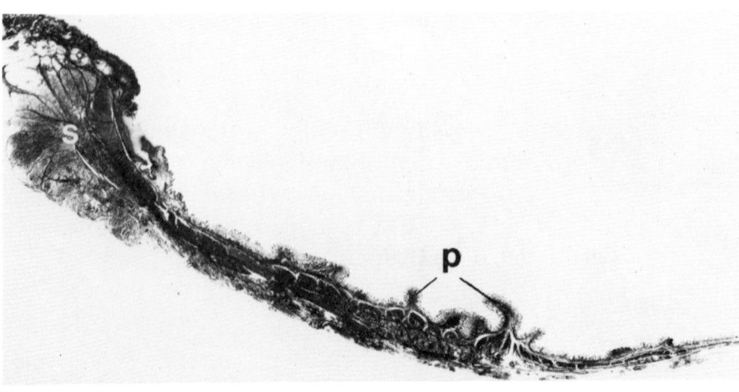

FIGURE 20-1 Macrophotograph of a longitudinal section through the small intestine. (p) Plicae circulares. (s) Pyloric sphincter.

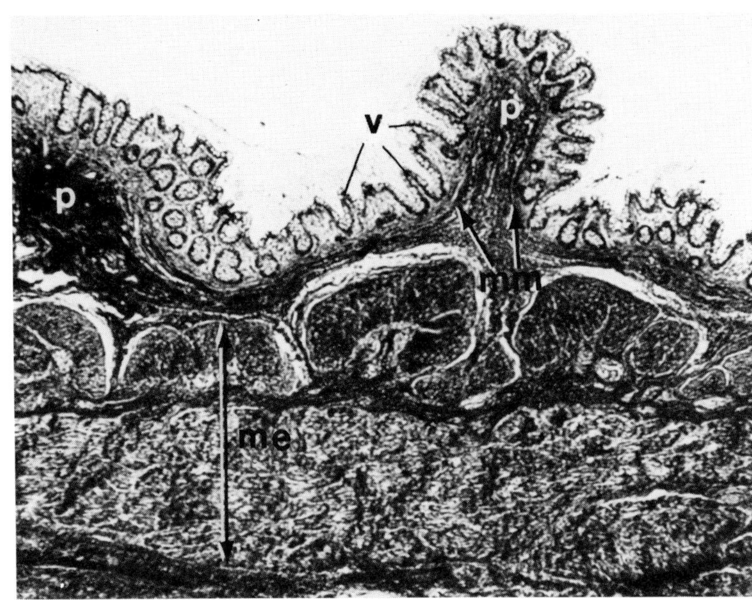

FIGURE 20-2 Light micrograph of a longitudinal section through the small intestine. (p) Plicae circulares. (v) Intestinal villi. (mm) Muscularis mucosae. (me) Muscularis externa.

influx of food material into the mucosa since it is present along the entire circumference of the epithelial cells at their apexes. Different amounts of mucopolysaccharide coat have been found at the surface of intestinal microvilli. The mucopolysaccharide coat corresponds to the glycocalyx discussed in Chap. 1.

The membrane of microvilli contains enzymes which aid in the hydrolysis of nutrient materials. Examples of these enzymes are alkaline phosphatase, ATPase, and peptidases. Materials hydrolyzed along the cell surface are taken through the plasma membrane between microvilli. Although small soluble molecules pass through the cell mem-

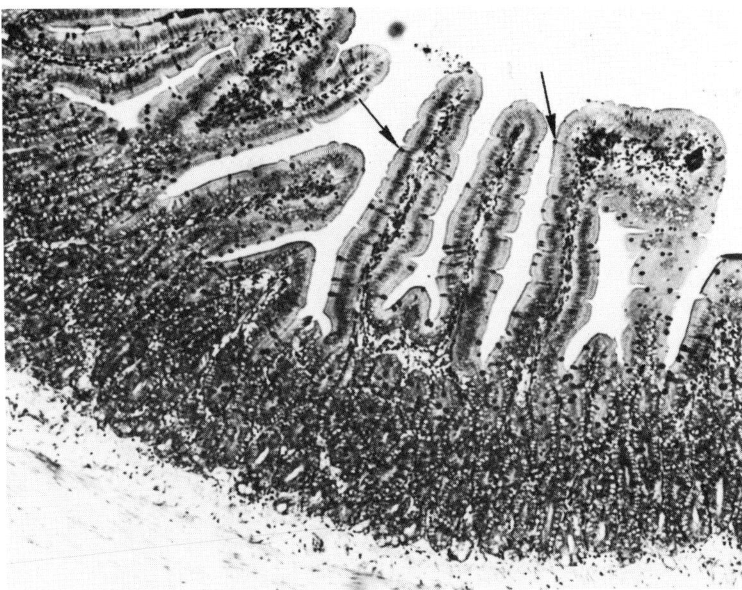

FIGURE 20-3 Light micrograph of the wall of the small intestine. Goblet cells (arrows). See text for explanation.

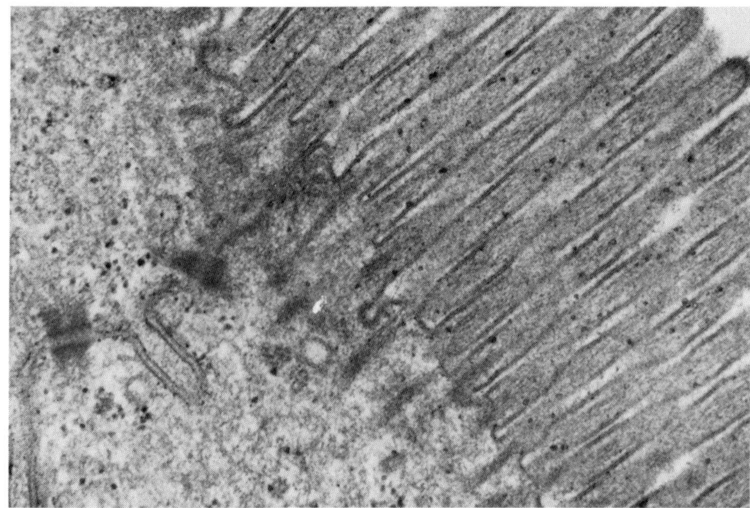

FIGURE 20-4 Electron micrograph of microvilli. Each microvillus has supporting axial microfilaments which form the terminal web (t) in the apical portion of the epithelial cells.

brane via diffusion and other transport mechanisms, it has been shown that larger molecules and marker protein particles are taken in by a process similar to pinocytosis. The process of pinocytosis appears to occur in the valley between microvilli, producing small vesicles that contain partially hydrolyzed products. They are brought to the supranuclear region of the Golgi complex where they are thought to be modified and resynthesized into larger, more complex molecules. Resynthesized nutrients contained in cytoplasmic vacuoles are exteriorized by a process akin to reverse pinocytosis along the lateral surface of the cell. The products penetrate the basal lamina and enter into the lamina propria.

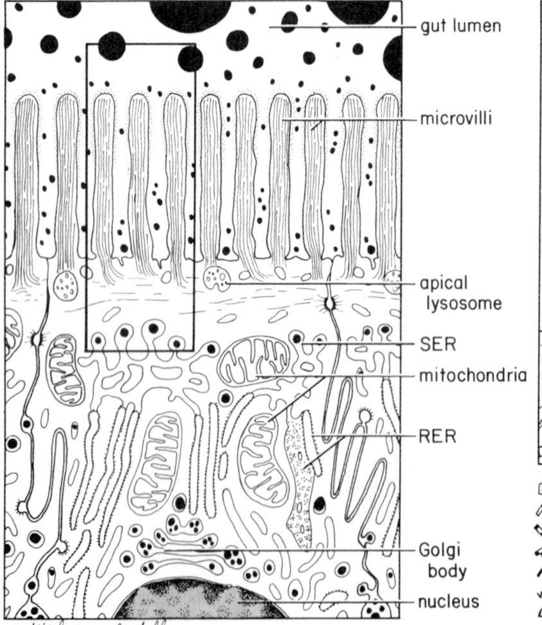

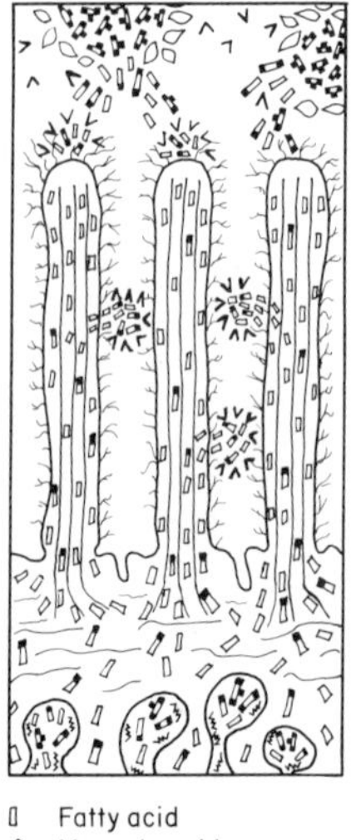

FIGURE 20-5 Diagram illustrating the absorption of lipids. (*Redrawn after R. R. Cardell.*)

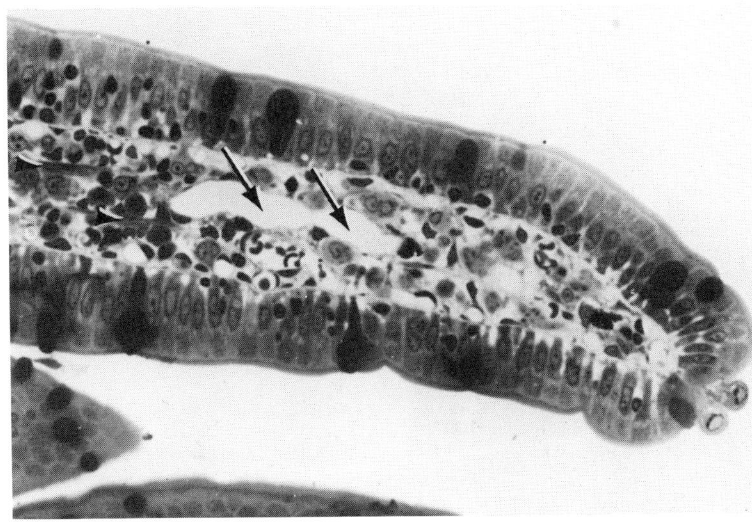

FIGURE 20-6 Light micrograph of intestinal villus. The center is occupied by the central lacteal (arrows). Occasional smooth muscle cells (arrowheads) are found in the wall of the lacteal.

The process of absorption and resynthesis has been particularly well demonstrated by lipid absorption studies which have shown that small lipid droplets are hydrolyzed to form free fatty acids and monoglycerides prior to entry into the cell, as shown in Fig. 20-5. The hydrolyzed products of lipid are thought to be resynthesized into more complex forms by the smooth and rough ER, which eventually produce visible droplets of a protein solution containing triglycerides, phospholipids and cholesterol. By the resynthesis and sequestration of triglycerides within the cell an inward diffusion gradient of monoglycerides and fatty acids is thought to be maintained.

LAMINA PROPRIA

The hydrolyzed products entering the lamina propria of the villus are drained by a centrally located collecting passage, the *central lacteal.* The central lacteal is a blind lymphatic duct

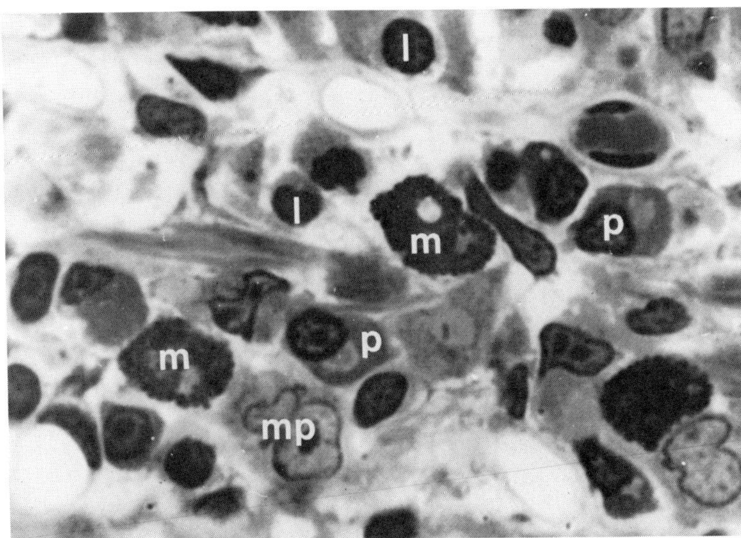

FIGURE 20-7 The lamina propria of an intestinal villus viewed at higher magnification. Numerous defensive cells are present. (p) Plasma cell. (mp) Macrophage. (m) Mast cell. (l) Lymphocyte.

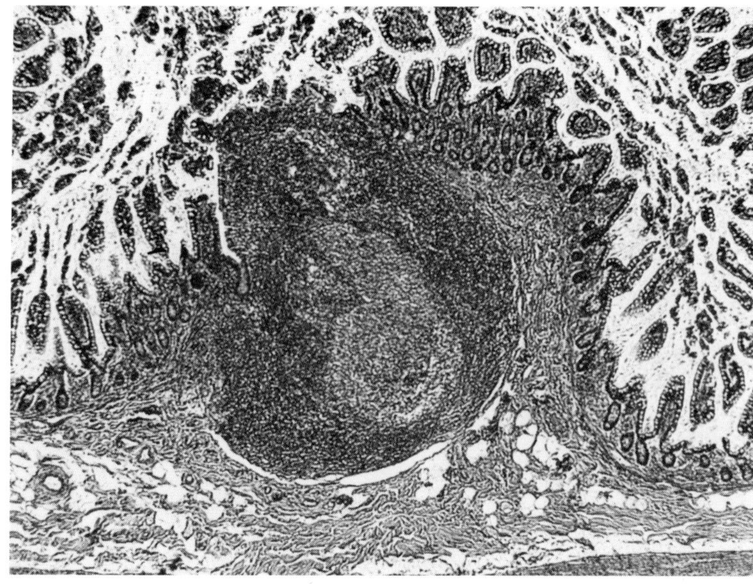

FIGURE 20-8 Light micrograph of a solitary lymphoid nodule in the lamina propria of the ileal mucosa. Notice the large lighter germinal center in the nodule.

which serves as the beginning of the lymphatics of the mesentery (Fig. 20-6). The wall of the central lacteal is supported by occasional smooth muscle cells (Figs. 20-6 and 20-7, arrowheads).

Numerous defensive cells, such as lymphocytes, plasma cells, and macrophages, are found in the lamina propria of the small intestine. They react against the numerous foreign elements introuced into the lamina propria via the GI tract in the form of digested food and microorganisms. Figure 20-7 represents the core of a villus in the jejunum in which a number of plasma cells (p), macrophages (mp), mast cells (m), and lymphocytes (l) are found within the lamina propria. Note

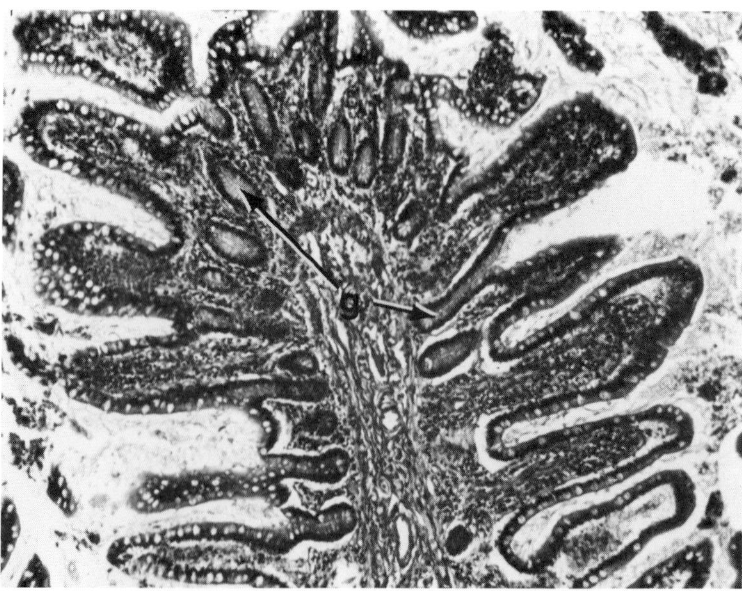

FIGURE 20-9 Light micrograph of the mucosa of the small intestine. Intestinal glands (g) extend from the bottom of the valleys between villi into the lamina propria.

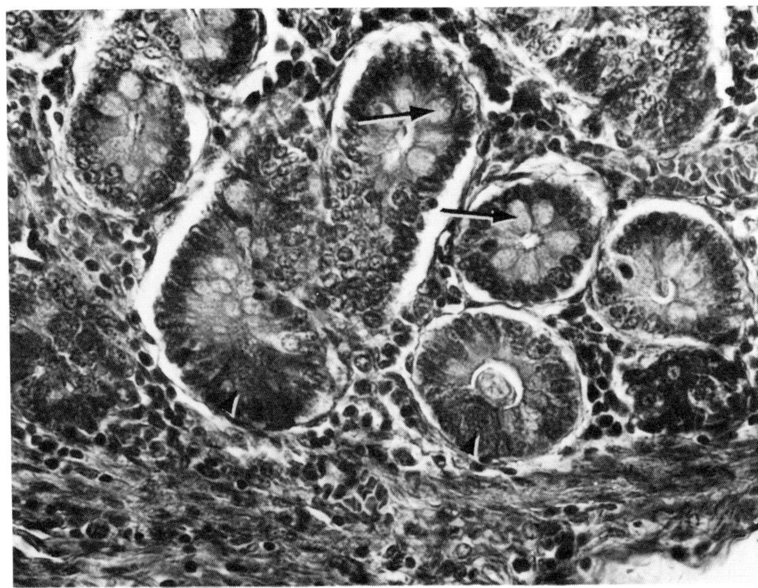

FIGURE 20-10 Cross sections of intestinal glands. The majority of the cells making up the crypts of Lieberkühn are goblet cells (arrows). In addition, there are Paneth cells (arrowheads) and argentaffin cells.

the eccentric nuclei with patchy and often cartwheel-shaped chromatin pattern of the plasma cells. The number of lymphoid cells increases steadily along the length of the small intestine. In the ileum, solitary lymphoid nodules and clusters of lymphoid nodules being to appear (Fig. 20-8). These clusters of lymphoid nodules are called *Peyer's patches.*

MUCOSAL GLANDS OF SMALL INTESTINE (CRYPTS OF LIEBERKÜHN)

Simple tubular glands are present throughout the small and large intestines. In the small intestine they begin from the valley between villi and extend into the mucosa. These glands are called the *crypts of Lieberkühn* (g in Fig. 20-9). A variety of cell types are found

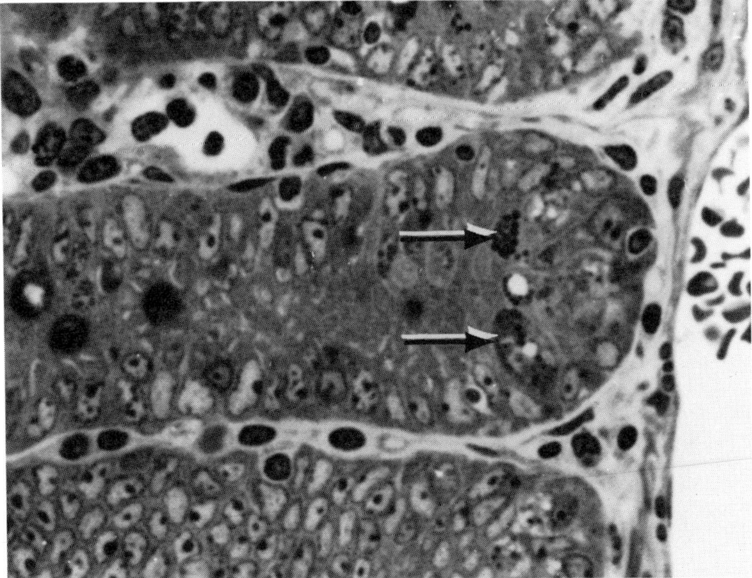

FIGURE 20-11 Light micrograph of intestinal glands. Paneth cells (arrows) are located at the fundus of the glands.

in the crypts of Lieberkühn (Fig. 20-10). The majority of cells making up the crypts of Lieberkühn are mucus-secreting cells that have the characteristic supranuclear cytoplasm filled with mucus and a basally located nucleus (Fig. 20-10, arrows).

In H and E preparation, the mucus content of these cells is often dissolved away and presents a foamy, clear appearance. In addition to the mucous cells, at least two other types of cells are found throughout the small intestine. *Paneth cells* are mainly found near the base of the crypts of Lieberkühn. They appear serous in nature, containing many bright red-stained serous granules in their cytoplasm (Fig. 20-10, arrowheads). A few well-defined Paneth cells are visible at the bottom of the gland in Fig. 20-11 (arrows). That the Paneth cells are clearly serous cells is seen in Fig. 20-12, in which the supranuclear Golgi's complex and granules of different densities (therefore in different stages of maturation) are depicted (arrow). Their function has not been clearly established.

Argentaffin cells, sometimes called *enterochromaffin cells,* are fewer and smaller in size than Paneth cells. As in the stomach, they are usually located in the depths of the crypts of Lieberkühn. Argentaffin cells are particularly common in the duodenum, but are found throughout the lower GI tract as well. The secretory granules are small and require special staining. Argentaffin cells are known to secrete serotonin into the underlying connective tissue, causing the constriction of blood vessels supplying the small intestine. Thus, they are often found between Paneth cells and the basement membrane. A few argentaffin cells whose granules are not stained are shown in Fig. 20-12 (arrowheads).

The lamina muscularis mucosae of the intestine has two layers, an inner circular layer and an outer longitudinal smooth muscle layer. Underlying the muscularis mucosae is a submucosa, with many blood vessels and lymphatics. Figure 20-13 shows the arrangement of lymphatics and blood vessels of the small intestine. Aggregations of parasym-

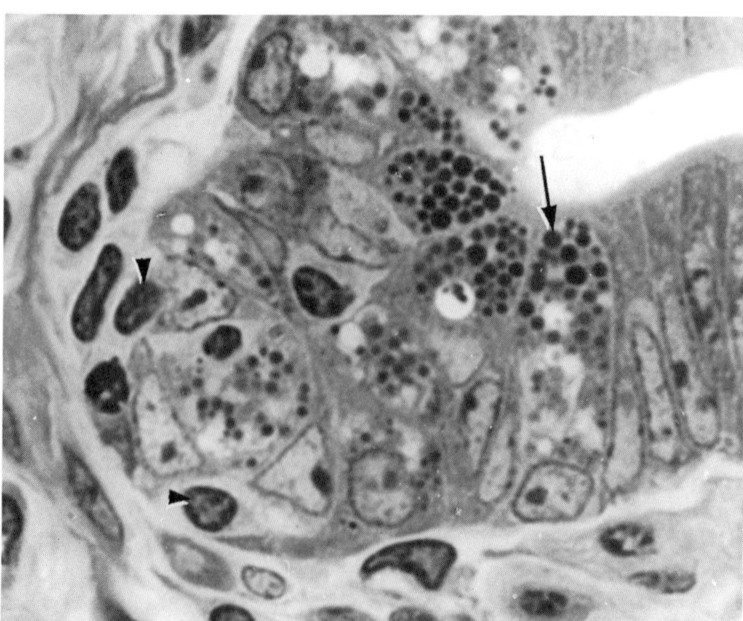

FIGURE 20-12 Light micrograph of the fundus of an intestinal gland. Several Paneth cells are seen containing zymogen granules (arrow) in the apical cytoplasm. A few argentaffin cells are located between the basal portions of the Paneth cells (arrowheads). Their granules are not apparent in this preparation.

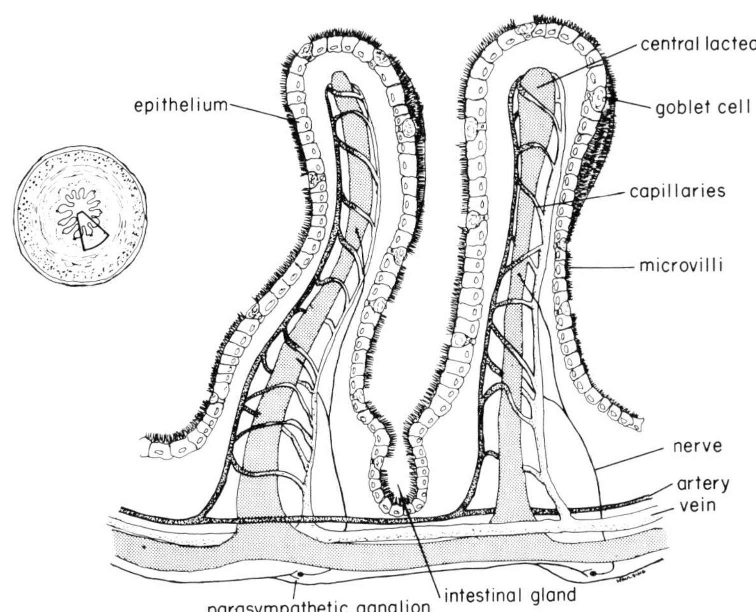

FIGURE 20-13 Diagram illustrating the arrangement of lymphatics and blood vessels in the wall of the small intestine.

pathetic elements, called *Meissner's plexuses,* are present in the submucosa of the small intestine.

STRUCTURAL CHANGES OCCURRING THROUGHOUT THE INTESTINE

Passing caudad through the small intestine, the number of lymphocytes and other defensive cells increases in the mucosa. Goblet cells also increase in number toward the more distal portions of the small intestine. Microvilli in the ileum are considerably longer than those in the duodenum, as the ileum absorbs more water. A greater number of lymphocytes, macrophages, and other defense-related cells are found in the lamina propria, and the number of Peyer's patches increase in the more distal segments.

The villi are also different in shape in different parts of the small intestine. They are leaf-shaped in the duodenum, become more rounded in the jejunum, and are club-shaped in the ileum.

SUBMUCOSA

One of the most distinctive histologic features of the duodenal portion of the small intestine is the presence of a well-developed and prominent submucosal gland called *Brunner's gland* (Fig. 20-14). Brunner's gland is a large, compound tubular gland which protects the duodenal wall with mucoid alkaline secretions. The secretion occurs when the duodenal mucosa is stimulated by the acidity of food being transmitted from the stomach.

Figure 20-15 shows a portion of the duodenal wall in which crypts of Lieberkühn (mucosal glands) are seen on the upper right. The lower left half of the field is occupied by the tortuous tubular gland of Brunner which presents randomly sectioned secretory portions that are stained lighter than the crypts of Lieberkühn. The mucous nature of the secretory portions of this gland is better depicted in Fig. 20-16. The common bile duct empties bile salts from the gallbladder and digestive enzymes from the pancreas into the duodenum. The addition of the bile products increases the pH necessary for activation of various enzymes in the small intestine.

MUSCULARIS EXTERNA

The muscle cells of this layer are well developed throughout the small intestine and

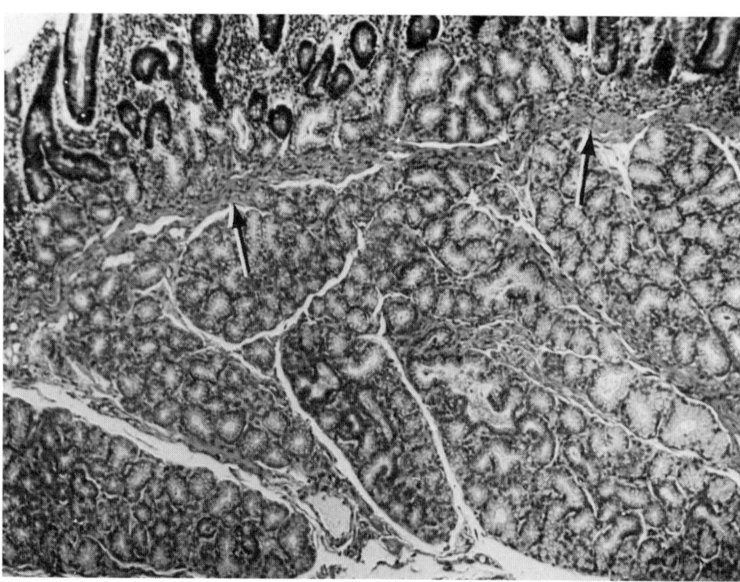

FIGURE 20-14 Light micrograph of Brunner's gland. Muscularis mucosae (arrows) demarcate the mucosa from the submucosa.

make up a clearly visible pattern of inner circular and outer longitudinal muscles. Auerbach's plexus of parasympathetic nerves is present between the two layers throughout the small intestine.

LARGE INTESTINE

The diameter of the large intestine, or *colon,* is much greater than that of the small intestine. Since little nutrient absorption occurs in the large intestine, it has a rather smooth inner surface. Crypts of Lieberkühn are still present, as they contribute to the lubrication of the epithelial cells. The cells of the large intestine maintain a striated border to increase the surface area for water absorption. As you may expect, goblet cells and lymphoid nodules are present in greater numbers than in the small intestine. Since intestinal villi are absent in the colon, the crypts of Lieber-

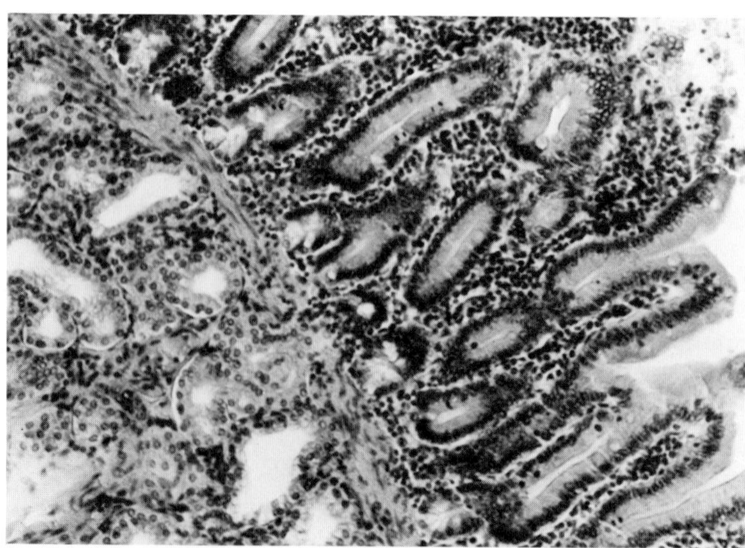

FIGURE 20-15 Portion of the duodenal wall. Secretory portions of Brunner's gland are seen in the lower left half.

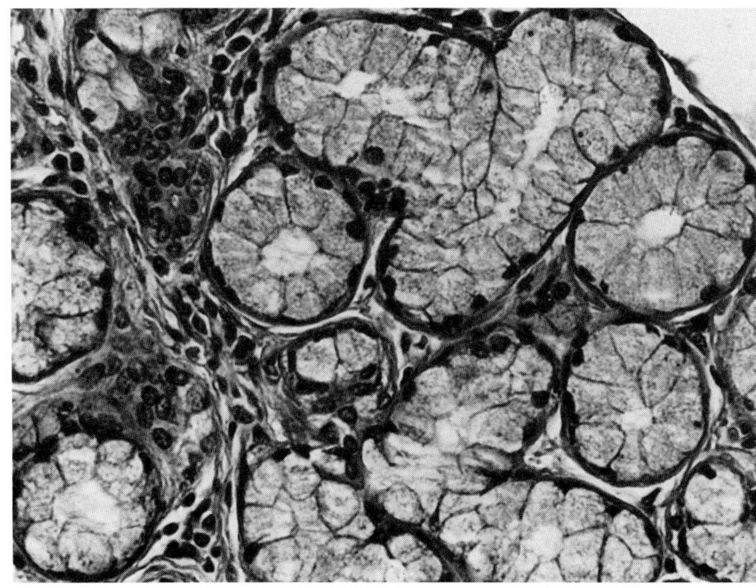

FIGURE 20-16 Secretory portions of the Brunner's gland viewed at higher magnification. They are pure mucous in nature.

kühn exclusively account for the invaginations of its epithelial surface, and therefore the muscularis mucosae clearly separates the mucosa from the submucosa (Fig. 20-17).

Goblet cells are abundant in intestinal glands (Figs. 20-18 and 20-19). Aggregates of lymphoid nodules in the lamina propria and the submucosa are more prominent. The lamina muscularis mucosae does not change appreciably from the small intestine to the large intestine, except that the muscularis mucosae appears as a straight circular structure without being thrown into the plicae, which are absent here. In the large intestine, the muscularis externa shows a unique specialization. The inner circular layer is prominent and easily seen. However, the muscle cells of the outer longitudinal layer become arranged into three bands of parallel muscles called *teniae coli*. Between these bands, the

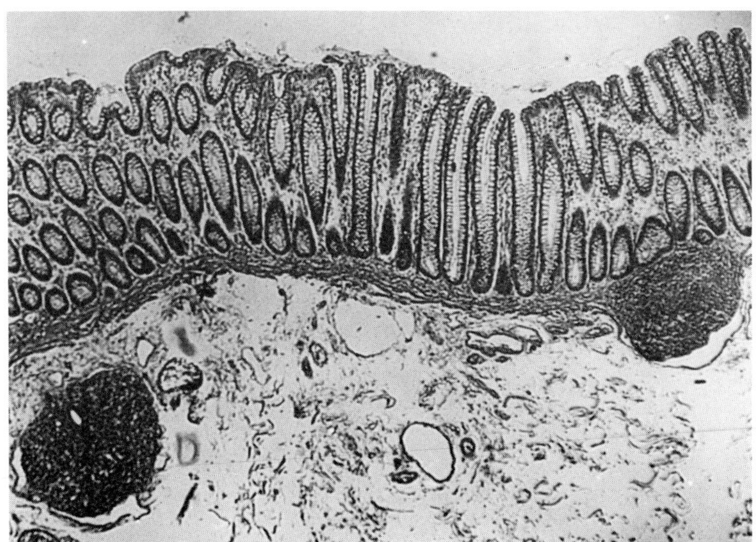

FIGURE 20-17 Light micrograph of the large intestine.

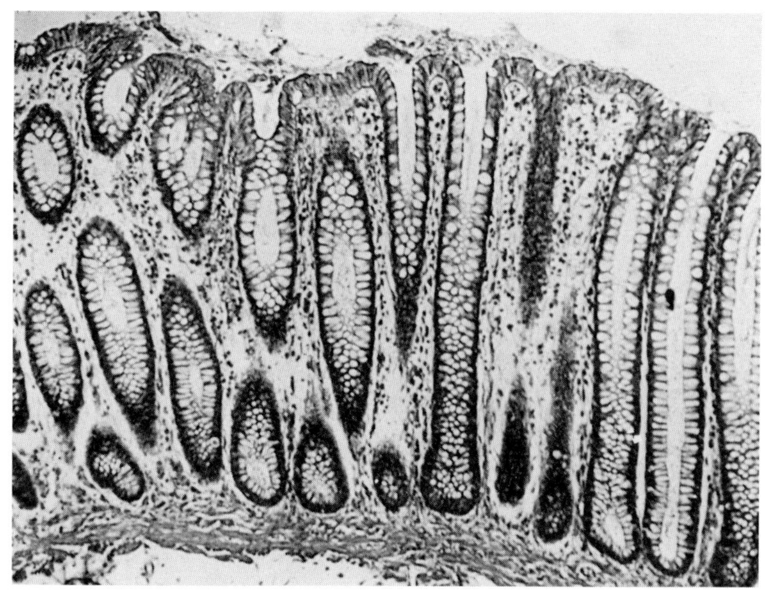

FIGURE 20-18 Crypts of Lieberkühn in the large intestine. They are lined primarily by goblet cells.

longitudinal muscle cells are sparse. Teniae coli function in the physical formation of the feces. As in other sections of the GI tube, aggregation of parasympathetic nerve cells and fibers, called *Auerbach's plexus,* occur in the muscularis externa of the large intestine.

APPENDIX

The appendix is a vestigial outpouching of the large intestine close to the ileocecal junction. It is short and of small diameter. The mucosa of the appendix is identical in appearance to that of the large intestine (Fig. 20-20).

Peyer's patches are prominent and are scattered throughout the entire submucosa. The muscularis mucosae may even be interrupted by lymphoid cells (Fig. 20-21). The small diameter and lack of teniae coli along with numerous Peyer's patches characterize the appendix.

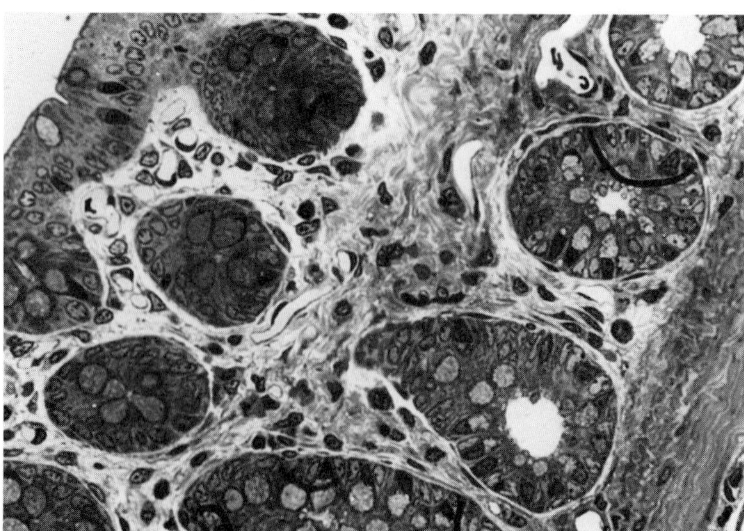

FIGURE 20-19 Cross sections of crypts of Lieberkühn in the large intestine. Note the abundance of goblet cells.

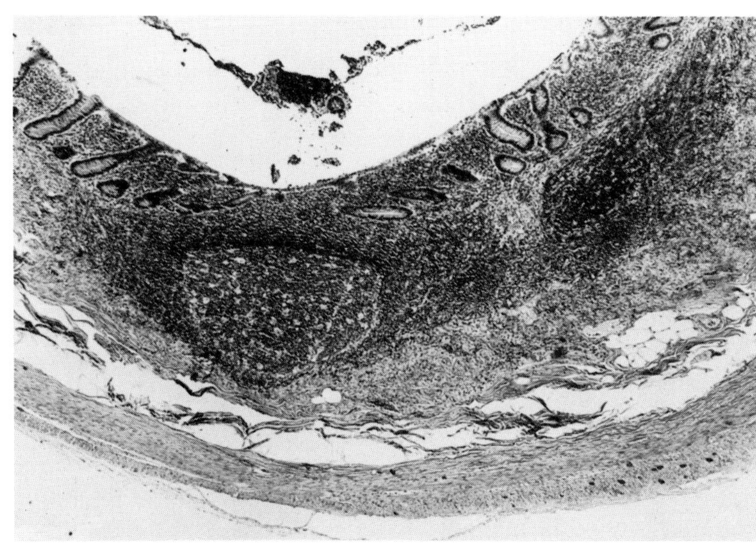

FIGURE 20-20 Light micrograph of a cross section of the appendix. Note the Peyer's patches and the dense accumulation of lymphoid cells.

RECTUM

The descending colon is continuous with the sigmoid colon, which in turn continues into the rectum. The rectum may be divided into two parts; an upper part, the rectum proper, and a lower part, the anal canal (Fig. 20-22). The *rectum proper* is similar in structure to the large intestine, with the exception that its crypts of Lieberkühn are slightly deeper. The crypts are lined by numerous goblet cells.

At the junction between the rectum proper and the anal canal, the simple columnar epithelium of the rectum changes abruptly into noncornified stratified squamous epithelium. Figure 20-23 shows a section from the rectum proper, while Fig. 20-24 shows the rectoanal junction. Note the abrupt change in epithelium (arrows).

Along the rectal wall are venous plexuses, dilations of which cause hemorrhoids. These are irregularly shaped sinusoidal spaces that occupy much of the submucosa (Fig. 20-23).

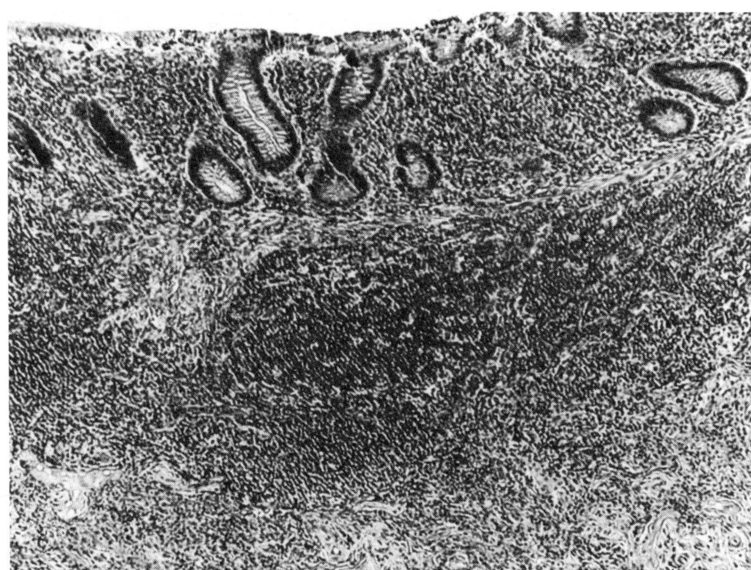

FIGURE 20-21 Portion of the appendix viewed at higher magnification. Muscularis mucosae is penetrated by lymphoid cell aggregations.

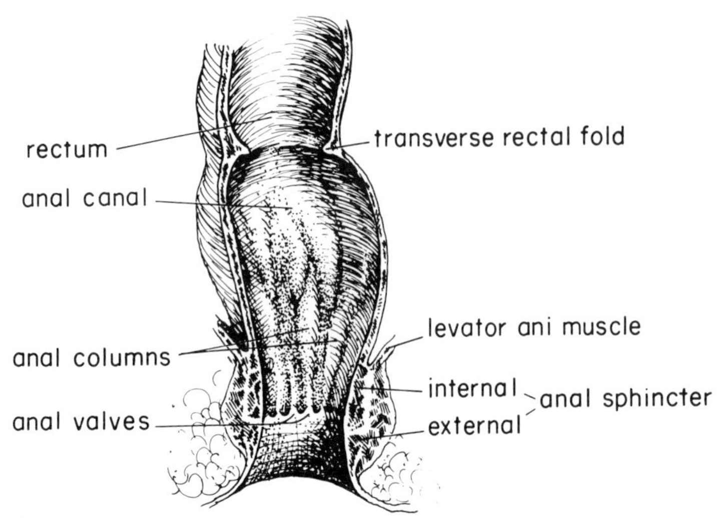

FIGURE 20-22 Diagram of the anal canal.

The distended vessels contain numerous red blood cells that are present in this stagnant area of circulation. The muscularis externa of the rectum is thick, forming an internal sphincter (Fig. 20-22). In addition to the involuntary internal sphincter, the anal region has an external sphincter muscle that is made up of skeletal musculature.

REVIEW SECTION

1. What kind of epithelium is present in Fig. 20-25? Name two structural adaptations for absorption seen in this picture.

This photomicrograph of intestinal villi shows simple columnar epithelium with striated border and goblet cells. The villi and central lacteals are

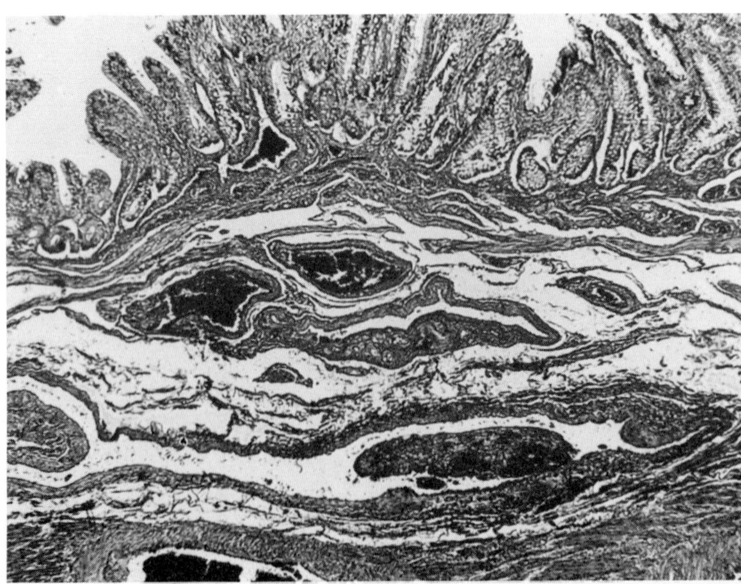

FIGURE 20-23 Light micrograph of the rectum proper. It is similar in structure to the large intestine.

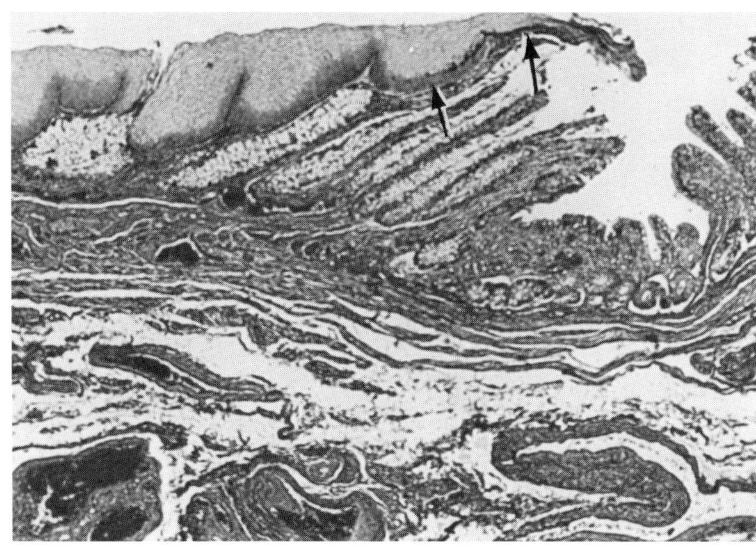

FIGURE 20-24 Light micrograph of the rectoanal junction. Note the abrupt change in epithelium at the junction (arrows).

structural adaptations for absorption. In an EM, the striated border will appear as a series of microvilli which represent another surface adaptation for absorption.

2. Two dense oval regions with a light center are shown in the lamina propria in Fig. 20-26. These lymphoid nodules are called __________. In what segment of the GI tract are they found?

Aggregations of lymph nodules (called Peyer's patches) are seen in the lamina propria. This section is taken from the ileum.

3. Name three defense-oriented cells found in the lamina propria of the small intestine. The number of these cells (increases/decreases) in the lower portion of the small intestine.

Lymphocytes, plasma cells, and macrophages are abundant in the lamina propria of the small intestine, and they increase in number in the lower portions.

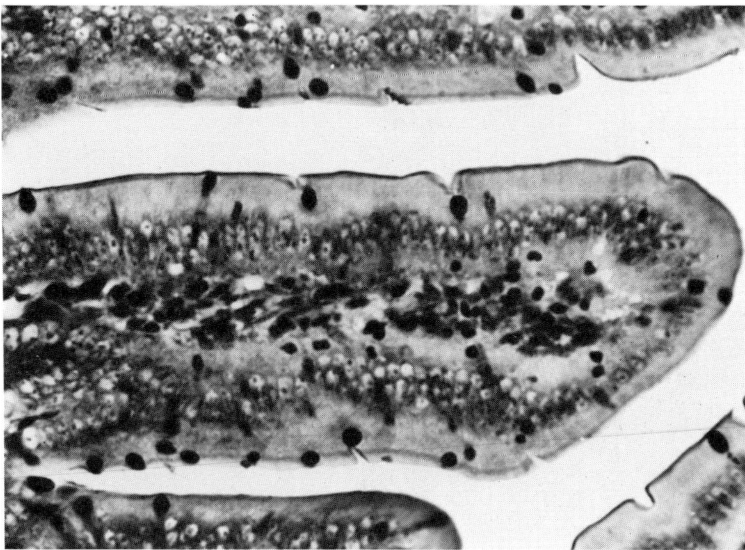

FIGURE 20-25

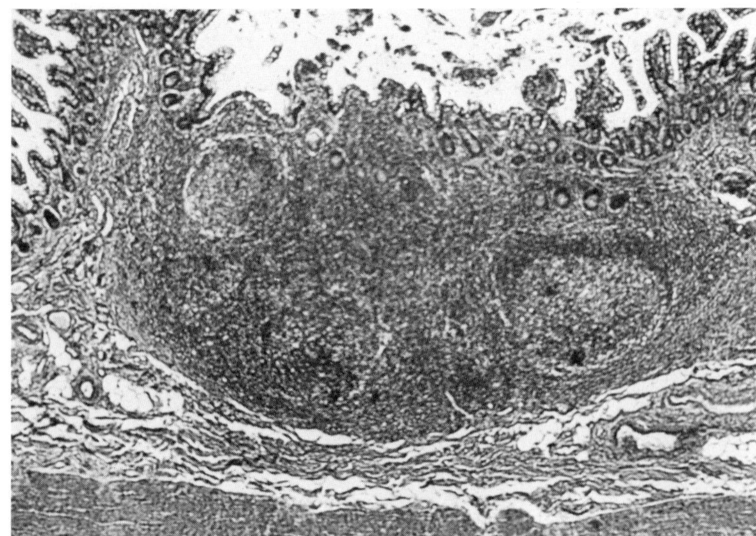

FIGURE 20-26

4. The microvillous border of the epithelium in Fig. 20-27 is stained dark due to the surface coating of the microvilli which is composed of __________. The dark-staining bodies are composed of __________. The dark-staining bodies labeled (a) are __________. The blind lymphatic duct (b) in the central portion of the villus is called the __________. A number of cells with dark-staining nuclei (c) have invaded the epithelium. What are they?

The special stain for indicating carbohydrates used in this slide causes the microvilli lining the epithelial cells of the intestinal villi to stain more intensely due to their mucopolysaccharide surface coating. The goblet cells (a) also stain intensely. The blind lymphatic (b) in the central portion of the villus is a central lacteal. Lymphocytes (c) are found between epithelial cells in this section.

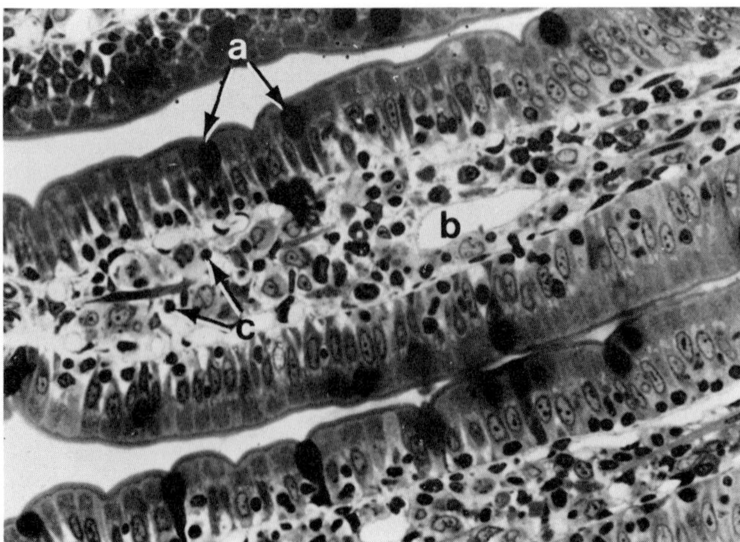

FIGURE 20-27

Lineback, P. E.: "Studies on the Musculature of the Human Colon, with Special Reference to the Taeniae," *Am. J. Anat.* **36:**357–383 (1925).

Rhodin, J. A. G.: *Histology: A Text and Atlas,* Oxford University Press, New York, 1974.

Strauss, E. W.: "Morphological Aspects of Triglyceride Absorption," in C. F. Code and W. Heidel (eds.), *Handbook of Physiology,* sect. 6, vol. 3, American Physiological Society, Washington, 1968, chap. 71, pp. 1377–1406.

Thier, J. S.: "Morphology of the Epithelium in the Small Intestine," in C. F. Code and W. Heidel (eds.), *Handbook of Physiology,* sect. 6, vol. 3, American Physiological Society, Washington, 1968, chap. 63, pp. 1125–1175.

Walker, W. A., and R. Hong: "Immunology of the Gastrointestinal Tract, I." *J. Pediatr.* **83:**517–530 (1973).

Weiss, L., and R. O. Greep: *Histology,* 4th ed., McGraw-Hill, New York, 1977.

ADDITIONAL READINGS

Bennett, G.: "Migration of Glycoprotein from the Golgi Apparatus to Cell Coat in the Columnar Cells of the Duodenal Epithelium," *J. Cell Biol.* **45:** 668–673 (1970).

Brunser, O., and J. H. Luft: "Fine Structure of the Apex of Absorptive Cells from Rat Small Intestine," *J. Ultrastruct. Res.* **31:**291–311 (1967).

Cardell, R. R., Jr., S. Badenhausen, and K. R. Porter: "Intestinal Triglyceride Absorption in the Rat: An Electron Microscopic Study," *J. Cell Biol.* **34:**123–155 (1967).

Dawson, I.: "The Endocrine Cells of the Gastrointestinal Tract," *Histochemical J.* **2:**527–549 (1970).

Deschner, E. E.: "Observations on the Paneth Cell in Human Ileum," *Exp. Cell Res.* **47:**624–628 (1967).

Freeman, J. A.: "Goblet Cell Fine Structure," *Anat. Rec.* **154:**121–148 (1966).

Friend, D.: "Fine Structure of Brunner's Gland in the Mouse," *J. Cell Biol.* **25:**563–576 (1965).

Ito, S.: "The Enteric Surface Coat on Cat Intestinal Microvilli," *J. Cell Biol.* **27:**475–491 (1966).

Laguens, R., and M. Briones: "Fine Structure of the Microvillus of Columnar Epithelial Cells of Human Intestine," *Lab. Invest.* **14:**1616–1623 (1965).

Leeson, T. S., and C. R. Leeson: "The Fine Structure of Brunner's Gland in Man," *J. Anat.* **103:**263–276 (1968).

Moe, H.: "The Goblet Cells, Paneth Cells and Basal Granular Cells of the Epithelium of the Mouse Small Intestine," *Int. Rev. Gen. Exp. Zool.* **3:** 241–287 (1968).

Mukherjee, T. M., and L. A. Staehlin: "The Fine Structural Organization of the Brush Border of Intestinal Epithelium," *J. Cell Sci.* **8:**573–599 (1971).

Ockner, R. K., and K. J. Isselbacher: "Recent Concepts of Intestinal Fat Absorption," *Rev. Physiol. Biochem. Pharmacol.* **71:**107 (1974).

Owen, R. L., and A. L. Jones: "Epithelial Cell Specializations within Human Peyer's Patches: An Ultrastructural Study of Intestinal Lymphoid Follicles," *Gastroenterology* **66:**189–203 (1974).

Palay, S. L., and J. O. Revel: "The Morphology of Rat Absorption," in H. C. Meng (ed.), *Lipid Transport,* Charles C Thomas, Springfield, Ill., 1964.

Toner, P. G.: "Cytology of Intestinal Epithelial Cells," *Int. Rev. Cytol.* **24:**223–243 (1968).

Walls, R. W.: "Anorectal Anatomy," *Sci. Basis Med. Ann. Rev.* 113–124 (1963).

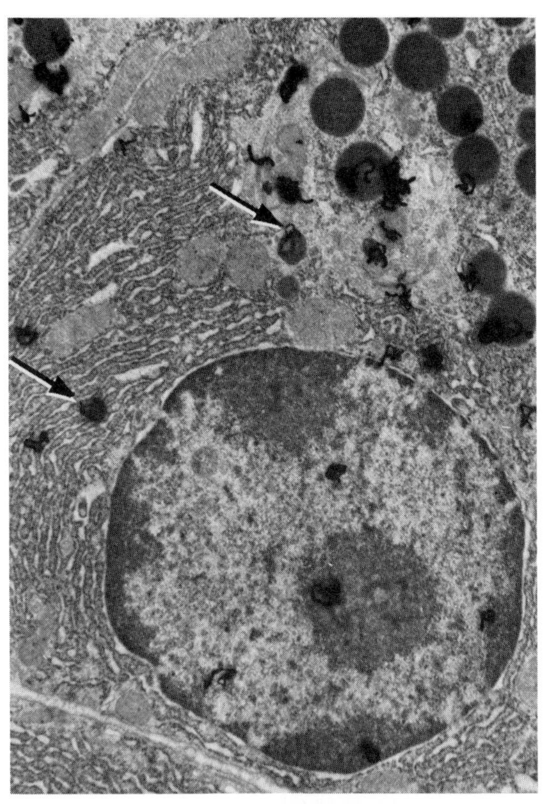

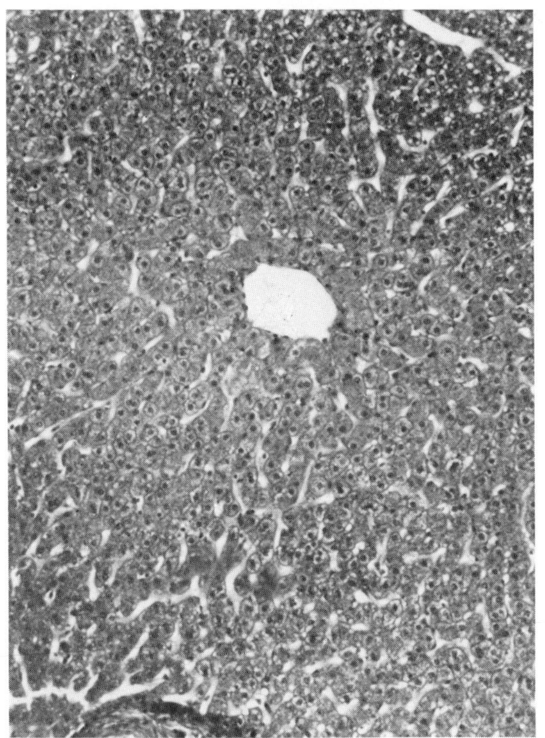

21 PANCREAS AND LIVER

OBJECTIVES

Upon completion of this chapter, the student will be able to:

1 Identify on a diagram of the pancreas the following structures:
(a) connective tissue capsule (b) lobules (c) area of exocrine acini (d) islets of Langerhans (e) ducts (main, interlobular, intercalated)

2 Identify on an LM of the pancreas the following:
(a) acinar cells (b) interlobular duct (c) intercalated ducts (d) centroacinar cells (e) alpha cells (f) beta cells

3 List the products produced by the following types of cells in the pancreas:
(a) exocrine cells (b) alpha cells (c) beta cells

4 List four major functions of the liver.

5 Identify the location in the liver of the following structures:
(a) hepatic lobe (b) hepatic lobule (c) connective tissue capsule (d) hepatic sinusoid (e) central vein (f) portal vein (g) hepatic artery (h) bile duct (i) portal triad (hepatic trinity) (j) sublobular vein

6 Identify the location, structure, and function of the following:
(a) von Kupffer cells (b) endothelial cells (c) space of Disse (d) hepatic sinusoids

7 Identify the structure, function, and location of the following structures for a typical liver cell:
(a) nucleus (b) RER (c) SER (d) Golgi apparatus (e) glycogen granules (f) mitochondria (g) beginning of the bile canaliculi

8 Identify the location and function of both alpha and beta granules of glycogen.

9 Trace the flow of blood through the liver.

10 Identify the location and function of the gallbladder.

11 Identify the histologic structure of the following three layers of the gallbladder:
(a) mucosa (b) muscularis (c) adventitia

PANCREAS

The pancreas is one of the three types of accessory digestive glands that have developed from the gut epithelium in early embryonic life. Over 90 percent of the organ is a compound alveolar gland responsible for exocrine secretion of a number of enzymes that, together with the bile, are poured into the duodenal portion of the small intestine. These enzymes include proteases, lipases, and amylases necessary for the breakdown of foodstuff in the GI tract. In addition to the primary functioning of the pancreas as an exocrine gland, it has islets of endocrine cells that secrete insulin and glucagon that are vital hormones regulating carbohydrate metabolism.

The large pinkish white organ stretches transversely from the duodenum to the spleen, behind the stomach. The gland is often arbitrarily divided into a head, neck, body, and tail. The head is an expanded portion located in the curve of the duodenum. The neck and body lie along the back of the stomach, and the tail tapers toward the spleen.

The two pancreatic ducts, a main duct and an accessory duct, begin in the tail by the union of small interlobular ducts. These ducts traverse the surface of the gland, collecting smaller ducts as they merge. Eventually the pancreatic duct joins with the common bile duct, which empties into the duodenum. The accessory pancreatic duct may open directly into the duodenum or connect with the main duct.

Embryologically, the pancreas develops from both dorsal and ventral invaginations of duodenal endoderm. The fusion of the two main ducts varies with the individual, as a result of their separate origin in the embryo.

As in other parenchymal organs, the pancreas is covered by a thin, dense, irregular fibrous elastic connective tissue (FECT) capsule which divides the organ into numerous lobules.

EXOCRINE PANCREAS

The pancreas is surrounded by a sparse connective tissue capsule, inward extensions of which (trabeculae) subdivide the gland into a series of lobes and lobules. Each lobule is composed of a number of secretory acini (Fig. 21-1). As in salivary glands, the acinus is a saclike structure composed of roughly spherically arranged, pyramidal-shaped cells that are supported by a basement membrane. These cells are polarized. The apical portion of acinar cells is filled with zymogen granules that are generally acidophilic in staining characteristics, whereas the basal portion is basophilic and contains the nucleus. The polarity of the cytoplasm is more dramatically demonstrated in Fig. 21-2, in which the acini stained for basophilic cytoplasm are depicted at a higher magnification.

The basophilia of acinar cells is confined to the basal half of the cell where a vesicular nucleus is located. Supranuclearly, light, irregularly shaped areas that are devoid of zymogen granules are present. These correspond to the regions of the Golgi complex

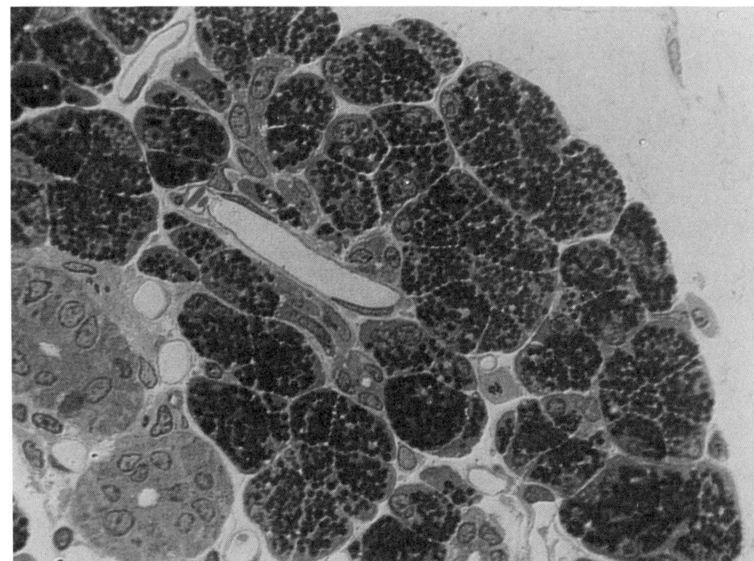

FIGURE 21-1 Light micrograph of a pancreatic lobule. It consists of several secretory acini.

where the production of secretory granules is known to take place. Numerous eosinophilic zymogen granules are found in the apical portion of these cells. Thus, each cell has a prominent basally located nucleus, vesicular nuclear chromatin, large nucleolus, and other features common to protein-secreting cells and a cytoplasm that has a typically polarized appearance.

Basal basophilia corresponds to the presence of RER and the supranuclear Golgi apparatus can be stained to show associated enzymes. The EM radioautograph of a pancreatic exocrine cell in Fig. 21-3 illustrates the

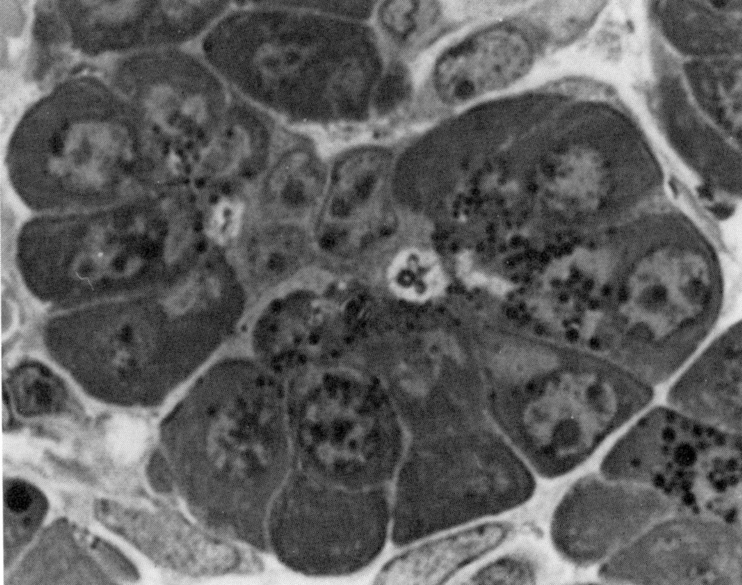

FIGURE 21-2 Pancreatic acini viewed at higher magnification. Silver grains are localized over RER, Golgi apparatus, and zymogen granules.

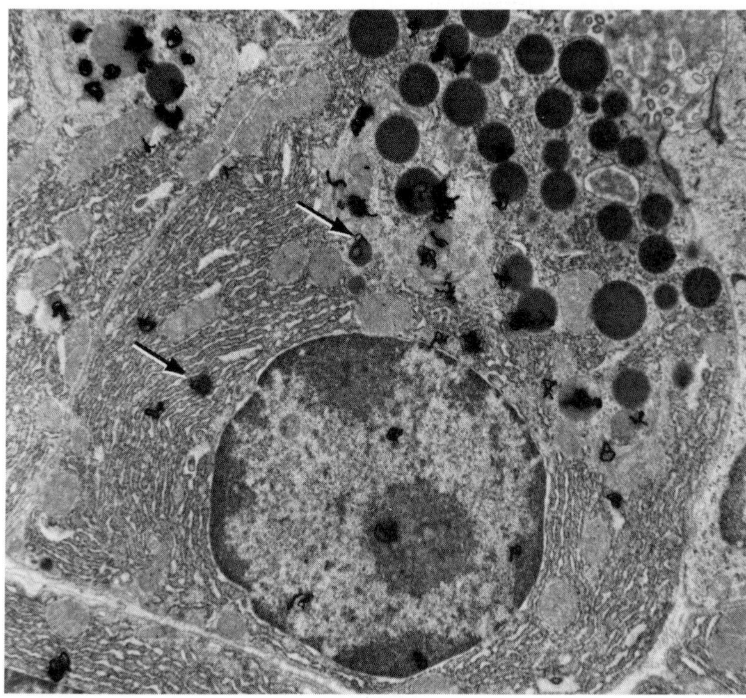

FIGURE 21-3 Electron microscopic radioautograph of a pancreatic exocrine cell. The silver grains (arrows) represent nuclear tracts on the emulsion of ^{3}H-labeled amino acids.

above, namely, a developed RER in the basal cytoplasm with a prominent Golgi and apical zymogen granules that are labeled with ^{3}H amino acids.

The supranuclear Golgi region has prozymogen or condensing vacuoles which aggregate and condense to form zymogen granules (Fig. 21-4). As in the previous figure, Fig. 21-4 is an electron microscopic radioautograph in which ^{3}H-labeled amino acids are localized (arrows) as they are incorporated into the prozymogen granules. Zymogen

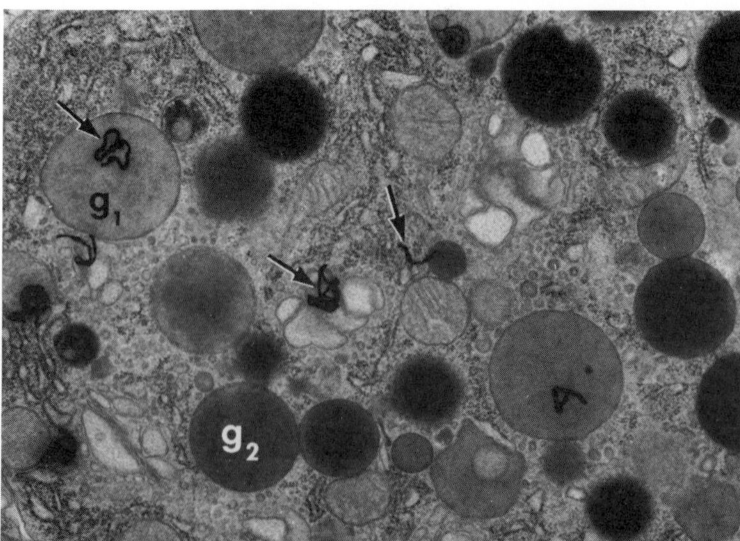

FIGURE 21-4 Electron micrograph of the Golgi region of a pancreatic acinar cell. It has prozymogen granules (g_1) that condense to form zymogen granules (g_2). Amino acids labeled with tritium (^{3}H) are incorporated into the prozymogen granules. The sites of incorporation are visualized as silver grains (arrows) in this radioautograph.

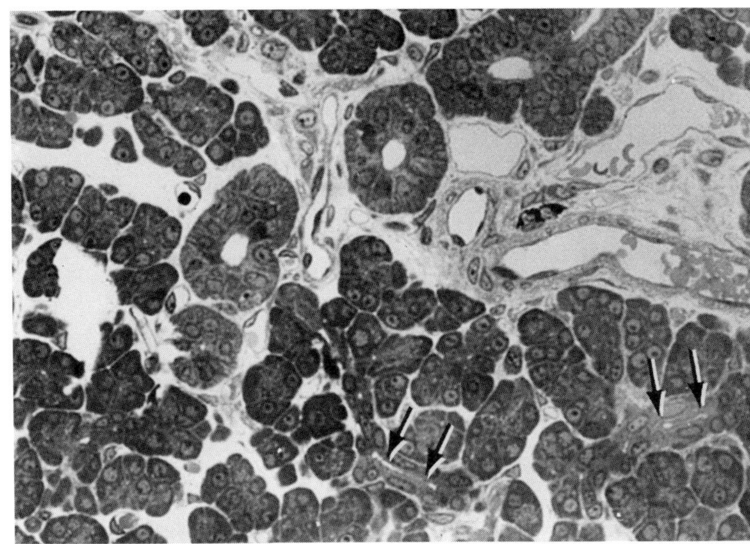

FIGURE 21-5 Light micrograph of a pancreatic lobule. The acinar lumen leads into a narrow duct, the intercalated duct (arrows).

granules contain several classes of enzyme products (lipases, amylase, proteases, and nucleases) to be released into the acinar lumen and then transported to the duodenum.

The acinar lumen leads into the narrow intercalated duct (Fig. 21-5, arrows). This duct becomes the *intralobular duct* which combines with others, becoming larger and passing through the connective tissue of the interacinar region. Intercalated ducts stain lighter than the acini and are located between the latter (Figs. 21-6 and 21-7, circles). As intercalated ducts join, groups of acini become lobules and are separated by connective tissue. At this point, the intercalated duct becomes an intralobular duct and joins similar ducts from other lobules. Several intralobular ducts combine and form an interlobular duct (Figs. 21-7 and 21-8). In turn, the

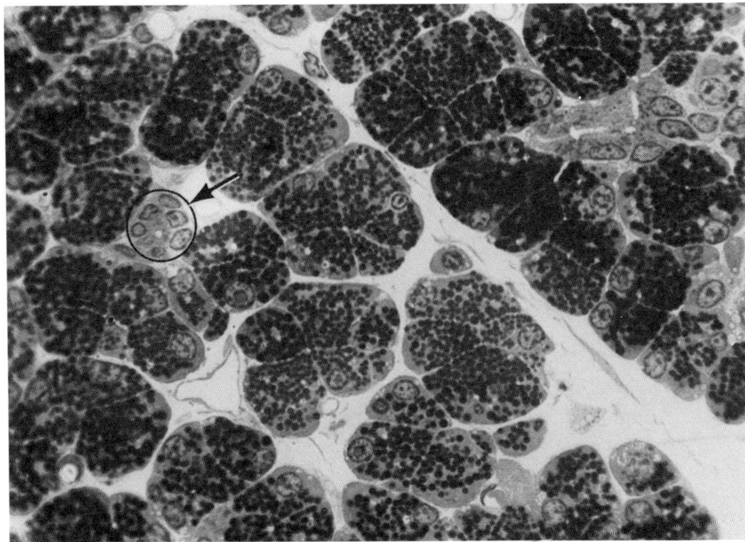

FIGURE 21-6 Light micrograph of pancreatic lobule. An intercalated duct (circle) is located in the loose FECT between the acini.

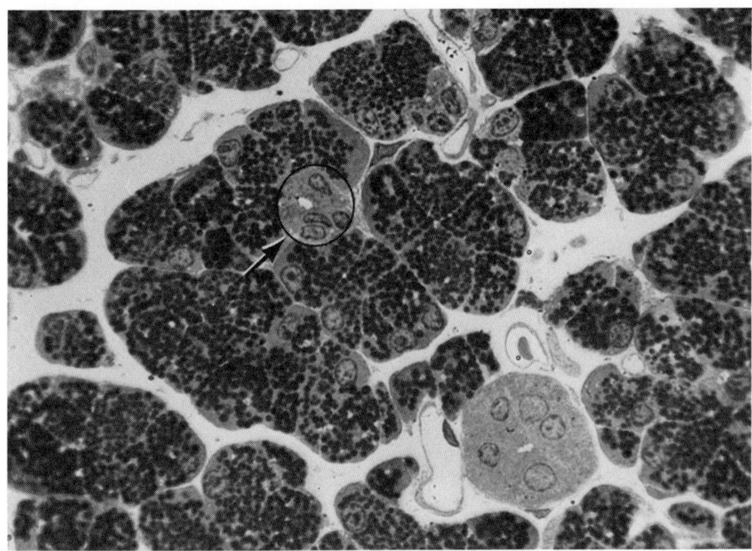

FIGURE 21-7 Pancreatic lobule. An intercalated duct (circle) is present within it, and an intralobular duct is seen in the FECT between the lobules in the lower right-hand corner.

interlobular ducts merge with the main excretory duct, which is lined with columnar epithelium. The main excretory duct, as already mentioned, carries the digestive enzymes to the common bile duct, to be discharged into the duodenum.

A pancreatic acinus often contains in its lumen a small, pale-staining epithelial cell adjacent to the apex of a secreting cell (Fig. 21-9, arrows). These *centroacinar cells* are thought to contribute to the regeneration of the acinus when necessary.

ENDOCRINE PANCREAS

Groups or clusters of endocrine cells that produce *insulin* and *glucagon* called *islets of Langerhans* are present within the pancreas and have a rich capillary supply which breaks up the secretory cells into cords (Fig. 21-10). In human beings more than 1 million islets

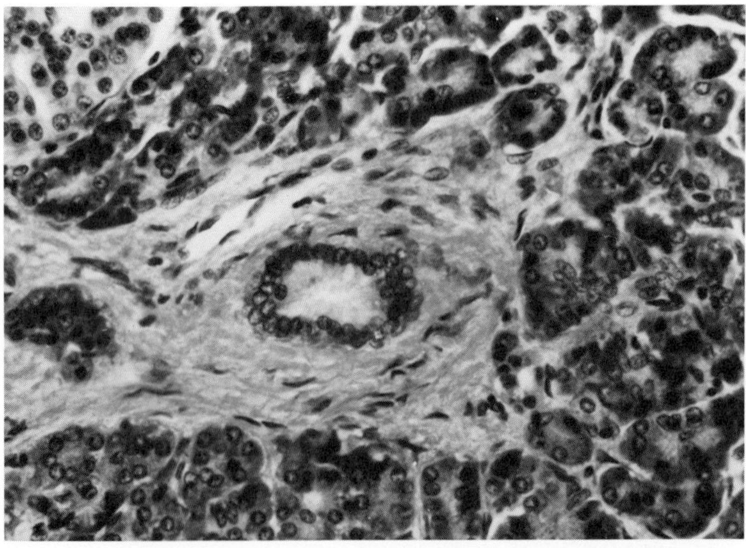

FIGURE 21-8 Intralobular duct seen in a conventionally prepared paraffin section. Previous micrographs were obtained from 1-μm-thick sections of plastic embedded tissue.

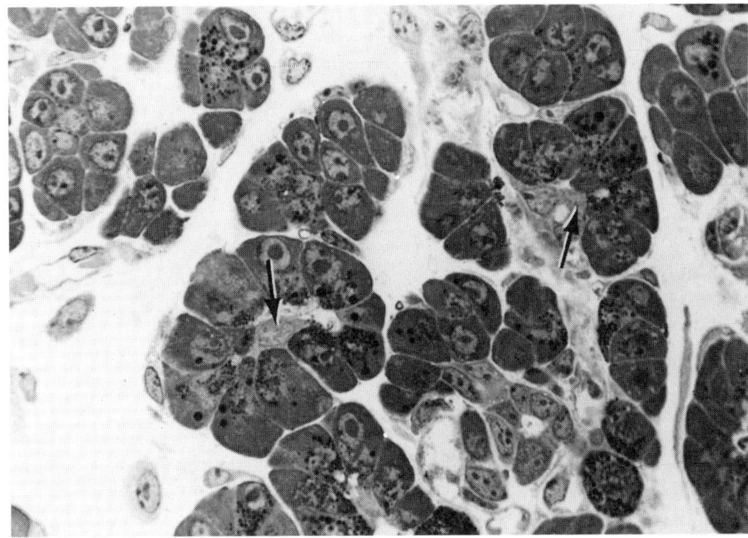

FIGURE 21-9 Light micrograph of pancreatic acini. Light-staining centroacinar cells (arrows) are associated with the apexes of the acinar cells.

of Langerhans have been reported. Cells in the islets contain rounded, centrally located nuclei and various small granules which can be discerned by special stains. Thus, the islet cells can be divided into at least two major subgroups which can be differentially stained.

Beta cells, or B cells, are more numerous, making up approximately 75 percent of islet cells. They are centrally located and produce *insulin.* The *alpha cells,* or A cells, constitute about 20 percent of islet cells, are more peripherally located, and produce *glucagon.* Since both of the hormones produced by the islets of Langerhans are peptide in nature, they are present in small granules within the cytoplasm of alpha and beta cells. Although beta-cell granules are larger than those of alpha cells, such difference in size is not evident in an LM, but becomes clear when

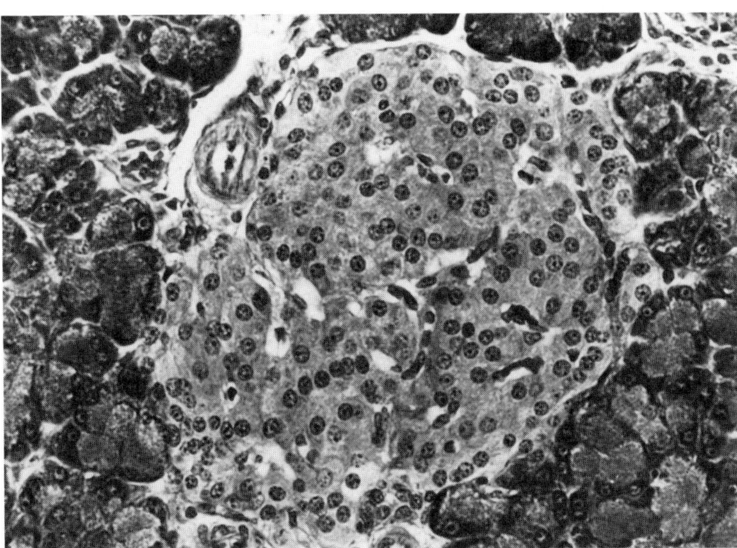

FIGURE 21-10 Light micrograph of an islet of Langerhans. The islet is demarcated from the surrounding acini by loose FECT. Its cells are arranged in irregular cords and are paler staining than the exocrine acinar cells. The islet is supplied by an extensive capillary network that surrounds the cords of islet cells.

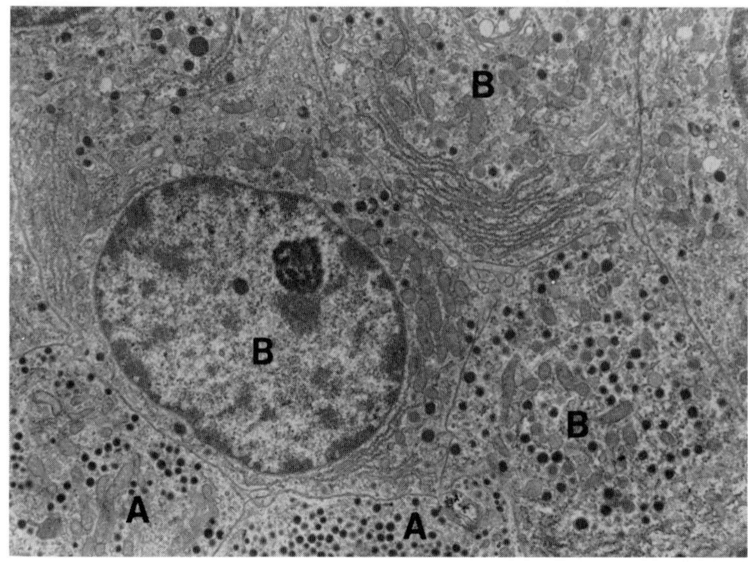

FIGURE 21-11 Electron micrograph of an islet of Langerhans. Both alpha cells (A) and beta cells (B) contain apparent secretory granules since the hormones secreted by both cell types are peptides.

viewed in an electron microscope. Beta-cell granules vary among different species. The EM in Fig. 21-11 shows portions of several beta cells (B), along with a couple of alpha cells (A) at the lower left corner. Both alpha and beta granules are bordered by a membrane and contain an electron-dense core.

Beta-cell granules are rounded and contain an electron-dense center which is separated from the membrane surrounding it (Fig. 21-12). There is a close positional relationship between capillaries (c) and peripherally located granules. The production of these granules is believed to involve the RER and condensation of the secretory material in the Golgi complex, as in most protein-secreting cells, since insulin and glucagon are peptides. Indeed, the basic structure of alpha and beta cells conforms to the usual structure of other protein-secreting cells, as shown in

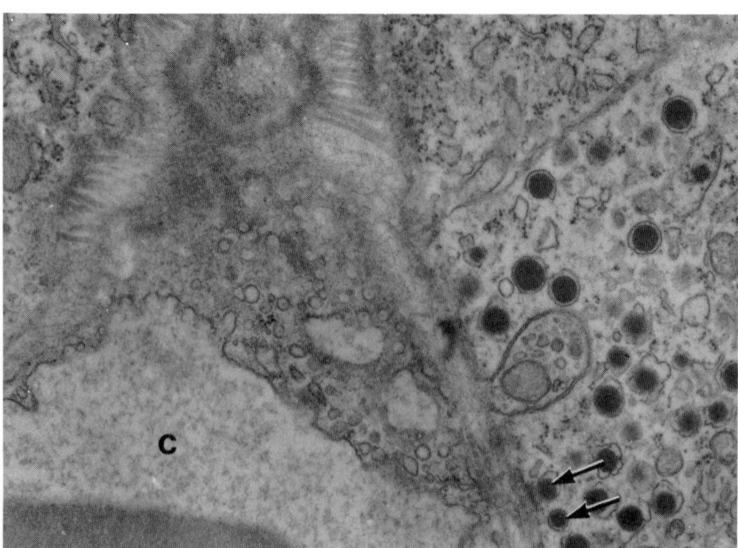

FIGURE 21-12 Portion of beta cell. The round membrane-bound granules have an electron-dense core. Note the close relationship between the peripherally located granules (arrows) and the capillary (c).

Fig. 21-11; that is, there are areas of RER, Golgi apparatus, and secretory granules. Endocrine secretory granules (approximately 200 μm) are much smaller than exocrine zymogen granules. They are exteriorized, or exocytosed, and passed into the capillary to join the bloodstream. A third cell type, called *delta,* or D cells, is also present in humans. Delta cells may make up 5 percent of the endocrine cell population. They do have granules that resemble those of alpha cells, but their function is not clearly understood.

LIVER

The liver, the largest internal organ of the body, is located in the right half of the abdominal cavity, below the diaphragm. It is an organ covered by a connective tissue capsule, and deep fissures divide it into four lobes. On the underside of the organ, the portal vein and hepatic vessels enter, and the bile duct exists via the hilus, or porta, of the liver. The liver is encapsulated by the *Glisson's capsule,* a dense but thin connective tissue covering of dense, irregular FECT mixture which is continuous with the adventitia of the vessels, forming a stromal framework for the radiating lobules within the four lobes.

The liver plays a central role in the many metabolic functions that are carried out by the blood. It regulates the amount of various useful metabolites by adding them to or removing them from the blood. It produces and regulates blood constituents such as serum albumin, fibrinogen, and cholesterol. Materials such as alcohol, drugs, and hormones are detoxified, or inactivated, then filtered out of the blood in the kidney. The liver also stores carbohydrate in the form of glycogen and releases its breakdown products into the blood at times when blood sugar levels are low. In addition, the liver has an exocrine function in that it synthesizes and excretes bile via the bile duct.

Because of these diverse functions, the liver is centrally located on the collective route of nutrient materials. Blood from the abdominal viscera, the stomach and intestines, is collected in the portal veins and passed through the liver. Portal blood comprises 75 percent of the blood passing through the liver, while the remaining 25 percent enters via the hepatic artery to nourish the parenchymal cells of the organ. The microscopic organization of the liver reflects the functional requirements mentioned above. The liver is composed of *hepatic lobules* which are made up of many cords of cells, or *hepatic plates,* radiating out from the branches of the portal vein, the *central veins* (c; Fig. 21-13).

Because of the close metabolic relationship between blood and parenchymal cells of the liver it is important to have a clear understanding of the blood supply to comprehend the structure and function of the liver. Two blood vessels bring blood to the liver: the portal vein from the intestines and the hepatic artery, which nourishes the hepatic cells. Branches of these vessels travel in delicate connective tissue septa between the hepatic lobules, called portal spaces. The diagram in Fig. 21-14 shows the interrelationships of these structures.

Blood from branches of the portal vein drains into the fenestrated sinusoids of a lobule and travels toward the central vein. The fenestrated lining of the sinusoids allows the blood to come into direct contact with parenchymal cells. The hepatic artery branches into the capillary systems invested in the connective tissue supplying the parenchyma.

The blood collected in the central veins is passed to the sublobular veins. The sublobular veins merge to form the inferior vena cava, which returns the blood to the heart. Branches of the portal vein, hepatic artery, and bile duct, as they feed into and receive secretory products from the hepatic lobule, are present in connective tissue spaces be-

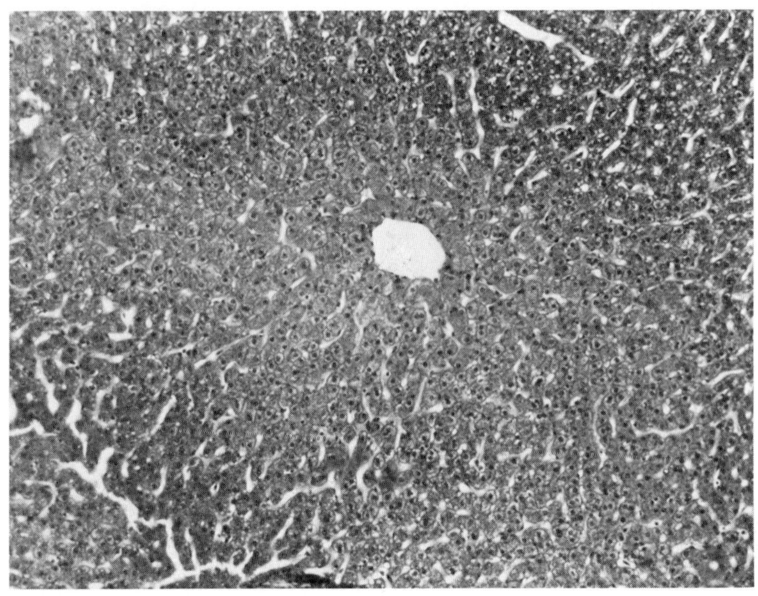

FIGURE 21-13 Light micrograph of a hepatic lobule. It has a polyhedral shape, and cords of cells, hepatic plates, radiate from the central vein (c), a branch of the portal vein.

tween adjacent hepatic lobules. Since they travel in a group, they are called a *hepatic triad* or *trinity* (Fig. 21-15).

LIGHT MICROSCOPIC VIEW OF HEPATIC LOBULES

Hepatic cells are structurally similar throughout the liver despite their multiple functions. They are arranged in somewhat regular rows, called *hepatic cords,* as seen in Fig. 21-16, which represents the appearance of hepatic cords delineated by a dye which has been injected into the portal vein, thereby filling the sinusoids between the cords.

Most liver cells have a centrally located nucleus, although a small number of them

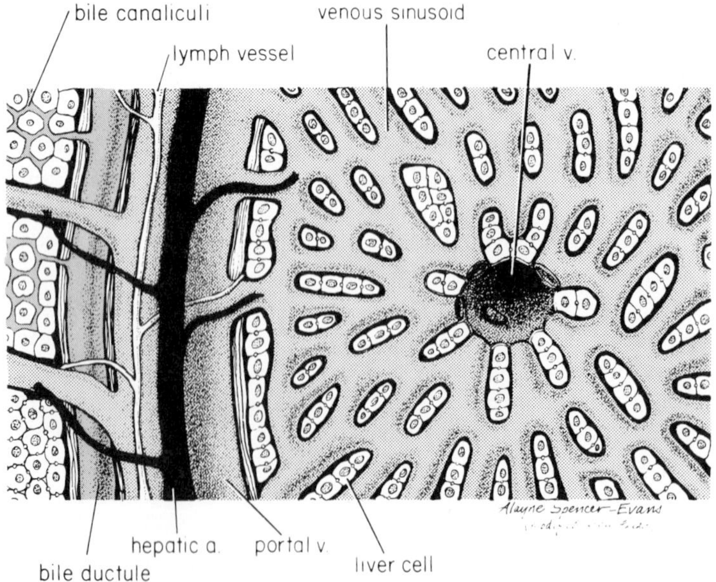

FIGURE 21-14 Diagram of the hepatic structure. (*Modified from Elias.*)

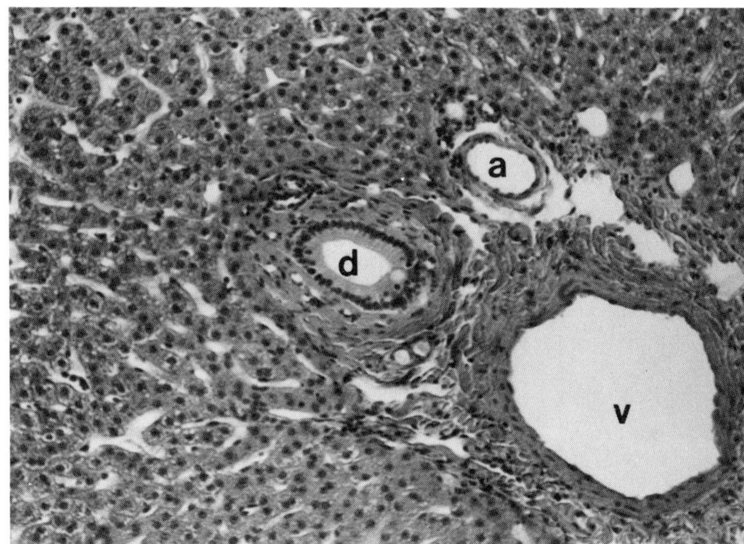

FIGURE 21-15 Light micrograph of a portion of the liver. It shows a region of connective tissue between lobules containing a hepatic triad. The triad consists of a branch of the portal vein (v), a branch of the hepatic artery (a), and a bile duct (d).

are binucleated (Fig. 21-17). The nucleus contains light vesicular material, indicating the relatively high activity of the cell. Nucleoli are often prominent. Elsewhere in the cytoplasm, lightly stained areas correspond to locations of glycogen storage (Fig. 21-18).

In addition, the hepatic cell has a number of basophilic regions which represent the RER responsible for bile and albumin synthesis. Within the cytoplasm, other areas of differentiation are present, but they can only be seen with special stains and in the electron microscope. By making very thin sections of liver cells prepared for electron microscope studies, some of the structural details may be made out in LMs. Figure 21-19 is an LM taken from such a section and stained for basophilia and lipid. It demonstrates the moderately dense cytoplasmic patches where RER is present (arrows). Lipid droplets of

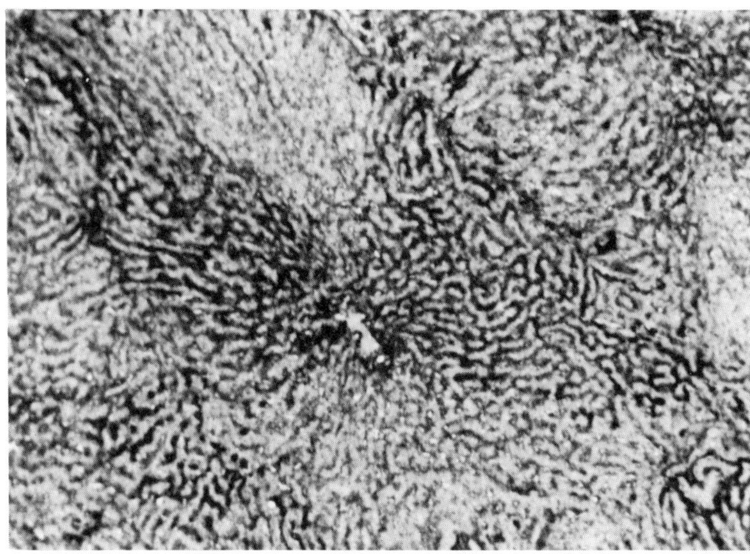

FIGURE 21-16 Portion of dye-injected liver. Hepatic cords are delineated by the dye, which fills the sinusoids between them.

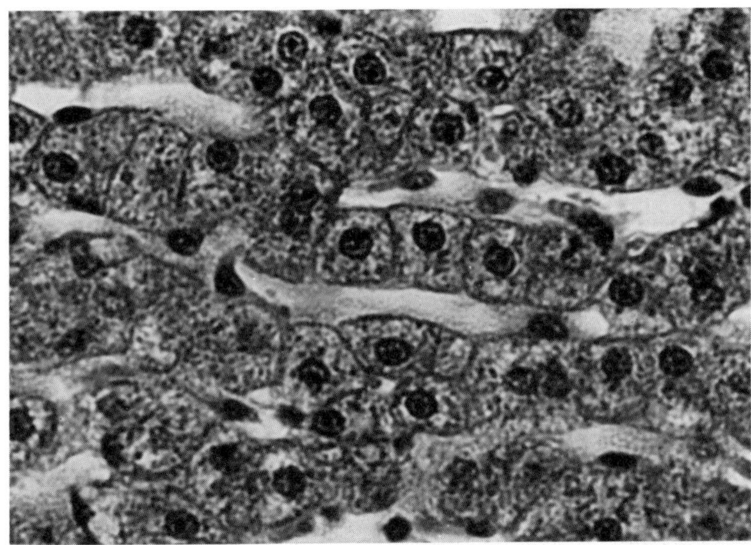

FIGURE 21-17 Hepatic cords viewed at higher magnification showing the appearance of individual cells.

different sizes are present in black globular forms of varying size.

Hepatic sinusoids maintain a population of flattened endothelioid cells which form an incomplete lining of the sinusoid (Fig. 21-20, arrows). These cells have a dense and somewhat flattened nuclei. They can differentiate into the phagocytic cells of von Kupffer (arrowhead) which frequently contain foreign particles and occasional bile pigments. When properly stained, reticular fibers can be demonstrated along the sinusoidal walls, closely associated with these von Kupffer cells (Fig. 21-21). There is a small patent space between the parenchymal cells and the sinusoidal lining where lymph originates. This is called the space of Disse, from which the lymph is conducted opposite to the direction of blood flow, draining into the interlobular lymphatic vessels.

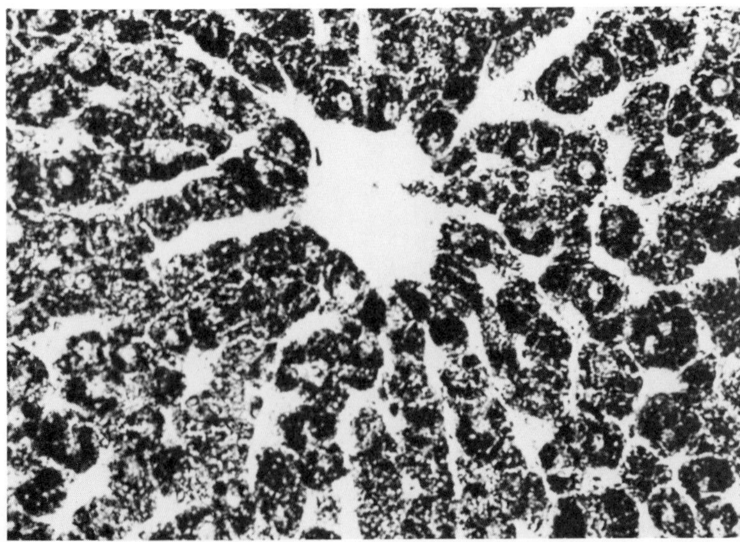

FIGURE 21-18 Light micrograph of a liver lobule that has been stained for the demonstration of glycogen.

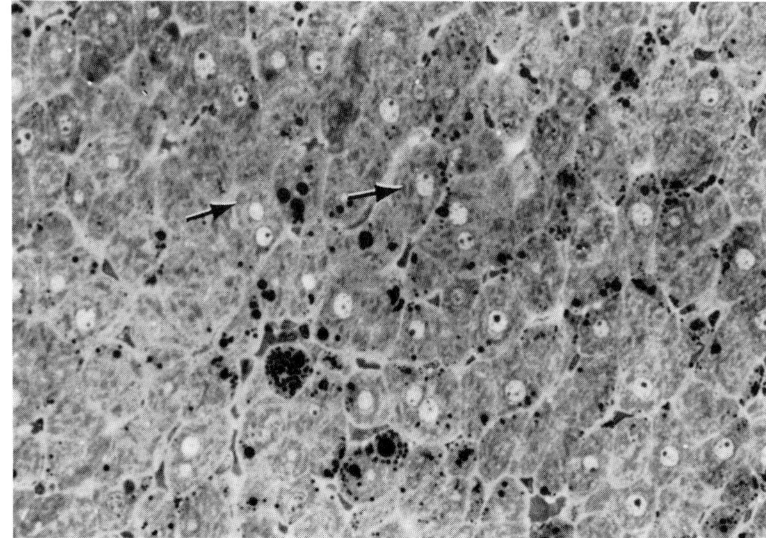

FIGURE 21-19 Light micrograph of a 1-μm-thick section of plastic embedded liver tissue. It shows more clearly structural details of the polyhedral hepatic cells. The cells have a large, round vesicular nucleus with prominent nucleoli. A few cells are binucleate. The basophilic stain (toluidine blue) demonstrates patches of RER (arrows). Lipid droplets appear as black globules of varying size.

ELECTRON MICROSCOPIC VIEW OF HEPATIC LOBULES

An electron microscopic view presents more details of the structure of liver cells. The composite diagram in Fig. 21-22 demonstrates salient structural details of the cell. The hepatic cell has two active surfaces: one that faces the sinusoids and another adjacent to the bile canaliculi. Note the difference in size between these two types of passages.

As pointed out earlier, RER is present in the basophilic portion of the cytoplasm. Frequently, a large amount of smooth-surfaced endoplasmic reticulum (SER), often called agranular reticulum, is found in association with the glycogen granules, which appear as clear areas next to the SER (Figs. 21-23 and 21-24). Glycogen granules can be stained for electron microscopy (Fig. 21-24). Granules, stained with lead, accumulate among the

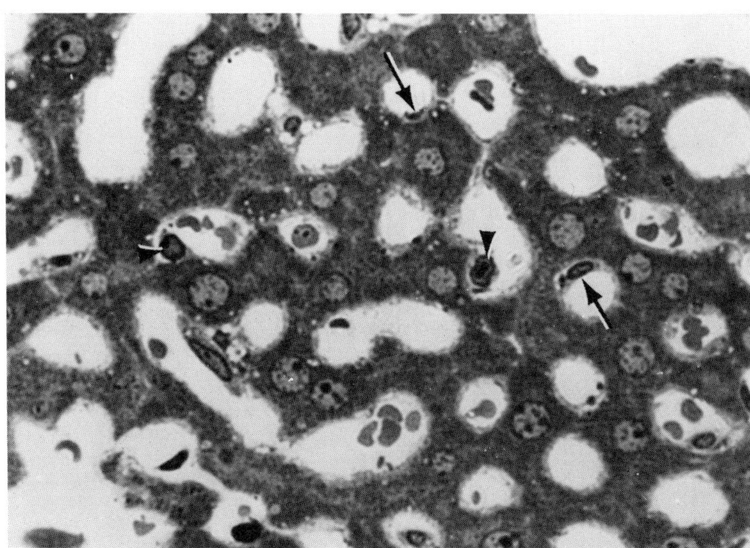

FIGURE 21-20 Light micrograph of hepatic sinusoids. The sinusoids have an incomplete lining of flattened endothelial cells (arrows). Among these cells are phagocytic cells, von Kupffer cells, that contain foreign ingested material (arrowheads).

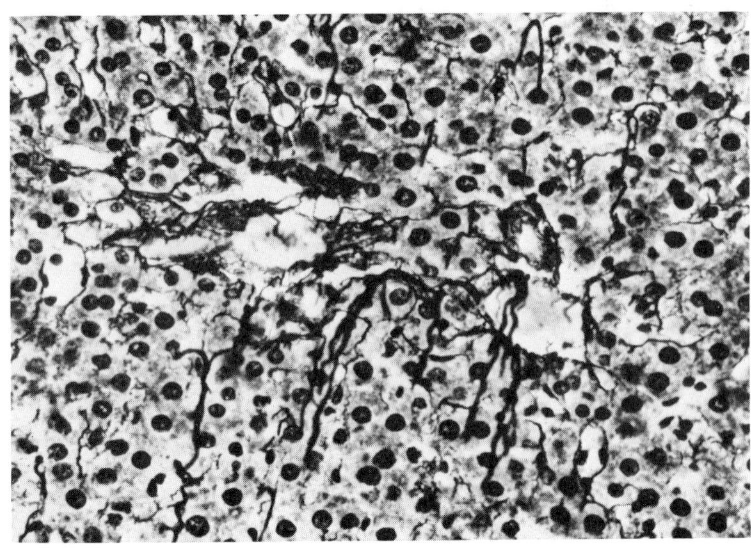

FIGURE 21-21 Light micrograph of a liver section stained for demonstration of reticular fibers. The fibers appear black and are closely associated with the walls of the sinusoids.

SER, and they may be present in aggregates (alpha form) or as individual granules (beta form).

Mitochondria in the liver cells (Figs. 21-23 and 21-24) are usually round to ovoid, with cristae that seldom go across the entire mitochondrial matrix. The matrix is moderately dense, with occasional intramitochondrial granules. Experiments have associated the moderately dense matrix of the liver mitochondria with their numerous intermediary metabolic involvements in addition to the generation of ATP.

Bile canaliculi are small passages pro-

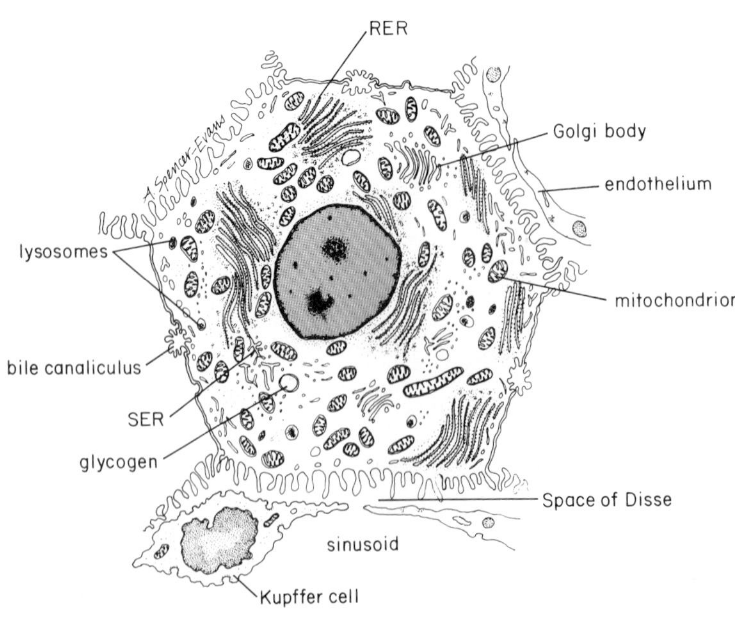

FIGURE 21-22 Composite diagram illustrating the fine structure of hepatic cells.

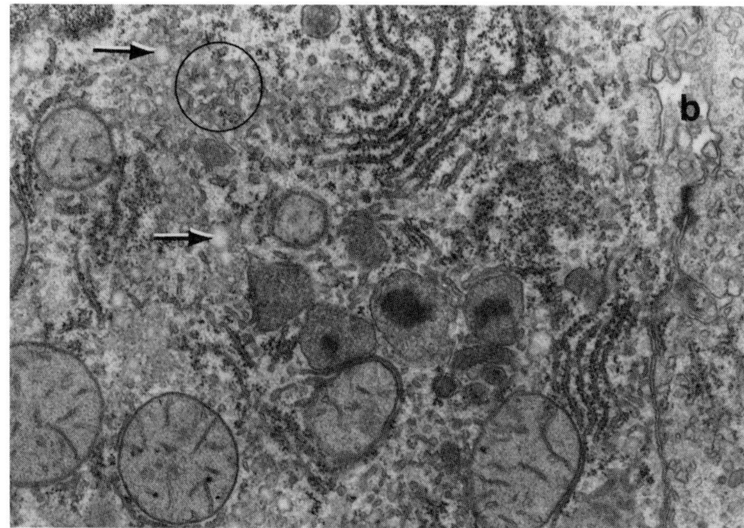

FIGURE 21-23 Portion of a liver cell. Glycogen deposits (arrows) are closely associated with SER (circle). A bile canaliculus is separated from the intercellular space by tight junctions and occasional desmosomes.

duced by irregular membranes between adjacent cells (Figs. 21-22 and 21-23). Along the lumen of bile canaliculi, a number of microvillous projections are seen. Desmosomes and occasional tight junctions are present around the canaliculi.

The sinusoidal walls between the hepatic parenchymal cells and von Kupffer cells are highly irregular. Numerous short processes project out of the parenchymal cells (Fig. 21-25). Small bundles of collagen fibrils (arrow) are often present in association with von Kupffer cells. These correspond to the reticular fibers observed in Fig. 21-21. The cytoplasm of von Kupffer cells, although not seen in this figure, often have the usual characteristics of a phagocytic cell, namely, vacuoles, vesicles, and ruffling surface mem-

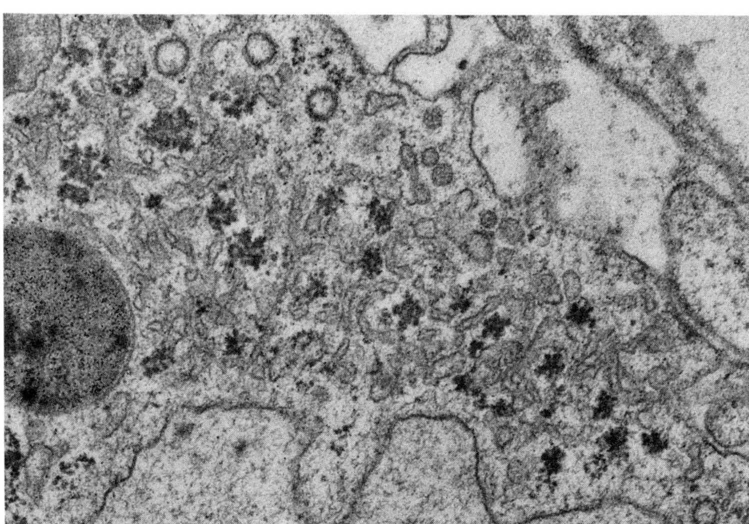

FIGURE 21-24 Portion of liver cell cytoplasm containing high concentrations of glycogen in aggregations of varying size (alpha particles). The glycogen is closely associated with profiles of agranular endoplasmic reticulum.

branes, along with occasional foreign particles. Note the irregular-shaped space of Disse (s) between the Kupffer cell (K) and hepatic parenchymal cell to the right.

GALLBLADDER

The gallbladder is a hollow, pear-shaped organ lying along the inferior surface of the liver. It functions as a storage depot for bile and also helps concentrate it through reabsorption of water and ions. The gallbladder is connected to both the liver and the duodenum by the bile duct system.

The wall of the gallbladder is composed of three layers: mucosa, muscularis, and ad-

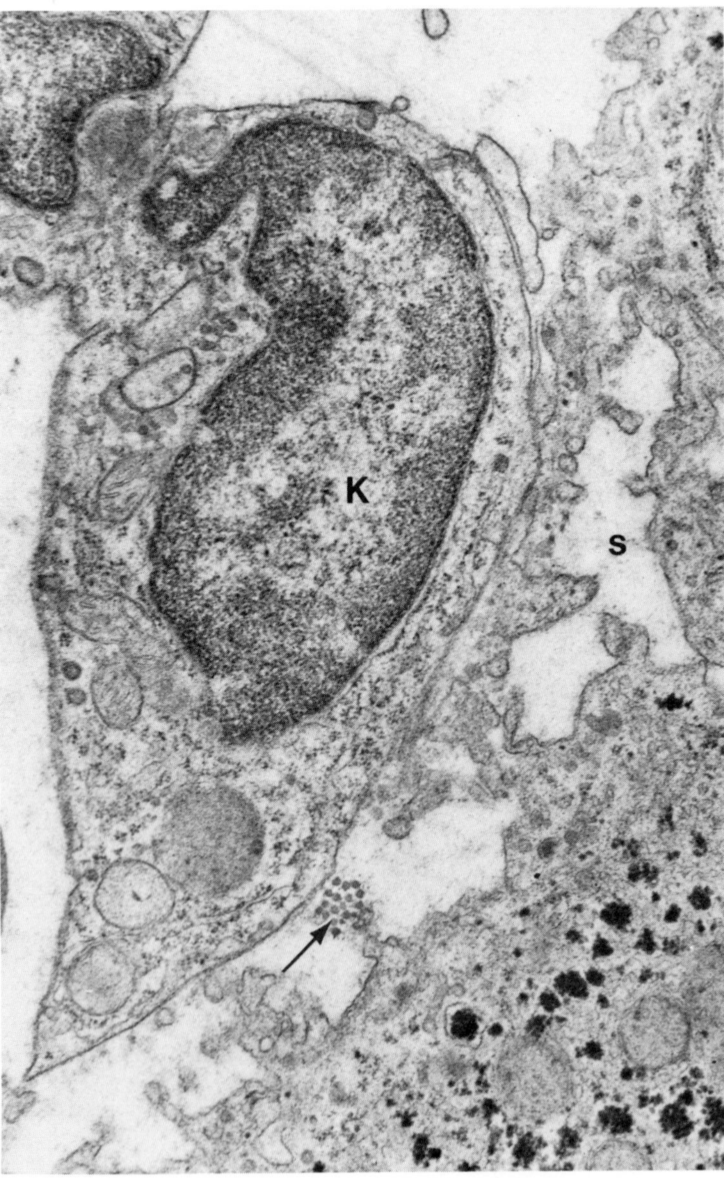

FIGURE 21-25 Electron micrograph of von Kupffer cell. Collagen fibrils (arrow) are associated with the cell. The irregular space of Disse (s) is present between the von Kupffer cell (K) and the liver cells.

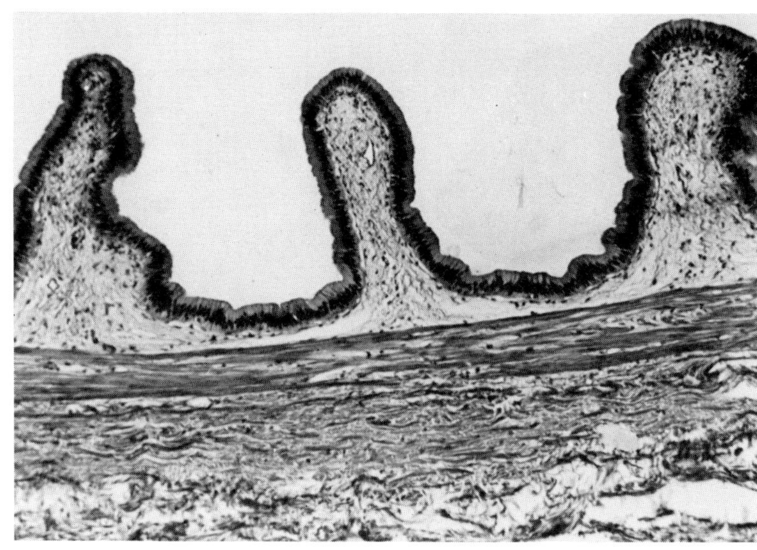

FIGURE 21-26 Light micrograph of the wall of the gallbladder. The appearance of its three layers—mucosa, muscularis, and adventitia—is shown.

ventitia (Fig. 21-26). The thinnest layer, the mucosa, folds into numerous ridges. Its epithelium is tall and columnar, with ovoid nuclei in the basal cytoplasm (Fig. 21-27). In EM, numerous microvilli are seen along the apical surface. Underlying the epithelium is a lamina propria of connective tissue and a few scattered smooth muscle cells from the muscularis layer.

The muscular layer consists of alternating smooth muscle fibers with interlacing layers of FECT. Bundles run longitudinally near the lamina propria, covering the length of the gallbladder and traveling over its fundus (Fig. 21-27). The remaining muscle fibers are circularly arranged.

The adventitia is usually a thick, fibrous connective tissue and contains blood vessels, lymphatics, and nerves. This FECT is covered by mesothelium, thus forming a serosa.

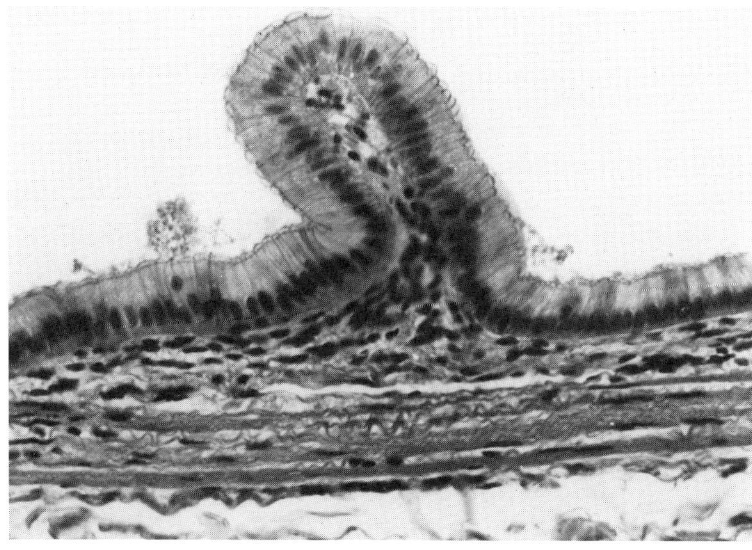

FIGURE 21-27 Light micrograph of gallbladder mucosa. Smooth muscle fiber bundles run longitudinally near the lamina propria.

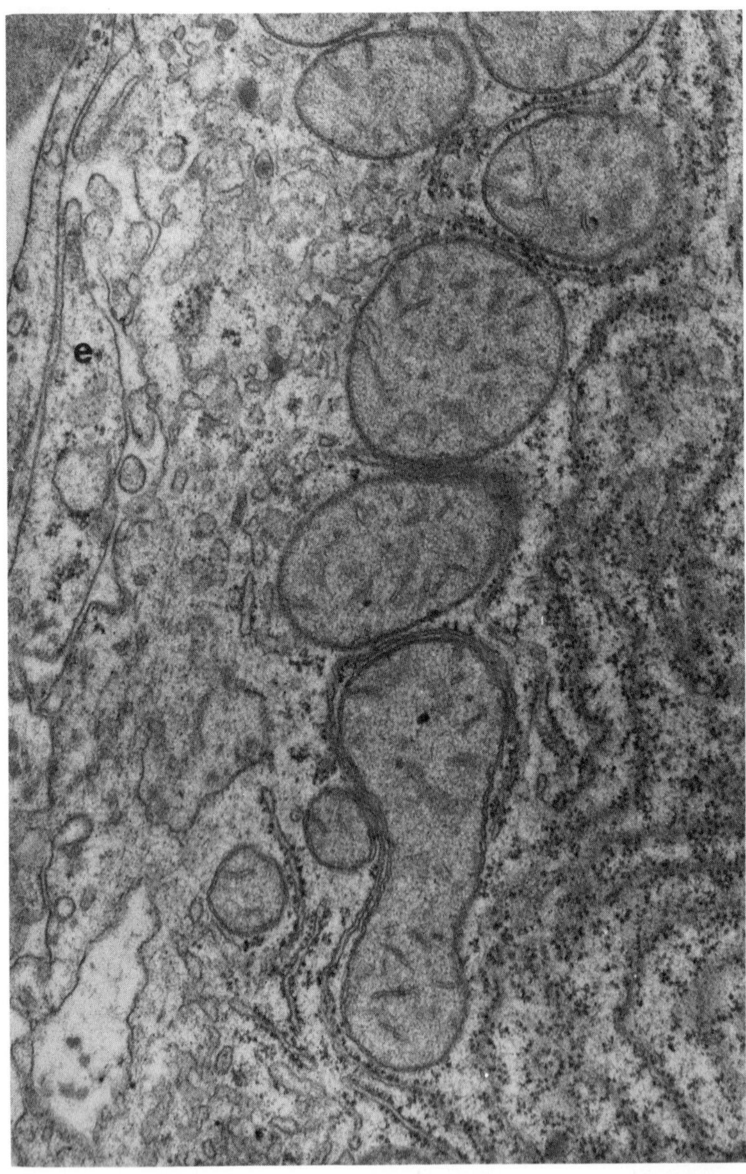

FIGURE 21-28

REVIEW SECTION

1. The EM in Fig. 21-28 is a section through a liver cell. What is the name given to the specialized endothelium appearing along the left margin? The boundaries of the space of Disse are ________ and ________. What is formed in the space of Disse? The structures in the lower right corner of the micrograph function to ________.

The endothelial cells lining the sinusoids of the liver are von Kupffer cells. Between the lining of endothelial and liver cells, extensions of the liver cells' plasma membranes form small villous projections. In this space (of Disse), lymph from the hepatic parenchyma is formed. The large amounts of RER found in liver cells are indicative of their protein synthetic activities.

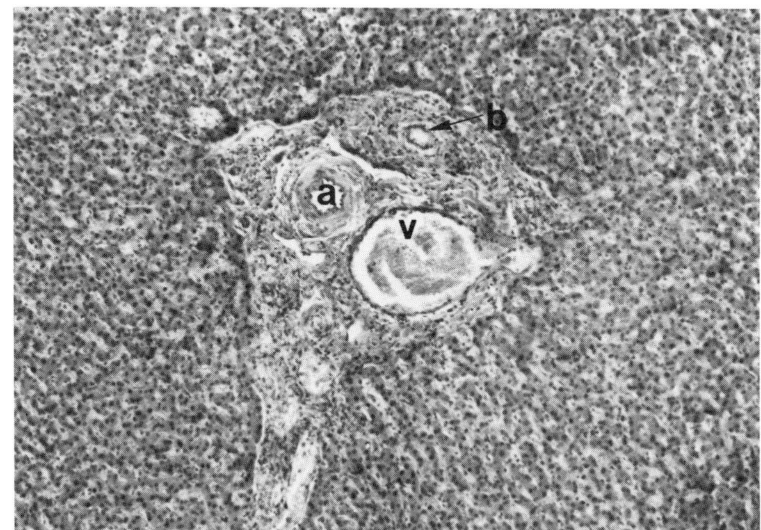

FIGURE 21-29

2. Figure 21-29 is a section through the liver. Name the various circular profiles indicated by the letters.

This picture shows a portal vein (v), hepatic artery (a), and bile duct (b). The hepatic triads are formed by branches of the portal vein, hepatic artery, and bile duct next to each other. Note the parenchymal cells lined up in cords.

3. How many layers of tissue can be seen in Fig. 21-30? What type of cell lines the luminal surface? How are the muscle cells arranged in this picture? This section is taken from a __________.

This section through the gallbladder reveals the mucosa, muscularis, and part of the adventitia. The tall columnar epithelium on the inner surface aids in the concentration of the bile. The microvilli and the irregular inner surface increase the effective area of interaction between the gallbladder and the bile.

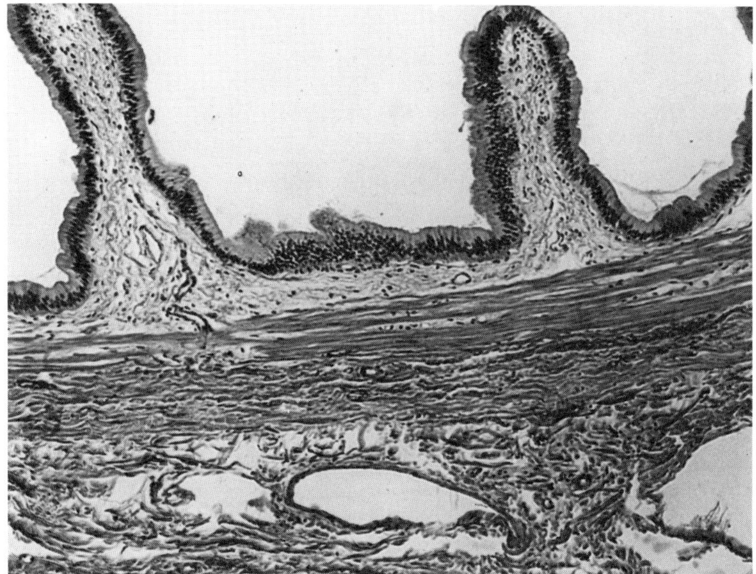

FIGURE 21-30

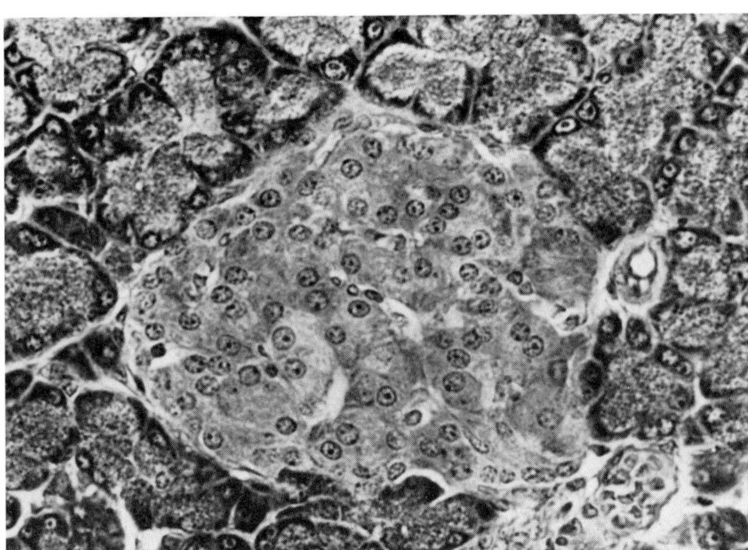

FIGURE 21-31

4. Identify the organ in Fig. 21-31. It is subdivided into ________. The light-staining groups of cells are the ________. The dark cells located at the periphery of the field form ________. The two major types of products formed by this organ are ________ and ________. These products are produced in ________ and ________.

The pancreas consists of exocrine glands (the dark, peripherally located acini) which produce digestive juices and the lighter round islets of Langerhans which produce endocrine hormones (insulin and glucagon) that help regulate carbohydrate metabolism.

BIBLIOGRAPHY

Brauer, R. W.: "Liver Circulation and Function," *Physiol. Rev.* **43:**115–213 (1963).

Elias, H., and J. C. Sherrick: *Morphology of the Liver,* Academic, New York, 1969.

Lacy, P. E., and M. H. Greider: "Ultrastructural Organization of Mammalian Pancreatic Islets," in R. O. Greep and E. B. Astwood (eds.), *Handbook of Physiology,* sect. 7, vol. 1: *Endocrine Pancreas,* American Physiological Society, Washington, 1970, pp. 77–100.

Ma, M., and L. Biempica: "The Normal Human Liver Cell," *Am. J. Pathol.* **63:**353–376 (1971).

Palade, G. E., P. Siekevitz, and L. G. Caro: "Structure, Chemistry, and Function of the Pancreatic Exocrine Cell," in A. V. S. de Reuck and M. P. Cameron (eds.), Ciba Foundation Symposium, *The Exocrine Pancreas,* Little, Brown, Boston, 1962, pp. 32–55.

Rhodin, J. A. G.: *Histology: A Text and Atlas,* Oxford University Press, New York, 1974.

Weiss, L., and R. P. Greep: *Histology,* 4th ed., McGraw-Hill, New York, 1977.

ADDITIONAL READINGS

Babcock, M. B., and R. R. Cardell, Jr.: "Hepatic Glycogen Patterns in Fasted and Fed Rats," *Am. J. Anat.* **140:**299–337 (1974).

Bruni, C., and K. R. Porter: "The Fine Structure of the Parenchymal Cell of Normal Rat Liver," *Am. J. Pathol.* **46:**691–755 (1965).

Burkel, W. E., and F. N. Low: "The Fine Structure of Rat Liver Sinusoids, Space of Disse and Associated Tissue," *Am. J. Anat.* **118:**769–784 (1966).

Caro, L. G., and G. E. Palade: "Protein Synthesis, Storage, and Discharge in the Pancreatic Exocrine Cell: An Autoradiographic Study," *J. Cell Biol.* **20:**473–495 (1964).

Ekholm, R., T. Zelander, and Y. Edlund: "The Ultrastructural Organization of the Rat Exocrine Pancreas. I. Acinar Cells," *J. Ultrastruct. Res.* **7:** 61–72 (1962).

———, ———, and ———: "The Ultrastructural Organization of the Rat Exocrine Pancreas. II. Centroacinar Cells, Intercalary and Intralobular Ducts," *J. Ultrastruct. Res.* **7:**73–83 (1962).

Elving, G.: "Crypts and Ducts in the Gall Bladder Wall," *Acta Pathol. Microbiol. Scand.,* vol. 49, suppl. 135, 1960.

Greenway, C. V., and R. D. Stark: "Hepatic Vascular Bed," *Physiol. Rev.* **51:**23–65 (1971).

Greider, M. H., S. A. Benscome, and J. Lechago: "The Human Pancreatic Islet Cells and Their Tumors. I. The Normal Pancreatic Islet," *Lab. Invest.* **22:**344–354 (1970).

Hayward, A. F.: "The Fine Structure of the Gall Bladder Epithelium of the Sheep," *Z. Zellforsch.* **65:**331–339 (1965).

Kaye, G. I., H. O. Wheeler, R. T. Whitlock, and N. Lane: "Fluid Transport in Rabbit Gall Bladder," *J. Cell Biol.* **30:**237–268 (1966).

Lacy, P. E.: "The Pancreatic Beta Cell: Structure and Function," *New Engl. Med. J.* **276:**187 (1967).

Land, A. V.: "A Quantitative Stereological Description of the Ultrastructure of Normal Rat Liver Parenchymal Cells," *J. Cell Biol.* **37:**27–46 (1968).

Lazarow, A.: "Cell Types of the Islet of Langerhans and the Hormones They Produce," *Diabetes* **6:**222 (1957).

Like, A. A.: "The Ultrastructure of the Secretory Cells of the Islet of Langerhans in Man," *Lab. Invest.* **16:**937–951 (1967).

Renold, A. E.: "Insulin Biosynthesis and Secretion. A Still Unsettled Topic," *New Engl. Med. J.* **283:** 173–182 (1970).

Widman, J. J., R. S. Cotran, and H. D. Fahimi: "Mononuclear Phagocytes (Kupffer Cells) and Endothelial Cells: Identification of Two Functional Cell Types in Rat Liver Sinusoids by Endogenous Peroxidase Activity," *J. Cell Biol.* **52:**159–170 (1971).

Wisse, E.: "An Electron Microscopic Study of the Fenestrated Endothelial Lining of Rat Liver Sinusoids," *J. Ultrastruct. Res.* **31:**125–150 (1970).

———: "An Ultrastructural Characterization of the Endothelial Cell in the Rat Liver Sinusoids under Normal and Various Experimental Conditions, as a Contribution to the Distinction between Endothelial and Kupffer Cells," *J. Ultrastruct. Res.* **38:** 528–562 (1972).

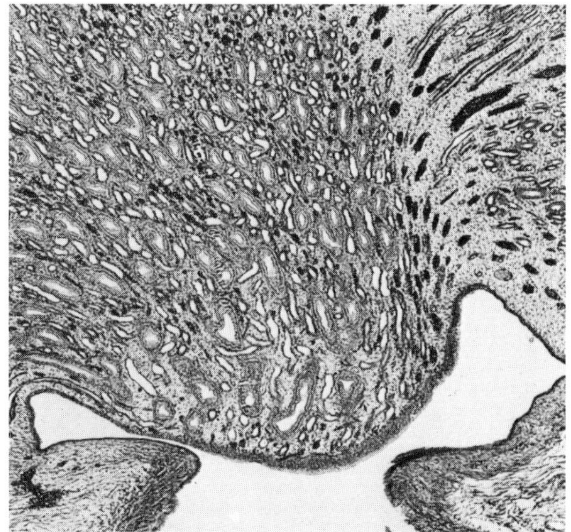

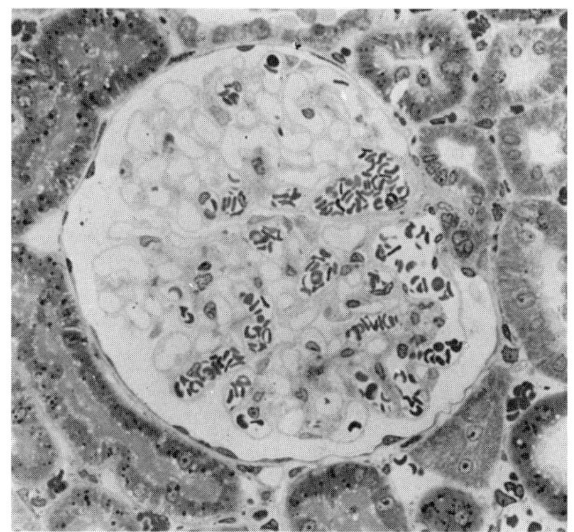

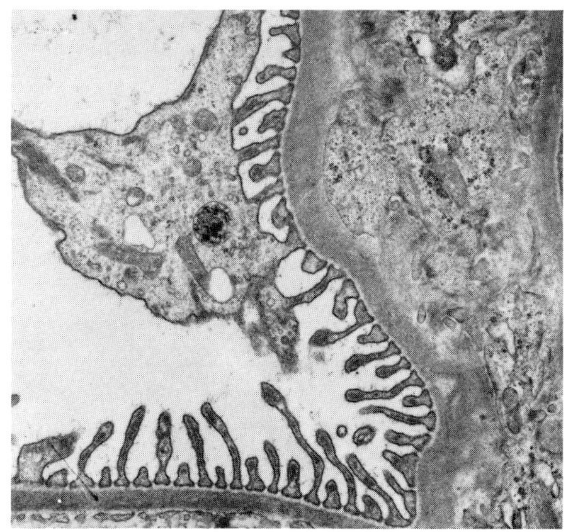

22
KIDNEY AND URINARY PASSAGES

OBJECTIVES

Upon completion of this chapter, the student will be able to:

1 Identify the location, structure, and function of the following:
(a) major and minor calyces (b) renal pelvis (c) ureter (d) bladder (e) urethra

2 Identify the histologic characteristics, location, and function of each of the following:
(a) renal corpuscle (b) proximal convoluted tubule (c) loop of Henle (d) distal convoluted tubule (e) collecting duct

3 Identify the structural characteristics of each of the following at the electron microscopic level:
(a) renal corpuscle (b) podocyte (c) mesangial cell (c) filtration barrier (e) proximal convoluted tubule (f) distal convoluted tubule

4 Trace the course of blood through the kidney.

5 Identify the location, structure, and function of the following:
(a) macula densa (b) juxtaglomerular apparatus

6 Describe the procedure by which the countercurrent multiplier system produces a hypertonic urine.

The kidneys are paired bean-shaped retroperitoneal organs approximately 4 in long located on each side of the vertebral column in the wall of the back. They comprise the functional organs of the urinary system because of the filtering and regulating activities they carry out in the formation of urine. Therefore, the kidney is composed of a blood vascular system, from which the waste products are removed, and a tubular system, which collects the urine. Anatomically, these two systems are in very close contact. The urine is carried away from each kidney by a tube called the *ureter* and deposited in the bladder, which stores the urine until it is voided periodically through the urethra to the exterior. Figure 22-1 shows a simple diagram of these structures.

The kidneys are responsible for the elimination of metabolic waste products and for the regulation of fluid volume to maintain homeostasis of the body. They are capable of separating nitrogenous by-products including urea from the blood, concentrating them, and then excreting them via a series of tubules. The kidney is also involved in a multitude of complex activities that are vital in maintaining the fluid-and-salt balance of the body. Depending upon the body's needs, varying amounts of fluids and electrolytes are excreted. Similarly, the hydrogen ion concentration is closely regulated, thus stabilizing the acid-base balance of the body fluids. Toxic products, including drugs broken down in the liver, may also be removed from the bloodstream by the kidney.

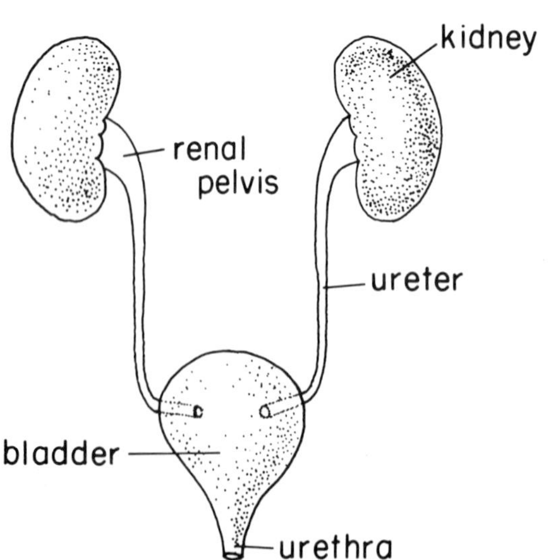

FIGURE 22-1 Diagram of the urinary system.

GENERAL STRUCTURE

The kidney is covered by a well-developed renal capsule which is penetrated by the renal artery, renal vein, and ureter, which enter and leave through a depression in the medial border called the *hilus.* Extending inward from the hilus is a large cavity, the *renal sinus.* This area contains the expanded upper part of the ureter, the *renal pelvis,* as well as the blood vessels and nerves, all of which are embedded in loose FECT and adipose tissue. The renal pelvis is divided to form two or three cuplike structures, the *major calyces,* which are further subdivided into eight to twelve *minor calyces.* The minor calyces fit like a funnel over the tips of pyramidal-shaped structures, the *renal papillae* of the *medullary pyramids.* The medullary pyramid is made up of collecting tubules which transport the urine to the minor calyx.

In a hemisection through the kidney (Fig. 22-2), the parenchyma consists of an outer cortex and inner medulla, partially surrounding the renal sinus. The cortical tissue sends projections into the medulla, called *renal columns,* which separate the *medullary pyramids.* The bases of the medullary pyramids send fine extensions, *medullary rays,* into the outer cortex, while the apexes of the pyramids project into the calyces in the renal sinus area. Each pyramid along with its overlying cortex is considered to be a renal lobe, al-

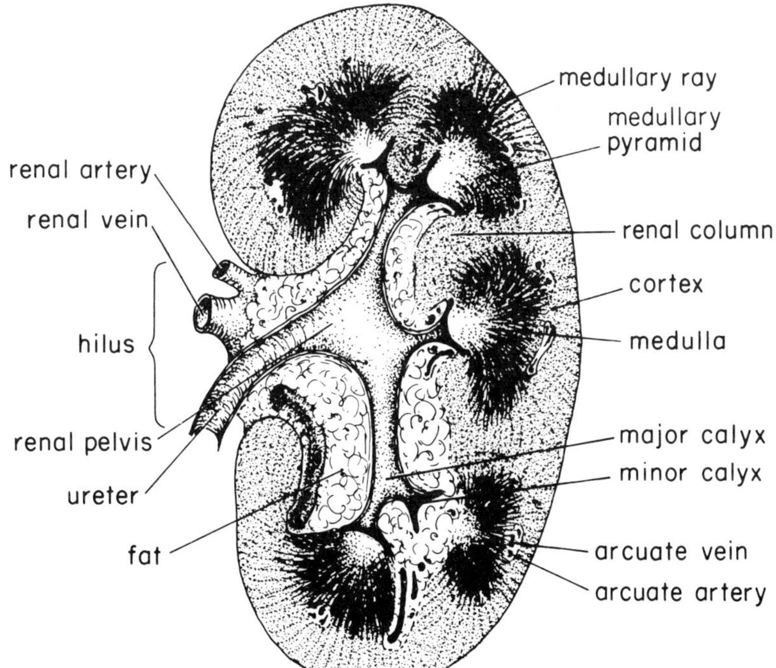

FIGURE 22-2 Diagram of a semisection through the kidney.

though there are no connective tissue septa separating it from the rest of the medullary substance. The base (cortex) of a lobe is seen in Fig. 22-3, and the apex (medullary pyramid) of a lobe is shown in Fig. 22-4. A *renal lobule* is a smaller unit consisting of a medullary ray and its associated cortical tissue.

The medulla has a striated appearance due to its straight tubular and vascular components being arranged in parallel rows. The "granular appearance" of the cortex is due to the presence of numerous round bodies, the renal corpuscles, and convoluted tubules (Fig. 22-5).

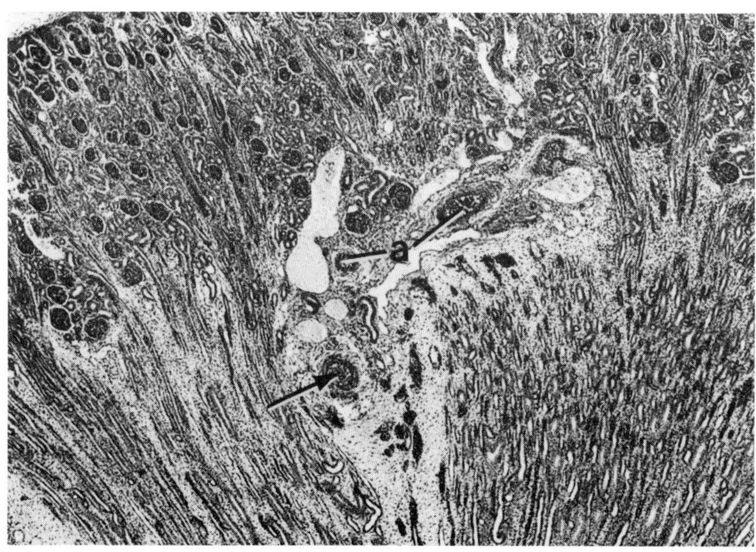

FIGURE 22-3 Light micrograph of a portion of the kidney showing the base, the cortical part of a renal lobe. An interlobar artery (arrow) passes through the renal column. It branches into arcuate arteries (a) at the corticomedullary junction.

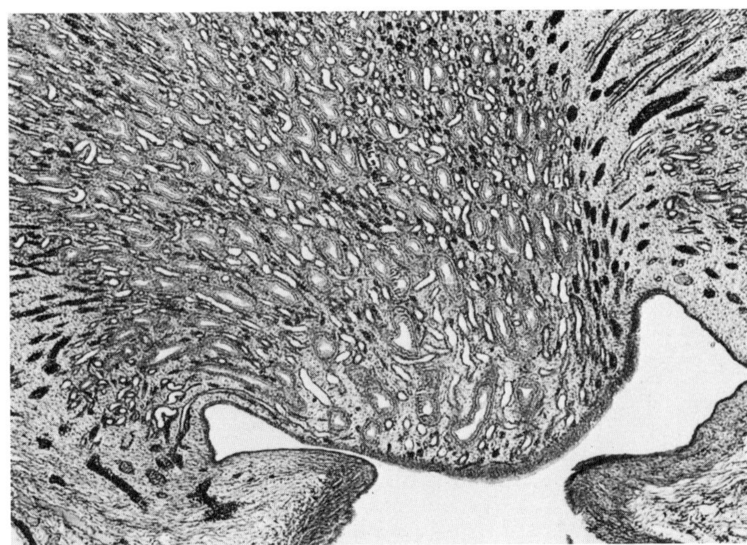

FIGURE 22-4 Light micrograph of the apex of a renal lobe, the medullary pyramid.

BLOOD SUPPLY

The kidneys have an extremely rich blood supply, with the total volume of circulating blood passing through the kidneys once every 5 min. Each kidney receives a direct branch of the abdominal aorta called the *renal artery,* which enters the hilus and divides into several branches. These branches give rise to *interlobar arteries* (Figs. 22-2 and 22-3, arrows) which pass through the renal columns (between lobes) to the base of the medullary pyramids. Here the interlobar arteries bend sharply, forming *arcuate arteries* (a in Fig. 22-3) which arch over the pyramids and run parallel to the kidney surface along the corticomedullary junction. Finer branches, *interlobular arteries,* arise from arcuate arteries and are located between medullary rays. Interlobular arteries approach the glomerular

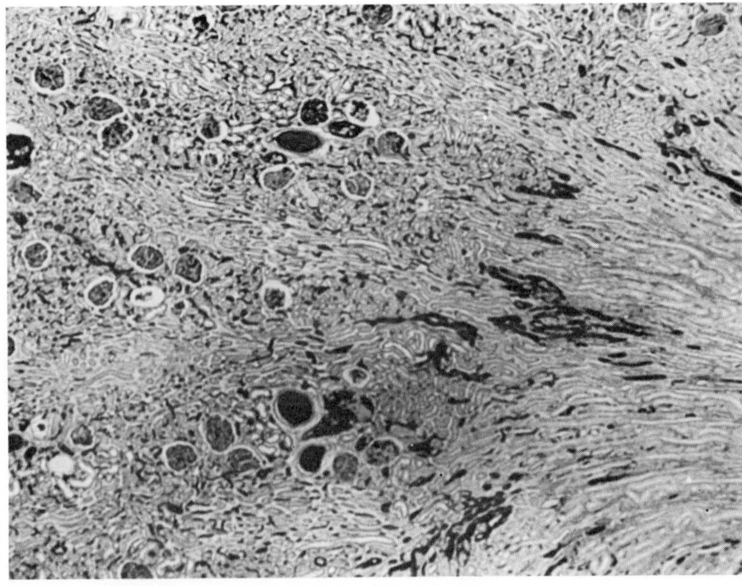

FIGURE 22-5 Light micrograph of the renal cortex (at left) and the renal medulla (at right).

region where they enter a renal corpuscle as the *afferent glomerular arteriole.* The afferent arteriole divides into a network of capillaries which reunite as the *efferent arteriole.* Immediately upon leaving the renal corpuscle, the efferent arteriole forms a capillary network around the cortical tubules called the *peritubular plexus.* Efferent arterioles lying close to the medulla provide an arterial supply to it by sending *arteriolae rectae spuriae* (straight, false arterioles) into the pyramids. The arteriolae rectae spuriae, together with their capillary network and *venae rectae,* are called the *vasa recta.* The vasa recta are thin-walled, fenestrated vessels that facilitate rapid movement of diffusible substances (the countercurrent exchange system).

The venous drainage is similar to the arterial supply in arrangement, except that there is no venous component in the glomerulus. Cortical capillaries collect into *stellate veins* which join in a starlike pattern forming *interlobular veins.* These vessels pass toward the medulla to join *arcuate veins,* which also receive venae rectae from the medullary pyramids. Arcuate veins drain into *interlobar veins,* which join to form the *renal vein,* finally emptying into the inferior vena cava.

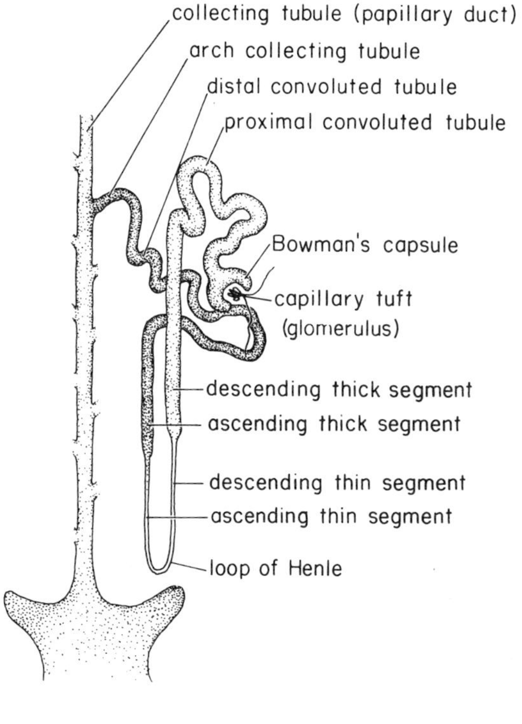

FIGURE 22-6 Diagram of a nephron.

NEPHRON

The nephron is the functional unit of the uriniferous tubules concerned with the formation of urine. Basically, the nephron is a long epithelium-lined tube with several morphologically distinct sections, each occupying a specific location in the cortex or medulla. The tube starts blindly at the glomerulus, winds tortuously, and ends by joining the system of collecting ducts. The first portion of the nephron is a thin-walled, cup-shaped expansion, *Bowman's capsule,* which is deeply indented by a tuft of capillaries, the *glomerulus.* These components together comprise the nearly spherical *renal corpuscle,* which is found exclusively in the cortical tissue.

From Bowman's capsule, the nephron continues as the *proximal tubule* which becomes the thin segment of the tubule (loop of Henle), and finally the *distal tubule,* which connects to the collecting duct. Both the proximal and distal tubules have convoluted portions that lie adjacent to the renal corpuscle in the cortex, while the *loop of Henle* extends into the medulla. The basic structure of the nephron is shown in Fig. 22-6.

RENAL CORPUSCLE

Bowman's capsule is composed of a double-walled cup of squamous epithelium surrounding the glomerulus. The inner, or *visceral layer* (glomerular epithelium), which closely invests the capillary tuft and the outer,

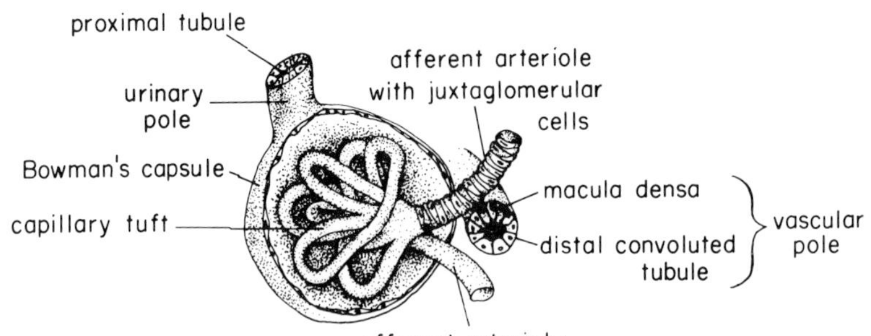

FIGURE 22-7 Diagram of renal corpuscle.

or *parietal layer* (capsular epithelium), are separated by the space of the capsule. There is a *vascular pole* where afferent and efferent arterioles enter and leave the glomerulus (Fig. 22-7).

The LM in Fig. 22-8 shows the glomerular arterioles as they enter and leave the glomerulus (arrows). Notice the visceral and parietal layers of the Bowman's capsule. The moderately dense deposits located in the cavity of the Bowman's capsule (arrowhead) represent coagulated urinary substances of the provisional urine which were fixed during preparation of this tissue specimen. Opposite the vascular pole is the *urinary pole* where the capsular space is continuous with the proximal convoluted tubule. Here, the parietal epithelium of the capsule becomes continuous with the cuboidal epithelium of the tubule (Fig. 22-9).

The parietal layer of Bowman's capsule is a typical simple squamous epithelium resting on a thin basal lamina with nuclei protruding slightly into the capsular space (Fig. 22-10). The space between the two layers of the Bowman's capsule is clearly visible. However, the nuclei of the parietal layer of the capsule protrude into the space only slightly (arrows). The visceral layer is composed of highly modified cells, *podocytes,* which are

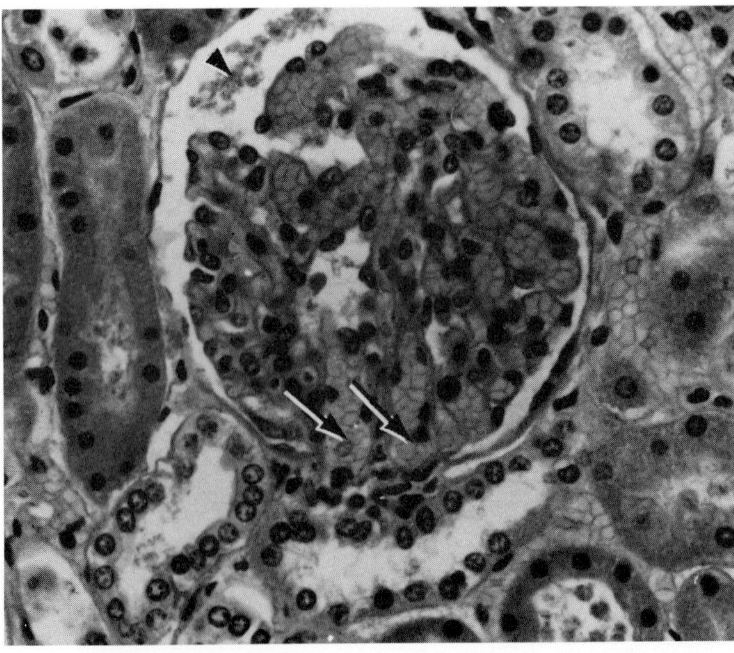

FIGURE 22-8 Light micrograph of renal corpuscle. Blood vessels enter and leave the glomerulus at the vascular pole (arrows). Coagulated urinary substances are seen in the lumen of Bowman's capsule (arrowhead).

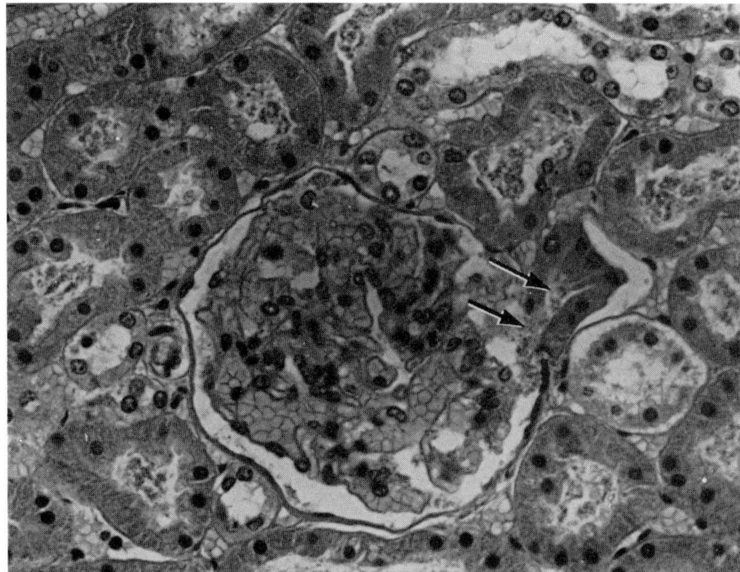

FIGURE 22-9 Light micrograph of a renal corpuscle. The parietal epithelium of the capsule is continuous with the cuboidal epithelium of the proximal tubule at the urinary pole (arrows). Opposite the urinary pole is the vascular pole, although it cannot be seen clearly since the two poles are usually oriented at an angle to one another.

so closely applied to the glomerulus that they are difficult to delineate by light microscopy. However, podocytes have a larger cell body which often protrudes into the space of the Bowman's capsule (Fig. 22-10, arrowheads).

In an electron micrograph, a number of radiating processes give off numerous footlike extensions called *pedicels* (Fig. 22-11). The three-dimensional appearance of podocytes is illustrated in Fig. 22-12. Note the interdigitations of pedicels from different branches of different podocytes. The pedicels are attached to the basal lamina of the outer surface of the capillaries (b_1) and along the surface (b_2) of mesangial cells (m) which are mesodermal cells located within the connec-

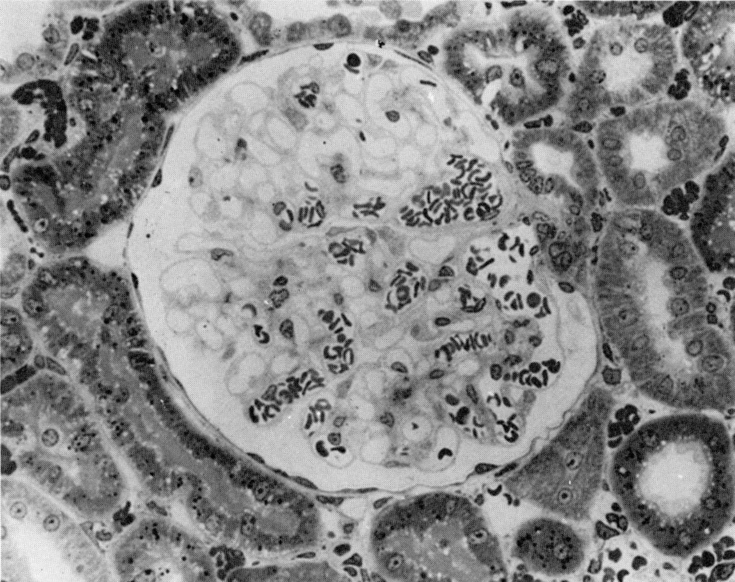

FIGURE 22-10 Light micrograph of renal corpuscle. Arrows indicate nuclei of the cells in the parietal layer of Bowman's capsule and arrowheads point to the nuclei of podocytes.

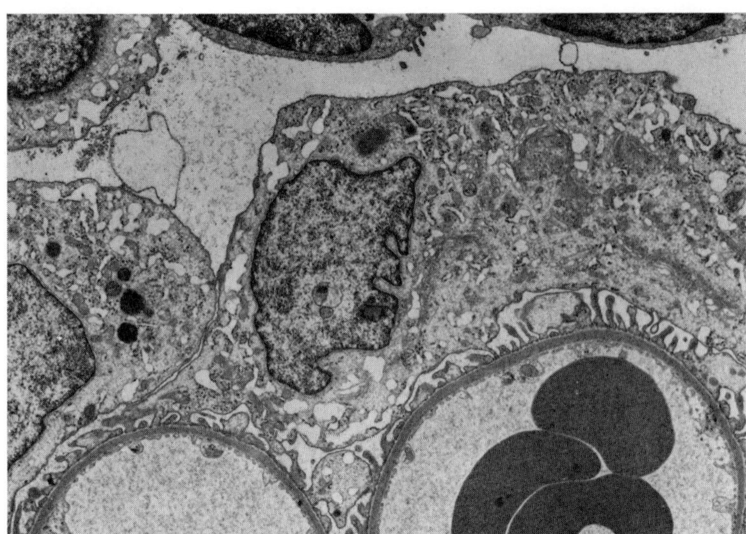

FIGURE 22-11 Electron micrograph of podocytes. Note the numerous footlike extensions, pedicels, that are closely apposed to the capillary wall.

tive tissue between the glomerular capillary and the visceral layer of the renal capsule (Fig. 22-13). Mesangial cells maintain an embryonic appearance (Fig. 22-14) and can be induced to become phagocytic. Thus, the mesangial cell is not unlike a pericyte or septal cell of the lung.

Adjacent pedicels are separated by narrow gaps about 25 nm wide, called *filtration slits* or pores. These filtration slits are bridged by 60-nm-thick *slit membranes* which are comparable to diaphragms closing the pores of fenestrated capillary endothelium. When observed from above, pedicels between adjacent cells show a series of interlocking pedicels (Fig. 22-15).

The endothelium of glomerular capillaries is greatly attenuated and discontinuous, being perforated by pores or fenestrae, 50 to 100 nm (with an average of 80 nm) in diameter. The pores are large and numerous and possibly are not traversed by a thin diaphragm, thereby differing from fenestrae of capillaries elsewhere in the body.

The structures which separate the blood in the capillaries from the filtrate in the capsular space are together termed the *filtration membrane* or *barrier.* The barrier is composed of the fenestrated capillary endothelium, the basal lamina, and the pedicels of the podocytes. The basal lamina is the only continuous layer and has been thought to be the main barrier preventing the passage of large molecules. Although larger particulate tracers, such as ferritin, pass easily through endothelial pores, they do not pass through the basal lamina. However, smaller particles [horseradish peroxidase, molecular weight

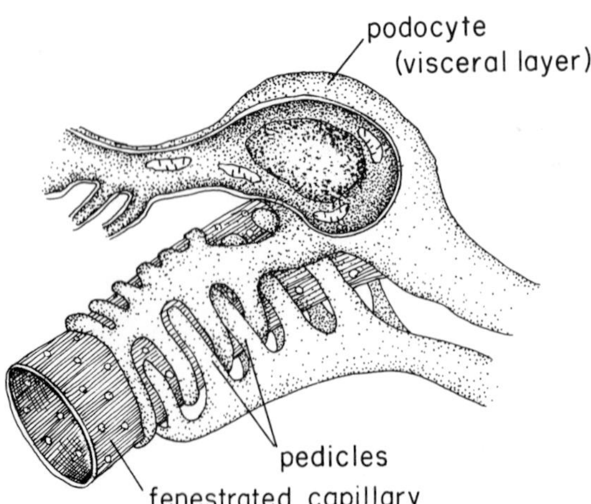

FIGURE 22-12 Illustration of the relationship between glomerular capillary and podocyte. Note the three-dimensional arrangement of podocytes.

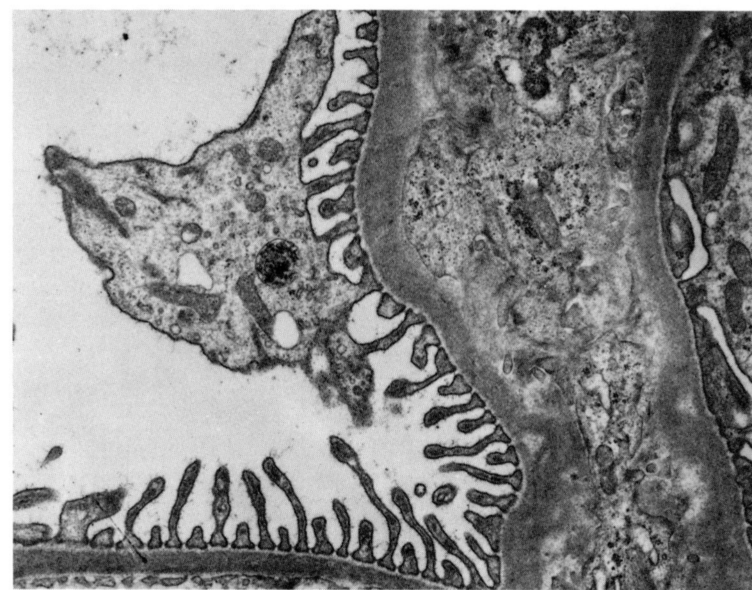

FIGURE 22-13 Electron micrograph of podocyte pedicels. The pedicels are attached either to the outer aspect of the basal lamina of capillaries (b_1) or along the surface (b_2) of a mesangial cell (m).

(MW) 40,000] readily pass through all components to appear in the capsular space. Slightly larger particles (myoperoxidase, MW 160,000) are held up by the endothelial pores and basal lamina. Thus, the filtration slits appear to be the ultimate barrier responsible for differential glomerular permeability to proteins of varying molecular size.

In the renal corpuscle, provisional urine is formed. The provisional urine undergoes substantial alterations as it passes through the tubular segments of the nephron, either by reabsorption of filtered products including water or by secretions from the tubule cells. Provisional urine is an ultrafiltrate of blood plasma which contains small molecules such

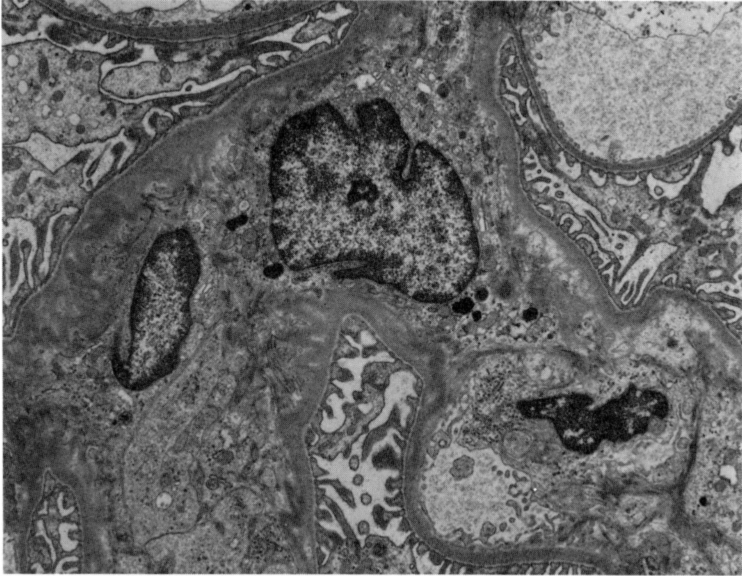

FIGURE 22-14 Electron micrograph of mesangial cells.

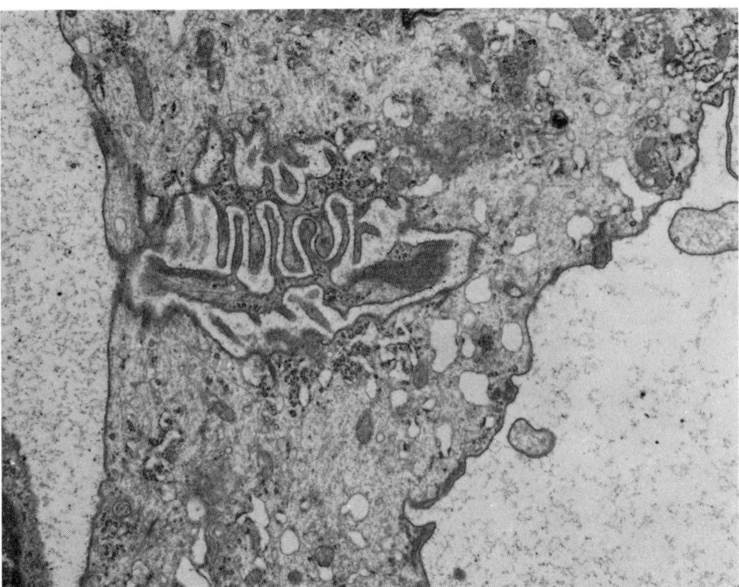

FIGURE 22-15 Electron micrograph of podocytes illustrating the interlocking of the pedicels of two adjacent cells.

as phosphates, creatinine, uric acid, urea, and albumin. It is free of larger protein molecules which are held back by the filtration barrier. The glomerulus, as it is a specialized region of the arterial tree, provides a high-pressure system which tends to push the blood fluid through the endothelial pores, across the basal lamina, and through the filtration slits into Bowman's capsule. In human beings, approximately 125 mL of glomerular filtrate is produced per minute from the 1200 mL of blood that flows through. Eventually 124 mL of filtrate are reabsorbed, leaving a final volume of 1 mL to be excreted as urine. Therefore, in 24 h, the total amount of glomerular filtrate is reduced from 170 to 200 L to about 1.5 L of concentrated urine.

PROXIMAL CONVOLUTED TUBULE

Being the longest and most convoluted part of the nephron, the proximal convoluted tubule forms the major portion of the cortical substance. It consists of a single layer of cuboidal epithelium surrounding a large lumen. The apical surface has a prominent brush border which undergoes rapid postmortem change, thereby giving it a ragged appearance in most histologic preparations. Each cell contains a single spherical nucleus in a deeply staining eosinophilic, granular cytoplasm. Faint striations may be seen in the basal portion of the cells. Figure 22-16 is a portion of the cortex in which a number of proximal convoluted tubules (p) surrounding the renal corpuscle (c) are depicted. Note the presence of a thick brush border in the apical (luminal) portion of the proximal convoluted tubules.

Electron micrographs show a prominent supranuclear Golgi complex, but poorly developed RER in the epithelial cells of the proximal convoluted tubule. The cytoplasm has closely packed microvillous extensions at its apical end (Fig. 22-17). Unlike microvilli of the GI tube, these projections do not have supporting axial filaments and, therefore, tend to collapse upon fixation. Elsewhere, the cytoplasm contains numerous vesicles and vacuoles which are thought to be formed by micropinocytosis during tubular resorption (Fig. 22-18).

The basal plasma membrane is thrown into deep folds, between which are elongated mitochondria oriented in the axial direction of the cells. This resembles the orientation of mitochondria that is seen in striated ducts of

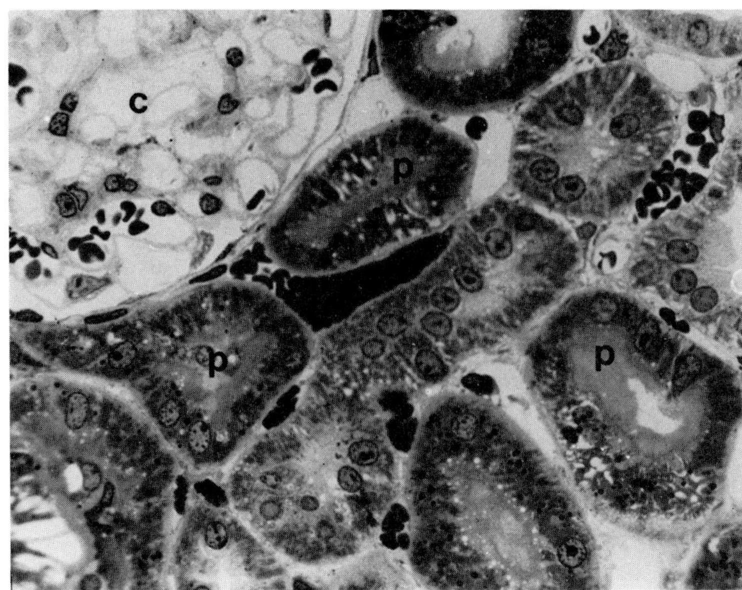

FIGURE 22-16 Light micrograph of renal cortex. Several proximal convoluted tubules (p) are adjacent to the renal corpuscle (c).

salivary glands. This is shown in the diagram of the proximal convoluted tubule appearing in Fig. 22-19. Surrounding the tubule is the basal lamina, separating the epithelial cells from the capillaries. In this diagram, the lumen is above (note the brush border) and the capillaries would be below (not shown).

As much as 85 percent of the initial filtrate is resorbed in the proximal tubule by selective reabsorption. Sodium is actively transported by the cells, with chloride and water following passively to maintain osmotic equilibrium. Sodium is moved out at the base of the cell by the sodium pump to be retrieved by the capillaries and is drawn into the cell at the apical end by diffusion.

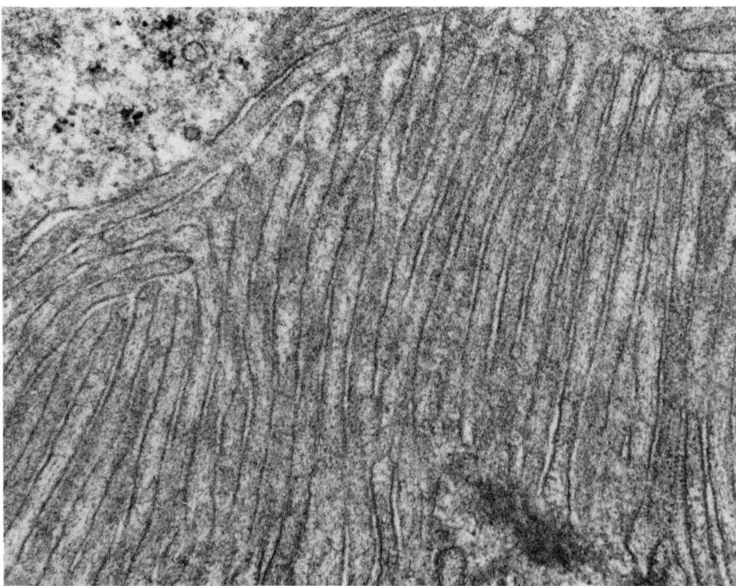

FIGURE 22-17 Electron micrograph of the apical cytoplasm of cells in the proximal convoluted tubule. Note the appearance of the closely packed microvillous extensions.

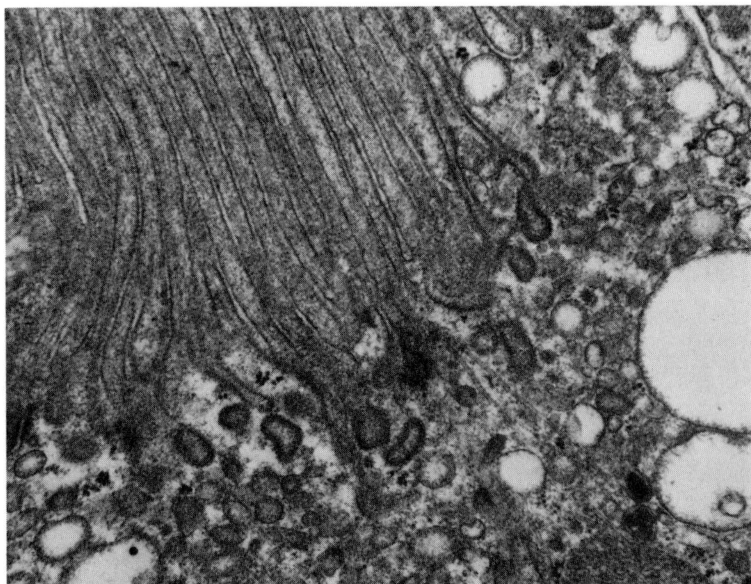

FIGURE 22-18 Apical portion of a cell in the proximal convoluted tubule. Note the numerous vacuoles and vesicles in the cytoplasm subjacent to the microvillous border. They are thought to be formed during the tubular resorption by micropinocytosis.

Both the basal infoldings and microvillous projections increase the surface area. The numerous mitochondria inserted between basal infoldings provide energy, (ATP molecules) for driving the metabolic pump. Glucose is totally resorbed in the proximal convoluted tubule, as well as small proteins and amino acids by active transport.

LOOP OF HENLE

The loop of Henle has a *descending limb* which is composed of the straight proximal tubule and a thin segment along with an *ascending limb* made up of a thin segment and straight portion of the distal tubule (see Fig. 22-6). The two limbs lie close together, oriented radially in the medulla, sometimes extending

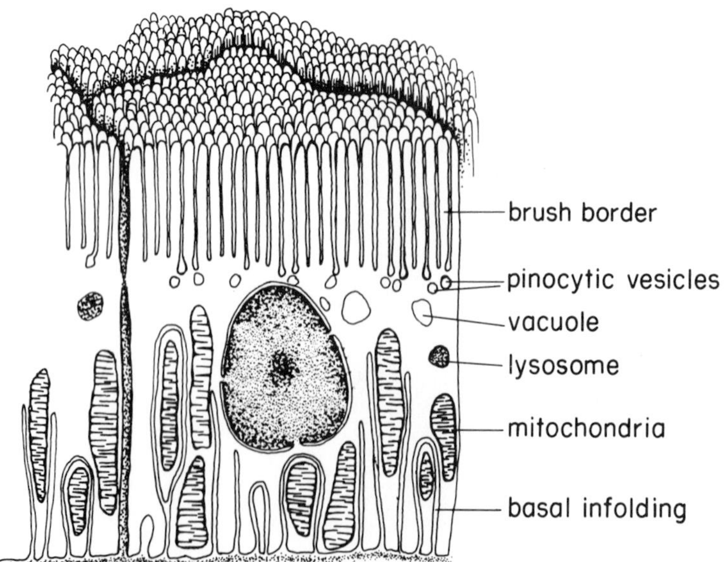

FIGURE 22-19 Diagram of proximal convoluted tubule.

nearly to the apex of the papilla. The transition between the straight proximal tubule (descending thick limb), which has an epithelium similar to that of the proximal convoluted tubule, and the thin segment is rather abrupt. The epithelium changes from cuboidal to squamous with a sudden termination of the brush border. The cells of the thin segment resemble those of a capillary endothelium except that their nuclei protrude more into the lumen. Electron microscopy shows that the cells have rare, short apical microvilli, few cytoplasmic organelles, and a thick basal lamina.

The transition from thin segment to straight distal tubule (ascending thick limb) is also abrupt, with the epithelium becoming cuboidal and staining slightly darker. The basal plasma has elaborate infoldings with complex interdigitations between cells (Fig. 22-20). Numerous elongated mitochondria have a complex internal membrane structure and many granules in the matrix. There are numerous free ribosomes as well as some profiles of RER.

Figure 22-21 is a section of a medullary ray showing cross sections of the various tubular elements. Note the similarity in structure of the thick descending limb (d) to that of the proximal convoluted tubule. The thick ascending limb (a) resembles the distal convoluted tubule, to be discussed next. Collecting tubules (c) are also present in this field.

The straight part of the distal tubule ascends to the glomerulus where it becomes convoluted and lies between afferent and efferent arterioles. Here the tubular epithelium forms an elliptical disk of tall, dark cells, the *macula densa* (dense spot) (Fig. 22-22, arrow). These cells are closely related, structurally as well as functionally, to specialized cells in the wall of the afferent arteriole called *juxtaglomerular cells* (Fig. 22-23). The two types of cells are separated only by a thin basal lamina. The juxtaglomerular cells contain conspicuous cytoplasmic granules thought to form a vasopressor substance, renin, which may be important in regulating the blood flow to the glomerulus.

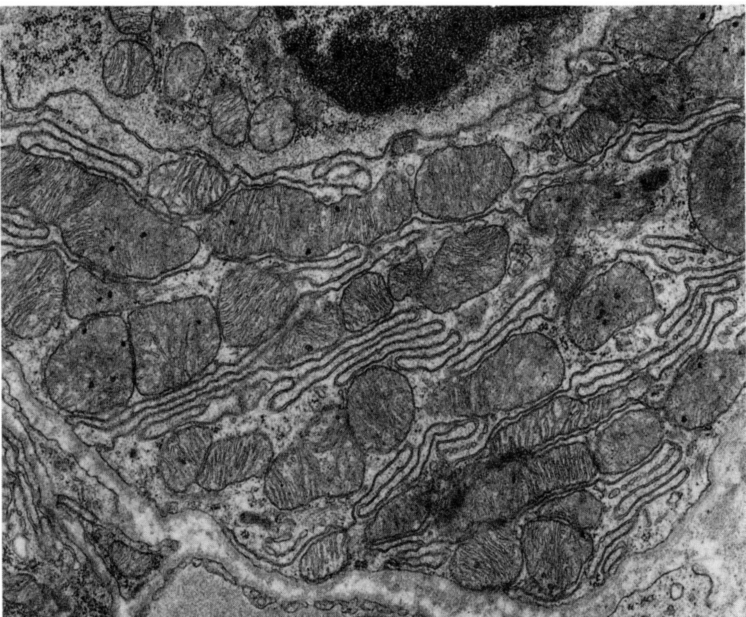

FIGURE 22-20 Electron micrograph of the straight distal tubule showing the infoldings of the basal plasma membranes. Numerous elongated mitochondria are seen in the cytoplasm.

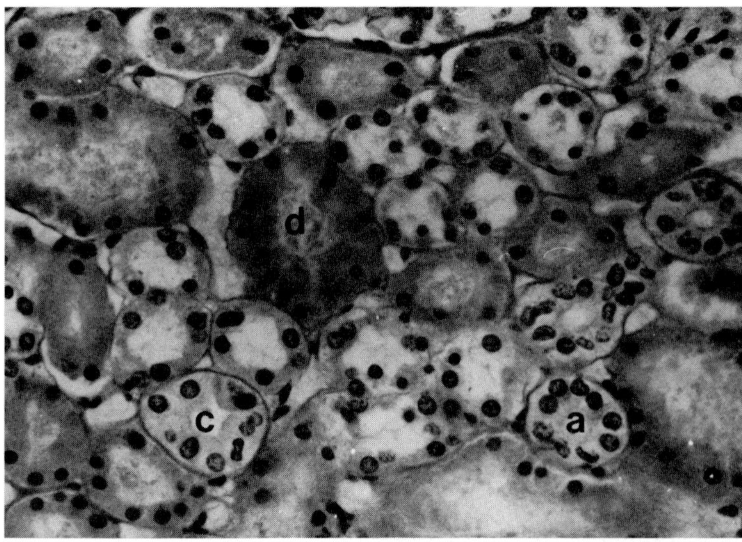

FIGURE 22-21 Light micrograph of a medullary ray. Various tubular elements are seen in cross section. (d) Thick descending limb. (a) Thick ascending limb. (c) Collecting tubule.

The essential function of the loop of Henle is to provide a hypertonic environment in the medullary interstitium for the eventual final concentration of urine in the collecting ducts. The hypertonic environment is accomplished by the arrangement of the loop of Henle and by properties in different segments of the tubular epithelium. Upon leaving the proximal tubule, provisional urine is isoosmotic with blood plasma and surrounding connective tissue. There is a progressive increase in tonicity of the provisional urine as it passes down the descending limb and a progressive decrease in tonicity as it travels up the ascending limb. It is somewhat hypotonic as it enters the distal convolution. These changes are in correlation with differences in osmolarity of the medullary tissue which becomes more hypertonic toward the papillae of the pyramids.

The epithelium of the ascending thick limb actively pumps sodium (without a cor-

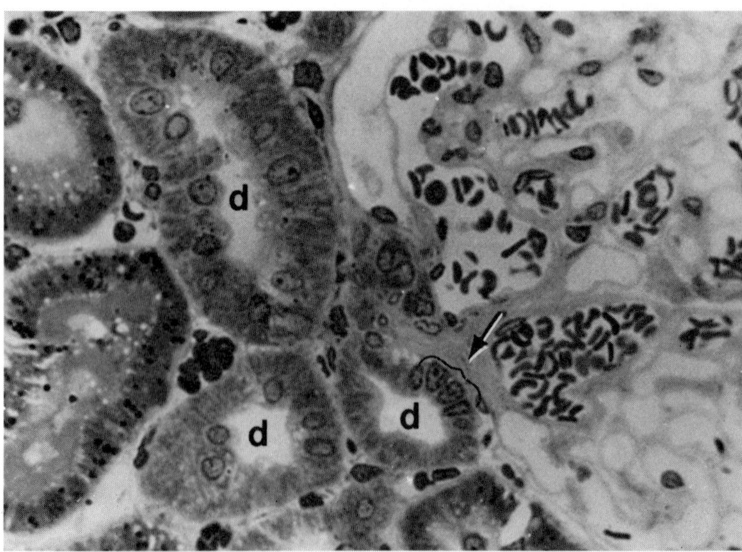

FIGURE 22-22 Cross section of distal convoluted tubules (d) situated between the glomerulus and proximal convoluted tubules also present. Where the distal tubule contacts the afferent arteriole of the glomerulus, the tubular epithelium forms an elliptical disk of tall, dark cells, the macula densa (arrow).

responding loss of water) from the urine, thus making the intertubular tissue hypertonic. The descending thin limb, acting as a dialyzing membrane, takes up some of this sodium by diffusion and loses some water as well, thereby making the urine more hypertonic. The sodium circulates through the loop to be pumped out again without water in the ascending thick limb. This process provides a *countercurrent multiplier effect*. There is also a countercurrent exchange between ascending and descending limbs of the straight blood vessels (vasa recta) of the medulla, permitting equilibration of blood concentration and thus preserving the osmotic gradient in the medulla.

DISTAL CONVOLUTED TUBULE

The macula densa region marks the beginning of the distal convoluted tubule. These tubules end by joining a collecting duct. The distal convolutions are shorter and less tor-

FIGURE 22-23 Electron micrograph showing juxtaglomerular cells from rat kidney. The juxtaglomerular granules (Gr) can be seen within the modified smooth muscle cells.

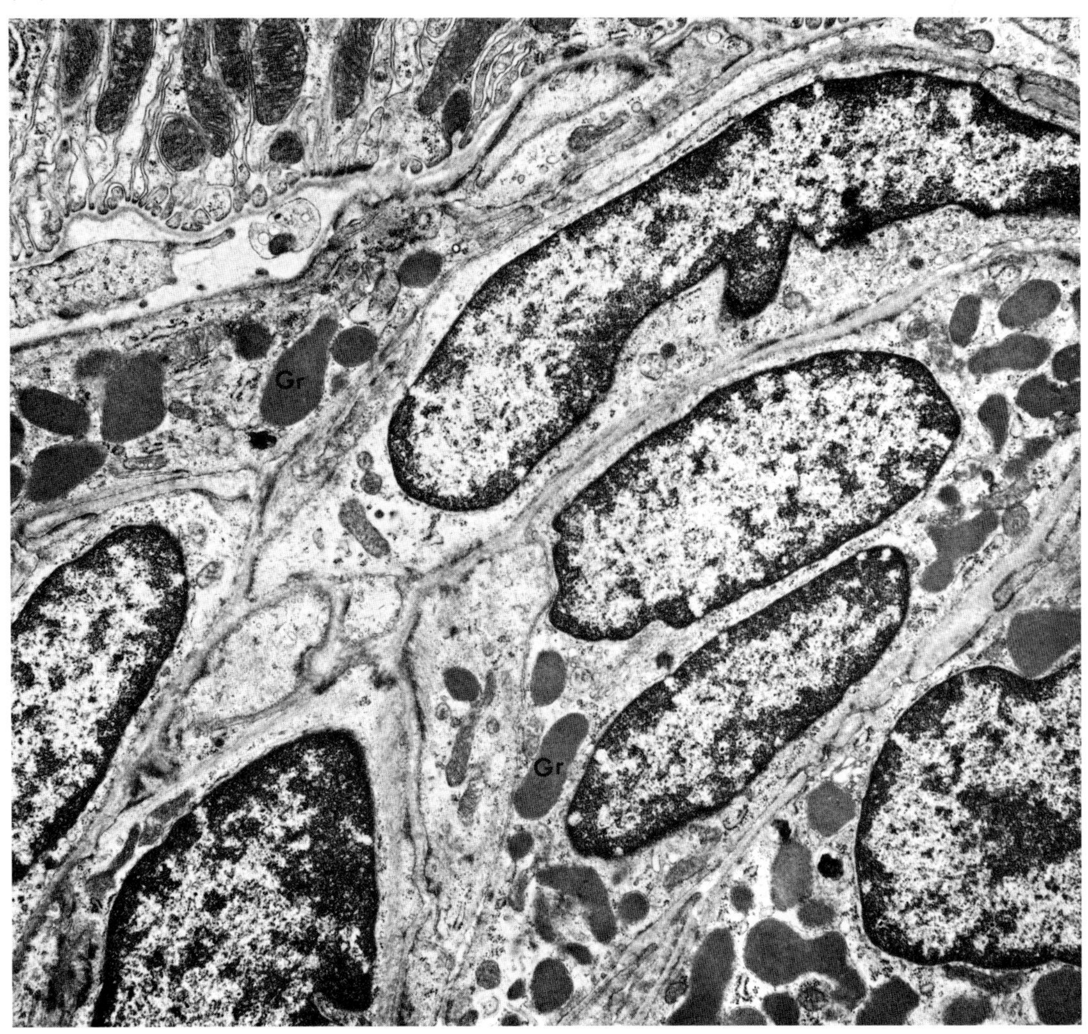

tuous than the proximal and are therefore seen in fewer numbers in a section (see Fig. 22-22). The lower epithelial cells create a larger-appearing lumen even though the overall diameter of the tubule is less. The granular cytoplasm stains less intensely than that of the proximal tubules, and there is no brush border. Electron microscopy shows that, as in the ascending thick limb, the basal infoldings filled with elongate mitochondria are more highly developed than in the proximal tubule (see Fig. 22-20, proximal convoluted tubule).

The distal tubule is the principal site of acidification of urine. Sodium is actively removed, being replaced by potassium, hydrogen, and ammonia. The efficiency of sodium resorption is under the control of the hormone *aldosterone* (from the adrenal cortex). The permeability to water of the distal convolution and collecting ducts is controlled by the *antidiuretic hormone* (ADH) from the posterior hypophysis. In the presence of ADH, the tubular epithelium becomes highly permeable to water, which is then readily absorbed due to the high osmolarity of the medulla. This process results in a highly concentrated urine.

COLLECTING DUCTS

The collecting tubules begin where the distal convoluted tubule passes into a medullary ray, then down into the medulla, joining with several others to form large ducts, *papillary ducts* which open into the calyces at the apex of the papilla. The most proximal part of the collecting duct is called the *arched collecting duct* as it represents the arched beginning of the collecting duct system. The cells lining these ducts vary in size, depending on location; they start as cuboidal cells and become tall columnar cells in the papillary ducts. Figure 22-24 demonstrates an arched collecting duct (c) which is surrounded by proximal (p) and distal (d) convoluted tubules. A similar duct is depicted in Fig. 22-25 at higher magnification. A cross section of several collecting ducts in the papilla is presented in Fig. 22-26. (The smaller tubules are thin segments of the loop of Henle, H.) The cells of collecting ducts characteristically show intercellular boundaries, darkly staining spherical nuclei, and a pale cytoplasm with few

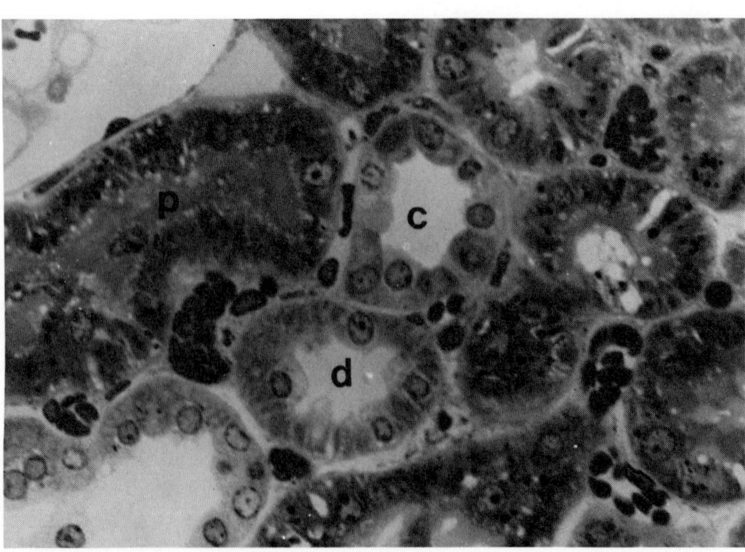

FIGURE 22-24 Light micrograph of a medullary ray. An arched collecting duct (c) is surrounded by proximal (p) and distal (d) convoluted tubules.

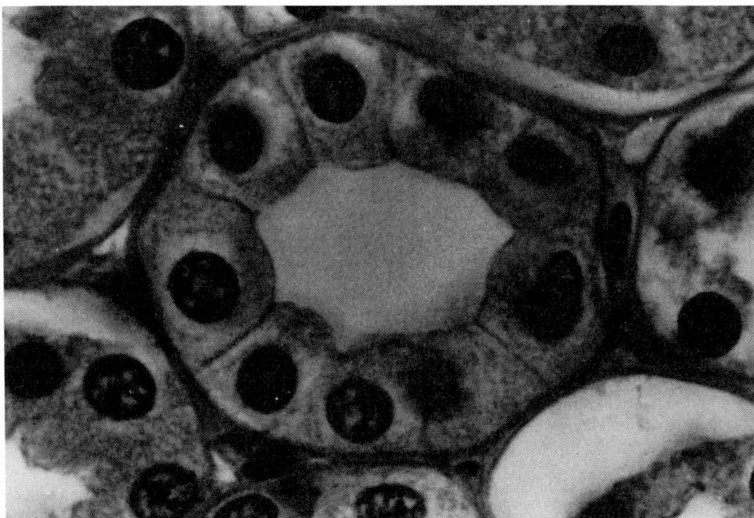

FIGURE 22-25 Collecting duct.

organelles. The collecting ducts conduct urine from the nephron to the renal pelvis, concentrating it under the influence of ADH.

EXCRETORY PASSAGES

The excretory passages are simple ducts which are relatively impermeable to urinary substances. The calyx, pelvis, ureter, and bladder are all excretory passages with similar structures, although the wall increases in thickness from the calyx to the bladder. The inner surface is lined by a mucous membrane supported by a lamina propria. This is attached to a smooth muscle coat which is surrounded by an adventitia of connective tissue. Near the end of the papillary duct the lining epithelium becomes transitional (Fig. 22-27). The transitional epithelium is present throughout the entire excretory passages, is two to three layers thick in the contracted calyces (Fig. 22-28), and increases to six to

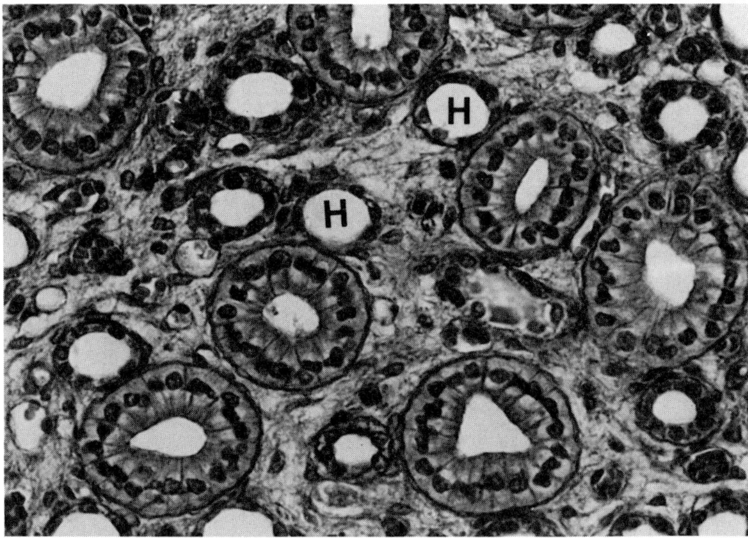

FIGURE 22-26 Cross section of several collecting ducts. They are seen between thin loops of Henle (H) in a renal papilla.

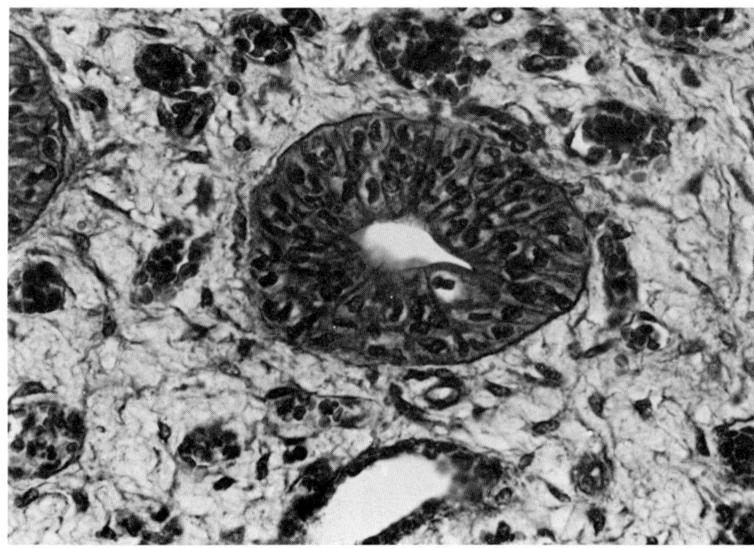

FIGURE 22-27 Light micrograph of cross section of papillary duct near its opening at the apex of the papilla. It is lined by transitional epithelium.

eight layers in the bladder (Fig. 22-29).

The thickness of epithelium in the bladder varies with the amount of distension. The muscular coat consists of an *inner longitudinal* layer and an *outer circular* layer. A third and outer longitudinal layer is added in the lower third of the ureter, becoming more prominent in the bladder. Waves of muscular contraction aid in the expulsion of urine.

The terminal passage to the exterior is the urethra, which differs depending on sex. The male urethra is long, being 15 to 20 cm in length and also serves as the pathway for semen in the reproductive system. The lining epithelium begins as transitional, changes to stratified or pseudostratified columnar, and finally becomes stratified squamous. The female urethra is short, 3 to 4 cm in length and lined mainly by moist stratified squamous epithelium. There is a muscularis with two layers of smooth muscle, reinforced by a striated muscle sphincter at the outer orifice in the female.

FIGURE 22-28 Light micrograph of the wall of a calyx. The mucosa is composed of a transitional epithelium that is two to three layers thick and a lamina propria.

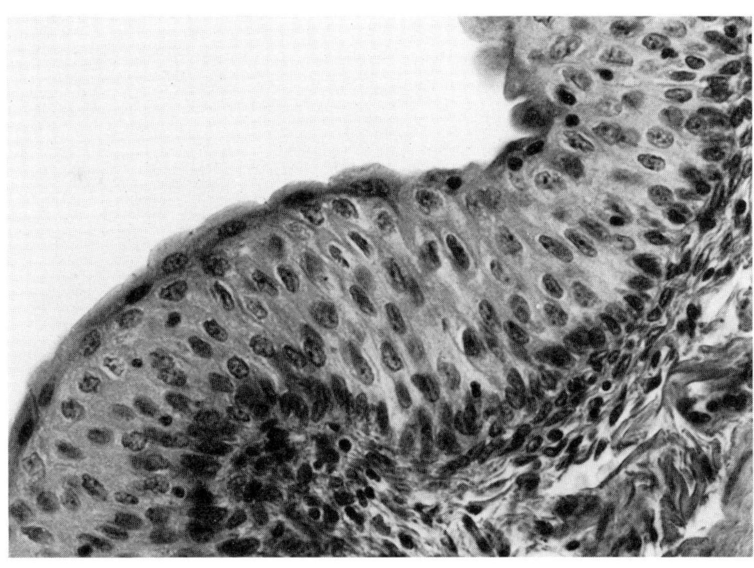

FIGURE 22-29 Bladder mucosa. The transitional epithelium is six to eight layers thick and is supported by loose FECT, the lamina propria.

REVIEW SECTION

1. What are the rounded bodies shown in Fig. 22-30? What part of the kidney are they in? Identify the part of the kidney shown at the top lower half of the figure.

The renal corpuscles are the spherical structures located among the convoluted tubules of the renal cortex. The cortical labyrinth is interrupted by medullary rays consisting of parallel rows of tubules which give it a striated appearance. The convergence of collecting ducts shown in the lower half indicates that this region is a portion of a medullary pyramid. Note also the arcuate vessels.

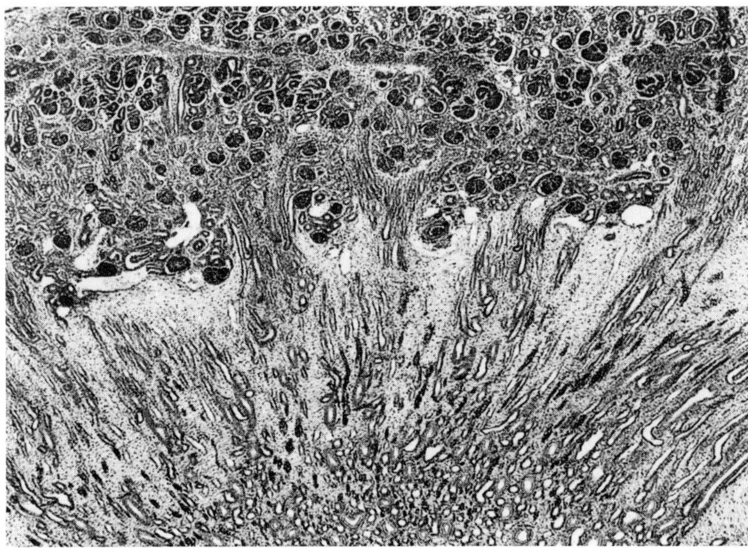

FIGURE 22-30

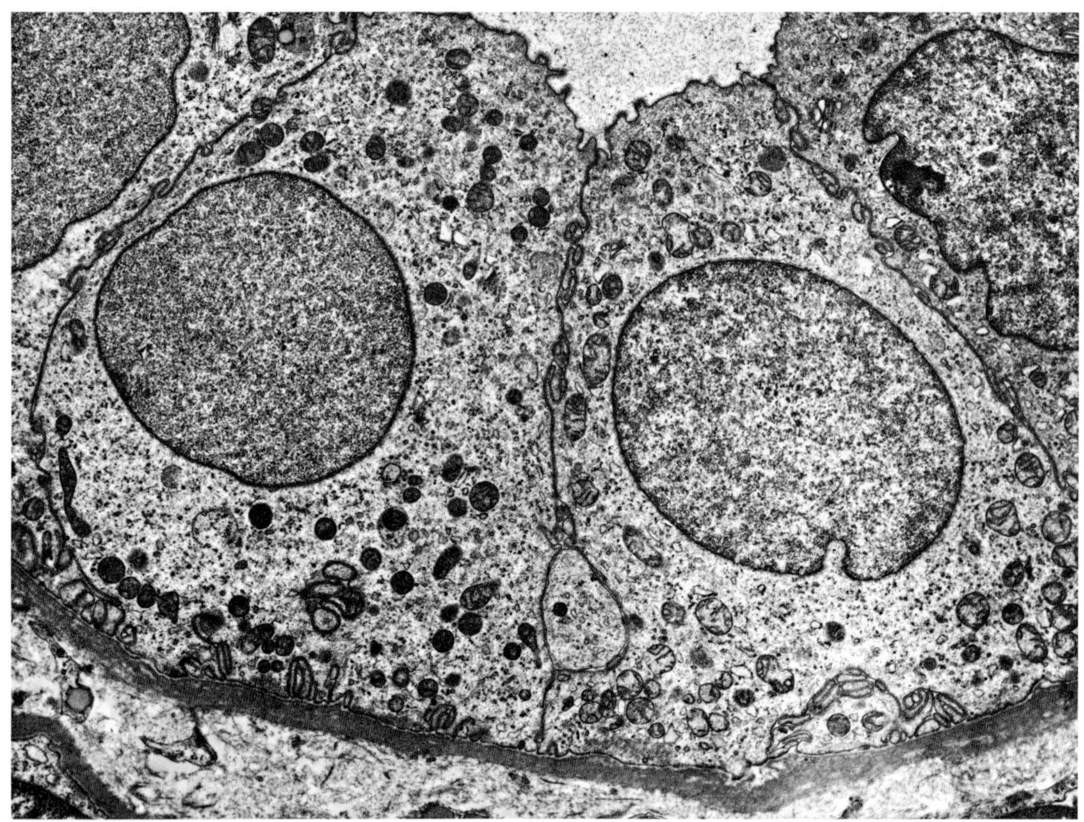

FIGURE 22-31

2. Identify the cells shown in the EM in Fig. 22-31. Where are these cells found?

A single layer of cuboidal cells lines the lumen of a collecting duct which is usually found in the medulla. The cells have round nuclei and pale cytoplasm which tends to bulge into the lumen and has few organelles. Also note the prominent appearance of the plasma membranes of these cells. This is quite apparent in light microscope preparations.

3. What is the large structure in the center of Fig. 22-32? This structure is filled with __________. What structure with three parts surrounds the irregular mass of the tissue contained in the circular area? What is the particular cell group indicated by the arrow and bracket? What happens in the structure identified above? The arrow is in the lumen of the __________.

The renal corpuscle is filled with glomerular capillaries surrounded by Bowman's capsule with its parietal and visceral layers and capsular space. Here an ultrafiltrate of plasma is formed, with the high hydrostatic glomerular pressure pushing fluid into the capsular space. The macula densa (bracket) of the distal convoluted tubule is in close association with the afferent arteriole in the vascular pole where it may have a function in controlling blood flow into the glomerulus.

4. Identify at least four tubular structures seen in Fig. 22-33. What portion of the kidney is this section from? What happens here?

This is a section of the renal medulla with collecting tubules, thick and thin portions of loop of Henle, and many capillaries. This is the location of the countercurrent exchange system which serves to increase the osmolarity of the medulla, thereby

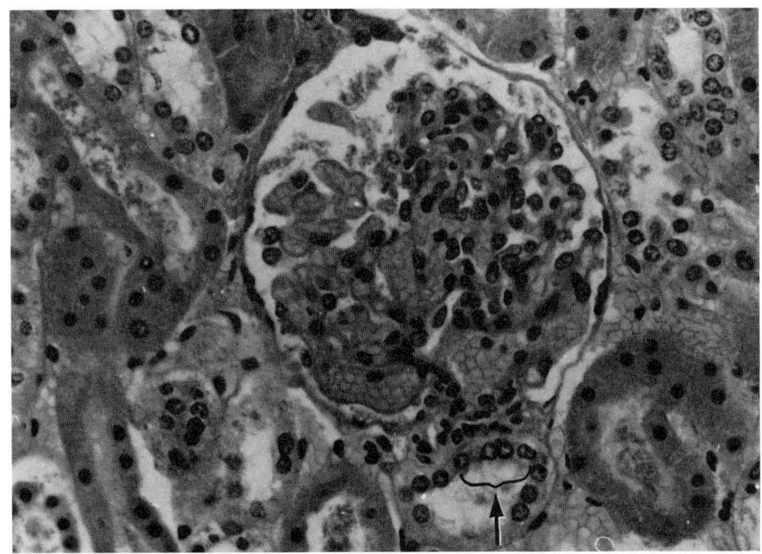
FIGURE 22-32

allowing for the final concentration of urine in the collecting ducts.

5. Name the general region which is surrounded by several glomeruli in Fig. 22-34, broken line. What tubular structures are expected to be present here?

Six glomeruli appearing in this figure delineate a medullary ray (broken line) which runs into the cortex. In the medullary ray are present ascending and descending thick limbs, parts of the loop of Henle, and collecting ducts.

BIBLIOGRAPHY

Jorgensen, F.: *The Ultrastructure of the Normal Human Glomerulus,* Munksgaard, Cophenhagen, 1966).

Pitts, R. F.: *Physiology of the Kidney and Body Fluids,* Year Book, Chicago, 1963.

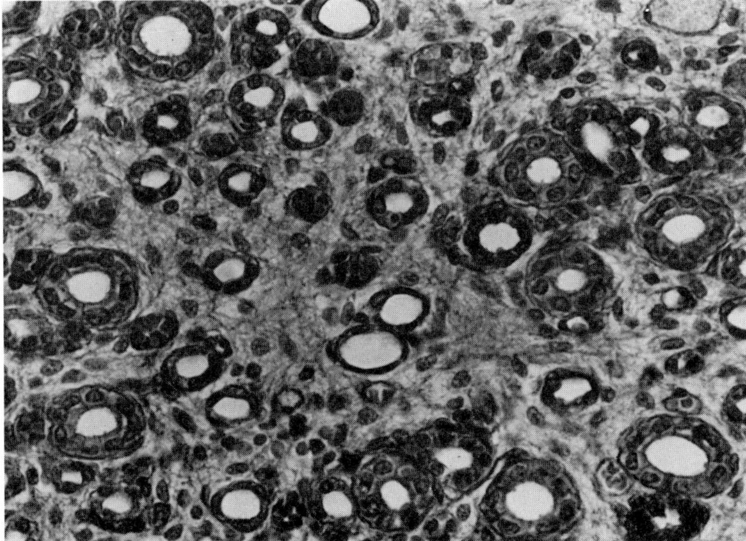
FIGURE 22-33

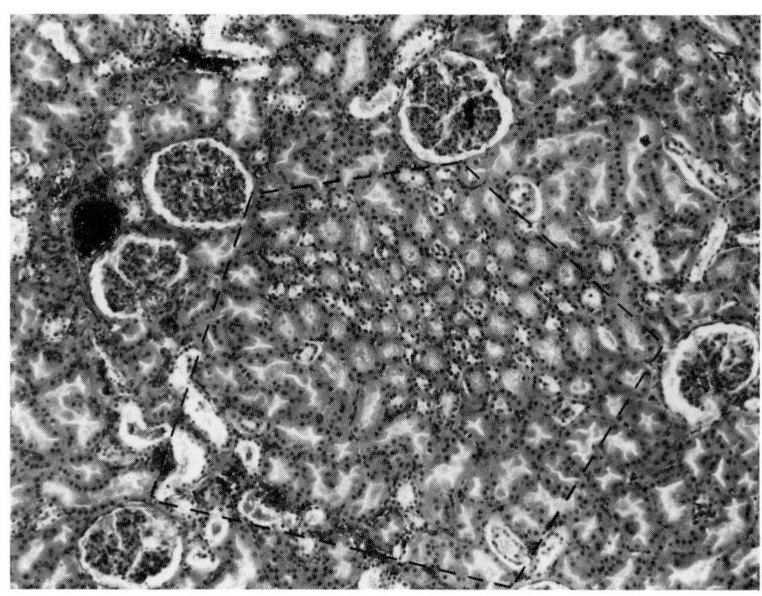

FIGURE 22-34

Rhodin, J. A. G.: "Anatomy of Kidney Tubules," *Int. Rev. Cytol.* **7:**485–534 (1958).

———: *Histology: A Text and Atlas,* Oxford University Press, New York, 1974.

Rouiller, C., and A. F. Muller (eds.): *The Kidney,* 3 vols., Academic, New York, 1969–1971.

Weiss, L., and R. O. Greep: *Histology,* 4th ed., McGraw-Hill, New York, 1977.

ADDITIONAL READINGS

Andrew, P. M., and K. R. Porter: "A Scanning Electron Microscopic Study of the Nephron," *Am. J. Anat.* **140:**81–116 (1974).

Arakawa, M.: "A Scanning Electron Microscopy of the Glomerulus of Normal and Nephrotic Rats," *Lab. Invest.* **23:**489–496 (1970).

Bulger, R. E., C. C. Tisher, C. H. Myers, and B. F. Trump: "Human Renal Ultrastructure. II. The Thin Limb of Henle's Loop and Interstitium in Healthy Individuals," *Lab. Invest.* **16:**124–141 (1967).

———, F. L. Siegel, and R. Pendengrass: "Scanning and Electron Microscopy of the Rat Kidney," *Am. J. Anat.* **139:**483–502 (1974).

Dixon, J. S., and J. A. Gosling: "Electron Microscopic Observations of the Renal Caliceal Wall in the Rat," *Z. Zellforsch.* **103:**328–340 (1970).

Fujita, T., J. Tokunuga, and M. Miyoshi: "Scanning Electron Microscopy of the Podocytes of Renal Glomerulus," *Arch. Histol. Jap.* **32:**99–113 (1978).

Hartroft, P. M.: "Juxtaglomerular Cells," *Circulation Res.* **12:**525–538 (1963).

Hicks, R. M.: "The Fine Structure of the Transitional Epithelium of Rat Ureter," *J. Cell Biol.* **26:**25–48 (1965).

Johnson, F. R., and S. J. Darnton: "Ultrastructural Observations on the Renal Papilla of the Rabbit," *Z. Zellforsch.* **81:**390–406 (1967).

Jorgensen, F.: "Electron Microscopic Studies of Normal Visceral Epithelial Cells (Human Glomerolus)," *Lab. Invest.* **17:**225–242 (1967).

Miyoshi, M., T. Fujita, and J. Tokunuga: "The Differentiation of Renal Podocytes: A Combined Scanning and Transmission Electron Microscope Study in the Rat," *Arch. Histol. Jap.* **33:**161–178 (1971).

Moffat, D. B., and J. Fourman: "The Vascular Pattern of the Rat Kidney," *J. Anat.* **97:**543–553 (1963).

Mueller, C. B.: "The Structure of the Renal Glomerulus," *Am. Heart J.* **55:**304–322 (1958).

Myers, C. H., R. E. Bulger, C. C. Tisher, and B. F. Trump: "Human Renal Ultrastructure. IV. Collecting Duct of Healthy Individuals," *Lab. Invest.* **15:** 1921–1950 (1966).

Oswaldo, L., and H. Latta: "The Thin Limbs of Henle," *J. Ultrastruct. Res.* **15:**144–168 (1966).

Richter, W. R., and S. M. Moize: "Electron Microscopic Observations on the Collapsed and Distended Mammalian Urinary Bladder," *J. Ultrastruct. Res.* **9:**1–9 (1963).

Schwartz, M. M., and M. A. Venkatachalam: "Structural Differences in Thin Limbs of Henle: Physiological Implications," *Kidney Internat.* **6:**193–208 (1974).

Tisher, C. C., R. E. Bulger, and B. F. Trump: "Human Renal Ultrastructure. I. Proximal Tubule of Healthy Individuals," *Lab. Invest.* **15:**1357–1394 (1966).

———, ———, and ———: "Human Renal Ultrastructure. III. The Distal Tubule in Healthy Individuals," *Lab. Invest.* **18:**655–668 (1968).

Tobian, L.: "Relationship of Juxtaglomerular Apparatus to Renin and Angiotensin," *Circulation* **25:** 189–195 (1962).

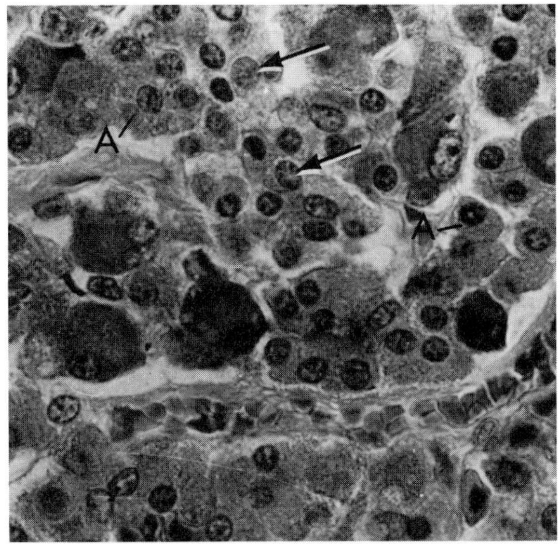

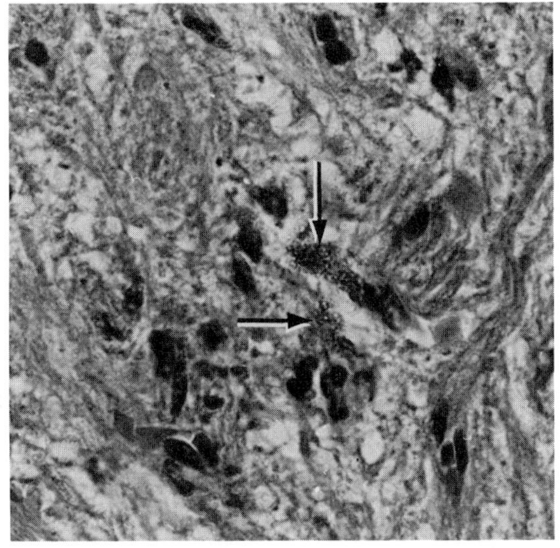

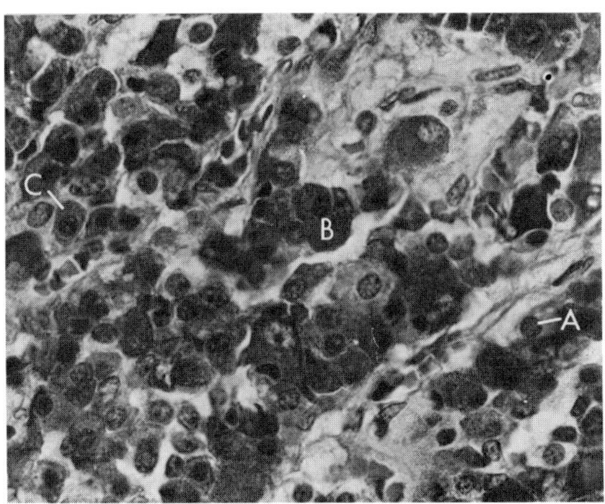

23 ENDOCRINE GLANDS I: HYPOPHYSIS AND HYPOTHALAMUS

OBJECTIVES

Upon completion of this chapter, the student will be able to:

1 Diagram the hypophysis, including the following components:
(a) hypophyseal stalk (b) infundibulum (c) pars distalis (d) pars nervosa (e) pars intermedia (f) vestigial lumen

2 Associate the above-listed components with their proper embryonic origin (neural ectoderm or oral ectoderm).

3 Identify the following in histologic section:
(a) pars distalis (b) pars nervosa

4 Identify the following pars distalis cells in a histologic section and state which hormones they produce:
(a) acidophils (b) basophils (c) chromophobes

5 Identify the function of each of the hormones produced by cells in the pars distalis.

6 Name and identify in section the cell type found in the neurohypophysis.

7 Name and identify the function of the two hormones stored in the neurohypophysis.

8 Define the hypophyseal portal system.

9 Define the following terms:
(a) Rathke's pouch (b) Herring bodies

10 Describe the means by which the hypothalamohypophyseal tract transports releasing factors from the hypothalamus to the hypophysis.

11 Identify one example of negative-feedback inhibition.

The endocrine system includes a variety of tissues and glands which produce hormones that are released directly into the circulatory system. Hormones are secretory products of endocrine cells and are capable of altering the metabolic activity of many cells some distances from the gland (target cells). Some hormones are rather specific in eliciting a stimulatory effect that may be limited to one type of cell, for example, the effect of follicle-stimulating hormone (FSH) on the ovarian follicle which induces the differentiation and maturation of the follicle. Others are more broadly effective in that many different types of cells are stimulated. The growth hormone produced by the anterior pituitary gland would be a good example.

The endocrine system includes a number of different glands such as the hypophysis (pituitary gland), thyroid, adrenals, parathyroids, ovaries, testes, and islets of Langerhans. Of these, the pituitary gland is often referred to as the master gland, as it produces a number of trophic hormones whose target cells are in other endocrine glands elsewhere in the body. For this reason, the pituitary gland will be discussed first. Since the pituitary gland, or hypophysis, is situated at the base of the brain and is closely regulated in its function by neural influence of the hypothalamus, considerations will be given to the hypothalamic control of pituitary function.

The islets of Langerhans, discussed in connection with the pancreas in Chap. 21, produce insulin and glucagon, vital factors in the regulation of glucose metabolism. The testes and ovaries produce a number of sex hormones including those that govern the development of secondary sexual characteristics. These sexual glands and their products are discussed in subsequent chapters on the reproductive organs.

HYPOPHYSIS (PITUITARY GLAND)

The hypophysis is a small gland located in the sella turcica of the sphenoid bone. It weighs about 0.5 g in men and slightly more in women. Since the gland is connected to the base of the brain, it is ensheathed by the *dura mater,* the fibrous connective tissue membrane which envelopes the central nervous system. The hypophysis can be divided into two sections which reflect the embryonic derivation of the gland and show histologic characteristics that are pertinent in understanding the structure and function of this gland. In the development of the hypophysis, oral ectoderm approaches neural ectoderm in the form of a small sac called *Rathke's pouch* (Fig. 23-1).

When these two ectoderms fuse, they induce each other to develop into the two major subdivisions of the hypophysis. Cells from the neural ectoderm develop into the *neurohypophysis,* and cells from the oral ectoderm develop into the *adenohypophysis.* The adenohypophysis of a fully mature gland is comprised of the *pars distalis, pars tuberalis* (*pars infundibularis*), and *pars intermedia.* The

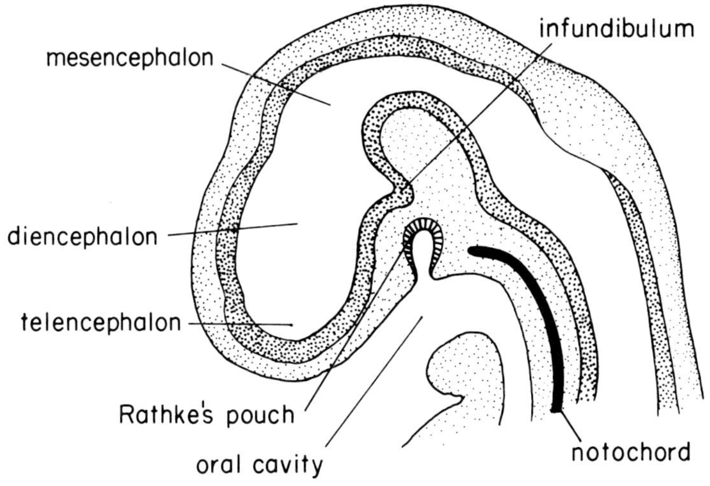

FIGURE 23-1 Schematic sagittal section of an embryo showing the relationship of the stomodeal and neural portions of the hypophysis during its development.

pars distalis is often called the *anterior lobe* of the pituitary. The pars intermedia and a portion of the pars nervosa, the infundibular process of the neurohypophysis, form the *posterior lobe* of the hypophysis. Table 23-1 summarizes the embryonic origins of major divisions and subdivisions of the hypophysis.

Figure 23-2 illustrates the topographic divisions as seen through the sagittal plane of the hypophysis. Figure 23-3 is a section through the hypophysis showing some of the parts mentioned above. The infundibulum is derived from neural ectoderm and is covered by a layer of orally derived ectoderm, the *pars tuberalis* (Fig. 23-2).

PARS DISTALIS (ANTERIOR LOBE)

The pars distalis or anterior lobe makes up about 75 percent of the hypophysis. It is the site of production of many *trophic hormones,* that is, hormones which stimulate activities of other endocrine organs. Since all of the trophic hormones are peptides or small proteins in nature, the cytoplasm of all secretory cells in the pars distalis show the features characteristic of protein-secreting cells, namely, abundant RER, Golgi complex, and secretory granules of different sizes. The anterior lobe is comprised of irregular cords of cells that are in close association with extensive capillary beds. Based on the nature of staining, the cells of the anterior lobe can be divided into several different types, each of which has been identified with one or more trophic hormones. In the discussion that follows, a description of different cell types based on the traditional staining will be made. This will be followed by comments on recent findings on pituitary immunocyto-

DERIVATIONS AND DIVISIONS OF THE HYPOPHYSIS

TABLE **23-1**

Neurohypophysis (neural ectoderm)	Infundibulum (neural stalk)	Median eminence Infundibular stalk	With pars infundibularis comprise the hypophyseal stalk
	Lobus nervus (neural lobe)	Infundibular process	Posterior lobe
Adenohypophysis (oral ectoderm)	Lobus glandularis	Pars intermedia	
		Pars distalis Pars infundibularis	Anterior lobe (pars tubularis)

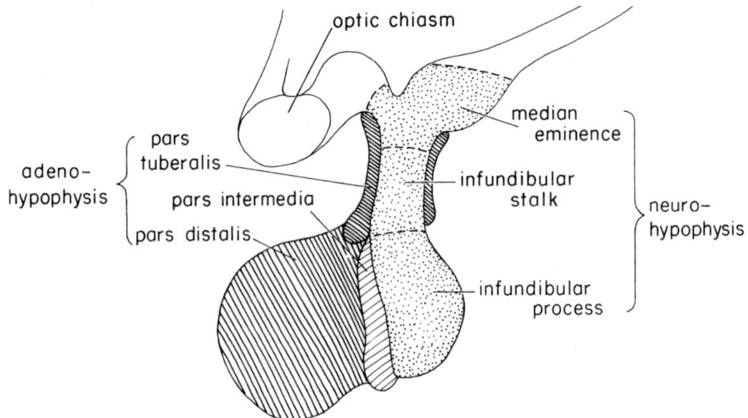

FIGURE 23-2 Diagram of a sagittal section through the hypophysis showing its topographic divisions.

chemistry which makes possible a more functional classification of the anterior pituitary cells.

Chromophobes

The chromophobes do not stain appreciably in conventional preparations (Fig. 23-4, arrows). They are the smallest, most numerous cell type in the pars distalis, comprising about 50 percent of the total cell population. These cells are believed to be reserve cells and therefore are not active in the production of hormones. However, a small proportion of chromophobes with secretory granules of 200 to 250 nm have been found to secrete adrenocorticosteroid-stimulating hormone, and hence are called *corticotrophs.*

ACIDOPHILS

Acidophils, sometimes called *alpha cells,* stain red with eosin. They are larger than chromophobes but smaller than the basophils (A in Figs. 23-4, 23-5, and 23-6). These cells are rounded and contain cytoplasmic granules of different sizes. Acidophils comprise approximately 35 percent of the total pars distalis cells. By application of special stains, two types of acidophils can be distinguished. *Somatotrophs* which produce somatotropin, or growth hormone, stain with orange G. In

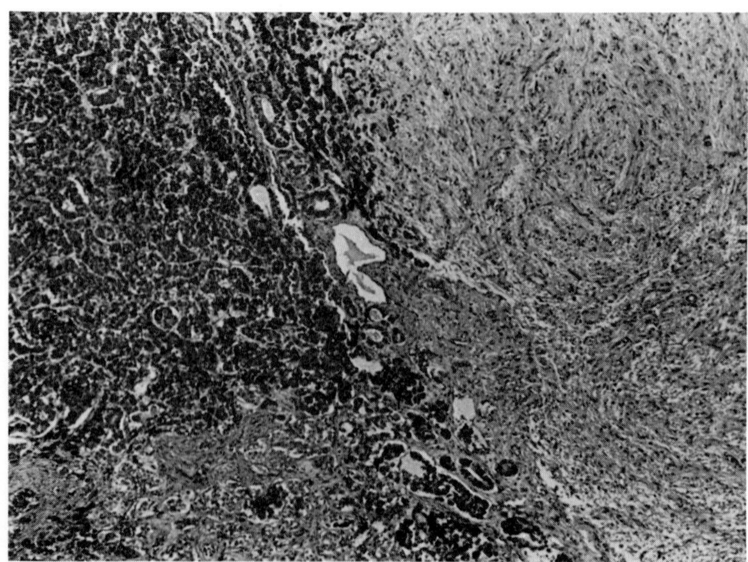

FIGURE 23-3 Light micrograph of the hypophysis. The anterior lobe is seen in the left part and the posterior lobe in the right part of the micrograph. Note the cysts in the pars intermedia of the posterior lobe.

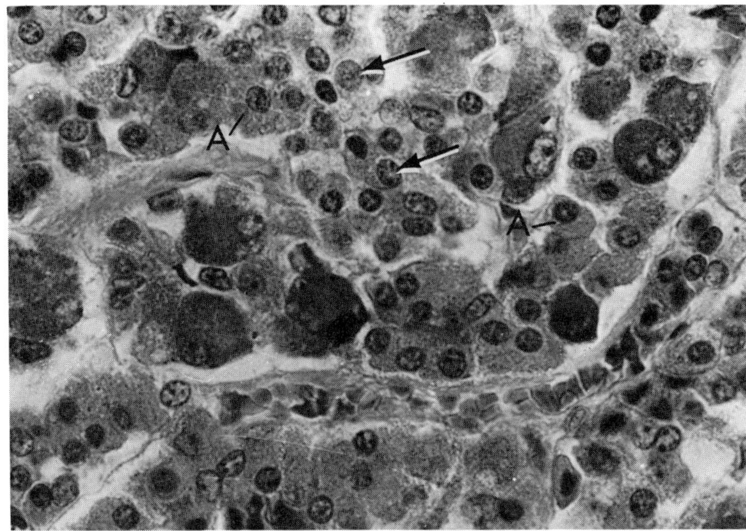

FIGURE 23-4 Light micrograph of the pars distalis. Chromophobes (arrows) are small, lightly stained cells that are present among the larger chromophilic cells. Acidophils (A) and basophils (B) are present among the more numerous chromophobes.

EM, somatotrophs measure 14 to 19 μm in diameter. Prolactin cells are also called *mammotrophs* because they stimulate the development of mammary glands.

BASOPHILS

Basophils stain with aldehyde fuchsin, PAS, or aniline blue and comprise approximately 15 percent of the pars distalis cells (B in Figs. 23-4, 23-5, and 23-6). Based on their response to aldehyde fuchsin, basophils have been subdivided into two categories. These are the beta cells (positive to aldehyde fuchsin) and the delta cells (staining with PAS).

Beta Cells (B Cells)

These basophils react positively to aldehyde fuchsin. They are more polyhedral than the rest of the basophils and are centrally located within the gland. In an EM they have been shown to contain small secretory granules measuring 140 to 160 nm in diameter. Beta

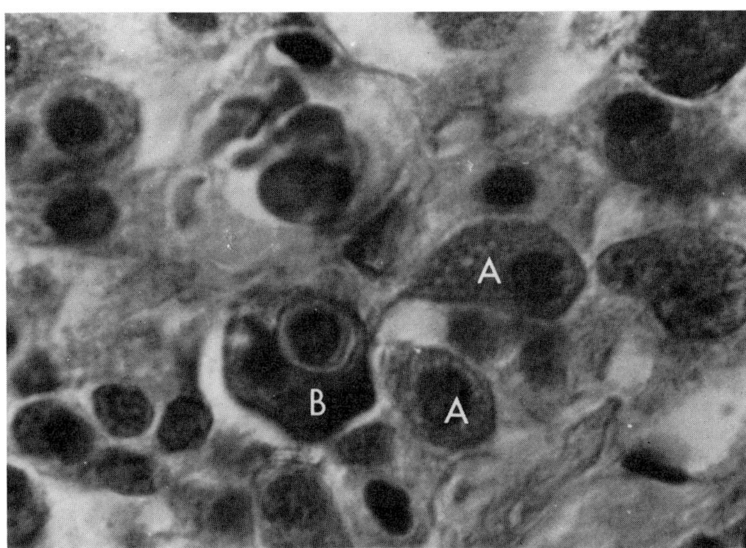

FIGURE 23-5 Light micrograph of the adenohypophysis. Acidophils (A) and basophils (B) are present among the smaller, more numerous chromophobes.

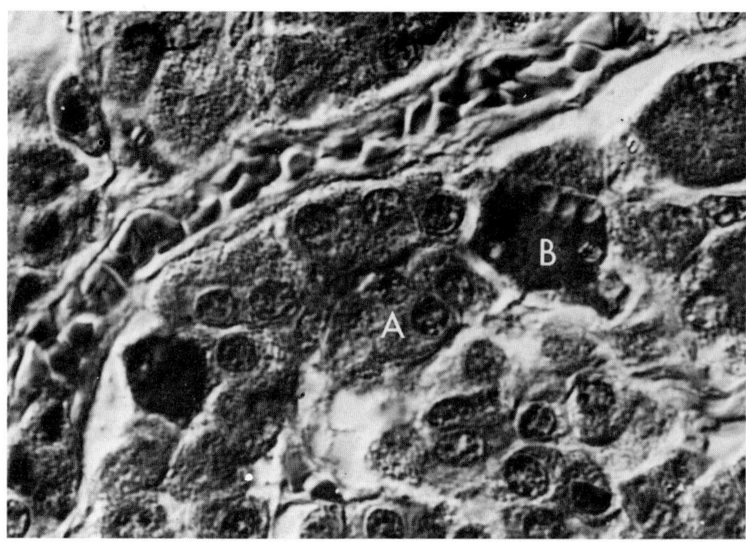

FIGURE 23-6 Phase-contrast micrograph of the pars distalis. Acidophils (A) and basophils (B) are seen among the chromophobes.

cells produce *thyroid-stimulating hormone* (*TSH*) that induces the thyroid to produce thyrosine.

Delta Cells (D Cells)

Delta cells are the larger of the two basophils. They stain with PAS, but do not react with aldehyde fuchsin. They are found mainly at the periphery of the pars distalis and are more rounded than the beta cells. Delta cells appear to be made up of two populations. There are D cells that produce *follicle-stimulating hormone* (*FSH*), which controls the growth of ovarian follicles and their production of estrogen. Other delta cells produce *luteinizing hormone* (*LH*), which maintains the ovarian cycle. In the male, this same hormone is called *interstitial cell-stimulating hormone* (*ICSH*) because it stimulates the interstitial cells of the testes, resulting in the production of testosterone. Delta-cell granules are rounded and measure 150 to 250 nm in diameter.

PARS INTERMEDIA

The pars intermedia is a relatively small segment of the hypophysis originating from the oral ectoderm which is located between the anterior lobe and the pars nervosa. The cleft between the infundibular process and oral ectoderm may persist after birth, resulting in a zone of cysts called *Rathke's cysts* (Figs. 23-7 and 23-8). The pars intermedia is composed of cuboidal cells which appear very similar to the cells of the pars distalis. Some of the cells are intensely basophilic, while others are smaller and lightly stained. The cytoplasm contains varying numbers of small granules (200 to 300 nm). It is thought that the basophilic cells of the pars intermedia are responsible for the production of *melanocyte-stimulating hormone* (*MSH*), which regulates the pigment-producing cells of the skin.

PARS TUBERALIS

The pars tuberalis is a layer of cuboidal cells covering the infundibulum. These cells originate from oral ectoderm and are slightly basophilic. Colloid-containing vesicles are frequently seen in the pars tuberalis cells, but no hormones are produced here.

NEUROHYPOPHYSIS

As pointed out earlier, the neurohypophysis is derived from neural ectoderm and includes the infundibulum and the pars nervosa. No hormones are produced in the neurohypophysis. Figure 23-7 shows the appearance of

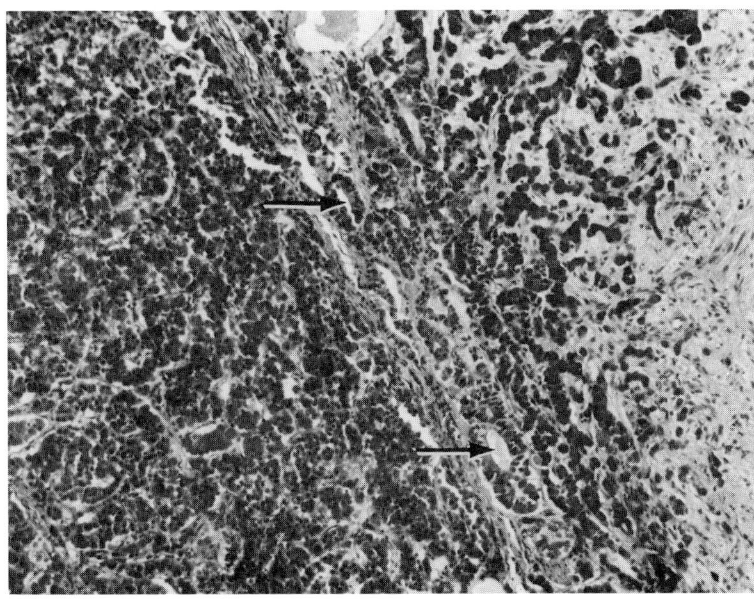

FIGURE 23-7 Light micrograph of the border zone, the pars intermedia, between the anterior lobe and pars nervosa. This zone contains cysts called Rathke's cysts (arrows).

the pars nervosa on the right. The pars distalis is seen on the left. The clear area in the center of the frame is the *vestigial lumen* (Rathke's cyst) which, as mentioned previously, is caused by the folding of Rathke's pouch, and is found between the pars intermedia and pars distalis. Figure 23-9 is a higher-power view of the pars nervosa. The most numerous cells of the neurohypophysis are *pituicytes* that are small with rounded nuclei, as seen in this figure. They are analogous to glial elements of the central nervous system.

The pars nervosa is a region of storage

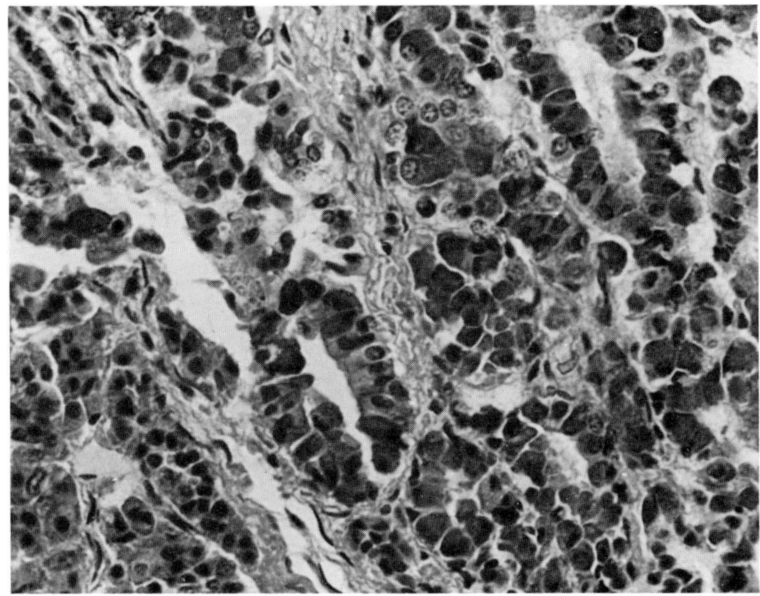

FIGURE 23-8 Pars intermedia viewed at higher magnification.

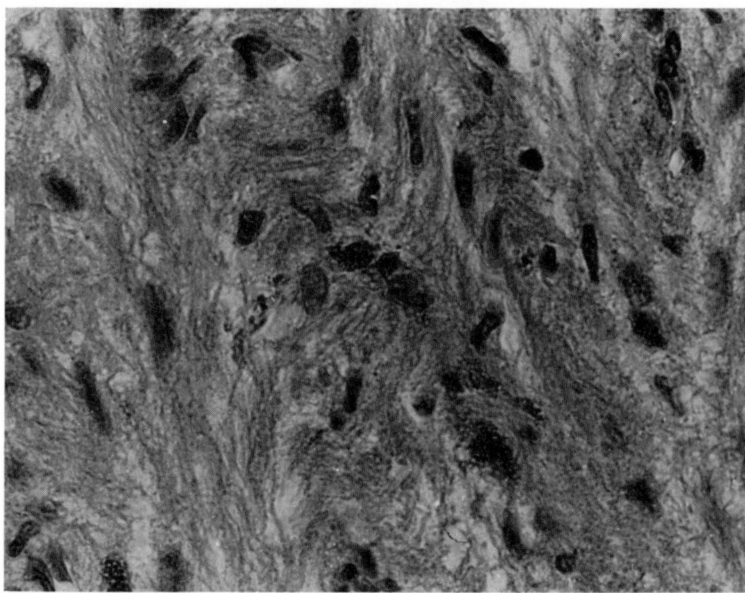

FIGURE 23-9 Pars nervosa viewed at higher magnification. Note the appearance of the pituicytes.

for two important hormones: *antidiuretic hormone* (*ADH*), which regulates fluid balance, and *oxytocin,* which stimulates the uterine muscles to contract. Both of these hormones are produced by nerve cells located in the *hypothalamus.* The relationship between the hypothalamus and pars nervosa of the hypophysis is seen in Fig. 23-10.

Nerve cell bodies making up the *paraventricular* and *supraoptic nuclei* send their axons to the pars nervosa. These axons make up the hypothalamohypophyseal tract through which the ADH and oxytocin are transported into the neurohypophysis. In other words, these two hormones are produced by the hypothalamic nuclei packaged in the form of small vesicles which migrate inside the axons, reaching the terminal ends in the pars nervosa. Some of these granules can be stained and visualized in LM (Fig. 23-

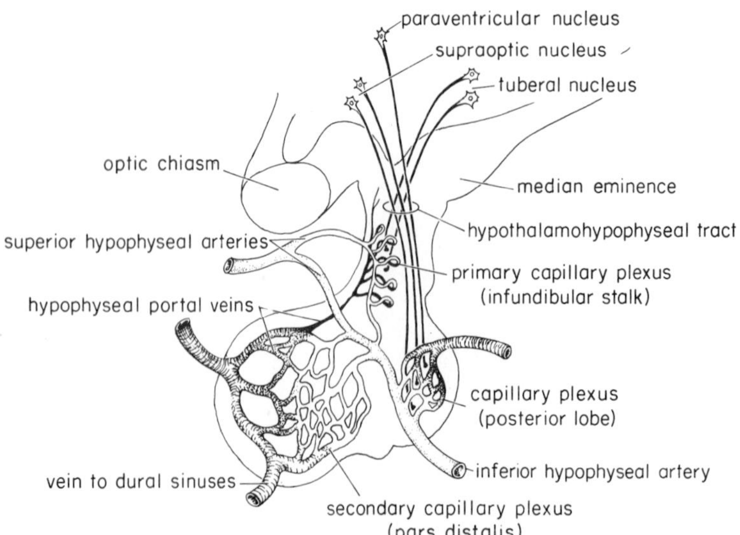

FIGURE 23-10 Diagram illustrating the blood supply to the hypophysis. Note also the connection between the hypothalamus and pars nervosa via the hypothalamohypophyseal tract.

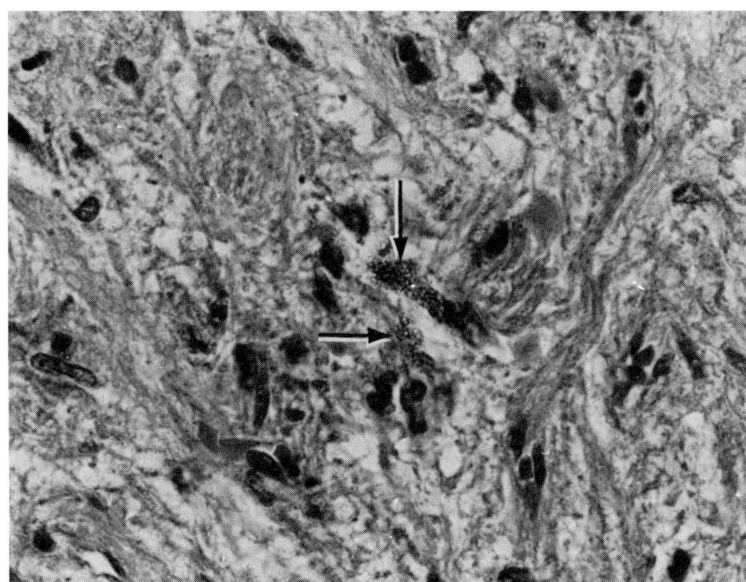

FIGURE 23-11 Light micrograph of the pars nervosa. Nerve terminals of the hypothalamohypophyseal tract contain basophilic granules, Herring bodies (arrows). These terminals are found among the pituicytes.

11). These granules are called *Herring bodies.* They are intensely basophilic and are scattered among the pituicytes.

RECENT DEVELOPMENTS IN PITUITARY CYTOLOGY

With the advent of electron microscopy, the detailed structural characteristics of the various cell types mentioned in this chapter came to be known. As already alluded to, all of the trophic hormones produced by the hypophysis are peptide in nature and, therefore, involve cytoplasmic structures characterized by an abundance of RER, a Golgi apparatus, and secretory granules. However, the extent of their development and the size and appearance of secretory granules are different among different cell types, thereby permitting an ultrastructural characterization of the different cell types. The problem in pituitary cytology at EM level has been the wide differences among different species in terms of relative percentage of cell types, as well as variable ultrastructural features in a given cell type.

Immunocytochemical techniques which are more specific have allowed the identification of immunologically specific sites of hormone production. The availability of these newer techniques has promoted intensive investigations on the subject, and much has been added to our understanding of the cytology of hypophyseal cells in human beings. Current understanding of cell types in terms of their products is summarized in Table 23-2.

BLOOD SUPPLY: THE HYPOPHYSEAL PORTAL SYSTEM

As with all endocrine glands, the hypophysis depends on the bloodstream to transport its products to their respective target cells and organs. The hypothalamus produces two hypothalamic hormones, ADH and oxytocin. In addition, it produces a number of factors that are responsible for initiating the release of trophic hormones produced by the pars distalis. The transport of these hypothalamic-releasing factors from the pars nervosa into the pars distalis is a vital function that is carried out by the local vascular system. For this reason, a hypophyseal portal system exists and links the two embryologically different portions of the gland. In Fig. 23-10 the blood supply to the hypophysis is illustrated diagrammatically.

The superior hypophyseal artery enters ventrally at the level of the stalk (pars tuberalis), and the inferior hypophyseal artery enters dorsally at the posterior lobe. The superior hypophyseal artery branches to form a network of capillaries which supply the infundibulum where the veins form the hypophyseal portal system. Portal veins, after passing into the anterior lobe, form a second capillary bed supplying the pars distalis. In this way, hypothalamic-releasing factors, as well as ADH and oxytocin, are picked up by the first capillary bed and carried into the pars distalis via the portal vein, which supplies the second capillary network where the release factors act upon the anterior lobe cells, discharging and stimulating the production of the various trophic hormones. The blood, now carrying hormones, leaves the anterior lobe through the two hypophyseal veins.

NEGATIVE-FEEDBACK INHIBITION

In order to maintain appropriate levels of metabolic activities in target cells or to stim-

HORMONES OF THE PITUITARY GLAND — TABLE 23-2

HORMONES	CELLULAR SOURCE [TRADITIONAL CLASS]	PRINCIPAL ACTIONS [GRANULE SIZE IN EM]
1. *Pars distalis* (anterior pituitary)		
Somatotropin (STH, growth hormone)	Somatotrophs [acidophil]	Growth; protein synthesis and modification of lipid and carbohydrate metabolism [350 nm]
Adrenocorticotropin (ACTH)	Corticotrophs [chromophobe]	Stimulation of secretion by adrenal cortex; certain extraadrenal actions [200–250 nm]
Thyrotropin (TSH)	Thyrotrophs [basophil]	Stimulation of thyroid gland to form and release its hormones [140 nm]
Gonadotropins		
(a) Luteinizing or interstitial cell–stimulating hormone (LH or ICSH)	Gonadotrophs (luteotrophs or interstitiotrophs) [basophil]	Ovary: formation of corpora lutea; secretion of progesterone probably in conjunction with FSH
(b) Follicle-stimulating hormone (FSH)	Gonadotrophs (folliculotrophs) [basophil]	Ovary: growth of follicles; ovulation and release of estrogens with LH [200 nm]
(c) Prolactin (lactogenic hormone, luteotropin)	Prolactin cells [acidophil]	Development of mammary glands during last trimester; initiation of milk secretion; certain other effects on many tissues [600 nm]
2. *Pars intermedia*		
Melanophore-stimulating hormone (intermedin, MSH)	Melanotrophs	Dispersion of pigment granules in the melanophores of lower vertebrates
3. *Neurohypophysis* (posterior pituitary) Vasopressin (ADH, antidiuretic hormone)	Hypothalamic neurons	Elevation of blood pressure
Oxytocin	Hypothalamic neurons	Contraction of uterus; facilitation of sperm transport in females and milking

ulate the secretion by target cells, it is important that hormone levels are maintained at an appropriate level. The stimulation of a target cell toward secretion and/or differentiation requires continued output of a hormone, which continues to amplify its effect. The continued functioning of the pituitary gland during growth of the various target organs is a good example of a positive-feedback mechanism. On the other hand, the maintenance of the differentiated state of a target organ is accomplished by modulation of hormone levels involving a negative-feedback mechanism. A good example of this process is found in the relationship between TSH production by the pituitary and thyroid functioning. Thyroid-stimulating hormone, produced by the B cells of the pars distalis, stimulates the thyroid gland to increase production of thyroxine. The increased amount of thyroid hormone produced and released into the blood has an inhibitory effect on the thyrotrophic cells of the pars distalis. Thus, the pars distalis responds to increasing levels of thyroxine by decreasing its production of TSH. This is called *negative-feedback inhibition,* which operates in a number of endocrine and other biologic systems. In addition to this direct negative feedback at the level of endocrine glands and target cells, a similar feedback mechanism operates through higher parts of the brain. Presently, a significant number of releasing hormones (or factors) has been identified in the hypothalamic nuclei and hypothalamohypophyseal tracts with respect to a host of pituitary trophic hormones. For example, an increase of the thyroid hormone level in blood induces the release of thyrotropin-releasing factor (TRF) by the hypothalamus which has an inhibitory effect on TSH release by the anterior pituitary.

REVIEW SECTION

1. Identify the three different types of cells labeled in Fig. 23-12. Are there any other cell types you might expect to see here? Identify the section.

This section from the pars distalis shows chromophobes (a), basophils (b), and acidophils (c). Endothelial cells and fibroblasts are also found in the pars distalis.

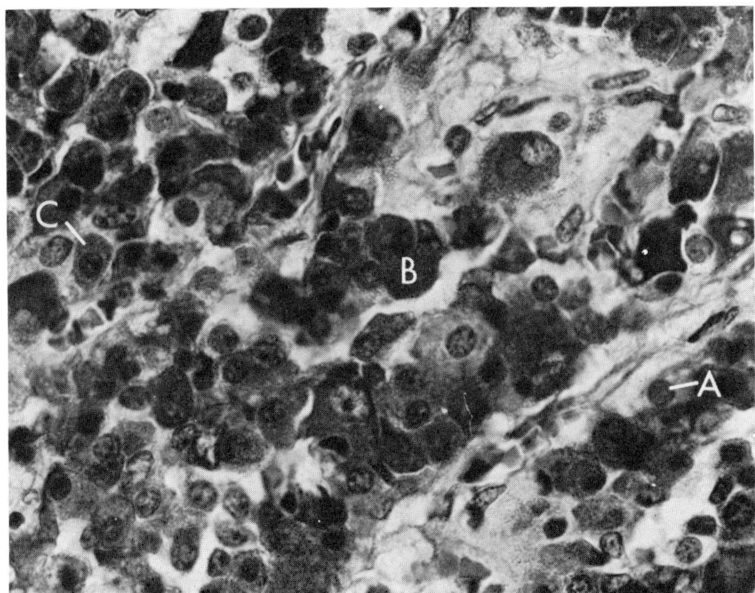

FIGURE 23-12

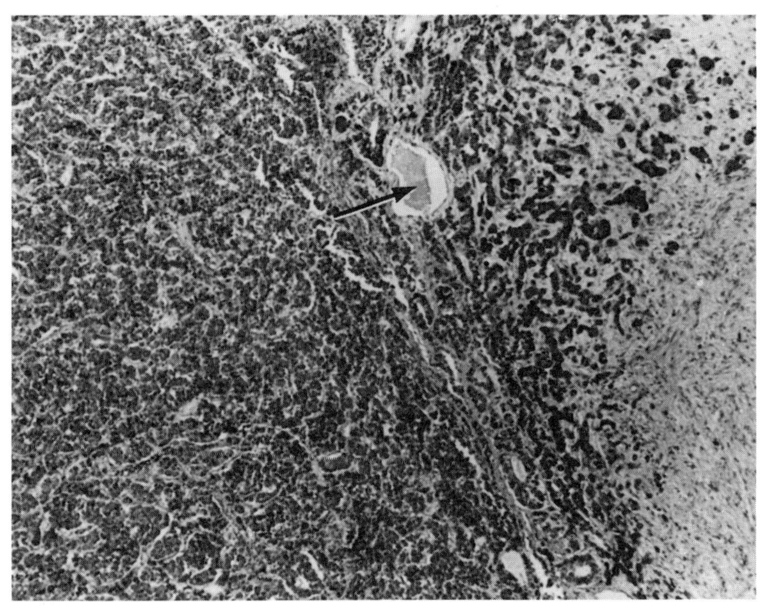

FIGURE 23-13

2. What is the clear region indicated by the arrow in Fig. 23-13? What regions lie on either side of the clear space? The anterior lobe appears on the (right/left) of the clear space, while the posterior lobe appears on the (right/left).

The vestigial lumen of the Rathke's cyst (arrow) in this figure divides the posterior lobe (pars nervosa) on the right and the anterior lobe on the left.

BIBLIOGRAPHY

Bargmann, W.: "Neurosecretion," *Int. Rev. Cytol.* **19:**183–201 (1966).

Bloom, W., and D. W. Fawcett: *A Textbook of Histology,* 10th ed., Saunders, Philadelphia, 1975.

Hall, R., and A. Gomez-Pan: "The Hypothalamic Regulatory Hormones and Their Clinical Applications," *Adv. Clin. Chem.* **18:**173–212 (1976).

Harris, G. W., and B. T. Donovan (eds.): *The Pituitary Gland,* 3 vols., University of California Press, Berkeley, 1966.

Heller, H. (ed.): *The Neurohypophysis,* Academic, New York, 1957.

Tixier-Vidal, A., and M. G. Farquhar (eds.): *The Anterior Pituitary,* Academic, New York, 1975.

Weiss, L., and R. O. Greep: *Histology,* 4th ed., McGraw-Hill, New York, 1977.

ADDITIONAL READINGS

Baker, B. L.: "Functional Cytology of the Hypophyseal Pars Distalis and Pars Intermedia," in R. O. Greep and E. B. Astwood (eds.), *Handbook of Physiology,* sect. 7, vol. 4, American Physiological Society, Washington, 1974, pp. 45–80.

Bodian, D.: "Herring Bodies and Neuroapocrine Secretion in the Monkey: An Electron Microscope Study of the Fate of the Neurosecretory Product," *Bull. Johns Hopkins Hosp.* **113:**57 (1963).

Fawcett, D. W., J. A. Long, and A. L. Jones: "The Ultrastructure of Endocrine Glands," *Recent Progr. Horm. Res.* **25:**315–380 (1969).

Green, J. D., and G. W. Harris: "The Neurovascular Link between the Neurohypophysis and Adenohypophysis," *J. Endocrinol.* **5:**136–146 (1947).

Halmi, N. S.: "The Current Status of Pituitary Cytophysiology," *N. Z. Med. J.* **80:**551–556 (1974).

Moriarty, G. C.: "Adenohypophysis: Ultrastructural Cytochemistry: A Review," *J. Histochem. Cytochem.* **21:**885–894 (1974).

Nakane, P.: "Classification of Anterior Pituitary Cell Types with Immunoenzyme Histochemistry," *J. Histochem. Cytochem.,* **18:**9–20 (1970).

Sachs, H., P. F. Takabatake, and R. Portanova: "Biosynthesis and Release of Vasopressin and Neurophysin," *Recent Progr. Horm. Res.* **25:** 447–491 (1969).

Stanfield, J. P.: "The Blood Supply of the Human Pituitary," *J. Anat.* **94:**259–273 (1960).

Wislock, G. B.: "The Vascular Supply of the Hypophysis Cerebri of the Rhesus Monkey and Man," in W. Timme, A. M. Fraintz, and C. C. Hare (eds.), *The Pituitary Gland,* Williams & Wilkins, Baltimore, 1938, pp. 48–68.

Zuereb, G. P., M. M. L. Prichard, and P. M. Daniel: "The Hypophyseal Portal System of Vessels in Man," *Quart. J. Exp. Physiol.* **39:**219–229 (1954).

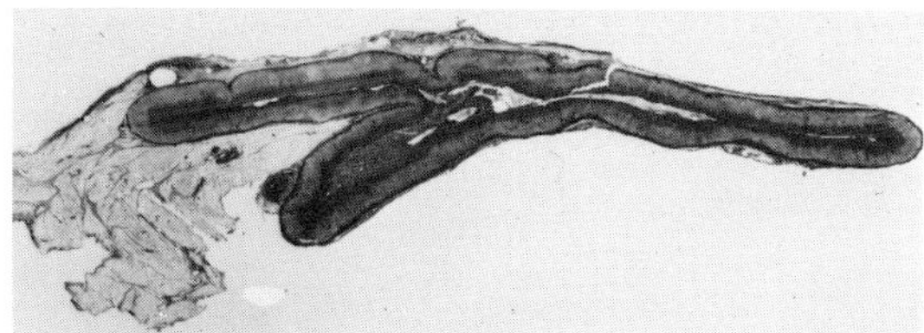

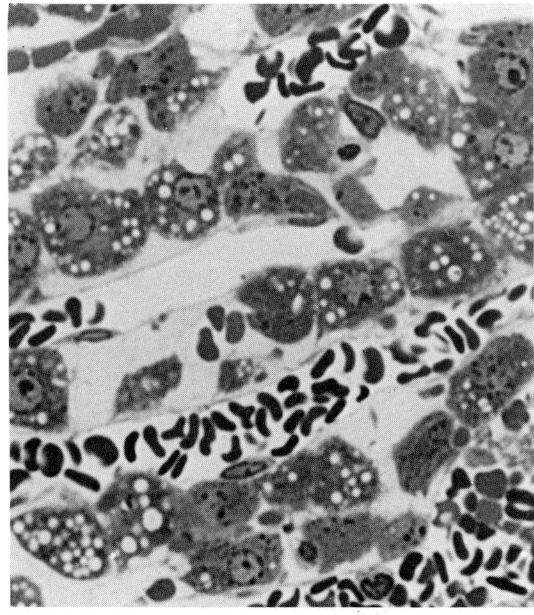

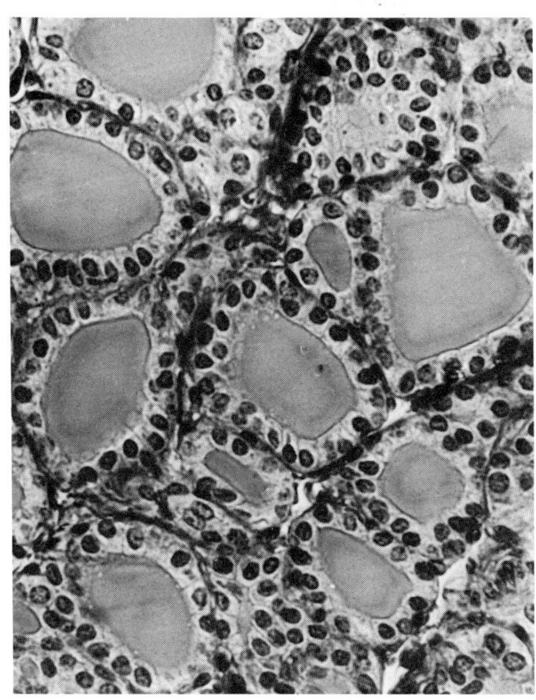

24 ENDOCRINE GLANDS II: ADRENALS, THYROID, AND PARATHYROID

OBJECTIVES

After completing this chapter, the student will be able to:

ADRENAL GLANDS

1 Identify the cellular regions of the adrenal gland in section:
(a) the three cortical zones (b) the adrenal medulla

2 State the names and functions of products from the different cellular regions.

3 Identify enterochromaffin cells.

4 List ultrastructural differences between a cell from the cortex and one from the medulla.

5 State the vascular supply of the gland (identify the central vein)

THYROID GLAND

1 Identify the cellular components of the thyroid gland in section:
(a) follicular cells (b) parafollicular cells

2 State the names and functions of products from the different types of cells.

3 Identify the mechanisms involved in the

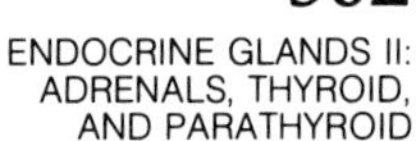

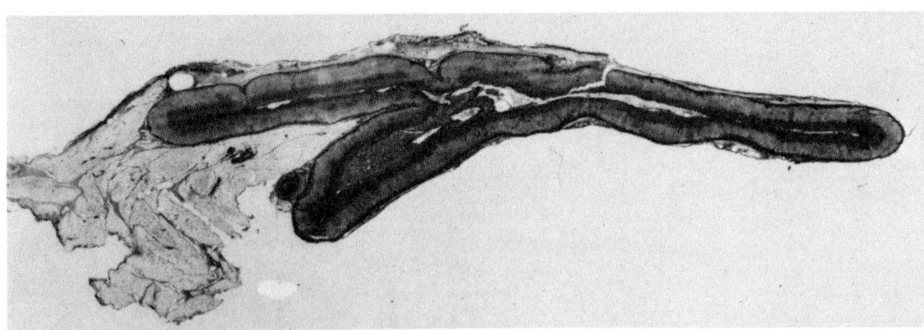

FIGURE 24-1 Cross section of an adrenal gland.

production and mobilization of the hormonal constituents of thyroid colloid.

PARATHYROID GLAND

1 Identify the location of the parathyroid gland.

2 Identify the types of cells found in the parathyroid:
(a) chief cells (b) oxyphils

3 State the name and function of the product of the parathyroid.

ADRENAL GLANDS

The paired adrenal glands, also called *suprarenal glands,* cap the upper pole of each kidney. Figure 24-1 is a cross section of an adrenal gland. An adrenal gland is divided into a cortex and a medulla. It is covered by a tough connective tissue capsule from which delicate trabeculae penetrate into the cortex. The *adrenal cortex* secretes about 40 different steroid hormones, or corticosteroids, into the bloodstream. They can be divided into two classes of steroid hormones: namely, mineralocorticoids, which maintain electrolyte and water balance, and glucocorticoids, which are involved in carbohydrate metabolism. The *adrenal medulla* secretes the peptide hormones epinephrine and norepinephrine.

ADRENAL CORTEX

The cells of the adrenal cortex are cuboidal in shape and are arranged in cords which are generally perpendicular to the capsule. These

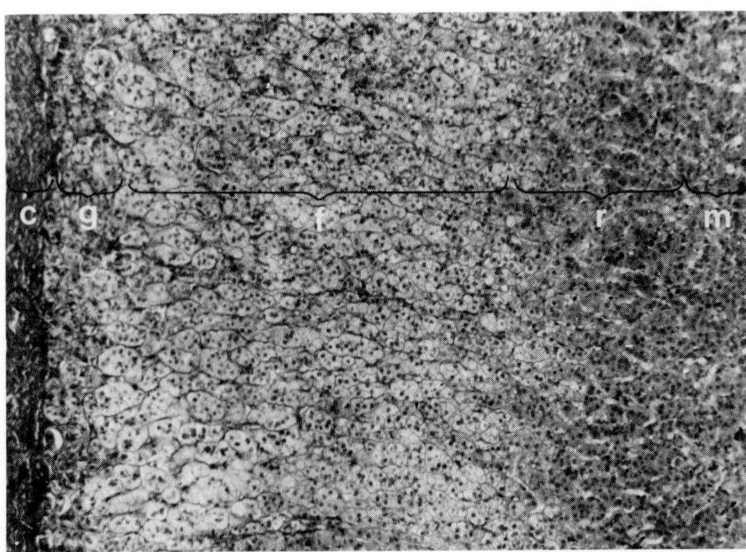

FIGURE 24-2 A portion of an adrenal gland viewed at higher magnification. (c) Capsule. (g) Zona glomerulosa. (f) Zona fasciculata. (r) Zona reticularis. (m) Medulla.

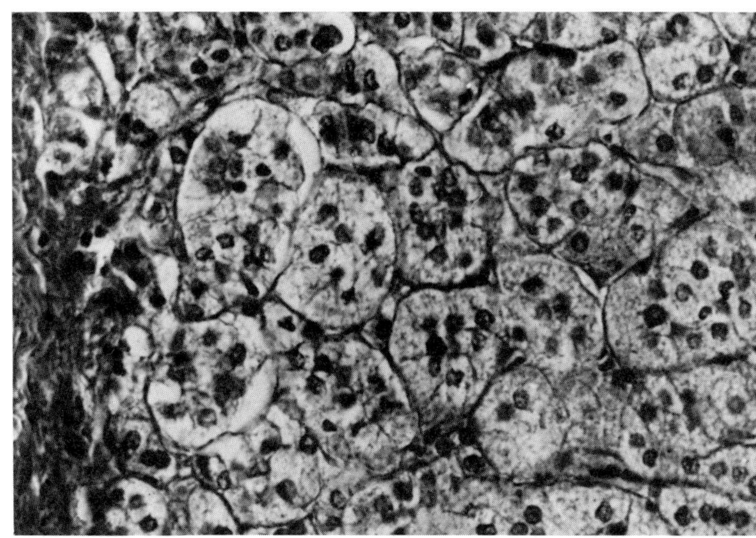

FIGURE 24-3 Zona glomerulosa viewed at higher magnification. Note the circular arrangement of the pale-staining cells.

cords of cells are interposed by the cortical sinusoids. They are straighter at the outer cortex and become irregularly arranged toward the medulla. Because of the changing pattern of organization of cortical cells, the cortex can be separated into three zones (Fig. 24-2). Immediately subjacent to the capsule (c) is the outer zone, the zona glomerulosa (g). The middle zone is the zona fasciculata (f) and the inner zone, zona reticularis (r), is nearest the medulla (m).

Cells in the *zona glomerulosa* are arranged in a circular fashion, presenting a glomerular appearance (Fig. 24-3). They are vacuolated since they are primarily concerned with production of steroid hormones and therefore contain a large number of lipid droplets (Fig. 24-4, arrows). In EM, the cells in the zona glomerulosa show SER and round mitochondria with tubular cristae typical of steroid-producing cells. They produce mineralocorticoids. *Mineralocorticoids* are steroids that

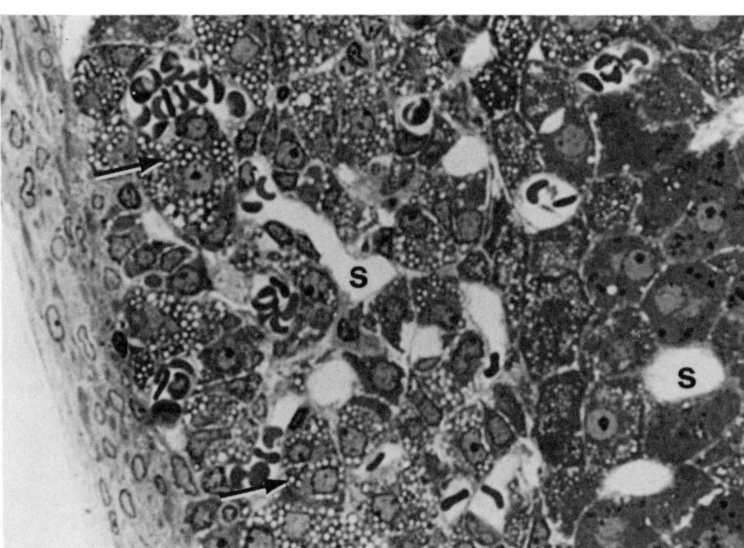

FIGURE 24-4 Light micrograph of 1-μm-thick section of plastic-embedded adrenal tissue. The cells of zona glomerulosa contain numerous lipid droplets (arrows). Cortical sinusoids (s) surround the cords of cells.

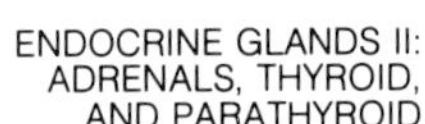

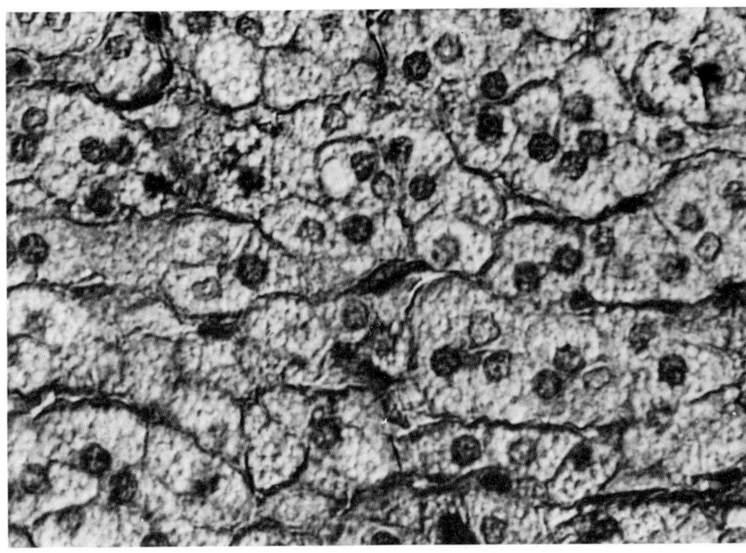

FIGURE 24-5 Light micrograph of zona fasciculata in a conventional preparation of the adrenal. The lightly stained cells are arranged in straight cords. Their cytoplasm is highly vacuolated.

regulate electrolyte concentrations in body fluids, thus regulating water balance. *Aldosterone* is one of the principal products of this region and regulates sodium-ion recovery in the kidney. The mineralocorticoids produced in the zona glomerulosa cells are secreted into the cortical sinusoids (s in Fig. 24-4).

The *zona fasciculata* is the thickest layer of the adrenal cortex. The cells are arranged in straight cords, one or two cells thick, and appear light in color because of their highly vacuolated cytoplasm. The cells are large and have vacuolated cytoplasm (Fig. 24-5). These features are better demonstrated in Fig. 24-6, which is a 1-μm section of the zona fasciculata. An EM reveals SER and mitochondria with tubular cristae along with numerous lipid droplets. Zona fasciculata cells produce *glucocorticoids,* that is, steroids which regulate carbohydrate metabolism. For

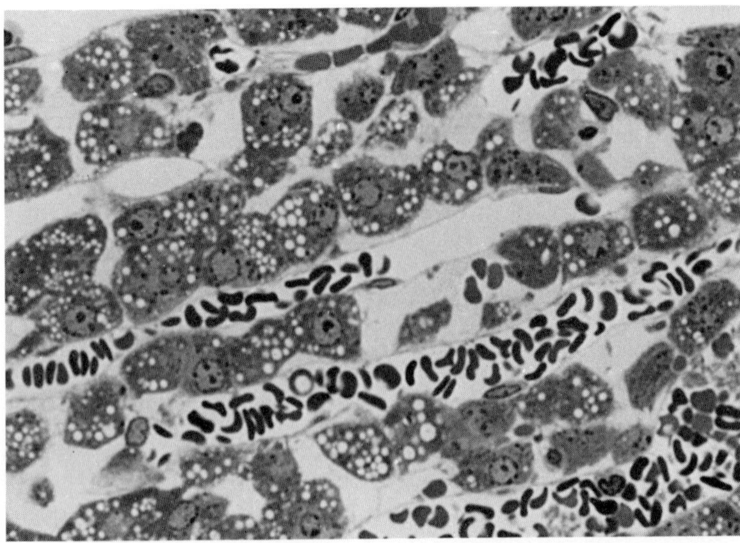

FIGURE 24-6 Light micrograph of 1-μm-thick section of plastic-embedded adrenal gland. The highly vacuolated cells of zona fasciculata are arranged in cords, one to two cells thick. Note their vesicular nuclei. A sinusoidal network surrounds the cell cords.

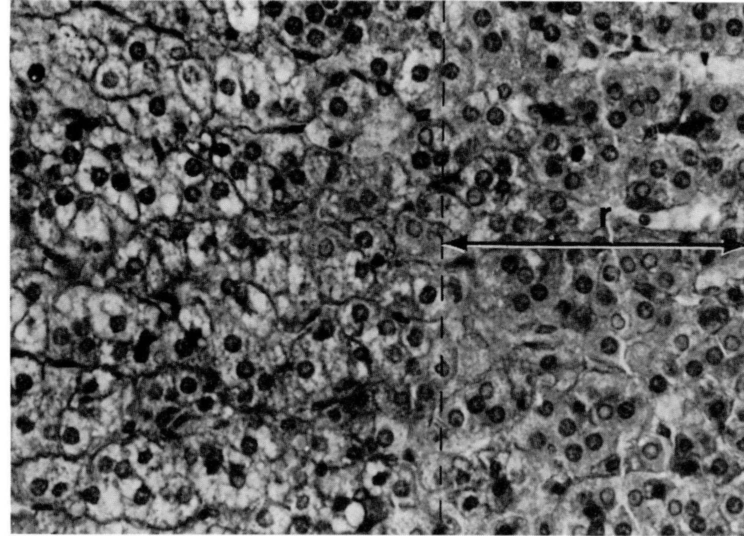

FIGURE 24-7 Light micrograph of the deeper portion of the adrenal cortex. Zona reticularis is located subjacent to zona fasciculata. Its cells have dark-staining nuclei and contain fewer lipid droplets than the other zones.

example, cortisone, one of the principal glucocorticoids, raises blood sugar levels.

The *zona reticularis* has the cells arranged in a branching network (Fig. 24-7). These cells are generally smaller and have denser cytoplasm than those of the zona fasciculata. They have dark-staining nuclei and contain few lipid droplets. Zona reticularis cells also secrete glucocorticoids. Compare the appearance of the zona reticularis (Fig. 24-8) with that of the zona fasciculata (Fig. 24-6) as seen in 1-μm-thick sections. Note the irregularly arranged cords of cells.

ADRENAL MEDULLA

The adrenal medulla differs from the adrenal cortex in both function and morphology. Within the substance of the medulla (Fig. 24-9) is a network of venous tributaries which form a large central vein (cv). The cells of the

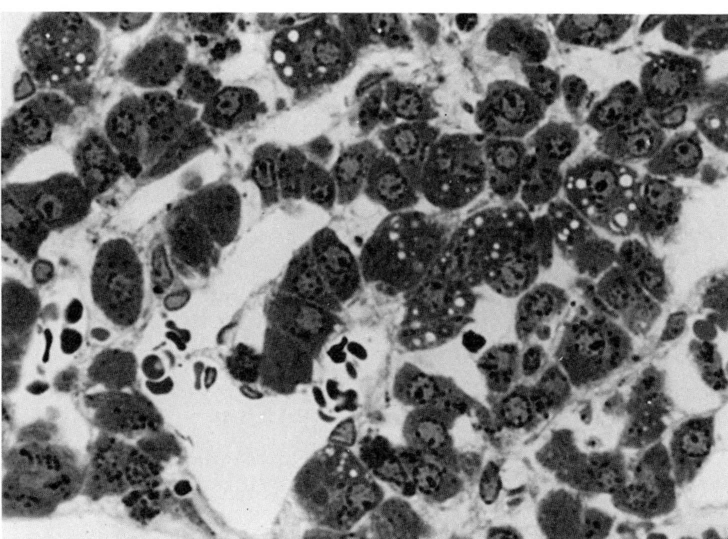

FIGURE 24-8 Light micrograph of the zona reticularis. The cells in this zone are smaller than those in zona fasciculata. They also have a darker cytoplasm that contains few lipid droplets.

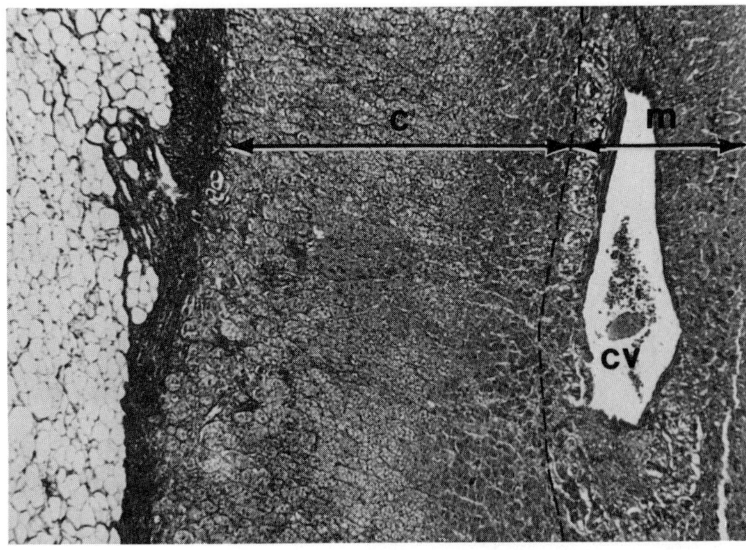

FIGURE 24-9 Portion of the adrenal gland. The cortex (c) surrounds the medulla (m). Small veins within the medulla merge and form a large central vein (cv).

medulla are interposed between the medullary capillaries and venules. They appear as irregular cords of cells surrounding the medullary sinusoids made up of distended capillaries (ms in Fig. 24-10). The medullary cells of the adrenals do not elaborate steroids and therefore are clearly different from the cortical cells. They become brown when they react with potassium bichromate. For this reason, they are called *chromaffin cells.* The Golgi regions and secretory granules of chromaffin cells are polarized toward a venule. The medullary cells (ms) are large and lighter than the cells of zona reticularis (r in Fig. 24-10).

Chromaffin cells secrete two catecholamines, namely, epinephrine (adrenaline) and norepinephrine (noradrenaline). These peptide hormones are responsible for modifying heart rate, blood pressure, and tonic control of smooth muscle throughout the

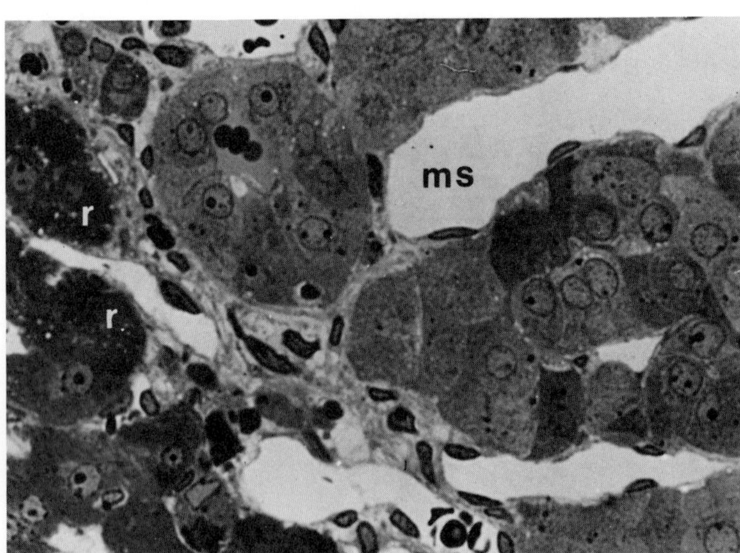

FIGURE 24-10 Light micrograph of the adrenal medulla. The cells of the medulla are arranged in irregular cords surrounding medullary sinusoids (ms). They are large and stain more lightly than the cells of zona reticularis (r).

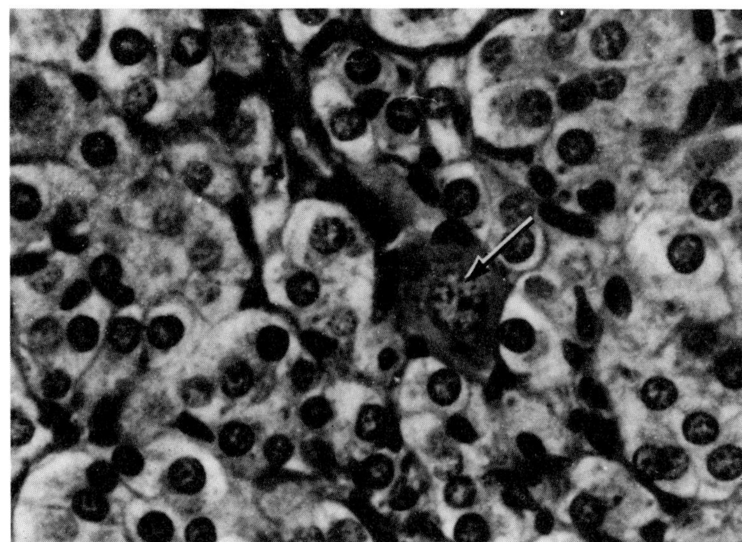

FIGURE 24-11 Light micrograph of the adrenal medulla. A large ganglion cell (arrow) that has a large vesicular nucleus with a prominent nucleolus is seen among the medullary cells. It has a rounded shape and highly basophilic cytoplasm.

body. In an EM, the chromaffin cells show catecholamine-containing granules. The epinephrine cells are more frequently found and have granules that are 200 nm in diameter. Norepinephrine cells are fewer in number than epinephrine cells and contain granules with a dense core. It is possible to distinguish the two types of cells that produce the two different hormones mentioned here. The chromaffin cells are often referred to as paraganglia or enterochromaffin cells.

A number of preganglionic sympathetic fibers enter the medulla where they synapse with a few ganglion cells that are found in the adrenal medulla. Ganglion cells are large, are basophilic, and have a rounded contour (Fig. 24-11). As in other ganglion cells, the nuclei are large and vesicular with prominent nucleoli. These cells regulate the release of epinephrine during stress situations or when blood sugar levels are lowered.

FLOW OF BLOOD THROUGH THE ADRENAL GLANDS

The suprarenal artery breaks into small branches as it penetrates the capsule, and its branches pour into the sinusoids of the cortex (Fig. 24-12). The blood flows through the

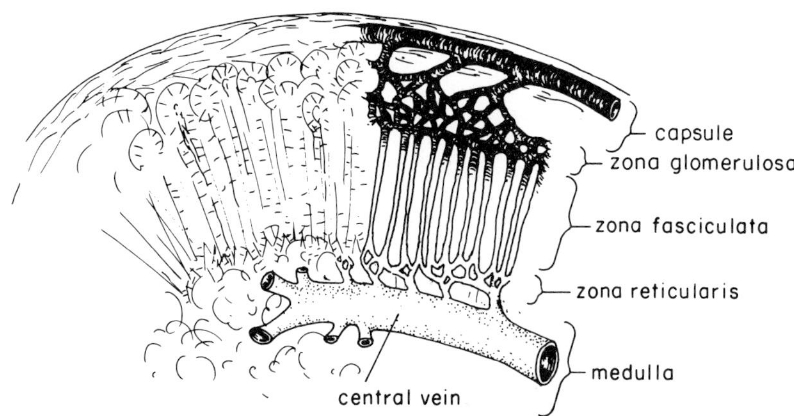

FIGURE 24-12 Diagram illustrating the flow of blood through the adrenal gland.

sinusoidal space of the cortex, receiving corticosteroids secreted by its three zones. The blood enters the medulla where irregular sinusoidal vessels join to form small veins which, in turn, merge into the *central vein* of the medulla, which leaves the gland as the suprarenal vein.

THYROID GLAND

The thyroid gland is bilobed and located in front of the larynx just below the thyroid cartilage. The two lateral lobes are connected by a bridge of thyroid tissue called the *isthmus.* The organ develops from an epithelial invagination from the portion of the oral ectoderm where the foramen cecum is found in the adult.

A thin capsule covers the entire gland which is separated by delicate trabeculae and septa. In human beings, the gland weighs from 25 to 40 g. It is primarily responsible for the elaboration of thyroid hormones, *thyroxine* and *tri-iodothyronine,* that are concerned with the regulation of the metabolic rate of all body cells. The thyroid gland is stimulated by the pituitary product thyrotrophic hormone. Unlike other endocrine glands, however, the thyroid gland has developed a unique storage space for its hormones in the form of *thyroid follicles.* In other words, the thyroid cells produce the hormones which are secreted out of the cell into the follicular lumen, where they are stored in the form of *thyroid colloid.* During the mobilization phase of the hormones, the luminal content of the follicle is broken down, transported across the thyroid follicular cells, and released into the blood vessels surrounding the follicles.

STRUCTURE OF THYROID FOLLICLES

The functional unit of the thyroid gland is the thyroid follicle. Each thyroid follicle consists of a layer of cuboidal cells surrounding a central lumen (Fig. 24-13). The follicles are generally spherical, are surrounded by a basement membrane, and contain a colloid substance in the follicular lumen. Some thyroid glands have thyroid follicles (Fig. 24-14) that are much more distended and thus contain more colloid. Due to the distension of the follicles, the epithelial cells appear flat-

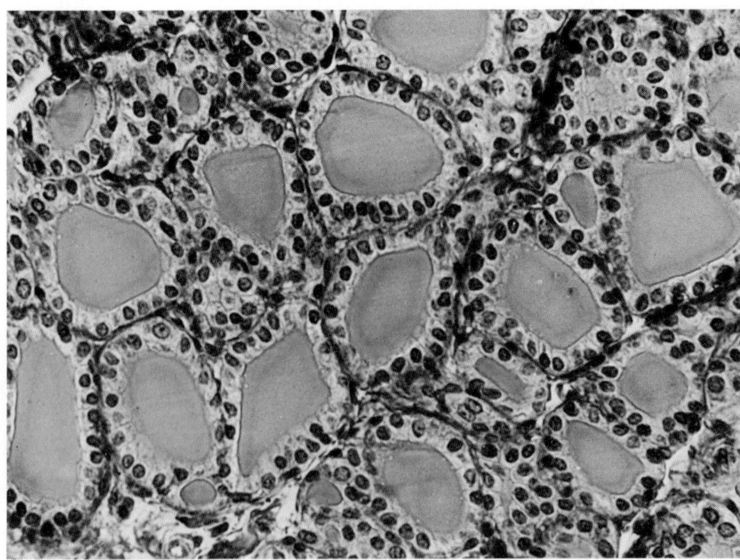

FIGURE 24-13 Light micrograph of thyroid follicles. Each follicle has a layer of cuboidal cells surrounding a central lumen and contains the thyroid colloid.

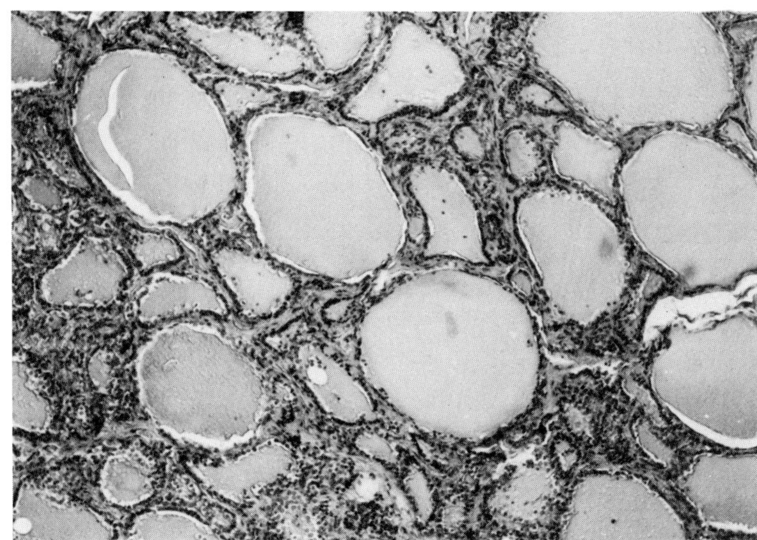

FIGURE 24-14 Light micrograph of a thyroid gland that has distended follicles containing a larger amount of colloid.

tened (Fig. 24-15). The wide variation in appearance of the follicular epithelium may reflect the difference in functional state of the gland.

The RER is abundant, and the Golgi complex is extensively developed during the active synthetic phase of the gland. Figure 24-16 shows the luminal aspect of a follicular cell which is rich in RER. Note the presence of small villous projections along the luminal surface facing the colloid.

SYNTHESIS AND MOBILIZATION OF THYROID HORMONES

The thyroxine molecule is produced when two tyrosine molecules are put together in the form of iodinated tyrosines. The first step in the production of thyroxine is the iodination of tyrosine to form monoiodotyrosine and di-iodotyrosine. These iodized tyrosines form tri-iodothyronine and tetraiodothyronine (thyroxine), which are the active hormones. The *tri-iodothyronine* is the more ac-

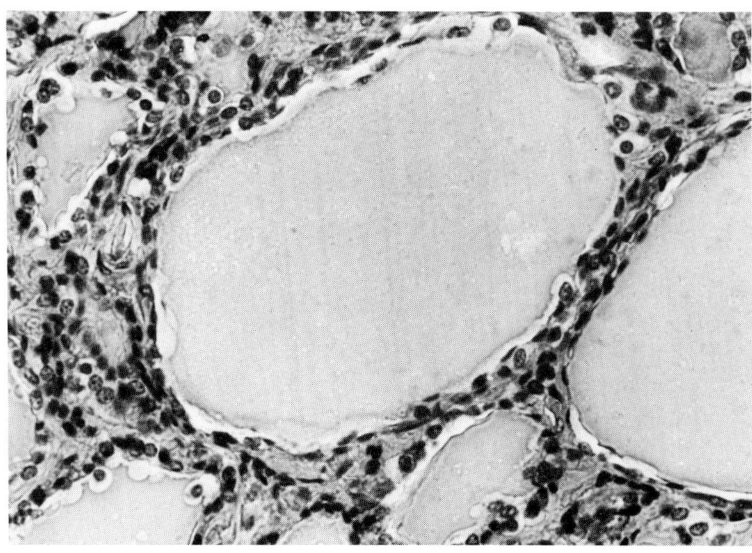

FIGURE 24-15 Distended thyroid follicles viewed at higher magnification. The flattened appearance of the epithelial cells is due to the large amount of colloid.

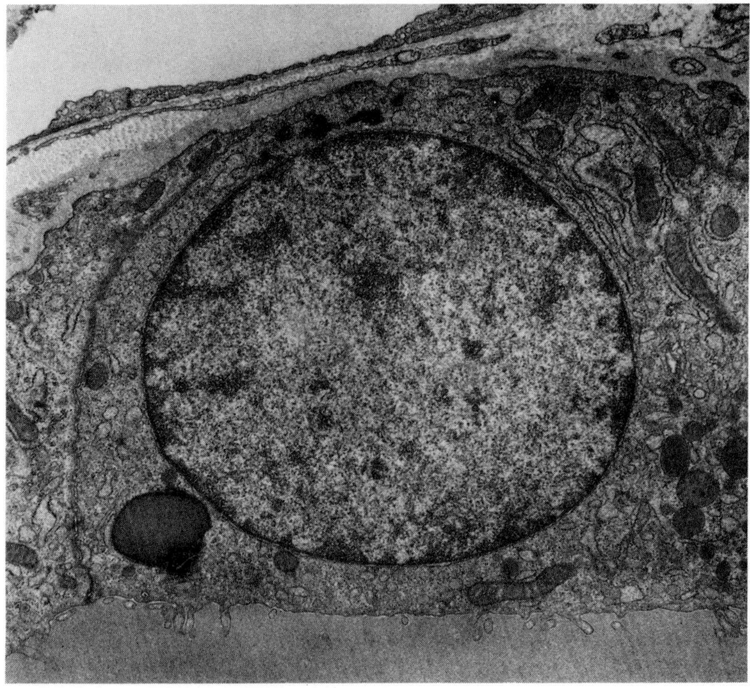

FIGURE 24-16 Electron micrograph of thyroid follicular cell. Note the presence of small villous projections along the luminal surface and the numerous profiles of RER in the apical cytoplasm. (*Courtesy of D. McDonald and S. Wissig.*)

tive of the two. However, because of the much larger quantity of tetraiodothyronine, or thyroxine, the latter is primarily responsible for thyroid hormone activity throughout the body. Following secretion into the follicular lumen, the active hormones become part of the thyroid colloid where they are stored as *thyroglobulins*. In addition to thyroid hormones, the colloid contains a number of enzymes, notably cathepsin, which catalyze the hydrolysis of a thyroglobulin during the mobilization phase. The precise mechanism by which cells become activated for mobilization is not known, although it is known

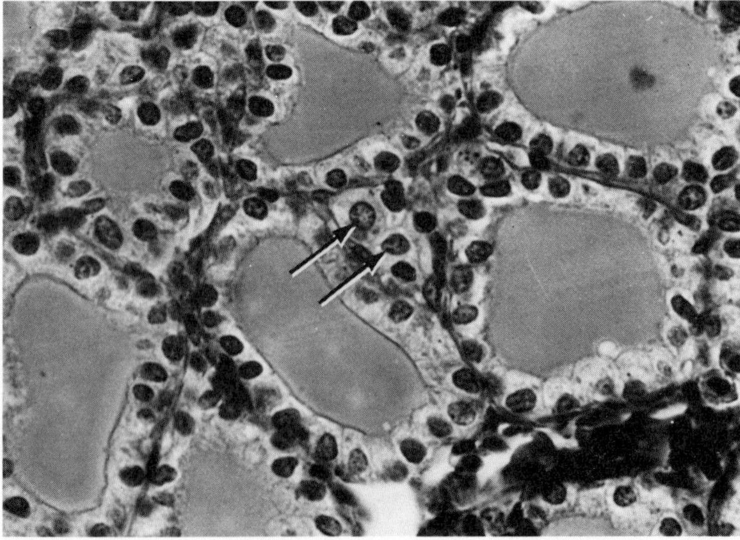

FIGURE 24-17 Light micrograph of thyroid follicles. Parafollicular cells (arrows) are located externally to the follicular cells.

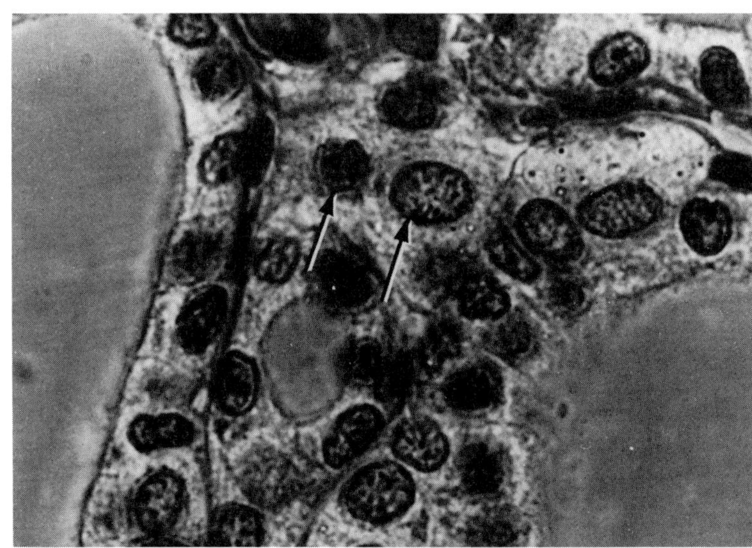

FIGURE 24-18 Parafollicular cells (arrows) seen at higher magnification. They are larger than the follicular cells and have less dense cytoplasm.

that TSH from the hypophysis has a direct effect upon discharge and synthesis of thyroxine. Once released, tri-iodothyronine and thyroxine are carried away by the bloodstream where they are present in association with carrier proteins.

PARAFOLLICULAR CELLS AND CALCITONIN

As the name implies, parafollicular cells, often called C or clear cells, are external to thyroid follicles (Figs. 24-17 and 24-18). They are somewhat larger in size and have a less dense cytoplasm than the follicular cells. Recently, it has been shown that parafollicular cells produce *calcitonin*, or thyrocalcitonin, a hormone which blocks osteoclastic activity and promotes osteoblastic functions, thereby decreasing the circulating calcium level. Since calcitonin is a peptide hormone, the parafollicular cells synthesizing it have a cytoplasm similar to other peptide-producing cells. In EM (Fig. 24-19), the parafollicular cells show a large number of secretory granules. Calci-

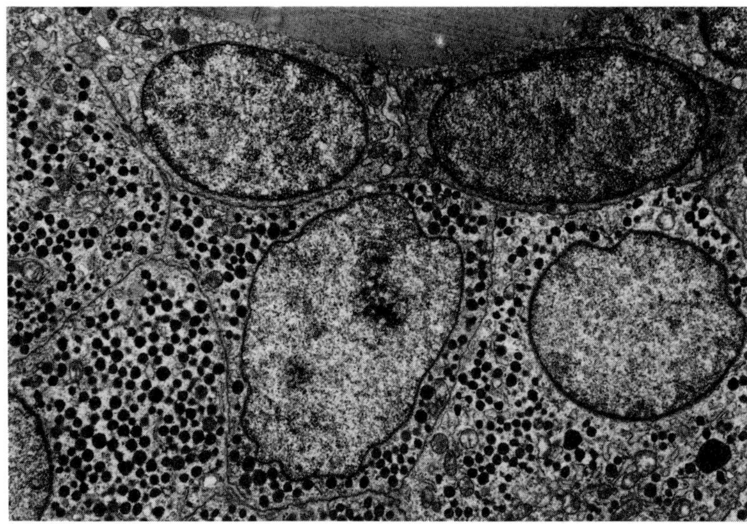

FIGURE 24-19 Electron micrograph of parafollicular cells. These cells contain a large number of secretory granules. (*Courtesy of D. McDonald and S. Wissig.*)

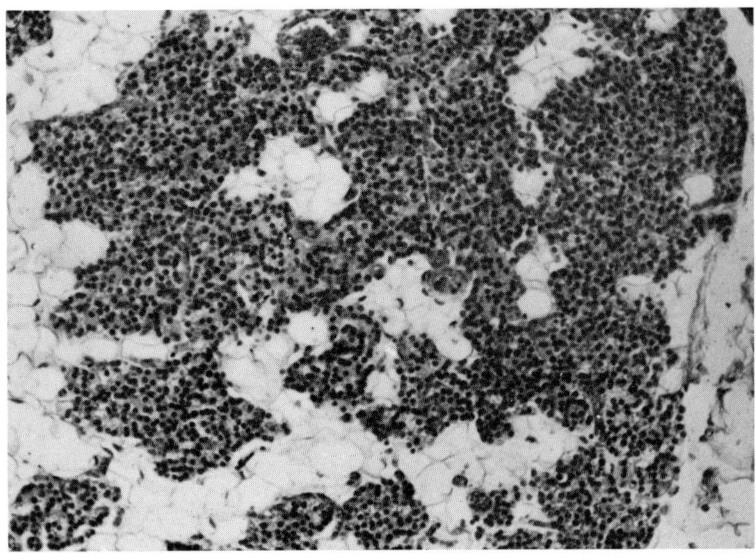

FIGURE 24-20 Light micrograph of the parathyroid gland.

tonin counters the effect of parathormone, which raises the blood calcium level. Both thyroxine and calcitonin gain entry into the bloodstream via the capillaries in the extrafollicular connective tissue. For this reason, the thyroid gland has, as have other endocrine organs, a rich supply of capillaries forming an extensive network around each follicle.

PARATHYROID GLAND

The parathyroid gland consists of two pairs of small ovoid bodies located on the posterior surface of the thyroid. This reflects the separate embryonic origins of the gland, which begins from the third and fourth branchial arches on each side. Their total weight might be as much as 0.3 g. Each parathyroid body

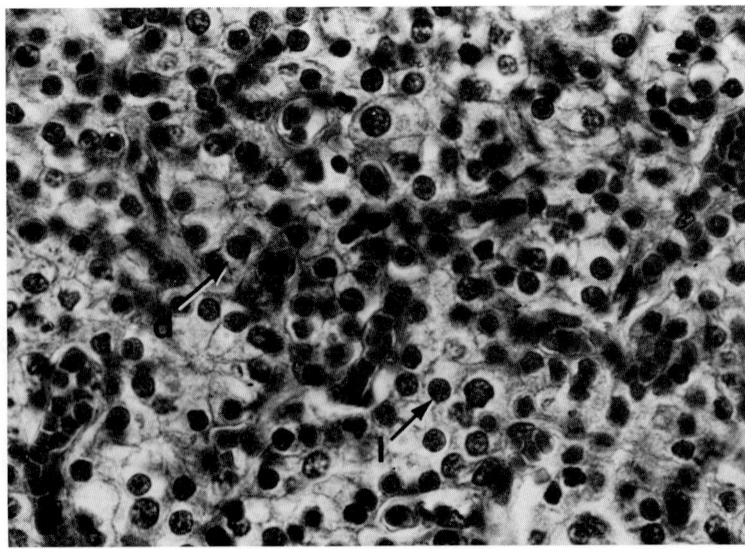

FIGURE 24-21 Parathyroid chief cells. Two subpopulations can be identified, namely, dark (d) and light (l) cells.

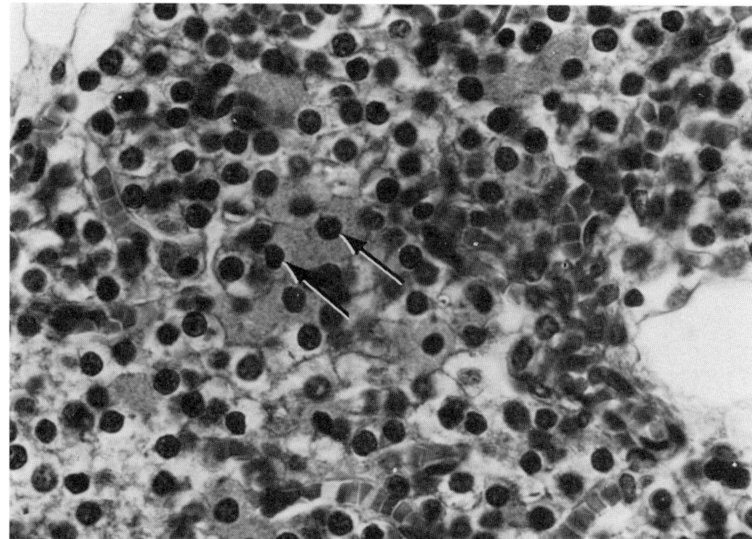

FIGURE 24-22 Light micrograph of the parathyroid. A few oxyphils (arrows) are present among the chief cells. Compare cell size, staining characteristics, and nuclear appearance of the chief cells and the oxyphil cells.

is surrounded by a connective tissue capsule which keeps it separate from the thyroid, yet firmly attached to it. Delicate connective tissue septa divide the parathyroid into obscure lobules, which are further subdivided into cords of cells. Figure 24-20 is a section of a parathyroid gland. The field shows an infiltration of adipose cells into the parenchyma, which is a common feature among older individuals.

Two types of cells make up the parathyroid parenchyma, namely, chief cells and oxyphils. *Chief cells* are more numerous, smaller, and more angular than oxyphils. They contain small, rounded nuclei often with vesicular chromatin. Two subpopulations of the chief cells have been recognized, namely, dark (d) and light (l) cells, as shown in Fig. 24-21. Such distinctions of the chief cells have been related to different levels of metabolism among them. The dark cells are regarded to be the active chief cells, as they contain numerous small argyrophilic granules, a large Golgi apparatus, and enlarged mitochondria. Few glycogen granules are found in the dark chief cells. Chief cells are PAS-positive.

In contrast, *oxyphils* are large and rounded with an eosinophilic cytoplasm and a small, dense nucleus. Oxyphils are readily distinguished even at medium magnification (Fig. 24-22). In actual slides, the cytoplasm is intensely eosinophilic. They may be present as islets of several oxyphils or singly among the chief cells. Oxyphils are usually found in patches along the periphery of the parathyroid glands. The electron microscope reveals an abundance of mitochondria. The function of oxyphils is unknown.

Parathormone, the product of active chief cells, induces the action of osteoclasts, involving the resorption of bone and the subsequent release of calcium stored in it. Although the calcium level of the blood and body fluids is influenced by such systemic factors as dietary intake and functioning of the intestine and kidneys, parathormone, in conjunction with calcitonin, is essential in the maintenance of calcium homeostasis in blood and body fluids. The maintenance of the calcium level by parathormone is extremely important with respect to the overall metabolic function of body cells since all cells require calcium.

REVIEW SECTION

1. Figures 24-23 and 24-24 represent sections of the same endocrine gland prepared in different ways. Identify the gland. Name the layers that are characteristic of the region of the gland shown.

The figures show the organization of the adrenal cortex. The cortex has three zones, namely, the zone glomerulosa, zona fasciculata, and zona reticularis. All cells have lipid granules in their cytoplasm.

2. Identify the portion of this endocrine organ appearing in Fig. 24-25. What is its function?

The portion of the organ shown here represents the adrenal medulla. The very last cells of the zona reticularis are partially seen at the upper right margins of the field. Note the irregular-appearing cords or large medullary cells and sinusoidal vessels. The parenchymal cells of the adrenal medulla produce epinephrine and norepinephrine.

3. Name the cytoplasm organelle filling much of the ground cytoplasm in Fig. 24-26. In what type of cell do you expect to find this structure?

This field shows a region of SER from a cell in the adrenal cortex. The rich SER is one of the cyto-plasmic characteristics of all steroid-secreting

FIGURE 24-23

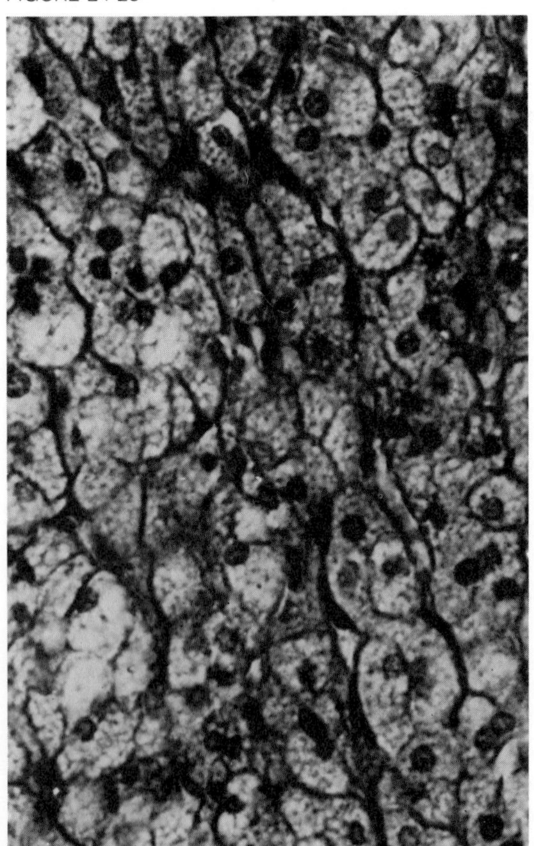

FIGURE 24-24

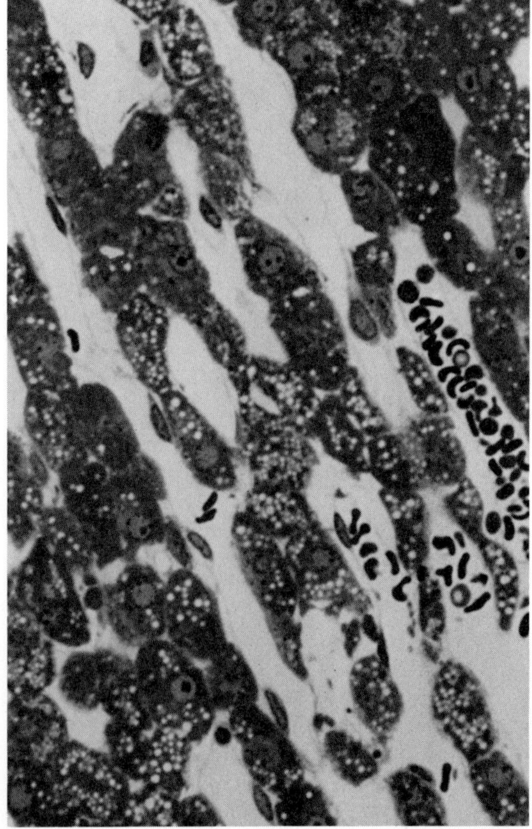

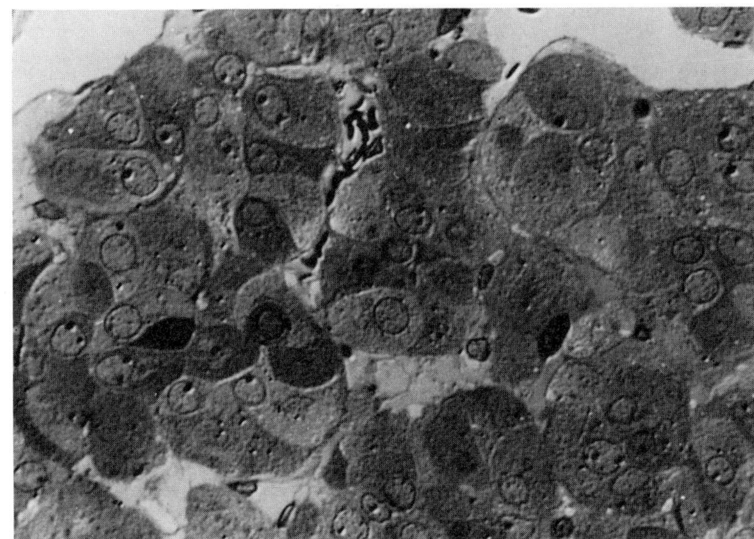

FIGURE 24-25

cells. Note that the cristae in the mitochondria are made up of fingerlike projections as shown by their cross-sectional appearance.

4. Figure 24-27 is a section of an endocrine organ. Identify the organ. What is the moderately dark material present in the circular areas?

The figure shows a section of thyroid follicles. The lumen of the follicles contains thyroid colloid, which stores the thyroid hormones thyroxine and tri-iodothyronine.

5. Figure 24-28 is a section of an endocrine gland. Numerous similar cells are seen throughout the field. The letter (a) indicates one of several larger cells with dark nuclei. Identify the gland. What are the names of the two principal cell populations making up the gland?

The parathyroid gland is composed of two cell types, the chief cell and the oxyphil (a). The chief cells produce parathormone. Chief cells are divided into active dark cells and less active light cells. The large eosinophilic cells with small, dark nuclei are oxyphils.

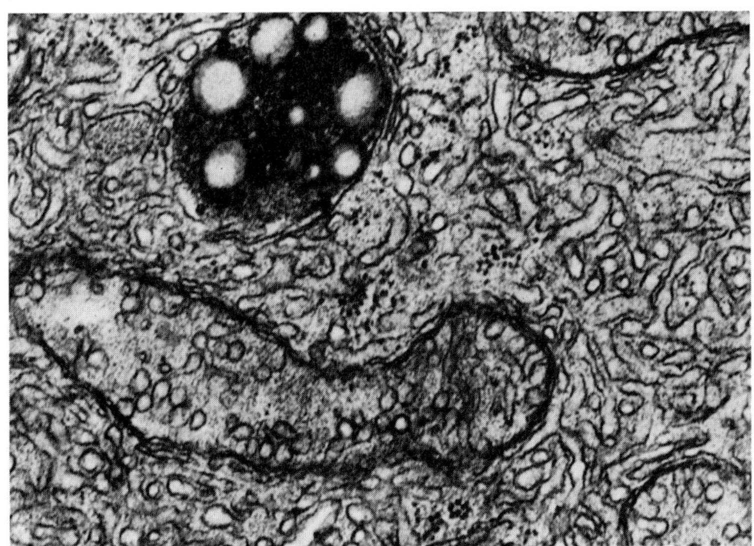

FIGURE 24-26

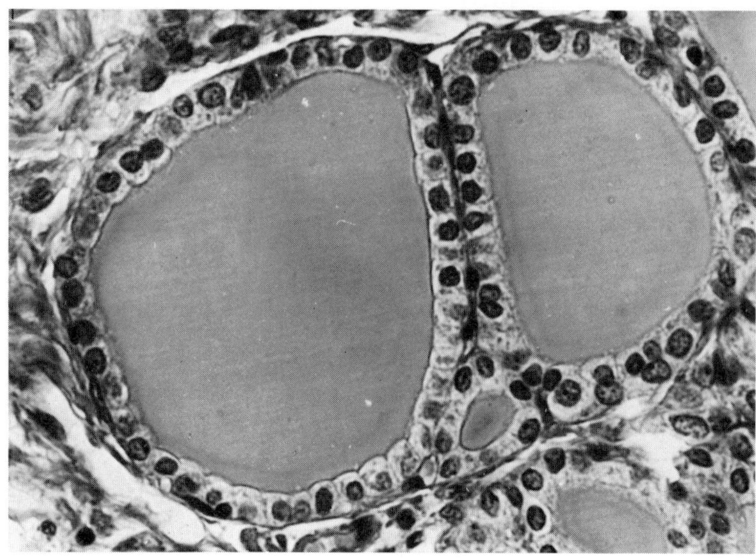

FIGURE 24-27

6. Figure 24-29 shows a portion of an endocrine gland. Name the portion shown. What does it produce?

The adrenal cortex is shown with the capsule at the left. The zona glomerulosa appears often as a circular arrangement of cells and hence its name. A portion of the zona fasciculata is present in the middle which continues as the zona reticularis on the right. Note the density of cells in the zona reticularis. The zona glomerulosa produces mineralocorticoids. Zonae fasciculata and reticularis produce glucocorticoids.

BIBLIOGRAPHY

Bloom, W., and D. W. Fawcett: *A Textbook of Histology,* 10th ed., Saunders, Philadelphia, 1975.

Christy, N. P. (ed.): *The Human Adrenal Cortex,* Harper & Row, New York, 1971.

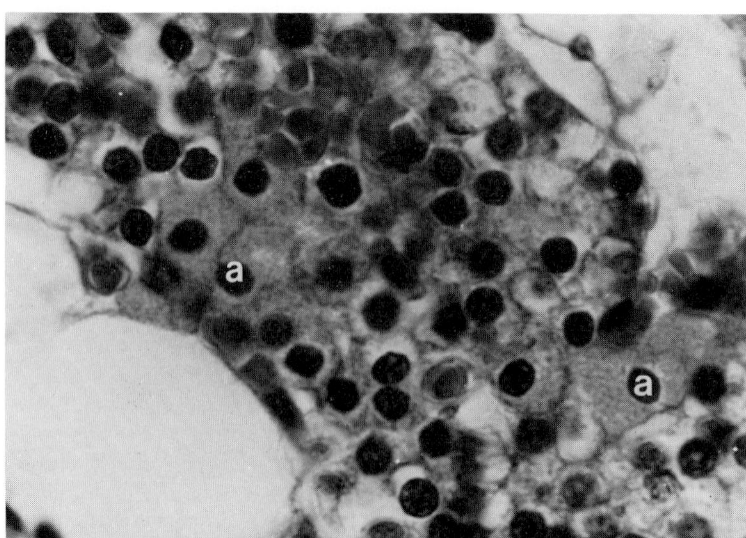

FIGURE 24-28

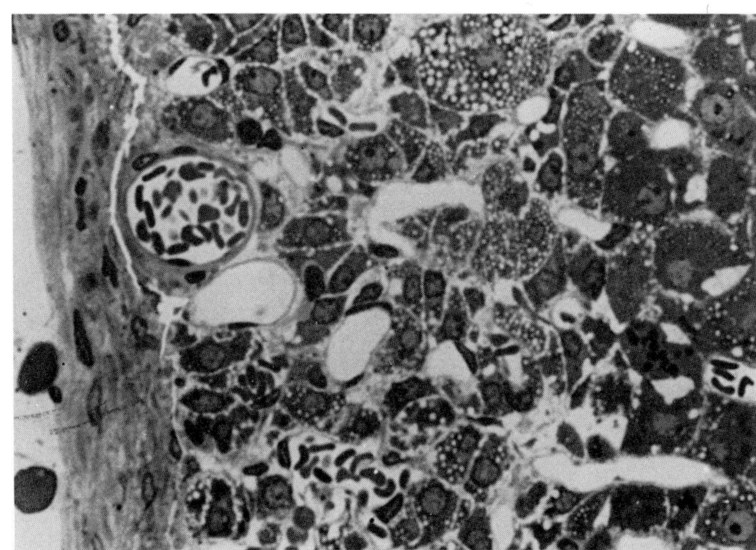

FIGURE 24-29

Coupland, R. E.: *The Natural History of the Chromaffin Cell,* Longmans, London, 1965.

Greep, R. O., and R. V. Talmage (eds.): *The Parathyroids,* Charles C Thomas, Springfield, Ill., 1961.

Hazard, J. B., and D. E. Smith (eds.): *The Thyroid,* Williams & Wilkins, Baltimore, 1964.

Lupulescu, A., and A. Petrovici: *Ultrastructure of the Thyroid Gland,* Williams & Wilkins, Baltimore, 1968.

Taylor, S. (ed.): "Calcinotin," *Proceedings of the Symposium on Thyro-Calcitonin and the C Cells,* Heinemann, London, 1968.

Weiss, L., and R. O. Greep: *Histology,* 4th ed., McGraw-Hill, New York, 1977.

ADDITIONAL READINGS

Al-Lami, E.: "Light and Electron Microscopy of the Adrenal Medulla of Macaca Mulatta Monkey," *Anat. Rec.* **164:**317–322 (1969).

Andros, G., and S. H. Wollman: "Autoradiographic Localization of Radio-Iodide in the Thyroid Gland of the Mouse," *Am. J. Physiol.* **213:**198–208 (1967).

Bennett, H. S.: "Cytological Manifestations of Secretion of the Renal Medulla in the Cat." *Am. J. Anat.* **69:**333–387 (1941).

Brown, W. J., L. Barajas, and H. Latta: "The Ultrastructure of the Human Adrenal Medulla: With Comparative Studies of White Rat," *Anat. Rec.* **169:** 173–184 (1971).

Coupland, R. E., and B. S. Weakley: "Electron Microscopic Observation on the Adrenal Medulla and Extra-Adrenal Chromaffin Tissue of the Postnatal Rabbit," *J. Anat.* **106:**213–231 (1970).

Ekholm, R., and V. Strandberg: "Thyroglobulin Biosynthesis in the Rat Thyroid," *J. Ultrastruct. Res.* **20:**103–110 (1967).

———, and L. E. Ericson: "The Ultrastructure of the Parafollicular Cells of the Rat," *J. Ultrastruct. Res.* **23:**378–402 (1968).

Idelman, S.: "Ultrastructure of the Mammalian Adrenal Cortex," *Int. Rev. Cytol.* **27:**181–281 (1970).

McCann, S. M. (ed.): *Endocrine Physiology,* Butterworth, London, 1974.

Munger, B. L., and S. I. Roth: "The Cytology of the Normal Parathyroid Glands of Man and Virginia Deer," *J. Cell Biol.* **16:**379–400 (1963).

Nadler, N. J., B. A. Young, and C. P. Leblond: "Elaboration of Thyroglobulin in the Thyroid Follicle," *Endocrinology* **74:**333–354 (1964).

Nunez, E. A., J. P. Whalen, and L. Krook: "An Ultrastructural Study of the Natural Secretory Cycle of the Parathyroid Gland of the Bat," *Am. J. Anat.* **134:**458–480 (1972).

Whur, P., A. Herscovics, and C. P. Leblond: "Radioautographic Visualization of the Incorporation of Galactose-^{3}H and Mannose-^{3}H by Rat Thyroids *in vitro* in Relation to the Stages of Thyroglobulin Synthesis," *J. Cell Biol.* **43:**289–311 (1969).

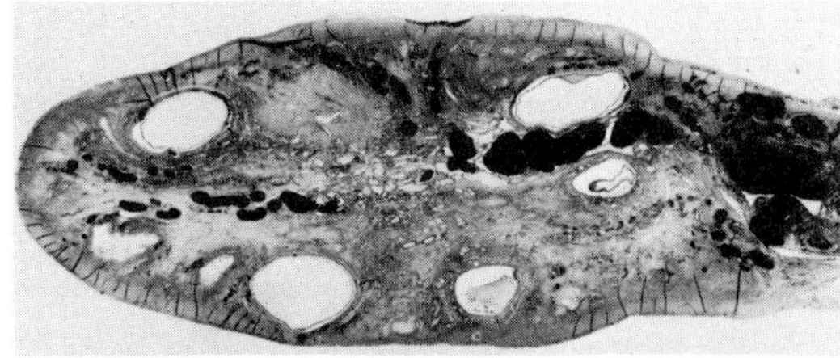

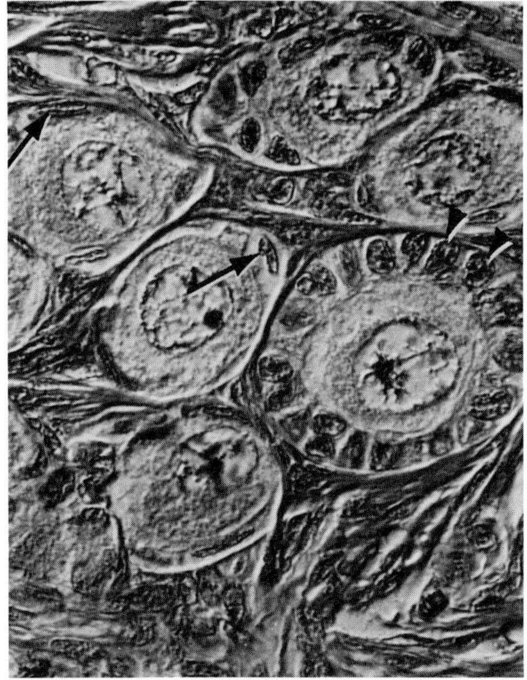

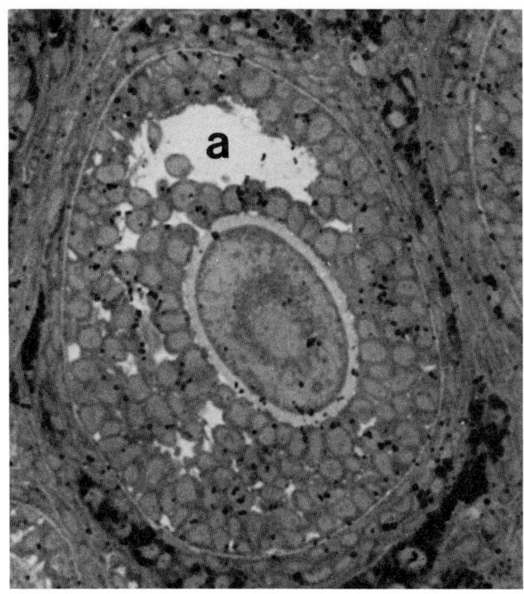

25 OVARY AND OVARIAN CYCLE

OBJECTIVES

Upon completion of this chapter, the student will be able to:

1 Identify the location and structure of the following parts of the ovary:
(a) cortex (b) medulla (c) mesovarium (germinal epithelium) (d) tunica albuginea (e) follicles

2 Describe the sequence of events in the differentiation of a primary follicle into a mature graafian follicle by identifying the:
(a) proliferation of follicular cells (b) production of the antrum and its contents (c) size of the ovum (d) formation of the connective tissue capsule

3 Identify the structure and function (where applicable) of the major parts of a developing follicle as follows:
(a) antrum (b) liquor folliculi (c) cumulus oophorus (d) corona radiata (e) theca interna (f) theca externa

4 List the hormones involved in the ovarian cycle and tell where they are produced, explain how these hormones influence the developing follicle, and describe how hormones produced by the developing follicle influence the level of hormones secreted by the anterior pituitary.

5 Identify and define the following structures:
(a) atretic follicle (b) corpus luteum (c) corpus albicans (d) interstitial cells of the ovary

6 Identify the structure and function of:
(a) theca externa (b) theca interna (c) granulosa lutein cell (d) theca lutein cell

7 Identify in an electron micrograph the fine structure of granulosa and luteal cells.

8 Define the molecular mechanism of secretion of gonadotropic hormones in general terms.

The female reproductive system (Figs. 25-1 and 25-2) consists of a pair of ovaries, a pair of uterine tubes (oviducts, also named fallopian tubes), a uterus, a vagina, and external genitalia. During the menstrual cycle and pregnancy of a sexually mature female, structural and functional changes occur, primarily in the ovary and uterus, that are regulated by complex hormonal interactions among several related hormones. They include pituitary gonadotropic hormones and steroid hormones produced within the gonads. Gonadotropic hormones are peptides in their chemical composition.

STRUCTURE OF THE OVARY

The bilateral ovaries are similar in development to the testes and contain the germ cells or ova and produce endocrine secretions under the influence of hypophyseal gonadotropic hormones. The ovary is covered by a special kind of cuboidal epithelium, *germinal epithelium,* which is continuous with the peritoneal fold (*mesovarium*) that supports the ovary as part of a ligament system. Cells

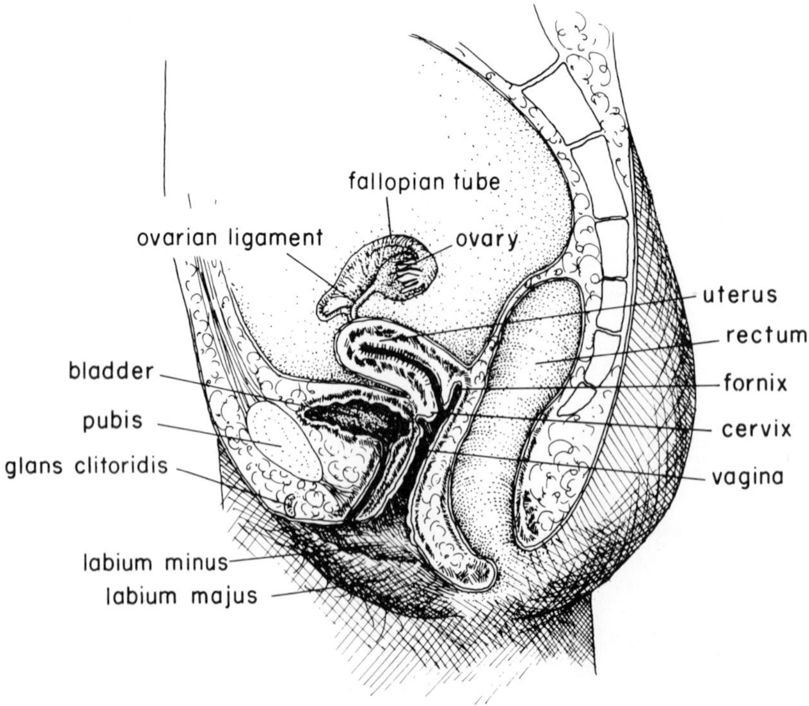

FIGURE 25-1 Schematic median section of the female pelvis.

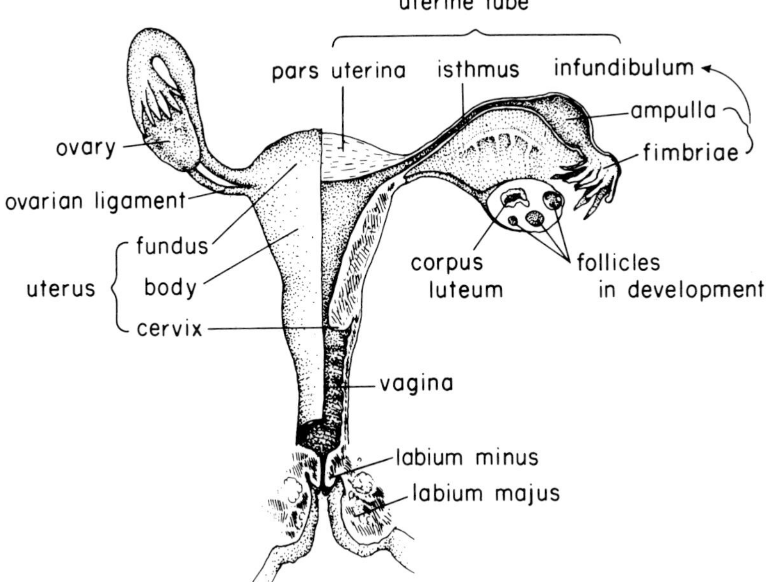

FIGURE 25-2 Diagram of the female reproductive system.

from the *germinal epithelium* are thought to sink into the parenchyma of the ovary to form primary ova during development. The epithelium covering the outer surface of the ovary in the postnatal female is best described as the ovarian surface epithelium. The ovary (Figs. 25-3 and 25-4) consists of two zones which are not clearly demarcated, the medulla and the cortex.

The *medulla* carries large blood vessels, lymphatics, and nerves in a loose connective tissue. The ovary appearing in Fig. 25-3 has been injected with a dye which delineates the vascular architecture in black. The broad outer *cortex* contains numerous primary ova, or oocytes, and ovarian follicles in various stages of development, all within a cellular connective tissue stroma. Surrounding the cortex just underneath the surface epithelium (e), there is a distinct fibrous connective tissue layer, *tunica albuginea* (ta in Fig. 25-5).

PRIMARY FOLLICLES

As many as 400,000 primary ova "sink" into the cortical stroma from the germinal epithelium in development. Cells of this surface epithelium are cuboidal in appearance and have fine microvillous projections (Fig. 25-6, arrow). Figure 25-7 is a scanning EM of the surface epithelium showing the numerous cytoplasmic projections. In the sexually mature female, there is a chance to ovulate only

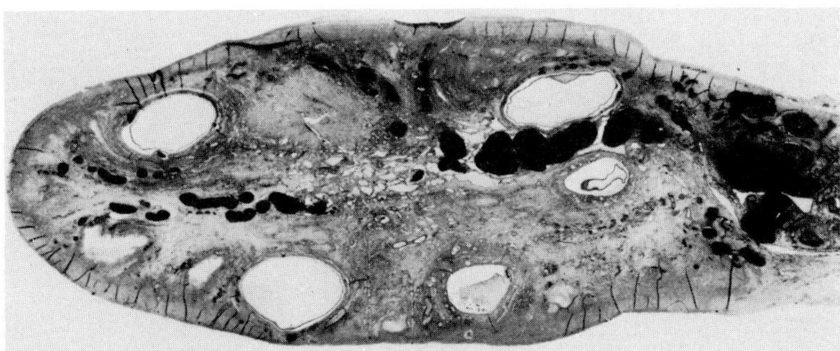

FIGURE 25-3 Macrophotograph of a section through an ovary.

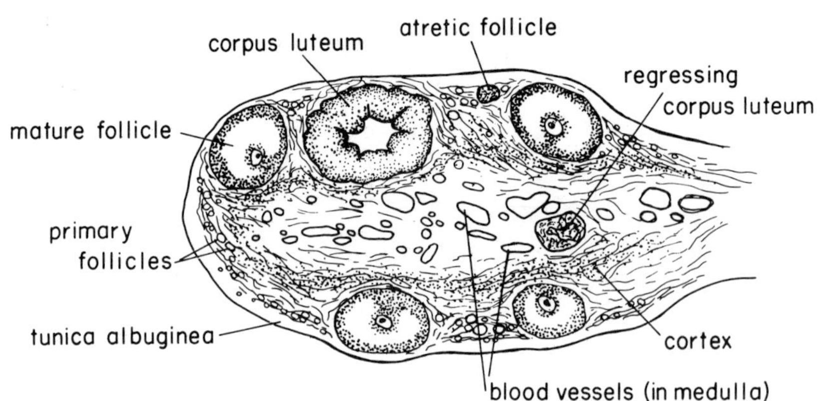

FIGURE 25-4 Schematic section through an ovary. Compare with Fig. 25-3.

about 400 times between puberty and menopause, thus creating a huge biological backup system. The primary ova undergo degeneration throughout life. Even prior to puberty, follicular *atresia* occurs as well as a continuous follicular development of cells surrounding certain ova, thus providing a rather smooth transition into sexual maturity.

Primary follicles form a thick layer immediately beneath the tunica albuginea in the periphery of the cortex (Fig. 25-5). Each is about 40 μm in diameter and consists of a large, round oocyte surrounded by a single layer of flattened epithelioid cells (Fig. 25-8, arrows). The oocyte has a large, somewhat eccentric vesicular nucleus with a prominent nucleolus.

OVARIAN CYCLE

MATURATION OF FOLLICLES

Corresponding to the monthly rhythmic changes in the rates of secretion of pituitary gonadotropic hormones, changes occur in the ovary and associated organs. These changes result in the release of a single mature ovum from the ovary and the preparation of the lining of the uterus for implantation of the

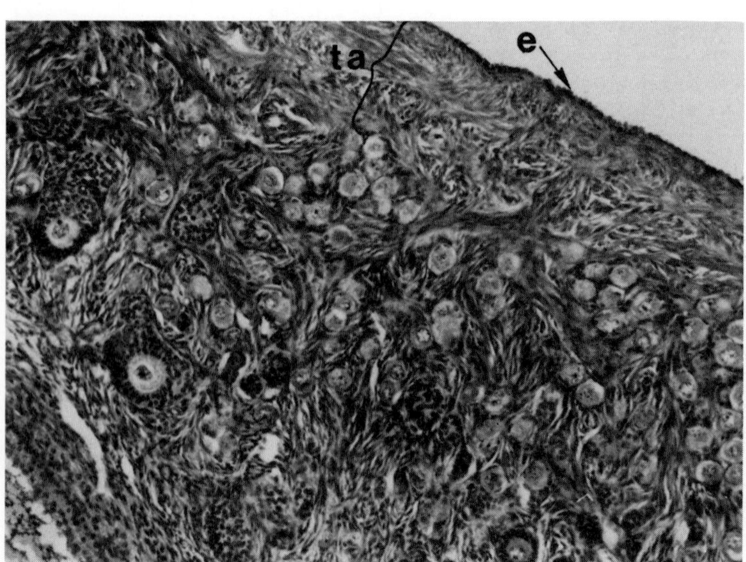

FIGURE 25-5 Light micrograph of the ovarian cortex. A dense connective tissue layer, tunica albuginea (ta), that is subjacent to the surface epithelium (e) encloses the cortex. Note the numerous primary follicles immediately beneath the tunica albuginea.

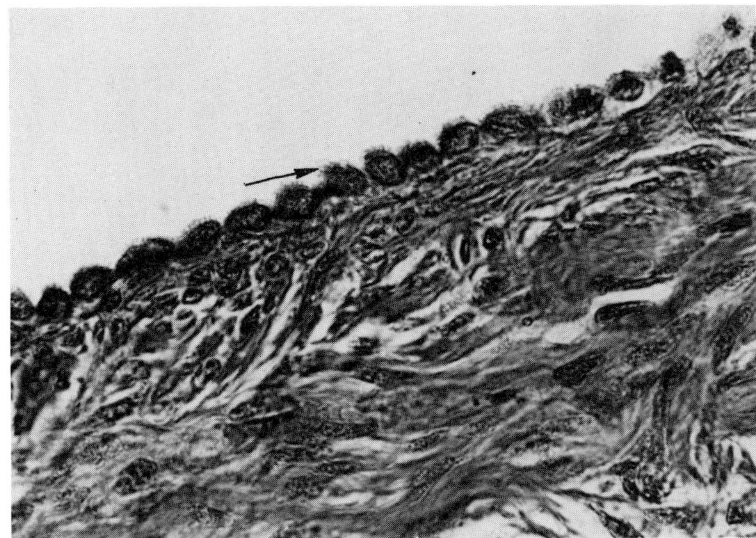

FIGURE 25-6 Light micrograph of the surface epithelium. The cells are cuboidal and have fine microvillous projections (arrow).

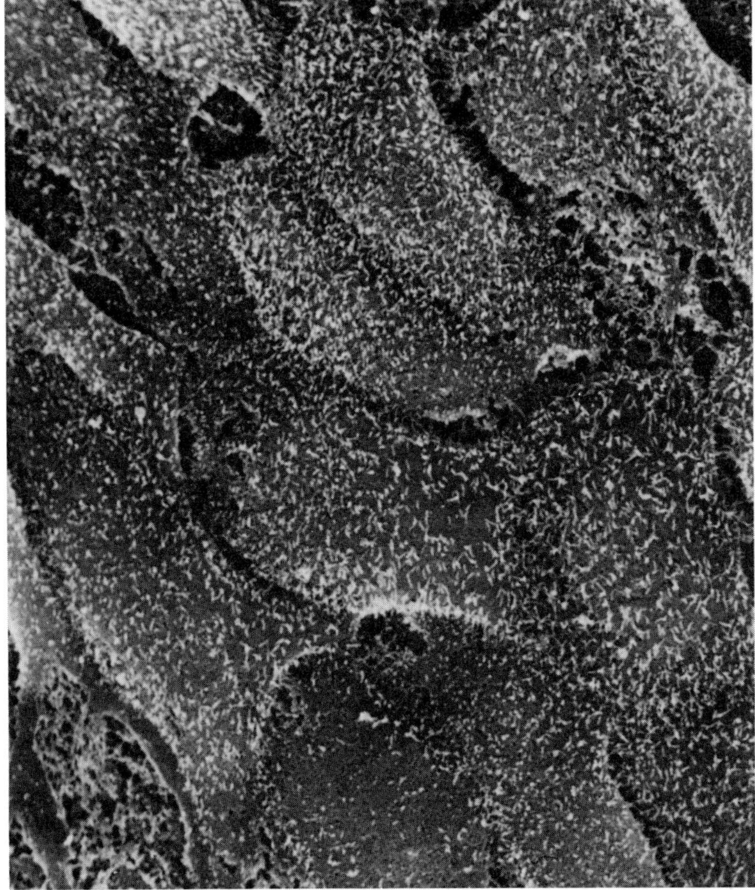

FIGURE 25-7 Scanning electron micrograph of ovarian surface epithelium.

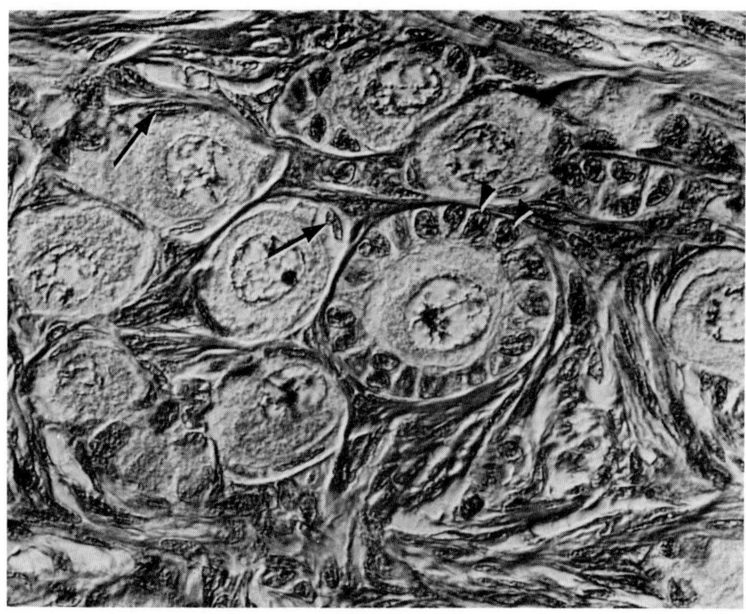

FIGURE 25-8 Phase micrograph of the ovarian cortex. The primary follicles present are surrounded by a single layer of flattened epithelioid cells (arrows). As the follicles mature, these cells become cuboidal (arrowheads).

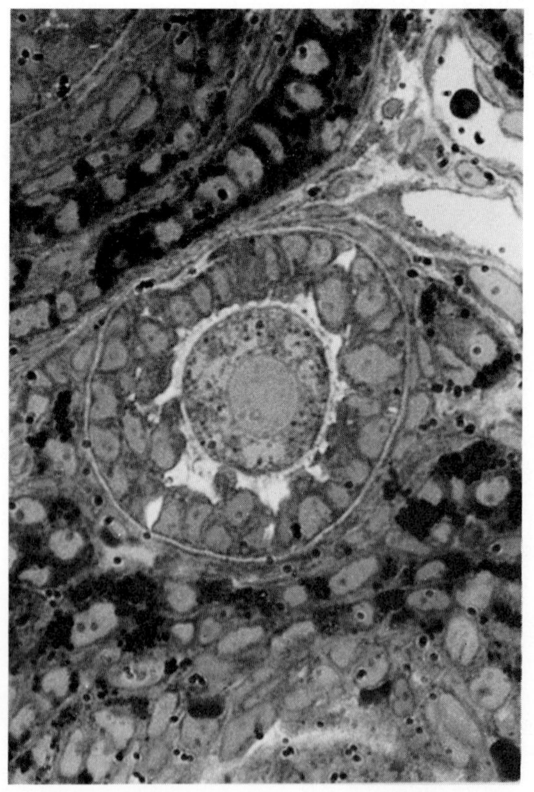

prospective fertilized ovum. The ovarian changes involve the growth and differentiation of a primary follicle into a mature or graafian follicle. This process is characterized (1) by the proliferation of follicular or *granulosa* cells, (2) by formation of a capsule of connective tissue cells surrounding the growing follicle, and (3) by the growth in size and maturation of the ovum.

Initially there is an increase in size of the follicle as the flattened follicular cells become cuboidal (Fig. 25-8, arrowheads) and then divide actively to form stratified layers around the ovum (Fig. 25-9). With continuing proliferation, irregular fluid-filled spaces appear among the granulosa cells, eventually fusing to form a single crescent-shaped cavity, the *antrum* (a in Fig. 25-10). The fluid filling the antrum, *liquor folliculi,* is produced

FIGURE 25-9 A more mature follicle. The epithelioid cells form stratified layers around the ovum.

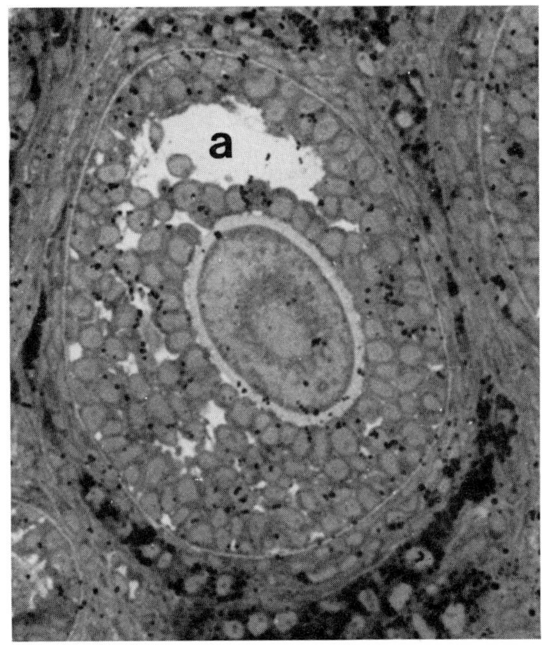

FIGURE 25-10 A further developed follicle. Fluid-filled spaces appear between the follicular cells and an antrum (a) is forming.

by the follicular cells and physically sustains the developing ovum. At this point, the follicle is sometimes referred to as a *secondary follicle*. With further cellular proliferation, the ovum, surrounded by follicular cells, is displaced to one side, forming a projection into the antrum called *cumulus oophorus*, or germ hill (Fig. 25-11).

Figure 25-12 is a diagram summarizing the rapid growth of the follicle. It should be noted that a thick glycoprotein coat called *zona pellucida* is produced by developing granulosa cells (zp in Fig. 25-13). In an EM the granulosa cells appear to be cuboidal to columnar at the periphery of the follicles (Fig. 25-14). The basement membrane separating the follicle and the surrounding connective tissue is prominent and often called *membrana granulosa* (mg in Fig. 25-13). The fine structure of granulosa cells demonstrates a large number of free ribosomes, a fair amount of RER profiles, small but numerous mitochondria, and occasional lipid droplets (l in Fig. 25-14).

A number of junctional specializations including gap junctions have been found between granulosa cells. Granulosa cells located near the antrum surface are more rounded in appearance and contain a cyto-

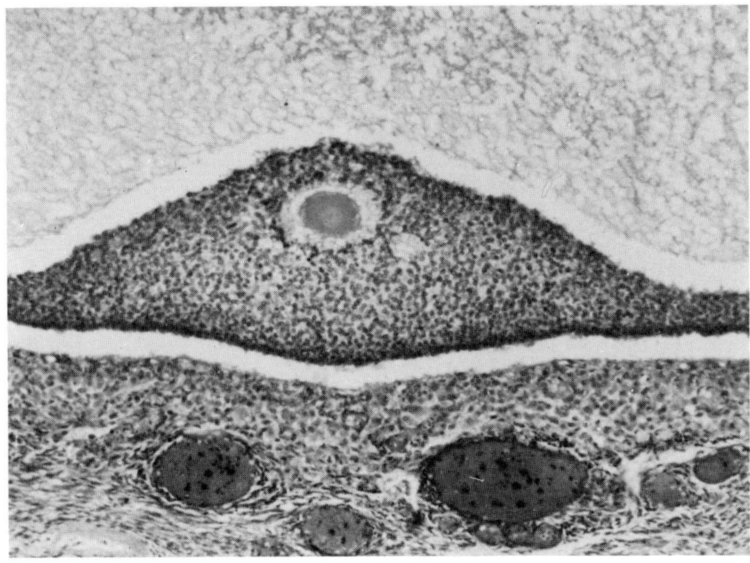

FIGURE 25-11 Light micrograph of cumulus oophorus.

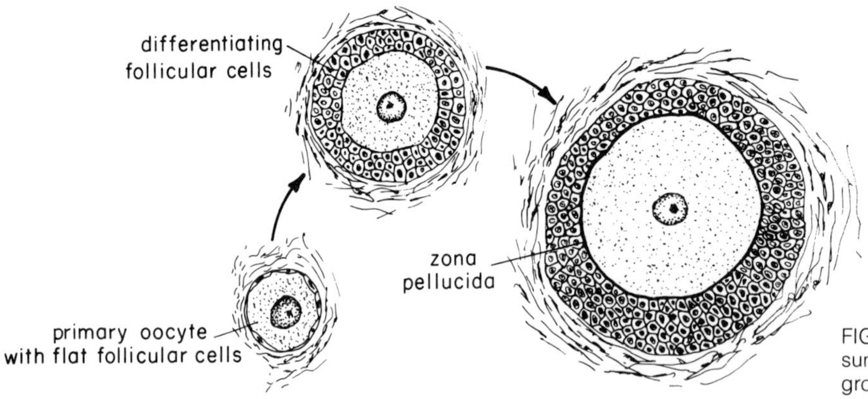

differentiating follicular cells

zona pellucida

primary oocyte with flat follicular cells

FIGURE 25-12 Diagram summarizing early follicular growth.

plasm which is full of free polyribosomes. Many of them show one or two prominent nucleoli (Figs. 25-14 and 25-15). These cells are known to divide during follicular maturation, whereas the aforementioned peripheral cells are known to respond to pituitary gonadotropins. The cells of the follicular layer directly surrounding the ovum become radially arranged to form the *corona radiata*. The follicle as a whole continues to enlarge until it reaches a diameter of 10 mm or more.

Concurrently, the connective tissue stroma around the follicle develops into a follicular sheath, *theca folliculi,* which is separated from the granulosa cells by a prominent basement membrane *membrana granulosa* (mg in Fig. 25-13). The theca folliculi is composed of two layers (Fig. 25-16): an inner vascular layer, *theca interna* (i), and an outer fibrous layer, *theca externa* (e). Although both layers are derived from the connective tissue, the cells of theca interna are epithelioid and have cytologic characteristics resembling cells in other steroid-secreting glands. As shown in the EM in Fig. 25-17, thecal cells (t) are large and spindle-shaped, or polyhedral, with a flat-to-ovoid nucleus and a cytoplasm containing lipid droplets.

The cristae of their mitochondria are tubular, rather than shelflike, as in other

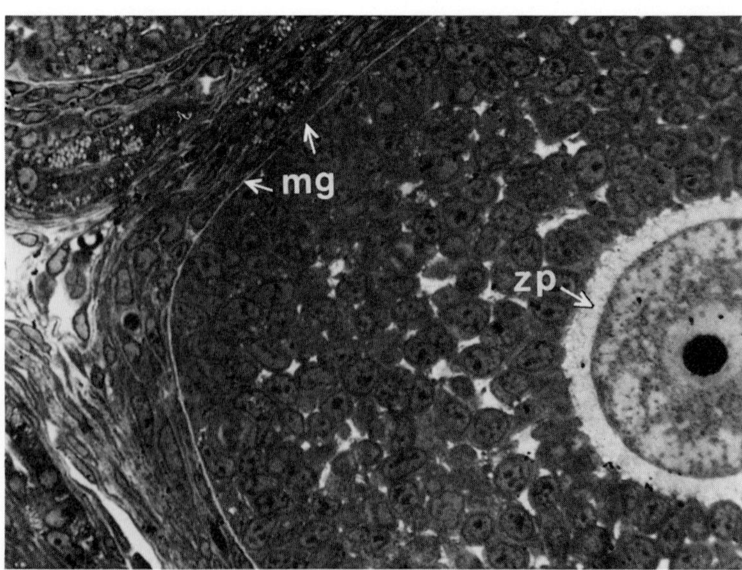

FIGURE 25-13 Portion of a growing follicle. A thick glycoprotein coat on the ovum, the zona pellucida (zp), is produced by developing granulosa cells. The basement membrane surrounding the follicle is thick and is called *membrana granulosa* (mg).

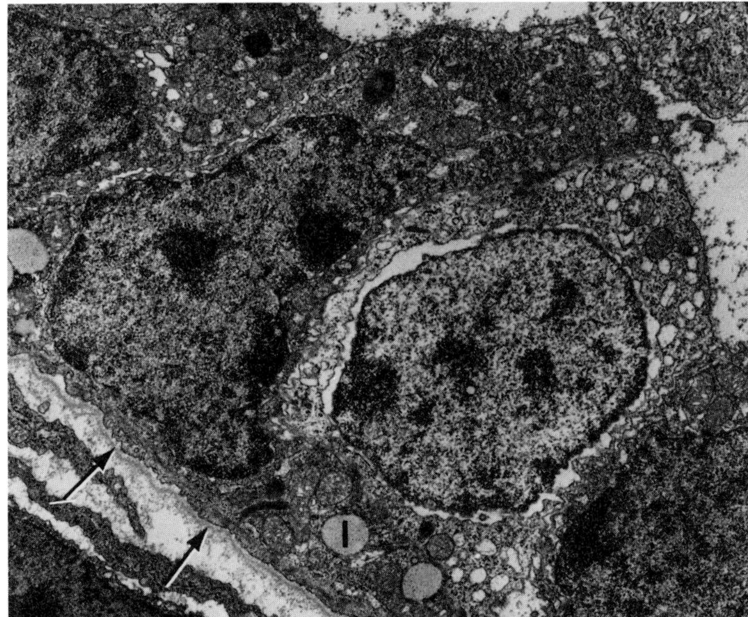

FIGURE 25-14 Electron micrograph of peripheral granulosa cells. These cells are separated from the surrounding connective tissue by a distinct basal lamina (arrows). They have a large number of free ribosomes, profiles of RER, small but numerous mitochondria, and lipid droplets (l) in their cytoplasm. Note the prominent nucleoli in the nuclei.

steroid-producing cells (Fig. 25-18). Thus, it is likely these cells are involved in the production of the female sex hormones, estrogens, which are necessary for the development and maintenance of secondary sexual characteristics. This endocrine function is consistent with the presence of a rich capillary supply in this layer.

The theca interna merges without a distinct junction into the theca externa which

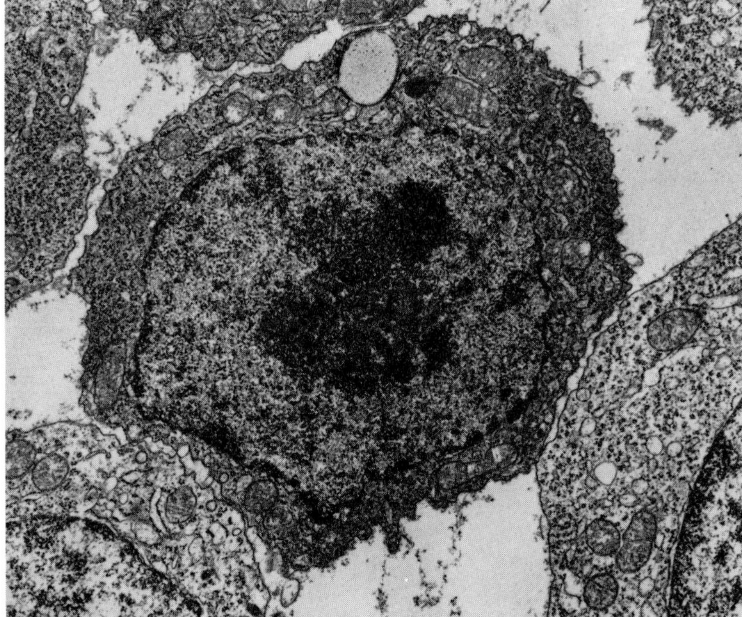

FIGURE 25-15 Electron micrograph of granulosa cells located near the antrum.

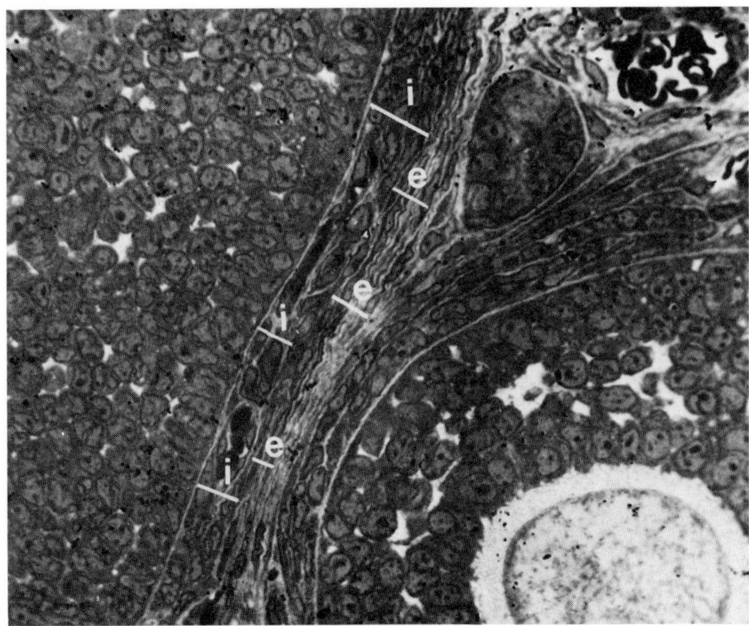

FIGURE 25-16 Growing follicles. The surrounding connective tissue sheath consists of two layers: an inner vascular layer, theca interna (i), and an outer fibrous layer, theca externa (e).

provides physical support for the growing follicle. It is composed of concentrically arranged connective tissue fibers and fusiform cells which merge peripherally with the surrounding ovarian stroma. The cells of the theca externa show a cytoplasm which resembles that of fibroblasts (f) or smooth muscle cells (m), as indicated in Fig. 25-19.

The primary oocyte, which results from a series of mitotic divisions of primordial ova

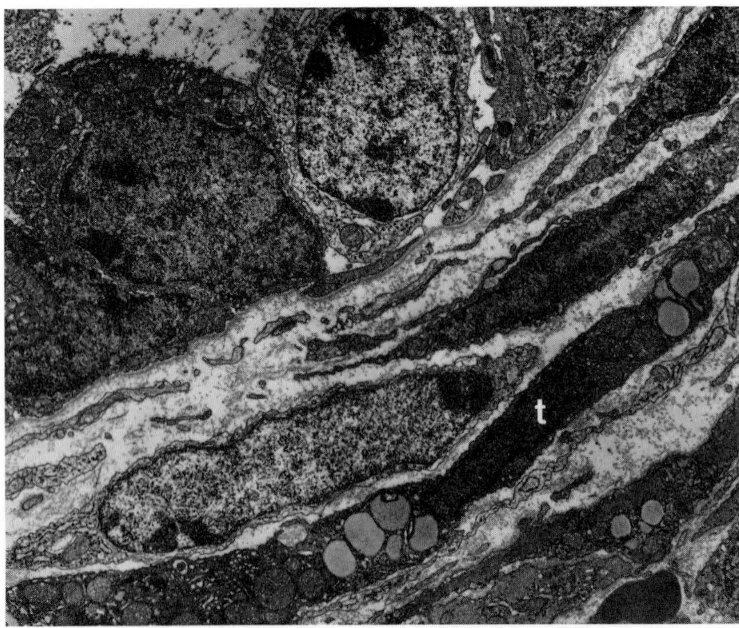

FIGURE 25-17 Electron micrograph of theca interna cells (t).

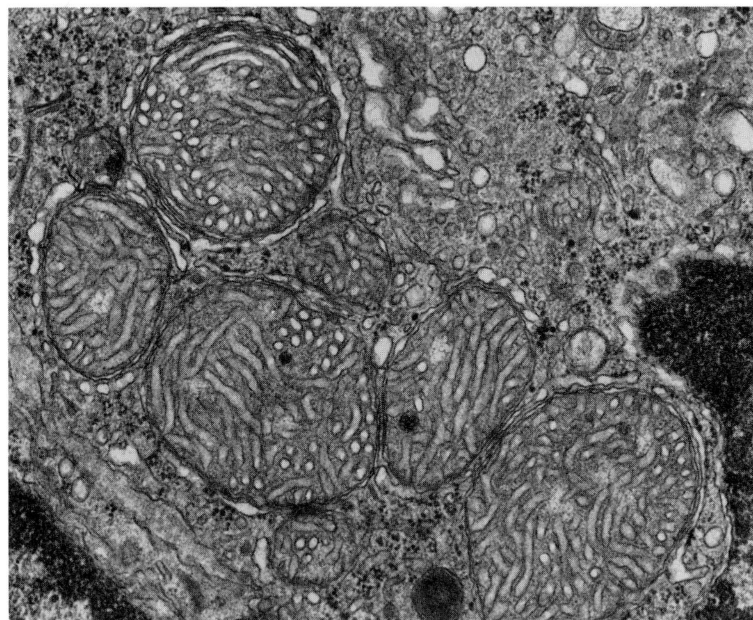

FIGURE 25-18 Mitochondria of theca interna cells. The cristae are tubular.

called *oogonia,* increases in size with beginning follicular growth and reaches full size at the time of antrum formation. It becomes surrounded by a refractile, deeply staining membrane, the *zona pellucida,* clearly seen in Fig. 25-13. It is rich in polysaccharides elaborated by both the ovum and surrounding follicular cells. In an EM (Fig. 25-20) it appears as a fine fibrillar structure between granulosa cells (g) and the ovum (o). Fine villous proc-

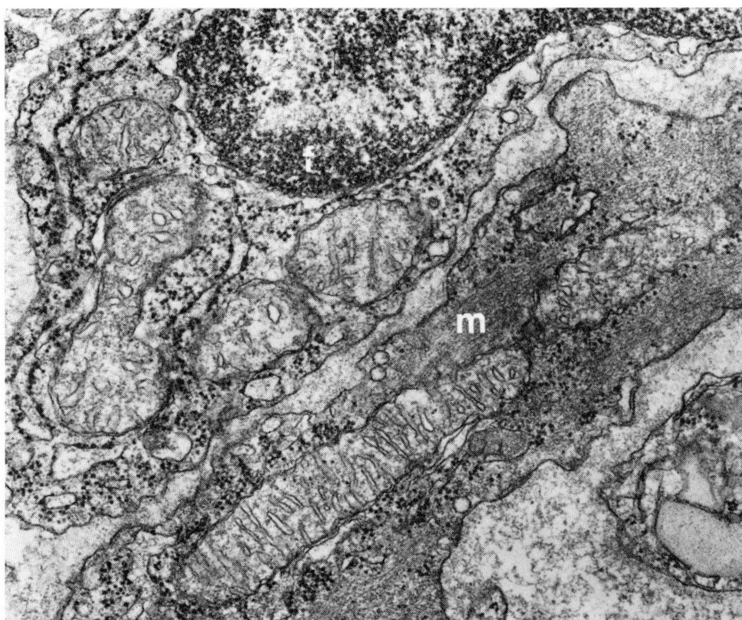

FIGURE 25-19 Electron micrograph of theca externa cells. These cells resemble either fibroblasts (f) or smooth muscle cells (m).

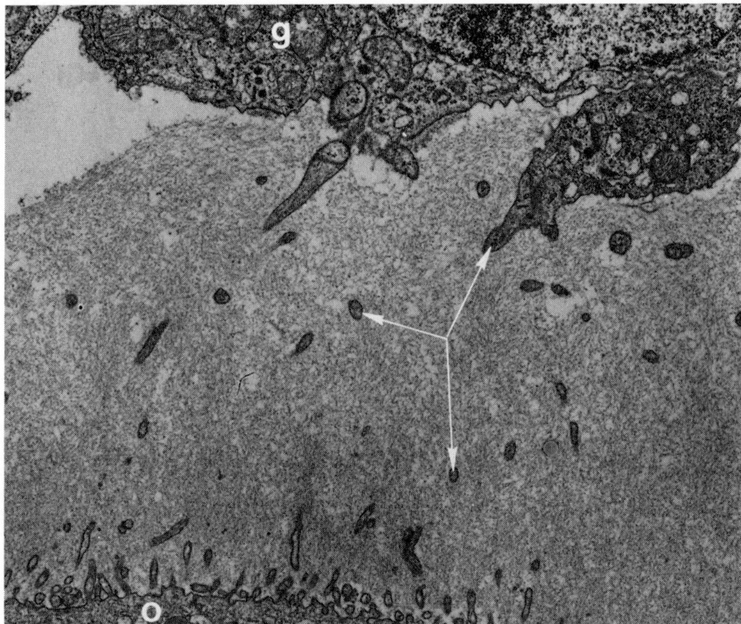

FIGURE 25-20 Electron micrograph of the zona pellucida. Zona pellucida has a fine fibrillar structure situated between the ovum (o) and the granulosa cells (g).

esses (arrowheads) have been seen to originate from both the ovum and those granulosa cells immediately surrounding the zona pellucida. Points of contact between cytoplasmic processes between the two cells have cellular communications between them. Figure 25-21 is an EM showing the cytoplasmic processes of granulosa cells (arrows) as they project themselves into the zona pellucida in which a number of slender processes (arrowheads) appear to run in random directions.

During development, the ovum under-

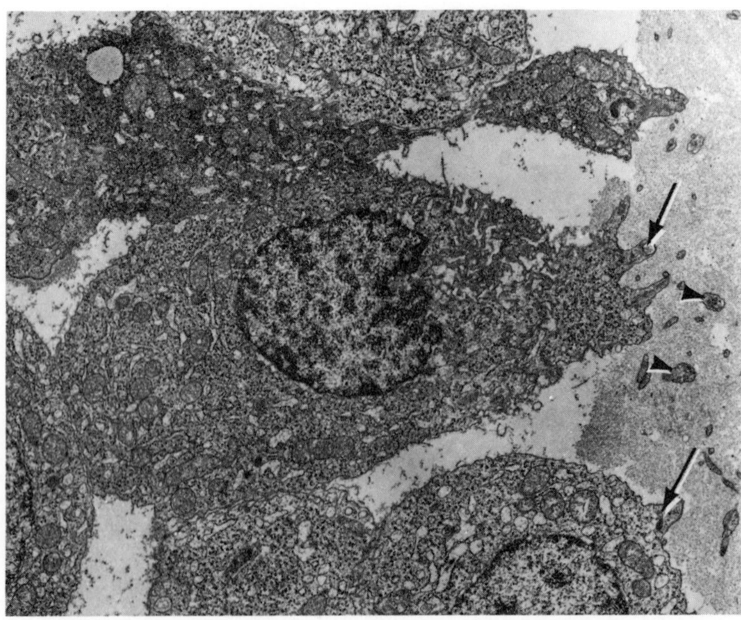

FIGURE 25-21 Granulosa cells adjacent to the zona pellucida. They have villous projections (arrows) that penetrate into the zona pellucida where slender processes (arrowheads) appear to run in random directions.

goes a series of divisions in order to reduce its chromosomes to the haploid number. They include two *reduction* or *maturation divisions* (also called *meiosis*), similar to those occurring in spermatogenesis. However, there is a difference in that only one ovum results as a consequence of mitotic division of oogonia. Each primary oocyte enters the prophase of the meiotic division, where further meiotic processes are arrested. Thus, practically all primary oocytes remain in postprophase or premetaphase conditions throughout the development of the ovary. In the adult, one oocyte per ovarian cycle is induced to differentiate leading to its ovulation. The particular oocyte so recruited undergoes maturation, completing the meiotic division shortly before ovulation. The long suppression of the first meiotic division, therefore, lasts anywhere from 12 to 50 years or so. The recruitment of a selected follicle and its subsequent differentiation is known to require a pituitary gonadotropic hormone, follicle-stimulating hormone (FSH), which is recognized by selected follicular cells. Under the stimulatory influence of FSH, further development of the secondary follicles to the mature ones then takes place. It is known that the entire complement of DNA necessary for the two maturation divisions, that is two sets of the diploid number of chromosomes, is synthesized prior to entering the prophase of the first divisions. Although the chromatin is divided equally between the daughter cells, the division of cytoplasm is unequal and thus produces a *secondary oocyte* with the bulk of the cytoplasm and the *first polar body,* a small package of chromosomes. The second division, occurring at the time of ovulation, is similar to the first, producing a mature ovum retaining the entire cytoplasm with its nutritive contents and the second polar body. The second division proceeds to metaphase, where another arrest of the division process occurs. The second division occurs just prior to or shortly after ovulation, but is completed by the time fertilization has occurred.

The first polar body may also divide, resulting in the formation of a total of three polar bodies, all of which eventually degenerate. Thus, only one daughter cell of the primary oocyte becomes functional. With respect to the reduction in chromosome numbers, it should be emphasized that the first meiosis reduces two sets of diploid chromosomes to a single diploid set per secondary oocyte, and that the second maturation division results in a haploid chromosome in each of the four daughter cells.

As mentioned previously, relatively few follicles reach full maturity with release of an ovum, since only one ovum per month is needed during the up to 40 years of procreativity in human beings. The remaining hundreds of thousands of follicles undergo degeneration or *atresia* throughout the life of the individual, either as primary follicles or after a varying period of follicular development. In atresia of a primary follicle, the ovum and follicular cells are resorbed, leaving a space filled with stroma. Atresia of more mature follicles is a more complicated process.

A mature *graafian follice* results after a maturation period of 10 to 14 days in the human being. It fills the width of the cortex and often produces a bulge on the free surface of the ovary. Fluid-filled spaces begin to appear among the cells of the cumulus oophorus, thus loosening the attachment of the ovum and its closely associated cells from the follicular wall.

OVULATION

At intervals of 28 days, rupture of the mature follicle occurs in the process of ovulation. Just prior to ovulation, there is a fusion of the follicular cells with the surface epithelium. With increasing pressure produced by the more rapid accumulation of liquor folliculi and the contractile properties of the ovarian stroma, the bulging wall ruptures, liberating the fluid and the ovum with its corona radiata. Recent EM studies have shown that there are a fair number of smooth muscle–like cells in the stroma which probably contribute to the process of ovulation with their contractile properties. Recent experimental evidence indicates that the luteinizing hormone (LH) from the anterior pituitary gland is required to trigger the resumption of the second meiotic division. The peripherally located granulosa cells of the preovulatory follicle have receptors for this gonadotropic hormone which binds to specific receptor molecules associated with the plasma membrane of granulosa cells. Although the precise molecular mechanism of ensuing responses leading to the ovulation is not clearly known, it has been shown that a sharp rise in plasma LH level precedes the ovulation, and that the subcellular reaction following the LH surge involves the activation of LH-dependent systems in the granulosa cells.

The ovum, entering the peritoneal cavity briefly, is then literally captured by the *fimbriated end* of the oviduct (fallopian tube). In a short time, cells of the corona radiata are freed, and the ovum is ready for fertilization, which must occur in a human being within 24 h. The location of fertilization can be anywhere from the ovulation site through the passage of the oviduct. Following fertilization, the journey to the uterus takes about 3 days and implantation occurs about 6 days after ovulation.

CORPUS LUTEUM

Upon completion of ovulation, the ruptured follicle collapses, forming an irregularly folded mass (Fig. 25-22). The corpus luteum becomes much larger during pregnancy, as shown in Fig. 25-23, cl, which is a corpus luteum from a pregnant woman taken at identical magnification as the one in Fig. 25-22. Tissue fluids and blood may fill the luminal area which otherwise becomes filled by a loose connective tissue.

The collapsed follicle then becomes transformed temporarily into a glandular structure, *corpus luteum* (yellow body), which secretes progesterone. Both the granulosa and theca interna cells contribute to the formation of corpus luteum by differentiation and are known as *granulosa lutein* and *theca lutein* cells, respectively (Fig. 25-24). Theca lutein cells are smaller and more peripherally located than granulosa lutein cells. Luteal cells have a vacuolated cytoplasm (Fig. 25-25). In ovaries stained for lipids, the cytoplasm of luteal cells is filled with dark granules that represent lipid droplets in vivo (Fig. 25-26).

Electron micrographs demonstrate varying amounts of smooth ER and mitochondria with tubular cristae, as evidence of their steroid-producing capability in the formation of progesterone (Figs. 25-27 and 25-28). The cytoplasm also contains pleomorphic lipid

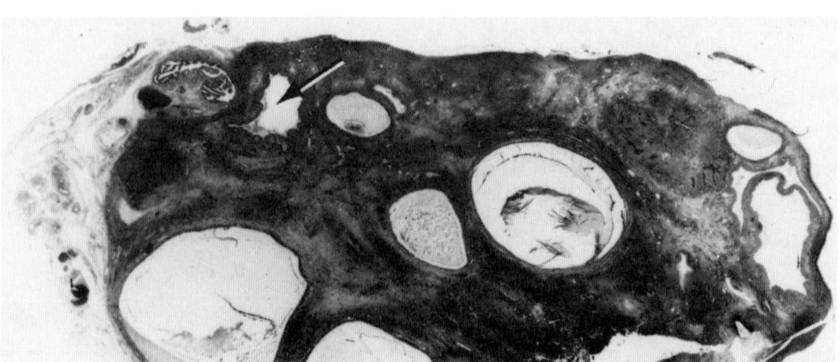

FIGURE 25-22 Light micrograph of an ovary. After ovulation, the follicle forms an irregular mass, a corpus luteum (arrow). The lumen is filled with tissue fluid.

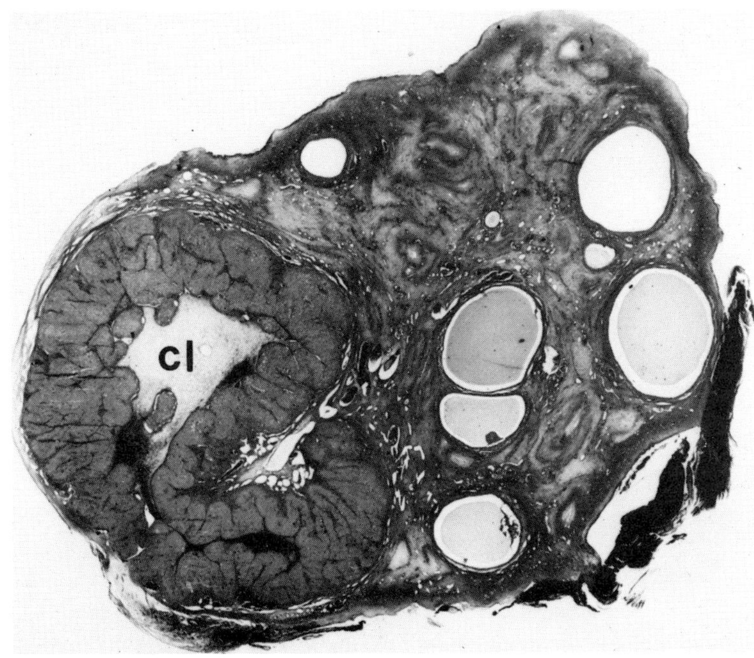

FIGURE 25-23 Corpus luteum (cl) in an ovary from a pregnant woman.

droplets. Numerous capillaries and connective tissue septa from the theca invade the lutein cellular mass (Fig. 25-27).

If the ovum is not fertilized, the corpus luteum begins to degenerate approximately 1 week after ovulation. The cells become smallar and are eventually resorbed. Loose connective tissue fills the space, finally producing a white scar, *corpus albicans* (Fig. 25-29).

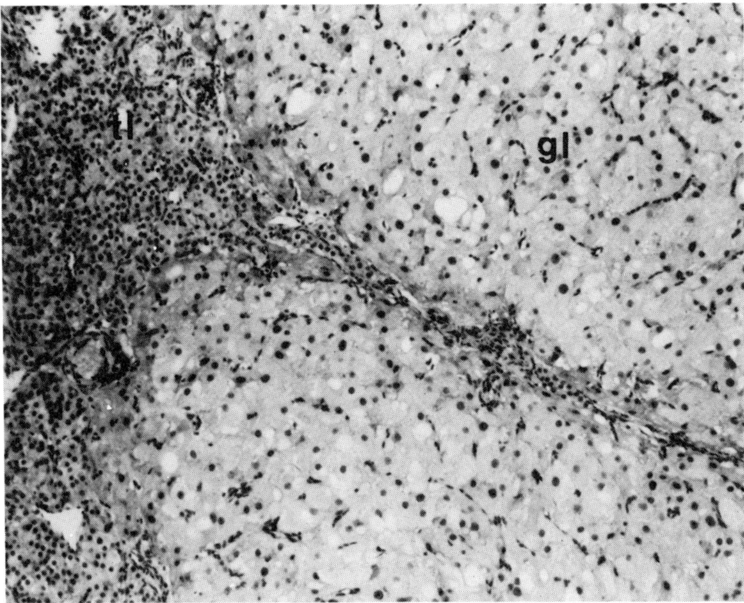

FIGURE 25-24 Light micrograph of a peripheral area of human corpus luteum. The smaller darker cells at the left are theca lutein cells (tl), and the larger paler cells are granulosa lutein cells (gl).

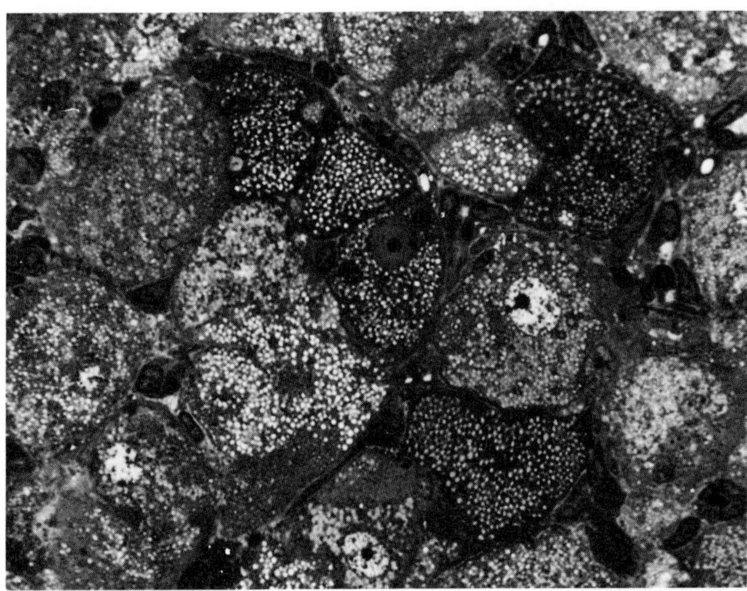

FIGURE 25-25 High-power light micrograph of luteal cells. The cytoplasm of the cells appears vacuolated.

If the ovum is fertilized, the corpus luteum increases in size, continues to differentiate, and functions as a major source of estrogen and progesterone during the early phase of pregnancy. These hormones are required for the development of the placenta and associated structures. The corpus luteum persists until the end of pregnancy when it undergoes involution, producing a corpus albicans similar to that produced in a normal menstrual cycle. The luteal involution involves autophagic processes within luteal cells as well as a mobilization of phagocytic cells of the connective tissue. A macrophage filled with light-stained lipid is seen in Fig. 25-30 which represents a portion of a regress-

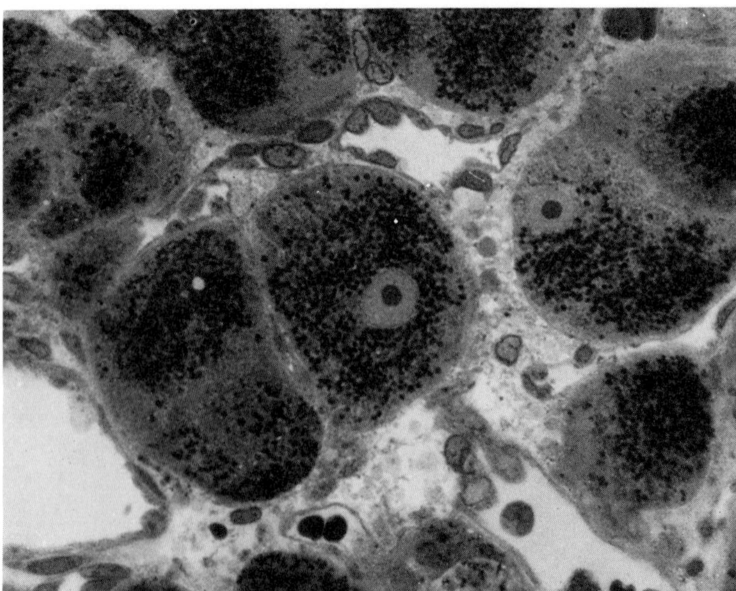

FIGURE 25-26 Section from an ovary stained for lipids. The cytoplasm of the luteal cells is filled with dark granules. Note the rich network of capillaries surrounding the cords of luteal cells.

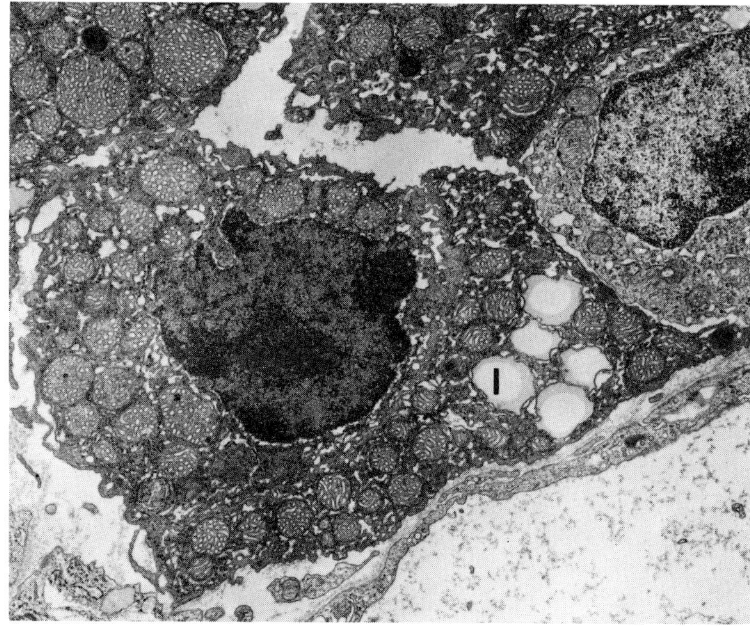

FIGURE 25-27 Electron micrograph of luteal cells. The cytoplasm contains numerous mitochondria with tubular cristae, varying amounts of SER, and lipid droplets (l).

ing corpus luteum. In many mammals, but poorly developed in human beings, the ovarian stroma contains epithelioid cells called *interstitial cells*. They are rich in lipids and are thought to secrete steroid hormones.

Figure 25-31 summarizes the sequence of morphological changes in the developing ovarian follicles from the primordial ovum that is recruited from the germinal epithelium (1) to the corpus albicans (10). The primordial ovum forms the primary follicle (2 to 4). A follicular antrum (5) is produced, and it con-

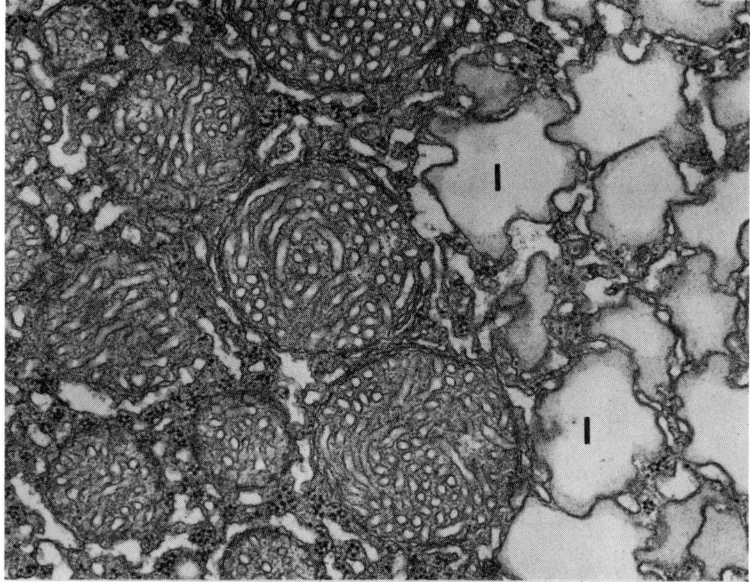

FIGURE 25-28 Mitochondria in luteal cells viewed at higher magnification. Note their tubular cristae. Lipid droplets (l).

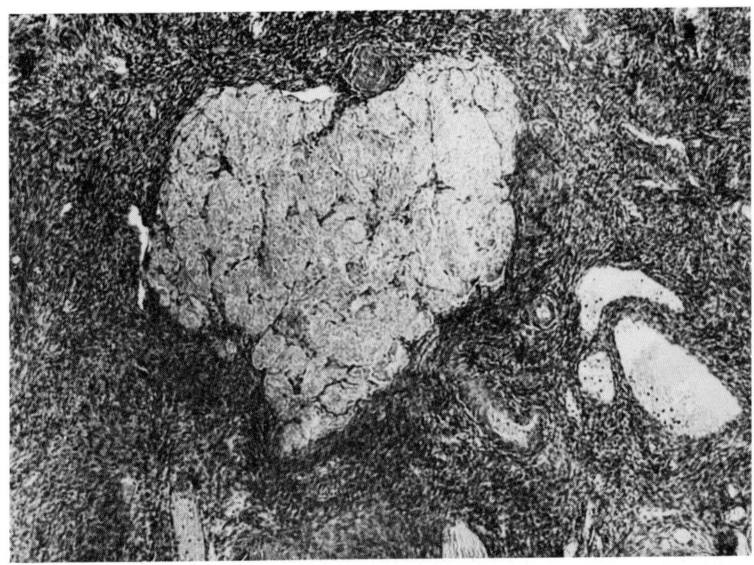

FIGURE 25-29 Light micrograph of a corpus albicans. The corpus albicans consists of white scar tissue replacing a degenerating corpus luteum.

tinues to enlarge (6) as the follicle differentiates into a mature graafian follicle (7). Following ovulation (8), the remaining follicular cells form the corpus luteum (9) and corpus albicans (10).

HORMONAL INFLUENCE ON THE OVARIAN CYCLE

The sexual cycle is dependent on the gonadotropic hormones secreted by the anterior

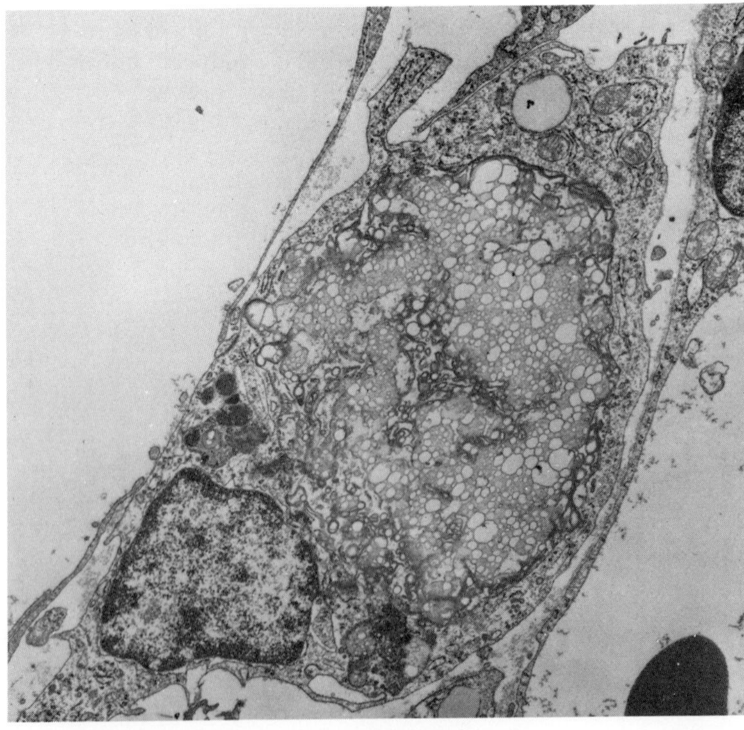

FIGURE 25-30 Electron micrograph of a macrophage in a regressing corpus luteum.

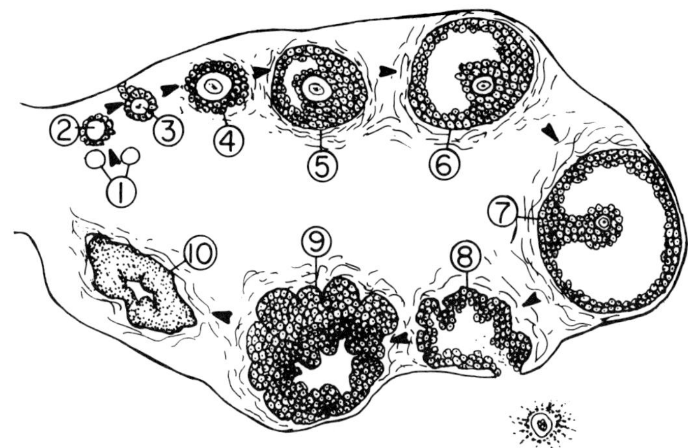

FIGURE 25-31 Diagram summarizing the morphologic changes occurring during the development of ovarian follicles.

hypophysis which, in turn, is under control of the hypothalamus. Ovaries not receiving this hormonal stimulus remain inactive, as is the case throughout childhood. Production of gonadotropic hormones in small amounts begins at the age of about 8 years and increases until the initiation of the monthly menstrual cycle at puberty between the ages of 11 and 15 years. Subsequently, the cyclic variations in these hormones control the maturation of follicles and the formation of corpora lutea. In recent years, it has been shown that a specific gonadotropin-releasing hormone (GnRH) for each gonadotropin is produced by hypothalamic nuclei which carry GnRH via the hypothalamohypophyseal tract and secrete it in the posterior hypophysis into the hypophyseal portal system.

The cycle starts with increased secretion of FSH which stimulates the ovarian follicle to grow. At the time of antrum formation the secretion of LH begins to increase, acting synergistically with FSH to promote further follicular development. Luteinizing hormone is required for final follicular maturation and ovulation. As mentioned earlier, an especially large amount of LH is released just prior to ovulation. The beginning development of lutein cells from follicular cells is also dependent on LH. Prolactin, or luteotropic hormone (LTH), which increases at the same time as LH, is responsible for continued development of the corpus luteum, along with production and release of its hormonal secretions.

The ovary in turn produces hormones with cyclic variations which not only affect such accessory reproductive organs as the uterus, but also regulate the hormones produced by the anterior hypophysis via a negative-feedback mechanism. Estrogen is secreted by the growing follicles and to a lesser degree by the corpus luteum. It has an inhibitory effect on FSH production as well as a stimulating one on production of LH. With the formation of corpus luteum, progesterone secretion rises, which in turn inhibits the pituitary secretion of LH. As the corpus luteum degenerates, estrogen and progesterone levels decline, thus stimulating FSH production and the start of another cycle (Fig. 25-32).

In summary, FSH leads to follicle growth and estrogen secretion. An increase in circulating estrogen leads to FSH decline and LH rise. The increase in LH level leads to ovulation and progesterone secretion. Increasing progesterone level leads to LH decline, which promotes corpus luteum degeneration with subsequent decline of estrogen and progesterone levels, resulting in menstruation. Estrogen decline leads to FSH rise and the beginning of a new cycle.

Birth control pills, consisting of synthetic

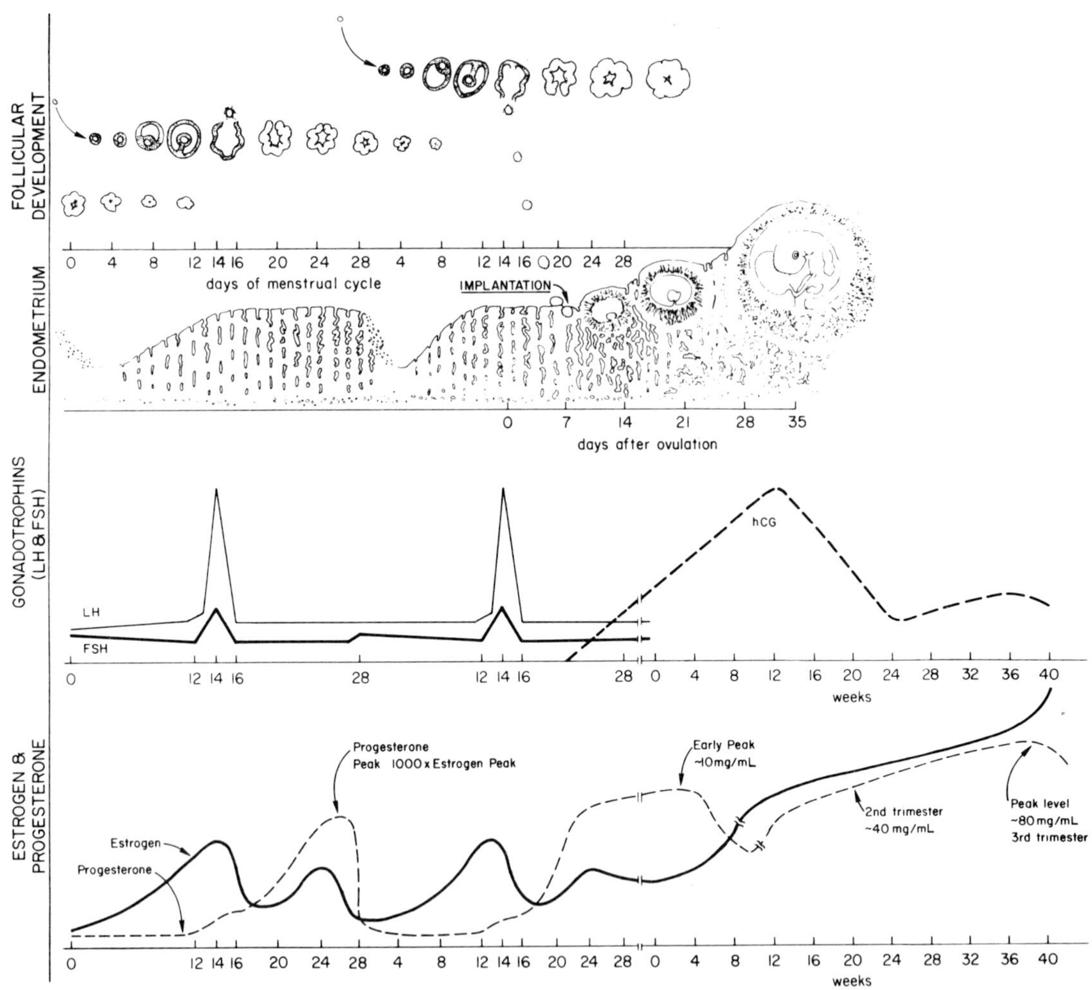

FIGURE 25-32 Diagram depicting the morphologic changes occurring in the ovary and the endometrium in the course of a menstrual cycle and the changes in the levels of the hormones controlling the morphologic changes.

hormones, prevent pregnancy primarily by inhibiting development of ova and subsequent ovulation. For example, synthetic progesterone inhibits LH secretion, including the sharp preovulation surge, and thus prevents ovulation. With continued suppression of LH, follicle development is unable to progress and release a mature ovum. Small amounts of synthetic estrogen are also included to prevent side effects such as the occurrence of spotty bleeding from the uterine lining. Once a month, the pill is withdrawn (or replaced by a placebo), resulting in the sudden drop of estrogen and progesterone and causing the onset of menstruation.

REVIEW SECTION

1. Figure 25-33 shows a developing ovarian follicle. Identify the various components of this developing follicle. Identify the layer of cells encir-

This figure shows a follicle growing within the theca interna. This particular section is from a rat. The nucleus of the ovum, zona pellucida, stratified

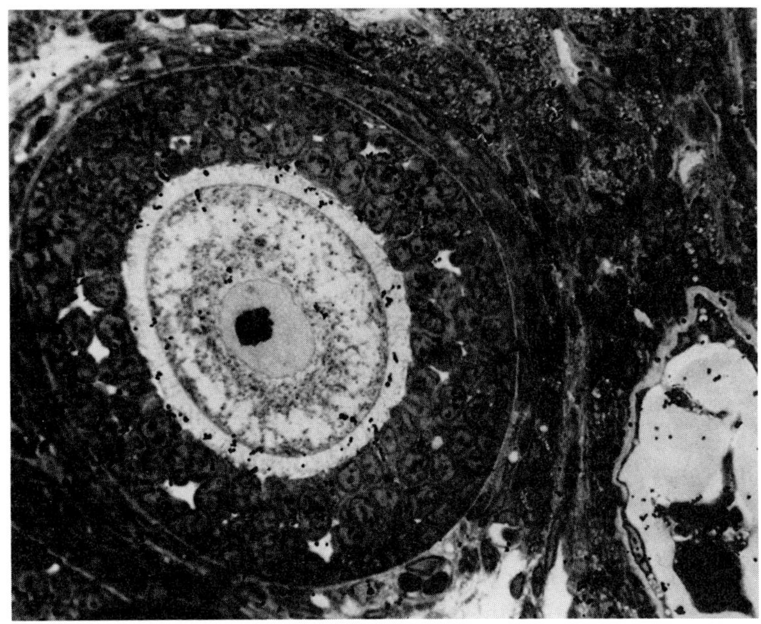

FIGURE 25-33

cling the structure. What is the eventual hormonal function of these cells?

granulosa cells, and theca can all be identified. The production of estrogen is the eventual function of these cells and is vital to maintaining other structures of the reproductive system.

2. Name the structure indicated by the arrow in Fig. 25-34. What hormone is being produced here? What fills the space?

The graafian follicle in Fig. 25-34 is producing estrogen. The follicular lumen is filled with liquor folliculi.

3. Name the structure shown in Fig. 25-35. What fills the internal space indicated by arrows? What hormones are beginning to be produced? What will happen to this structure if pregnancy occurs? What will happen to this structure if pregnancy does not occur?

Figure 25-35 depicts a corpus luteum of a human being filled with connective tissue. Following ovulation, this structure maintains some estrogen production, but primarily begins to produce progesterone. The corpus luteum will continue to grow and produce hormones if a pregnancy occurs, or it will degenerate to form the corpus albicans.

BIBLIOGRAPHY

Austin, C. R.: *The Mammalian Egg,* Blackwell, Oxford, 1961.

Bloom W., and D. W. Fawcett: *A Textbook of Histology,* 10th ed., Saunders, Philadelphia, 1975.

Grandy, H. G., and D. E. Smith (eds.): *The Ovary,* Williams & Wilkins, Baltimore, 1962.

Weiss, L., and R. O. Greep: *Histology,* 4th ed., McGraw-Hill, New York, 1977.

FIGURE 25-34

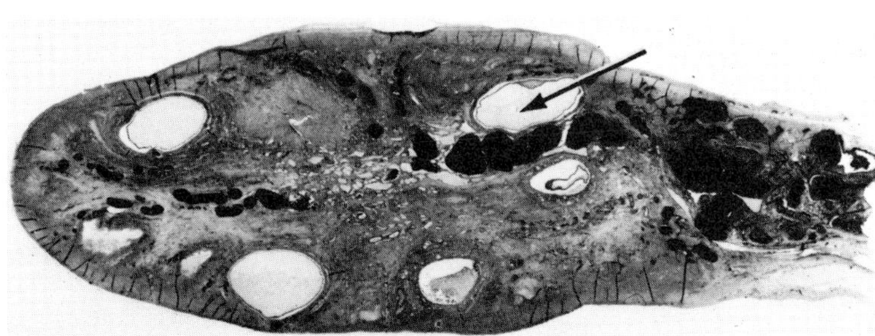

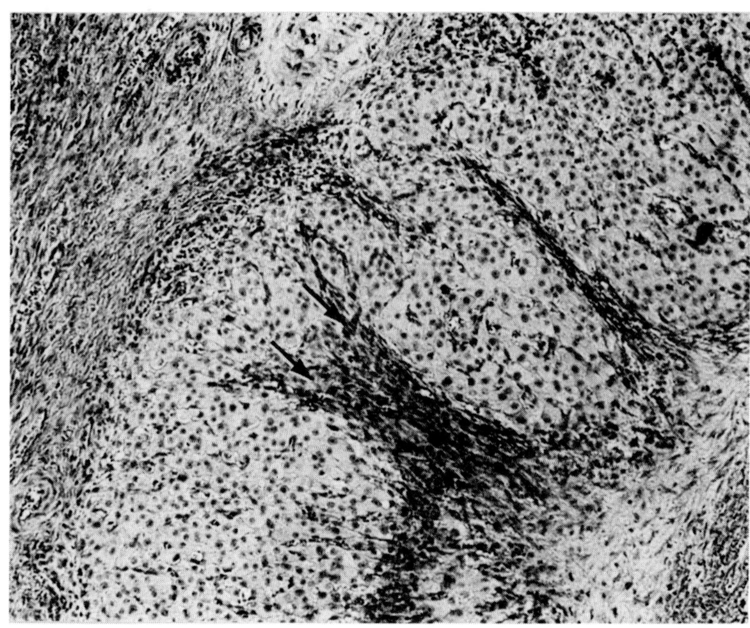

FIGURE 25-35

Zuckerman, S. (ed.): *The Ovary,* Academic, New York, 1962.

ADDITIONAL READINGS

Adams, E. C., and A. T. Hertig: "Studies on Guinea Pig Oocytes. I. Electron Microscopic Observations on the Development of Cytoplasmic Organelles in Oocytes of Primordial and Primary Follicles," *J. Cell Biol.* **21:**397–427 (1964).

———, and ———: "Studies on the Human Corpus Luteum. I. Observations on the Ultrastructure of Development and Regression of the Luteal Cells during the Menstrual Cycle," *J. Cell Biol.* **41:** 696–715 (1969).

———, and ———: "Studies on the Human Corpus Luteum. II. Observations on the Ultrastructure of Luteal Cell during Pregnancy," *J. Cell Biol.* **41:** 716–735 (1969).

Baca, M., and L. Zamboni: "The Fine Structure of Human Follicular Oocytes," *J. Ultrastruct. Res.* **19:** 354–381 (1967).

Baker, T. G. and L. L. Franchi: "The Fine Structure of Oogonia and Oocytes in Human Ovaries," *J. Cell Sci.* **2:**213–224 (1967).

Bjersing, L.: "On the Morphology and Endocrine Function of Granulosa Cells in Ovarian Follicles and Corpora Lutea," *Acta. Endocrinol.,* suppl. **125:** 1–23 (1967).

———: "On the Ultrastructure of Granulosa Lutein Cells in Porcine Corpus Luteum with Special Reference to Endoplasmic Reticulum and Steroid Hormone Synthesis," *Z. Zellforsch.* **82:**187–211 (1967).

Blanchette, E. J.: "Ovarian Steroid Cells. II. The Lutein Cell," *J. Cell Biol.* **31:**517–542 (1972).

Blandau, R. J.: "Ovulation in the Living Albino Rat," *Fertil. Steril.* **6:**391–404 (1955).

Corner, G. W.: "The Histological Dating of Corpus Luteum of Menstruation," *Am. J. Anat.* **98:**377–401 (1956).

Crisp, T. M., D. A. Dessouky, and F. R. Denys: "The Fine Structure of the Human Corpus Luteum of Early Pregnancy and during the Progestational Phase of the Menstrual Cycle," *Am. J. Anat.,* **127:** 37–69 (1970).

Hertig, A. T., and E. C. Adams: "Studies on the Human Oocyte and Its Follicle. I. Ultrastructural and Histochemical Observations on the Primordial Follicle Stage," *J. Cell Biol.* **34:**647–675 (1967).

Richardson, G. S.: "Ovarian Physiology," *New Engl. Med. J.* **274:**1008–1015, 1064–1075, 1121–1134, 1184–1194 (1966).

Ryan, K. J., and R. V. Short: "Formation of Estradiol by Granulosa and Theca Cells of the Equine Ovarian Follicle," *Endocrinology* **76:**108–114 (1965).

Young, W. C. (ed.): *Sex and Internal Secretions,* 2 vols., Williams & Wilkins, Baltimore, 1961.

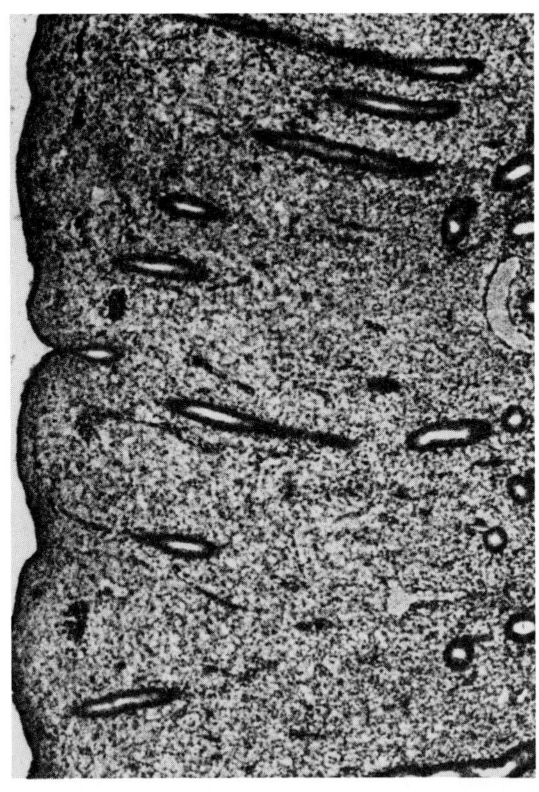

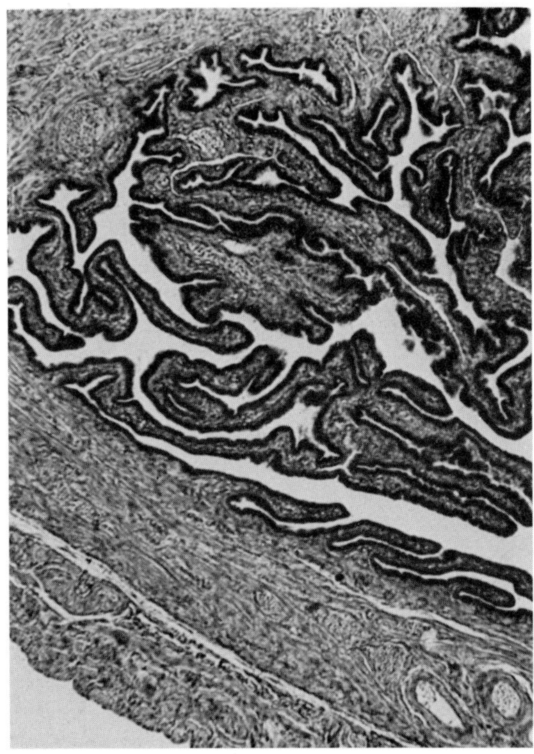

26 UTERUS, OVIDUCT, PLACENTA, AND VAGINA

OBJECTIVES

Upon completion of this chapter, the student will be able to:

1 Identify the layers of the uterus, their functions, and their changes throughout the menstrual cycle.

2 Identify the oviduct including the fimbriated end.

3 Relate the endometrial cycle to the ovarian cycle including the endocrine aspects.

4 Define *decidua* and list its parts.

5 Describe the structure and function of the following layers of chorionic villi:
(a) syncytiotrophoblastic layer (b) cytotrophoblastic layer

6 List the hormones produced by the placenta and their functions.

UTERUS

The uterus is a large, muscular organ made up of three layers: the endometrium, myometrium, and perimetrium (Figs. 25-1 and 26-1). The endometrium is made up of simple columnar epithelium supported by a lamina propria with endometrial glands and varies in thickness according to the menstrual cycle. The myometrium or muscle layer constitutes the bulk of the organ, while the perimetrium is the thin connective tissue and peritoneal covering of the organ.

ENDOMETRIUM

The uterus is lined by a mucosa, the endometrium, which is closely attached to the myometrium. The epithelial lining is simple columnar and is invaginated to form numerous tubular endometrial glands. During childbearing years, the endometrium undergoes cyclic changes which are closely related to the ovulatory cycle. These endometrial changes consist of hypertrophy of the glandular, vascular, and interstitial elements of the mucosa in preparation for pregnancy. If implantation of the fertilized ovum occurs, the hypertrophy continues; if not, there is a breakdown of the prepared layer and it is discharged as menstrual flow.

MYOMETRIUM

The muscle cells of the myometrium appear on cursory examination to be randomly organized; however, they actually form bundles that are separated into three layers. Next to the endometrium is the inner layer of muscles which is thin and is called *stratum subvasculare* (Fig. 26-2). It consists primarily of longitudinal fibers that run parallel to the long axis of the uterus. The middle layer (Figs. 26-2 and 26-3) is the thickest layer, called *stratum vasculare*, and contains circular muscle fiber bundles, oriented to function efficiently in parturition. The tissue spaces here contain large blood vessels, especially veins; hence the name *stratum vasculare* (vascular layer). The blood vessels are mostly veins (Fig. 26-4). The outermost layer directly beneath the perimetrium contains both circularly and longitudinally oriented bundles running in various directions. This layer,

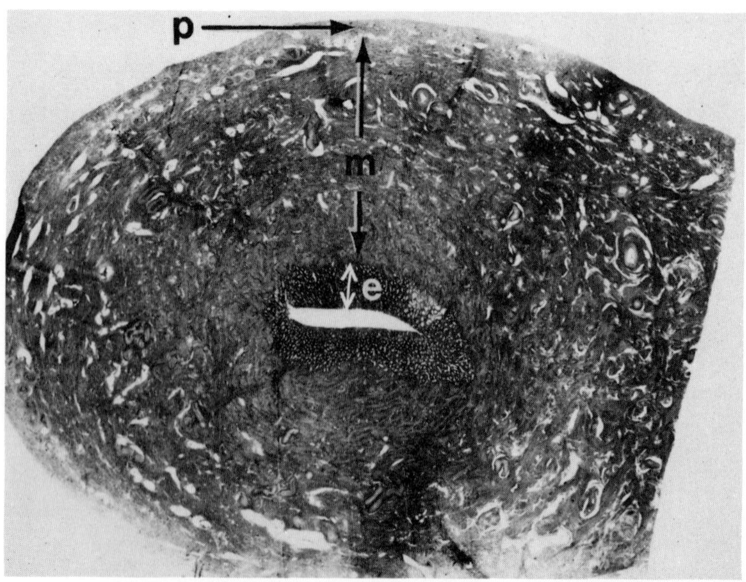

FIGURE 26-1 Macrophotograph of a section through the uterus. The wall of the uterus consists of three layers: endometrium (e), myometrium (m), and perimetrium (p).

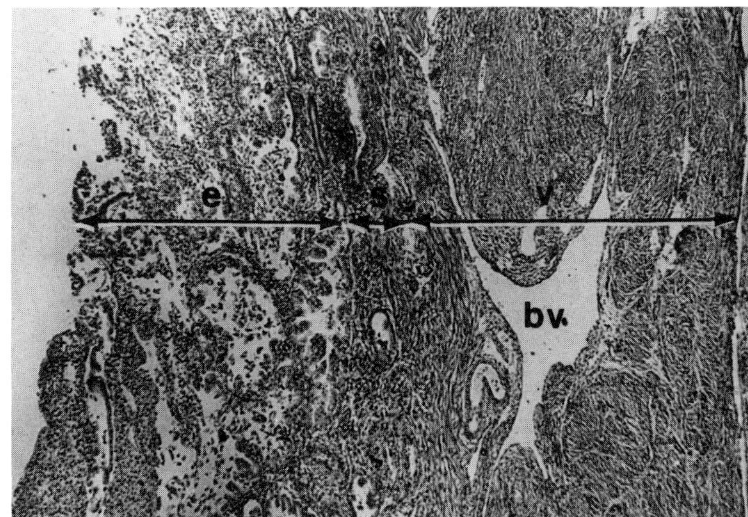

FIGURE 26-2 Light micrograph of the uterus. The endometrium (e) lines the lumen. Subjacent to the endometrium is the inner muscle layer, stratum subvasculare (s). Next to the inner muscle layer is the middle muscle layer, called stratum vasculare (v). The latter contains large blood vessels (bv).

called *stratum supravasculare*, is also thin (Fig. 26-3). The longitudinal fibers seem to predominate and are continuous with the longitudinal muscle coat of the vagina.

The smooth muscle cells of the myometrium periodically vary in size, corresponding to the normal menstrual cycle. Since there is little division of muscle cells, the increase in size is due to the enlargement of existing cells. The fiber size increases tremendously to accommodate a growing embryo, growing from about 40 μm to more than 500 μm in length during the later stages of pregnancy.

PERIMETRIUM

The perimetrium is a continuation of the peritoneal covering from the adjoining areas. This serous membrane is made up of a thin layer of loose connective tissue covered by mesothelium. Since the perimetrium is continuous with the covering of the bladder and the rectum, it is found only on the fundus

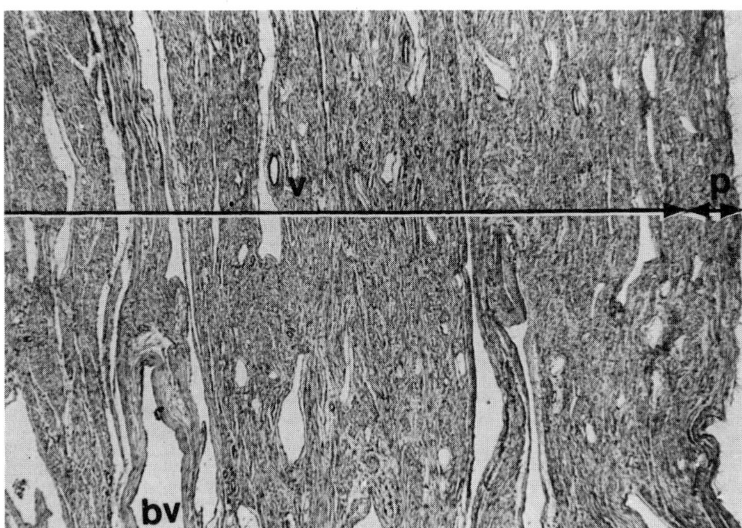

FIGURE 26-3 The outer part of the uterine wall. Stratum vasculare (v) which contains large blood vessels (bv) and the perimetrium (p) are shown.

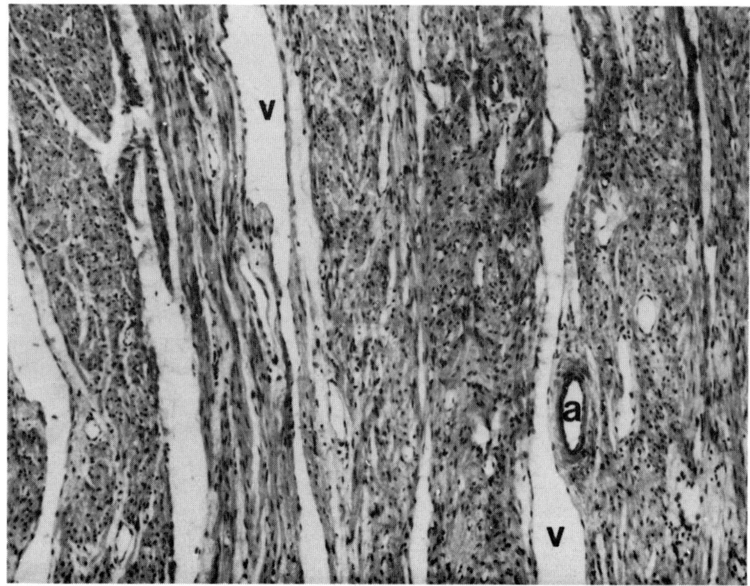

FIGURE 26-4 Stratum vasculare seen at higher magnification. It contains large veins (v) and arteries (a) of smaller caliber than veins.

(see Fig. 25-1) and posterior surfaces of the uterus, making the latter retroperitoneal.

ENDOMETRIAL CYCLE

The endometrial cycle consists of four characteristic phases being regulated by the cyclic hormonal levels discussed in Chap. 25. The first day of menstruation is regarded as the first day of the endometrial cycle. Lasting 3 to 5 days, it is called the menstrual phase. Following this is the proliferative or estrogenic phase of about 10-days' duration in which there is a buildup of tissue. Next is the secretory phase (progestational or progravid) lasting about 10 days and leading to a 1- or 2-day period of premenstrual phase in which certain basic changes occur in the mucosa prior to menstruation and a repeat of the cycle. The histologic description of the endometrium best begins with the proliferative phase, since this is where the structural buildup of the mucosa begins.

PROLIFERATIVE PHASE

The proliferative phase, also called the *follicular phase,* is characterized by increasing thickness of the endometrium from 0.5 mm just following menstruation to 2 or 3 mm by the end of the phase. The thin basal layer of endometrium remaining after the preceding menstrual phase begins to proliferate rapidly. Epithelial cells from remnants of glands show numerous mitoses and migrate to cover the exposed surfaces of the endometrium, regenerating the layer of simple columnar epithelium. After reepithelialization, the gland cells continue to increase in number and produce long tubular endometrial glands that are closely packed. Figure 26-5 demonstrates the reepithelialized endometrium in the beginning stages of proliferation. It has not yet become very thick; however, numerous glands are quite evident. The presecretory columnar cells of the endometrial glands are tall (Fig. 26-6).

Endometrial glands remain relatively straight and uniform in diameter, although they may have occasional branched endings. The glandular cells begin to show a collection of glycogen droplets and mucinous secretory granules in the basal region which displace the nucleus toward the apex. Blood vessels in the form of *coiled arteries* (Fig. 26-7) grow into the developing tissue, maintaining a high degree of vascularization.

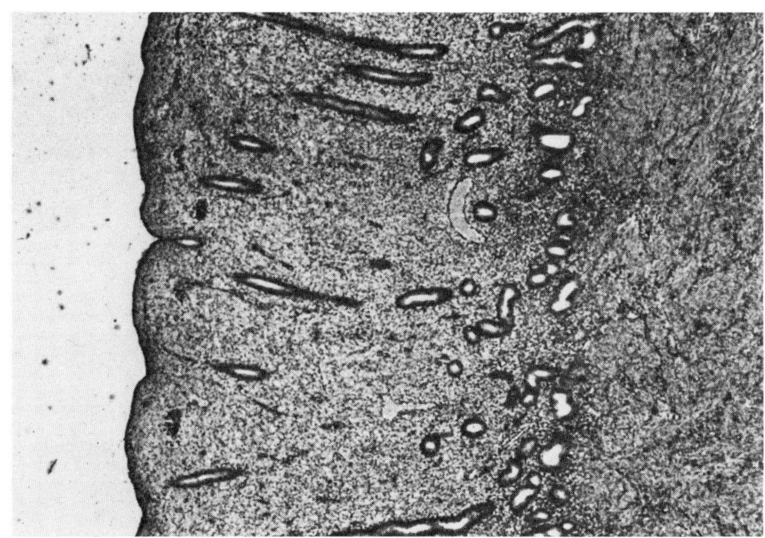

FIGURE 26-5 Light micrograph of the endometrium at the beginning of the proliferative stage.

SECRETORY PHASE

During the secretory phase, which is often referred to as the *luteal phase,* several layers can be distinguished in the endometrium (Fig. 26-8). The deepest layer is *basalis* which changes very little with the cycle and is not lost at menstruation (Fig. 26-8). The glandular cells at the base of the endometrial glands appear dense and less differentiated than the remainder of the gland cells (Figs. 26-8 and 26-9). The *functionalis* is superficial to the basalis and is that portion of the endometrium which is receptive to hormonal influence and, hence, is sloughed off during the course of menstruation. The functionalis is divided into two layers: a thin superficial *compacta* layer and a deep *spongiosa* layer which makes up the bulk of the endometrium (Fig. 26-8). The spongiosa is named for its "spongy" appearance due to the irregular convolutions of the glands and the edematous changes in the stroma.

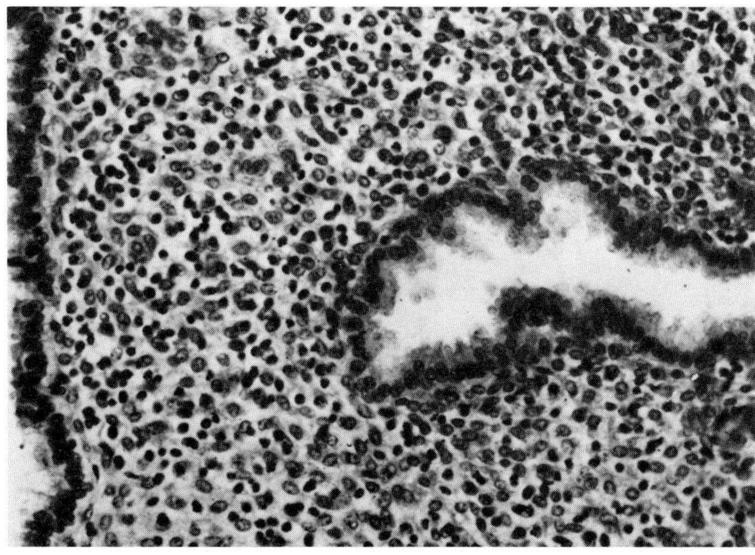

FIGURE 26-6 Endometrial gland in the proliferative stage. The lining cells are tall columnar cells with basally located nuclei. Note the similar appearance of the surface cells.

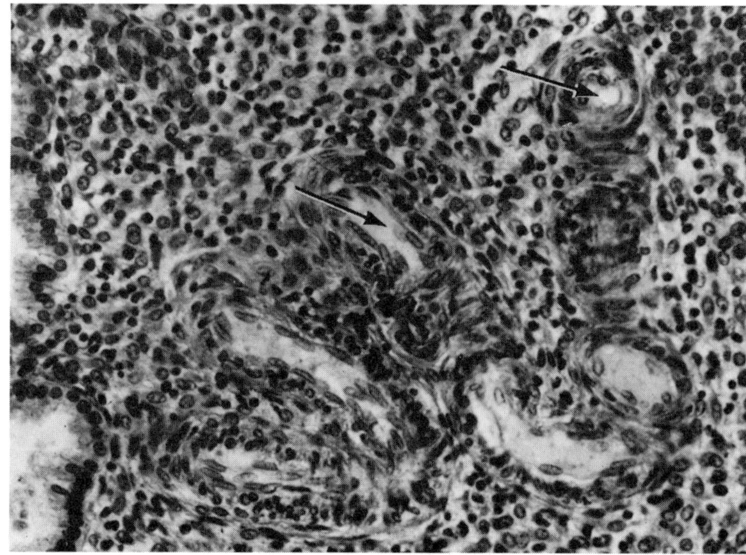

FIGURE 26-7 Portion of the endometrium in the proliferative phase. In addition to developing glands, it contains growing coiled arteries (arrows).

The secretory phase is marked by an even greater increase in thickness of the endometrial layer, largely due to glandular swelling, presence of edema, and increased vascularity. As the glands become large due to their secretory activity, the endometrial layer is unable to accommodate them, which results in the coiling of the glands like corkscrews. Edema appears in the deeper functionalis region. Coiled arteries form near the surface and become highly tortuous. By the time the endometrial glands are highly convoluted, their cells are fully differentiated, and they can be characterized as secretory (Fig. 26-10). The cells displace their nucleus toward the free surface owing to the increased synthesis and accumulation of glycogen. The secretory products, including some glycogen and mucin as well as a small amount of lipid, increase rapidly and are released into the

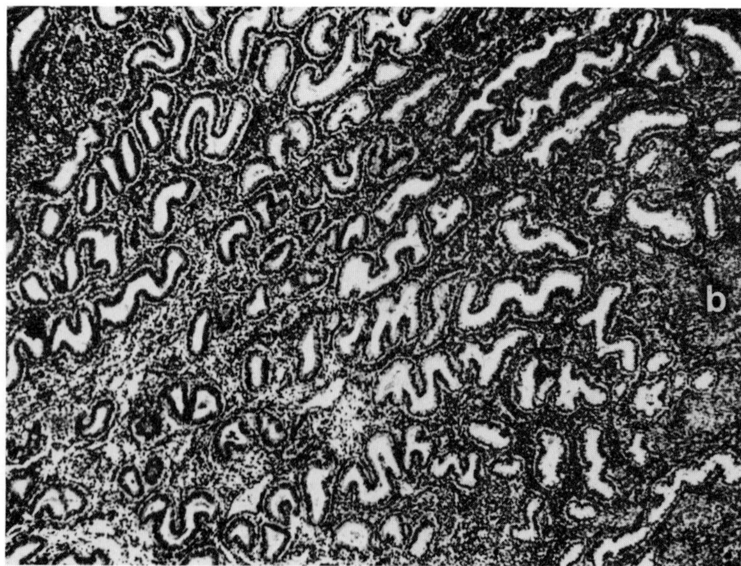

FIGURE 26-8 Light micrograph of the endometrium in the luteal phase. The deepest layer, basalis (b), undergoes very little change. Functionalis is superficial to basalis and consists of two layers: a superficial layer, the compacta, and a deeper layer, the spongiosa.

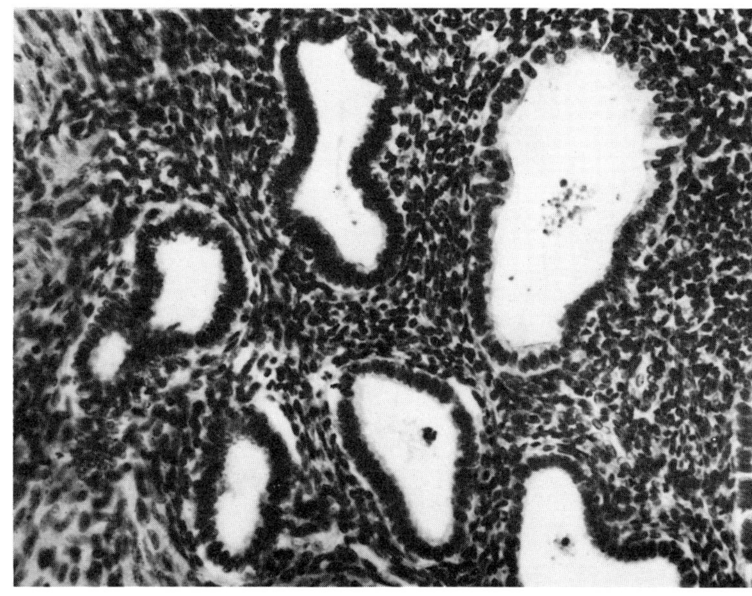

FIGURE 26-9 Endometrial glands in the basalis layer. Note the dark and less differentiated appearance of the epithelial cells and the irregular shape of the gland lumina.

widened lumina of the glands (Figs. 26-9 and 26-10).

PREMENSTRUAL PHASE

If fertilization fails to occur, the ovarian hormones decline, inducing a marked vascular change in the endometrium. The coiled arteries, which are sensitive to this, constrict intermittently, causing ischemia (lack of blood and oxygen) in the region. Finally, the coiled arteries (Fig. 26-11, arrows) close down completely, producing a rather massive and quick degeneration of the functionalis portion of the endometrium.

MENSTRUAL PHASE

During menstruation, the degenerated functionalis, including broken-down glandular

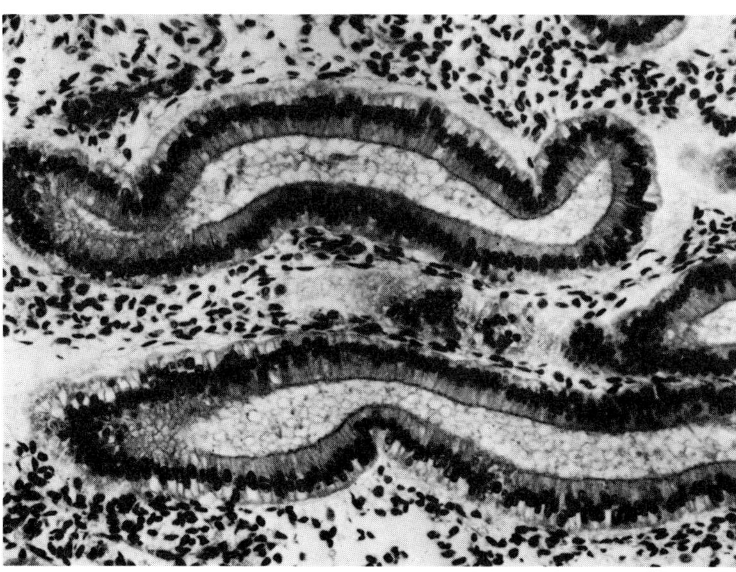

FIGURE 26.10 Light micrograph of fully differentiated endometrial glands in the secretory phase. The nuclei of the columnar epithelial cells are displaced toward the free surface as the result of accumulation of glycogen in the basal cytoplasm. Note the irregular shape of the lumina.

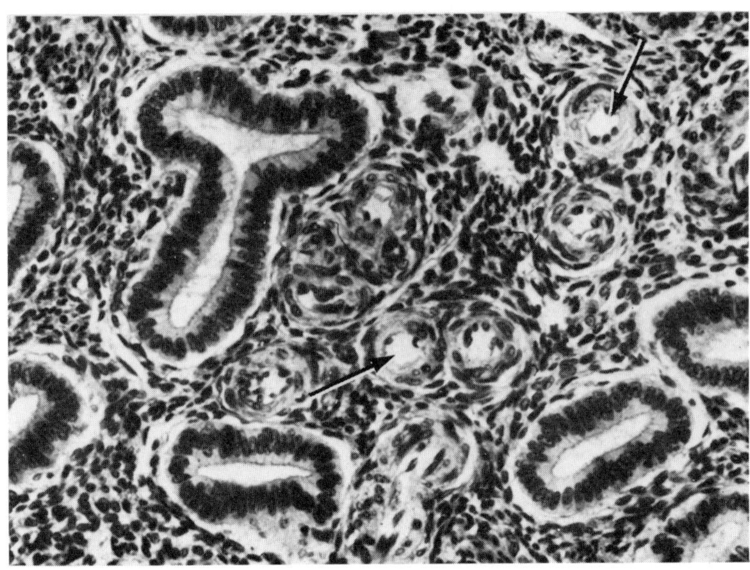

FIGURE 26-11 Light micrograph of the endometrium in the premenstrual phase. Note the constricted lumina of the coiled arteries (arrows).

tissue, secretions, and blood oozing from veins and coiled arteries, is sloughed off leaving a uterine surface without an intact epithelial covering (Fig. 26-12). The basal layer remains intact and induces regeneration of the epithelium at the termination of menstruation. Figure 26-13 demonstrates the basalis layer remaining with its denuded surface (s). Note the remnants of endometrial glands (arrows) and coiled arteries (arrowheads). Following menstruation, the cycle repeats continuously until menopause unless conditions are altered such as an ovum becoming fertilized or hormonal levels being changed. The diagram appearing in Fig. 26-14 provides a three-dimensional summary of the histologic structures pertinent to the endometrial cycle.

The intimate relation between the endometrial and ovarian cycles is shown in Fig.

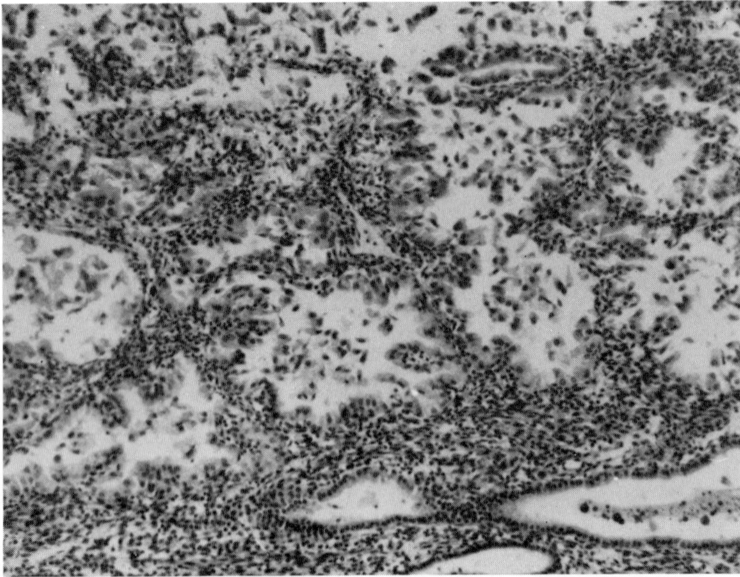

FIGURE 26-12 Light micrograph of the endometrium in the menstrual phase. Note that the surface lacks epithelial covering.

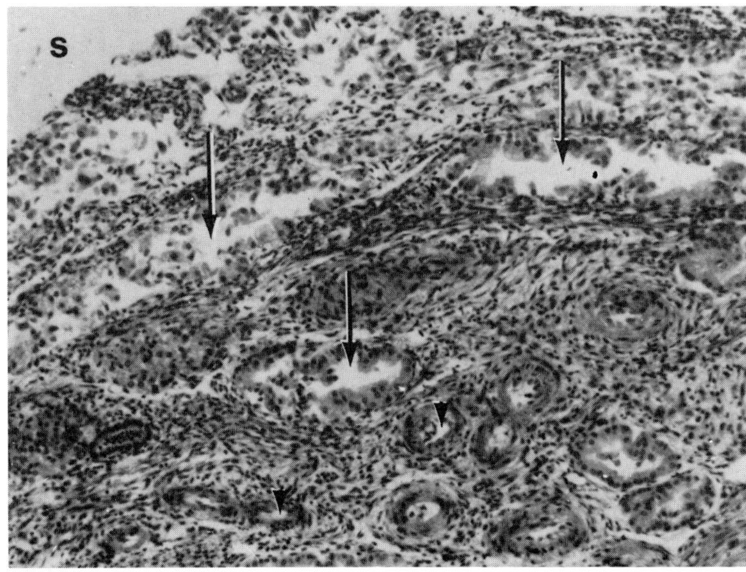

FIGURE 26-13 The basal layer of the endometrium during the menstrual phase. The bases of glands (arrows) and coiled arteries (arrowheads) remain.

25-32, along with hormonal levels. Cyclic uterine changes are induced by estrogen, a product of the developing ovarian follicle, and progesterone, a product of the corpus luteum. If fertilization occurs, the hormonal levels remain high in order to maintain the endometrium and prevent further ovulation; if not, there is a sudden drop in estrogen and progesterone levels, resulting in menstruation.

OVIDUCT

The oviduct, or fallopian tube, transports the ovum from the ovary to the uterus. Its open end, the *infundibulum,* is funnel shaped with a number of fingerlike processes, the *fimbriae* (Fig. 26-15). Its rather extensive irregular surface is lined by ciliated simple columnar epithelium (Fig. 26-16). The simple columnar epithelium, due to its irregular surface topography, often appears pseudostratified. The ciliated epithelium captures the ovum and facilitates its movement down through the oviduct. A second cell type, a nonciliated secretory cell, is present among the ciliated variety (Figs. 26-16 and 26-17). The oviduct then expands into a thin-walled segment, the *ampulla* (see Fig. 25-2) which leads to a short, narrowed region, the *isthmus.* The funnel-shaped ampulla contains a tall columnar epithelial layer and a thin, clearly visible muscle layer (Fig. 26-17). The isthmus is surrounded by a heavy circular muscle layer which aids in ovum propulsion by a wave of contractions. The *uterine portion* (Fig. 26-18) of the oviduct then enters the substance of the uterus.

The epithelium of the oviduct gradually becomes nonciliated as it nears the uterus, becoming similar to the epithelial lining of the uterus. The epithelium of the oviduct is also influenced by the menstrual cycle. During the preovulatory period or the following phase, the cells are taller with more cilia in preparation for the coming ovum. During the luteal phase, or second half of the menstrual cycle, after ovulation the cells become low columnar and devoid of cilia, as there is no need for the transport of ova.

FERTILIZATION, PLACENTA, AND PREGNANCY

FERTILIZATION

Fertilization of the ovum takes place in the oviduct when a sperm penetrates into the

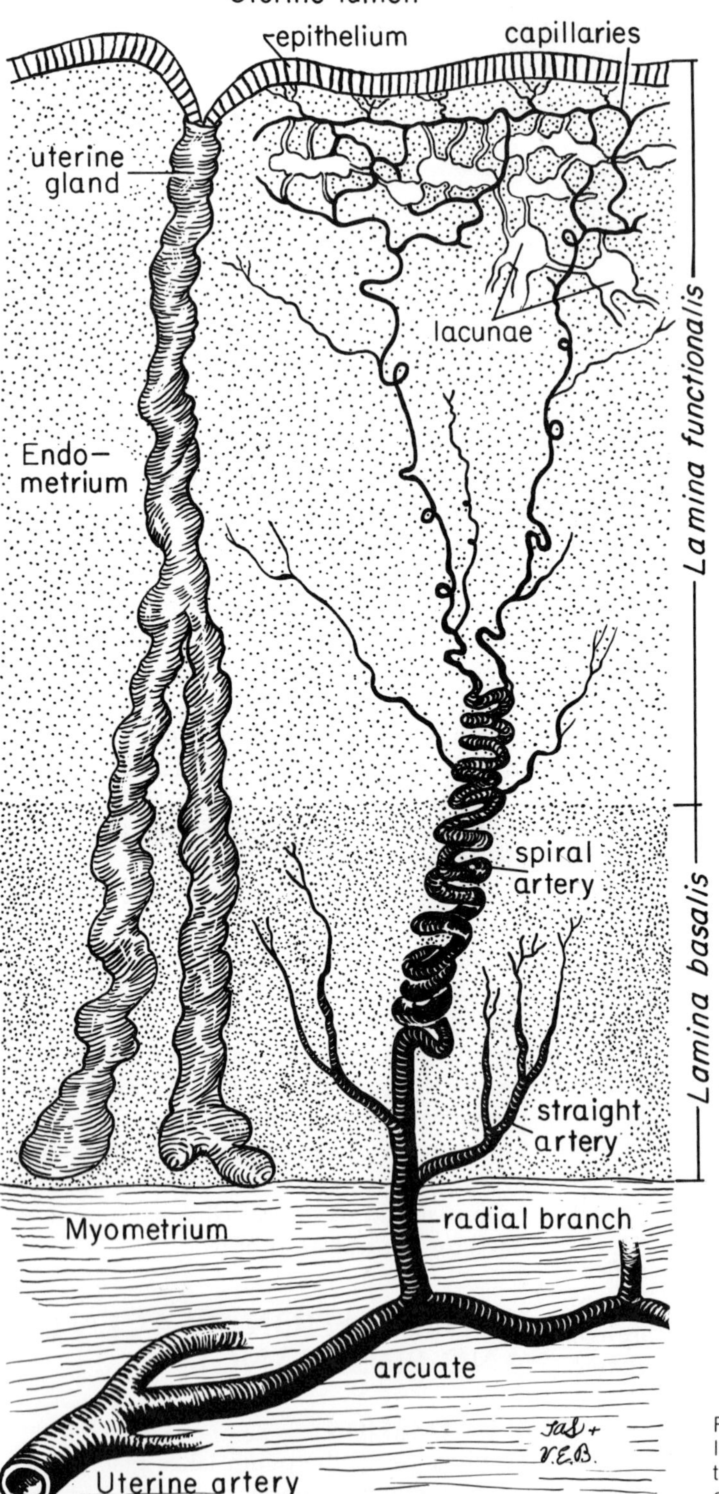

FIGURE 26-14 Diagram illustrating histologic structures participating in the endometrial cycle.

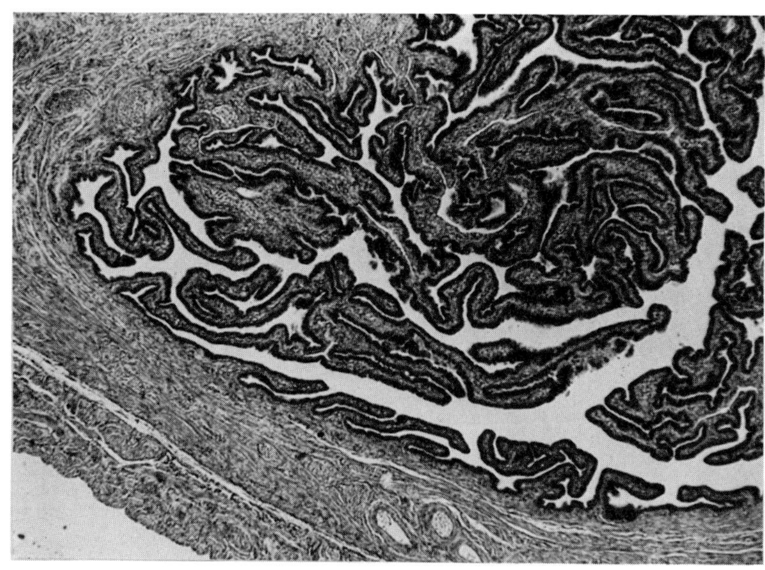

FIGURE 26-15 Low-power light micrograph of the fimbriated end of the oviduct. Note the numerous fingerlike processes, the fimbriae.

ovum. The joining of these two cells results in a single cell with a full genetic complement, and thus development is initiated. It takes approximately 3 days for the fertilized ovum to reach the uterus. During this time a number of divisions occur, transforming the one single-celled ovum into a sphere of cells with a fluid-filled lumen known as a *blastocyst*. By the seventh or eighth day after ovulation the blastocyst rests against the endometrial lining of the uterus which at this time is at the height of its secretory or progestational phase. The blastocyst now consists of an inner cell mass, destined to become the embryo, and an outer cell layer, which is called the *trophoblast layer* or *trophectoderm* (Fig. 26-19).

IMPLANTATION

The trophoblast cells are regarded as highly invasive cells, thus aiding implantation. Somehow, there is a separation of cells on

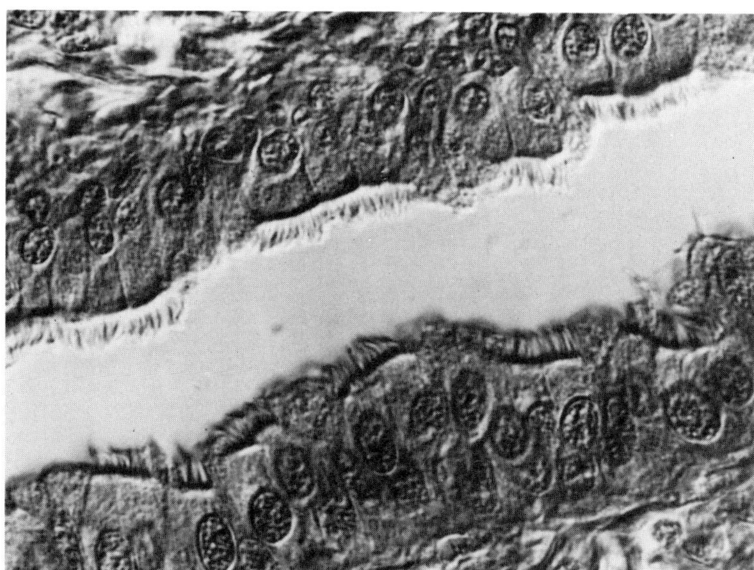

FIGURE 26-16 Light micrograph of the surface epithelium on the fimbriae. The epithelium consists of tall ciliated columnar cells. Among the ciliated cells is a second type of cell, a nonciliated one.

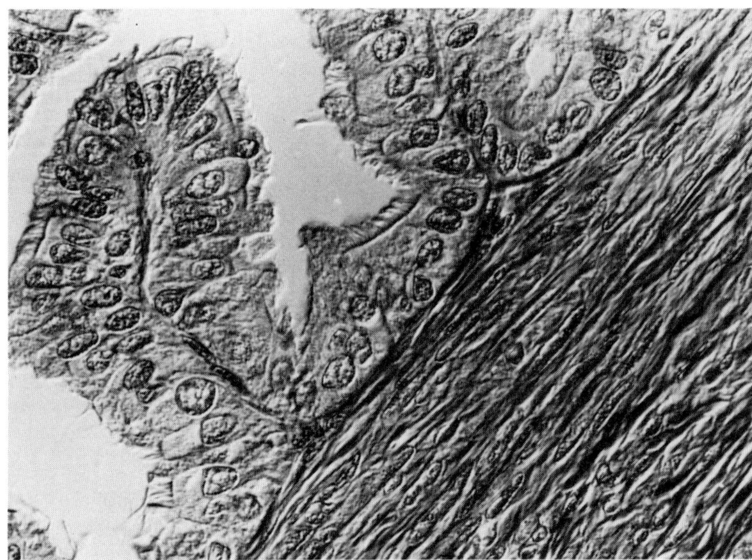

FIGURE 26-17 Phase micrograph of the ampulla. Many folds that are branched extend into the lumen. They are lined by tall columnar ciliated cells. Also note the thin muscle layer.

the endometrial surface that allows the blastocyst to sink into the lining tissue. The area of implantation is then quickly reepithelialized; the blastocyst becoming enclosed by the endometrial tissue. The inner cell mass (embryoblast) of the blastocyst becomes the developing embryo, and the outer trophoblast layer becomes transformed into the fetal portion of the *placenta,* which develops into a nutritional link between mother and fetus (Fig. 26-20).

ORGANIZATION OF THE PLACENTA

The placenta is a combination of interlocking maternal tissues consisting of the modified endometrium and fetal tissues consisting of the chorionic plate with branching processes or villi. For the duration of gestation, ex-

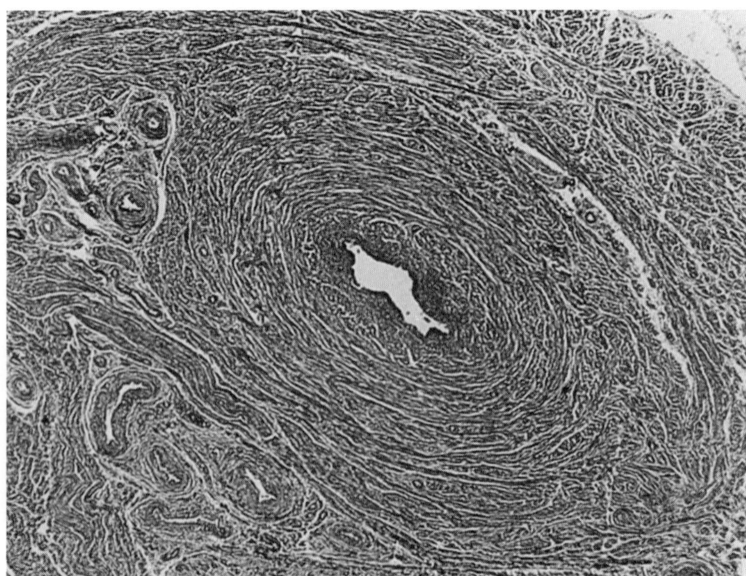

FIGURE 26-18 Light micrograph of the uterine portion of the oviduct. The epithelium is nonciliated and gradually becomes similar to the epithelium lining the uterus.

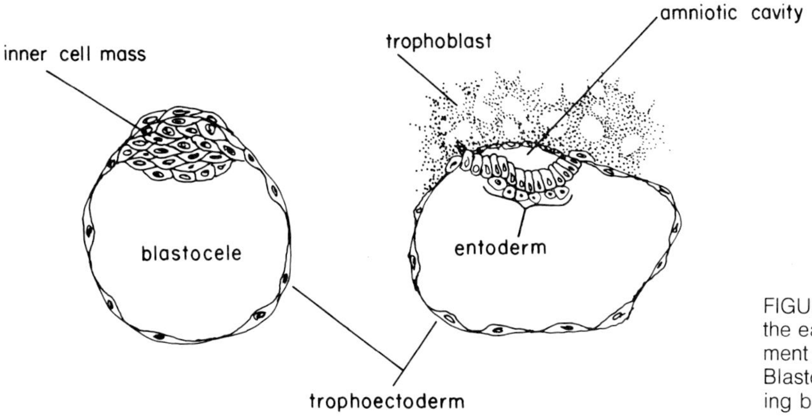

inner cell mass
trophoblast
amniotic cavity
blastocele
entoderm
trophoectoderm

FIGURE 26-19 Diagrams of the early stages of development of human embryos. Blastocyst (left) and implanting blastocyst (right).

change of selected materials between fetal and maternal blood via the placenta occurs (Fig. 26-21).

During parturition the placenta is discharged, leaving the deepest of basal layers of the endometrium to restore the uterine lining. Thus it is very similar in principle to menstruation, although the process of parturition involves much more profound physiologic and histologic changes. The portion of the endometrium that is sloughed off is called the *decidua* and can be divided into three parts. Surrounding the developing fetus like a capsule and adjacent to the uterine lumen is the *decidua capsularis.* The *decidua basalis* is the tissue located between the developing fetus and the myometrium. This is the maternal component of the placenta and is the area where nutritional exchange occurs. It is in intimate contact with the chorionic plates and villi of the fetal placenta. The decidua which lines the remainder of the uterine lumen is called *decidua parietalis.* It has nothing to do with the fetal development,

endometrial connective tissue
(maternal)
trophoblastic lacunae
trophoblast layer
yolk sac
endodermal
cells

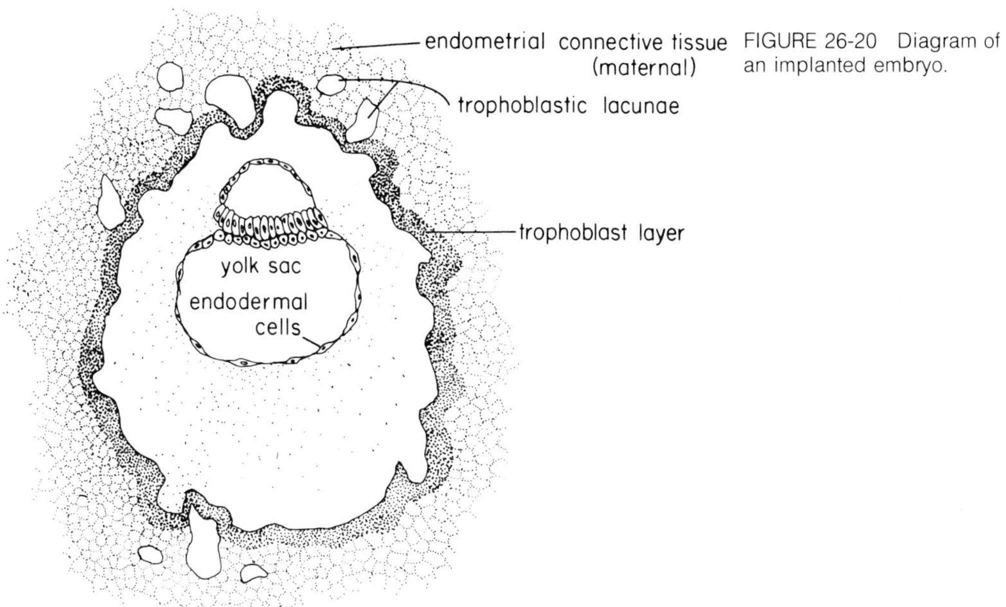

FIGURE 26-20 Diagram of an implanted embryo.

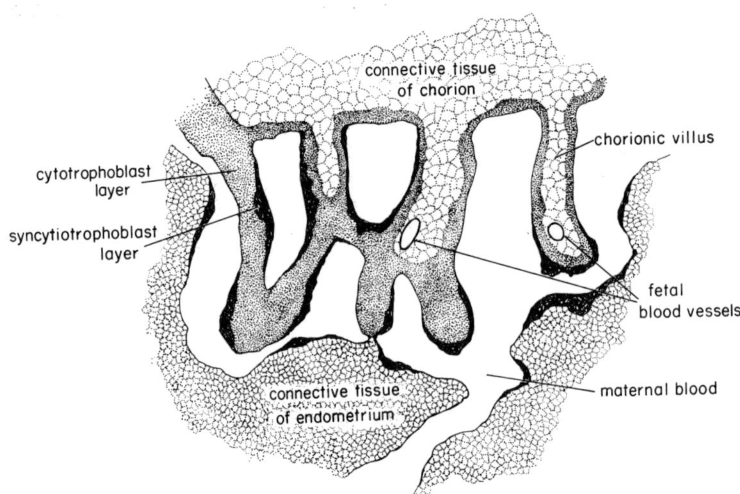

FIGURE 26-21 Diagram of the interrelationship between maternal and fetal tissues in the placenta.

but follows other placental portions in the sloughing process. As the fetus grows to fill the entire luminal space of the uterus, the decidua capsularis comes in contact with the decidua parietalis and they fuse. These features are diagramed in Fig. 26-22. The fetal portion of the placenta develops as the trophoblast cells invade the decidua basalis. They produce irregular fingerlike projections, *chorionic villi,* which serve to anchor the chorionic plate of the fetus to the decidua (anchoring villi) and to provide a large surface area in close contct with maternal blood for nutritional exchange (Fig. 26-21).

Trophoblasts which begin to fuse along the surface eventually produce two different layers of cells. The syncytial layer with no distinguishable cell boundaries is called the *syncytiotrophoblast layer.* It is adjacent to the spaces filled with maternal blood. The inner layer, consisting of a simple cuboidal epithelium, is called the *cytotrophoblast layer* since the cells are well defined. The layer is more prominent during the early phase of pregnancy.

All of the nutritional requirements for the developing fetus must be met by transplacental transport involving the two troph-

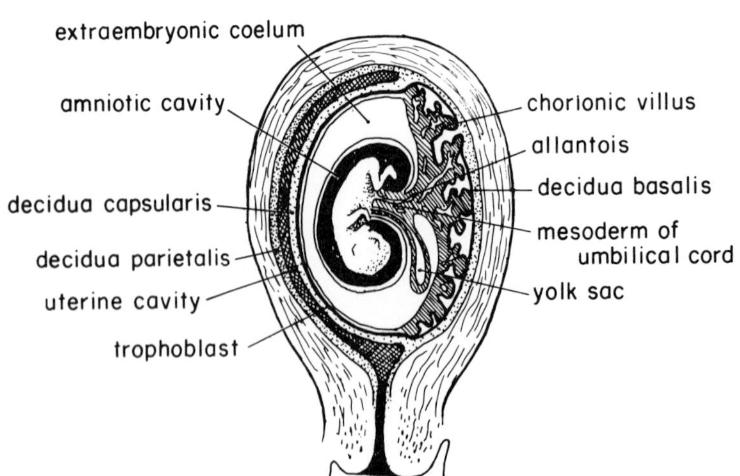

FIGURE 26-22 Drawing of human pregnancy, showing the obliterated uterine lumen, along with deciduae capsularis, parietalis, and basalis. Note that the decidua basalis is largely responsible for the establishment of the definitive discoid placenta.

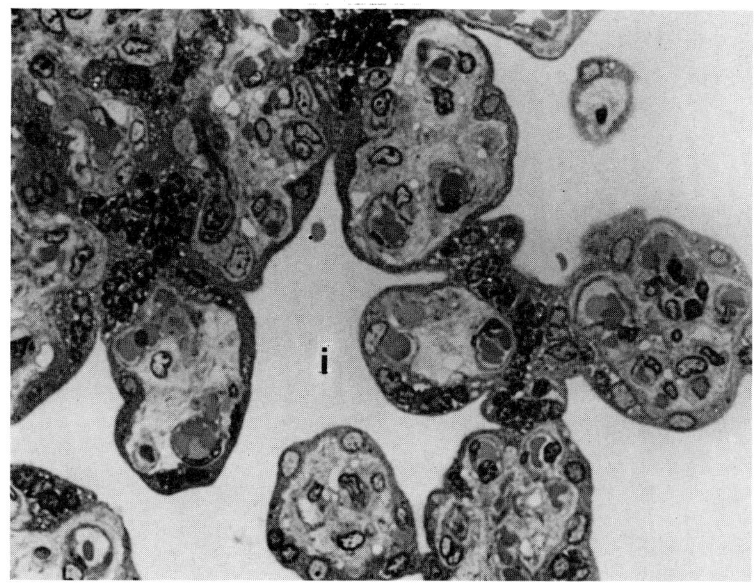

FIGURE 26-23 Light micrograph of chorionic villi from a placenta at term. (i) Intervillous space.

oblast layers. These trophoblast cells change continuously throughout pregnancy, not only in number but also in cytology. Toward the latter half of pregnancy when the fetus starts to grow more rapidly, the trophoblast layer becomes extremely attenuated. The cytotrophoblasts decrease in number, and the syncytiotrophoblasts show less complicated intracellular structures. Figure 26-23 is an LM of a number of human chorionic villi from a placenta at term in which the *intervillous space* of the maternal circulation shows a couple of red blood cells (RBCs). The individual villi contain fetal vessels and connective tissue which is surrounded by vacuolated syncytiotrophoblasts (Fig. 26-24, arrows).

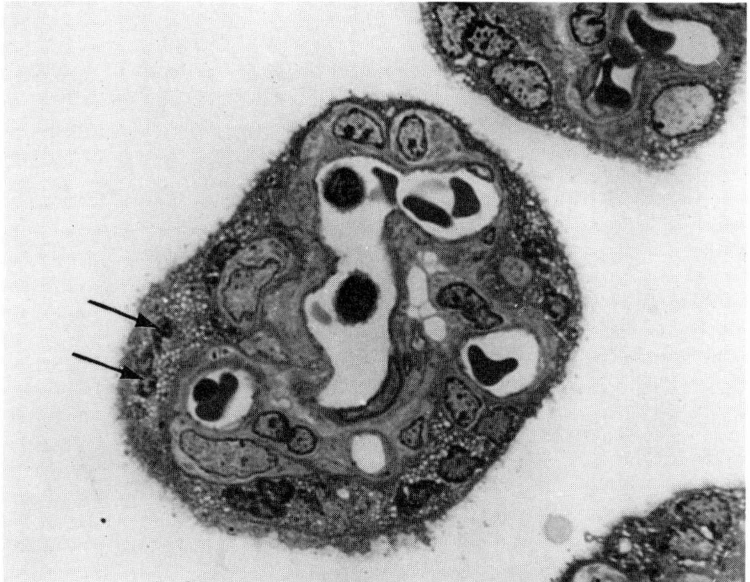

FIGURE 26-24 A chorionic villus seen at high magnification. It has a center of fetal connective tissue containing blood vessels. The cytoplasm of the surrounding syncytiotrophoblasts is vacuolated (arrows).

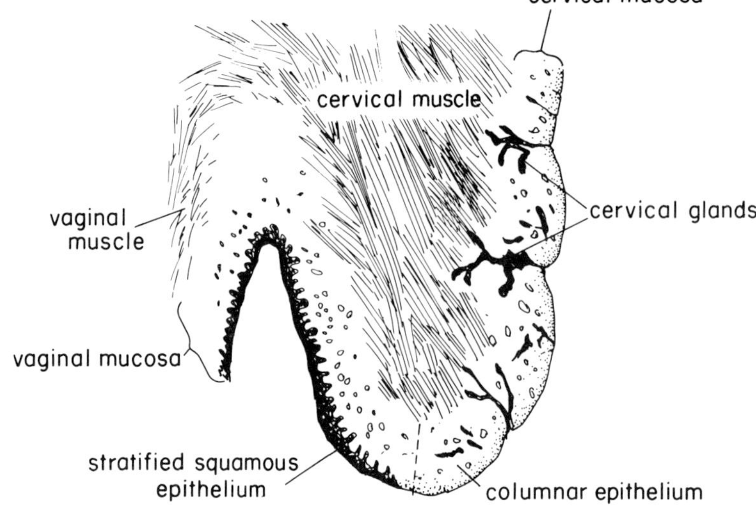

FIGURE 26-25 A drawing of a sagittal section through the cervical region of the uterus and the upper part of the vagina.

The trophoblast cells are also responsible for the production of *chorionic gonadotropin* as well as the two steroid hormones necessary for the maintenance of pregnancy, estrogen and progesterone. In fact, there is a point at which the placenta is capable of maintaining pregnancy independent of the ovary. This point is reached toward the end of the first trimester in human beings, coinciding with the maximum rise of chorionic gonadotropin which is needed prior to the onset of estrogen and progesterone production by the trophoblast cells, notably by syncytiotrophoblasts.

VAGINA

The vagina, which extends from the uterine cervix to the external genitalia, is basically a hollow tube with three layers in its wall; the mucosa, muscularis, and fibrosa layers. The muscular layer is continuous with the uterine myometrium (see Fig. 26-25). The mucosa consists of a stratified squamous epithelial lining, rich in glycogen, and the underlying lamina propria. There is an abrupt change in epithelium from the cervix of the uterus (simple columnar) to the vagina. The juxtapositioning of these two types of epithelium seems to predispose this area to malignant transformation. A similar situation was noted in the cardioesophageal junction discussed in Chap. 19. The vaginal epithelium is also sensitive to endocrine changes. During the preovulatory or estrogenic phase there is a thickening of the cellular layers, whereas during the postovulatory phase several cell layers slough off, thus showing a thinned appearance.

The lamina propria is directly continuous with the muscularis and is rich in lymphocytes and polymorphonuclear neutrophil leukocytes. Few glands are present in the vagina. Therefore, the mucus discharged by the vagina originates from the cervical glands of the uterus. However, the vaginal epithelium is rich in glycogen which tends to support the microbial flora, making the vaginal flow acidic.

The vaginal muscularis consists of longitudinally oriented smooth muscle bundles that are continuous with the myometrium. A few circularly oriented smooth muscle fibers can be found, and many elastic fibers are also seen in the connective tissue of this region. Striated muscle cells can be located in the wall near the vaginal orifice.

The fibrosa, or outer layer, of the vagina consists of dense FECT, containing abundant elastic fibers. This fibrous layer supports the vagina and serves to connect it to surrounding structures.

REVIEW SECTION

1. Figure 26-26 is a section through the uterus. What are the irregular structures occupying the left half of the tissue? What type of tissue is in the right half of the picture? What two layers of the uterus are shown? When during the menstrual cycle will this view be found?

The cross sections of arteries and glands in the tissue next to the muscular myometrium identify the layer on the left as the basalis portion of the endometrium. The degenerating tissue on the right indicates that the functionalis being lost in the menstrual flow, making this the middle of the menstrual phase.

2. Figure 26-27 is another section of the uterus. In comparison with Fig. 26-26, what changes have occurred? Which phase of the endometrial cycle is this? What phase in the ovarian cycle occurs continuously?

When ovulation occurs, the corpus luteum secretes progesterone and estrogen. This changes the late proliferative phase into the early secretory phase, with a corresponding increase in mucosal thickness and glandular development.

3. Identify the two types of epithelial cells shown in Fig. 26-28. Where in the female reproductive system are they found? How will this tissue be affected by hormonal changes?

This figure shows a portion of the oviduct which is lined by a simple columnar ciliated epithelium. Occasional nonciliated cells that are secretory in function are present between ciliated cells. Since the ciliated epithelium facilitates the movement of the fertilized ovum along the oviduct, the ciliated cells are taller and more numerous during the follicular phase of the cycle.

BIBLIOGRAPHY

Amoroso, E. C.: "Histology of the Placenta," *Brit. Med. Bull.* **17**:81–90 (1961).

Blandau, R. J., and K. Moghissi: *The Biology of the Cervix,* University of Chicago Press, Chicago, 1973.

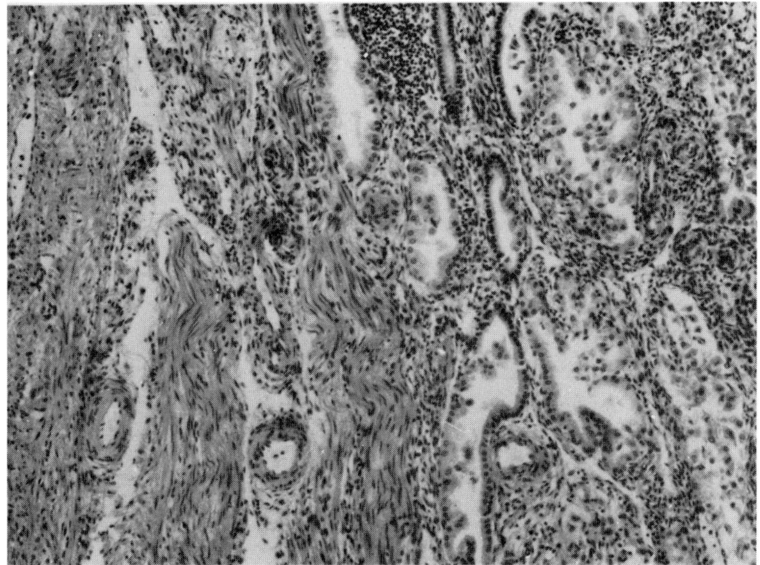

FIGURE 26-26

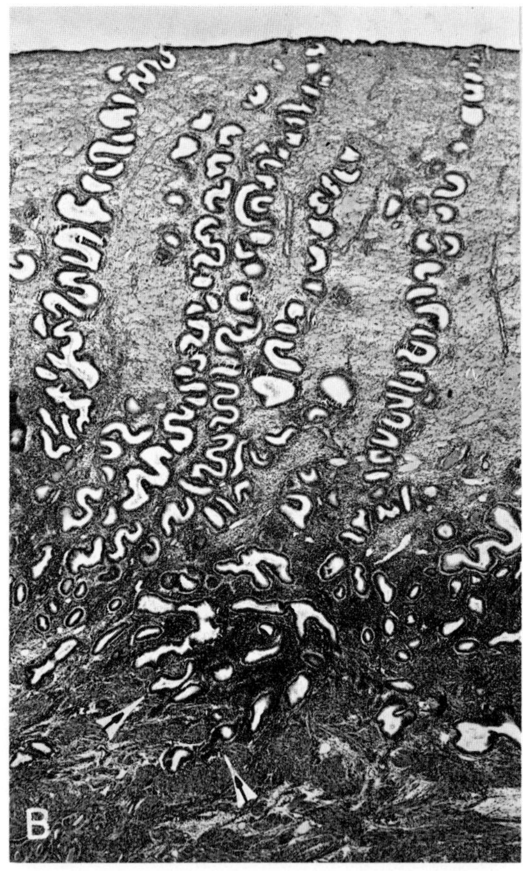

FIGURE 26-27

Bloom, W., and D. W. Fawcett: *A Textbook of Histology,* 10th ed., Saunders, Philadelphia, 1975.

Finn, C. A., and D. G. Porter: *The Uterus.* Handbooks of Reproductive Biology, Elek, London, 1974.

Hafez, E., and R. J. Blandau: *The Mammalian Oviduct,* University of Chicago Press, Chicago, 1969.

Hertig, A. T.: *Human Trophoblast,* Charles C Thomas, Springfield, Ill., 1968.

Schmidt-Mattiessen, H.: *The Normal Human Endometrium,* McGraw-Hill, New York, 1963.

Weiss, L., and R. O. Greep: *Histology,* 4th ed., McGraw-Hill, New York, 1977.

Young, W. D. (ed.): *Sex and Internal Secretions,* vols. 1 and 2, Williams & Wilkins, Baltimore, 1961.

ADDITIONAL READINGS

Baker, B. L., A. Hook, and A. E. Severinghaus: "The Cytological Structure of Human Chorionic Villus and Decidua Parietalis," *Am. J. Anat.* **74:**297–327 (1944).

Blandy, J. R. (ed.): *The Biology of the Blastocyst,* University of Chicago Press, Chicago, 1971.

Borell, V., O. Nilsson, and A. Westman: "The Cyclical Changes Occurring in the Epithelium Lining of the Endometrial Glands: An Electron Microscopic Study in the Human Being," *Acta. Obstet. Gynecol. Scand.* **38:**364–377 (1959).

Boyd, J. D., and W. J. Hamilton: "Development of

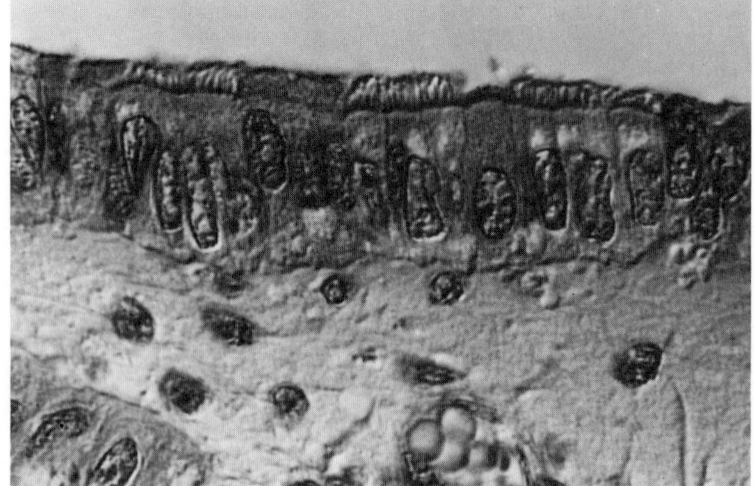

FIGURE 26-28

the Human Placenta in the First Three Months of Gestation," *J. Anat.* **94:**297–328 (1960).

———, and ———: *The Human Placenta,* Haffer, Cambridge, Eng., 1970.

Brenner, R. M.: "Renewal of Oviduct Cilia during the Menstrual Cycle of the Rhesus Monkey," *Fertil. Steril.* **20:**599–611 (1969).

Crawford, J. M.: "The Fetal Placental Circulation," *J. Obstet. Gynaecol. Brit. Europ.* **63:**542–547 (1956).

Cregioire, A. T., O. Kandil, and W. J. Ledger: "The Glycogen Content of Human Vaginal Epithelium," *Fertil. Steril.* **22:**64–68 (1971).

Enders, A. C.: "Formation of the Syncytium from Cytotrophoblasts in the Human Placenta," *Obstet. Gynecol.* **25:**378–386 (1965).

———: "Fertilization, Cleavage and Implantation," in E. S. E. Hafez (ed.), *Reproduction and Breeding Techniques for Laboratory Animals,* Lea & Febiger, Philadelphia, 1970, pp. 137–156.

———, and S. Schlafke: "Cytological Aspects of Trophoblast-Uterine Interaction in Early Implantation," *Am. J. Anat.* **125:**1–30 (1969).

———, and ———: "Penetration of the Uterine Epithelium during Implantation in the Rabbit," *Am. J. Anat.,* **132:**219–240 (1971).

Gay, V. L., A. R. Midgley, and G. D. Niswender: "Pattern of Gonadotropin Secretion Associated with Ovulation," *Fed. Proc.* **29:**1880–1887 (1964).

Hertig, A. T.: "Gastrointestinal Hyperplasia of the Endometrium," *Lab. Invest.* **13:**1153–1191 (1964).

Markee, J. E.: "The Morphological and Endocrine Basis for Menstrual Bleeding," *Prog. Gynecol.* **2:** 63–100 (1950).

Papanicolou, G. N.: "The Sexual Cycle in the Human as Revealed by Vaginal Smears," *Am. J. Anat.* **53:**519–637 (1933).

Rhodin, J. A. G., and J. Terzakis: "The Ultrastructure of the Human Full-term Placenta," *J. Ultrastruct. Res.* **6:**88–106 (1962).

Salvatore, C. A.: "The Growth of Human Myometrium and Endometrium: Studies of Cytological Aspects," *Anat. Rec.* **108:**245–259 (1950).

Schlafke, S., and A. C. Enders: "Cellular Basis of Interaction between Trophoblast and Uterus at Implantation," *Biol. Reprod.* **12:**41–65 (1975).

Smith, B. G., and E. K. Brunner: "The Structure of the Human Vaginal Mucosa in Relation to the Menstrual Cycle and to Pregnancy," *Am. J. Anat.* **54:**27–86 (1934).

Terzakis, J. A.: "The Ultrastructure of Normal Human First Trimester Placenta," *J. Ultrastruct. Res.* **9:**268–284 (1963).

Vickery, B. H., and J. P. Bennett: "The Cervix and Its Secretions in Mammals," *Physiol. Rev.* **48:** 135–154 (1968).

Wislocki, G. B., and H. S. Bennett: "The Histology and Cytology of the Human and Monkey Placenta, with Special Reference to the Trophoblast," *Am. J. Anat.* **73:**335–449 (1943).

Wynn, R. M., and J. A. Harris: "Ultrastructural Cyclic Changes in the Human Endometrium. I. Normal Preovulatory Phase," *Fertil. Steril.* **18:** 623–648 (1967).

———, ———, and R. S. Wooley: "Ultrastructural Cyclic Changes in the Human Endometrium. II. Normal Postovulatory Phase," *Fertil. Steril.* **18:** 721–738 (1967).

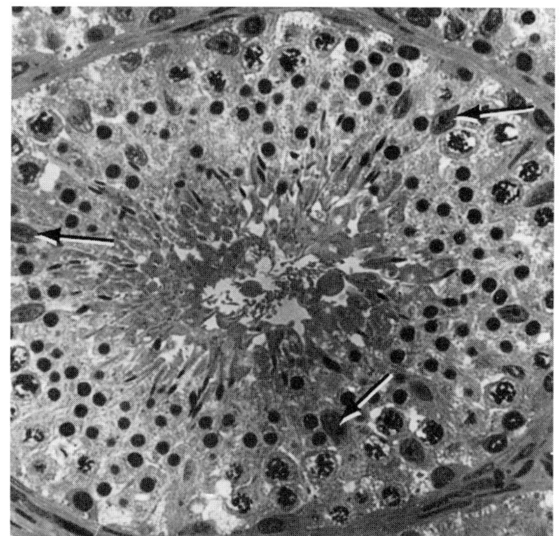

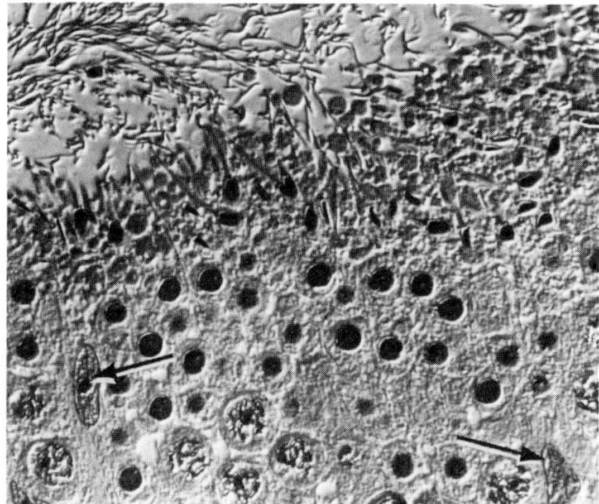

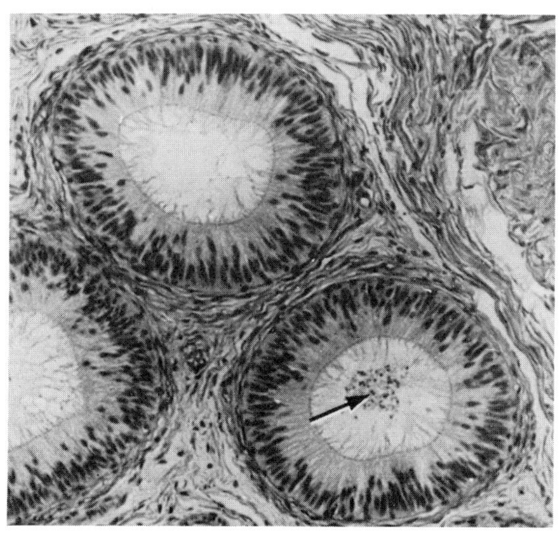

27 MALE REPRODUCTIVE SYSTEM

OBJECTIVES

Upon completion of this chapter, the student will be able to:

1 Identify on a diagram of the testis the following structures:
(a) septum (b) tunica albuginea (c) seminiferous tubules (d) straight tubules (e) rete testis (f) ductuli efferentes (g) duct of the epididymis (h) prostate gland (i) seminal vesicles

2 Identify the location and structure of the following:
(a) spermatogonia (b) primary spermatocytes (c) secondary spermatocytes (d) spermatids

3 Define the following terms:
(a) blood-testes barrier (b) meiotic (reduction or maturation) division I and II (c) spermiogenesis

4 Describe the cellular changes that occur during spermatogenesis (production of spermatids).

5 Identify the location, structure, and function of the following:
(a) Sertoli cell (b) interstitial cell (Leydig cell)

6 Identify the histologic structure of the following:

(a) straight tubules (b) rete testis (c) ductuli efferentes (d) ductus epididymidis (e) ductus deferens

The male reproductive system is composed of a series of structures concerned with the production and delivery of spermatozoa and an endocrine portion responsible for the production of androgenic hormones which enable the maintenance of secondary sexual characteristics of the male (Fig. 27-1). The paired testes, various ducts for transport of spermatozoa, associated glands, and the penile portion of the urethra make up the male reproductive system as shown in Fig. 27-2.

TESTIS

Since the testis produces spermatozoa, it could be regarded as a cytogenic gland. The parenchyma of the testis is made up of numerous *seminiferous tubules* where the sperm production takes place. Because the seminiferous tubules join to form various segments of a duct system, the testis is a cytogenic compound tubular gland. Within the scrotum, supported by associated connective tissue and cremaster muscle layers, the testis is enclosed in a dense fibroelastic capsule, the *tunica albuginea,* which sends out *septa* to subdivide the testis into many lobules (Fig. 27-2). Under the tunica albuginea is a less dense, richly vascularized tissue, the *tunica vasculosa*. The third and outermost coat, *tunica vaginalis,* is found covering the anterior and lateral aspects of the testis. This covering is a closed serous sac originating during development as a diverticulum from the peritoneal cavity. The visceral layer of the tunica vaginalis adheres to the tunica albuginea, while the parietal layer lines the scrotum. Capping each testis is the epididymis containing a convoluted tube of about 4 m in length. The convoluted tube of the epididymis, *ductus epididymidis,* connects the testis with the *ductus deferens* (or *vas deferens*). The ductus deferens in turn transports the sperm

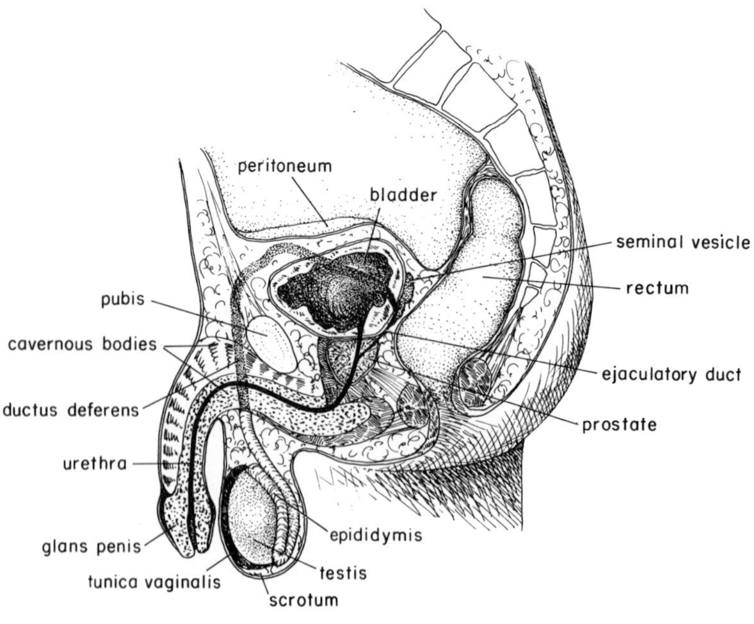

FIGURE 27-1 Schematic median section of the male pelvis.

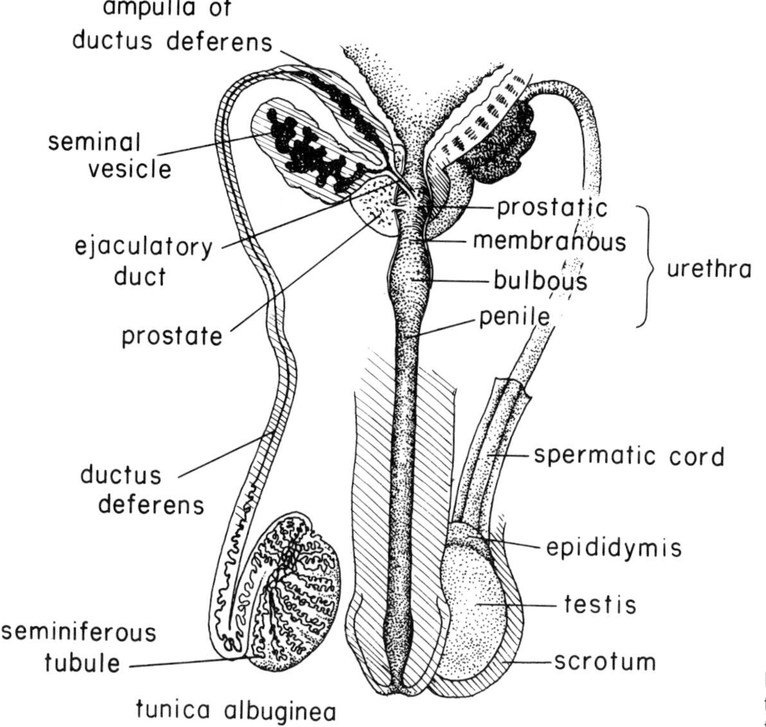

FIGURE 27-2 Diagram of the male reproductive system.

to the prostate. At the level of the prostate, the vas deferens becomes the ejaculatory duct, passing through the prostate and in turn joining the urethra. Posterior to the prostate where the vas deferens becomes the ejaculatory duct is a small pouchlike gland, the seminal vesicle, which provides alkaline fluid and glucose to the semen.

The penile urethra carries sperm through the penis. A cross section of the penis reveals the urethra and a specialized, highly vascular erectile tissue that make up its body.

SEMINIFEROUS TUBULE

The testis is divided by septa into about 40 lobules. Each lobule contains a number of continuous and coiled seminiferous tubules where sperm is produced. Figure 27-3 is a

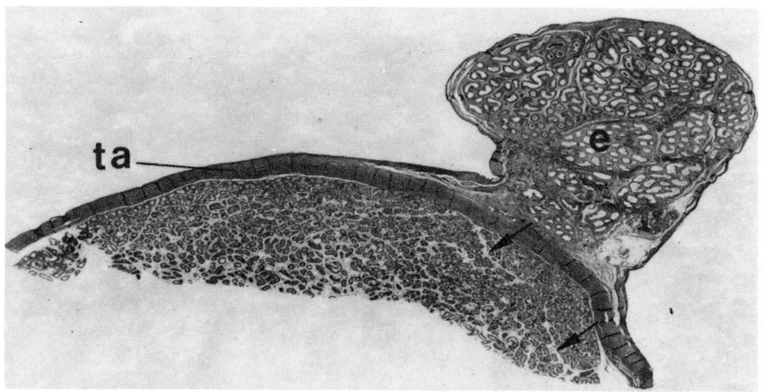

FIGURE 27-3 Light micrograph of a section through the posterior portion of the testis including the epididymis. (ta) Tunica albuginea. (e) Epididymis. (arrows) Septa.

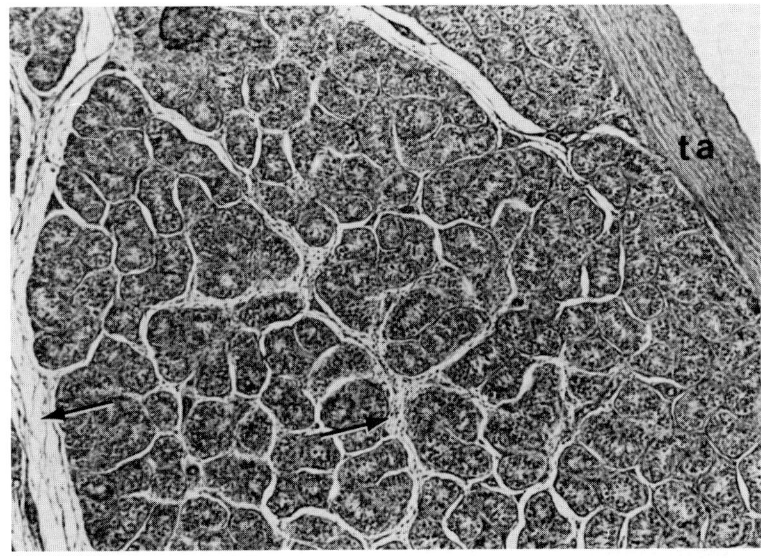

FIGURE 27-4 Light micrograph of an immature testis showing the appearance of the seminiferous tubules. (ta) Tunica albuginea. (arrows) Septa.

section through the posterior aspect of a testes depicting the tunica albuginea (ta), a number of ill-defined septa (arrows), and numerous seminiferous tubules. At the upper right-hand region is the epididymis (e) which is surrounded by a connective tissue capsule outside the tunica albuginea.

The seminiferous tubules may form single, double, or triple arches. Different arches of the same tubule may be in different lobules, passing through the incomplete interlobular septa. The seminiferous tubule is surrounded by a layer of loose FECT and a basement membrane upon which the lining epithelial cells rest. Figure 27-4 shows the appearance of seminiferous tubules in an immature testis of a child and the arrangement of the connective tissue that invests them. The tubules lack a lumen. Fine strands of connective tissue surround individual tubules, while heavier bundles of FECT make up the septa (arrows) which are continuous with the thick tunica albuginea (ta).

In the mature testis (Fig. 27-5), the sem-

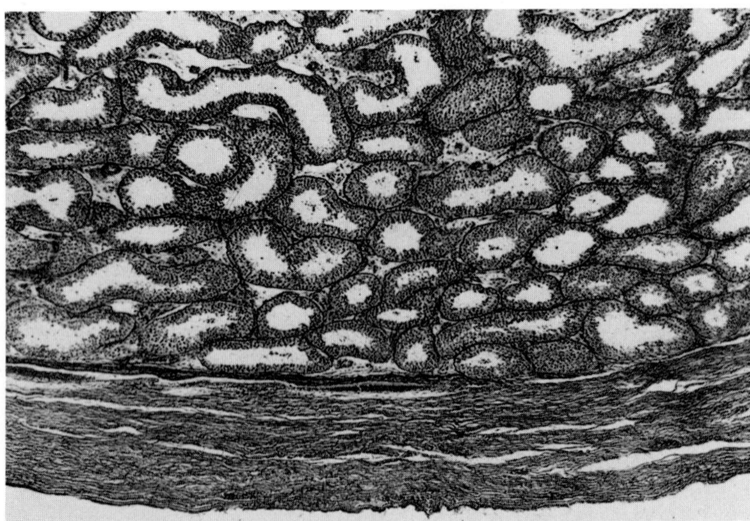

FIGURE 27-5 Light micrograph of a mature testis showing the appearance of the seminiferous tubules.

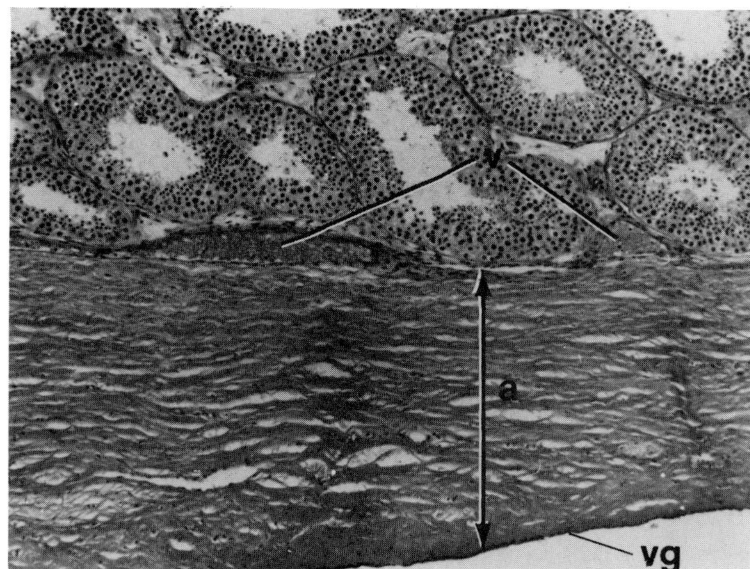

FIGURE 27-6 The testicular capsule viewed at higher magnification. (v) Tunica vasculosa. (a) Tunica albuginea. (vg) Tunica vaginalis.

iniferous tubules appear as large, tortuous tubular structures with a clear lumen. The connective tissue surrounding the testis is thick and shows differentiation of the three layers (Fig. 27-6): that is, the inner tunica vasculosa (v), middle tunica albuginea (a), and the outer tunica vaginalis (vg.) As mentioned earlier, the tunica vaginalis is covered by a mesothelium, but is present only along the anterior and lateral aspects of the testis.

The seminiferous epithelium is stratified and consists of two types of cells, spermatogenic and Sertoli. Spermatogenic cells with round nuclei of different chromatin patterns are situated between Sertoli cells with oblong nuclei (Fig. 27-7). The Sertoli cells extend

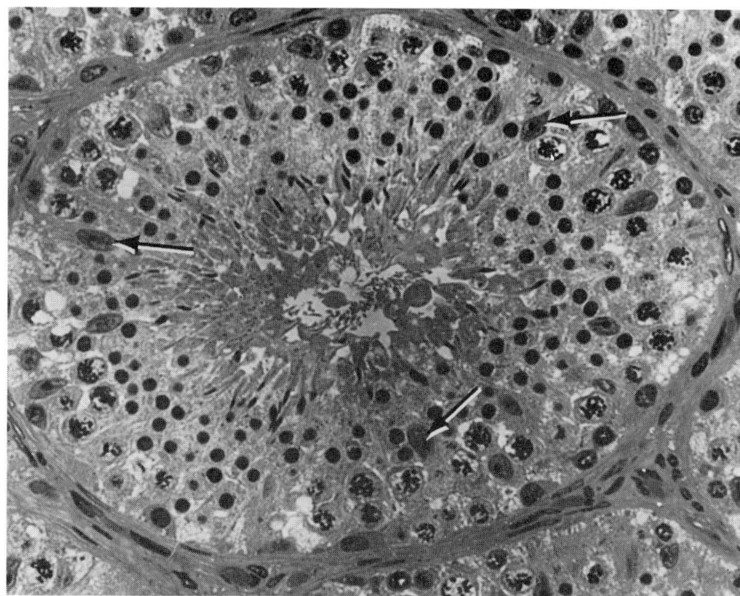

FIGURE 27-7 Light micrograph of a seminiferous tubule. Sertoli cells (arrows) are situated among the spermatogenic cells. They extend from the basement membrane to the lumen.

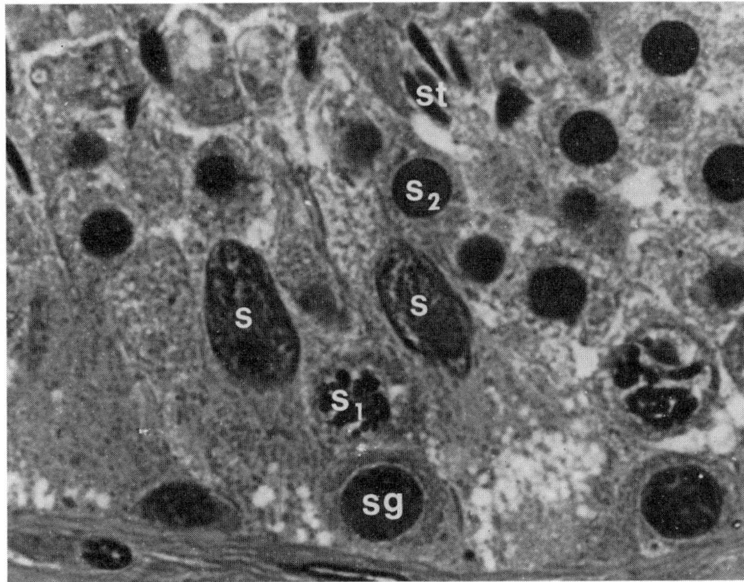

FIGURE 27-8 Portion of seminiferous tubule viewed at higher magnification. Sertoli cells (s) are interspersed among spermatogenic cells and extend from the basement membrane to the lumen. They are thought to provide nutrition to the developing sperm. Beginning at the basement membrane going toward the lumen, cells in different stages of spermatogenesis (sperm development) can be distinguished. (sg) Spermatogonia. (s_1) Primary spermatocytes. (s_2) Secondary spermatocytes. (st) Spermatids.

from the basal lamina to the lumen. In a sexually mature male, spermatogenic cells are layered between the tall cells of Sertoli in four to eight levels representing different stages of sperm development (Figs. 27-8 and 27-9).

Several structurally different cell types can be distinguished among the spermatogenic cells; they are *spermatogonia, primary spermatocytes, secondary spermatocytes,* and *spermatids.* Adjacent to the basement membrane of the seminiferous tubule are the *spermatogonia* (sg in Fig. 27-8) which represent the initial stage of the developing sperm. They are made up of two populations, namely, type A cells which maintain the stem cell population and type B cells which differentiate into other spermatogenic cells. Spermatogonia are spherical or cuboidal in shape, with a diameter of about 12 μm. Their nuclei are rounded with dense granular chromatin. Spermatogonia divide by mitosis to maintain their base level population, while allowing some to develop into primary spermatocytes (Fig. 27-9). Following the last mitosis, those spermatogonia destined to become primary spermatocytes undergo the S phase of the cell cycle. Thus, each primary spermatocyte has twice the normal number of chromosomes (that is, $2 \times 2n$).

Primary spermatocytes grow to a diameter of 17 to 19 μm and lie closer to the lumen than spermatogonia (s_1 in Fig. 27-8). The chromatin pattern of the large vesicular nuclei varies, depending on the stage of prophase the cell is in. In human beings, the prophase of primary spermatocytes lasts more than 3 weeks. It is for this reason that many primary spermatocytes show the chromosomal structure characteristic of a prophase nucleus. Each primary spermatocyte divides by meiosis into two smaller secondary spermatocytes that are displaced toward the lumen (s_2 in Figs. 27-8, and 27-9). The secondary spermatocytes are characterized by a small, dense nucleus. Soon after formation, each secondary spermatocyte divides again by meiosis to form two spermatids.

Spermatids are small and are the innermost cells in the seminiferous tubule (st in Figs. 27-8 and 27-9). They undergo structural changes as they mature into spermatozoa, without further divisions. The maturation of spermatids into spermatozoa occurs while the spermatids are invested in the apical cytoplasm of Sertoli cells. The growing tails

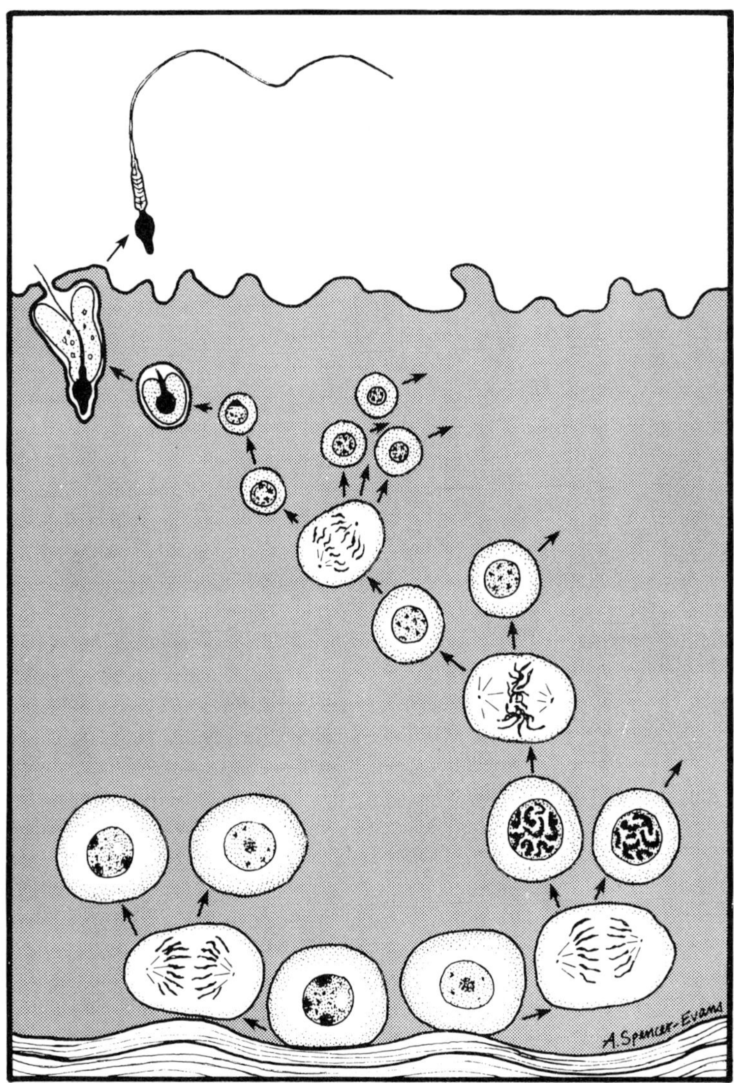

FIGURE 27-9 Diagram illustrating the various stages of sperm development.

of spermatozoa always extend into the lumen. Figure 27-10 depicts the slender nuclei of growing spermatids and spermatozoa clustered around the apexes of the tall Sertoli cells. Free spermatozoa are rarely seen in seminiferous tubules, as they rapidly pass to the epididymis upon maturation. With the exception of a small number of spermatogonia, type A spermatogonia, most spermatogenic cells are incomplete with respect to cytoplasmic division. Therefore, cytoplasmic connections through these cells have been found on the electron micrograph (EM).

Sertoli cells are tall, irregularly shaped columnar cells that extend from the basal lamina to the lumen of the seminiferous tubule (Figs. 27-7, 27-8, and 27-10). Spermatogenic cells located between Sertoli cells cause depressions along the lateral aspects of the Sertoli cells. The nucleus of the Sertoli cell is ovoid and pale staining, having a finely dispersed chromatin pattern and one or more prominent nucleoli. Occasionally, crystalloid structures appear in the cytoplasm. Sertoli cells are believed to provide nutritional support for the developing sperm and contribute

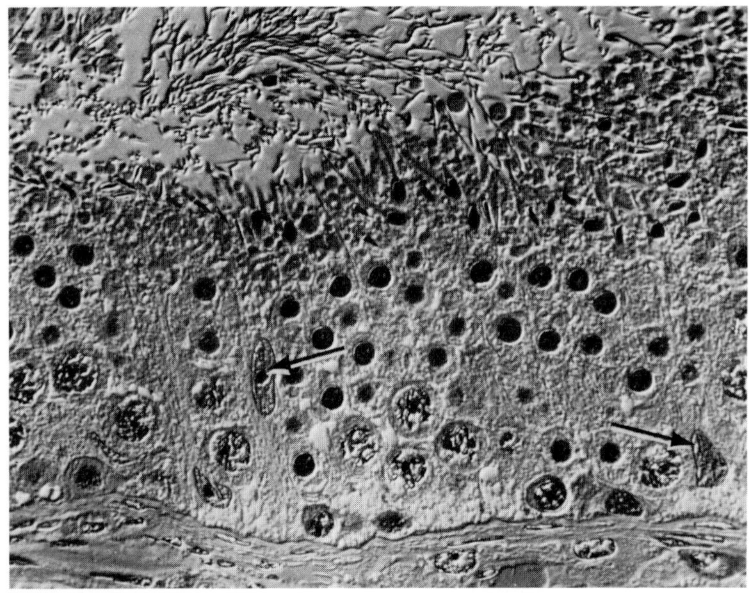

FIGURE 27-10 Phase light micrograph of seminiferous tubule. Sertoli cells are labeled with arrows.

to the organization of the seminiferous tubule.

BLOOD-TESTIS BARRIER

During the past decade the structure of Sertoli cells has been studied in great detail. In the electron microscope, the Sertoli cell is seen to have a cytoplasm that contains many intracellular fibrils and microtubules that are longitudinally oriented (Fig. 27-11). The surface plasma membrane shows numerous indentations which provide space for developing spermatogenic cells at both the lateral walls and the apical surface where differen-

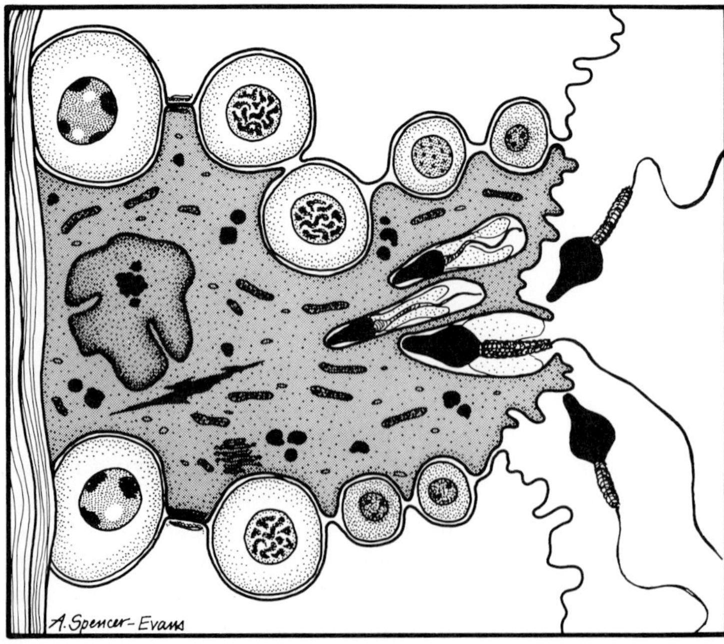

FIGURE 27-11 Diagram illustrating the blood-testis barrier.

tiating spermatids are present. In addition, the lateral surface of adjacent Sertoli cells have spacialized junctional complexes that are characterized by many parallel lines of fusion of the plasma membranes. In fact, these junctional areas appear to make up a belt around the entire circumference of individual Sertoli cells. The use of electron-dense marker molecules has shown that such markers can reach up only to the junctional belt, but not beyond. Thus, a small *basal compartment* and a large *adluminal compartment* are produced (Fig. 27-11). Therefore, bloodborne substances are allowed to reach only the basal compartment, making the adluminal compartment a privileged space. The junction belt acts in a manner similar to that of the blood-brain or the blood-thymus barrier. The functional significance of this blood-testis barrier has been speculated in different ways, but its role as an immunologic shield enjoys greater popularity than other theories. What is clear is that spermatogonia which have the same diploid quantity ($2n$) of DNA as other somatic cells reside in the basal compartment, whereas the primary spermatocytes and other further advanced cells having tetraploid ($4n$) or haploid ($1n$) complements of recombined genes are segregated in the adluminal space. Coincidental with this nuclear change is the development of antigenicity at the surface of differentiating spermatocytes.

The supportive role of Sertoli cells noted above, therefore, includes metabolic and possible immunologic functions, as well as the physical support that has been known since the nineteenth century. Furthermore, recent studies have shown that Sertoli cells respond to follicle-stimulating hormone (FSH) from the anterior pituitary. A class of protein called androgen-binding protein (ABP) is known to be produced by the Sertoli cell. The ABP recognizes and binds androgenic steroids such as testosterone that are essential for maintenance of spermatogenesis.

In human beings, spermatogenesis is a continuous process that proceeds in a wave-like sequence, and therefore various stages of spermatid production may be seen at any given point along a seminiferous tubule. Several different stages of spermatogenesis in the human being, as well as in other animals, have been found. Figure 27-12 depicts examples of several different stages of spermatogenesis.

CHROMOSOMAL CHANGES DURING SPERMATOGENESIS

Since a sperm is to unit with an ovum in the production of a fertilized ovum, both the spermatid and the ovum should have half the number of chromosomes present in other cells of the body. The chromosome number in somatic cells is 44 + XY or 44 + XX, and therefore a mature sperm and ovum have 22 + Y or 22 + X chromosomes each.[1] This reduction in chromosome number is accomplished in two reduction divisions (or meioses) that occur following the primary spermatocyte. The last DNA synthesis has already occurred before the last mitosis. The first meiotic division is characterized by a long prophase which can be subdivided into four substages. During this prolonged prophase, homologous chromosomal strands pair closely and exchange parts of their genome. This recombination of hereditary material is what makes the resulting chromosomes in the spermatozoa to consist of alternating segments of maternal and paternal chromosomes. The interchange of DNA segments is known to occur during the final stage of the prophase. Points where interchange of chromosome segments take place are called *chiasma*.

[1] The female chromosome is designated X and the male, Y. There are two sex chromosomes in somatic cells (diploid, 46 in number.) The mature sperm or ovum has 23 chromosomes (haploid); one of these is a sex chromosome.

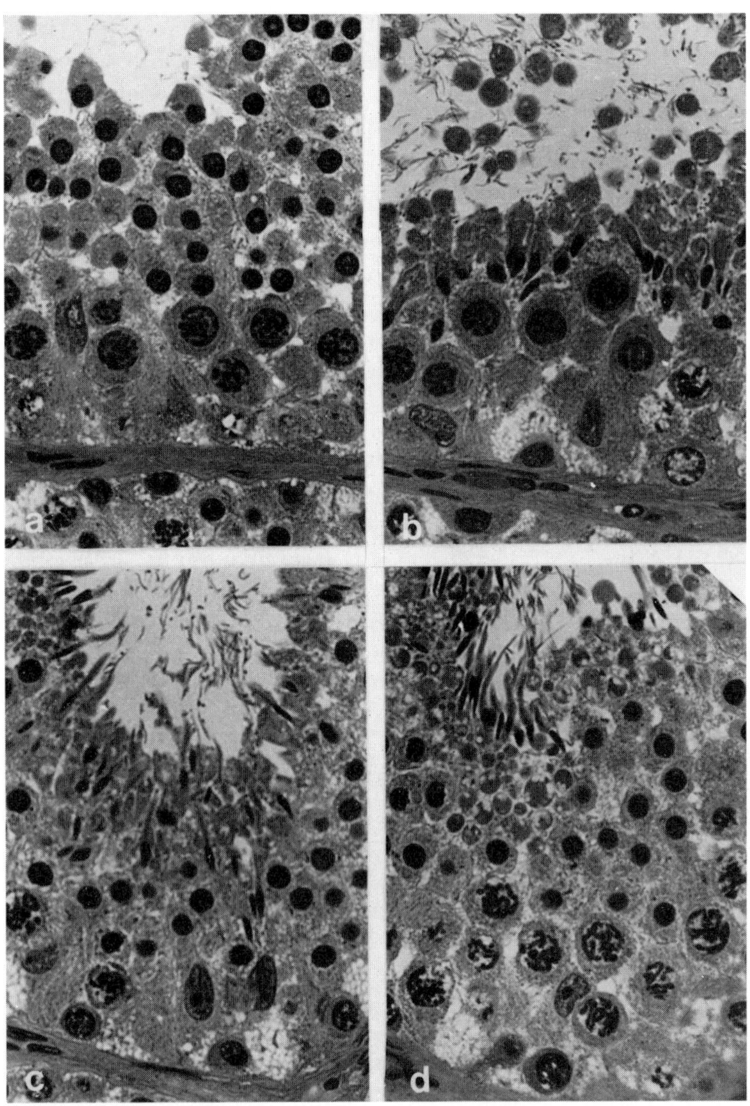

FIGURE 27-12*a* to *d* Different fields of seminiferous tubules from monkey testis illustrating various stages in the cycle of the seminiferous epithelium during spermatogenesis.

STRUCTURE OF SPERMATOZOA

The last step in the formation of mature spermatozoa is called *spermiogenesis* during which sequential developmental events characterizing the transformation of spermatids into spermatozoa take place. The entire process of spermiogenesis occurs in the adluminal space. Briefly, this involves differentiation of various parts recognizable in mature spermatozoa from rounded spermatids. Thus, there is an orderly differentiation of head, neck, middle piece, principal piece, and end piece. Along the way, much of the cytoplasm which initially wraps around the spermatid nucleus becomes "squeezed" to one side and is eventually discharged as a piece of residual cytoplasm (Fig. 27-13).

The role that the cytoplasm of cells plays during spermatogenesis has not been firmly established. It is of interest that the cytoplasm of committed spermatogonia (i.e., committed to become primary spermatocytes) does not completely divide throughout the mitotic and meiotic divisions. This would mean that the

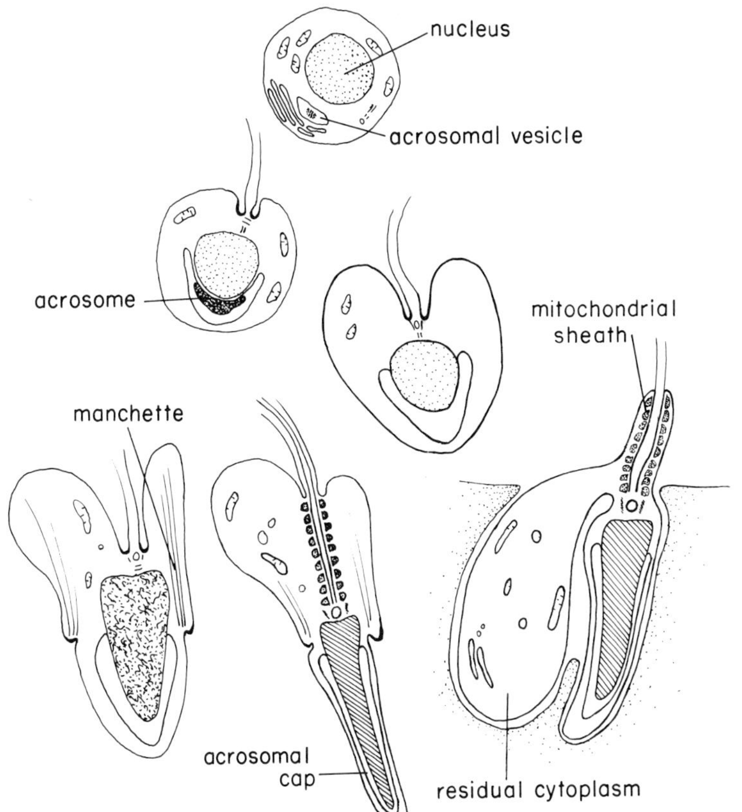

FIGURE 27-13 Diagram of successive stages in spermatogenesis including nuclear condensation, appearance of the manchette, elongation of the cell, appearance of acrosomal cap, and formation of mitochondrial sheath and residual cytoplasm.

cytoplasm of the several generations of spermatids originating from a single committed spermatogonium is sharing a common ground cytoplasm until the last step in spermiogenesis when the residual cytoplasm is released.

The mature spermatozoon is an elongated and motile cell that is 55 to 65 μm long. Approximately 60,000 spermatozoa are contained in 1 mm^3 of seminal fluid. In the favorable environment of the male genital tract, the sperm may remain alive for some time (up to several weeks) after leaving the testes. As diagramed in Fig. 27-14, the sperm consists of a head and a tail. The junction between the head and tail is known as the neck. The head is a conical, rather ovoid structure 4 to 5 μm long. A nucleus, with compact, deeply staining chromatin enclosed within the nuclear envelope, comprises the bulk of the head. The cephalic end of the head is covered by a cell membrane or plasmalemma.

A pair of centrioles are located in the neck near the basal end of the nucleus where it continues with the proximal end of the tail. This tail portion is comprised of three parts: the middle piece (close to the neck), the principal piece, and the end piece (Fig. 27-14). The middle piece is about 5 to 9 μm long and 1 μm in diameter. In a cross-sectional EM, it shows the typical structural arrangement of microtubules that make up cilia. There are nine doublets of peripheral microtubules surrounding a central pair. These microtubules are surrounded by tightly packed, perpendicularly oriented mitochondria. The principal piece, the central segment of the tail, is 40 to 45 μm in length and contains the central pair of microtubules con-

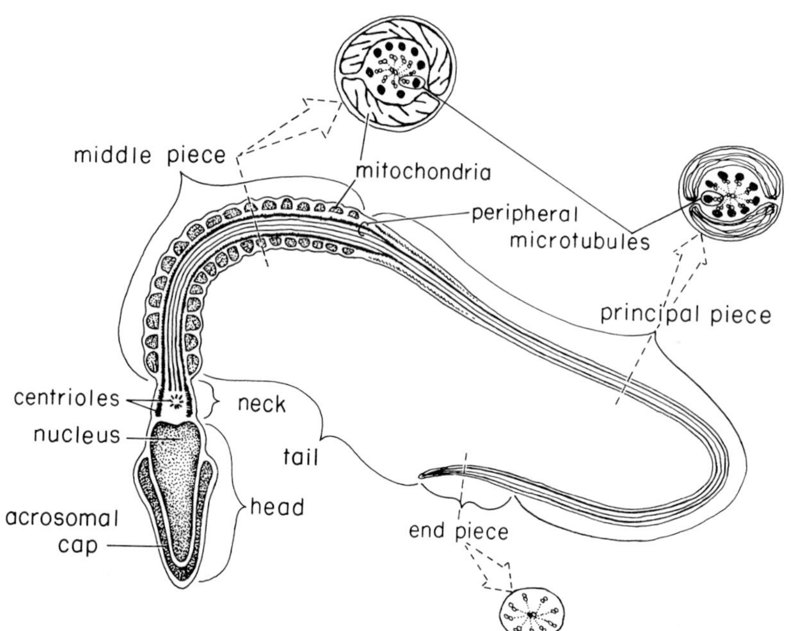

FIGURE 27-14 Drawing illustrating the fine structure of the human spermatozoon.

tinuing from the middle piece. The mitochondria are lacking in the middle piece. The microtubules are surrounded by a fibrous, circumferentially oriented riblike sheath (Fig. 27-14). The end piece of the tail, 5 to 10 μm in length, consists of a pair of microtubules and a cell membrane.

INTERSTITIAL CELLS

The endocrine function of the testes is performed by interstitial cells of Leydig. They occupy the interstitial spaces (between seminiferous tubules) along with connective tissue elements and a rich capillary network. In human beings, the interstitial connective tis-

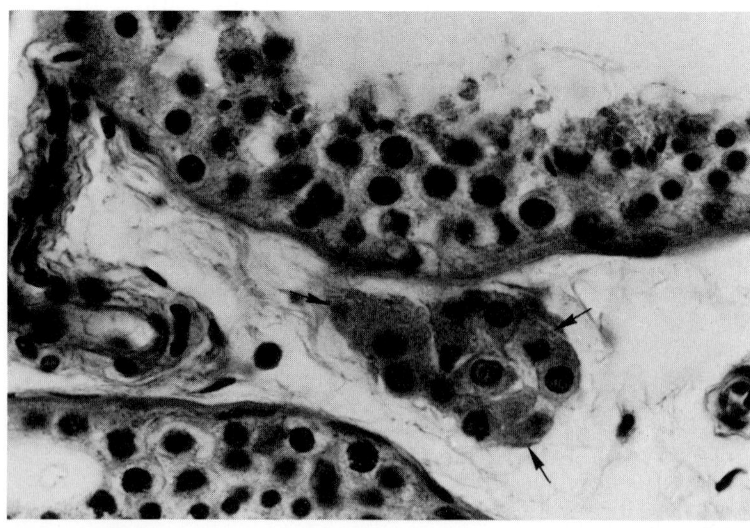

FIGURE 27-15 Light micrograph of an interstitial region of the testis. A group of Leydig cells (arrows) is seen between the seminiferous tubules.

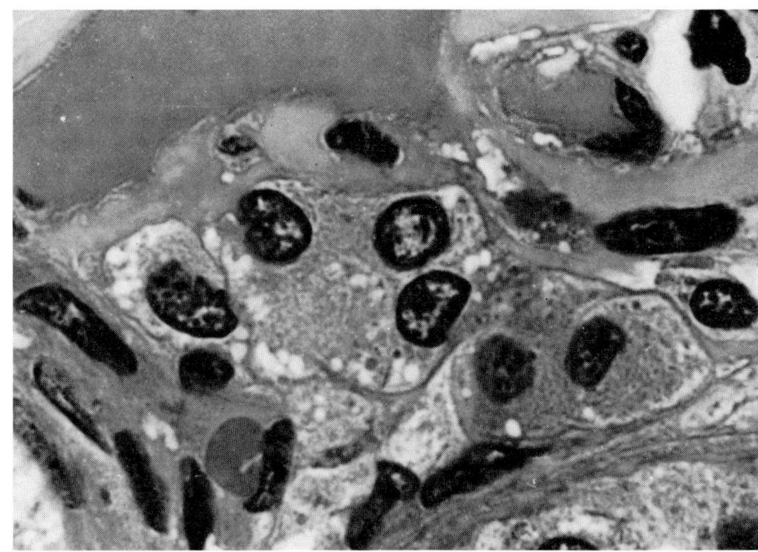

FIGURE 27-16 Leydig cells viewed at higher magnification. Their peripheral cytoplasm appears vacuolated and is lightly stained.

sue is made up of fine collagen fibrils. In certain lower forms, the interstitium is much more heavily developed. Leydig cells produce *testosterone,* the male sex hormone, and may occur in groups of varying sizes. They are large and often polygonal in shape (Fig. 27-15, arrows). Their large nucleus is frequently eccentric. Fairly dense eosinophilic cytoplasm is seen near the nucleus, but it is vacuolated in the periphery and stains lightly (Fig. 27-16). These vacuoles are a common feature to many endocrine cells producing steroid hormones.

In the electron microscope, Leydig cells reveal many profiles of SER and many mitochondria with tubular cristae (Fig. 27-17).

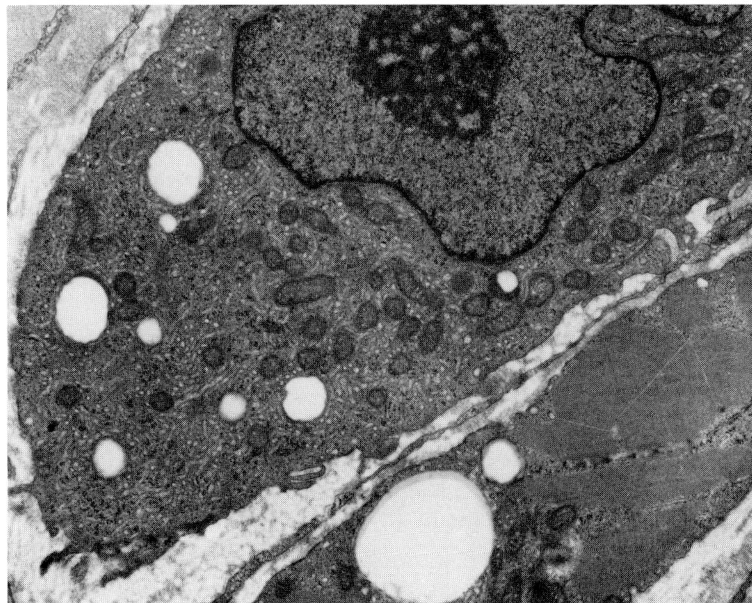

FIGURE 27-17 Electron micrograph of a portion of a Leydig cell. The cytoplasm contains extensive SER and numerous mitochondria. Crystals are seen in an adjacent cell in the lower left-hand corner. (*Courtesy of A. K. Christensen.*)

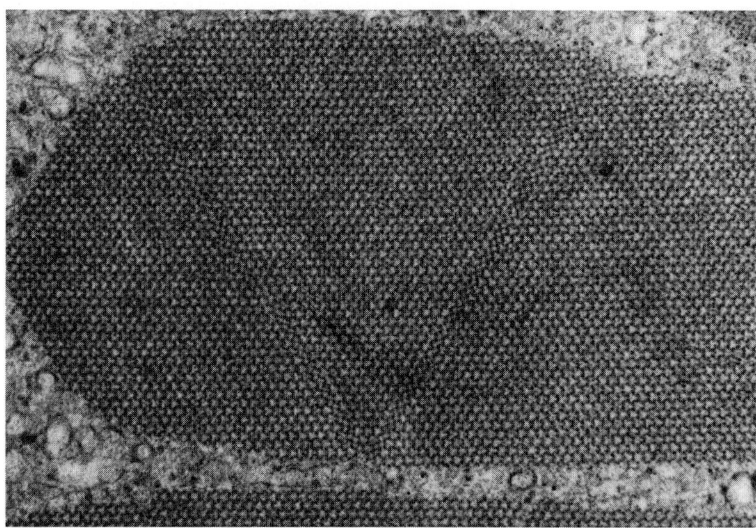

FIGURE 27-18 Electron micrograph of a crystal found in a Leydig cell of a human testis. (*Courtesy of A. K. Christensen.*)

Both of these characteristics are typical of steroid-secreting cells. Two types of Leydig cells can be distinguished. One type, a fusiform cell, has relatively few organelles, and the other, a large cell, has many small membrane-bound vesicles, granules, lipid droplets, and large crystalloid inclusions. These cells may reflect different functional states of the Leydig cell populations. Leydig cell granules are rod-shaped and are made of crystalloid substance. The fine structure of such a crystal is depicted in Fig. 27-18.

The dependence of Leydig cells on interstitial cell–stimulating hormone (ICSH or LH) from the pituitary gland has been clearly shown by numerous experiments in which retrograde degeneration of Leydig cells after hypophysectomy has been established. Testosterone produced by Leydig cells under the trophic influence of ICSH is essential to the later stages of spermatogenesis. In addition, it has been found that Leydig cells also influence other androgen-dependent gonadal organs, including the duct system through which the sperms travel. It should be emphasized that both FSH and LH are required for the maintenance of normal seminiferous tubule function. The difference appears to be that LH exerts an indirect effect through the synthesis and release of testosterone, while FSH acts upon Sertoli cells to stimulate the production of ABP which in turn binds androgens produced by Leydig cells. Thus, the two independent cell populations of the testis are responsive to the two different gonadotropic hormones which mutually augment their trophic functions.

DUCT SYSTEM

Toward the apex of a testicular lobule, the seminiferous tubules abruptly become thin and merge to form the *straight tubules,* which in turn converge into the rete testis, located at the hilar region of the testis. Figure 27-19 shows the junction of a straight tubule (s) with a seminiferous tubule on the right. Note the sudden changes in the wall structure. Nearing this junction, the developing spermatogenic cells in the walls of the seminiferous tubule decrease in number and eventually disappear. Thus, most of the cells seen at the end of the seminiferous tubule are Sertoli cells. The Sertoli cells in this region become modified and continue as the flat or cuboidal epithelium of the straight tubule. This transition is diagramed in Fig. 27-20.

Occasional cells in this region have a single cilium with a basal body. The tubules

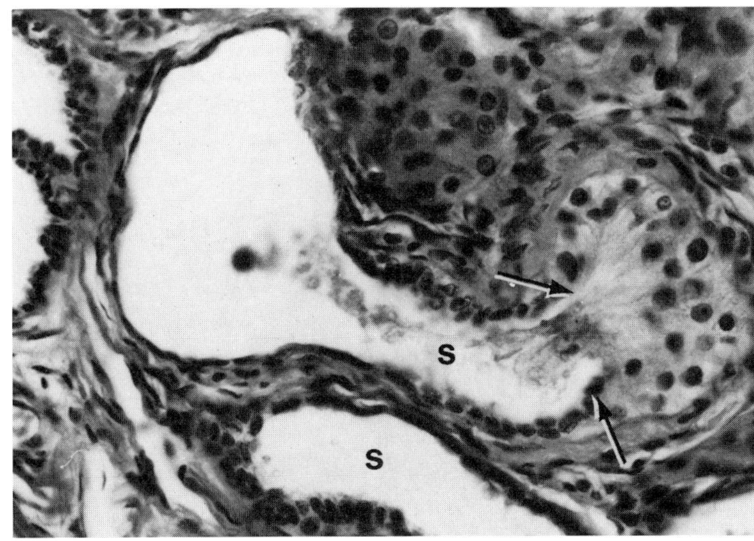

FIGURE 27-19 Light micrograph of the junction of a seminiferous tubule with a straight tubule (s). Arrows indicate the junction between the two types of tubules.

of the *rete testis* (Figs. 27-21 and 27-22), the next segment of the duct system, have a cuboidal epithelium similar to that of the straight tubule. These tubules make up an extensive anastomosing network and therefore serve to collect spermatozoa from the many straight tubules and pass them on to the less numerous ductuli efferentes.

The *ductuli efferentes,* composed of 8 to 15 tubules, carry the sperm from the rete testis to the superior aspect of the epididymis. Here, the most cephalic duct becomes continuous with the ductus epididymidis, and the remaining ducts join at a lower level. The ductuli efferentes have a lining epithelium of columnar cells alternating with groups of cuboidal cells (Fig. 27-23). This gives the lumen an irregular contour. The epithelium rests on a distinct basement membrane, surrounded by a thin connective tissue layer which contains some circularly oriented smooth muscle fibers and many capillaries. Many of the tall cells have cilia on their luminal surface. Unlike stereocilia found elsewhere, these are motile cilia and aid in the movement of spermatozoa. Each ductule is about 7 to 8 cm in length and joins the single winding duct of the epididymis individually. The transition between the efferent ductules and epididymis is gradual.

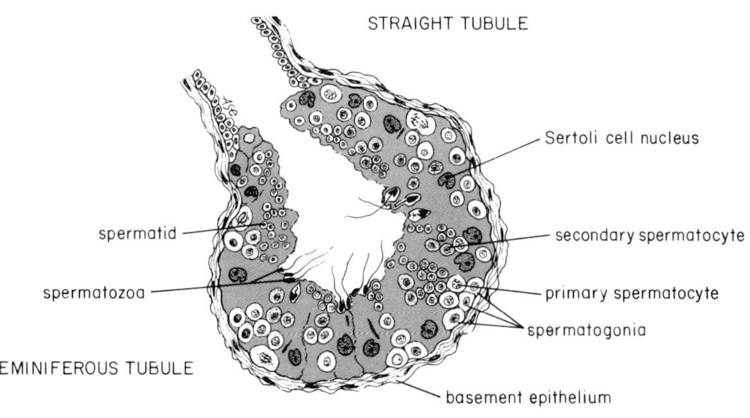

FIGURE 27-20 Diagram of the junction of a seminiferous tubule with a straight tubule.

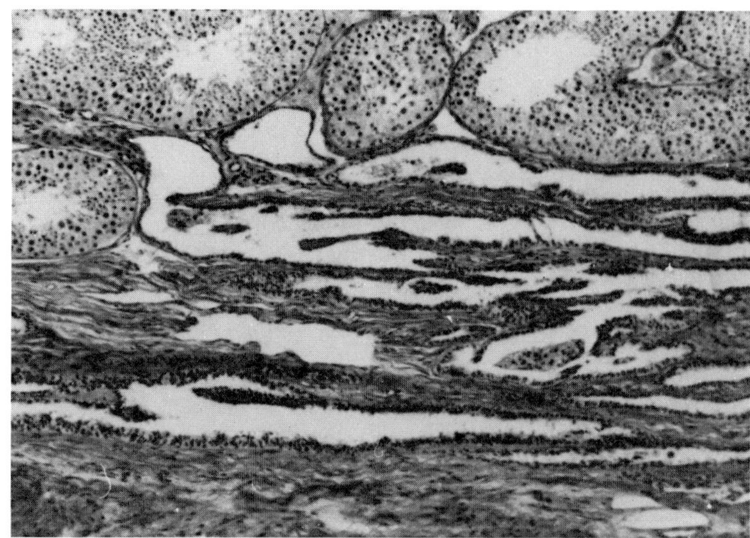

FIGURE 27-21 Light micrograph of a rete testis.

EPIDIDYMIS

The body of the epididymis contains the long, coiled tubule of the *ductus epididymidis*. The total length of the ductus in the human being is approximately 6.5 m. This tubule has its origin at the cephalic pole of the epididymis where a single tubule is formed by merger of the ductuli efferentes. In the middle and caudal portions the duct follows a convoluted course to the tail of the epididymis. At the caudal pole it folds upon itself and becomes the *ductus deferens*. The external and luminal contours of the ductus epididymidis are quite even (Fig. 27-24). Ductus epididymidis is lined with narrow, pseudostratified columnar cells with stereocilia. A basement membrane and lamina propria are present, as well as a small amount of circular smooth muscle. Upon approaching the juncture with the *ductus* (*vas*) *deferens*, the muscle content increases and tiny longitudinal muscle bundles begin to appear. Storage of sperm takes place in the lumen of the ductus epididymidis, as indicated in Fig. 27-24 (arrow).

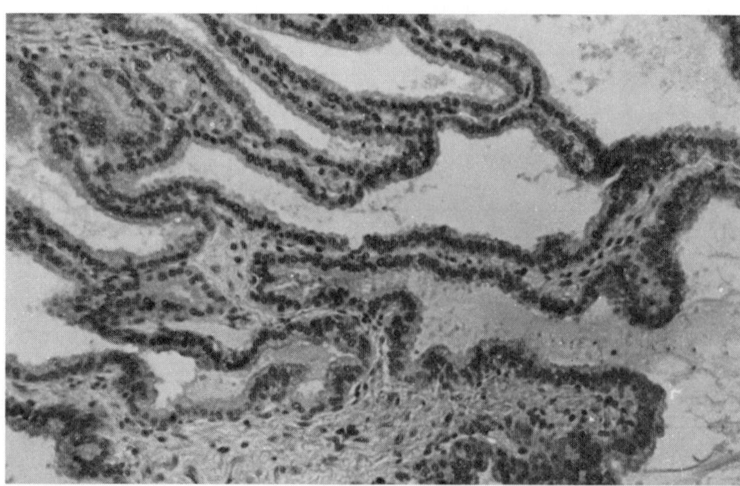

FIGURE 27-22 Rete testis viewed at higher magnification. The irregular lumina of the rete are lined by cuboidal epithelium.

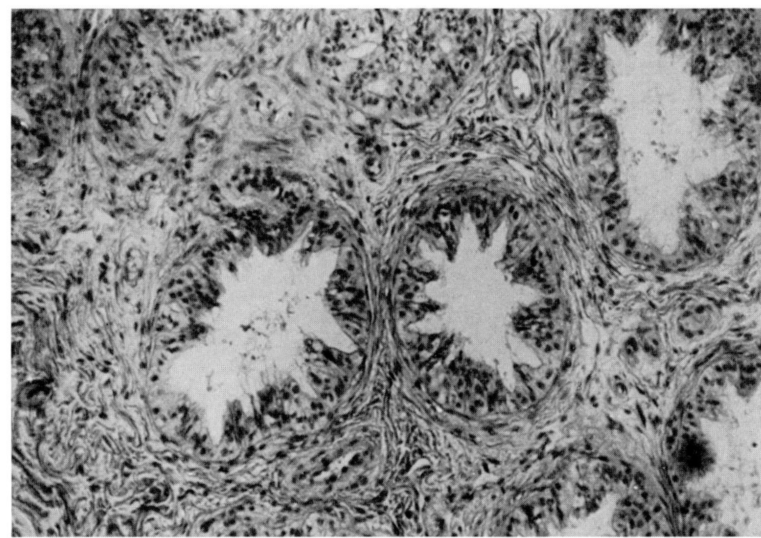

FIGURE 27-23 Light micrograph of transverse section through the ductuli efferentes.

DUCTUS DEFERENS

The ductus deferens consists of the mucosa, muscularis, and fibrosa (Fig. 27-25). The mucosa is lined by a pseudostratified columnar epithelium similar to that of the epididymis except that a corrugated surface exists due to heavy development of its wall (Fig. 27-26). The epithelium is surrounded by elastic fibers which are condensed around the basal lamina. The muscularis has three smooth muscle layers: inner longitudinal, middle circular, and outer longitudinal. The longitudinal layer is better developed than are the circular layers. The fibrosa, which is external

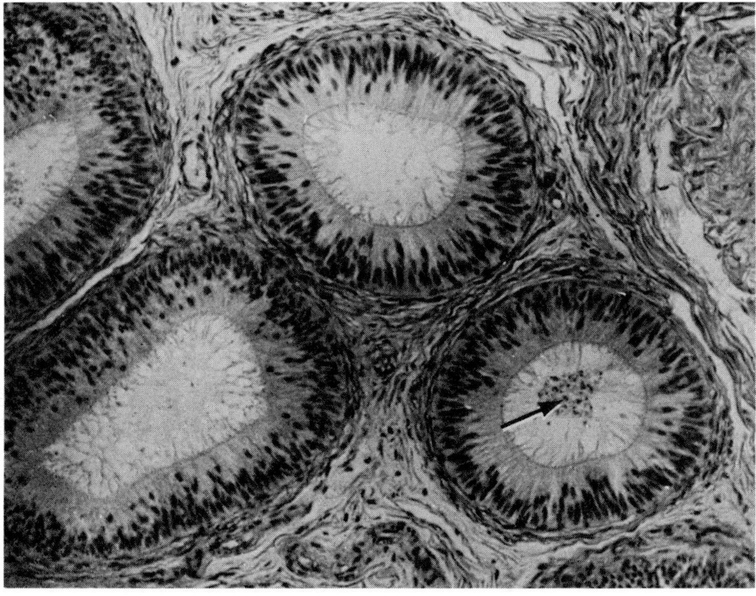

FIGURE 27-24 Light micrograph of transverse section through the ductus epididymidis. Spermatozoa are stored in the lumen (arrow) of the ductus epididymidis.

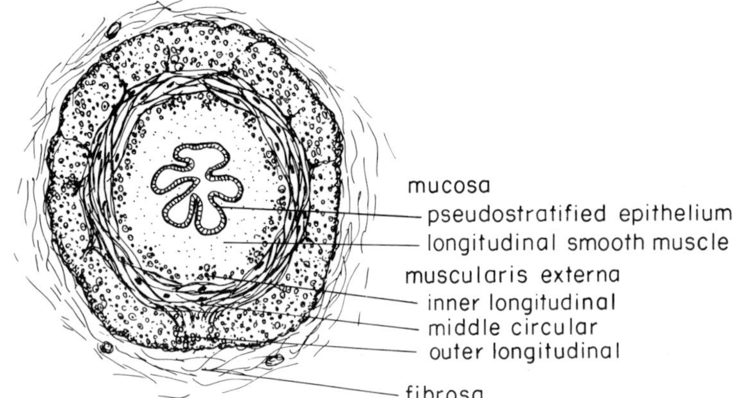

FIGURE 27-25 Drawing of a cross section through the vas deferens.

to the muscle layers, contains numerous blood vessels, nerves, and scattered bundles of smooth muscle fibers.

GLANDS ASSOCIATED WITH THE MALE GENITAL TRACT

SEMINAL VESICLES

The pouchlike seminal vesicles open into the genital tract at the junction of the two ductus deferentes and the ejaculatory duct. The convoluted walls of the pouchlike glands contain cells which secrete a slightly alkaline viscous fluid. The secretion contains fructose which is used as an energy source by the spermatozoa.

The mucosa of the seminal vesicles is folded into numerous irregular chambers or crypts, as shown in Fig. 27-27. The epithelial lining is composed of short pseudostratified columnar cells containing yellowish lipoid pigment granules. Epithelial characteristics resemble those present in the genital tract. The lamina propria forms a continuous fibroelastic layer around the vesicle. Inner and

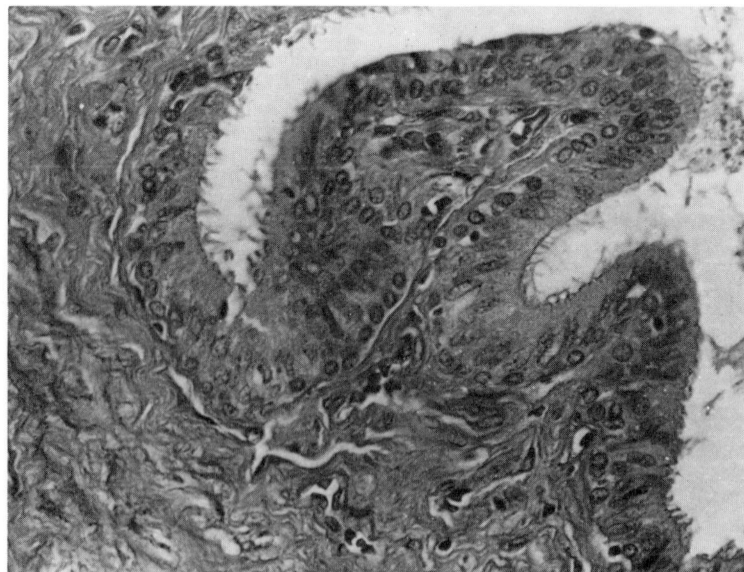

FIGURE 27-26 Light micrograph of the mucosa of the vas deferens.

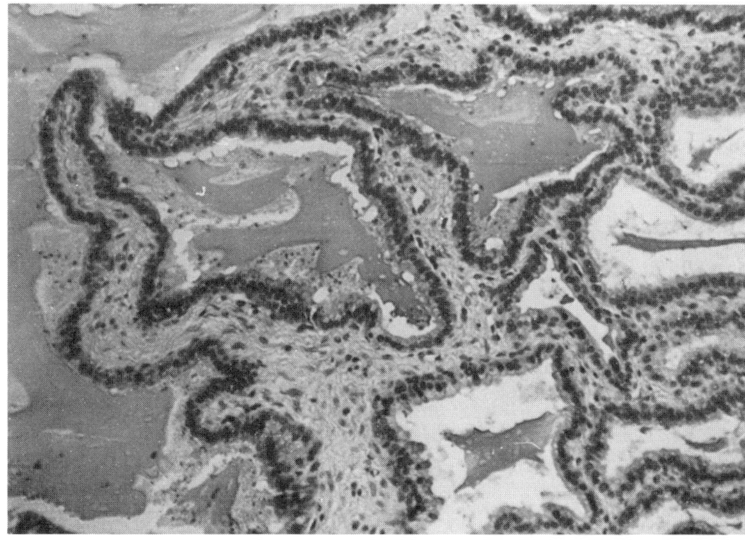

FIGURE 27-27 Light micrograph of the mucosa of the seminal vesicles.

outer smooth muscle layers are present, surrounding the lamina propria. However, the muscle is less developed than in the ductus deferens.

PROSTATE GLAND

The prostate is an aggregate of 30 to 50 masses of tubuloalveolar glands opening into the penile urethra via 20 or more ducts. A portion of the prostate is shown in Fig. 27-28. A fibroelastic capsule containing smooth muscle fibers surrounds the gland. Separating the glandular masses are numerous connective tissue septa arising from the capsule. Citric acid, lipids, and acid phosphatase are major components of prostate secretion. Acid phosphatase has become an important clinical indicator of prostate carcinoma, as it increases markedly in the blood of patients with prostate cancers.

The epithelium of the prostate shows a wide variation between glands and even

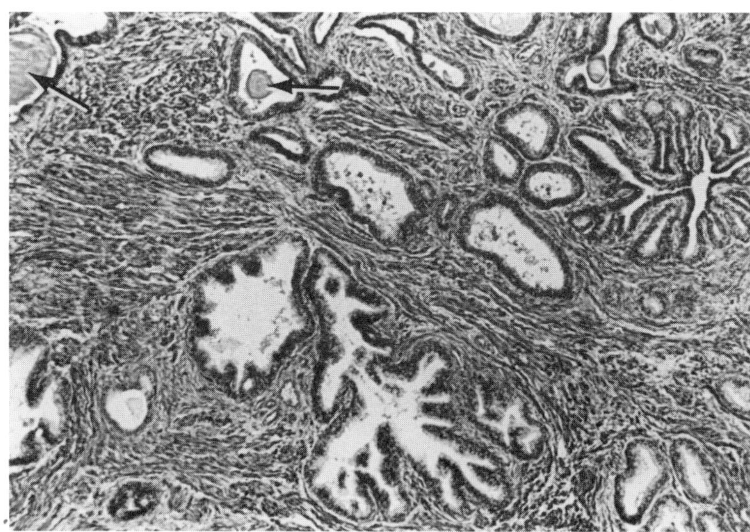

FIGURE 27-28 Light micrograph of the prostate gland. Some of the glandular lumina contain amyloid bodies (arrows).

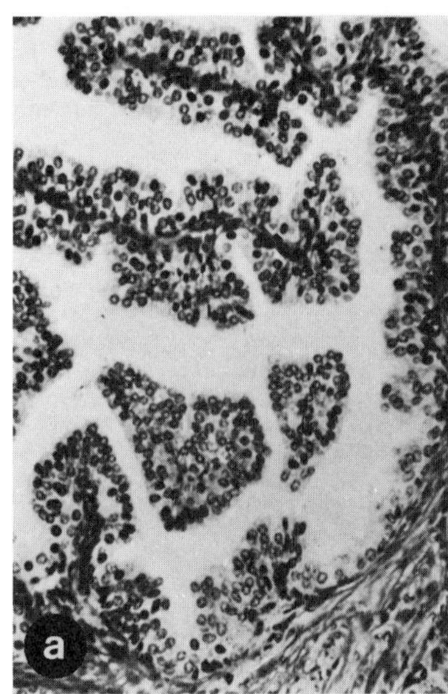

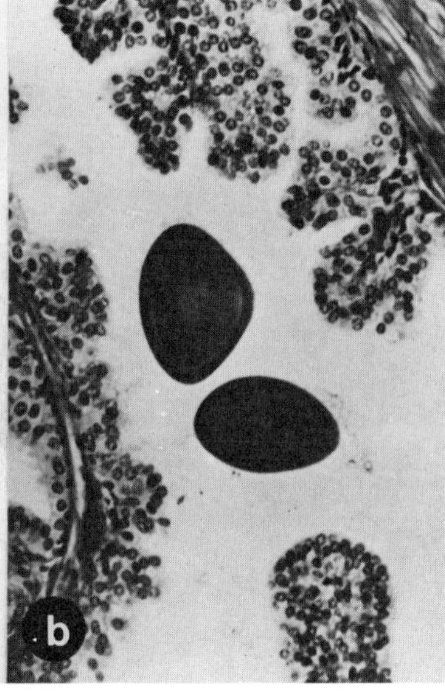

FIGURE 27-29 (*a*) Glandular epithelium of the prostate viewed at higher magnification. (*b*) Corpora amylacea in the glands of the prostate.

within a gland (Fig. 27-29*a*). It is usually of a cuboidal or columnar type and contains secretory granules and occasional lipid droplets. A common finding in many of the prostates is *corpora amylacea* (or amyloid bodies) (Figs. 27-28 and 27-29*b*). These spherical bodies are composed of protein and carbohydrates. In the normal state they are small and soft, but with advancing age may calcify and hypertrophy.

PENIS

The penis consists of a core of erectile tissue surrounded by a fascia of loose irregularly arranged elastic connective tissue and a thick layer of skin (Fig. 27-30). The erectile tissue is divided into three parallel compartments by connective tissue septa, as diagramed in Fig. 27-31. The two larger compartments, *corpora cavernosa*, are joined at the distal end.

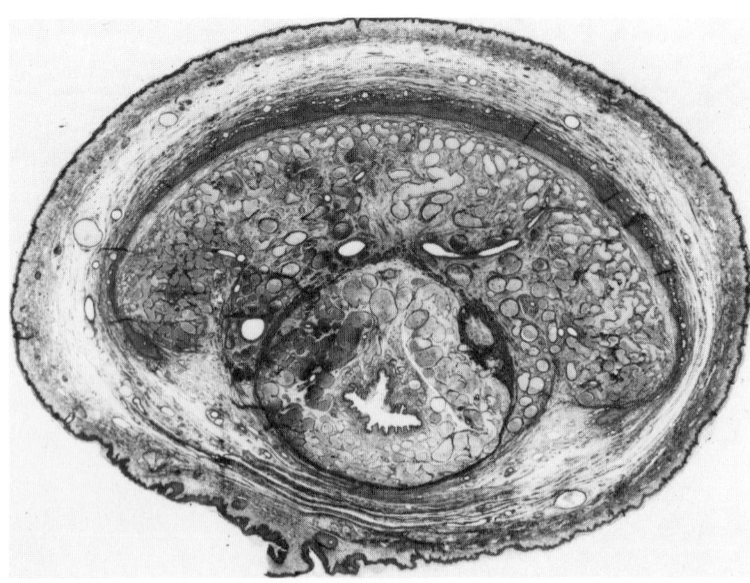

FIGURE 27-30 Light micrograph of a transverse section through the penis.

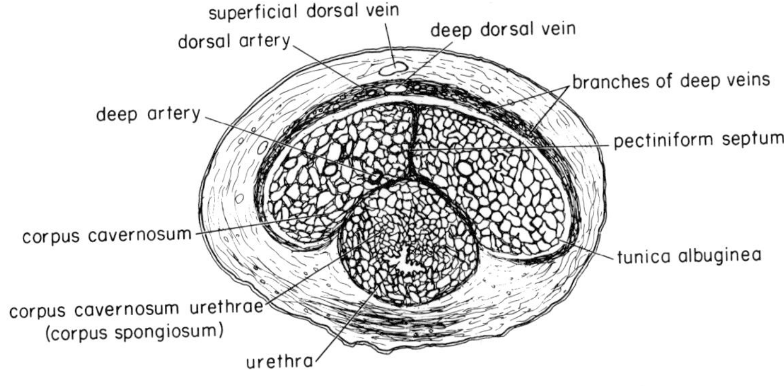

FIGURE 27-31 Schematic drawing of the transverse section of the penis seen in Fig. 27-30.

The third and smaller *corpus spongiosum urethrae* is centered between and caudal to the other two. The spongiosum contains the penile urethra and ends in an enlargement, the *glans penis.*

The tissue in the corpus cavernosum consists of a network of lacunae (spaces in erectile tissue) lined by endothelium. Between the endothelium are fibrous trabeculae rich in smooth muscle fibers. Each corpus cavernosum is surrounded by a dense collagenous tunica albuginea capsule.

The blood supply to erectile tissues depends on the varying functional states of the penis. During the flaccid state, the peripheral dorsal arteries supply blood to the organ, which is drained by peripheral veins in the albuginea. In addition, an arteriovenous shunt, which is present in the deep artery and draining vein, is open throughout the flaccid period. During erection, the shunt is closed, and the deep arteries that run lengthwise in each of the cavernous bodies force blood through the arterial branches that feed erectile tissue. These branches, called *helicine arteries,* are normally constricted during the flaccid state, but dilate during erection. The sudden filling of the tissue spaces with blood blocks the veins which drain the compartments. At the termination of the erection, the arterial branches feeding the erectile tissue constrict, allowing the pressure to lower. Subsequently, the veins open and the blood leaves the tissue. The corpus spongiosum fills similarly, but does not become as rigid as the other compartments.

REVIEW SECTION

1. The tubular structure in Fig. 27-32 is __________. Two primary cell types found within this tubule are __________ and __________.

This figure represents a cross section of seminiferous tubules. The epithelium of these tubules consists of spermatogenic and supportive Sertoli cells. Mature sperm are not stored in the lumen of seminiferous tubules and therefore would not be expected to be found in the lumen of these tubules. Seminiferous tubules contain columns of maturing spermatogenic cells.

2. Refer to Fig. 27-33. What stage do the large epithelial cells with thick chromosomes in the nuclei (arrows) represent? Which stage is next represented toward the lumen?

The large nuclei belong to the primary spermatocytes in prophase, which is extremely long in the human being (about 3 weeks). The dense smaller nuclei located over the primary spermatocytes belong to secondary spermatocytes or spermatids.

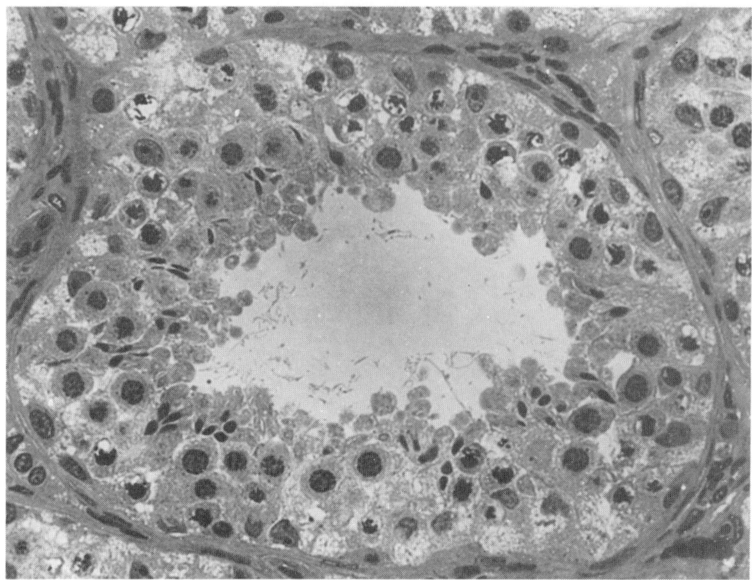

FIGURE 27-32

3. The large cell with light, oblong nuclei extending from basement membrane to luminal surface in Fig. 27-34 (arrows) is called ________. What is its function?

In this field Sertoli cells can be seen extending from the basement membrane to the lumen. Many spermatids are clustered at the apex of these cells, presumably deriving nourishment from them.

4. The sperm is composed of two principal parts, the head and the tail. What is each part primarily composed of? What primary function does each part play?

The nucleus comprises the major portion of the sperm head and functions as a carrier of genetic material. The tail is concerned with locomotion and as such is composed of axial microtubules and mitochondria.

5. Refer to Fig. 27-35. The large cells shown in

Leydig cells are shown here. Testosterone, the

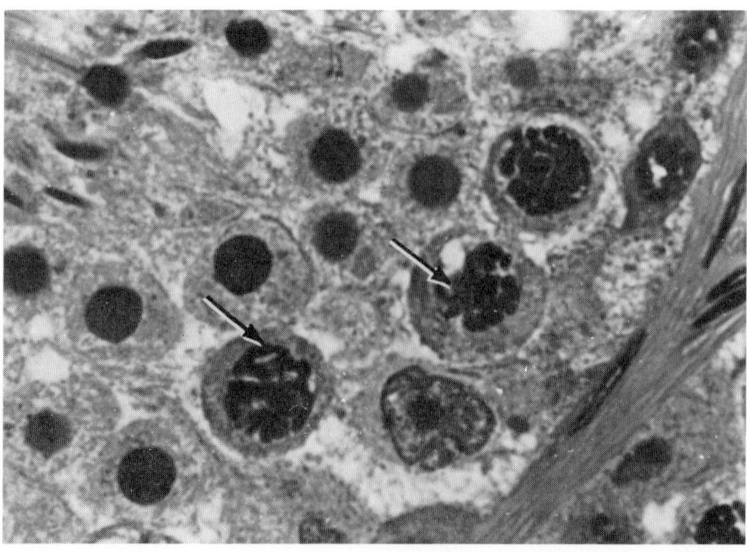

FIGURE 27-33

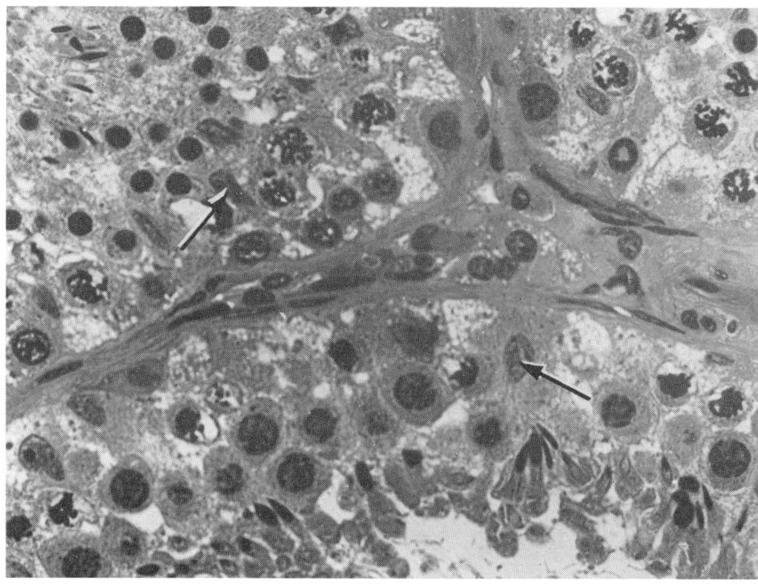
FIGURE 27-34

the central portion of the field are called __________. What is their function?

male sex hormone, is produced by these cells and distributed via the rich capillary network.

6. The duct in Fig. 27-36 is representative of what part of the genital tract?

This figure presents a section of the epididymis. The even contour of pseudostratified, stereociliated columnar cells helps one to make this identification. Sperm present in the lumen indicate the storage function of this duct.

7. Figure 27-37 represents a section from the wall of the __________.

A section of seminal vesicle is shown in this figure. It can be identified by the highly convoluted walls of often pseudostratified columnar epithelium. A slightly alkaline viscous fluid is produced by the seminal vesicle and provides a source of energy for the sperm.

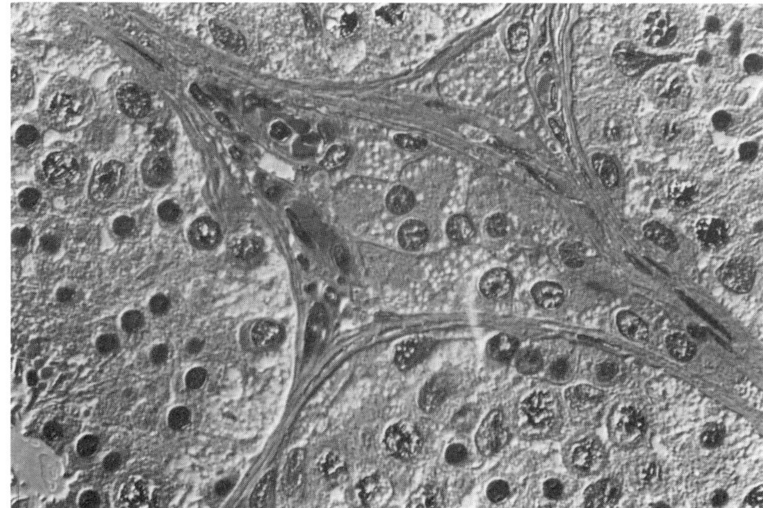
FIGURE 27-35

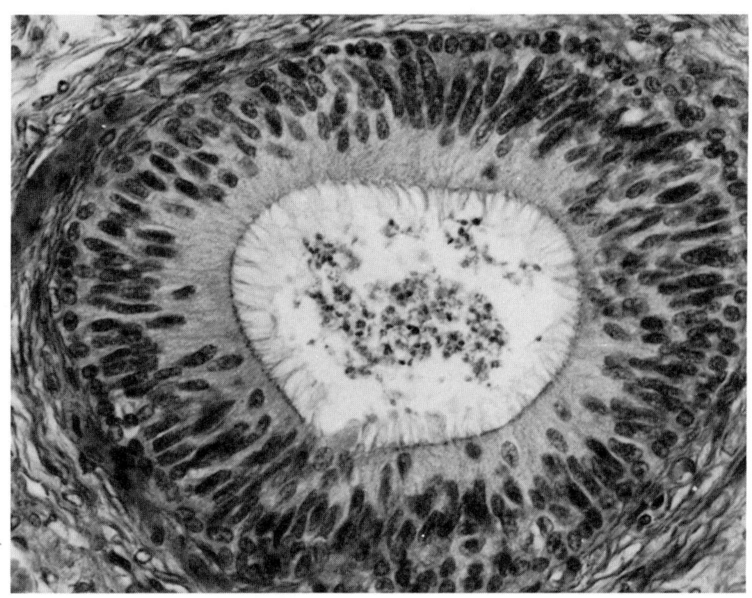

FIGURE 27-36

BIBLIOGRAPHY

Bloom, W., and D. W. Fawcett (eds.): *A Textbook of Histology,* 10th ed., Saunders, Philadelphia, 1975.

Brandes, D.: *Male Accessory Sex Organs: Structure and Function in Mammals,* Academic Press, New York, 1974.

Hamilton, D. W.: "The Mammalian Epididymis," in H. Bahn and S. Glasser (eds.), *Reproductive Biology,* Excerpta Medica, Amsterdam, 1972.

———, and R. O. Greep (eds.): "Male Reproductive System," sect. 7, *Endocrinology,* vol. 5, in *Handbook of Physiology,* American Physiological Society, 1975.

Johnson, A. D., W. R. Gomes, and N. L. VanDemark (eds.): *The Testis,* Academic Press, New York, 1970.

Steinberger, E.: "Hormonal Control of Mammalian Spermatogenesis," *Physiol. Rev.* **51:**1–22 (1971).

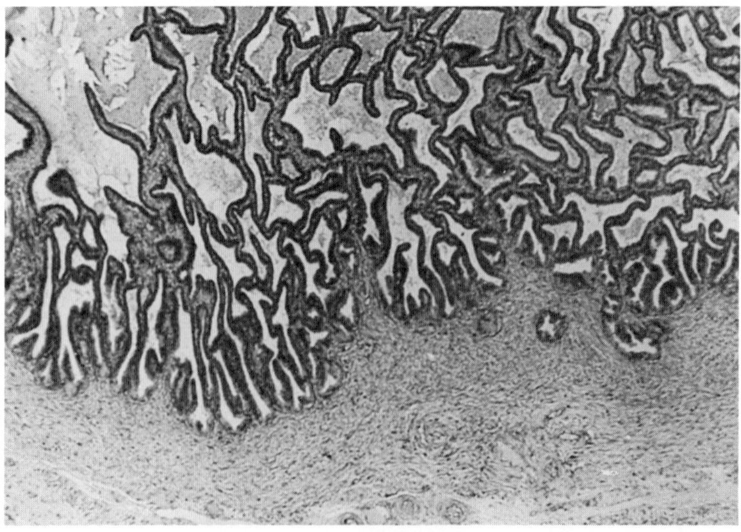

FIGURE 27-37

Weiss, L., and R. O. Greep: *Histology,* 4th ed., McGraw-Hill, New York, 1977.

Young, W. C. (ed.): *Sex and Internal Secretions,* vols. 1 and 2, Williams & Wilkins, Baltimore, 1961.

ADDITIONAL READINGS

Bawa, S. R.: "The Fine Structure of the Sertoli Cell of the Human Testis," *J. Ultrastruct. Res.* **9:**459–474 (1963).

Bishop, D.: "Sperm Motility," *Physiol. Rev.* **42:**1–59 (1962).

Brandes, D.: "The Fine Structure and Histochemistry of Prostatic Glands in Relation to Sex Hormones," *Int. Rev. Cytol.* **20:**207–276 (1966).

Clermont, Y.: "The Cycle of the Seminiferous Epithelium in Man," *Am. J. Anat.* **112:**35–52 (1963).

———: "Renewal of Spermatogonia in Man," *J. Anat.* **118:**509–524 (1966).

———: "Kinetics of Spermatogenesis in Mammals: Seminiferous Epithelium Cycle and Spermatogonial Renewal," *Physiol. Rev.* **52:**198–236 (1972).

deKretser, D. M.: "The Fine Structure of the Testicular Interstitial Cells in Man of Normal Androgenic Status," *Z. Zellforsch.* **80:**594–609 (1967).

Dym, M.: "The Mammalian Rete Testis: A Morphological Examination," *Anat. Rec.* **186:**493–524 (1976).

———, and D. W. Fawcett: "The Blood-Testis Barrier in Rat and the Physiological Compartmentation of the Seminiferous Epithelium," *Biol. Reprod.* **3:**308–326 (1970).

Fawcett, D. W.: "The Mammalian Spermatozoan," *Devel. Biol.* **44:**395–436 (1975).

Flickinger, C., and D. W. Fawcett: "The Junctional Specialization of Sertoli Cells in the Seminiferous Epithelium," *Anat. Rec.* **158:**207–222 (1967).

Heller, C. H., and Y. Clermont: "Kinetics of the Germinal Epithelium in Man," *Recent Progr. Horm. Res.* **20:**545–575 (1964).

Ladman, A. J., and W. C. Young: "An Electron Microscopic Study of the Ductuli Efferentes and Rete Testis of the Guinea Pig," *J. Biophys. Biochem. Cytol.* **4:**219–226 (1958).

Leeson, T. S., and C. R. Leeson: "The Fine Structure of Cavernous Tissue in the Adult Rat Penis," *Invest. Urol.* **3:**144–154 (1965).

McNeal, J. E.: "The Prostate and Prostatic Urethra: A Morphological Synthesis," *J. Urol.* **107:** 1008–1016 (1972).

Morita, I.: "Some Observations on the Fine Structure of the Human Ductuli Efferentes Testis," *Arch. Histol. Jap.* **26:**341–365 (1966).

Orgebrin-Crist, M. C.: "Studies on the Function of the Epididymis," *Biol. Reprod.,* suppl., **1:**155–175 (1969).

Riva, A.: "Fine Structure of Human Seminal Vesicle Epithelium," *J. Anat.* **102:**71–86 (1967).

Roosen-Rungo, E. C.: "The Process of Spermatogenesis in Mammals," *Biol. Rev.* **37:**343–377 (1962).

Zamboni, L., R. Zemjanis, and M. Stefanini: "The Fine Structure of Monkey and Human Spermatozoa," *Anat. Rec.* **169:**129–154 (1971).

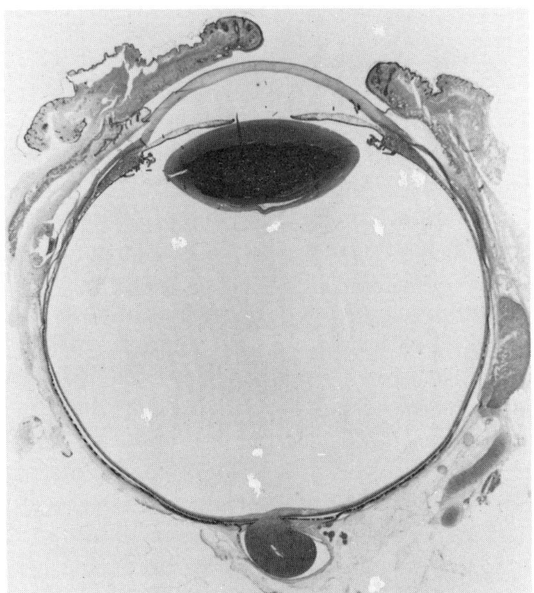

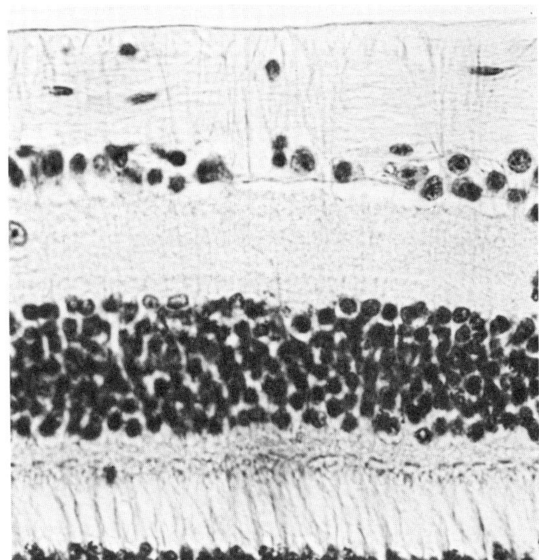

28 EYE

OBJECTIVES

Upon completion of this chapter, the student will demonstrate an understanding of the histology of the eye by being able to:

1 Define the general structural plan of the eye.

2 List the functions of the three layers of the eye:
(a) tunica fibrosa (b) uvea (c) retina

3 Locate the sites where pigmented cells are present.

4 List the layers of the retina and identify the three functional groups of cells in the chain of neurons.

5 Identify the structure of rod and cone cells in an electron micrograph.

6 Locate the sites where photosensitive pigments (rhodopsin and iodopsin) are located.

7 Name and give reasons for the two physically vulnerable sites of the retina.

8 Cite the layers of the cornea.

9 List the structural components of the eyeball and suspensory structures.

10 Identify the anterior and posterior chambers and their boundaries.

11 Follow the aqueous humor from its formation to drainage from the orbit.

12 State the chemical composition of the:
(a) lens (b) vitreous body

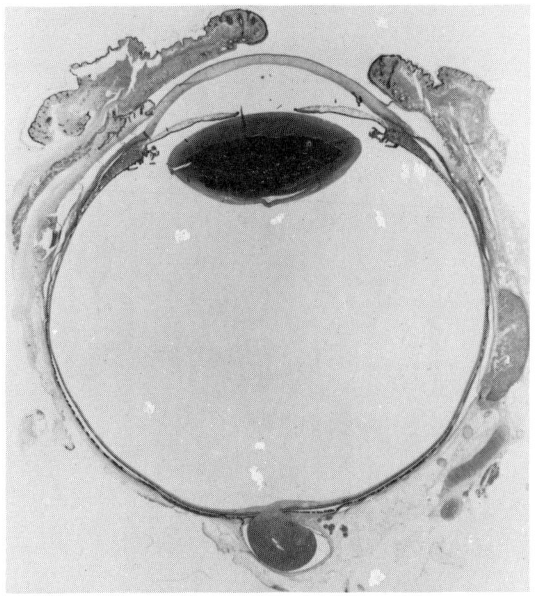

FIGURE 28-1 Median section of monkey eye showing the eyelids and the globe.

13 List the functions of the lacrimal and tarsal glands.

14 Define the conjunctiva.

15 List the involuntary muscles of the eye and give their functions.

The eye as a visual organ is made up of the eyeball (*bulbus oculi*) and the optic nerve. In addition, the eyeball is supported by a set of extrinsic muscles and the lacrimal gland. These components of the visual organ are protected by the bony orbit. In human beings the eyeball is a spherical structure measuring approximately 2.6 cm in diameter. It is a light-tight box except for the cornea which represents its transparent anterior surface. In a way, it is comparable, both in function and structure, to a camera. The incoming light from the cornea passes through certain refractive media, a diaphragm and lens, which form the focused image on the photosensitive posterior wall, that is, the retina. Figure 28-1 is a low-power LM of the eyeball. The diagram in Fig. 28-2 shows the eyeball as visualized in a horizontal plane through the center.

The wall of the eyeball proper consists of three layers. The outer layer is the tough *tunica fibrosa* which is made up of the opaque sclera and the transparent cornea. The middle layer is a vascular layer which is responsible for nutritive support. This layer contains an abundant blood supply and therefore is called *tunica vasculosa*. The tunica vasculosa is also called *uvea* and has a pigmented epithelium to prevent the reflection of light, as in a camera. Most of the uvea is made up of choroid, whereas the anterior region becomes specialized in the form of a ciliary body and an iris. The inner layer is a photosensitive neural tissue called *retina*. The retina becomes insensitive where it lines the ciliary body and the iris.

The refractive tissue filling the eyeball from the anterior surface to the retina is represented by the cornea, the lens, and two transparent media: aqueous humor in front of the lens and vitreous humor behind the lens (Fig. 28-2). The lens is held in position by suspensory ligaments which change its surface curvature, facilitating the variable refraction of the organ. The change in the refractive index of the lens is achieved by the smooth muscle fibers present in the ciliary body. The term *ciliary zone* (*zonula ciliaris*) is used to designate the radiating ligaments and the muscular ciliary body.

The ciliary body secretes the watery aqueous humor into the large *posterior chamber* which is located behind the lens and is occupied by the jellylike vitreous body. The turnover of intraocular fluids involves the fluid movement from the posterior chamber through the circular slit between the ciliary zone and iris (Fig. 28-3). Upon entering the *anterior chamber,* the aqueous humor is drained out of the eye through *Schlemm's canal* located at the periphery of the anterior chamber.

In development, the beginning of the eyeball formation is seen as early as in the

optical axis
visual axis
posterior chamber
anterior chamber
cornea
conjunctiva
canal of Schlemm
lens
ciliary body
ciliary zone
ora serrata
rectus lateralis
rectus medialis
vitreous body
sclera
uvea
fovea
retina
lamina cribrosa
optic nerve

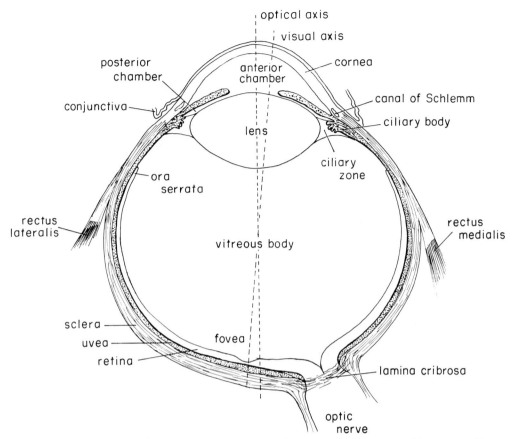

FIGURE 28-2 Diagram of a horizontal section of the eye through the optic nerve and fovea centralis.

FIGURE 28-3 Drawing of the limbus area.

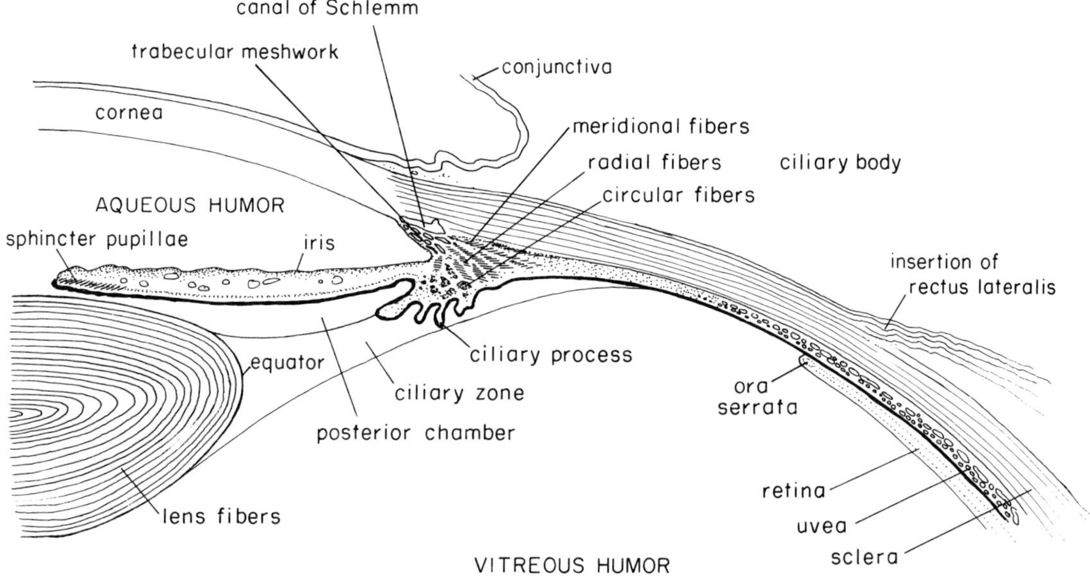

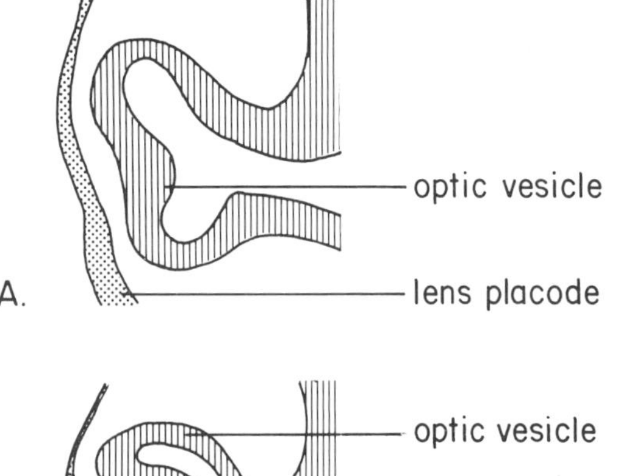

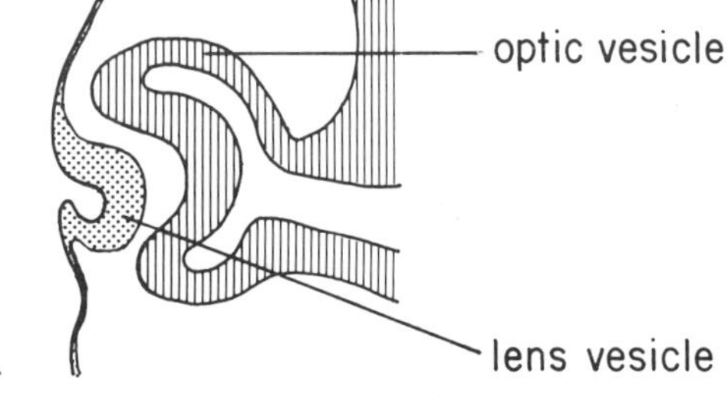

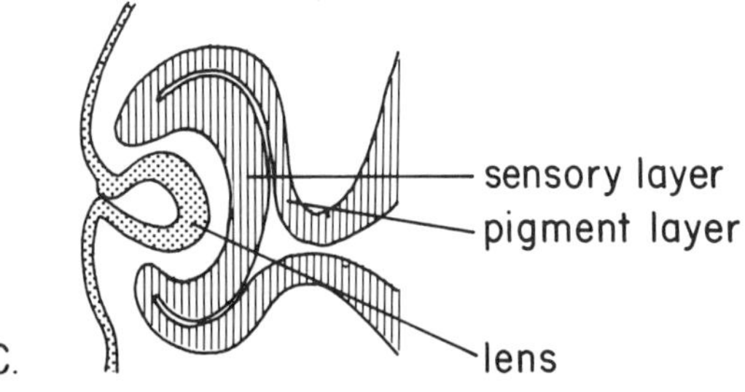

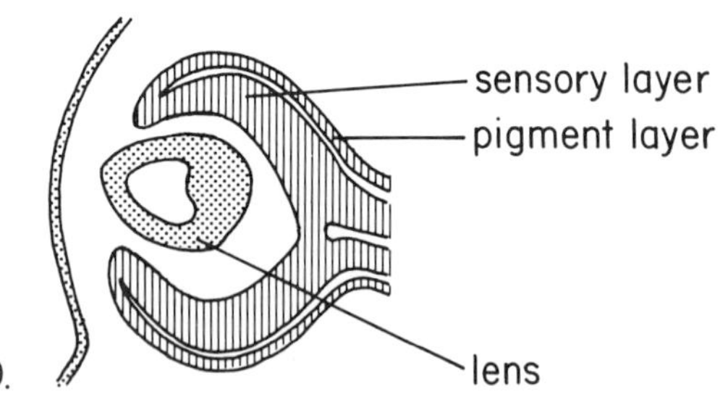

FIGURE 28-4 Early development of the optic cup and lens during the fourth to the seventh week in utero. (*Modified from B. M. Patten.*) (*A*) The ectoderm overlying the optic cup begins to develop a local thickening, the lens placode, toward the end of the fourth week. (*B*) The cavity of the optic cup then deepens and the lens placode becomes invaginated. (*C*) During the fifth week, the lens vesicle becomes closed. (*D*) The lens vesicle then breaks away from the surface and by the seventh week constitutes a rounded epithelial body lying in the opening of the optic cup.

5-week-old embryo. It appears as a pair of lateral outpouchings from the diencephalon. When the pouchings approach the surface ectoderm, each pouch produces an invagination called an *optic cup*. The inner layer of the optic cup eventually develops into retina, while the outer layer from the brain forms the pigmented epithelium of the choroid (Fig. 28-4). The ectodermal surface of the optic cup thickens and differentiates into the lens.

TUNICA FIBROSA

As mentioned earlier, the tunica fibrosa is divided into the transparent *cornea* which delineates the anterior aspect of the eyeball and the *sclera* which occupies the remainder (about 80 percent) of the tunica fibrosa. Thus, the cornea makes up approximately 20 percent of the fibrous capsule. The transitional area between the two regions is called *limbus*.

CORNEA

As shown in Figs. 28-1 and 28-2, the cornea bulges out at the anterior pole and is nonvascular. From its anterior surface five layers can be identified in the cornea (Fig. 28-5). These are the corneal epithelium, Bowman's membrane, stroma, Descemet's membrane, and endothelium.

The *corneal epithelium* is a thin stratified squamous epithelium of approximately five to six cell layers lining the anterior surface of the cornea (Figs. 28-5 and 28-6). This accounts for one-tenth of the thickness of the cornea. The epithelium has more or less columnar basal cells that rest upon a prominent base-

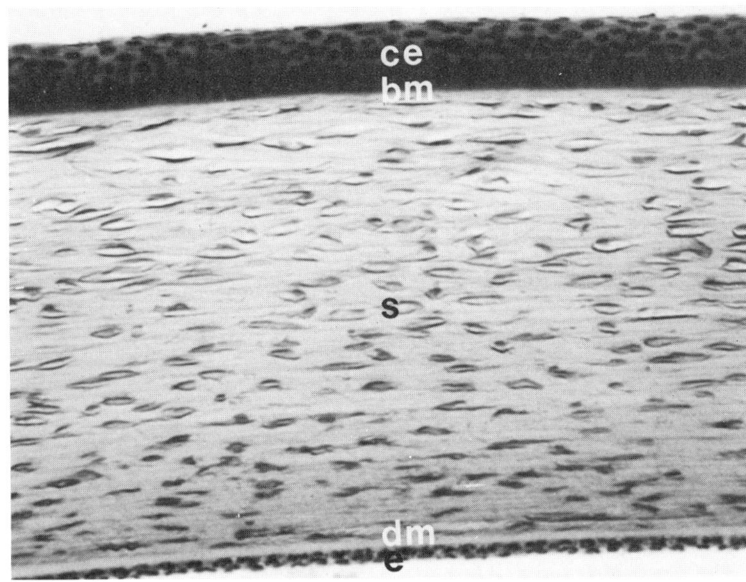

FIGURE 28-5 Light micrograph of the cornea. (ce) Corneal epithelium. (bm) Bowman's membrane. (s) Stroma. (dm) Descemet's membrane. (e) Endothelium.

ment membrane, which in turn is continuous with the Bowman's membrane. Its outer surface has a smooth contour.

Advances in electron microscopy have shown that the *Bowman's membrane* (Fig. 28-6), also called the *anterior limiting membrane,* is made up of collagen fibrils that are packed in a manner that produces the optical smoothness of the anterior layers of the cornea. Bowman's membrane ends abruptly at the limbus region where the corneal epithelium terminates.

The *stroma,* which is also called *substantia propria,* consists of up to 60 layers of regularly arranged collagen fibrils along with fibroblasts trapped between layers of collagen fibrils. In an electron micrograph (EM), the corneal stroma resembles the sclera, but is chemically different in that it contains a large amount of chondroitin sulfates which may contribute to the reversible swelling and transparency of the cornea.

Descemet's membrane (Fig. 28-7), often called the *posterior limiting membrane,* has a

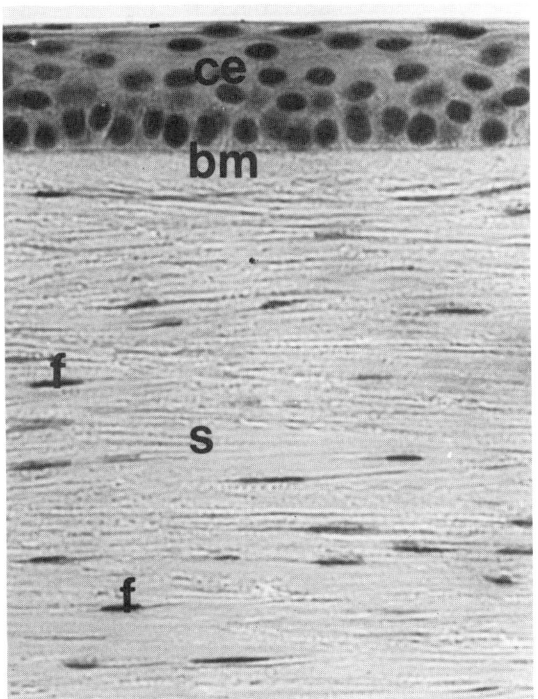

FIGURE 28-6 Corneal epithelium viewed at higher magnification. (ce) Corneal epithelium. (bm) Bowman's membrane. (s) Stroma. Fibroblasts (f) are squeezed between the bundles of collagen fibers.

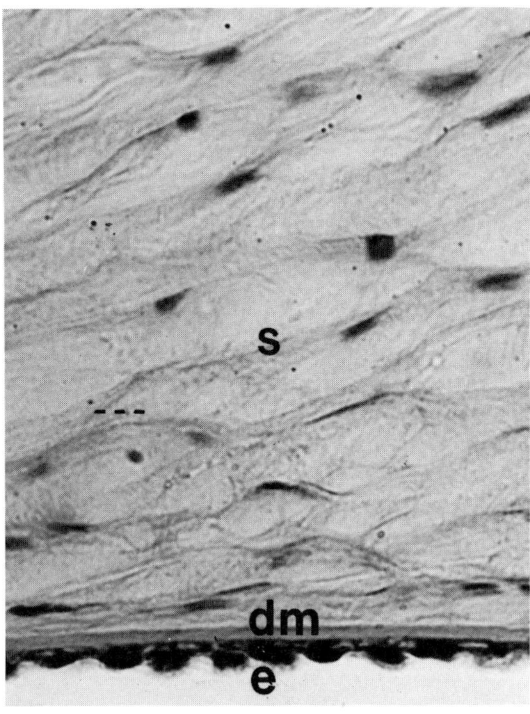

FIGURE 28-7 Corneal stroma (s) viewed at higher magnification. (dm) Descemet's membrane. (e) Endothelium.

degree of elasticity which accommodates the surface changes resulting from the swelling of the corneal matrix. The anterior surface of Descemet's membrane is firmly attached to the stroma, while its posterior surface is lined by a layer of endothelial cells (Fig. 28-7). The thickness of Descemet's membrane is approximately 10 μm.

As in other regions of the body, *endothelial cells* covering the posterior surface of the cornea are single layered and become extremely tenuous along the periphery covering the pores of the trabecular meshwork, as visualized in Fig. 28-8. It should be pointed out that, contrary to common belief, the refraction of light depends more on the cornea than on the lens.

Limbus, the boundary between the cornea and sclera, shows certain specializations. The limbus is approximately 1 mm in width. The corneal epithelium becomes gradually thicker, reaching a maximum of 10 layers or so of squamous cells (Fig. 28-8). Thus, the transition between the corneal and conjunctival epithelia takes place at the limbus. Both Bowman's and Descemet's membranes terminate at the limbus. The *trabecular meshwork*

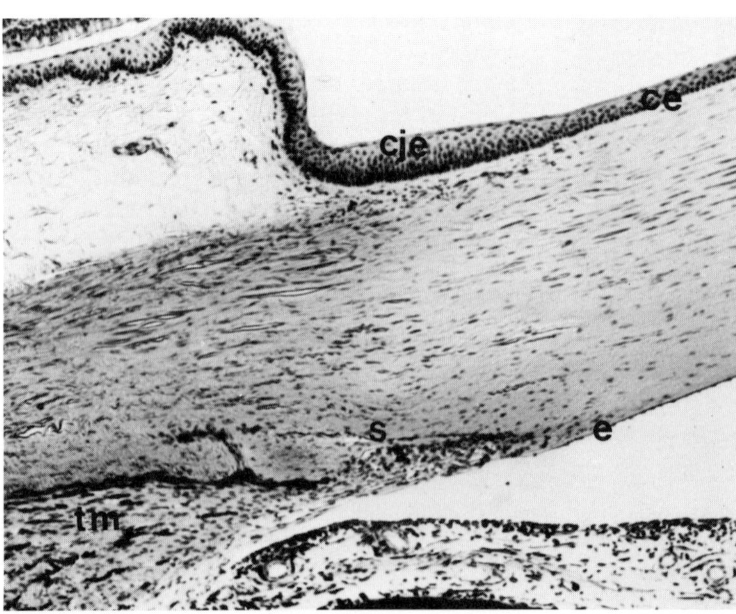

FIGURE 28-8 Light micrograph of the limbus region. It shows the transition between the cornea (ce) and conjunctival epithelium (cje). The endothelium (e) becomes extremely attenuated in the region of the trabecular meshwork (tm). The canal of Schlemm (s) is situated anteriorly and laterally to the trabecular meshwork.

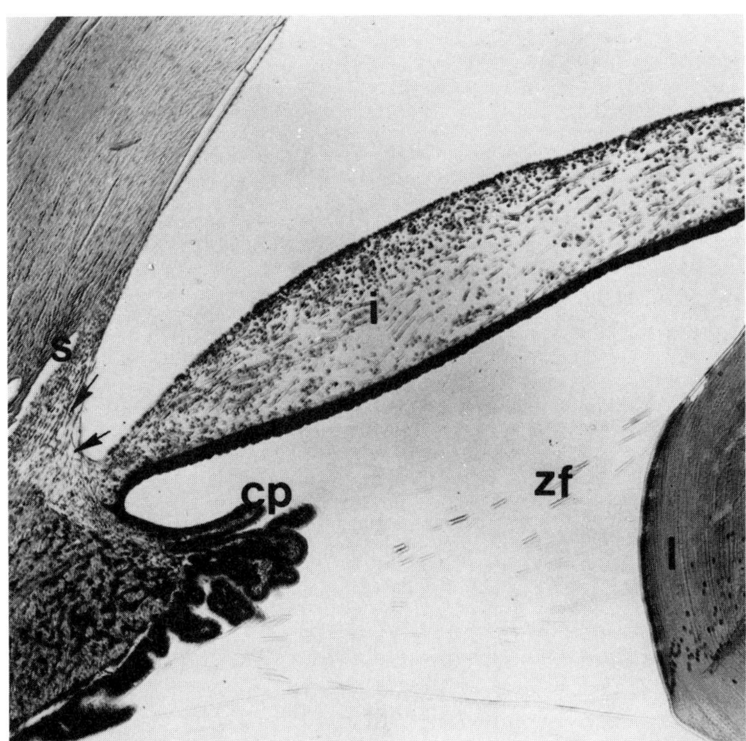

FIGURE 28-9 Light micrograph of the iris angle region. The canal of Schlemm (s) is situated laterally and anteriorly to the trabecular meshwork (arrows). The lens (l) is suspended by zonular fibers (zf) from the ciliary processes (cp). (i) Iris.

(Fig. 28-9) is continuous with Descemet's membrane and is covered by the endothelial cells that cover the cornea (Fig. 28-8). The space produced by the trabecular meshwork is called the *space of Fontana* which is in direct communication with the anterior chamber.

Anterior and lateral to the trabecular meshwork is the *canal of Schlemm* which courses circumferentially about the cornea (Figs. 28-8 and 28-9). The canal of Schlemm is made up of one or more channels and serves to collect the aqueous humor as it penetrates through the space of Fontana. Since the connective tissue supporting the trabecular meshwork and cornea tends to become heavier with age, an obstruction of the draining canal in the elderly may result in the higher intraocular pressure which is one of the main causes of glaucoma. The aqueous humor in a healthy individual is drained through episcleral vessels.

SCLERA

This external coat represents the "white" of the eye and is thick, having an average thickness of 0.5 mm. Sclera is the site where tendons of the extraocular muscles attach. It is largely made up of fibrous connective tissue and is continuous with the cornea. However, the collagen fibrils of the sclera are not as regularly arranged as those in the cornea. Blood supply is rich, particularly near the limbus. The four rectus muscles of the orbit have their insertions in the superficial layers of the sclera at 5 to 8 mm from the limbus. The two oblique muscles insert more posteriorly on the sclera. The sclera becomes thin and fenestrated at the site where the optic nerve begins. This region, called *lamina cribrosa*, has numerous holes (1 mm in diameter) through which the optical nerve fibers exit in small bundles. For this reason, the lamina cribrosa is the weakest point of the sclera.

UVEA (TUNICA VASCULOSA)

The uvea is primarily vascular, is pigmented, and has a muscular region. With the exception of the lamina cribrosa and ciliary body, it is tightly bound to the retina. The association of this vascular tunic to the sclera is a loose one. The uvea has two openings; one is the *pupil* and the other is about the optic nerve. The uvea consists of the choroid, ciliary body, and iris.

CHOROID

That portion of the uvea which is present between the sclera and retina is called *choroid.* The choroid, therefore, terminates at the *ora serrata* (Figs. 28-1, 28-2, and 28-10) where the retina ends. Four different layers may be recognized from outside inward. These are suprachoroid, lamina vasculosa, lamina choriocapillaris, and lamina vitrae.

The *suprachoroid* is often called *epichoroid.* It is made up of loose FECT with its fibers coursing obliquely from choroid to sclera (Fig. 28-11). This layer contains a large number of chromatophores.

The *lamina vasculosa* is characterized by the abundance of blood vessels present in it. The *stratum vasculosa* has numerous veins that form large whorllike "vortices." Four vortex veins drain blood from the choroid. Slender chromatophores of stellate shapes are found in this layer.

The *lamina choriocapillaris* represents the main capillary network supporting the retina in terms of nutrients and oxygen. As such, the capillary layer of the choroid shows a broad-based network of capillary vessels against the retina with the thin lamina vitrae interposed between the two (Fig. 28-11).

Also called Bruch's membrane, the *lamina vitrae* is a cellular glossy membrane which is composed of an outer collagenous layer and an inner layer which represents the basement of the pigmented epithelium of the retina.

CILIARY BODY

As mentioned earlier, the choroid and retina extend as far as the ora serrata. From this point on, the uvea thickens progressively toward the iris where it gives rise to some 70 ridges. Each ridge can now be called a *ciliary process* which runs in a meridional plane (a plane going through the anteroposterior axis of the eyeball). The ciliary processes are present in a concentric manner, and when viewed from an anteroposterior direction, they produce a wheellike structure which is called the *ciliary crown.*

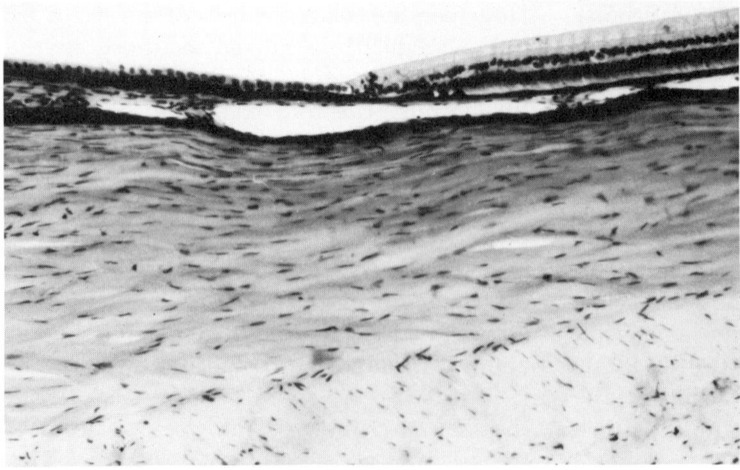

FIGURE 28-10 Light micrograph of the ora serrata region. The retina and the choroid end here.

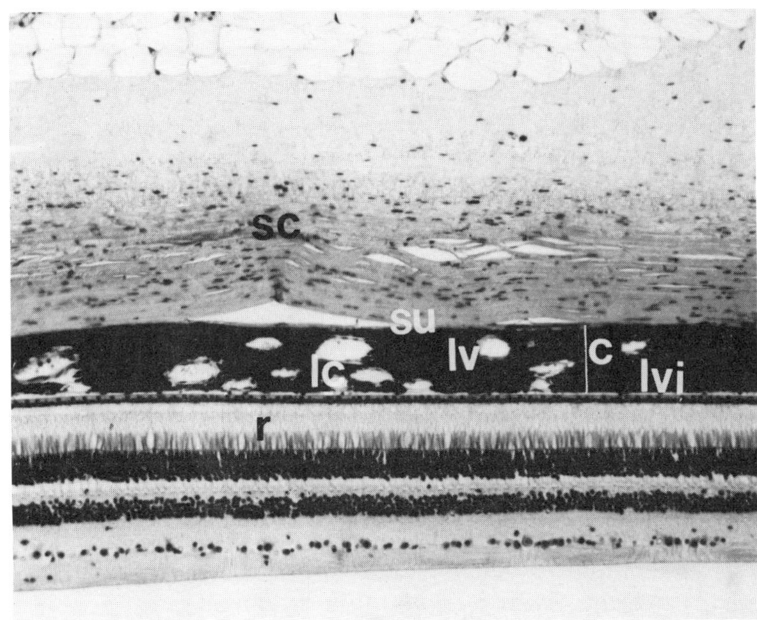

FIGURE 28-11 Light micrograph of the wall of the eyeball. The choroid (c) consists of four different layers: suprachoroid (su), lamina vasculosa (lv), lamina choriocapillaris (lc), and lamina vitrae (lvi). (sc) Sclera. (r) Retina.

Essentially the same layers observed in the choroid are present in the ciliary body. In addition, the ciliary body is covered by a continuation of the sensory epithelium of the retina. The modified retinal epithelium covering the ciliary body is called *ciliary epithelium* and consists of single-layered columnar epithelium without pigmentation.

The ciliary body is unique in that it contains a mass of smooth muscle cells which run radially, meridionally, and circularly. They are present in the suprachoroid layer. The *meridional* (longitudinal) *fibers* form the outermost fibers, while the *radial fibers* lie internal to the meridional fibers. The radial fibers continue to the angle of the iris where they terminate in a loose meshwork. The *circular fibers* are the heaviest and occupy the inner edge of the ciliary body. The circular fibers, called Müller's muscle, relax tension on the lens upon contraction and thereby permit the lens to accommodate for near vision. They are supplied by the parasympathetic postganglionic fibers from the ciliary ganglion. The function of the radial and meridional fibers is not clearly understood at this time.

IRIS

The *iris* is a washer-shaped diaphragm placed directly in front of the lens (Figs. 28-1, 28-2, 28-3, 28-12, and 28-13). The *pupil*, a circular aperture, is located slightly to the nasal side of the center. The root of the iris is attached to the ciliary body, while the free pupillary end is suspended in the aqueous humor between the cornea and lens. Although the iris is seen to cover the lens, there is no real attachment between the two, thereby allowing a continuous flow of the aqueous humor from the posterior chamber to the anterior one.

As in the ciliary body, several layers can be recognized in the iris. They are the endothelium, anterior border layer, vessel (stromal) layer, sphincter and dilator pupillae muscles and pigmented epithelium (Figs. 28-9, 28-12, and 28-13). The first three layers represent the uveal portion of the iris, whereas the last two constitute a forward continuation of the tunica intima (i.e., retina). The *endothelial layer* of the iris is often indistinct. It is continuous with the endothelium of the cornea at the angle of the iris (Fig. 28-9).

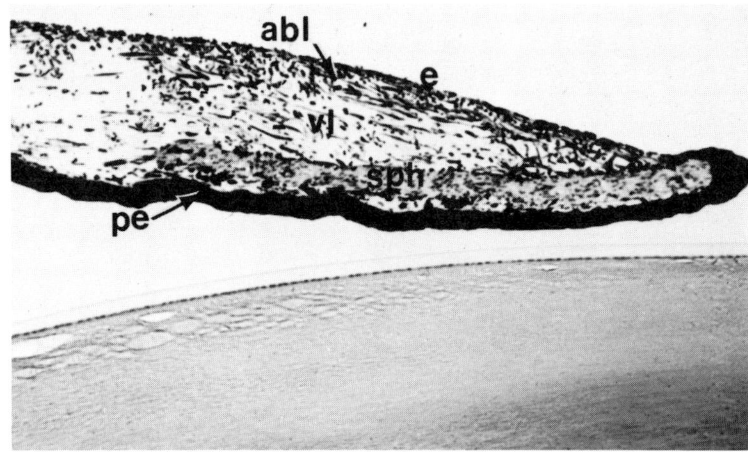

FIGURE 28-12 Light micrograph of the iris. The different layers are readily identified. (e) Endothelium. (abl) Anterior border layer. (vl) Vessel layer. (sph) Sphincter pupillae muscle. (pe) Pigmented epithelium.

The *anterior border layer,* also called the anterior stromal layer, immediately under the endothelium is avascular and rich in chromatophores (Figs. 28-12 and 28-13). This is the layer which determines the color of the eye. The color of the eye is a function of the thickness and the concentration of chromatophores in the anterior stromal layer. If the thickness of this layer is thin and few chromatophores are present, the light will penetrate the various layers of the iris and become reflected in its pigmented epithelium located at the posterior surface. Depending on the amount and arrangement of pigmented cells, the color of the eye in such an individual may vary from blue, to green, to gray.

The deeper *vessel layer* is often called the *stromal layer* and is made up of spongy FECT (Figs. 28-12 and 28-13). Fibers are delicate, and most of the stromal cells are pigmented. Branches from the anterior and posterior ciliary arteries make extensive anastomoses, producing the greater arterial circle of the iris which supplies the stromal layer.

The *sphincter pupillae* muscle, although located in the stromal layer, is know to have been derived from the ectodermal pigmented cells (Figs. 28-12 and 28-13). They are circularly arranged near the papillary margin. Upon contraction, the sphincter muscle cells reduce the diameter of the pupil. In addition, the *dilator pupillae* muscle is formed within the basement membrane of the pigment epithelium. Being radially arranged, the contraction of the dilator pupillae results in an increased pupillary diameter. The sphincter pupillae muscle is innervated by parasympathetic fibers by way of long ciliary nerves. The dilator pupillae muscle receives sympathetic innervation via the superior cervical ganglion. Together with ciliary muscles, the sphincter and dilator pupillae muscles make up the *intrinsic muscles* of the eye.

The *pigmented epithelium* of the posterior surface of the iris is a double layer of heavily pigmented cells (Figs. 28-12 and 28-13). At the margin of the pupil these pigmented cells extend centrally more than does the stroma, producing a collarette of pigmented cells.

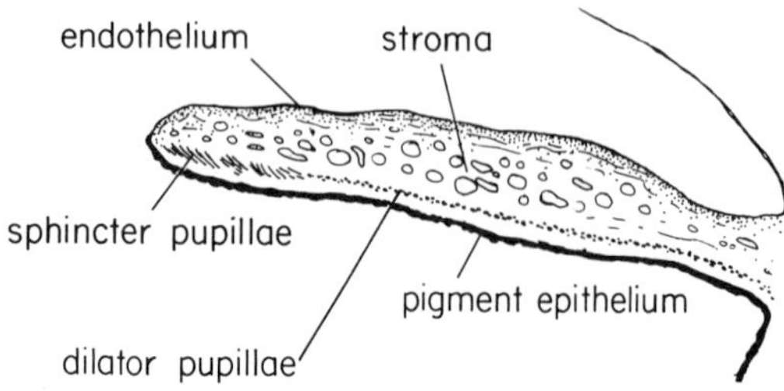

FIGURE 28-13 Diagram of the iris.

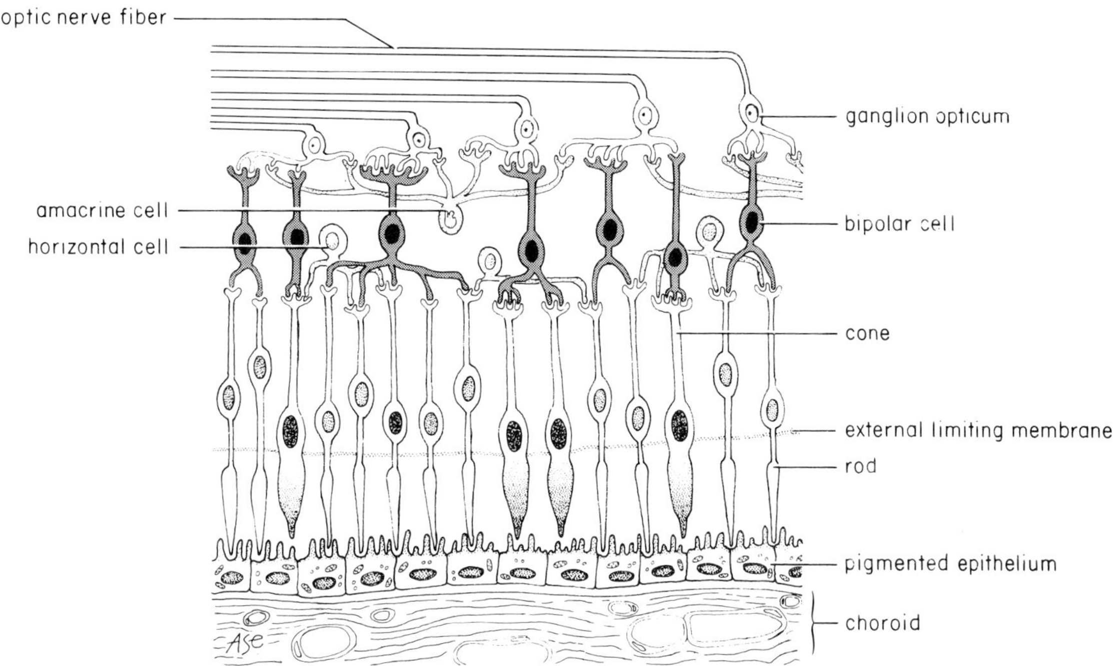

FIGURE 28-14 Diagram of the retina.

Iris Angle. As the iris meets the sclerocorneal junction at an acute angle, this relationship creates an angle between the cornea and the iris (Figs. 28-3 and 28-9). Here, the space of Fontana continues to the anterior chamber on one side and approaches the canal of Schlemm encircling the iris on the other side. Thus, the canal of Schlemm, together with the space of Fontana (of iris angle), forms the draining passage of the intraocular fluid from the eye. As it circles the iris, the canal of Schlemm communicates peripherally with numerous tributaries to the anterior ciliary veins.

RETINA (TUNICA INTIMA OCULI)

The innermost layer lines the tunica vasculosa throughout its extent. As described previously, the retina arose as a protrusion from the diencephalon which later became a double-walled optic cup. The outer layer of this cup became the nonsensory pigment epithelium, while its inner layer differentiated into the retina proper which contains several sublayers. The *retina* is a delicate membrane of no more than 0.5 mm in thickness. It is here that the visual stimulus is perceived, integrated to some extent locally, and then transmitted via the optic nerve to the brain for further processing. The retina extends from the optic nerve to the posterior scalloped margin of the ciliary body, *ora serrata.* Except for certain minor modifications, the retina is made up of recognizable layers as follows (Fig. 28-14):

1. Pigment epithelium
2. Inner and outer segments of rods and cones
3. Outer limiting membrane
4. Outer nuclear layer
5. Outer plexiform layer
6. Inner nuclear layer
7. Inner plexiform layer
8. Ganglion cell layer
9. Nerve fiber layer
10. Internal limiting membrane

Pigment Epithelium. The pigment epithelial layer is represented by cuboidal epithelium resting on a basal lamina that is continuous with the inner lamella of the lamina vitrae of the choroid. The spherical nuclei are located along the basal (outer) portion (Figs. 28-11, 28-14, and 28-17). The apical (inner) portion of the cells contains different amounts of melanin known as *fuscin*. The slender processes of pigment cells insinuate between the outer segments of rods and cones.

Inner and Outer Segments of Rods and Cones. Rods and cones are the photosensitive receptors. Rods are cylindrical and cones are pyramidal in shape. Rods outnumber cones 130 million to 7 million. The outer segments of both rods and cones can be stained with periodic acid Schiff (PAS) and other agents for phospholipid staining. They contain photosensitive substances which are called *rhodopsin* (or visual purple) in rods and *iodopsin* in cones. The outer segments have been shown in an EM to be composed of a series of laminated membranes (Fig. 28-15) that are responsible for the absorption of light that triggers the visual stimulus. The inner segment of each photoreceptor contains a concentration of mitochondria in a body called *ellipsoid* (Fig. 28-16).

FIGURE 28-15 Schematic drawing of rod and cone. Arrows indicate level of external limiting membrane.

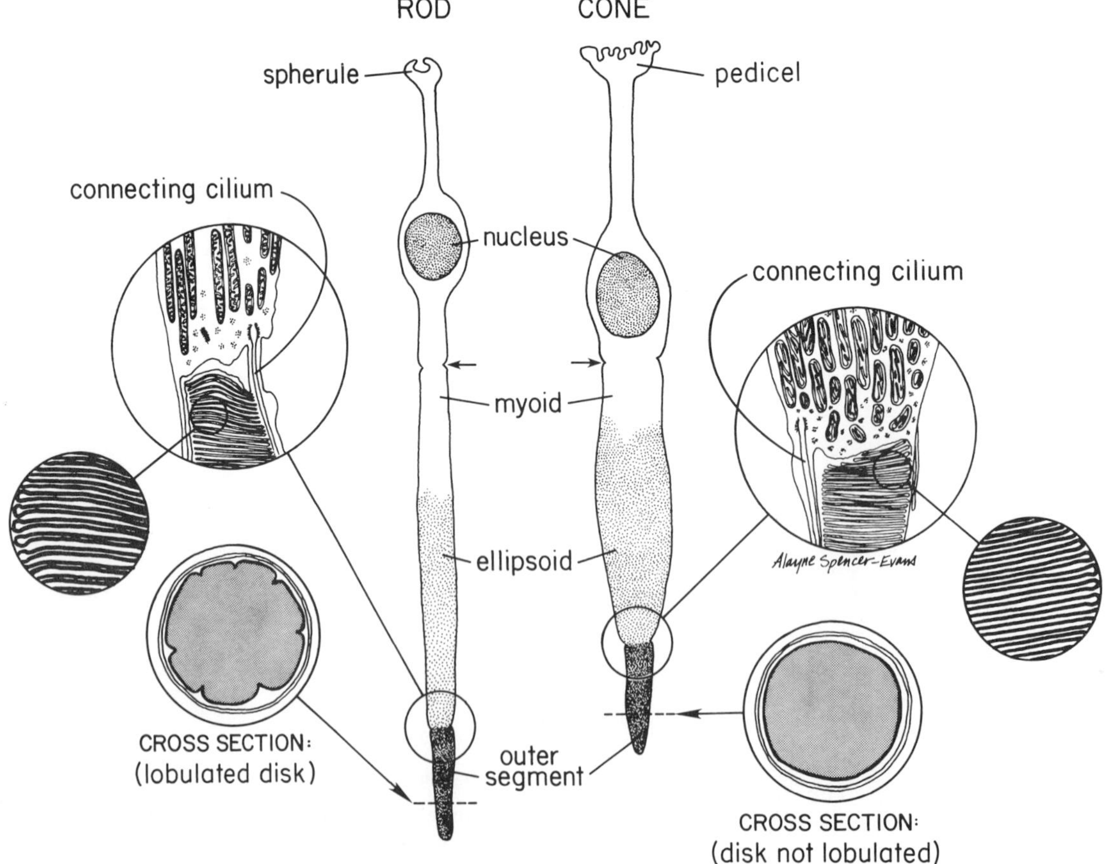

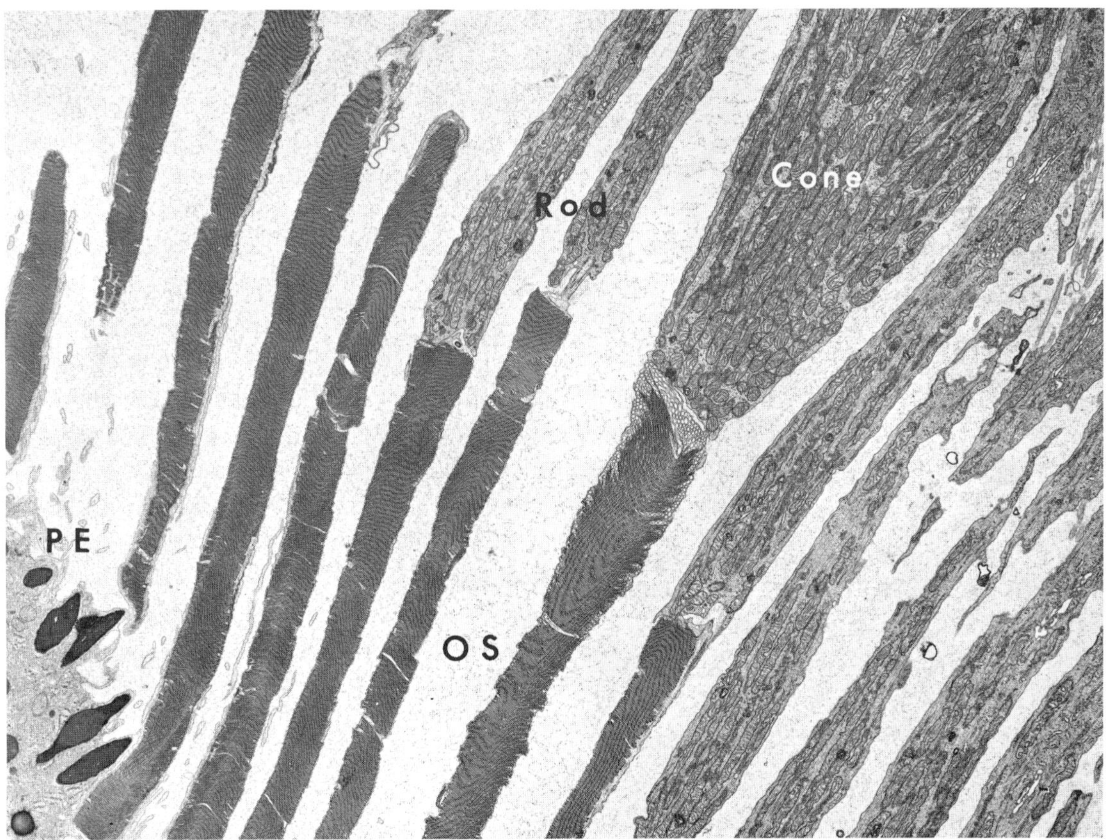

FIGURE 28-16 Photoreceptor elements of the retina. Inner segments of rods (Rod) and cones (Cone) contain packs of mitochondria. Outer segments (OS) consist of lamellar membranes. Pigment epithelium (PE) touches lightly at the tips of the outer segments.

The rods have more photosensitive segments and are therefore capable of responding to low levels of illumination. Thus, rods provide a greater acuity for night vision. The laminated plates of the rods number approximately 1300, with a turnover time of approximately 10 days. In addition, the rods have more extensive neural connections.

The cones, having less photosensitive pigments and less summation at the retina, provide a greater resolution of images and therefore better visual acuity in daylight.

Outer Limiting Membrane. Electron microscopy has rendered that the outer limiting membrane is made up of junctional complexes between the supporting Müller's cells and adjoining photoreceptor cells. As indicated in the diagram appearing in Fig. 28-14, the outer limiting membrane is *not* a membrane.

Outer Nuclear Layer. The cell bodies and nuclei of the rods and cones constitute the outer nuclear layer. Nuclei of the cone cells are pale, are ovoid, and lie just beneath the outer limiting membrane. Nuclei of rod cells are rounded and dark and situated at deeper levels (Figs. 28-14 and 28-17). Anterior to the outer nuclear layer is the outer plexiform layer.

Outer Plexiform Layer. Both rod and cone cells terminate at the outer plexiform level. The rod cell processes end in a knob, while the cone cells end in branching expansions.

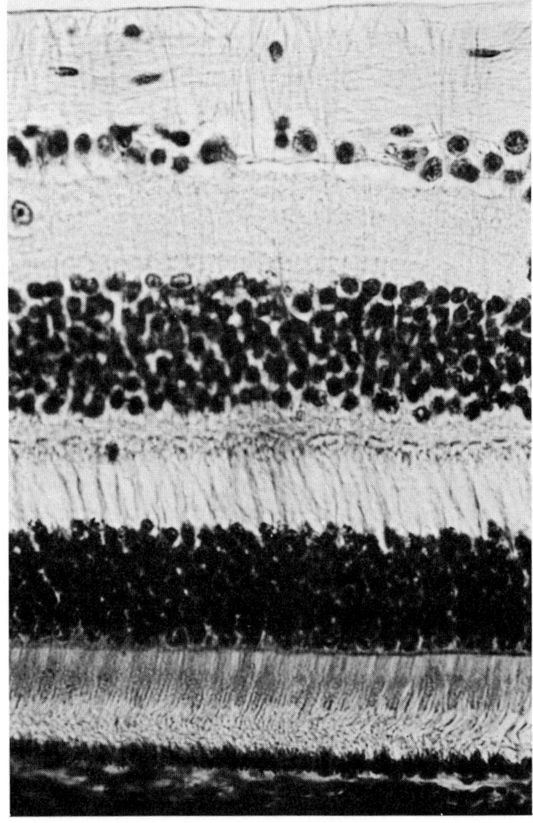

FIGURE 28-17 Light micrograph of the retina. Compare with the diagram in Fig. 28-14.

Synapses between the visual cells and the bipolar cells are made.

Inner Nuclear Layer. Cell bodies of bipolar cells, association neurons, *horizontal cells,* and supportive (Müller's) cells reside in the inner nuclear layer. Of these, the bipolar cells are most numerous, and therefore this layer is often called the *bipolar layer.* This layer is important in summation of impulses originated in rod cells. It also affects some integration of rods and cones. In addition, large monopolar amacrine cells may be found (Figs. 28-14 and 28-17). They connect with each other, with axonal endings of bipolar cells, and with dendrites of ganglion cells.

Inner Plexiform Layer. Located interior to the inner nuclear layer, the inner plexiform layer is the site where synapses between the axons of bipolar cells and the dendrites of the next order of cells, that is, ganglion cells, occur.

Ganglion Cell Layer. Ganglion cells are the most interior and therefore the last layer of cells in the retina. They are large cells containing Nissl bodies and represent the third link in the chain of neurons. Dendrites of the ganglion cells make synapses with axons of the bipolar cells, while axons of these cells form the optic nerve.

Nerve Fiber Layer. As indicated in the preceding paragraph, this layer represents the collection of axons of the ganglion cells. These axons converge toward the lamina cribrosa. The layer becomes thicker as it approaches the optic nerve. Along with nerve fibers, this layer also contains the large blood vessels entering from the optic nerve.

Internal Limiting Membrane. The inner boundary of the retina, the internal limiting membrane, represents the basement membrane formed by the expanding ends of the supportive (Müller's) cells that have penetrated the nerve fiber layer. Accordingly, the internal lining membrane can be stained with PAS.

SPECIALIZATIONS IN THE RETINA

Macula (Fovea). Located at the posterior pole, the macula is the region of the highest visual acuity (Fig. 28-18). Here, the retina is greatly reduced in thickness. This is accomplished by a progressive disappearance of rods. The remaining cones are more slender in appearance, resembling rods in shape. The inner retinal layers spread aside, creating a pitlike depression called *foveola* or *fovea centralis.* The absence of outer neurons and vessels and the tight packing of slender cones increase the resolution. Only ganglion cells surround the margin of the foveola.

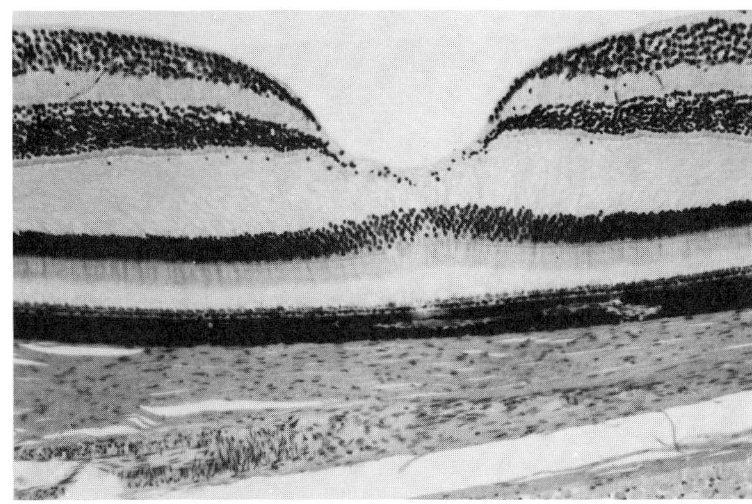

FIGURE 28-18 The macular region of the retina.

Nerve Head. The nerve head, also called the *papilla,* is a circular area (1 mm in diameter) through which nerve fibers leave the retina to become the optic nerve (Fig. 28-19). The head is situated approximately 3 mm nasal to the center of the macula.

Ora Serrata. The anterior end of the retina is scalloped and terminates abruptly (Fig. 28-1). Approaching this region, visual cells become shorter and thicker, and the ganglion cells and nerve fibers disappear. The two nuclear layers (i.e., visual cells and bipolar cells) blend.

OPTIC NERVE

Since the retina has its origin in the developing diencephalon, the optic nerve is a central nerve. It is covered by the same meningeal layers that surround the brain. At the posterior surface of the eyeball, subdural and subarachnoid spaces cease, and the meningeal sheaths join the sclera. Pia mater ex-

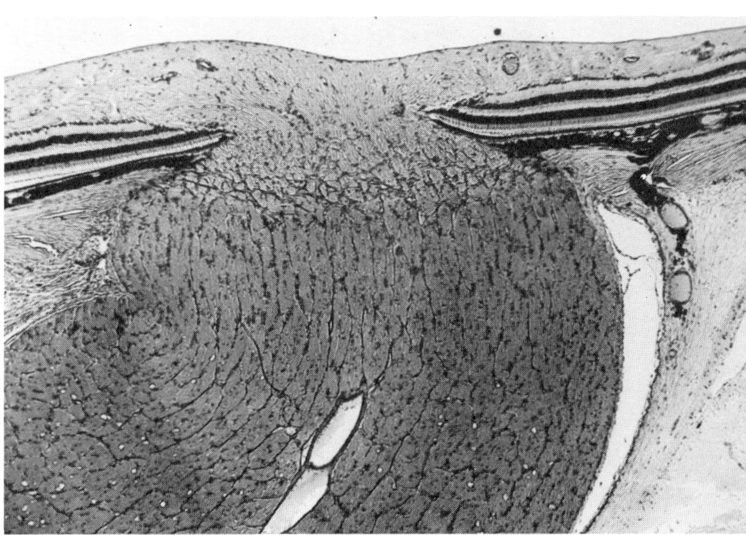

FIGURE 28-19 Light micrograph of the optic nerve entering the eyeball through the sclera.

tends into the optic nerve, dividing the nerve into small bundles. These small nerve fiber bundles create the lamina cribrosa of the sclera (Fig. 28-19). The optic nerve is a collection of axons from the ganglion cells. They acquire myelin sheaths as soon as they exit from the sclera. The connective tissue trabeculae from the pia and the glial cells form a system of partitions which unites in the center of the optic nerve. Through this central connective tissue matrix, the central artery and vein provide blood supply to the retina.

LENS AND ADNEXA

Posterior to the cornea, the eyeball has a series of refractive elements that contribute to the functioning of the eye. They are the lens, ciliary zone, vitreous body, and aqueous humor. That the cornea is the strongest refractive structure has already been discussed.

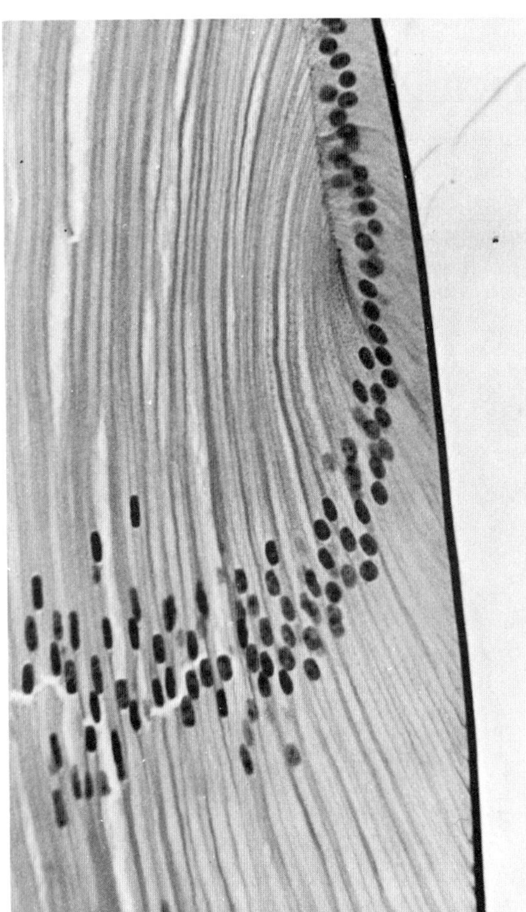

FIGURE 28-20 Equator of the lens of a child. The lens epithelial cells are differentiating into recognizable lens fibers. The nuclei are found at considerable depth in the cortex of the young lens. The capsule is heavily stained with PAS. (× 150)

LENS

The lens is a transparent, plastic, and biconvex disk situated behind the iris and held in place by the zonular fibers. Its posterior surface is more convex than the anterior surface (Fig. 28-1). The two surfaces meet each other at a rounded edge called the *equator.* In human beings, its diameter is about 10 mm, with the thickness at its center being approximately 5 mm.

The lens is covered by a homogenous hyaline capsule that has the same physical property as the Descemet's membrane. The suspensory annular fibers insert in the *lens capsule* (Fig. 28-20). Under the anterior portion of the capsule is a layer of cuboidal cells with uniform height. The *lens substance* is made up of differentiated lens cells which are flattened. These cells, also called *lens fibers,* are arranged in concentric lamellae. Lens fibers are continuously added throughout life from the epithelial cells near the equator.

Recent studies have shown that the capsule is made up of the basement membrane of the lens epithelium. The lens fibers are closely packed without obvious intercellular space. The lens does not receive a blood supply or additional cells. The initial collection of ectodermal cells continues to provide additional lens fibers. Thus, the lens is an immunologically protected area which is free from cellular traffic. This permits the

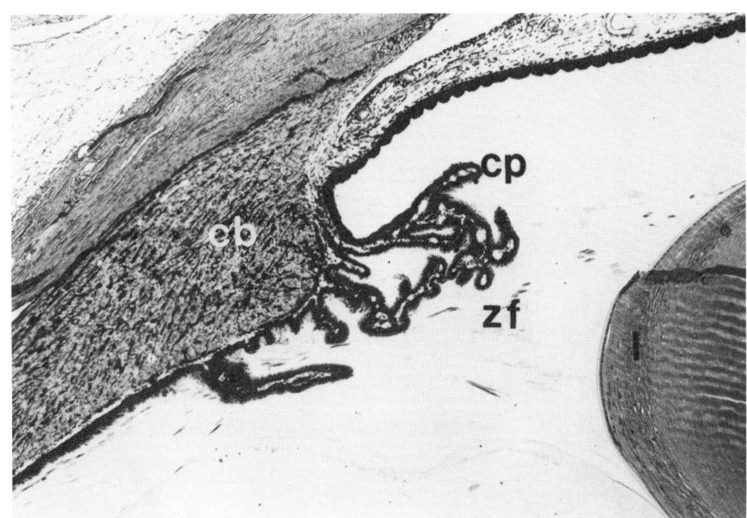

FIGURE 28-21 The ciliary zonule. Suspensory fibers extend from the ciliary processes to the lens capsule near the equator. (cb) Ciliary body. (cp) Ciliary process. (l) Lens. (zf) Zonular fibers.

maintenance of the remarkable transparency the lens shows throughout life.

CILIARY ZONULE

As indicated earlier, the suspensory fibers of the *zonule* maintain the lens in place. The zonular filaments originate from the inner surface of the ciliary body and insert into the lens capsule near the equator at both the anterior and posterior surfaces (Fig. 28-21). Zonular filaments resemble the filamentous structure of the vitreous body. The zonules form the posterior limit of the posterior chamber and the anterior demarcation of the vitreous space.

VITREOUS BODY

The viscous, yet transparent, fluid filling the large space between the lens and retina is called the *vitreous* or *vitreous body.* The vitreous is composed of a small quantity of collagen suspended in a hyaluronic acid–rich polysaccharide matrix. The vitreous borders on the basement membranes of adjacent tissues (viz, lens capsule and internal limiting membrane of the retina). In addition to the refractive function, the vitreous body physically supports the retina.

AQUEOUS HUMOR

Aqueous humor, a fluid similar to other tissue fluids in composition, fills the *ocular chamber.* The ocular chamber is divided into anterior and posterior chambers by the iris. The production, flow, and drainage of the aqueous humor have been discussed previously.

EYELIDS

The eyelids are movable skin folds which have certain specializations that protect the eye from physical injury and excessive light. They have a thin skin which has a specialized mucosa on the posterior surface, the *conjunctiva.* That portion of the conjunctiva lining the posterior surface of the eyelids is called *palpebral conjunctiva,* while that over the ocular surface is called *bulbar conjunctiva.* The bulbar conjunctiva ends at the cornea. The fold between the two portions of the conjunctiva is called the *fornix.* These features are diagramed in Fig. 28-22.

The free margin between the exterior skin and palpebral conjunctiva has eyelashes and the openings of the simple branched alveolar glands (Fig. 28-23). These *tarsal glands* are sebaceous in nature and function in lubricating and forming a watertight seal

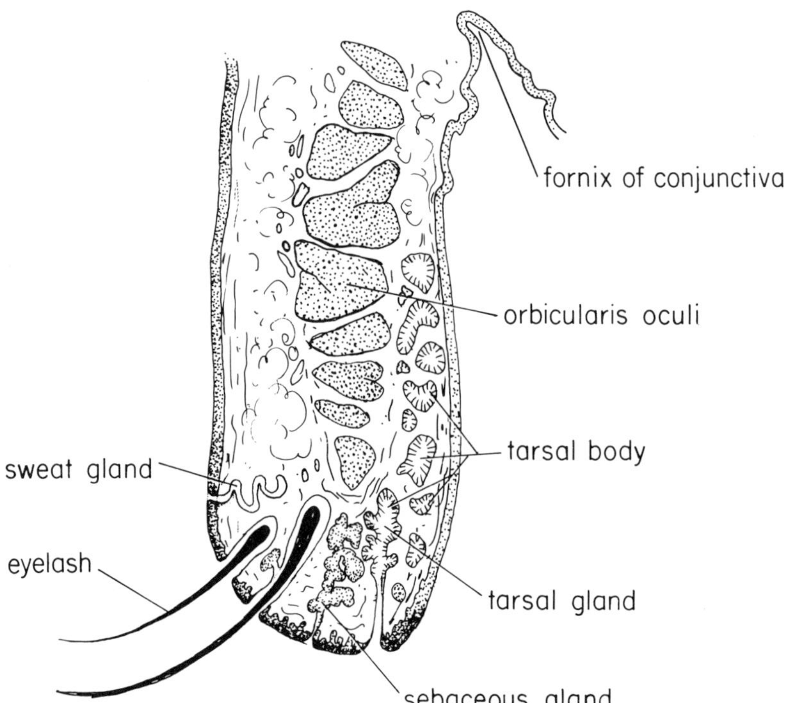

FIGURE 28-22 Drawing of the eyelid.

while the lids are closed. They are also called meibomian glands.

In addition, sebaceous glands are present in relation to the *eyelashes* and are much like those found in the hair shafts throughout the body. They used to be called the glands of Zeis. Large spiraling sweat glands in the skin near the edge of the eyelid open their ducts between eyelashes. They are different from other sweat glands in that they do not coil, but present a loose spiral form. They are often called the glands of Moll. The shape of the lid is maintained by a tough fibrous plate, the *tarsus*.

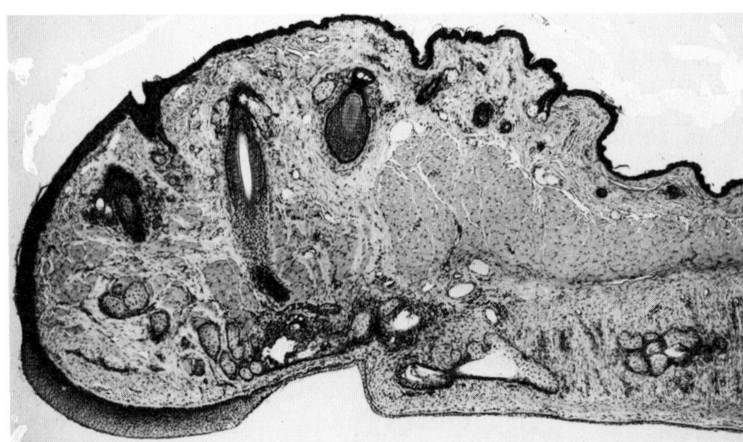

FIGURE 28-23 Light micrograph of the eyelid. Meibomian glands and eyelashes are present.

LACRIMAL GLANDS

The lacrimal glands are serous, compound tubuloalveolar glands resembling the parotid. They are situated in the superolateral corner of the orbit. The acinar cells are somewhat taller than those of the parotid. Between the acinar cells and the basement membrane are numerous basketlike myoepithelial cells. As in parotid glands, the stroma of lacrimal glands has a number of lymphoid elements infiltrating the region. This suggests the possibility that secretory IgA might be produced in the lacrimal glands.

The secretory product is a clear and salty tear which flushes the conjunctival sac, flows toward the medial angle of the eye, and is collected by the lacrimal canaliculus in the eyelid. From this point, each tear eventually reaches the nasal cavity via the lacrimal sac and the nasolacrimal duct.

REVIEW SECTION

1. Figure 28-24 shows a section through the wall of the eyeball. How many layers can be seen? Name the various layers. What are their functions?

The wall of the eyeball has three layers. They are the tunica fibrosa, the uvea, and the retina. Tunica fibrosa consists of the cornea which is transparent and the sclera. The sclera is the site of attachment for the extrinsic eye muscles. The uvea is highly vascularized and supplies nutrition to the retina, the ciliary body, and the iris. In the retina, the visual stimuli are perceived.

2. Figure 28-25 shows a section of retina. How many layers can you identify? Name the layers. Where are the nuclei of the photoreceptors located? Where are the photosensitive substances located? Where does the summation of impulses originated in the rods and cells occur? What takes place in the ganglion cell layer?

The retina is made up of 10 recognizable layers, namely, pigment epithelium, inner and outer segments of rods and cones, outer limiting membrane, outer nuclear layer, outer plexiform layer, inner nuclear layer, inner plexiform layer, ganglion cell layer, nerve fiber layer, and internal limiting membrane. Nuclei of the photosensitive receptors, the

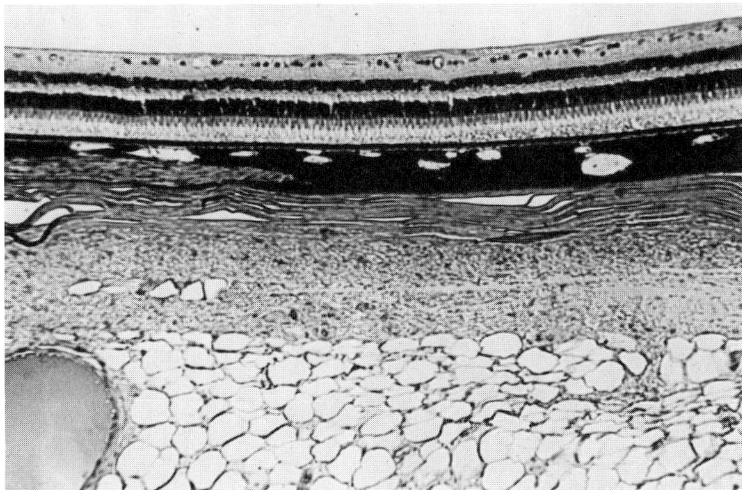

FIGURE 28-24

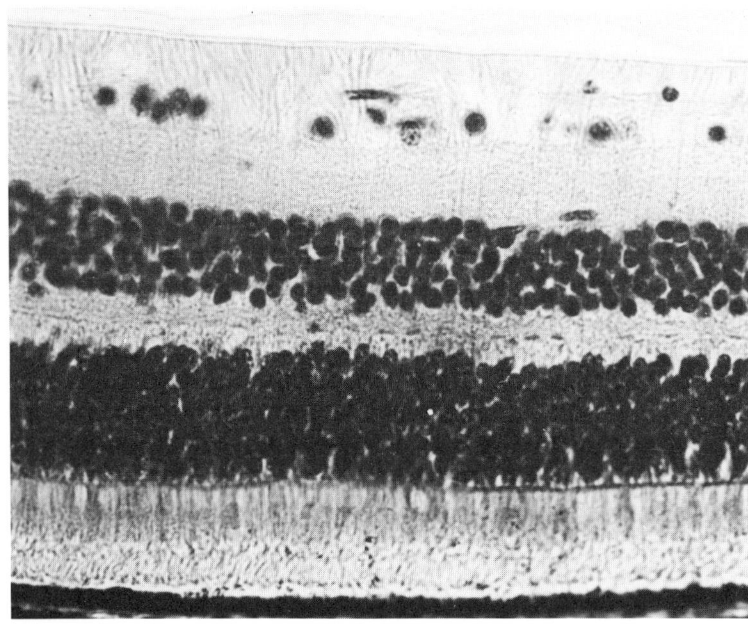

FIGURE 28-25

rods and cones, are located in the outer nuclear layer. The photosensitive substances of the retina are located in the outer segments of rods and cones. Impulses originated in the rods and cones undergo summation in the inner nuclear layer. Dendrites of the ganglion cells make synapses with axons of bipolar cells, while axons of these cells form the optic nerve.

3. Figure 28-26 shows a section of a portion of the iris and the lens. Name the space indicated by (a). Name the space indicated by (b). How many layers are present in the iris? Name them. What determines the color of the eye? What structures can be recognized in the lens?

The triangular area (a) bounded by the iris anteriorly and the lens posteriorly is the posterior chamber. Anterior to the iris and the lens, space (b) is part of the anterior chamber. Its anterior boundary is the cornea. Five layers are recognizable in the iris. They are the endothelium, the

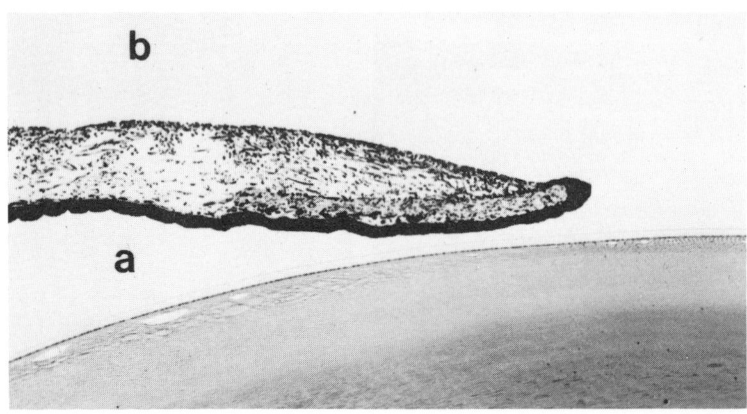

FIGURE 28-26

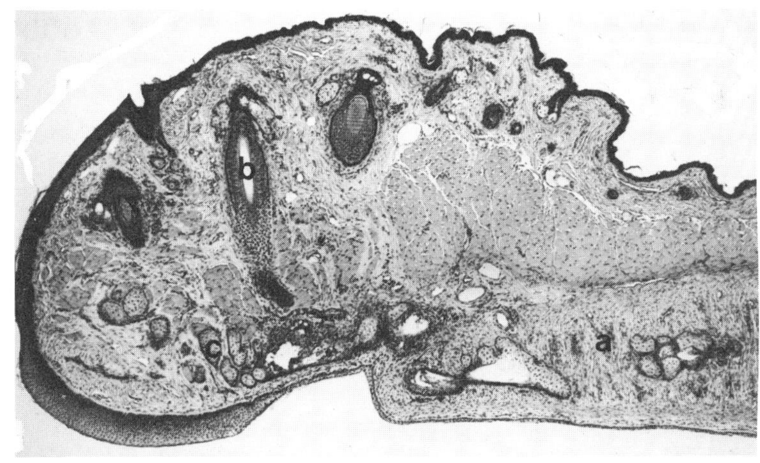

FIGURE 28-27

anterior border layer, the vessel layer, the sphincter and dilator pupillae, and the pigmented epithelium. Light will penetrate the various layers of the iris and become reflected in the pigment epithelium. The color of the eye thus depends on the number and arrangement of pigmented cells. The lens capsule, the lens epithelium, and the lens fibers are apparent in the lens.

4. Figure 28-27 shows a section of the eylid. Identify:
(a)________ (b)________ (c)________

The shape of the eyelid is maintained by the tarsus (a), a tough fibrous plate. Tarsal glands (c), or meibomian glands, are sebaceous in nature and function in lubricating and forming a watertight seal when the lids are closed. The lids have eyelashes (b).

BIBLIOGRAPHY

Dawson, H. (ed.): *The Eye,* vols. 1–4, Academic Press, New York, 1962.

Fine, B. S., and M. Yanoff: *Ocular Histology,* Harper & Row, New York, 1972.

Hogan, M. J., J. A. Alvaredo, and J. E. Wedell: *Histology of the Human Eye,* Saunders, Philadelphia, 1971.

Rhodin, J. A. G.: *Histology: A Text and Atlas,* Oxford, New York, 1974.

Weiss, L., and R. O. Greep: *Histology,* 4th ed., McGraw-Hill, New York, 1977.

ADDITIONAL READINGS

Boycott, B. B., and J. E. Dowling: "Organization of the Primate Retina: Light Microscopy," *Phil. Trans. Roy. Soc., London,* **B255:**109 (1969).

Cohen, A. I.: "Vertebrate Retinal Cells and Their Organization," *Biol. Rev.* **8:**427–459 (1963).

Dowling, J. E.: "Organization of Vertebrate Retinas," *Invest. Ophthalmol.* **9:**665–680 (1970).

———, and B. B. Boycott: "Organization of the Primate Retina: Electron Microscopy," *Proc. Roy. Soc.,* **B116:**80–111 (1966).

Hect, S.: "Rods, Cones, and the Chemical Basis of Vision," *Physiol. Rev.* **17:**239–290 (1937).

Leeson, T. S.: "Rat Retinal Rods: Freeze Fracture Replication of Outer Segments," *Canad. J. Ophthalmol.* **5:**91–107 (1970).

———: "Lens of the Rat Eye: An Electron Microscope and Freeze Etch Study," *Exp. Eye Res.* **11:** 78–82 (1971).

———, and C. R. Leeson: "Choriocapillaries and

Lamina Elastica (Vitrea) of the Rat Eye," *Brit. J. Ophthalmol.* **51:**599–616 (1967).

Sjostrand, F. S.: "An Electron Microscope Study of the Retinal Rods of the Guinea Pig Eye," *J. Cell Comp. Physiol.* **33:**383–403 (1949).

———: "Ultrastructure of the Outer Segments of Rods and Cones of the Eye as Revealed by the Electron Microscope," *J. Cell Comp. Physiol.* **42:** 15–44 (1953).

Villegas, G. M.: "Ultrastructure of the Human Retina," *J. Anat.* **98:**501–513 (1964).

Wald, G.: "The Photoreceptor Process in Vision," *Am. J. Ophthalmol.* **49:**18–41 (1955).

———: "The Receptors of Human Color Vision," *Science* **145:**1007–1016 (1964).

Wanko, T., and M. A. Gavin: "Electron Microscopic Study of Lens Fibers," *J. Biophys. Biochem. Cytol.* **6:**97–102 (1959).

Young, R. W.: "Visual Cells and the Concept of Renewal," *Invest. Ophthalmol.* **15:**700–725 (1976).

———, and D. Bok: "Participation of the Retinal Pigment Epithelium in the Rod Outer Segment Renewal Process," *J. Cell Biol.* **42:**392–403 (1969).

———, and ———: "Autoradiographic Studies on the Metabolism of the Retinal Pigment Epithelium," *Invest. Ophthalmol.* **9:**524–536 (1970).

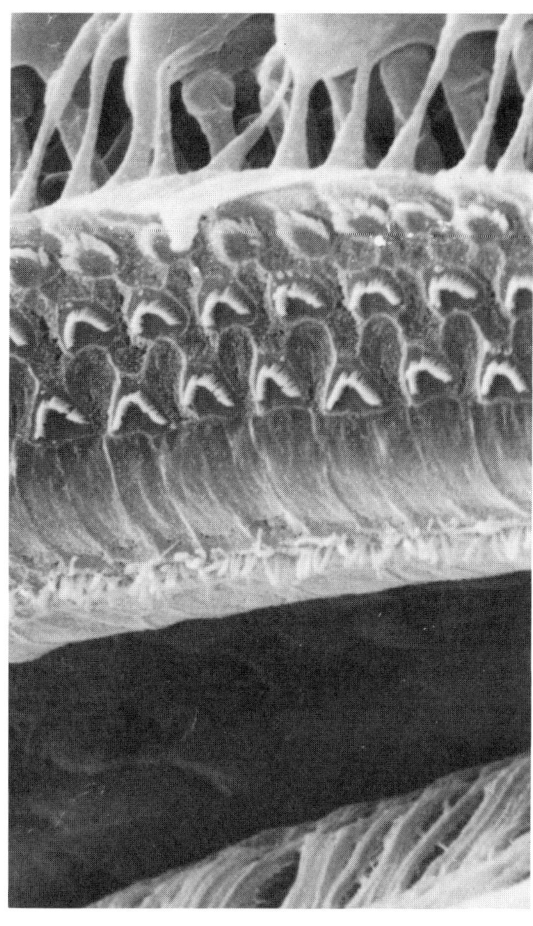

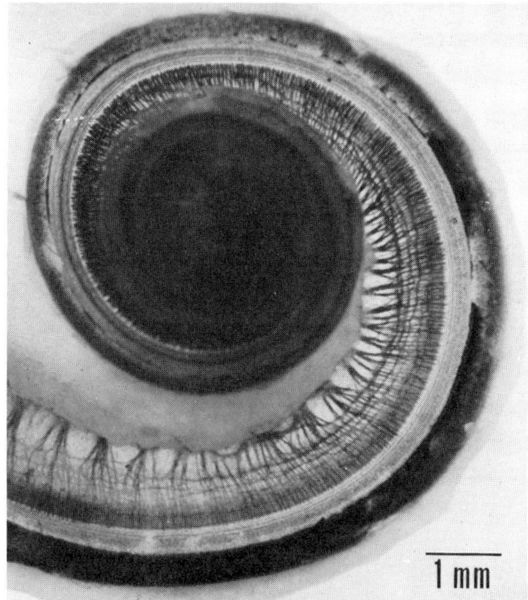

29
EAR

OBJECTIVES

Upon completion of this chapter, the student will demonstrate an understanding of the structure of the ear by the ability to:

1 Define the anatomical boundaries and gross structural compositions of the external ear, middle ear, and inner ear including the: (a) auricle (b) external auditory meatus (c) tympanic membrane (d) eustachian tube (auditory tube) (e) auditory ossicles (f) membranous labyrinth (g) vestibular labyrinth (h) semicircular canal (i) cochlea (j) tympanum

2 Identify the histologic regions of the tympanic membrane:
(a) anulus tympanicus (b) stratum cutaneum (c) stratum radiatum (d) stratum circulare (e) stratum mucosum

3 Define and identify the:
(a) oval window (vestibular) (b) round window (cochlear) (c) relationship between stapes and oval window

4 State the functional significance of the eustachian tube.

5 Name the intrinsic muscles of the ear.

6 Identify the functioning of the:
(a) semicircular canal (b) saccule (c) utricle

7 Define the spaces where endolymph and perilymph are present.

8 State the differences in chemical composition of endolymph and perilymph.

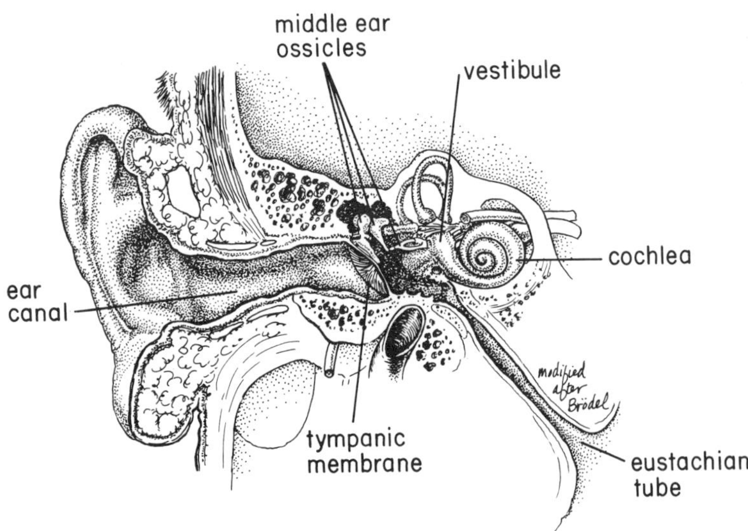

FIGURE 29-1 Schematic drawing of the human ear, illustrating the anatomical relations of the various parts of the ear and the surrounding and connecting structures.

9 Identify and state the similarities in chemical composition between cupula, otolithic membrane, and tectorial membrane.

10 Identify the following histologic structures of the cochlea:
(a) scala vestibuli (b) scala tympani (c) scala media (d) Reissner's membrane (e) stria vascularis (f) inner and outer hair cells (EM also) (g) inner and outer phalangeal cells (h) cochlear nerve (i) bony spiral limbus (j) spiral ganglion (k) modiolus (l) helicotrema

The ear is composed of two functionally different portions: one that receives sound and transmits it to the brain and another which registers the gravity and motion of the head. Accordingly, the ear is supplied by both the acoustic and the vestibular nerves. Anatomically, the ear can be divided into three parts (Fig. 29-1). The first part is the *external ear* which includes the pinna or auricle and the external auditory meatus leading inward to the tympanic membrane. The second portion is the *middle ear* which is a membranous chamber and includes the tympanic membrane or drum and a chain of three ossicles extending from the drum to the *inner ear* (Fig. 29-1). The tympanic cavity is in communication with the nasopharynx via the auditory (eustachian) tube. The third and innermost component, the inner ear, is a labyrinthine structure which includes the organs of hearing and equilibrium, the surrounding bones, and associated nerves.

EXTERNAL EAR

AURICLE

The auricle, also called *pinna,* is made up of a thin cartilage plate which is covered on both sides by a thin skin. On the anterior surface, the skin is tightly attached to the perichondrium, while at the dorsal surface a subcutaneous FECT is interposed between the skin and the cartilage. The cartilage of the auricle is elastic in nature, containing an unusually abundant number of cartilage cells. The extrinsic muscles of the ear insert into the fibrous perichondrium. The skin over the auricle is supplied with fine hairs and many large sebaceous and sweat glands. In the lobule, the subcutaneous connective tissue contains a large number of adipose cells.

EXTERNAL AUDITORY MEATUS

The external auditory meatus (or auditory canal) can be divided into an outer cartilagenous portion (one-third) and inner bony (two-thirds) portion (Fig. 29-1). It is lined with epithelium containing hairs, sebaceous glands, and ceruminous glands. *Ceruminous glands* are branched tubuloalveolar glands which resemble large sweat glands. Their secretory portions elaborate and secrete the waxy *cerumen*. The secretory portion differs from sweat glands in that a very large lumen is present. A distinct cuticular border is often seen on the gland cells, and many pigment granules and lipid droplets are present in their cytoplasm. The duct of ceruminous glands is lined with a stratified cuboidal epithelium. Ceruminous gland ducts end directly at the skin or open together with the sebaceous glands into the neck of hair follicles.

MIDDLE EAR

TYMPANIC MEMBRANE

The middle ear is separated from the external ear by the *tympanic membrane* or the *eardrum* which consists of four distinct strata. They are the outermost cutaneum, the radiatum, the circulare, and the innermost mucosum. The margin of the tympanic membrane is attached to a fibrocartilaginous ring which in turn is connected to the *anulus tympanicus*, a bony ridge on the temporal bone (Fig. 29-2). The *stratum cutaneum* is a thin skin without papillae in its dermis. Cutaneum is thicker near the center and contains nerves and vessels which descend along the manubrium of the malleus and spread radially from it.

FIGURE 29-2 (*a*) Schematic drawing of a longitudinal section through the eardrum and its attachments. (*b*) Schematic drawing of the eardrum illustrating the different layers.

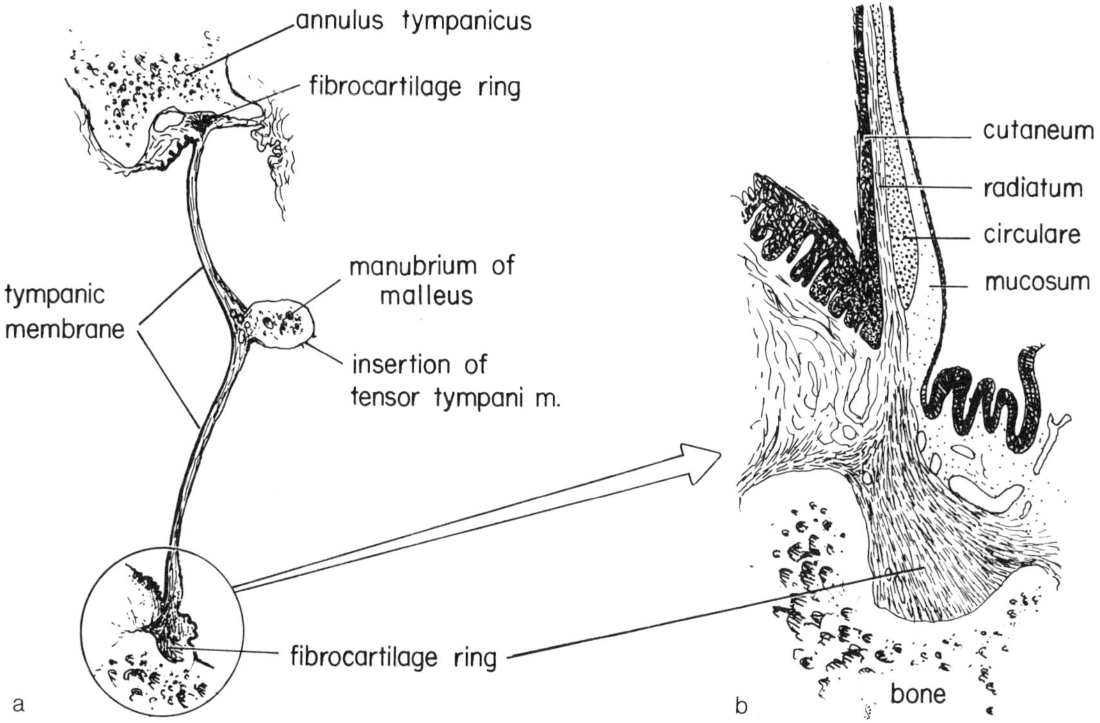

The *strata radiatum* and *circulare* consist of compact FECT fibers that are so arranged as to suggest a tendon (Fig. 29-2). In the radial layer, the fibers blend with the perichondrium of the hyaline cartilage that covers the manubrium. Peripherally, they contribute to the formation of the fibrocartilaginous ring. The *stratum mucosum* is a thin layer of connective tissue covered by a simple nonciliated cuboidal epithelium continuous with the lining of the tympanic cavity. The tympanic membrane can be divided into tense and flaccid portions. The latter is a small upper part which lacks a fibrous layer. The tendon of the *tensor tympani muscle* is attached to the center of the tympanic membrane.

TYMPANIC CAVITY

The *tympanic cavity,* also called *tympanum,* is air filled and lined with a mucous membrane which is closely attached to the surrounding periosteum. The connective tissue is generally thin and is covered by a simple cuboidal epithelium of varying heights. Cilia may be present in a sporadic manner except on the floor of the tympanum where they are always found. Along the anterior border, small mucous glands are often present. The contour of the tympanic cavity is highly irregular (Fig. 29-3). Laterally, the tympanic membrane marks the border, while the orifice of the eustachian tube marks the anterior boundary. Superiorly and posteriorly, the tympanic cavity forms a deep recess called the *epitympanic cavity;* the mastoid cells open here. The upper portion also contains the rounded heads of the malleus and incus. Anteriorly on the medial wall, the cavity presents a bony prominence called the *promontory* which corresponds to the first (broadest) turn of the spiral canal of the cochlea. Inferior to this prominence is a small recess leading to the *round window* which is covered by a delicate membrane (the *second tympanic membrane*). Behind the promontory and slightly superior to it, a deep recess leads to the *oval window* (or *vestibular window*) which is closed by the base of the stapes.

AUDITORY OSSICLES

The function of transmitting sound waves from the tympanic membrane to the oval window is borne by a chain of three ossicles: the *malleus,* the *incus,* and the *stapes* (Fig. 29-4). The *manubrium* of the malleus is firmly attached to the tympanic membrane, while the head of the bone articulates with the head of the incus in the epitympanic space (Figs.

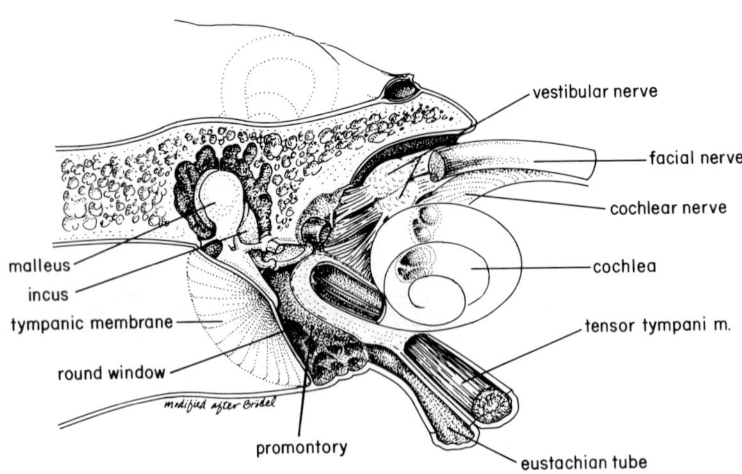

FIGURE 29-3 Schematic drawing of the middle and inner ear.

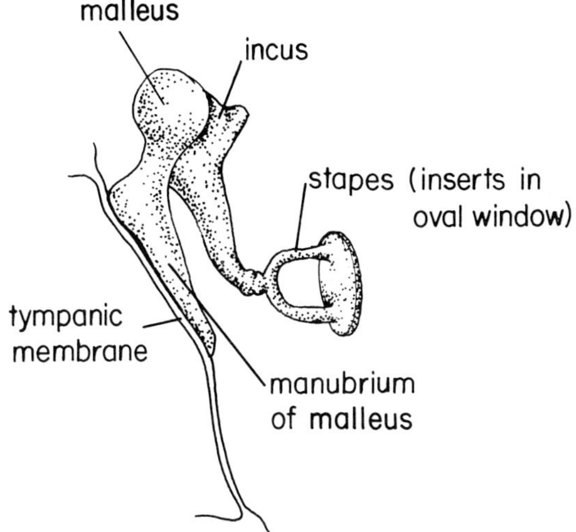

FIGURE 29-4 The ossicles of the middle ear.

29-1, 29-3, and 29-4). The long process of the incus extends downward along the tympanic wall, in a course nearly parallel to that of the manubrium mallei, and makes a sharp turn to articulate with the head of the stapes. The foot process of the stapes is held by an annular fibrous ligament in the oval window (*fenestra vestibuli*) which opens into the inner ear.

The joints articulating the ossicles are regular joints and are supported by connective tissue strands. Should an infection of the middle ear occur, joints often become involved, resulting in a fibrous ankylosis. Hearing is thereby impaired due to deterioration of the mechanical conduction of sound from the external ear to the neurosensory elements of the inner ear.

AUDITORY TUBE (EUSTACHIAN TUBE)

The auditory tube connects the tympanic cavity with the nasopharynx (Fig. 29-1). It includes a bony part (about one-third) near the tympanum and a cartilaginous part toward the pharynx. Its mucosa consists of a ciliated columnar epithelium of endodermal origin supported by a fibrous lamina propria directly attached to the bony cartilaginous or muscular surface. The stroke of the cilia is toward the pharyngeal orifice. A series of pockets containing air, *cellulae pneumaticae*, are present along the floor of the auditory tube. The mucosa is thicker in the cartilaginous part than in the osseous portion. Near the pharynx, many seromucous glands are found in the mucosa. As in other regions of the respiratory mucosa, many lymphocytes are present and form solitary nodules which blend with nodules of the pharyngeal tonsils near the pharyngeal opening.

INNER EAR

BONY AND MEMBRANOUS LABYRINTHS

The internal ear is situated within the petrous portion of the temporal bone (Fig. 29-1). A series of tortuous canals make up the *bony labyrinth* which is shaped to accommodate the *membranous labyrinth*. Thus, the membranous labyrinth constitutes the internal lining of the bony labyrinth. The membranous labyrinth is surrounded by and bathed in *perilymph*, a fluid similar in composition to other extracellular fluids. A narrow canal, the *vestibular aqueduct*, connects the bony labyrinth with the cerebrospinal space.

The membranous labyrinth has two parts (Fig. 29-5): the vestibular labyrinth and the cochlea. The *vestibular labyrinth* contains the semicircular canals, the utricle, and the saccule, which collectively make up the organ of equilibrium. Within the *cochlea* is the *organ of Corti*, which is the organ of hearing. The membranous labyrinth is filled by *endolymph*, a fluid that differs from perilymph in that it is rich in potassium and therefore resembles the cytosol of most cells.

VESTIBULAR LABYRINTH

The *vestibule* of the inner ear is the ovoid central portion of the bony labyrinth where the saccule and the utricle are housed (Fig. 29-1).

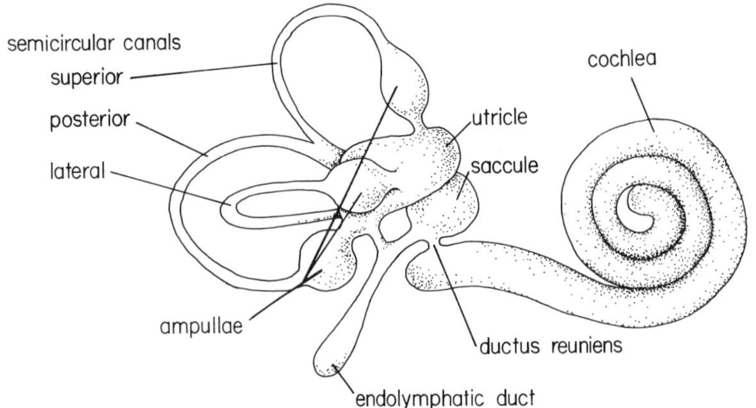

FIGURE 29-5 Schematic drawing of the membranous labyrinth.

Saccule

The saccule is a spherical membranous bag which is connected with the *scala media* of the cochlea at one end via the ductus reuniens and with the utricle and endolymphatic duct at the other end via a common duct (Fig. 29-5). The epithelium that lines the inner surface is of ectodermal origin. It consists of squamous to cuboid cells and is supported by a fibrous lamina propria. On the anterior and inferior surfaces of the saccule is a small (3 × 2 mm) ovoid patch of neuroepithelium called the *macula sacculi* (Fig. 29-7). It lies in a parasagittal plane.

Utricle

The utricle is slightly larger than the saccule and is situated behind and slightly above the saccule. It is connected to the latter and also communicates with the semicircular canals (Fig. 29-5). In common with the saccule, the utricle has a small round patch of neuroepithelium (2 × 2 mm). This area, called *macula utriculi* (Fig. 29-6), is located on the anterosuperior surface of the utricle and lies in a horizontal plane which parallels that of the cranial base and the lateral (horizontal) semicircular canal.

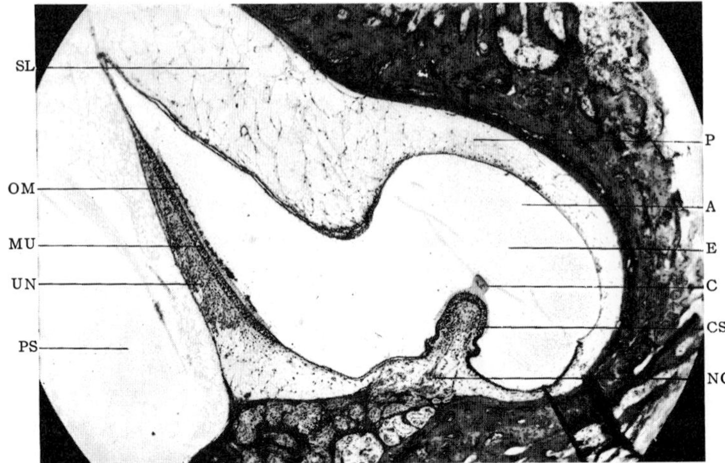

FIGURE 29-6 Macula utriculi and the ampulla of a semicircular canal of a guinea pig. Supporting ligaments run from the membranous semicircular canal and the utricle to the bony wall of the cavity. The utricle is suspended in the vestibule. (PS) Perilymphatic space. (SL) Supporting ligaments. (MU) Macula utriculi. (UN) Utricular nerve. (OM) Otolithic membrane. (A) Ampulla. (CSC) Crista and semicircular canal. (C) Cupula. (NC) Nerve to crista. (E) Endolymph. (P) Perilymph. (× 40). (*Courtesy of Lurie.*)

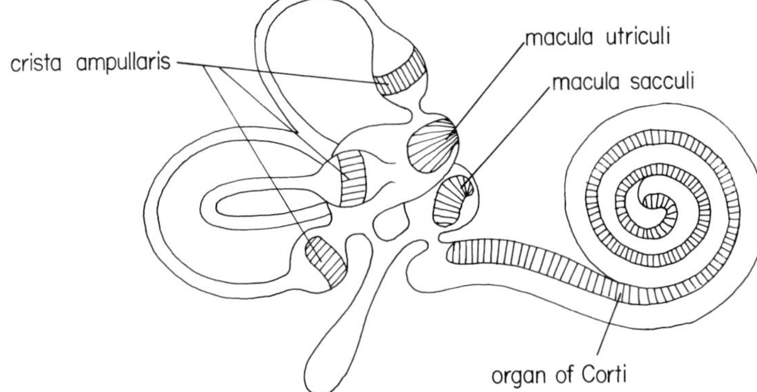

FIGURE 29-7 Schematic drawing of the membranous labyrinth illustrating the orientation of the sensory cells.

Semicircular ducts

Three semicircular ducts are positioned at right angles to each other in the bony semicircular canals. They conform in shape with the canals, and each has an ampullar end (Fig. 29-5). The ampullar ends of the anterior (superior), posterior (vertical) and the lateral (horizontal) ducts have separate openings, whereas their nonampullar ends form a common opening into the superior aspect of the utricle. The posterior (vertical) duct is perpendicular to the anterior (superior) duct. Each semicircular canal has a sensory organ that responds to angular movement in the plane of the canal (Fig. 29-7). This is the *crista ampullaris* which is composed of a ridge of connective tissue protruding into the ampulla which is lined with neuroepithelium. At both ends where the crista joins the inner wall of the ampulla, the epithelium lining the crista becomes tall columnar cells (Fig. 29-8). The neuroepithelium again has both sustentacular cells and sensory hair cells. The hairs of the sensory cells project into a gelatinous structure called *cupula* (Fig. 29-8). The cupula moves back and forth, conforming to the movement of the endolymph, and excites the sensory cells by displacement of the hairs of the hair cells.

Neurosensory epithelium

In the neuroepithelia of the semicircular ducts, the cells become tall columnar in shape and contain hair cells. Electron microscopy has revealed clearly that there are two types of hair cells (Fig. 29-9). Type I is a rounded cell with a constricted neck. Its cell body is surrounded by a chalice-like afferent nerve ending which in turn has synapses with efferent fibers containing numerous vesicles. Type II hair cell is columnar in shape and makes synapses with both afferent and efferent terminals. In both cell types, each cell has up to 80 stereocilia. Their length gradually increases toward one pole of the cell where a single cilium is present (Fig. 29-9). This polarization appears to have a functional significance in that the movement of the hair toward the single cilium increases the firing frequency of hair cells, while a movement away from the cilium decreases the firing frequency of the hair cells. The supporting cells have a more basally located nucleus and contribute to the formation of the cupula. The movement of the head is recorded by a simultaneous unidirectional displacement of all hair cells in a given crista. Thus, in the horizontal crista, they move toward the utricle, while in the two vertically oriented cristae (posterior and superior) they move away from it.

The neurosensory epithelium of the maculae utriculi and sacculi has the same basic structure as that of the crista ampullaris. In the maculae, the gelatinous substance of the cupula is replaced by the *otolithic membrane* which has a similar glycoprotein matrix into which the sensory cells penetrate (Fig. 29-

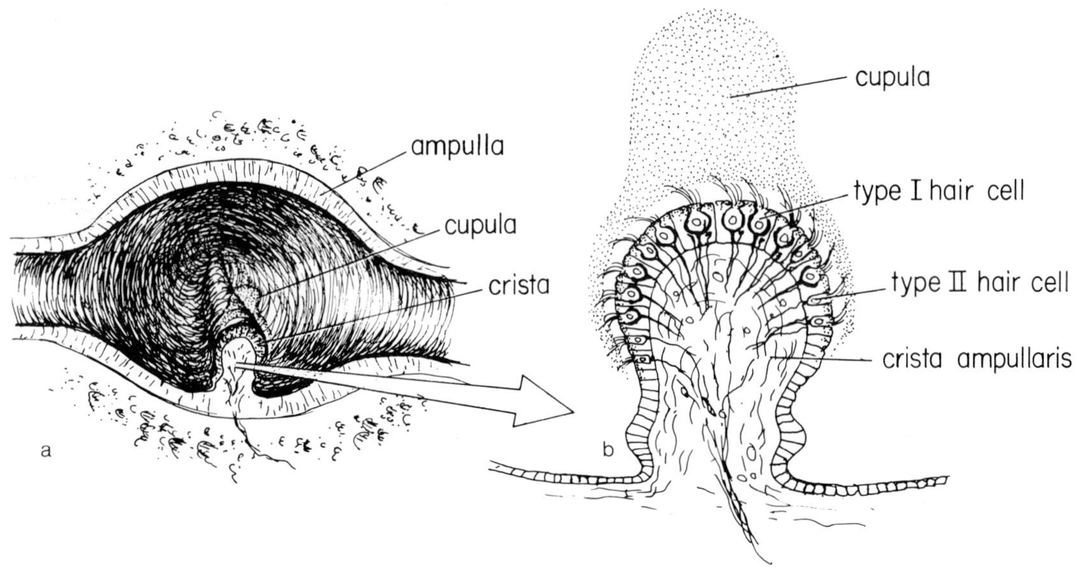

FIGURE 29-8 (*a*) Diagram of the ampullary end of a semicircular duct. (*b*) Diagram of crista ampullaris. The crista contains hair cells with apical sensory hairs joined to the cupula.

10). Unlike the cupula, the otolithic membrane has on its surface numerous crystals of calcium carbonate called *otoconia*. The weight of the otoconia creates a sheer motion of the otolithic membrane when a linear acceleration or positional change of the head occurs. The neurosensory elements, embedded in the gelatinous otolithic membrane, are either excited or inhibited by this relative motion.

The scanning EM in Fig. 29-11 shows otoconial crystals found in a healthy individual. Degenerative changes in otoconial structure occur during the process of normal aging and under certain pathologic conditions.

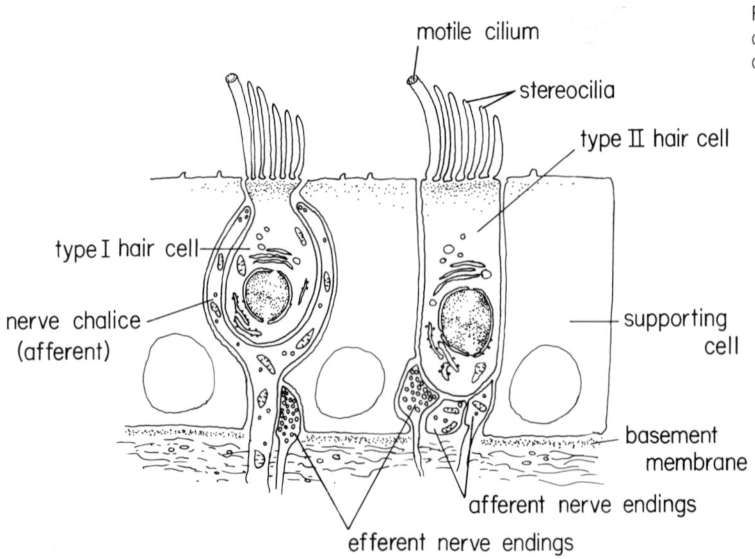

FIGURE 29-9 Schematic drawing of vestibular sensory cells.

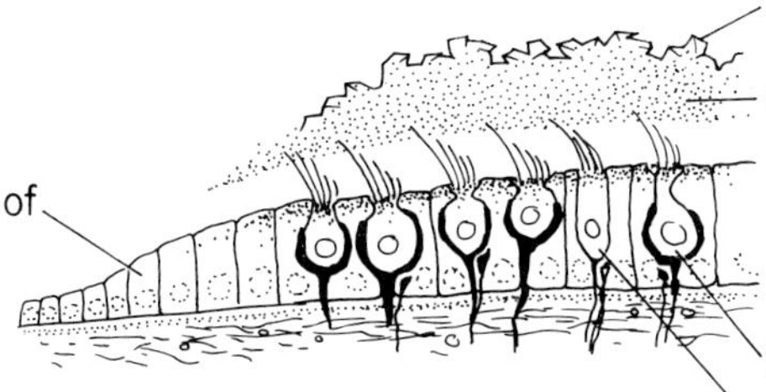

FIGURE 29-10 Sensory epithelia of the maculae utriculi and sacculi have identical structures.

COCHLEA

The critical structure essential to hearing is the membranous cochlea which is invested in a rigid osseous shell, the *osseus cochlea.* The term *cochlea* comes from the fact that its shape resembles a snail, having a broad base and a pointed apex. The spiraling cochlea makes two and one-half turns around a central pillar, called *modiolus,* which is made up of spongy bone (Figs. 29-12 and 29-13). A thin shelflike osseous projection originating in the modiolus partially divides the bony canal. It is called the *spiral lamina.*

Perilymphatic space

The spiral duct lining the bony cochlea is divided into two partitions by the *basilar membrane* which parallels the spiral course of the duct, bisecting it horizontally (Figs. 29-14 and 29-15). The space above the basilar membrane is called the *scala vestibuli* because it opens into the vestibule, whereas the compartment below is named the *scala tympani* and ends basally at the round window. Both of these compartments are filled with perilymph. The wedge-shaped compartment between them is called the *scala media,* which is delimited toward the scala vestibuli and scala tympani by the *Reissner's* and basilar membranes, respectively. The membranous cochlea is filled with endolymph and ends as a blind sac at the apex. At this apicalmost end of the cochlear duct the two perilymphatic spaces of the scala vestibuli and scala tympani become confluent. The union of the two parts is called the *helicotrema* (Figs. 29-12 and 25-13). Near the round window at the base of the cochlea a small duct (aqueductus cochlae) originates in the scala tympani and reaches the subarachnoidal space in the vicinity of the jugular fossa.

Membranous cochlea (endolymph space)

The membranous cochlea limiting the scala media is triangular in histologic sections (Figs. 29-14 and 29-15). At the base it is bordered by the basilar membrane and bony spiral lamina. Laterally, it is delimited by the *spiral ligament* which is a fibrous thickening

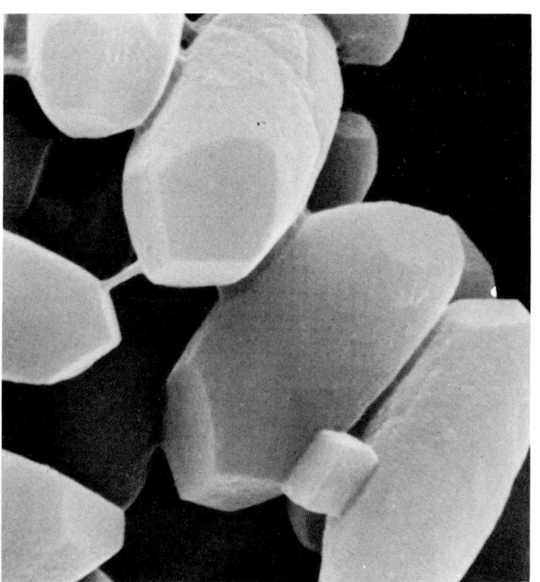

FIGURE 29-11 Scanning electron micrograph of otoconia found in a healthy individual. (*Courtesy of M. D. Ross.*)

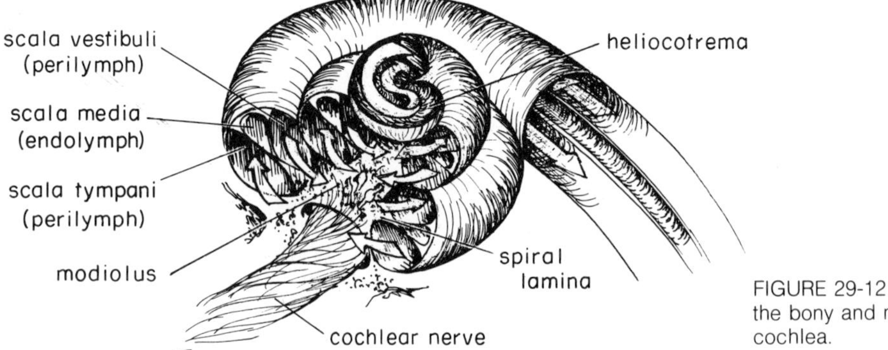

FIGURE 29-12 Diagram of the bony and membranous cochlea.

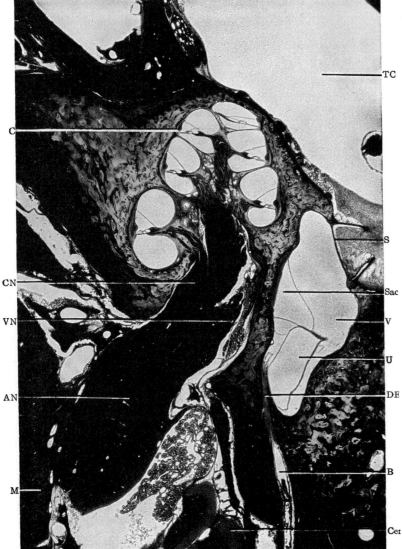

FIGURE 29-13 Cochlea and vestibule of a cat. Horizontal section shows the foot plate of the stapes in the oval window, the vestibule showing the relationship of saccule and utricle to the oval window, and the large perilymphatic space around the foot plate of the stapes. The ductus endolymphaticus can be followed in the aqueductus vestibuli. The bulb of the ductus lymphaticus can be seen. The cochlea is cut to show three turns. The cochlear nerve and the vestibular nerve can be seen in their course from the spiral ganglia and Scarpa's ganglion into the medulla. (TC) Tympanic cavity. (S) Stapes. (V) Vestibule. (Sac) Saccule. (U) Utricle. (DE) Ductus endolymphaticus. (B) Bulb (sarcus) of ductus endolymphaticus. (C) Cochlea. (CN) Cochlear nerve and spiral ganglia. (VN) Vestibular nerve. (M) Scarpa's ganglion. (AN) Auditory nerve. (Cer) Cerebellum. (× 12) (*Courtesy of Lurie.*)

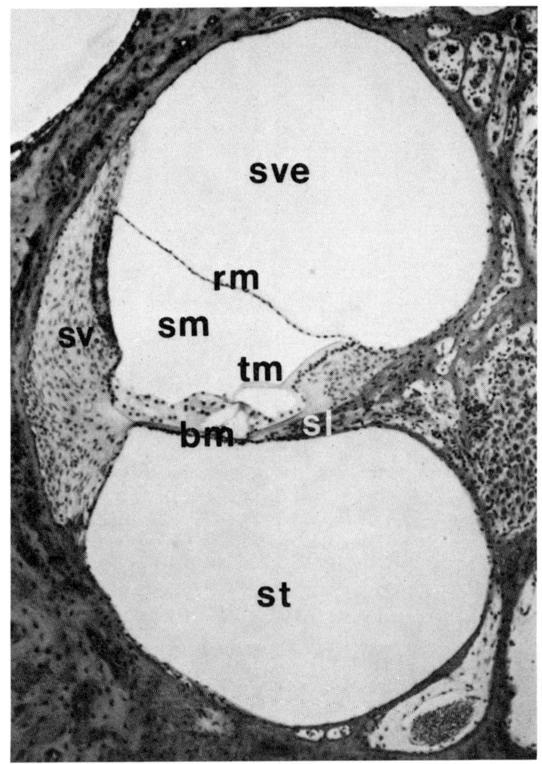

FIGURE 29-14 Organ of Corti. (sve) Scala vestibuli. (sm) Scala media. (st) Scala tympani. (sl) Spiral lamina. (sv) Stria vascularis. (bm) Basilar membrane. (rm) Reissner's membrane. (tm) Tectorial membrane.

of the periosteum (Figs. 29-14 and 29-15). A pseudostratified columnar epithelium covers the surface of the spiral ligament. The subjacent connective tissue is rich in blood vessels. Together they are called the *stria vascularis* (Figs. 29-14 and 29-15) which is believed to be the main source of the endolymph. Above, the Reissner's membrane separates the scala vestibuli from the scala media.

The basilar membrane varies in width from the base of the cochlea to its apex. It is the narrowest at the round window (0.16 mm) and widest at the helicotrema (0.52 mm). The length of the basilar membrane in human beings is approximately 31 mm. On the surface facing the scala tympani, the basilar membrane is lined by mesothelial cells resting upon a thin fibrous connective tissue layer, whereas on the side toward the scala media, the membrane has a homogenous ground substance supporting the organ of Corti. The Reissner's membrane is attached to the medial superior lip of the bony spiral lamina and reaches the outer wall of the bony cochlea at the upper end of the stria vascularis (Figs. 29-14 and 29-15).

Organ of Corti

As the neurosensory receptor apparatus for hearing, the organ of Corti is strategically situated on the basilar membrane. It is, as such, present throughout the entire stretch of the basilar membrane between the round window and the helicotrema of the cochlea. A small portion of the organ lies over the tympanic lip of the bony spiral lamina (Fig. 29-15). That portion of the organ of Corti which rests on the basilar membrane has three rows of outer hair cells lodged between associated elements, while the region which lies over the bony spiral lamina contains the inner hair cells and supportive structures. The *inner* and *outer pillar cells* supporting respective hair cells make a triangular space called the *tunnel*. The pillar cells have a broad base in which the nucleus lies. They are filled with numerous tonofilaments which make a cuticular plate at the free surface. Altogether, there are approximately 6000 inner pillar cells and 4000 outer pillar cells, resulting in a ratio of 3:2 interlocking of the inner and outer pillar cells as they meet at the roof of the tunnel.

Sensory cells of the organ are arranged in two groups. *Inner hair cells* totaling about 3500 in number make a single row near the inner pillar cells. Some 20,000 or so of *outer hair cells* make up three to four rows. The inner and outer hair cells slant toward each other as they lie against their respective pillar cells (Fig. 29-15). Investing the hair cells are tall supportive cells called *inner* and *outer phalangeal cells* which in nature resemble the sustentacular cells found in other neurosen-

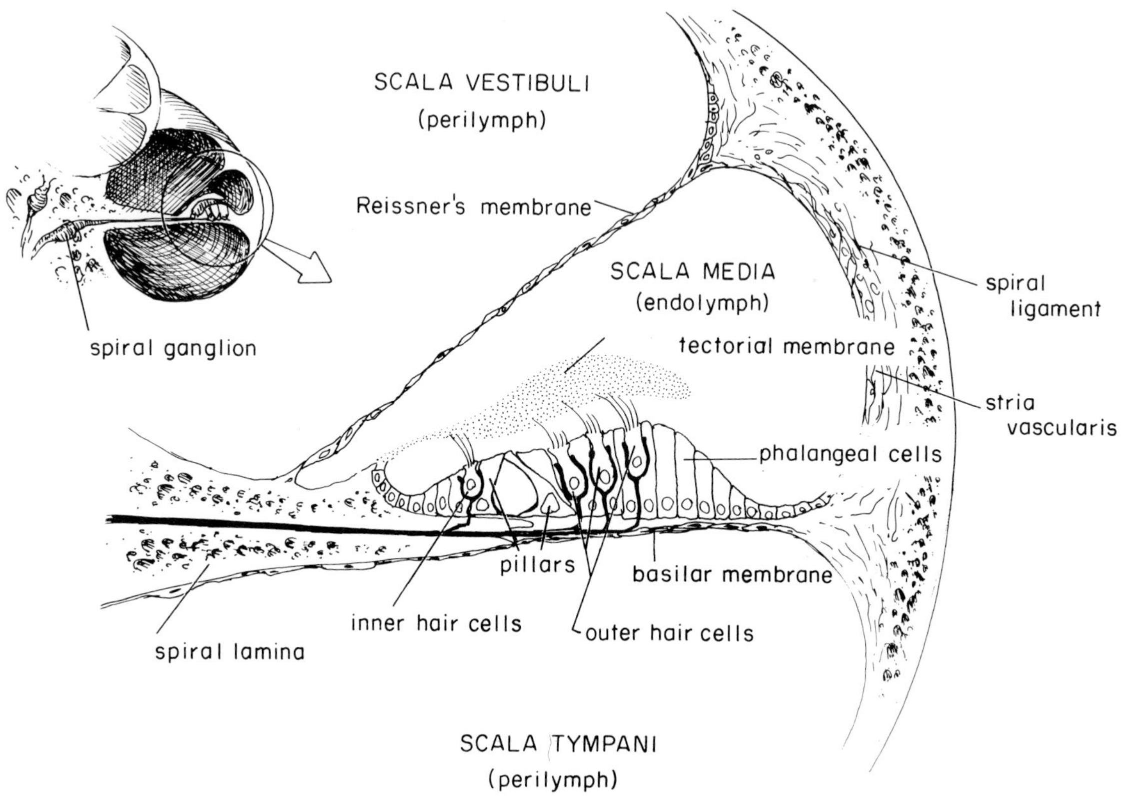

FIGURE 29-15 Diagram of a cross section of the membranous labyrinth illustrating the morphology of the organ of Corti.

sory end organs (viz, taste buds). The inner hair cells have a bulbous cell body with their sensory "hairs" lining up in straight rows (Fig. 29-16). The scanning EM in Fig. 29-17 shows the appearance of the hairs of inner hair cells.

The outer hair cells have a columnar shape, with their base showing a rounded appearance. From the apex of each outer hair cell, a bundle of stereocilia project in a V-shaped pattern (Figs. 29-16 and 29-17). A gelatinous *tectorial membrane* lies over the organ of Corti (Figs. 29-14 and 29-15). It is attached to the spiral limbus which contains the interdental cells that secrete the substance of the membrane. It rests upon the tallest row of stereocilia in each hair cell.

With respect to *innervation* both inner and outer hair cells have a few afferent and several efferent nerve endings per cell. The efferent nerve terminals in both cell types contain a large number of vesicles.

NERVE SUPPLY OF THE INNER EAR

The sensory areas of the membranous labyrinth is supplied by the *VIIIth cranial nerve* which, in accordance with the anatomical division of the inner ear itself, is made up of two major components, namely, the vestibular nerve and the cochlear nerve. The *vestibular nerve* has a superior division supplying the macula of the utricle and the cristae of the anterior vertical and horizontal canals. The macula of the saccule and the crista of the posterior vertical canal are innervated by the inferior branch. Cell bodies of the vestibular nerve are bipolar and reside in the *vestibular* ganglion (also called Scarpa's gan-

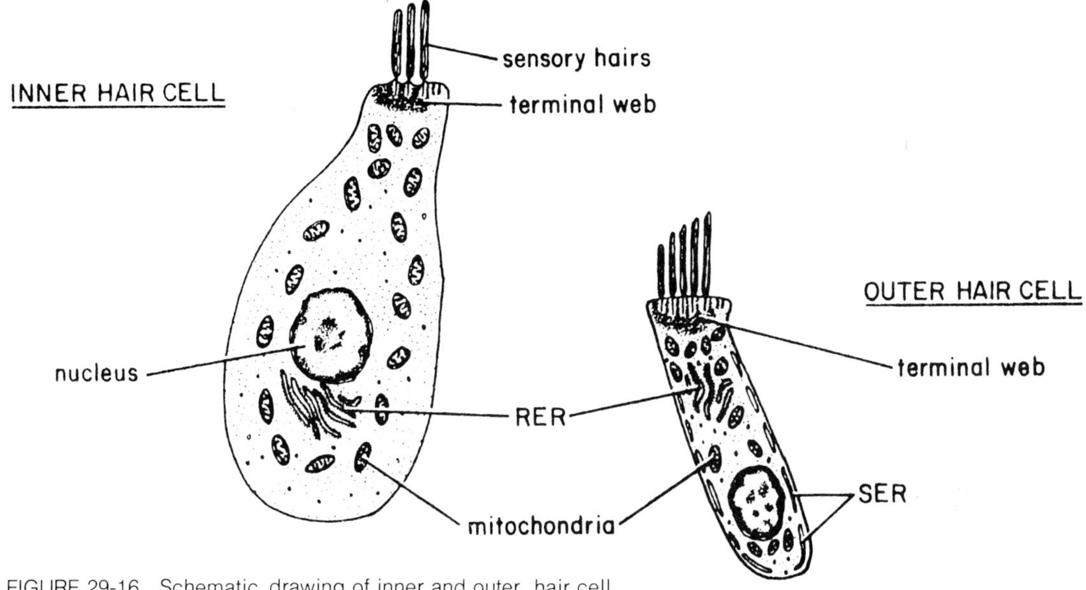

FIGURE 29-16 Schematic drawing of inner and outer hair cell.

glion). Their axons enter the medulla oblongata, ending in the vestibular nuclei near the fourth ventricle. From there, the nerve makes a connection with the cerebellum and the nuclei of cranial nerves III, IV, and VI.

Fibers of the *cochlear nerve* exit the cochlea through the modiolus and reach the *spiral ganglion* where the bipolar cell bodies occupy a prominent position in the bony spiral lamina (Figs. 29-12 and 29-15). The orifices through which cochlear nerve fibers emerge are called *foramina nervosa*. All cochlear nerve fibers are myelinated beyond the foramina nervosa. The cochlear nerve enters the medulla and terminates in the cochlear nuclei where its fibers make connections with other

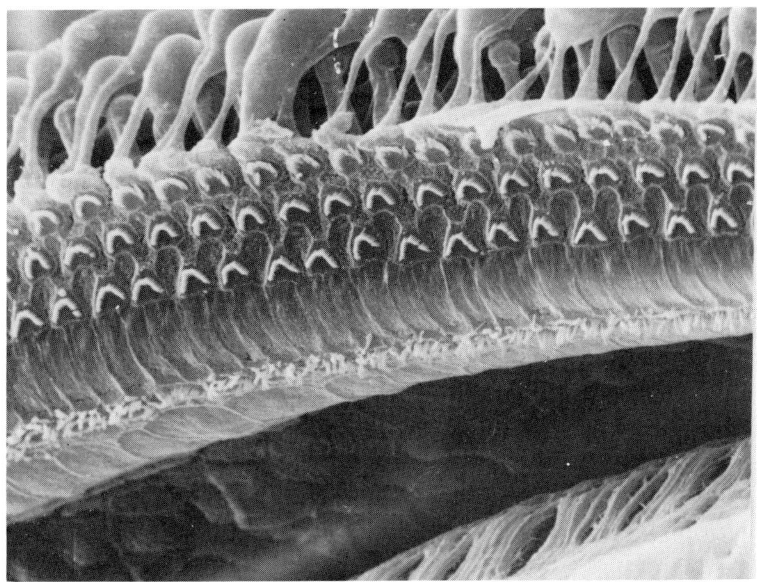

FIGURE 29-17 Sensory hair bundles of inner and outer hair cells project from the surface of the organ of Corti. There are three rows of outer hair cells, while there is only a single row of inner hair cells. Also note the phalangeal cells supporting the outer hair cells. *(Courtesy of M. D. Ross.)*

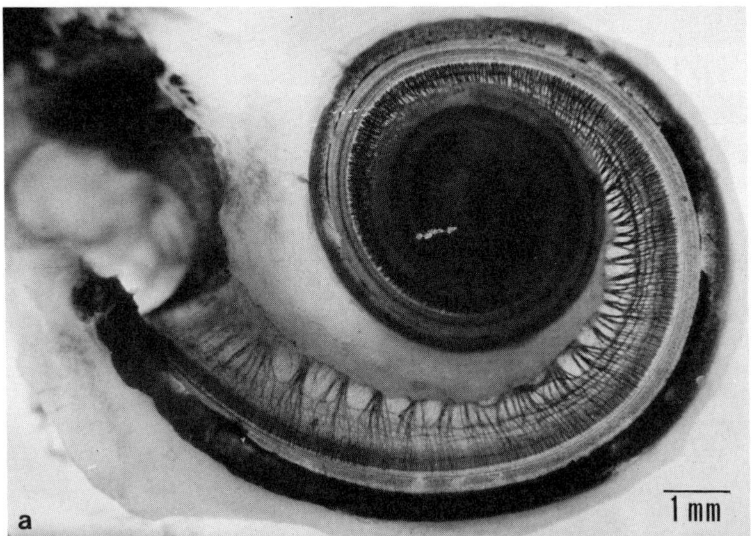

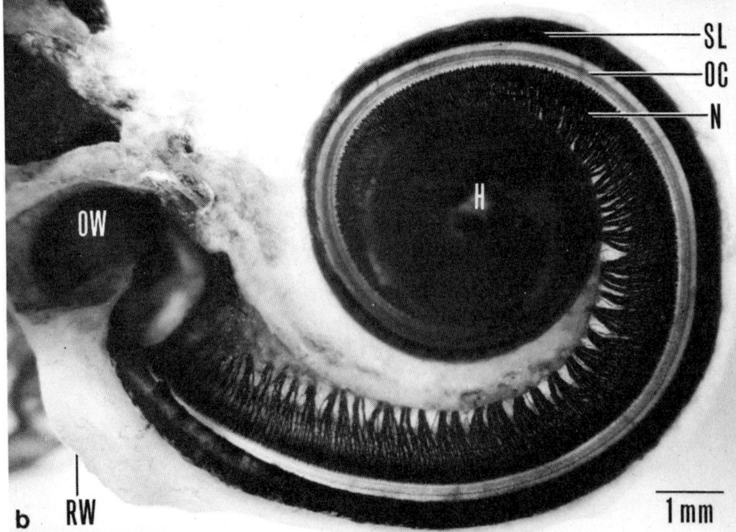

FIGURE 29-18(*a*) Dissected cochlea in an old individual. Several areas of degeneration in the organ of Corti can be seen. The innervation of the organ of Corti is sparse in comparison to that of the young individual seen in (*b*). (*Courtesy of L. E. Johnsson and J. E. Hawkins.*) (*b*) Dissected cochlea in a young individual. (H) Helicotrema. (N) Nerves. (OC) Organ of Corti. (OW) Oval window. (RW) Round window. (SL) Spiral ligament. (*Courtesy of L. E. Johnsson and J. E. Hawkins.*)

nuclei located at higher levels. The main portion of the cochlear fibers goes to the medial geniculate nucleus of the thalamus. From there, the fibers radiate to the temporal lobe of the cerebrum where the auditory centers are located.

It has been known that sympathetic components of the autonomic nervous system from the superior cervical ganglion innervate the blood vessels of the inner ear. They also form a delicate plexus which terminates in both the vestibular and the spiral ganglia. The functional significance of the sympathetic innervation is not clearly understood.

Certain changes take place in the structure of the inner ear as a function of time. In general, the neurosensory elements of the cochlea show varying degrees of degeneration including nerve cell death. These degenerative changes appear to be age-dependent phenomena and largely involve the basal portion of the cochlea which is responsible for hearing high-frequency sounds. Therefore, an old person with hearing difficulties (*presbycusis*) characteristically has problems in hearing high-frequency notes. The question of when the first organic changes occur in the inner ear is not clearly understood, although degenerating nerve fibers of the cochlea have been observed even in neonatal subjects. Figures 29-18*a* and *b* show neural degeneration at the cochlear base in an old individual (*a*) as compared with a nearly intact cochlear nerve in an infant (*b*).

It is generally thought that some of the neurosensory degradation may indeed occur as a function of time. However, recent studies involving veterans of the Vietnam war and hunters clearly indicate that noise pollution may significantly contribute to an early degeneration of the cochlear nerve. Thus, industrial noise pollution as well as such "pleasurable" life experiences as loud music from electronic equipment may result in permanent damage to the ability to perceive and discriminate sounds of subtle qualitative differences.

REVIEW SECTION

1. Figure 29-20 is a section through the cochlea and the spiral ganglion. What are the spaces labeled (a), (b), and (c) called? Identify the structure labeled (d). Name the structure labeled (e). Identify the structures labeled (f) and (g).

Scala vestibuli (a) and scala tympani (c) are parts of the membranous cochlea and contain the perilymph. The wedge-shaped space between them (b) and limited by the Reissner's membrane (f) and the basilar membrane (g) is scala media. It contains endolymph. The organ of Corti (e) is the neurosensory apparatus for hearing.

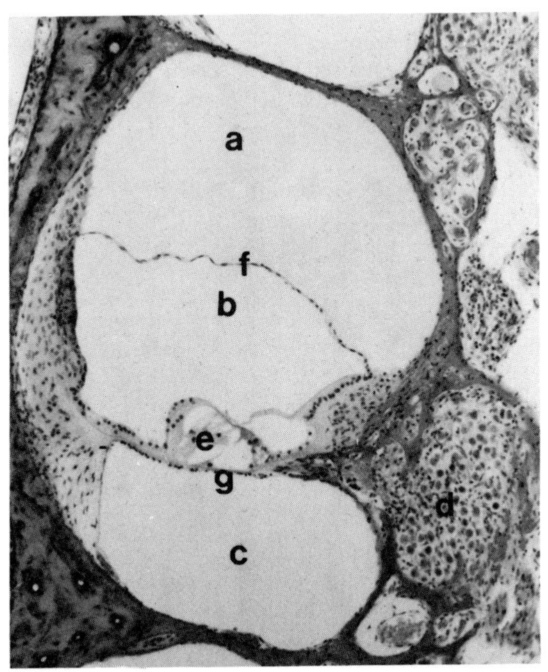

FIGURE 29-19

2. The auditory tube connects the tympanic cavity with the ________. What is its function?

The eustachian tube connects the tympanic cavity with the nasopharynx. It functions in equilibrating the air pressure in the tympanic cavity with the surroundings.

3. The cristae of the semicircular ducts and the maculae of the utricle and sacculi are end organs of what system?

They are all end organs of the vestibular system and play an important role in the regulation and coordination of the movements of equilibrium and locomotion.

4. Where are the auditory ossicles located? What is their function? What is the result of ankylosis of the joints between them?

The three auditory ossicles, malleus, incus, and stapes, are found in the air-filled tympanic cavity. They function in transmitting sound waves from the eardrum to the oval window. If for any reason the joints between them are ankylosed, hearing is impaired due to deterioration of the mechanical conduction of sound.

5. What do the cristae and the maculae of the vestibular labyrinth have in common?

They are all lined with neuroepithelium.

BIBLIOGRAPHY

Bloom, W., and D. W. Fawcett: *A Textbook of Histology,* 10th ed., Saunders, Philadelphia, 1975.

Engstrom, H., J. Ader, and A. Anderson: *Structural Pattern of the Organ of Corti,* Almquist & Wiksell, Stockholm, 1966.

———, and H. W. Ades (eds.): "Inner Ear Studies," *Acta Otolaryngol.,* suppl. 301 (1972).

Iurato, S. (ed.): *Submicroscopic Structure of the Inner Ear,* Pergamon, New York, 1967.

ADDITIONAL READINGS

Bredberg, G.: "Cellular Pattern and Nerve Supply of the Human Organ of Corti," *Acta Otolaryngol.,* suppl. **236:**1–135 (1968).

Davis, H.: "Mechanisms of the Inner Ear," *Am. Otol.* **56:**84–95 (1968).

Engstrom, H., H. Ades, and J. Hawkins: "Structure and Function of the Sensory Hairs of the Inner Ear," *J. Acoust. Soc. Amer.* **34:**1356 (1962).

———, and J. Wersall: "Structure and Innervation of the Inner Ear Sensory Epithelia," *Int. Rev. Cytol.* **7:**535–585 (1958).

Friedman, I.: "The Cytology of the Ear," *Brit. Med. Bull.* **18:**209–213 (1962).

Graves, G. O., and L. F. Edwards: "The Eustachian Tube: A Review of Its Descriptive, Microscopic, Topographic and Clinical Anatomy," *Arch. Otolaryngol.* **39:**359–397 (1944).

Kimura, R. S.: "Hairs of the Cochlear Sensory Cells and Their Attachment to the Tectorial Membrane," *Acta Otolaryngol.* **51:**55–72 (1966).

———, and H. F. Schuknecht: "The Ultrastructure of the Human Stria Vascularis," part I, *Acta Otolaryngol.* **69:**415–426 (1970).

———, ———, and I. Sundo: "Fine Morphology of the Sensory Cells in the Organ of Corti in Man," *Acta Otolaryngol.* **58:**390–408 (1965).

Lindeman, H. H.: "Studies on the Morphology of the Sensory Regions of the Vestibular Apparatus (with 45 figures)," *Ergeb. der Anat. u. Entwicklgesch.* **42:**1–113 (1969).

Smith, C. A.: "Ultrastructure of the Organ of Corti," *Adv. Sci.* **24:**419 (1968).

———: "Electron Microscopy of the Inner Ear," *Ann. Otol. Rhin. Laryngol.* **77:**629–643 (1968).

INDEX

Page numbers in *italic* indicate illustrations or tables.